Reproductive Processes and Contraception

Reproductive Processes and Contraception

BIOCHEMICAL ENDOCRINOLOGY

Series Editor: Kenneth W. McKerns

STRUCTURE AND FUNCTION OF THE GONADOTROPINS
Edited by Kenneth W. McKerns

SYNTHESIS AND RELEASE OF ADENOHYPOPHYSEAL HORMONES
Edited by Marian Jutisz and Kenneth W. McKerns

REPRODUCTIVE PROCESSES AND CONTRACEPTION
Edited by Kenneth W. McKerns

Reproductive Processes and Contraception

Edited by

Kenneth W. McKerns
The International Society for Biochemical Endocrinology
Blue Hill Falls, Maine

PLENUM PRESS • NEW YORK AND LONDON

Library of Congress Cataloging in Publication Data

Main entry under title:

Reproductive processes and contraception.

(Biochemical endocrinology)
Sponsored by the International Society for Biochemical Endocrinology.
Includes index.
1. Reproduction–Congresses. 2. Contraception–Congresses.
3. Hormones, Sex–Congresses.
I. McKerns, Kenneth W. II. International Society for Biochemical Endocrinology.
QP251.R4446
ISBN 0-306-40534-2 599.01′6 80-20744

A Division of Plenum Publishing Corporation
233 Spring Street, New York, N.Y. 10013

Printed in the United States of America

In Memoriam

A few weeks after the Asticou conference, we were shocked to hear from Dr. Janet Nolin of the accidental death of Professor E. M. Bogdanove, her husband. He was a vital physical force, with a provocative intellectual capacity. His death was a sad loss to his family – a sad loss to us all.

Contributors

A. Aakvaag, Hormone and Isotope Laboratory, University of Oslo Hospitals, Aker Hospital, Oslo 5, Norway

Maria Alexandrova, Institute of Experimental Endocrinology, Slovak Academy of Sciences, Bratislava, Czechoslovakia

L. D. Anderson, Department of Anatomy, University of Maryland School of Medicine, Baltimore, Maryland 21201

Claude Auclair, Department of Molecular Endocrinology, Le Centre Hospitalier de l'Université Laval, Quebec G1V 4G2, Canada

Om P. Bahl, Department of Biological Sciences, Division of Cell and Molecular Biology, State University of New York at Buffalo, Buffalo, New York 14260

U. K. Banik, Department of Endocrinology and Immunochemistry, Ayerst Research Laboratories, P.O. Box 6115, Montreal H3C 3J1, Canada

Satish Batta, Department of Physiology, University of Maryland School of Medicine, Baltimore, Maryland 21201

Etienne-Emile Baulieu, Unité de Recherches sur le Métabolisme Moléculaire et la Physio-Pathologie des Stéroides de l'Institut National de la Santé et de la Recherche Médicale (U 33 INSERM) and ER 125 CNRS, 94270 Bicêtre, France

F. Bayard, Laboratoire d'Endocrinologie Expérimentale, INSERM U 168, CHU Toulouse Rangueil, 31054 Toulouse Cedex, France

Fuller W. Bazer, Department of Animal Science, University of Florida, Gainesville, Florida 32611

Alain Bélanger, Department of Molecular Endocrinology, Le Centre Hospitalier de l'Université Laval, Quebec G1V 4G2, Canada

K. K. Bergstrom, Fertility Research, The Upjohn Company, Kalamazoo, Michigan 49001

D. C. Beuving, Fertility Research, The Upjohn Company, Kalamazoo, Michigan 49001

Frederick J. Bex, Endocrinology Section, Wyeth Laboratories, Inc., Philadelphia, Pennsylvania 19101

Steven Birken, Department of Medicine, Columbia University College of Physicians & Surgeons, New York, New York 10032

B. Bizzini-Kouznetzova, INSERM FRA 8, Institut Pasteur, 75724 Paris, France

Xuan-Hoa Bui, Unité de Chimie Hormonologique et de Pharmacognosie, Université de Louvain, B-1200 Brussels, Belgium

Simon Caron, Department of Molecular Endocrinology, Le Centre Hospitalier de l'Université Laval, Quebec G1V 4G2, Canada

Cornelia P. Channing, Department of Physiology, University of Maryland School of Medicine, Baltimore, Maryland 21201

Scott C. Chappel, Department of Obstetrics and Gynecology, University of Pennsylvania, Philadelphia, Pennsylvania 19104

Herman Cohen, Carter-Wallace, Inc., Cranbury, New Jersey 08512

Alan Corbin, Endocrinology Section, Wyeth Laboratories, Inc., Philadelphia, Pennsylvania 19101

Nicole Crozet, INRA, Station Centrale de Physiologie Animale, 78350 Jouy-en-Josas, France

Lionel Cusan, Department of Molecular Endocrinology, Le Centre Hospitalier de l'Université Laval, Quebec G1V 4G2, Canada

Raymond Devis, Unité de Chimie Hormonologique et de Pharmacognosie, Université de Louvain, B-1200 Brussels, Belgium

F. Dray, INSERM FRA 8, Institut Pasteur, 75724 Paris, France

L. Duguet, Laboratoire d'Endocrinologie Expérimentale, INSERM U 168, CHU Toulouse Rangueil, 31054 Toulouse Cedex, France

A. Esquifino, Chair for Experimental Endocrinology, Department of Physiology, University Complutensis Medical School, Madrid-3, Spain

J. C. Faye, Laboratoire d'Endocrinologie Expérimentale, INSERM U 168, CHU Toulouse Rangueil, 31054 Toulouse Cedex, France

Mark Feldman, Department of Pediatrics and Laboratories for Reproductive Biology, University of North Carolina, Chapel Hill, North Carolina 27514

Frank S. French, Department of Pediatrics and Laboratories for Reproductive Biology, University of North Carolina, Chapel Hill, North Carolina 27514

M. L. Givner, Department of Endocrinology and Immunochemistry, Ayerst Research Laboratories, P.O. Box 6115, Montreal H3C 3J1, Canada

Roy H. Hammerstedt, Paul M. Althouse Laboratory, Biochemistry Graduate Program, Department of Microbiology, Cell Biology, Biochemistry and Biophysics, The Pennsylvania State University, University Park, Pennsylvania 16802

F. Haour, INSERM U 162, Hôpital Debrousse, 69322 Lyon, France

M. T. Hochereau-de-Reviers, INRA, Station de Physiologie de la Reproduction, 37380 Nouzilly, France

Pushpa S. Kalra, Department of Obstetrics and Gynecology, University of Florida College of Medicine, Gainesville, Florida 32610

Satya P. Kalra, Department of Obstetrics and Gynecology, University of Florida College of Medicine, Gainesville, Florida 32610

Paul A. Kelly, Department of Molecular Endocrinology, Le Centre Hospitalier de l'Université Laval, Quebec G1V 4G2, Canada

S. S. Koide, Center for Biomedical Research, The Population Council, The Rockefeller University, New York, New York 10021

Fernand Labrie, Department of Molecular Endocrinology, Le Centre Hospitalier de l'Université Laval, Quebec G1V 4G2, Canada

Oscar A. Lea, Department of Pharmacology, University of Bergen, Bergen, Norway

A. Lemay, Laboratory of Endocrinology of Reproduction, Department of Obstetrics and Gynecology, Hôpital St. François d'Assise, Quebec G1L 3L5, Canada

Bonnie K. Loeser, Department of Obstetrics and Gynecology and Department of Biochemistry, Medical College of Ohio, Toledo, Ohio 43699

Janet M. Loring, Department of Biological Chemistry and Laboratory of Human Reproduction and Reproductive Biology, Harvard Medical School, Boston, Massachusetts 02115

Takeshi Maruo, Center for Biomedical Research, The Population Council, The Rockefeller University, New York, New York 10021

J. Mather, Center for Biomedical Research, The Population Council, The Rockefeller University, New York, New York 10021

M. Mazzuca, Laboratoire d'Endocrinologie Expérimentale, INSERM U 168, CHU Toulouse Rangueil, 31054 Toulouse Cedex, France

Rodrigue Mortel, Unité de Recherches sur le Métabolisme Moléculaire et la Physio-Pathologie des Stéroides de l'Institut National de la Santé et de la Recherche Médicale (U 33 INSERM) and ER 125 CNRS, 94270 Bicêtre, France

Janet M. Nolin, Department of Physiology, Medical College of Virginia, Richmond, Virginia 23298

Tiiu Ojasoo, Centre de Recherches Roussel-Uclaf, 93230 Romainville, France

Paul Ordronneau, Department of Anatomy, University of North Carolina, Chapel Hill, North Carolina 27514

A. Oriol-Bosch, Chair for Experimental Endocrinology, Department of Physiology, University Complutensis Medical School, Madrid-3, Spain

Vladimir R. Pantić, Serbian Academy of Science and Arts, Belgrade, Yugoslavia

Peter Petrusz, Department of Anatomy and Laboratories for Reproductive Biology, University of North Carolina, Chapel Hill, North Carolina 27514

Richard J. Pietras, Department of Biology and the Molecular Biology Institute, University of California, Los Angeles, California 90024

Jacques I. Quivy, International Institute of Molecular and Cellular Pathology, B-1200 Brussels, Belgium; Unité de Chimie Hormonologique et de Pharmacognosie, Université de Louvain, B-1200 Brussels, Belgium

Jean-Pierre Raynaud, Centre de Recherches Roussel-Uclaf, 93230 Romainville, France

Paul Robel, Unité de Recherches sur le Métabolisme Moléculaire et la Physio-Pathologie des Stéroides de l'Institut National de la Santé et de la Recherche Médicale (U 33 INSERM) and ER 125 CNRS, 94270 Bicêtre, France

R. Michael Roberts, Department of Biochemistry and Molecular Biology, University of Florida, Gainesville, Florida 32611

Guy G. Rousseau, International Institute of Molecular and Cellular Pathology, B-1200 Brussels, Belgium

Judith Saffran, Department of Obstetrics and Gynecology and Department of Biochemistry, Medical College of Ohio, Toledo, Ohio 43699

M. R. Sairam, Institut de Recherches Cliniques, Montreal H2W 1R7, Canada

Jean-Pierre Schmit, Département de Chimie, Université de Sherbrooke, Sherbrooke (Québec) J1K 2R1, Canada

Sheldon J. Segal, The Rockefeller Foundation, New York, New York 10036

Carl Séguin, Department of Molecular Endocrinology, Le Centre Hospitalier de l'Université Laval, Quebec G1V 4G2, Canada

D. C. Sharp III, Department of Animal Science, University of Florida, Gainesville, Florida 32611

J. W. Simpkins, Department of Obstetrics and Gynecology, University of Florida College of Medicine, Gainesville, Florida 32610

Melvyn S. Soloff, Department of Biochemistry, Medical College of Ohio, Toledo, Ohio 43699

Harold G. Spies, Oregon Regional Primate Research Center, Beaverton, Oregon 97005

C. H. Spilman, Fertility Research, The Upjohn Company, Kalamazoo, Michigan 49001

Sarah Lipford Stone, Department of Physiology, University of Maryland School of Medicine, Baltimore, Maryland 21201

Clara M. Szego, Department of Biology and the Molecular Biology Institute, University of California, Los Angeles, California 90024

Daniel Szöllösi, INRA, Station Centrale de Physiologie Animale, 78350 Jouy-en-Josas, France

W. W. Thatcher, Department of Dairy Science, University of Florida, Gainesville, Florida 32611

P. A. Torjesen, Hormone and Isotope Laboratory, University of Oslo Hospitals, Aker Hospital, Oslo 5, Norway

J. A. F. Tresguerres, Chair for Experimental Endocrinology, Department of Physiology, University Complutensis Medical School, Madrid-3, Spain

Viviane Vaché, Centre de Recherches Roussel-Uclaf, 93230 Romainville, France

Claude A. Villee, Department of Biological Chemistry and Laboratory of Human Reproduction and Reproductive Biology, Harvard Medical School, Boston, Massachusetts 02115

Preface

This monograph represents the eighth sponsored by the International Society for Biochemical Endocrinology. The topics should be of interest to basic research scientists, medical practitioners, and students of reproductive biology. It complements our monograph published in 1979 on *Structure and Function of the Gonadotropins*.

The monograph is organized in ten topic areas relative to the general theme of reproduction and contraception. There are several chapters in each area. Obviously, all aspects of each area could not be covered. An attempt was made to seek interesting basic research ideas and concepts that might in the future be applicable to fertility regulation. The topics are: interactions in gonadotropin regulation; GnRH analogues as contraceptive agents; receptors in cellular localization of hormones; uterine and mammary receptors; germ-cell regulation and secretory proteins; control mechanisms and metabolic regulations; hCG peptides and antisera as antifertility agents; leutinization, oocyte maturation, and early pregnancy; steroids and cell growth; and finally, prostaglandins and cell function. The studies encompass many disciplines and techniques in anatomy, physiology, biochemistry, and endocrinology in animals and humans, both *in vitro* and *in vivo*.

A conference of contributors was held in Maine at the Asticou Inn in Northeast Harbor during the week of September 9–13, 1979. The chapters as written for the monograph were presented for discussion by the participants, who were selected for their knowledge of, and contributions to, this area of scientific investigation.

The format and procedures involved in the development of the Society for Biochemical Endocrinology monographs are different from those involved in publication of symposia proceedings where these provide a record of individual contributions. Participation has always been by invitation and is extended on the recommendation of ISBE Research Council members to researchers deemed to be making outstanding, unique, and creative contributions to the furthering of relevant concepts and data. Chapters are written for the monograph and are then presented informally for the members assembled in order that the information can be assessed and integrated within theoretical unifying themes. Some of the discussion following the presentations is included.

Several months were allowed after the meeting for revision of manuscripts in accordance with direction given by the interaction of the conference participants and for editing. Copyediting was begun by the publisher at the end of February 1980.

I wish to thank the following participants who presided over the various sessions of the conference: Dorothy Villee, E. M. Bogdanove, Asbjørn Aakvaag, Jennie Mather, France Haour, Cornelia Channing, Danielle Szöllösi, Claude A. Villee, Vladimir Pantić, Paul A. Kelley, A. Oriol-Bosch, Janet M. Nolin, M. T. Hochereau-de-Reviers, and Donald L. Wilbur. I am also indebted to the Research Council members of the society for the many suggestions for speakers and topics.

The next in the continuing series of cultural and scientific interactions sponsored by the International Society for Biochemical Endocrinology will be entitled *Hormonally Active Brain Peptides: Structure and Function*. The conference for the participants in the monograph was held in Dubrovnik, Yugoslavia, in late September 1980.

Kenneth W. McKerns

Blue Hill Falls, Maine

Contents

I INTERACTIONS IN GONADOTROPIN REGULATION

UTERINE AND MAMMARY RECEPTORS

VII hCG PEPTIDES AND ANTISERA AS ANTIFERTILITY AGENTS

VIII LUTEINIZATION, OOCYTE MATURATION, AND EARLY PREGNANCY

Reproductive Processes and Contraception

I

Interactions in Gonadotropin Regulation

1

Control of Gonadotropin and Prolactin Secretion in Rhesus Monkeys and Rodents

Scott C. Chappel and Harold G. Spies

1. Introduction

Since the pioneering work of Moore and Price (1932), it has been well documented that the ovary and anterior pituitary gland (AP) exist in a dynamic equilibrium. Simply stated, the AP secretes the gonadotropins, luteinizing hormone (LH) and follicle-stimulating hormone (FSH). These glycoproteins bind to the ovary at specific receptor sites and stimulate the synthesis and secretion of ovarian steroids. These steroids travel from the ovary, through the peripheral circulation, to the AP and the hypothalamus (the area of the brain that regulates the secretions of the AP) to stimulate or inhibit further gonadotropin secretion.

In the female, LH and FSH secreted during the follicular phase of the menstrual or estrous cycle stimulate the synthesis and secretion of estradiol. When serum concentrations of estradiol become maximally elevated during the late follicular phase, a neuroendocrine event occurs that causes the release of pituitary LH and FSH at midcycle. Ovulation occurs shortly thereafter. Low levels of estradiol and elevated levels of progesterone (from the corpus luteum) inhibit further gonadotropin

Scott C. Chappel • Department of Obstetrics and Gynecology, University of Pennsylvania, Philadelphia, Pennsylvania 19104. *Harold G. Spies* • Oregon Regional Primate Research Center, Beaverton, Oregon 97005.

secretion. With the demise of the corpus luteum, levels of both steroids decline and a release of FSH occurs that is not accompanied by an elevation in serum LH concentrations (Resko *et al,* 1974; Schwartz, 1969). This elevation in serum FSH acts on immature follicles to initiate their growth, maturation, and hormone production. This new crop of follicles will be ovulated after the next midcycle gonadotropin release (Greenwald, 1974).

It is clear that in the two mammalian species to be discussed in this chapter, the golden hamster and the rhesus monkey, ovarian estradiol is the principal signal for gonadotropin release or inhibition and therefore ovarian cyclicity (Knobil, 1974; Norman *et al.,* 1973). The hypothalamus and AP of both species specifically bind estradiol (Krieger *et al.,* 1976; Pfaff *et al.,* 1976). Removal of the ovaries causes a cessation of cyclicity and the occurrence of an elevated, pulsatile release of both gonadotropins. Administration of estradiol to ovariectomized hamsters or monkeys will lower elevated gonadotropin levels and, under the proper circumstances, elicit a preovulatorylike discharge of these pituitary hormones (Knobil, 1974; Norman *et al.,* 1973).

It has recently been demonstrated that a nonsteroidal substance, inhibin, exists within the ovaries of many mammalian species and exerts an inhibitory effect on pituitary FSH secretion. This material may play a role in the regulation of FSH secretion at a specific time during the ovarian cycle.

This chapter discusses the work in our laboratory directed toward a thorough understanding of the mechanisms of action, interactions, and physiological roles of estradiol and inhibin, in their ability to suppress the secretion of pituitary LH and FSH in the hamster and rhesus monkey.

In addition, we will present evidence for the modulatory role of estradiol in the control of prolactin secretion. This gonadal steroid appears to influence central aminergic and perhaps peptidergic neurons that regulate the release of pituitary prolactin in the female rhesus monkey.

2. Demonstration of the Presence of "Inhibin" within Proestrous Hamster Ovaries

To demonstrate the presence of an FSH-inhibiting material within hamster ovaries, proestrous hamster ovaries were collected before the onset of the endogenous gonadotropin release in cyclic animals. Each pair of ovaries was homogenized in 1 ml phosphate-buffered saline (PBS). Each ovarian homogenate was incubated with 50 mg charcoal/ml for 30 min at 4°C. This procedure removed all radioimmunoassayable estradiol and progesterone. Charcoal-treated ovarian extract demonstrated FSH-suppressing ability when injected into cyclic female hamsters during proestrus and estrus (Fig. 1). The equivalent of two steroid-free ovaries was injected into each animal at 3-hr intervals beginning at 18:00 hr of proestrus and continuing until 06:00 hr of estrus. Thus, these studies suggest the presence of a nonsteroidal substance capable of inhibiting pituitary FSH secretion (inhibin) in proestrous ovaries. This substance has been observed in proestrous ovaries of other species (Welschen *et al.*, 1977), and secretion of inhibin from the ovary into the ovarian vein occurs during proestrus in rats (DePaulo *et al.*, 1979).

An assumption may be made that the existence of inhibin in proestrous ovaries is due to a greater rate of synthesis and storage than of secretion. With a reliable bioassay system to assess relative amounts of inhibin within a sample, changes in the concentrations of inhibin within ovaries during the estrous cycle would allow us to hypothesize a time at which inhibin exerts an effect at the level of the AP. Presumably, a decrease in ovarian inhibin concentrations at one time during the estrous cycle would suggest that secretion had occurred; this phenomenon would be similar to that observed when gonadotropic-hormone concentrations are compared within pituitary tissue during proestrus (Schwartz and Bartosik, 1962; Chappel *et al.*, 1979). A decrease in pituitary concentrations of gonadotropin occurs at the same time as elevations in peripheral LH and FSH levels. Presumably, depletion of ovarian inhibin would occur at the time of its release and the occurrence of elevated serum levels of this inhibitory substance. To determine the

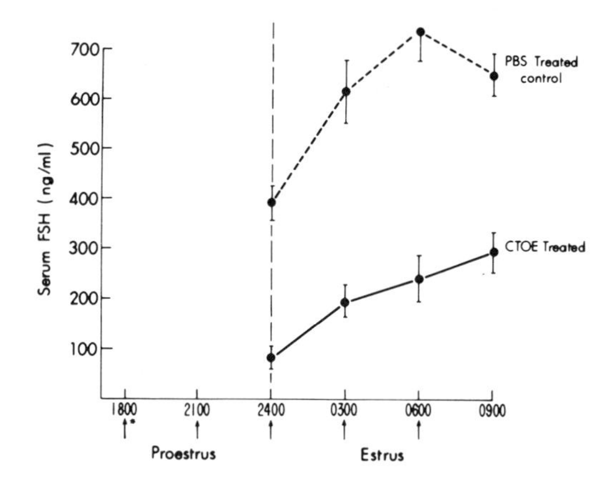

Figure 1. Effect of repeated injections of charcoal-treated ovarian extract (CTOE) or phosphate-buffered saline (PBS) on the estrous release of FSH in hamsters.* Arrow denotes administration of CTOE (2 ovaries/injection).

changes in ovarian inhibin content during the estrous cycle of the golden hamster, groups of hamsters were bilaterally ovariectomized between 09:00 and 10:00 hr on each of the four days of the estrous cycle. Steroid-free ovarian extracts were prepared (2 ovaries/ml) and measured for relative changes in inhibin by the following *in vivo* system. Long-term ovariectomized hamsters were lightly anesthetized with ether and bled (0.6 ml via cardiac puncture). The ovarian homogenates to be analyzed for inhibin content were injected (intraperitoneally) immediately thereafter. Blood collections (0.6 ml) were obtained at 2, 4, and 8 hr after the injection. The criterion for the presence of inhibin was a significant decline in circulating levels of FSH with no change in elevated levels of LH. The injection of 1 ml PBS or 1 ml charcoal-treated adrenal extract (4 adrenals/ml) failed to affect the secretion of FSH or LH in this model. As shown in Figs. 2 and 3, the injection of 2 or 4 steroid-free proestrous ovaries into ovariectomized hamsters elicited a significant decline in serum levels of FSH within 2 hr. A second injection of 4 steroid-free proestrous ovaries (Fig. 3) 3 hr after the first sustained this suppression. Neither injection affected circulating levels of LH in these test animals. The injection of ovarian extracts collected from any one of the remaining three days of the estrous cycle (estrus, diestrus day 1, and diestrus day 2) elicited no significant decline in circulating levels of FSH in the *in vivo* model. As mentioned above, others have demonstrated increased levels of ovarian-vein inhibin concentrations during proestrus that reach very low levels by estrus morning (DePaulo *et al.*, 1979). Thus, these data

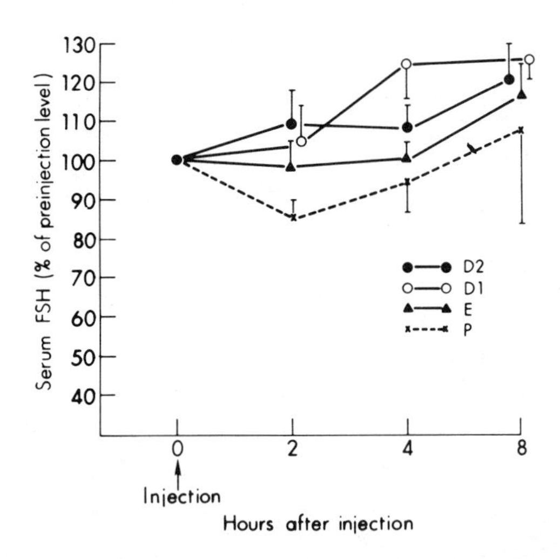

Figure 2. Percentage change in serum FSH concentrations in ovariectomized hamsters after the injection of 2 steriod-free ovaries collected from each day of the estrous cycle.

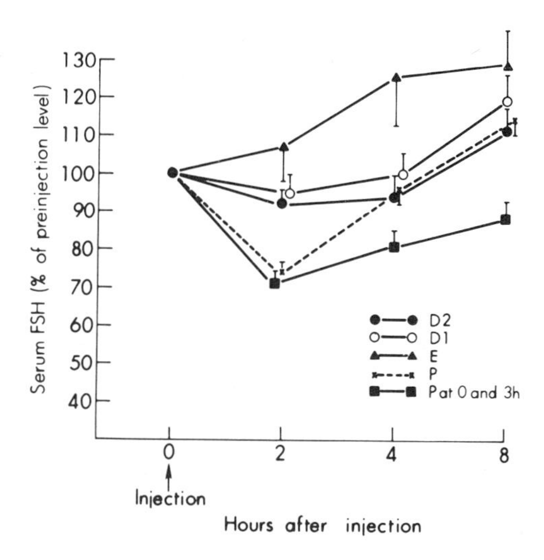

Figure 3. Percentage change in serum FSH concentrations in ovariectomized hamsters following the injection of 4 steroid-free ovaries collected from each day of the estrous cycle. (■) Effect of repeated administration of proestrous ovaries.

suggest that increased secretion and therefore ovarian depletion of inhibin occurs between proestrus and estrus. The importance of proper peripheral levels of inhibin during the periovulatory period, with regard to the estrous FSH release, is shown in the following experiments. If proestrous hamsters are injected at 3-hr intervals, beginning at 18:00 hr, with the equivalent of 2 steroid-free proestrous ovaries/injection, peripheral levels of FSH during estrus are depressed vs. PBS-treated controls (see Fig. 1). These observations have been reported previously (Schwartz and Channing, 1977). Further, bilateral ovariectomy of hamsters at 13:00 hr during proestrus does not prevent the preovulatory rise of serum LH and FSH; however, a premature secondary rise in serum FSH is observed in these animals (Fig. 4). Normally, the estrous FSH rise begins at 24:00 hr of proestrus (Chappel *et al.,* 1978). In these acutely ovariectomized animals, a release was observed by 18:00 hr of that day, 6 hr in advance of the normal estrous rise. These results suggest that ovarian inhibin may be exerting a regulatory effect on the timing of the estrous release of FSH in this species. It has been demonstrated that the neural event responsible for the estrous release of FSH in hamsters has occurred by 18:00 hr of proestrus (Chappel *et al.,* 1979). However, the expression of this neural event (the estrous elevation in serum FSH levels) does not occur until 24:00 hr (Chappel *et al.,* 1978, 1979). Perhaps the secretion of ovarian inhibin before 24:00 hr places the AP in a refractory state. If

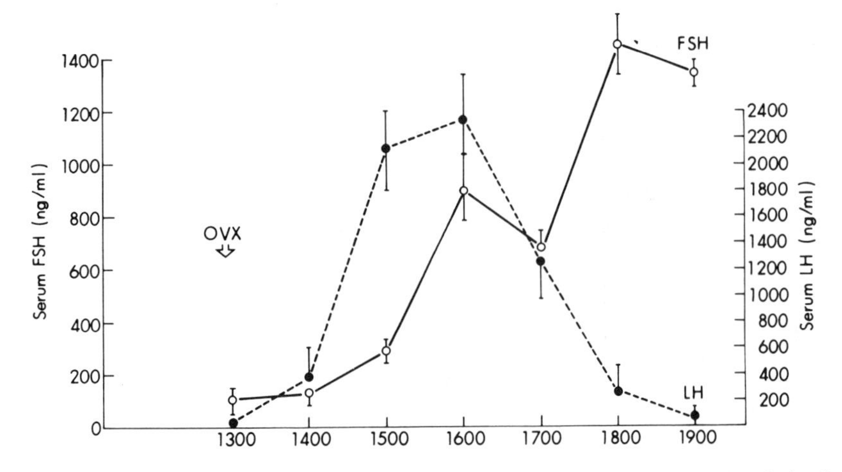

Figure 4. Effect of a bilateral ovariectomy (OVX) on serum LH and FSH levels in the hamster during proestrus.

this is the case, the estrous release of FSH may occur only after the inhibitory effects of inhibin have waned. As demonstrated above, ovariectomy during proestrus (and thus the removal of the source of inhibin) elicits a premature release of FSH normally observed during estrus. On the other hand, continued inhibin exposure during estrus depresses the elevations in serum FSH normally observed at this time (see Fig. 1). It has been reported that inhibin blocks FSH synthesis and secretion at the level of the AP (Chowdhury *et al.*, 1978). Interestingly, after the onset of the estrous FSH rise (perhaps due to the declining effects of inhibin) at 24:00 hr, an increase in pituitary tissue concentrations of FSH is observed, suggesting escape from an inhibitory influence (Chappel *et al.*, 1978).

Collectively, these data suggest that ovarian inhibin may be exerting an effect on pituitary FSH synthesis and secretion during the afternoon of proestrus. Perhaps as peripheral estradiol levels decline (as a result of luteinization) during proestrous afternoon, inhibin maintains an inhibitory influence on pituitary FSH secretion until 24:00 hr. This is the approximate time of ovulation in this species. Since the function of the estrous release of FSH is to initiate the growth and maturation of the next wave of ovarian follicles, it may be of evolutionary significance that mature follicles be ovulated before the next wave is stimulated. Thus, ovarian inhibin may provide this delay between the first and second FSH releases. Exogenous inhibin administration during estrus abolishes the estrous FSH rise in hamsters (Chappel and Selker, 1979). Proestrous elevations in estradiol and gonadotropins are not observed during the next cycle after treatment, and ovulation does not occur.

3. *Does Inhibin Play a Role in the Ovulatory Cycle of the Human?*

Chari *et al* (1979) have recently isolated and purified inhibin from human follicular fluid (hFF). We obtained samples of hFF collected from women at different times during the menstrual cycle, as determined by the measurement of peripheral estradiol and progesterone levels. Using two bioassay methods for the measurement of inhibin, and correlating the relative amounts with the stage of the menstrual cycle from which the fluid was collected, we have observed an interesting phenomenon, as demonstrated in the following experiments.

After six preinfusion peripheral blood collections, ovariectomized female rhesus monkeys with stainless steel cannulae placed in the AP were infused with 25 μl steroid-free hFF. Hourly blood collections were obtained to monitor the concentration changes in peripheral LH and FSH. Fluid collected from women during the follicular phase of the menstrual cycle contained amounts of inhibin adequate to exert an inhibitory effect on circulating levels of FSH, but not LH, in these monkeys (Fig. 5, bottom panel). Fluid collected from ovaries during the luteal phase of the menstrual cycle was ineffective in this regard (Fig. 5, top and middle panels). These observations are substantiated by the results of the following *in vitro* observations. AP glands were obtained from estradiol- and progesterone-pretreated hamsters and enzymatically dispersed. One half million cells were cultured per incubation dish for 4 days. On the fifth day, the cells were washed thrice and resuspended in medium containing 10 μl hFF. At 24 hr later, the cultures were washed thrice and resuspended in medium that contained 10^{-8} M luteinizing-hormone-releasing hormone (LH-RH). LH-RH administration to cultures pretreated with fetal calf serum (FCS) rather than hFF produced a significant increase in LH and FSH secretion (as reflected by the increase in medium concentrations of LH and FSH). Prior incubation with 10 μl hFF added to the culture medium failed to affect the LH-RH-induced LH rise in any of the pituitary culture dishes (Fig. 6). However, 24 hr of incubation with 10 μl of fluid collected from women during the follicular phase (patients SU–CA), but not the luteal phase (patients BO–JA), of the menstrual cycle inhibited the LH-RH-induced FSH rise (Fig. 7).

These data suggest that in women, as in the hamster, the greatest amounts of ovarian inhibin are present within preovulatory ovaries. This

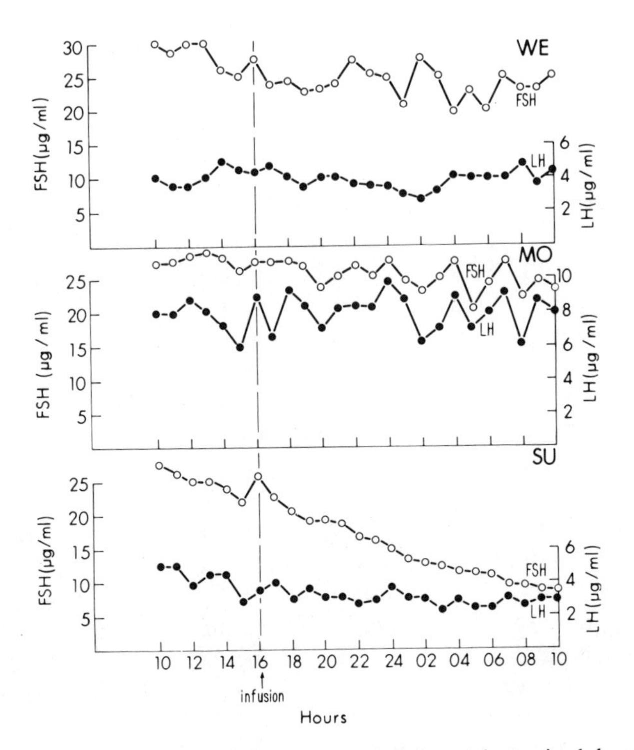

Figure 5. Changes in serum LH and FSH concentrations in ovariectomized rhesus monkeys following the infusion of 25 µl steroid-free hFF into the AP at 16:00 hr. Follicular fluid was collected from patient SU during the follicular phase and from patients WE and MO during the luteal phase of the menstrual cycle.

finding agrees well with the findings of Welschen *et al.* (1977), who have observed that the greatest amounts of inhibin are present in bovine follicular fluid collected from large preovulatory follicles. The exact role of inhibin in the human is at present unknown. However, it is assumed that this material exerts a negative feedback effect on FSH synthesis and secretion during the follicular phase of the menstrual cycle.

4. Determination of Estradiol's Site of Negative Feedback Action in the Rhesus Monkey

In addition to the nonsteroidal substance inhibin, estradiol plays a major role in the feedback regulation of LH and FSH secretion. In the rhesus monkey, estradiol is the major hormone that inhibits or stimulates gonadotropin secretion.

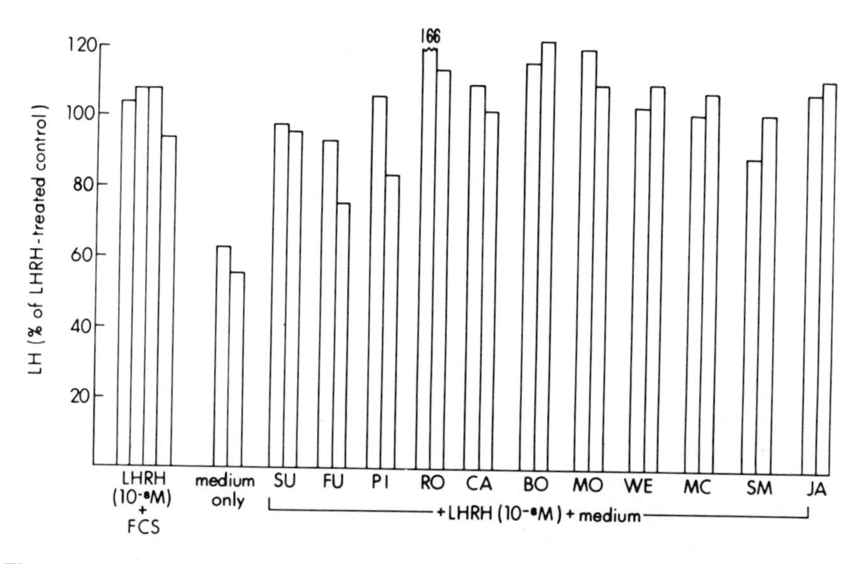

Figure 6. Percentage rise in LH from cultured pituitary cells following 4 hr of incubation with or without LH-RH and with the addition of hFF to the culture medium. Samples of hFF were collected from patients SU, FU, PI, RO, and CA during the follicular phase and from patients BO, MO, WE, MC, SM, and JA during the luteal phase of the menstrual cycle.

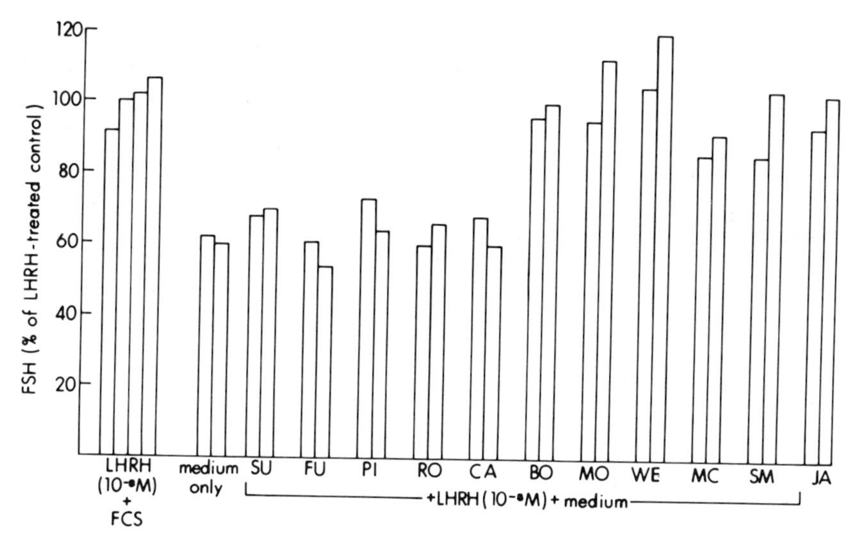

Figure 7. Percentage rise in FSH from cultured pituitary cells following 4 hr of incubation with or without LH-RH and with the addition of hFF to the culture medium. Samples of hFF were collected from patients SU, FU PI, RO, and CA during the follicular phase and from patients BO, MO, WE, MC, SM, and JA during the luteal phase of the menstrual cycle.

It has been demonstrated that treatment of ovariectomized female rhesus monkeys with estradiol reduces the elevated levels of serum LH and FSH as well as inhibits the occurrence of pulsatile releases of these gonadotropins (Knobil, 1974). Estradiol appears to exert its inhibitory effects at the level of the hypothalamus, presumably by inhibiting the secretion of the hypothalamic hormone LH-RH (Ferin *et al.*, 1974). A pituitary site of action has also been postulated, since *in vivo* (Spies and Norman, 1975) and *in vitro* (Tang and Spies, 1975) evidence has shown that estradiol treatment decreases the sensitivity of the AP to LH-RH treatment.

Investigators have attempted to determine whether estradiol exerts an effect at the level of the CNS, by implanting crystals of the steroid hormone into various areas of the hypothalamus. However, these studies have been difficult to interpret due to the possible diffusion of the hormone from its implantation site to the pituitary portal vasculature and therefore directly to the AP [implantation paradox (Bogdanove, 1964)]. We have attempted to reassess this problem in ovariectomized rhesus monkeys with cannulae placed in either the third ventricle of the hypothalamus or the AP. Initially, we attempted to elevate hypothalamic levels of estradiol by an acute infusion of 1 μg or 100, 50, 10, or 5 ng of an estradiol solution in 25 μl sterile saline. However, only when serum estradiol concentrations were elevated as a result of the third-ventricle infusion did peripheral gonadotropin levels decline. These results suggested only a pituitary site of action for estradiol in the inhibition of gonadotropin secretion. However, with this experimental protocol, an assessment of actual estradiol diffusion from the third ventricle to the AP was impossible. Therefore, the following experimental protocol was developed to deliver small quantities of estradiol directly to the third ventricle or the AP and to regulate the duration of exposure. This experimental design was made possible with the use of the Alzet osmotic minipump. The minipump was filled with 170 μl of a (5:95) solution of ethyl alcohol and sterile saline. The first concentration of estradiol tested was 100 pg/μl. The delivery rate of the minimpump was l μl/hr. Following six hourly preinfusion blood collections, the minipump was connected via Teflon tubing to a stainless steel cannula and inserted directly to the level of the third ventricle or the AP (Fig. 8). Hourly blood collections were obtained to monitor changes in peripheral gonadotropin levels. Treatment with this concentration of estradiol exerted no effect on peripheral gonadotropin levels. In the next series of experiments, the dosage of estradiol was increased to 1 ng/μl per hr. Again, no effect was observed. Finally, the dose was increased to 10 ng/μl per hr, and as shown below, an immediate inhibitory effect of estradiol was observed when infused into the third ventricle.

Four rhesus monkeys with indwelling third-ventricle cannulae were

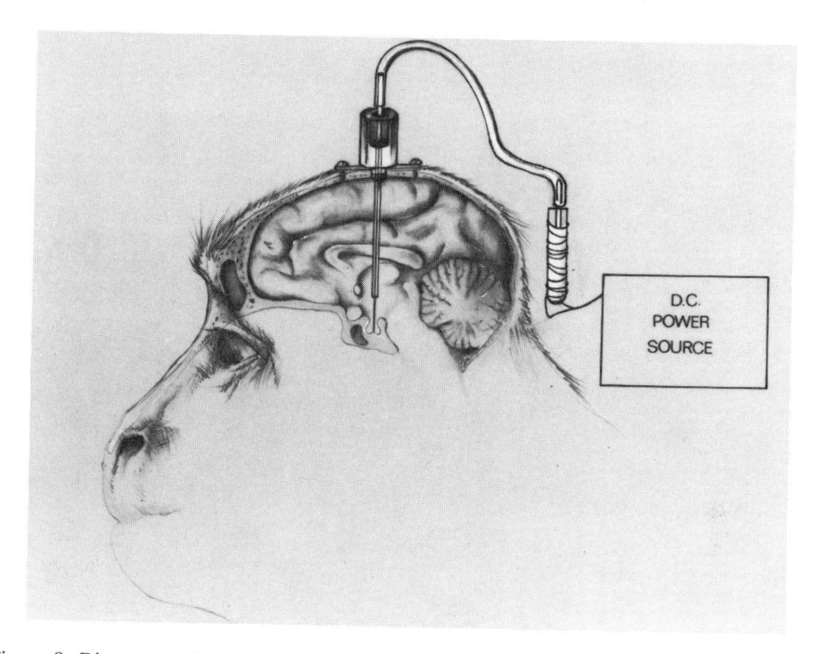

Figure 8. Diagrammatic representation of the minipump–stainless steel cannula apparatus used to deliver microliter volumes of an estradiol solution to either the third ventricle or the AP of rhesus monkeys. An AP cannula is depicted in this figure.

then tested. As before, following six preimplantation blood collections, each animal was implanted with a minipump system that delivered 10 ng estradiol/μl per hr. As shown in Fig. 9, this treatment elicited an immediate inhibitory effect on serum levels of LH and FSH. Gonadotropin levels continued to decline in a linear fashion throughout the blood-collection period (until 08:00 the next day). Interestingly, in these animals, treatment with 25 μg LH-RH [intravenously (i.v.)] at the termination of the experiment caused an unusually large gonadotropin release in the third-ventricle-infused animals (Fig. 10).

The same procedure was repeated in three rhesus monkeys with cannulae placed in the AP. Each animal received the same dose of estradiol as the group described above (10 ng/μl per hr). Unlike the third-ventricle-infused response, the secretory pattern of gonadotropins remained pulsatile at hourly intervals after the infusion began (Fig. 11). However, a significant inhibition of gonadotropin secretion was observed. Treatment of three animals with 25 μg LH-RH (i.v.) at the end of the pituitary estradiol infusion elicited an elevation in both LH and FSH that was significantly ($P < 0.01$) less than that observed in the third-ventricle-infused group (see Fig. 10).

To ensure that the estradiol infused into the third ventricle was not

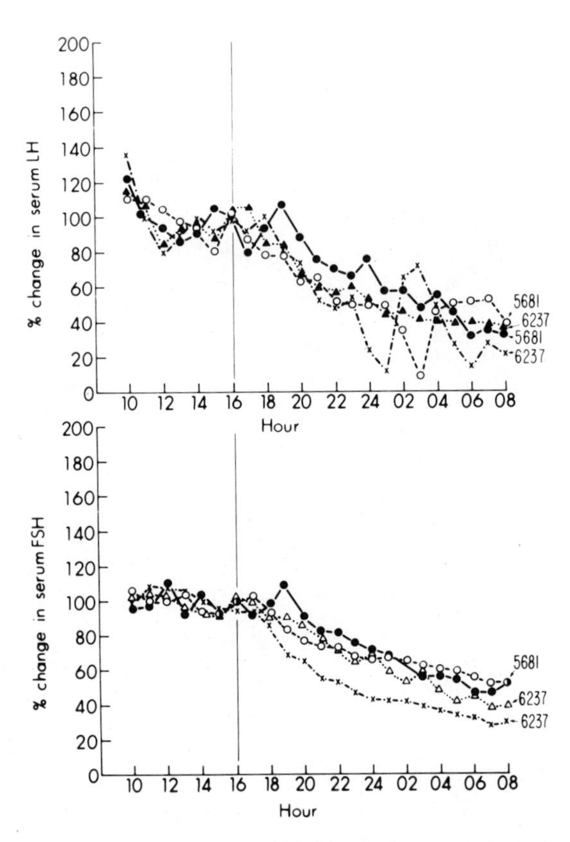

Figure 9. Percentage change in serum LH and FSH concentrations following the continuous infusion of 10 ng/μl per hr of estradiol into the third ventricle of ovariectomized rhesus monkeys beginning at 16:00 hr.

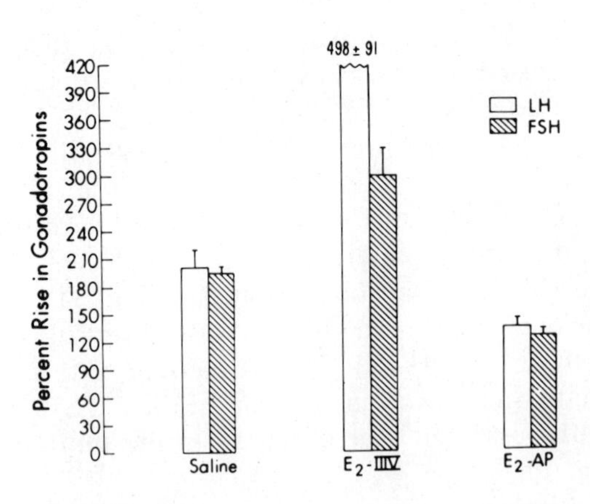

Figure 10. Percentage rise in serum LH and FSH in saline-treated control rhesus monkeys and anterior pituitary (AP)- or third ventricle (IIIV)-estradiol (E_2)-infused ovariectomized rhesus monkeys after the administration of 25 μg LH-RH at 08:00 hr (16 hr after the onset of the infusion).

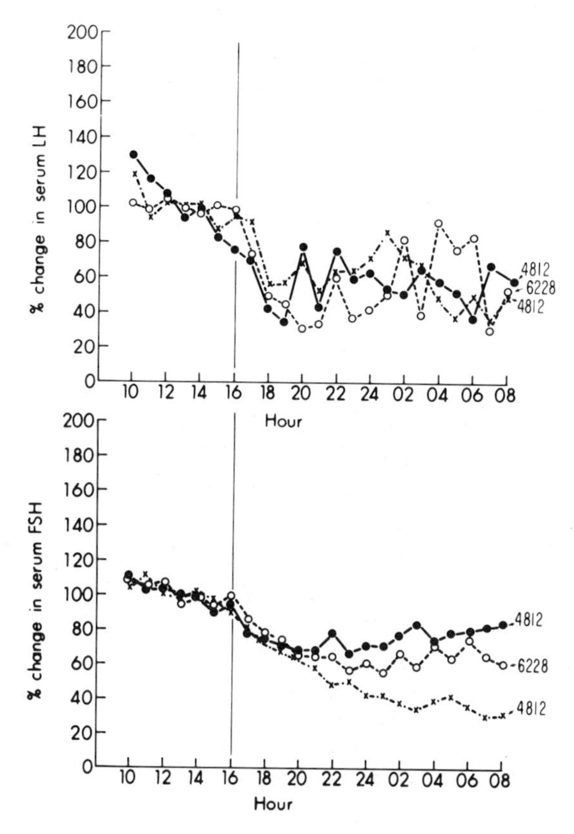

Figure 11. Percentage change in serum LH and FSH concentrations following the continuous infusion of 10 ng/μl per hr of estradiol into the AP of ovariectomized rhesus monkeys beginning at 16:00 hr.

diffusing to the AP, the third-ventricle infusion was repeated in three animals. In these experiments, a portion of the estradiol was replaced with [^{3}H]estradiol such that 1 μCi estradiol would be delivered each hour. At the termination of the experiment, the animals were sacrificed, and neural and pituitary tissue was removed, processed, and quantitated for [^{3}H]estradiol activity. As demonstrated in Table I, very little estradiol diffusion occurred, as reflected by the very low amount of [^{3}H]estradiol recovered from the AP.

Infusion of saline into either the third ventricle or the AP of ovariectomized rhesus monkeys at the same rate (1 μl/hr) elicited no effect on the pulsatile patterns of gonadotropin secretion (Figs. 12 and 13). Further, the gonadotropin responses to LH-RH in these groups were

Table I. Amount of Tritiated Estradiol Present in Various Tissues at the Termination of Third-Ventricle Infusion[a]

	Monkey No.		
Tissue[b]	2495	5737	6324
AP	138	324	762
PP	195	113	318
FC	78	762	245
CSF	18	32	39
Plasma	44	76	488
ME	23,506	11,791	6,520
RH	104,296	11,736	87,256

[a] Amounts are expressed in dpm/100 mg tissue or 1 ml.
[b] (AP) Anterior pituitary gland; (PP) posterior pituitary gland; (FC) frontal cortex; (CSF) cerebrospinal fluid; (ME) median eminence; (RH) rostral hypothalamus.

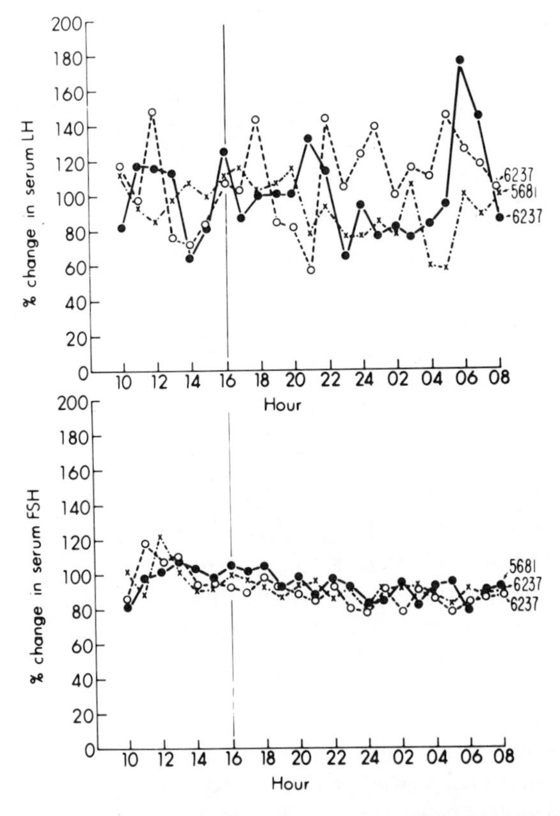

Figure 12. Percentage change in serum LH and FSH concentrations following the continuous infusion of saline (1 μl/hr) into the third ventricle of ovariectomized rhesus monkeys beginning at 16:00 hr.

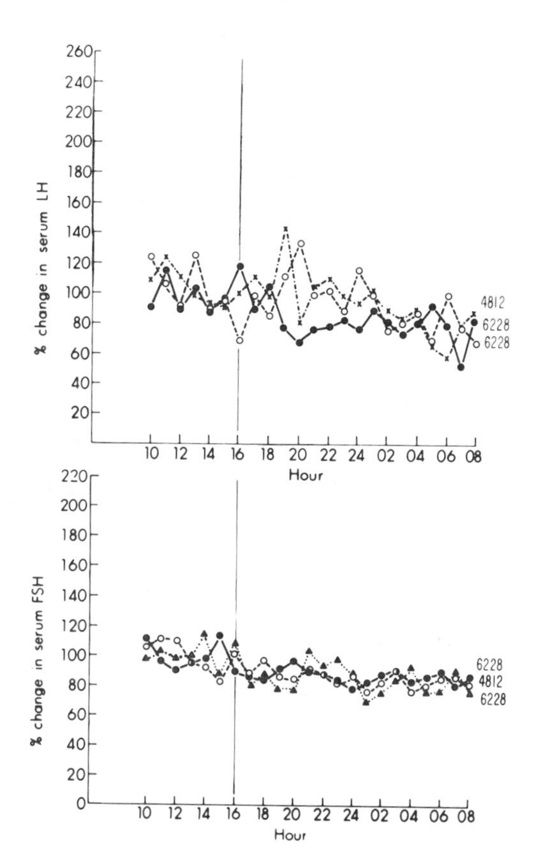

Figure 13. Percentage change in serum LH and FSH concentrations following the continuous infusion of saline (1 μl/hr) into the AP of ovariectomized rhesus monkeys beginning at 16:00 hr.

intermediate between those observed for the two estradiol-treated groups (see Fig. 10).

The results of these studies suggest that estradiol exerts its inhibitory effect at the level of both the medial basal hypothalamus and the AP. After the infusion of estradiol into the third ventricle, the pulsatile pattern of gonadotropin secretion was not observed. In addition, a decline of approximately 50% in circulating levels of both gonadotropins was observed at the end of the infusion period. The AP in this preparation is hyperresponsive to the stimulatory effects of LH-RH compared with saline-treated control animals. We feel that this observation indicates that within the hypothalamus, estradiol inhibits the neurosecretory elements responsible for the pulsatile discharges of LH-RH. Peripheral levels of LH and FSH decline and pituitary LH-RH receptors are vacant. A hyperresponsiveness to exogenous LH-RH ensues. This hyperresponsiveness also suggests that the AP is not affected by the inhibitory actions of estradiol administered to the hypothalamus within the time course of this experiment.

When estradiol is delivered directly to the AP, the LH-RH neurosecretory unit remains operative, as evidenced by the continued gonadotropin pulses. However, the responsiveness of the gland to the stimulatory effects of LH-RH is greatly diminished.

Thus, these results suggest that the negative feedback effects of estradiol influence gonadotropin secretion by two separate mechanisms. When serum concentrations increase, this gonadal steroid binds to preopticohypothalamic tissue and elicits a decreased release of LH-RH. In addition, estradiol appears to directly affect the sensitivity of the AP to LH-RH, perhaps by causing an internalization of the LH-RH receptor on the gonadotrope. The overall effect is a decline in the release of gonadotropins from the AP.

5. *How Does the Hypothalamus Regulate the Secretion of Prolactin and What Is the Role of Estradiol in This Phenomenon in the Female Rhesus Monkey?*

It has been well documented that prolactin secretion from the AP is maintained under an inhibitory influence by the medial basal hypothalamus (MBH). Lesions placed in this neural region or disconnection of the AP from the CNS by transection of the pituitary stalk elicits an increase in the secretion of prolactin as evidenced by an elevation in serum prolactin concentrations (Diefenbach *et al.*, 1976; Norman *et al.*, 1980). The catecholamine dopamine (DA), which is found in high concentrations within the MBH and is secreted into the pituitary portal vasculature (Neill *et al.*, 1978), inhibits the secretion of prolactin. Since DA receptors are contained on the plasma membrane of the AP in rhesus monkeys (Brown *et al.*, 1976), DA is considered to be a hypothalamic prolactin-inhibiting factor (PIF).

The following studies were designed to determine the neural mechanisms involved in the hypothalamic regulation of prolactin release and to test the involvement of estradiol in this phenomenon.

Electrical stimulation of the arcuate nucleus within the MBH of ovariectomized female rhesus monkeys elicits a prompt elevation in serum prolactin concentrations that returns to prestimulation levels within 2 hr (Fig. 14). Electrical activation of this region may cause the release of a hypothalamic prolactin-releasing factor (PRF), inhibit the release of a PIF, or both. In this experimental model, the infusion of DA shortly before and during the electrical stimulation of the MBH completely abolishes the elevation in serum prolactin. These results suggest that maintenance of elevations in systemic DA by exogenous administration prevents the AP's recognition of the electrically stimulated reduction

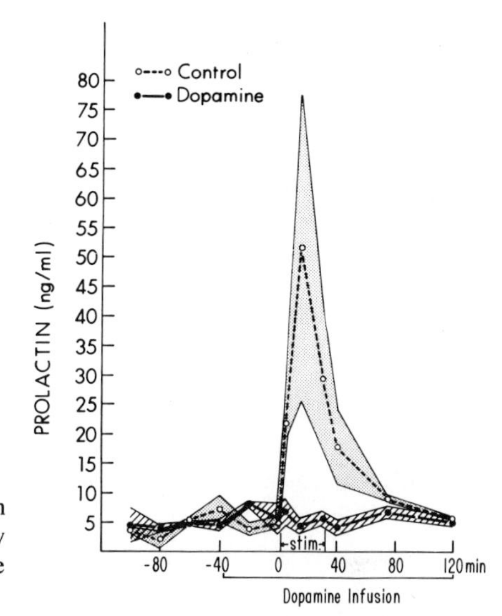

Figure 14. Effects of electrical stimulation of the MBH of the female rhesus monkey on serum levels of prolactin following saline or DA pretreatment.

in hypothalamic DA secretion. Alternately, DA may inhibit the activation of hypothalamic PRF-containing neurons during stimulation or prevent the stimulatory effects of PRF at the level of the AP.

It has recently been demonstrated that the administration of the endogenous opiate endorphin elicits a rise in serum levels of prolactin in rhesus monkeys (Spies *et al.*, 1980). Thus, it is possible that electrical stimulation of the MBH may increase the secretion of endorphins from the hypothalamus to the AP via the pituitary portal vasculature. Thus, endorphins may represent a PRF. If rhesus monkeys are treated with the opiate antagonist Naloxone (3.2 or 6.4 mg/kg body weight) before and during electrical stimulation of the MBH, no inhibition in the rise of serum levels of prolactin or LH is observed (Fig. 15). These results suggest that endogenous opiates do not play a role in the release of prolactin induced by electrical stimulation of the MBH. However, Naloxone pretreatment does abolish the rise in serum prolactin observed after electrical stimulation of the rostral hypothalamus (Spies *et al.*, 1980). Collectively, these data suggest that an opiate-dependent neural network may be involved in the extrahypothalamic regulation of prolactin secretion.

If ovariectomized rhesus monkeys are pretreated with Silastic capsules containing crystalline estradiol (which delivers follicular-phase levels of this hormone) and the MBH or rostral hypothalamus (RH) is electrically stimulated, a significantly greater release of prolactin is observed vs. untreated stimulated controls (Fig. 16). Estradiol may act

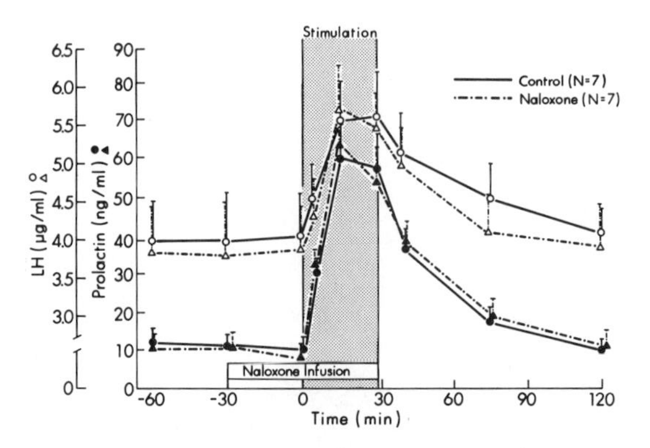

Figure 15. Failure of Naloxone (a narcotic antagonist) pretreatment to affect the rise in serum LH or prolactin concentrations following electrical stimulation of the MBH in female rhesus monkeys.

at the level of the MBH to cause a greater release of PRF or a greater suppression of PIF following electrical stimulation. Alternatively, these results suggest that estradiol acts at the level of the AP to sensitize that gland to the stimulatory effects of PRF. In rats, acute administration of estradiol-17β decreases pituitary portal-blood levels of DA (Cramer *et al.*, 1979). Thus, a decrease in the DA-induced inhibition of prolactin secretion caused by estradiol exposure may explain our results in rhesus monkeys.

Although these studies are not yet conclusive, we postulate the existence of a DA-sensitive, nonopiate factor within the MBH that acts at the level of the AP to stimulate the release of prolactin. Estradiol may increase both the response of PRF-containing neurons to electrical stimulation and the sensitivity of the pituitary mammotrophs to this hypothalamic substance as well as decrease tonic DA release from the MBH.

6. Conclusions

These experiments have demonstrated that the nonsteroidal ovarian substance inhibin exerts an effect on FSH secretion during the follicular phase of the ovarian cycle. Similarly, estradiol influences gonadotropin secretion at this time, exerting its effect at both the hypothalamic and the pituitary level (Fig. 17). Although the exact physiological role of inhibin in the feedback regulation of FSH has yet to be determined, it appears that this substance exerts its effect at the level of the AP as the

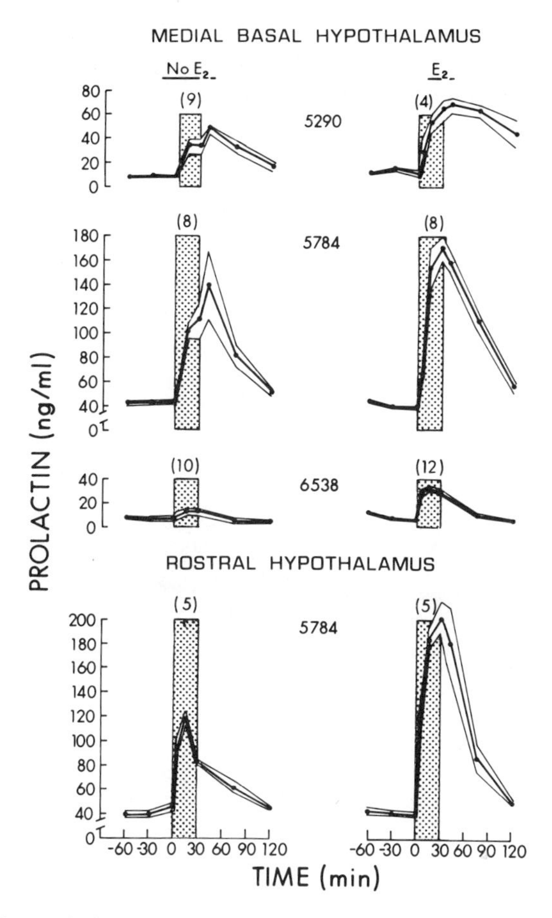

Figure 16. Effects of electrical stimulation of the MBH or RH on serum prolactin concentrations in female rhesus monkeys with or without pretreatment with estradiol (E_2).

follicle grows. Serum levels of FSH have been observed to decline during this time. The source of inhibin production within the ovary is the granulosa cells (Erickson and Hsueh, 1978), and these cells increase in number as the follicle grows. Thus, large preovulatory follicles contain the greatest amounts of inhibin. In addition, rising levels of estradiol exert a negative feedback effect on the secretion of both gonadotropins during this period. The secretion of LH appears to be regulated by an estradiol interaction at the level of the hypothalamus and AP. The regulation of FSH secretion may require the joint efforts of both estradiol and inhibin. The complete suppression of FSH secretion may occur only after the follicle has signaled its maturation, through the elevated levels of both estradiol and inhibin. This dual control of estradiol and inhibin may ensure elevations in serum levels of FSH to guarantee the maturation and development of the preovulatory follicle, an event absolutely essential to reproduction in all species.

In addition to its effects on the regulation of gonadotropin secretion, estradiol appears to influence the secretion of prolactin (Fig. 17). Although the exact mechanism is unknown, we postulate that estradiol's facilitative effect is twofold: it acts to increase the sensitivity of hypothalamic PRF-containing neurons to endogenous or exogenous stimuli, effecting an increased secretion of pituitary prolactin, as well as its proposed inhibitory effects on hypothalamic DA secretion. Both mechanisms would result in increased prolactin secretion.

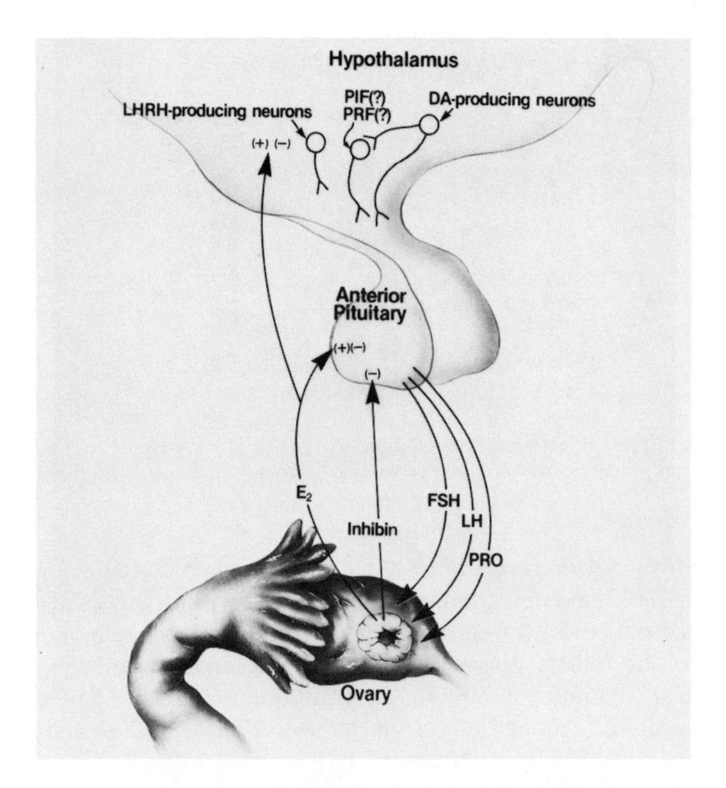

Figure 17. Diagrammatic representation of the sites of action of estradiol (E_2) and inhibin in the regulation of gonadotropin (LH and FSH) and prolactin (PRO) secretion. Also depicted are proposed stimulatory and inhibitory effects of E_2 on luteinizing-hormone-releasing hormone (LH-RH), prolactin-inhibiting (PIF) and -releasing factor (PRF), and dopamine (DA).

Discussion

Bogdanove: This was a nice and concise summary of the main features of the ovulating cycles in the hamster and the monkey. I think all of us would like to get some insight into the role(s) inhibin might play in the human menstrual cycle. I think I understood you to say that in the human you found much higher concentrations of inhibin in follicular fluid collected during the follicular phase than in fluid taken (from follicles or corpora lutea atretica?) during the luteal phase—as assayed *in vitro* using hamster pituitary tissue. Now, can we infer that the levels of inhibin in the blood parallel those in follicular fluid, and if so, would you feel that the physiological role of inhibin in the follicular phase of the menstrual cycle might be to preserve or spare FSH, protecting it from being exhausted during the preovulating surge so that it can be available to initiate folliculogenesis subsequently, after the luteal phase? Also, would you speculate that inhibin would not be needed during the luteal phase, when the luteal steroids become capable of taking over this FSH-sparing role?But is inhibin in ovarian venous blood higher in the follicular phase or the luteal phase?

Chappel: Until we have a sensitive and specific method for the measurement of inhibin in the peripheral serum, it is impossible to know whether blood levels parallel those changes observed within follicular fluid. Studies in rodents have demonstrated that following ovariectomy, elevated levels of LH may be suppressed by the administration of estradiol alone. Serum levels of FSH are also reduced, but not completely. Only when inhibin is administered with estradiol do these levels approach those of the intact animal. With this fact in mind, I would speculate that as the follicle matures, increased estradiol secretion occurs. In addition, the granulosa cells secrete inhibin. Only when estradiol and inhibin levels are elevated above a theoretical inhibitory threshold level is FSH secretion inhibited. Thus, folliculogenesis is ensured.

With the data presented here, it's difficult to speculate as to the role of inhibin during the luteal phase due to the questionable nature of the fluid collected at that time.

Channing: (1) You have done some elegant studies. I would like to caution about drawing rigorous physiological conclusions about changes in follicular fluid and ovarian content of inhibin. Ovarian-vein levels should be measured as well. There may be a situation where the rate of secretion of inhibin into the blood differs in the presence of a constant amount in the follicle.

We are measuring monkey ovarian-vein and follicular-fluid inhibin and should have data next year. Preliminary data in the rhesus monkey indicate that the ovarian vein draining the preovulatory-follicle-containing ovary contains inhibin.

(2) Additional evidence for inhibin being secreted by the primate ovary has been obtained by us. Removal of the preovulatory follicle can cause a rise in serum FSH (not LH) 2–4 days later. We have shown that fluid obtained from atretic human, mare, and monkey follicles contains less inhibin compared to that from viable follicles (unpublished observations).

(3) Human granulosa cells when cultured secrete inhibin in culture, whereas thecal tissue will not (Batta and Channing, 1980). Spent granulosa-cell media did not inhibit LH secretion. Spent media were assayed using basal rat pituitary cultures.

References

Batta, S.K., and Channing, C.P., 1980, Demonstración de una inhibina en el liquido follicular humano y su posible papel en la disfunción ovarica, in: *Advances en Obstetricia y Ginecologigia*, Vol. 6 (J. Gonzales-Merlo, J. Iglesiar-Guiu, and I. Burzaco, eds.), pp. 125–134, Salvat, Barcelona.

Bogdanove, E.M., 1964, The role of the brain in the regulation of pituitary gonadotropin secretion, *Vitam. Horm. (N.Y.)* **22:**205.

Brown, G.M., Seeman, P., and Lee, T., 1976, Dopamine/neuroleptic receptors in basal hypothalamus and pituitary, *Endocrinology* **99:**1407.

Chappel, S.C., and Selker, F., 1979, Relation between the secretion of FSH during the periovulatory period and ovulation during the next cycle, *Biol. Reprod.* **21:**347.

Chappel, S.C., Norman, R.L., and Spies, H.G., 1978, Effects of estradiol on serum and pituitary gonadotropin concentrations during selective elevations of follicle stimulating hormone, *Biol. Reprod.* **19:**159.

Chappel, S.C., Norman, R.L., and Spies, H.G., 1979, Evidence for a specific neural event that controls the estrous release of follicle-stimulating hormone (FSH) in golden hamsters, *Endocrinology* **104:**169.

Chari, S., Hopkinson, C.R.N., Duame, D., and Sturm, G., 1979, Purification of "inhibin" from human ovarian follicular fluid, *Acta Endocrinol.* **90:**157.

Chowdhury, M., Steinberger, A., and Steinberger, E., 1978, Inhibition of *de novo* synthesis of FSH by Sertoli cell factor (SCF), *Endocrinology* **103:**644.

Cramer, O.M., Parker, C.R., and Porter, J.C., 1979, Estrogen inhibition of dopamine release into hypophysial portal blood, *Endocrinology* **104:**419.

DePaulo, L.V., Shander, D., Wise, P.M., Channing, C.P., and Barraclough, C.A., 1979, Evidence for ovarian secretion of "inhibin," *Fed. Proc. Fed. Am. Soc. Exp. Biol.* (abstract 3985).

Diefenbach, W.P., Carmel, P.W., Frantz, A.G., and Ferin, M., 1976, Suppression of prolactin secretion by L-dopa in the stalk-sectioned rhesus monkey, *J. Clin. Endocrinol. Metab.* **43:**638.

Erickson, G.F., and Hsueh, J.W., 1978, Secretion of "inhibin" by rat granulosa cells *in vitro, Endocrinology* **103:**1960.

Ferin, M., Carmel, P.W., Zimmerman, E.A., Warren, M., Perez, L., and Vande Wiele, R.L., 1974, Localization of intrahypothalamic estrogen responsive sites influencing LH secretion in the female rhesus monkey, *Endocrinology* **95:**1059.

Greenwald, G.S., 1974, Role of follicle-stimulating hormone and luteinizing hormone in follicular development and ovulation, in: *Handbook of Physiology,* Section 7, Vol. IV (R.O. Greep and E.B. Astwood, eds.), pp. 293–323, Williams and Wilkins, Baltimore.

Knobil, E., 1974, On the control of gonadotropin secretion in the rhesus monkey, *Recent Prog. Horm. Res.* **30:**1.

Krieger, M.S., Morrell, J.I., and Pfaff, D.W., 1976, Autoradiographic localization of estradiol-containing cells in the female hamster brain, *Neuroendocrinology* **22:**193.

Moore, C.R., and Price, D., 1932, Gonad hormone functions and the reciprocal influence between gonads and hypophysis with its bearing on sex hormone antagonism, *Am. J. Anat.* **50:**13.

Neill, J.D., Frawley, L.S., Plotsky, P., and Tindall, G.T., 1978, Regulation of prolactin secretion by dopamine in the rhesus monkey, Society for the Study of Reproduction Annual Meeting, Abstract 113.

Norman, R.L., Blake, C.A., and Sawyer, C.H., 1973, Estrogen-dependent twenty-four-hour periodicity in pituitary LH release in the female hamster, *Endocrinology* **93:**965.

Norman, R.L., Quadri, S.K., and Spies, H.G., 1980, Differential sensitivity of prolactin release to dopamine and thyrotropin-releasing hormone in intact and pituitary stalk sectioned rhesus monkeys, *J. Endocrinol.* **84:**479.

Pfaff, D.W., Gerlach, J.L., McEwen, B.S., Ferin, M., Carmel, P., and Zimmerman, E.A., 1976, Autoradiographic localization of hormone-concentrating cells in the brain of the female rhesus monkey, *J. Comp. Neurol.* **170:**279.

Resko, J.A., Norman, R.L., Niswender, G.D., and Spies, H.G., 1974, The relationship between progestins and gonadotropins during the late luteal phase of the menstrual cycle in rhesus monkeys, *Endocrinology* **94:**128.

Schwartz, N.B., 1969, A model for the regulation of ovulation in the rat, *Recent Prog. Horm. Res.* **25:**1.

Schwartz, N.B., and Bartosik, D., 1962, Changes in pituitary LH content during the rat estrous cycle, *Endocrinology* **71:**756.

Schwartz, N.B., and Channing, C.P., 1977, Evidence for ovarian "inhibin": Suppression of the secondary rise in serum follicle stimulating hormone levels in proestrous rats by injection of porcine follicular fluid, *Proc. Natl. Acad. Sci. U.S.A.* **74:**5721.

Spies, H.G., and Norman, R.L., 1975, Interaction of estradiol and LHRH on LH release in rhesus females: Evidence for a neural site of action, *Endocrinology* **97:**685.

Spies, H.G., Quadri, S.K., Chappel, S.C., and Norman, R.L., 1980, Dopaminergic and opioid compounds: Effects on prolactin and LH release after electrical stimulation of the hypothalamus in ovariectomized rhesus monkeys, *Neuroendocrinology* **30:**249.

Tang, L.K., and Spies, H.G., 1975, Effects of gonadal steroids on the basal and LRF-induced gonadotropin secretion by cultures of rat pituicytes, *Endocrinology* **96:**349.

Welschen, R., Hermans, W.P., Dullart, J., and deJong, F.H., 1977, Effects of an inhibin-like factor present in bovine and porcine follicular fluid on gonadotropin levels in ovariectomized rats, *J. Reprod. Fertil.* **50:**129.

2

Regulation of LH-RH Secretion by Gonadal Steroids and Catecholamines

Satya P. Kalra, Pushpa S. Kalra, and J. W. Simpkins

1. Introduction

The presence of a hypothalamic factor that stimulates the release of pituitary luteinizing hormone (LH) was demonstrated two decades ago (McCann *et al.,* 1960). A great deal of effort was directed toward the purification and chemical identification of the hypothalamic LH-releasing hormone (LH-RH). Matsuo *et al.* (1971) characterized the chemical structure of the naturally occurring LH-RH and showed it to be biologically and chemically identical with the synthetic decapeptide. Antibodies against the synthetic decapeptide were produced for the radioimmunoassay of LH-RH in tissue and body fluids (Nett *et al.,* 1973). Recently, the distribution of the LH-RH neuronal network in the brain has been investigated by a variety of techniques including bioassay and radioimmunoassay and by immunocytochemical means (Naik, 1975; Elde and Hökfelt, 1978; Hoffman *et al.,* 1978; Sétáló *et al.,* 1978).

Abundant immunoassayable LH-RH is present in a restricted medial zone located basally in the hypothalamus of the rat. The highest concentrations of immunoreactive LH-RH are found in the nerve terminals that

Satya P. Kalra, Pushpa S. Kalra, and *J. W. Simpkins* • Department of Obstetrics and Gynecology, University of Florida College of Medicine, Gainesville, Florida 32610.

abut the hypophyseal portal vessels in the external layer of the median eminence (ME). The distribution of LH-RH terminals in the ME is uneven; they cover the rostral segment densely, but caudally these fibers are restricted to the lateral lip of the ME. Additionally, relatively small amounts of immunoreactive LH-RH are detectable in the nucleus arcuatus, periventricular nucleus, area retrochiasmatica, and suprachiasmatic nucleus in the basal hypothalamus of the rat. Of considerable interest is the dense network of LH-RH fibers (probably the projections of LH-RH neurons in the preoptic area) that lie close to the capillary network at the anterior boundary of the third ventricle in the organum vasculosum of the lamina terminalis (Sétáló *et al.*, 1978; Elde and Hökfelt, 1978).

The visualization of the LH-RH neuronal network in the brain has revealed a startling topographical relationship with sex-steroid-concentrating and aminergic neuronal systems, long known to influence pituitary gonadatropin release (S.P. Kalra, 1977; P.S. Kalra and S.P. Kalra, 1979; Sar and Stumpf, 1975; Stumpf and Sar, 1977; Ungerstedt, 1971; Swanson and Hartman, 1975). In the past, numerous attempts were made to study the effects of sex steroids on hypothalamic LH-RH levels (McCann, 1974). Since hypothalamic LH-RH levels were measured in these experiments by insensitive and generally unreliable bioassay procedures, no general consensus was reached on the role of sex steroids in modifying LH-RH secretion. The basic question of whether gonadal steroids modulate the secretion of the neurohormone into the portal circulation and whether this effect alone controls pituitary gonadotropin release is not yet firmly established.

Another major issue is whether the feedback-modulating action of gonadal steroids is exerted directly at the level of the peptidergic neurons or on some other neuronal system or systems that, in turn, regulate the secretion of LH-RH-producing cells. Although an array of evidence indicates that catecholaminergic systems are capable of influencing pituitary LH release (S.P. Kalra, 1977), it is unknown whether these system constitute an essential neural link in the steroid feedback regulation of LH-RH secretion. In this chapter, we will review our work on the effects of gonadal steroids on LH-RH secretions in male and female rats and analyze the participation of catecholamines in the central action of gonadal steroids.

2. *Effects of Gonadal Steroids on LH-RH Secretion in Male Rats*

With the availability of radioimmunoassay procedures to measure LH-RH, it was feasible to demonstrate temporal changes in the hypothalamic and preoptic area (POA) LH-RH levels during a 24-hr period (P.S.

Kalra and S.P. Kalra, 1977b). That androgens may influence hypothalamic LH-RH secretion was suggested by the observation that serum and hypothalamic LH-RH levels fluctuate in a circadian fashion as do serum androgen levels in male rats (P.S. Kalra and S.P. Kalra, 1977b). Further, it was found that following castration, hypothalamic LH-RH levels decreased and physiological replacement with testosterone (T) restored the levels to those found in intact rats (P.S. Kalra and S.P. Kalra, 1978, 1980). Evidently, androgens not only participate in the daily LH-RH secretion pattern but also exert a profound modulatory influence on the production of the neurohormone. Consequently, three important aspects of androgen action in the rat are considered in the following section: identification of the active metabolite of T, the intracranial site of action, and the mechanisms by which androgens modulate catecholamine and LH-RH activities in the hypothalamus.

2.1. *Central Metabolism of Testosterone*

Abundant evidence exists to show that one or more central target sites reduce T to dihydrotestosterone (DHT) or aromatize it to estrogens (Massa *et al.,* 1972; Naftolin *et al.,* 1975). In castrated male rats, administrations of estrogens or androgens (T and DHT) suppressed pituitary gonadotropin release. Our studies showed that DHT and estradiol (E_2) treatment promoted storage of LH-RH in the medial basal hypothalamus (MBH), and this response was indistinguishable from that observed after T treatment (P.S. Kalra and S.P. Kalra, 1980). In view of this equipotency of DHT and E_2, uncertainty existed concerning the nature of the product formed in the hypothalamus after T administration that actively modulated hypothalamic LH-RH secretion. Flutamide, an androgen antagonist (Neri, 1977), completely blocked the T- and DHT-induced LH-RH and serum gonadotropin responses. Seemingly, central aromatization is not an essential step, since an antiestrogen, Nafoxidine (Wade and Blanstein, 1978), was ineffective in counteracting the effects of T or DHT (P.S. Kalra and S.P. Kalra, 1980). Therefore, it is more likely that DHT may be the major metabolite of T involved in the regulation of hypothalamic LH-RH activity. In this respect, of significance are the findings that indicate that aromatization of T to E_2 may be a requisite step for the elicitation of male sex behavior (McEwen, 1976). These and our studies clearly point out that physiological and behavioral effects of T may be manifested through disparate cellular and chemical events in the hypothalamus. In support of this view are the observations of regional specificity of androgen action. The effects of DHT and T on LH-RH neurons were specifically exerted at the level of the MBH (P.S. Kalra and S.P. Kalra, 1979, 1980), where the highest concentrations of DHT receptors (Kato, 1975) and 5α-reductase activity were found (Massa *et al.*, 1972). More rostrally in the POA where the "sex center" is located, there are relatively large concentrations of aromatizing enzymes, and, importantly, discrete E_2 or T, but not DHT, implants in this region induced a typical male sexual response (McEwen *et al.*, 1979).

2.2. Where Do Gonadal Steroids Act in the Brain?

During the past decade, extensive investigations from several laboratories have shown that the gonadal steroids—androgens, E_2, and progesterone (P)—have target cells within and outside the hypothalamus (Stumpf and Sar, 1977; McEwen *et al.*, 1979). Specific cytoplasmic and nuclear receptors for each of the steroids have been documented in these neural regions (McEwen *et al.*, 1979). Within the hypothalamus E_2, T-, DHT-, and P-concentrating neurons have been found in the arcuate nucleus and dorsally and laterally to it in the ventromedial and dorsal medial nuclei. Outside the hypothalamus, E_2 target cells are present in the limbic system—medial cortical nucleus of the amygdala, stria terminalis, medial preoptic area (MPOA), and anterior hypothalamic area (AHA)—and are sparsely distributed in selected regions of the brainstem (Sar and Stumpf, 1975; Stumpf and Sar, 1977).

In view of the widespread distribution of sex-steroid-concentrating neurons in the brain and the fact that systemic administration of T, DHT, and E_2 influenced the MBH LH-RH levels, experiments were designed to identify the specific central steroid target sites that were capable of influencing the secretory activity of LH-RH neurons. Small amounts of T, DHT, or E_2 implanted in the MBH uniformly promoted storage of LH-RH locally; more rostrally in the POA, the LH-RH contents were unchanged (Fig. 1) (P.S. Kalra and S.P. Kalra, 1980). There was a clear dissociation in the effectiveness of steroids when implanted in the POA–AHA—a region also known to contain sex-steroid-concentrating neurons (Fig. 1). Androgens were ineffective in influencing the stores of LH-RH either locally or caudally in the MBH. However, although E_2 implants also failed to influence the POA LH-RH levels, they markedly raised the immunoreactive LH-RH levels in the basal hypothalamus (P.S. Kalra and S.P. Kalra, 1980). These differential responses indicated that the neuronal network, which may influence peptidergic neurons, was relatively more extensive for E_2 than for androgens (Fig. 1). If this were the case, it was of interest to identify estrogen-concentrating regions located outside the hypothalamus that may exercise control on the LH-RH neurons. Intracranial implantation of E_2 in the basal corticomedial nucleus of the amygdala and the medial septum in the diencephalon markedly raised the MBH LH-RH levels (Kalra and Kalra, unpublished). The responses were indistinguishable from those seen following E_2

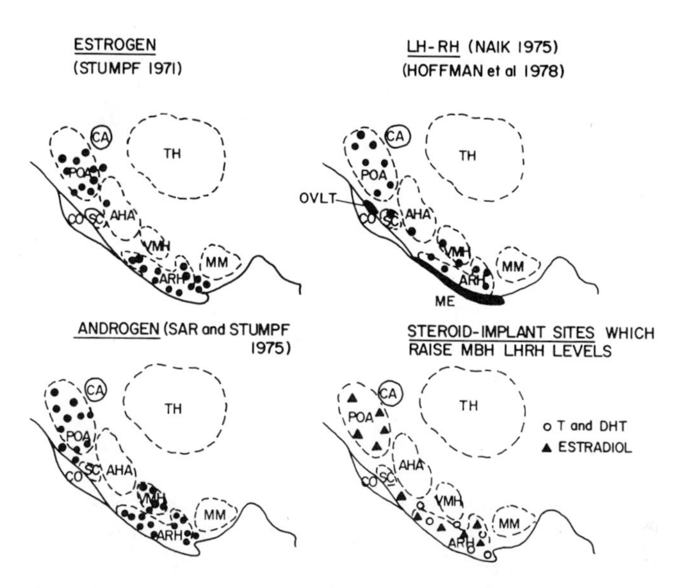

Figure 1. Diagrammatic comparison of sites containing estrogen- and androgen-concentrating neurons with the localization of LH-RH (●) in the hypothalamus and preoptic area (POA). Although LH-RH-containing perikarya and their projections are congruent with estrogen- and androgen-concentrating neurons, we were consistently unable to alter the POA LH-RH content [including LH-RH activity in the organum vasculosum of the lamina terminalis (OVLT)] by any steroid treatment. Estradiol implants (▲) in the POA as well as the medial basal hypothalamus (MBH) were equally effective in raising the MBH LH-RH stores, whereas androgens (○) were effective only when placed in the MBH. These results suggest the existence of a discrete population of LH-RH neurons that are steroid-sensitive and terminate in the median eminence (ME). (AHA) Anterior hypothalamic area; (ARH) arcuate region; (CA) anterior commissure; (CO) optic chiasm; (MM) mammillary body; (SC) suprachiasmatic area; (TH) thalamus; (VMH) ventromedial nucleus. Reprinted with permission from P.S. Kalra and S.P. Kalra, 1979, *J. Steroid Biochem.* **11**:981.

implantation in the POA–AHA or the MBH. Three possibilities can be raised to explain these intriguing observations of a limited field of androgen action compared to the rather extensive field of E_2 action. First, E_2 from the extrahypothalamic sites may find its way into the systemic circulation in concentrations sufficient to influence LH-RH neurons in the MBH. Second, diffusion of steroids in optimal concentrations to hypothalamus may occur within the brain. Third, the extrahypothalamic sites may have neural inputs to LH-RH neurons in the hypothalamus.

We failed to detect any significant diffusion of E_2 from the implants into the systemic circulation. However, administration of anti-E_2 serum in an attempt to neutralize E_2 before it reached the MBH produced interesting results. E_2 implants in the amygdala and septum were no longer effective in raising the MBH LH-RH levels and suppressing gonadotropin release. These results substantiated the suspicion that optimal levels of E_2, though undetectable by direct measurement, reached the hypothalamus by the systemic route. In contrast, the POA E_2 implants were still effective in E_2-antiserum-treated rats. The proximity of the POA–AHA to the MBH and the fact that the concentrating neurons of both androgens and estrogens have widespread distribution patterns and yet androgens were effective only when implanted in the MBH alone strongly suggested that only those sex-steroid target neurons that reside in the MBH are capable of influencing the secretion of LH-RH neurons (P.S. Kalra and S.P. Kalra, 1980).

2.3. *Identification of Sex-Steroid Target Neurons*

A more difficult and the least-studied aspect of steroid-hormone action in the brain is the identification of the target neurons or neuronal system or systems that regulate hypothalamic LH-RH secretion. Current concepts of hormone action indicate that these cells possess intracellular receptor sites that translocate steroids into the cell nucleus and initiate genetic information leading to modulation of neurohormone secretion. On the basis of the evidence available, there are two putative candidates that could serve as the target neurons for gonadal steroids: First, it is possible that sex-steroid cells are LH-RH-producing neurons. The action of these steroids in the nuclear compartment of the soma could directly alter the synthesis of the precursor hormone in the LH-RH neurons (Fig. 2).

There is another possible way that steroids could influence LH-RH levels in the peptidergic cells. In the absence of steroidal feedback, as in castrated rats, LH-RH secretion is apparently unimpaired. Treatment with sex steroids, E_2, T, or DHT, suppresses LH release (and possibly LH-RH release) and increases the MBH LH-RH concentrations (P.S. Kalra and S.P. Kalra, 1980). Seemingly, the action of these steroids in castrated rats is directed at modulating the intrinsic secretion pattern of the LH-RH neurons. This may well be manifested by steroids triggering the formation of proteins, which either reduce degradation of the neurohormone or drastically reduce LH-RH release from the nerve terminals in the ME or exert both effects. This mode of steroid action provides a suitable explanation for our results, but it does not necessarily exclude the possibility of steroid action on the nonpeptidergic neurons, which relate with LH-RH cells, and may consequently trigger similar responses in the LH-RH neurons (Fig. 2). There is a wealth of information implicating catecholamines as neurotransmitters mediating the central steroid feedback action (P.S. Kalra and McCann, 1973; S.P. Kalra, 1977).

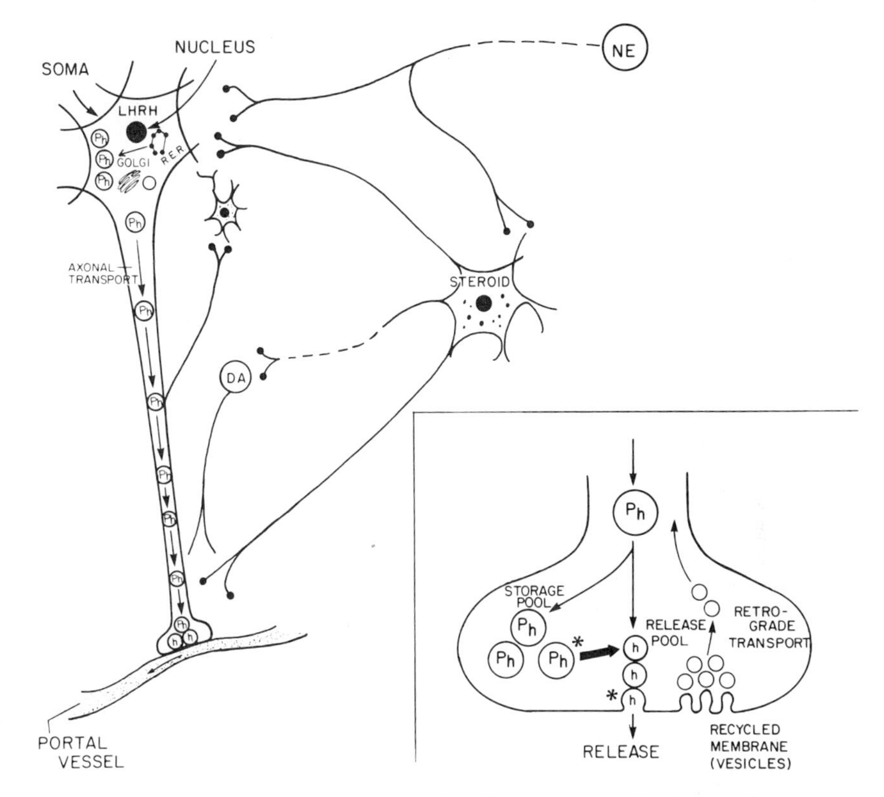

Figure 2. Diagrammatic representation of an LH-RH neuron terminating on a portal vessel and its relationship with sex-steroid-concentrating, dopamine (DA), and norepinephrine (NE) neurons. Our studies suggest that estradiol- and androgen-concentrating cells, may induce *de novo* synthesis of precursor hormone (P_h) at the level of LH-RH perikarya and may also activate the transformation of the stored precursor hormone into releasable immunoreative hormone pool (h) in the nerve terminal (shown by the heavy arrow in the inset). On the other hand, progesterone (P) may stimulate the processes involved in the transformation of P_h to h and, perhaps, augment LH-RH release into the portal vessels (P influence on these steps is indicated by asterisks in the nerve terminal). (R.E.R.) Rough endoplasmic reticulum. For further details, see the text.

As previously described, there is remarkable similarity in the distribution of catecholamine projections, LH-RH, and steroid neurons in the hypothalamus and POA. We found that T treatment, which inhibited gonadotropin release and raised hypothalamic LH-RH levels, decreased norepinephrine (NE) turnover in the MBH (Simpkins *et al.*, 1980b). Although evidence of this nature tended to support NE mediation of steroid action, a more direct experiment utilizing a neurotoxin, 6-hydroxydopamine (6-OH-DA) failed to support this view (Kalra *et al.*, unpublished). When implanted in the POA or MBH, 6-OH-DA reduced local NE levels by 80–90%. Despite this drastic depletion, T effects on the MBH LH-RH levels and serum gonadotropins were not impaired. Thus, it appears that NE may not be an important neural substrate to carry a T feedback message to LH-RH neurons. With respect to involvement of other hypothalamic monoamines, it is also evident that dopamine and serotonin may not mediate the regulatory influence of gonadal steroids over gonadotropin secretions (Nemeroff *et al.*, 1977; Hery *et al.*, 1978).

In view of these considerations, it seems quite likely that steroid target neurons may be a distinct group of cells that possess anatomical relationship with peptidergic and a host of other neurons in the brain. As depicted in Fig. 2, at the level of the hypothalamus they may regulate the secretions of LH-RH neurons and perhaps concurrently influence the activity of monaminergic neurons with which they presumably communicate.

2.4. Characterization of Gonadal-Steroid-Induced Effects on Hypothalamic LH-RH Levels

Although Fig. 2 depicts LH-RH and the steroid-concentrating cells as two distinct populations of neurons in the hypothalamus, it remains to be seen whether the steroid target neurons regulate the release or synthesis, or both, of the neurohormone in the LH-RH neurons. In this respect, our recent studies on the time course of steroid action to effectively raise the MBH LH-RH levels provide new insight.

We found that while T, DHT, or E_2 treatment suppressed LH release within hours, the effects on LH-RH levels were evident after 3 days (Fig. 3). Similar dissociation in serum LH and hypothalamic LH-RH responses was evident following withdrawal of T treatment (P.S. Kalra and S.P. Kalra, 1980). Two possibilities to explain this time lag between LH and the MBH LH-RH responses to T treatment can be advanced.

It is likely that in response to major shifts in T titers, LH-RH release into the portal circulation is altered, leading to accumulation of LH-RH in the MBH. However, numerous attempts to corroborate this view have not been successful (Eskay *et al.*, 1977). Another possibility that can be entertained to fit our data is that there may be more than one pool of LH-RH in the peptidergic neurons (see Fig. 2). One such pool, comprising a small portion of the total LH-RH in the neuron, is labile and is readily released into the portal vessel in response to acute changes in steroid titer. This pool of LH-RH in the nerve terminals (Fig. 2, h) may be derived from a more stable storage precursor hormone pool (Fig. 2,

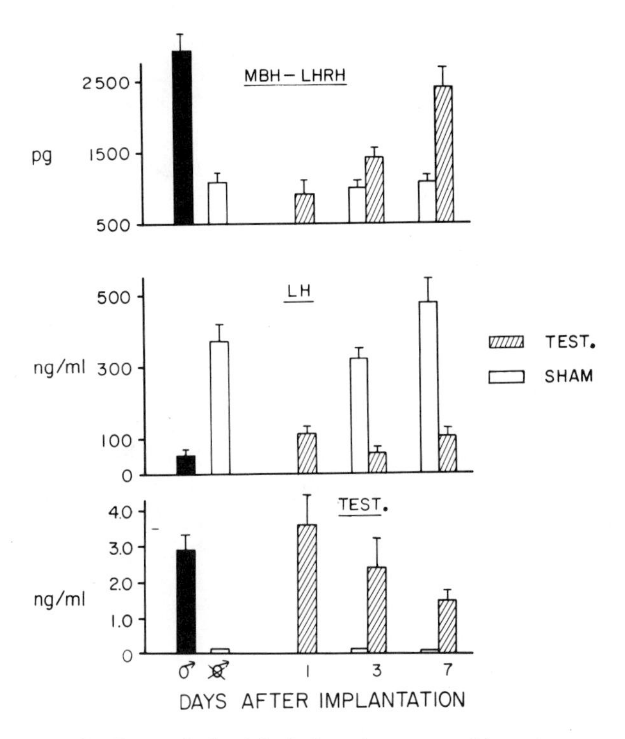

Figure 3. Temporal effects of physiological replacement with testosterone (TEST.) on serum LH and the medial basal hypothalamic (MBH) LH-RH levels in castrated rats. Serum LH levels decreased on day 1, but a significant increase in the MBH LH-RH levels was observed on day 7 after TEST. replacement (♂) Intact; (⚥)castrated. Reprinted with permission from P.S. Kalra and S.P. Kalra, 1980, *Endocrinology* **106:**390.

P_h) that has a relatively sluggish response to androgen and estrogen action. The androgen- and E_2-induced LH-RH accumulation that we have observed probably represents this population of LH-RH in the peptidergic neurons. Further, if the mechanism of biosynthesis of LH-RH in the peptidergic neurons is similar to the elaborate ribosomal-dependent synthesis in cell bodies postulated for other CNS peptides (Gainer *et al.,* 1977), it is conceivable that steroids stimulate *de novo* synthesis of LH-RH (Fig. 2, P_h) at the level of perikarya. On the basis of our studies, it appears that this stimulation of synthesis together with packaging and transport of the neurohormone for storage in the nerve terminals in the ME may require more than 3 days. It is noted, however, that in this respect P action is different from that observed for T (see Section 3.3).

3. Effects of Gonadal Steroids on LH-RH Secretion in Female Rats

The pattern of gonadotropin secretion in the female rat is distinctly different from that in the male rat. A sustained basal LH secretion pattern through most of the estrous cycle is interrupted briefly by a massive discharge of pituitary LH in the afternoon of proestrus (S.P. Kalra *et al.*, 1971; S.P. Kalra and P.S. Kalra, 1974). An intriguing and challenging problem for reproductive endocrinologists for the past 30 years has been the understanding of those integrating events—hormonal, neural, and environmental—that evoke pituitary LH release with remarkable regularity and precision in the afternoon of proestrus (Everett, 1977; P.S. Kalra and S.P. Kalra, 1977a). In this section, we present a brief review of our efforts to understand the modulating influence of steroids and neurotransmitters on the neural events that signal pituitary LH release. Special emphasis is placed on the temporal changes that occur in the activity of LH-RH neurons preceding and following this neural signal.

3.1. LH-RH Secretion in Proestrus

It is well known that LH release in proestrus occurs some time in the afternoon, and elevated levels are detectable until late evening (Everett, 1977; S.P. Kalra *et al.*, 1971). There is a lack of consensus on the precise time and duration of LH-RH discharge responsible for the LH surge. Some laboratories have detected a brief pulse of LH-RH in the peripheral and portal circulation prior to LH release (P.S. Kalra and S.P. Kalra, 1977a; Eskay *et al.*, 1977). Others have reported a slightly prolonged pattern of LH-RH secretion in the portal vessels in the afternoon of proestrus (Sarkar *et al.*, 1976). Recently, we have delineated these neural events in proestrus in considerable detail (P.S. Kalra and S.P. Kalra, 1977a). There were two notable sequences of events that characteristically occurred in proestrus—a decrease in the medial basal hypothalamus (MBH) LH-RH levels in the morning was followed by augmented levels in the afternoon. These temporal changes in the MBH LH-RH contents invariably occurred prior to LH discharge. A time lag between the rise in the MBH LH-RH and serum LH levels was also observed following electrochemical stimulation of the medial preoptic area (MPOA)–anterior hypothalamic area (AHA) in pentobarbital-blocked proestrous rats (S.P. Kalra *et al.*, 1973: Barr and Barraclough, 1978). Evidently, an enhancement of the bioactive (S.P. Kalra *et al.*, 1973), and immunoreactive LH-RH (P.S. Kalra and S.P. Kalra, 1977a; Barr and Barraclough, 1978) in the MBH is intimately linked with LH discharge in proestrus. Also, it is possible that these increments in LH-RH levels result from either *de novo* synthesis or activation of LH-RH from the preexisting precursor proteins in the hypothalamus (see Fig. 2), or from both processes.

Another intriguing outcome of these investigations was that despite the presence of LH-RH in the preoptic area (POA), the presumed site of origin of neurogenic stimuli responsible for LH release (S.P. Kalra, 1976a; S.P. Kalra *et al.,* 1977; Brownstein, 1977), temporally related changes in LH-RH levels in proestrus were observed in the MBH only. Since the POA LH-RH contents during the estrous cycle displayed an independent pattern (S.P. Kalra, 1976b), it would appear that the POA may participate in relaying the "trigger" signal to the MBH, but whether the LH-RH produced in the POA has any role in the discharge of LH is still uncertain.

3.2. Steroid Modulation of Hypothalamic LH-RH Levels Associated with Pituitary Gonadotropin Release

It has been known for a long time that the central action of ovarian steroids is responsible for the preovulatory discharge of gonadotropins (Everett 1977). This concept received unequivocal support when it was demonstrated that appropriate treatment of ovariectomized rats with estrogen alone or progesterone (P) in rats primed with low doses of estradiol benzoate (EBP) elicited a proestrous-type LH release (Caligaris *et al.,* 1968, 1971a, P.S. Kalra *et al.,* 1972).

We have recently found that the LH-RH secretion pattern is quite different following these two treatments (S.P. Kalra and P.S. Kalra, 1979). In fact, the pattern of LH-RH secretion in the EBP rats closely resembled the one observed in proestrus (P.S. Kalra and S.P. Kalra, 1977a). As in proestrus, in EBP rats there was a significant rise in the MBH LH-RH levels either in association with or preceding the serum LH and FSH rise (Fig. 4) (S.P. Kalra *et al.,* 1978; S.P. Kalra and P.S. Kalra, 1979). These remarkable similarities in the MBH LH-RH fluctuations on the day of proestrus and on the day of P injection (Fig. 4) prompted us to investigate further the site(s) at which the steroids act to modify LH-RH secretion and the nature of neurotransmitter participation in this stimulatory feedback action.

3.3. Site(s) of Progesterone Action

Due to the wide distribution of LH-RH pathways in the brain, it was of interest to identify the specific nuclei in the POA–AHA and the MBH that participate in the P-induced LH-RH response (Simpkins *et al.,*

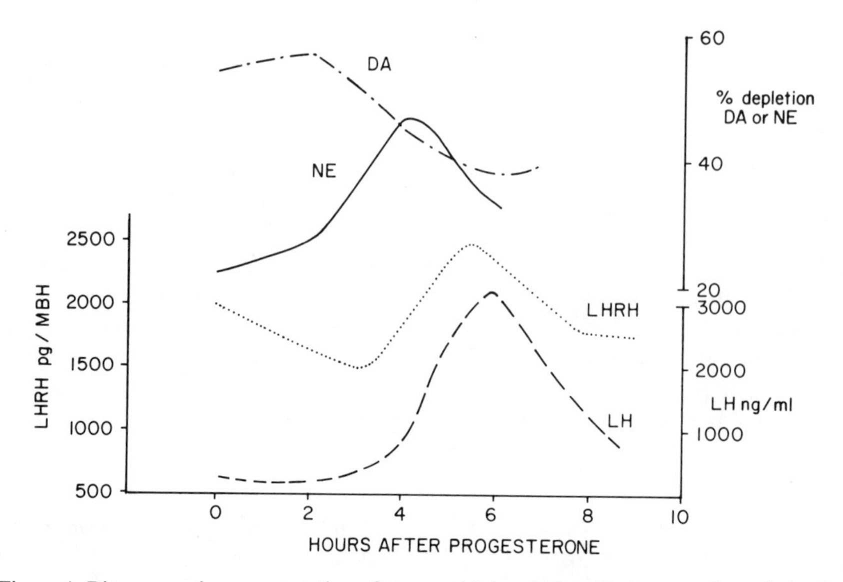

Figure 4. Diagrammatic representation of temporal interrelationship between hypothalamic dopamine (DA), norepinephrine (NE), and LH-RH activities and serum LH levels after progesterone injection to estrogen-primed rats. For details, see the text.

1980a). Analyses of six microdissected nuclei and regions of the brain of the EBP rats at 2-hr intervals indicated that only in the median eminence (ME) and nucleus arcuatus (NA) did LH-RH concentrations increase prior to LH discharge. Thereafter, LH-RH concentrations in these regions fell in coincidence with the peak serum LH levels. The other regions in the MBH, e.g., suprachiasmatic nucleus (SCN) and area retrochiasmatica, showed minimal decrease in LH-RH levels during the period of LH surge. Surprisingly, there were no alterations in the MPOA or AHA LH-RH concentrations prior to or through the LH surge period. These results raised a question as to the site(s) of the origin of LH-RH that accumulated in the NA–ME region in response to P treatment (see Fig. 2). Previous evidence from deafferentation studies suggested that the majority of LH-RH found in the MBH is probably produced in the perikarya located rostrally in the POA (S.P. Kalra, 1976a). In that case, it would seem that a significant amount of the POA LH-RH is transported to the NA–ME region within 4 hr after P treatment. However, our failure to observe any temporally related changes in LH-RH concentrations in the regions through which LH-RH fibers presumably course is not consonant with this view. Though there is no supportive evidence, it is likely that precursor LH-RH (Fig. 2, P_h) in the POA perikarya and fibers in passage is not immunoreactive but acquires this property in the NA–ME. Nevertheless, our consistent failure to alter POA LH-RH levels by steroid treatment (S.P. Kalra, 1976a; P.S. Kalra and S.P. Kalra, 1980), and conflicting findings on the localizations of LH-RH cell bodies in the brain (Sétáló *et al.,* 1978; Hoffman *et al.,* 1978), also are not compatible with the POA-origin-of-LH-RH hypothesis. In this context, the recent finding of heterogeneity of LH-RH employed to localize perikarya has raised considerable doubt concerning the earlier-reported visualization of cell bodies in the POA (Clayton and Hoffman, 1979).

An alternate explanation is that P stimulates the elaboration of LH-RH within the MBH either in the perikarya located in the NA or in their terminations in the ME (Fig. 2), or in both sites. The remarkable rapidity of LH-RH response strongly favors the NA–ME as a site of P action. Under these conditions, we still do not know whether P treatment stimulated the production of precursor hormone or activated the elaboration of LH-RH from it (Fig. 2). The rapid nature of LH-RH response is consistent with the latter possibility of P action.

3.4. *Neurotransmitters and LH-RH Secretion*

Evidence has accumulated to show that catecholaminergic systems in the brain may play a modulatory role in the induction of LH surge in proestrus or that induced experimentally by gonadal-steroid treatment of ovariectomized rats (P.S. Kalra *et al.,* 1972; S.P. Kalra and McCann, 1974). A careful analysis of the sequence of events in the EBP rats showed that hypothalamic dopamine (DA) and norepinephrine (NE) activities changed in opposite directions (see Fig. 4) (Simpkins *et al.,* 1979). This was illustrated by the fact that a significant rise in NE turnover was accompanied by decreased turnover in DA neurons. Coincident with these changes in neurotransmitter levels, the LH-RH levels also rose.

Recent evidence suggests that the tuberoinfundibular dopaminergic system may exercise an inhibitory influence on LH-RH secretion (S.P. Kalra, 1977; Drouva and Gallo, 1976; Sawyer, 1979; Füxe and Hökfelt, 1969). Our demonstration of concurrent alterations, though in opposite directions, in the DA and NE activities in the EBP rats concurs strongly with the view that these two neurotransmitters may function simultaneously to produce the characteristic increments in the MBH LH-RH and serum LH levels. This hypothesis was further supported by the results from two series of experiments. When central NE activity was suppressed by the administration of NE-synthesis inhibitors, the normally occurring increments in the MBH LH-RH and serum LH levels were blocked in EBP rats (S.P. Kalra *et al.*, 1978). On the other hand, when acceleration of DA-neuronal activity alone was achieved by pentobarbital administration, it resulted in suppression of afternoon LH release, but the increments in the MBH LH-RH levels occurred normally (S.P. Kalra *et al.*, unpublished). From these findings, it is apparent that excitation of NE neurons by P led to activation of LH-RH synthesis, whereas disinhibition of the DA influence on LH-RH neurons permitted the discharge of their product into capillaries in the ME (see Fig. 2).

Where do catecholamines and LH-RH neurons interact? On the basis of the anatomical distribution patterns of DA and LH-RH neurons, it is predictable that DA neurons may communicate with the peptidergic neurons in the basal hypothalamus either in the NA or in the ME. Studies showing that DA altered LH-RH secretion from the basal hypothalamus or ME tissue in the *in vitro* system tend to agree with this view (Negro-Villar *et al.*, 1979). With respect to the site(s) of interaction of NE and peptidergic neurons, the picture is not so clear. NE projections from the cell bodies in the brainstem are found in the hypothalamus and the POA (Ungerstedt, 1971; Swanson and Hartman, 1975). A few investigations indicated that NE receptors responsible for LH-RH release may be resident in the MBH (Krieg and Sawyer, 1976). On the contrary, evidence showed that the POA–SCN area may be a critical site of NE and LH-RH communication (Simpkins and S.P. Kalra, 1979). LH release following preoptic but not ME stimulation was suppressed in rats pretreated with NE-synthesis inhibitors (S.P. Kalra and McCann, 1973). More recently, implantation of 6-hydroxydopamine in the POA–SCN area, to acutely destroy NE terminals locally, blocked the P-induced increase in the MBH LH-RH contents and LH release (Simpkins and Kalra, 1979). Similar disruptions in the MBH NE system were ineffective in blocking P effects on LH-RH secretion. Collectively, these studies support the likelihood that P acts by transiently modifying the DA and NE activities at different central sites, leading to higher rates of LH-RH elaboration from the precursor pool in the NA–ME region (Fig. 2).

Although our knowledge of the precise nature of the neural links between NE and LH-RH neurons is still scanty, recent investigations have revealed that in the steroid feedback action, the communication between the two systems is dispensable. Destruction of NE inputs, either surgically (Clifton and Sawyer, 1979) or by pharmacological means (Nicholson *et al.*, 1978; Simpkins and Kalra, unpublished), to the hypothalamus or POA only transiently blocked the steroid feedback effects. Apparently, by reorganization of the neuronal circuitry in the hypothalamus, these rats are able to reinstate the positive feedback effects of steroids on gonadotropin secretion.

4. Summary

Continuing goals in the study of the central control of pituitary function have been to unravel the anatomical circuitry underlying gonadal feedback action. Gonadal steroids play a complex role in regulating the synthesis, release, and degradation of LH-RH in the peptidergic neurons. Our investigations over the past few years have delineated some of these in detail, although our understanding of them is still at the descriptive level. Nevertheless, morphological and physiological evidence supports the hypothesis that sex-steroid-concentrating neurons in the medial basal hypothalamus are a discrete group(s) of cells, the influence of which on LH-RH neurons is apparently local (see Fig. 2). Further, these steroids trigger a series of intraneuronal events leading to an increase in LH-RH in the nucleus arcuatus–median eminence region. In this respect, a mode of action of estrogen and androgen disparate from that of progesterone is afforded by the observation of a markedly different time course of effects. It is postulated that estrogens and androgens induce new synthesis of precursor hormone (Fig. 2, P_h) at the level of the LH-RH perikarya and possibly activate the transformation of precursor hormone into immunoreactive LH-RH in the terminals (Fig. 2, h). In contrast, progesterone appeared initially to accelerate this latter activational process and later facilitate LH-RH discharge from the nerve terminals into the hypophyseal portal circulation. Admittedly, many gaps have to be filled before the mechanism of steroid action on LH-RH secretion is fully understood.

The concept of central monoamine involvement in mediation of steroid feedback action is subject to considerable speculation. Evidence presented herein does not implicate monoamines as mediators of testosterone action on LH-RH neurons in male rats. There is compelling evidence, however, to support the hypothesis that the catecholaminergic link, though not indispensable, does play a modulatory role in cyclical or steroid-induced gonadotropin release in female rats.

ACKNOWLEDGMENTS. This work was supported by research grants from the Population Council and the National Institutes of Health (HD08634 and HD11362). We thank Penny Strother for excellent help in typing the manuscript.

DISCUSSION

PETRUTSZ: Immunocytochemical data seem to indicate that LRF-containing (synthesizing) cell bodies cannot be found in the medial basal hypothalamus (arcuate nucleus) region. Is it possible to reconcile your data with this situation, or do you expect that with the improvement of immunocytochemical methods and/or antisera, such cell bodies will eventually be found?

S.P. KALRA: Uncertainty continues to exist on localization of LH-RH-containing cell bodies not only in the arcuate nucleus but also elsewhere in the brain. Two observations suggest that some LH-RH-producing perikarya may be present in the basal hypothalamus. Partial or complete deafferentation reduces the hypothalamic LH-RH contents at the most by 80%. The remaining 20% or so of the LH-RH is probably derived from within the basal hypothalamus (S.P. Kalra, 1976a). The other observation is that following progesterone treatment in estrogen-primed rats, we have observed a rapid increase in LH-RH concentration in the arcuate–median eminence region only. The LH-RH response is rapid enough to justify a local origin. My view, therefore, is that with further improvement in immunocytochemical methods, it would be possible to observe LH-RH-containing perikarya in the arcuate nucleus and preoptic area as well.

REFERENCES

Barr, G.D., and Barraclough, C.A., 1978, Temporal changes in medial basal hypothalamic LH-RH correlated with plasma LH during the rat estrous cycle and following electrochemical stimulation of the medial preoptic area in pentobarbital-treated proestrous rats, *Brain Res.* **148:**413.

Brownstein, M., 1977, Biologically active peptides in the mammalian central nervous system, in: *Peptides in Neurobiology* (H. Gainer, ed.), p. 145, Plenum Press, New York.

Caligaris, L., Astrada, J., and Taleisnik, S., 1968, Stimulating and inhibiting effects of progesterone on release of luteinizing hormone, *Acta Endocrinol* **59:**177.

Caligaris, L., Astrada, J., and Taleisnik, S., 1971, Release of luteinizing hormone induced by estrogen injection into ovariectomized rats, *Endocrinology* **88:**810.

Clayton, C.T., and Hoffman, G.E., 1979, Immunocytochemical evidence for anti-LH-RH and anti-ACTH activity in the "F" antiserum, *Am. J. Anat.* **155:**139.

Clifton, D.K., and Sawyer, C.H., 1979, LH release and ovulation in the rat following depletion of hypothalamic norepinephrine: Chronic vs. acute effects, *Neuroendocrinology* **28:**442.

Drouva, S.V., and Gallo, R.V., 1976, Catecholamine involvement in episodic luteinizing hormone release in adult ovariectomized rat, *Endocrinology* **99:**651.

Elde, R., and Hökfelt, T., 1978, Distribution of hypothalamic hormones and other peptides in the brain, in: *Frontiers in Neuroendocrinology* (W.F. Ganong and L. Martini, eds.), p. 1, Raven Press, New York.

Eskay, R.L., Mical, R.S., and Porter, J.C., 1977, Relationship between luteinizing hormone releasing hormone concentration in hypophysial portal blood and luteinizing hormone release in intact, castrated, and electrochemically-stimulated rats, *Endocrinology* **100:**263.

Everett, J.W., 1977, The timing of ovulation, *J. Endocrinol.* **75:**1P.

Füxe, K., and Hökfelt, T., 1969, Catecholamines in the hypothalamus and the pituitary gland, in: *Frontiers in Neuroendocrinology* (W.F. Ganong and L. Martini, eds.), p. 47, Oxford University Press, New York.

Gainer, H., Pengloh, G., and Sarne, Y., 1977, Biosynthesis of neuronal peptides, in: *Peptides in Neurobiology* (H. Gainer, ed.), p. 183, Plenum Press, New York.

Hery, M., Laplante, E., and Kordon, C., 1978, Participation of serotonin in the phasic release of luteinizing hormone. II. Effects of lesions of serotonin containing pathways in the central nervous system, *Endocrinology* **102:**1019.

Hoffman, G.E., Melnyk, V., Hayes, T., Bennett-Clark, C., and Fowler, E., 1978, Immunocytology of LHRH neurons, in: *Neural Hormones and Reproduction* (D.E. Scott, G.P. Kozlowski, and L. Weindle, eds.), p. 67, S. Karger, Basel.

Kalra, P.S. and Kalra, S.P., 1977a, Temporal changes in the hypothalamic and serum luteinizing-hormone releasing hormone (LH-RH) levels and the circulating ovarian steroids during the rat estrous cycle, *Acta Endocrinol. (Copenhagen)* **85:**449.

Kalra, P.S., and Kalra, S.P., 1977b, Circadian periodicities of serum androgens, progesterone, gonadotropins and luteinizing hormone-releasing hormone in male rats: The effects of hypothalamic deafferentation, castration and adrenelectomy, *Endocrinology* **101:**1821.

Kalra, P.S., and Kalra, S.P., 1978, Effects of intrahypothalamic testosterone implants on LHRH levels in the preoptic area and the medial basal hypothalamus, *Life Sci.* **23:**65.

Kalra, P.S., and Kalra, S.P., 1979, Central regulation of gonadal steroid rhythms in rats, *J. Steroid Biochem.* **11:**981.

Kalra, P.S., and Kalra, S.P., 1980, Modulation of hypothalamic luteinizing hormone-releasing hormone levels by intracranial and subcutaneous implants of gonadal steroids in castrate rats: Effects of androgen and estrogen antagonists, *Endocrinology* **106:**390.

Kalra, P.S., and McCann, S.M., 1973, Involvement of catecholamines in feedback mechanisms, *Prog. Brain Res.* **39:**196.

Kalra, P.S., Kalra, S.P., Krulich, L., Fawcett, C.P., and McCann, S.M., 1972, Involvement of norepinephrine in transmission of the stimulatory influence of progesterone on gonadotropin release, *Endocrinology* **90:**1168.

Kalra, S.P., 1976a, Tissue levels of luteinizing hormone-releasing hormone in the preoptic area and hypothalamus, and serum concentration of gonadotropin following anterior hypothalamic deafferentation and estrogen treatment in the female rat, *Endocrinology* **99:**101.

Kalra, S.P., 1976b, Circadian rhythm in luteinizing hormone releasing hormone (LH-RH) content of preoptic area during the rat estrous cycle, *Brain Res.* **104:**254.

Kalra, S.P., 1977, Neuroamines in gonadotropin secretion, in: *Endocrinology,* Vol. 1, Proceedings of the International Congress of Endocrinology, *Excerpta Medica. Int. Congr. Series* **402** (V.H.T. James, ed.), p. 152, Excerpta Medica, Amsterdam.

Kalra, S.P., and Kalra, P.S., 1974, Temporal interrelationships among circulating levels of estradiol, progesterone and LH during the rat estrus cycle: Effects of progesterone, *Endocrinology* **95:**1711.

Kalra, S.P., and Kalra, P.S., 1979, Dynamic changes in hypothalamic LH-RH levels associated with the ovarian steroid-induced gonadotropin surge, *Acta Endocrinol. (Copenhagen)* **92:**1.

Kalra, S.P., and McCann, S.M., 1973, Effects of drugs modifying catecholamine synthesis on LH release induced by preoptic stimulation in the rat, *Endocrinology* **93:**356.

Kalra, S.P., and McCann, S.M., 1974, Effects of drugs modifying catecholamine synthesis on plasma LH and ovulation in the rat, *Neuroendocrinology* **15:**79.

Kalra, S.P., Ajika, K., Krulich, L., Fawcett, C.P., Quijada, M., and McCann, S.M., 1971, Effects of hypothalamic and preoptic electrochemical stimulation on gonadotropin and prolactin release on proestrous rats, *Endocrinology* **88:**1150.

Kalra, S.P., Krulich, L., and McCann, S.M., 1973, Changes in gonadotropin-releasing factor content in the rat hypothalamus following electrochemical stimulation of the anterior hypothalamic area and during the estrous cycle, *Neuroendocrinology* **12:**321.

Kalra, S.P., Kalra, P.S., and Mitchell, E.O., 1977, Differential response of luteinizing hormone releasing hormone in the basal hypothalamus and the preoptic area following anterior hypothalamic deafferentation, *Endocrinology* **100:**201.

Kalra, S.P., Kalra, P.S., Chen, C.L., and Clemens, J.A., 1978, Effects of norepinephrine synthesis inhibitors and a dopamine agonist on hypothalamic LHRH, serum gonadotropins and prolactin levels in gonadal steroid treated rats, *Acta Endocrinol. (Copenhagen)* **89:**1.

Kato, J., 1975, The role of hypothalamic and hypophyseal 5α-dihydrotestosterone, estradiol and progesterone receptors in the mechanisms of feedback action, *J. Steroid Biochem.* **6:**970.

Krieg, R.J., and Sawyer, C.H., 1976, Effects of intraventricular catecholamines on luteinizing-hormone release in ovariectomized steroid primed rats, *Endocrinology* **99:**411.

Massa, R., Stupnicka, Z., Kniewald, Z., and Martini, L., 1972, The transformation of testosterone into dihydrotestosterone by the brain and the anterior pituitary, *J. Steroid Biochem.* **3:**385.

Matsuo, H., Baba, Y., Nair, R.M.G., Arimura, A., and Schally, A.V., 1971, Structure of the porcine LH and FSH-releasing hormone, *Biochem. Biophys. Res. Commun.* **43:**1374.

McCann, S.M., 1974, Regulation of the secretion of follicle-stimulating (FSH) and luteinizing hormone (LH), in: *Handbook of Physiology,* Vol. IV (E. Knobil and W.H. Sawyer, eds.), p. 489, Williams and Wilkins, Baltimore.

McCann, S.M., Taleisnik, S., and Friedman, H.M., 1960, LH-releasing activity in hypothalamic extracts, *Proc. Soc. Exp. Biol. Med.* **104:**432.

McEwen, B., 1976, Steroid receptors in neuroendocrine tissues: Topography, subcellular distribution and functional implications, in: *Subcellular Mechanisms in Reproductive Neuroendocrinology* (G. Naftolin, K.J. Ryan, and J. Davis, eds.), p. 277, Elsevier, Amsterdam.

McEwen, B., Davis, P.G., Parsons, B., and Pfaff, D.W., 1979, The brain as a target for steroid hormone action, *Annu. Rev. Neurosci.* **2:**65.

Naftolin, F., Ryan, K.J., Davis, I.J., Reddy, V.V., Flores, F., Petro, Z., Kuh, M., White, R.J., Wolin, L., and Takaoka, Y., 1975, The formation of estrogens by central neuroendocrine tissue, *Recent Prog. Horm. Res.* **31:**295.

Naik, D.V., 1975, Immunoreactive LH-RH neurons in the hypothalamus identified by light and fluorescent microscopy, *Cell Tissue Res.* **157:**423.

Negro-Vilar, A., Ojeda, S.R., and McCann, S.M., 1979, Catecholaminergic modulation of luteinizing hormone-releasing hormone release by median eminence terminals *in vitro, Endocrinology* **104:**1749.

Nemeroff, C.B., Konkol, K.J., Bisette, G., Youngblood, W., Martin, J.B., Brazeau, P., Rone, M.S., Prange, A.J., Breese, G.R., and Kizer, J.S., 1977, Analysis of the disruption in hypothalamic pituitary regulation in rats treated neonatally with monosodium L-glutamate (MSG): Evidence for the involvement of tuberoinfundibular cholinergic and dopaminergic systems in neuroendocrine regulation, *Endocrinology* **101:**613.

Neri, R.O., 1977, Studies on biology and mechanism of action of nonsteroidal antiandrogens, in: *Androgen and Antiandrogens* (L. Martini and M. Motta, eds.), p. 179, Raven Press, New York.

Nett, T.M., Akbar, A.M., Niswender, G.D., Hedlund, M.T., and White, W.F., 1973, A radioimmunoassay for gonadotropin releasing hormone (GN-RH) in serum, *J. Clin. Endocrinol.* **36:**880.

Nicholson, G., Greely, G., Humm, J., Youngblood, W., and Kizer, J.S., 1978, Lack of effect of noradrenergic denervation of the hypothalamus and medial preoptic area on the feedback regulation of gonadotropin secretion and estrous cycle, *Endocrinology* **103:**559.

Sar, M., and Stumpf, W.E., 1975, Distribution of androgen concentrating neurons in rat brain, in: *Anatomical Neuroendocrinology* (W.E. Stumpf and L.D. Grant, eds.), p. 120, S. Karger, Basel.

Sarkar, D.K., Chiappa, S.A., Fink, G., and Sherwood, N.M., 1976, Gonadotropin-releasing hormone surge in proestrous rats, *Nature (London)* **264:**461.

Sawyer, C.H., 1979, Brain amines and pituitary gonadotropin secretion, *Can. J. Physiol. Pharmacol.* **57:**667.

Sétáló, G., Flerko, B., Arimura, A., and Schally, A.V., 1978, Brain cells as producers of releasing and inhibiting hormones, *Int. Rev. Cytol. Suppl.* **7:**1.

Simpkins, J.W., and Kalra, S.P., 1979, Blockade of progesterone-induced increase in hypothalamic luteinizing hormone-releasing hormone levels and serum gonadotropins by intrahypothalamic implantation of 6-hydroxydopamine, *Brain Res.* **170:**475.

Simpkins, J.W., Huang, H.H., Advis, J.P., and Meites, J., 1979, Changes in hypothalamic NE and DA turnover resulting from steroid-induced LH and prolactin surges in ovariectomized rats, *Biol. Reprod.* **60:**625.

Simpkins, J.W., Kalra, P.S., and Kalra, S.P., 1980a, Temporal alterations in LHRH concentrations in several brain nuclei of estrogen–progesterone treated rats: Effects of norepinephrine synthesis inhibition, *Endocrinology* **107:**12.

Simpkins, J.W., Kalra, P.S., and Kalra, S.P., 1980b, Effects of testosterone on catecholamine turnover and LHRH contents in the basal hypothalamus and preoptic area, *Neuroendocrinology* **30:**94.

Stumpf, W.E., 1971, Autoradiographic techniques and the localization of estrogen, androgen and glucocorticord in the pituitary and brain, *Am. Zool.* **11:**725.

Stumpf, W.E., and Sar, M., 1977, Steroid hormone target cells in the periventricular brain: Relationship to peptide hormone producing cells, *Fed. Proc. Fed. Am. Soc. Exp. Biol.* **36:**1976.

Swanson, L.W., and Hartman, B.K., 1975, The central adrenergic system: An immunofluorescence study of the localization of cell bodies and their efferent connections in the rat utilizing dopamine-β-hydroxylase as a marker, *J. Comp. Neurol.* **163:**467.

Ungerstedt, U., 1971, Stereotaxic mapping of the monoamine pathways in the rat brain, *Acta Physiol. Scand. Suppl.* **367:**1.

Wade, G.N., and Blanstein, J.D., 1978, Effects of an antiestrogen on neural estradiol binding and on behaviors in female rats, *Endocrinology* **102:**245.

3

Sensitivity of Pituitary Gonadotropic Cells and Gonads to Hormones

Vladimir R. Pantić

1. Introduction

Previous articles dealt briefly with the considerations of pituitary-cell specificity, the hypothalamic–pituitary–gonadal axis, and gonadal steroids (Pantić, 1974, 1975, 1980). There is no doubt that gonadotropic cells are one of the sites in the feedback effect of gonadal steroids and that estrogen (E) has the key role in the regulation of nervous and pituitary-cell activities. To contribute to further discussions related to the sensitivity of pituitary cells to gonadotropic-releasing hormone (GRH), gonadotropic hormone (GTH), and the gonadal steroids, an attempt is made also to summarize our data and recent other data closely connected with the role of various hormones in the differentiation and fate of genetically programmed oocytes in developing ovaries and of spermatogonia and Sertoli cells in developing testes.

1.1. General Properties of GTH Cells

It is now largely accepted that one type of pituitary basophil cell reacts with anti-follicle-stimulating hormone (FSH) and an anti-luteinizing hormone (LH) (Nakane, 1970), producing both FSH and LH (Tougard *et*

Vladimir R. Pantić • Serbian Academy of Science and Arts, Belgrade, Yugoslavia.

al., 1977; Herbert, 1975). Examining the subcellular organization of basophils, Soji (1978a–c) distinguished four cell types. There was a close interrelationship between loss of granules and LH, the latter being low in the pituitary and high in serum. The differences in the diameter of gonadotropic-cell granules are closely related to the cellular synthetic activities and suppression of hormone release (Kurosumi, 1968). In addition, the diameter is species-specific and depends on the dose of GTH or gonadal steroids, or both, ways of application, and other factors.

1.2. Receptors for GRH

Gonadotropic cells have receptors for gonadotropic-releasing hormone (GRH) and gonadal steroids, and are the sites of action of these hormones. The number of receptors for GRH present in the anterior pituitary was not different up to 20 hr after E injection; even LH levels were increased 100-fold 13 hr after treatment (Wagner *et al.*, 1979). Spona (1973) indicated the presence of two distinct binding sites in pituitary-cell plasma membranes.

LH-releasing hormone (LRH) acts on a series of basophils, not just on the cells analogous to the "LH cells," influencing the cells' transformation and release of granular content (Soji, 1978a–c). The repeated administration of this decapeptide greatly increased LH and FSH release in immature male rats (Horikoshi *et al.*, 1978). The effect of GRH on gonadotropin release was significantly affected by exogenous gonadal steroids (Debeljuk *et al.*, 1972). Gonadotropic cells are "target" cells for LRH as a decapeptide that influences the release of both FSH and LH (Schally *et al.*, 1973).

1.3. Mechanisms of Hormone Action

Mechanisms of hormone action through the membrane have been the subject of many papers. It is now accepted that in the mechanism of peptide and protein hormone effects, adenylate cyclase is involved. As a result of the binding of hormones to the plasma membranes, hormone–receptor complexes are formed, adenylate cyclase is activated, the production of cyclic AMP (cAMP) or cGMP is stimulated, and catalytic activity of target cells is increased. The role of dehydrogenase in the production of sequestered intramembranous protons, which are drivers of transport and membrane conformation, is mentioned by Grane *et al.* (1979). Modulators involved in the regulation of adenylate cyclase activity are dependent on the protein associated with the plasma-membrane target cells (Bradham, 1979). Recently, it was mentioned that enzyme activity can be regulated by hormones that mediate extracellular and intracellular signals. The role of K^+ in the release of GTH after treatment with synthetic GRH is mentioned by Kao and Weisz (1975).

2. Aim of the Investigation

In this chapter, data obtained by examining the response of pituitary GTH cells and gonadal tissue to peptides, glycoproteins, and gonadal steroids under various experimental conditions will be considered with respect to the answers they may suggest to the following questions:

- Which signs of brain-neuron reactions to gonadal steroids are pronounced?
- What are the main characteristics of the GTH-cell reaction to GRH?
- What is the difference in GTH-cell reaction during development and in adults up until old age, and is the reaction the result of direct or indirect effects?
- In which of the examined species are the pituitary GTH cells most sensitive to GRH?
- What is the fate of growing and maturing follicles in the ovaries of animals treated with female and male gonadal steroids?
- What is the difference in reaction of granulosa and theca cells in maturing follicles and interstitial cells in the ovaries of animals treated with human chorionic gonadotropin (hCG)?
- How can hCG or gonadal steroids or both be used in controlling reproductive properties of males and females of various animal species?
- What is the difference in GTH-cell reaction to female and male gonadal steroids?
- Is the response of Leydig and Sertoli cells in testes specific for the examined species, and is there any difference in reaction to female and male gonadal steroids?
- Is the long-term effect stimulatory or inhibitory to these steroid-producing cells up to puberty?

- At what level of spermatogenesis is the influence of the gonadal steroids most clearly pronounced, and is the reaction reversible?

3. Neonatal Rat as a Model for Studying the Endocrine-Cell Reaction to Gonadal Steroids

Neonatal female mice were used by Bern *et al.* (1975, 1976) as a model for examining the long-term consequences of exposure of the human fetus and neonate to hormones and other agents. These workers noted parallelism between the data obtained after treatment of mice with gonadal steroids and incidences of vaginal cancer in their adolescent and postadolescent daughters whose pregnant mothers were given diethylstilbestrol for treated abortion. Similar lesions may be induced not only with E, but also with progesterone (P) alone, E and P in combination (Jones and Bern, 1977), and testosterone (T) (Kimura and Nandi, 1967; Mori *et al.*, 1976). The first 3 days of mouse neonatal life apparently correspond to that at the end of the first trimester of human pregnancy (Bern *et al.*, 1975, 1976) and during the second trimester of pig pregnancy (Elsaesser *et al.*, 1978). The miniature pig was used as a model in neuroendocrine studies of reproduction (Ellendorff *et al.*, 1977).

Neonatal rats were used in our studies as a model for the investigation of the sequences of endocrine-cell reaction to gonadal steroids that were administered into the rats soon after birth. A single dose or repeated doses of E, E and P in combination, or testosterone propionate (TP) were injected into both female and male rats. Results of long-term effects of the gonadal steroids in some fishes, chicken, rats, and pigs, and the reaction of pituitary gonadotropic cells to LRH and gonadal steroids and of gonads to chorionic gonadotropins (hCG) or gonadal steroids are mainly the subject of this presentation.

The hypothalamic–pituitary–gonadal axis in newborn rats has been considered as sexually undifferentiated, and this time of life was known as the "critical" period. Recently, however, Corbier *et al.* (1979) reported that during the first hours of the life of newborn rats, the hypothalamic–pituitary–gonadal axis is differentiated and strongly stimulated.

Data obtained by examining neuroendocrine cells of immediately postnatally treated rats during development up to puberty were compared with the results obtained if the rats were treated with repeated doses during the prepubertal, peripubertal, adult, and old stages of life. With a view toward clarifying species specificities of the differentiation of neuroendocrine cells during periods of development, an attempt was made to follow the sensitivity of the brain–pituitary–gonadal axis in fishes treated briefly with steroids during early stages of development, in chickens treated soon after hatching, and in piglets neonatally gonadectomized or treated with gonadal steroids.

4. *Sensitivity of GTH Cells to GRH and Gonadal Steroids*

4.1. *Sensitivity of GTH Cells to GRH*

The gonadotropic-cell population was considered by Denef *et al.* (1978a,b) as a heterogeneous cell type in terms of ability to respond to LRH by releasing LH and to a lesser extent FSH. Both androgens and estrogens can modulate the release of FSH and LH selectively (Sétáló *et al.,* 1977).

An increased sensitivity of GTH cells to LRH was noted 6–10 hr after estrogen exposure (Vilchez-Martinez *et al.,* 1974). These cells are more sensitive to LRH, and LH release is maximal in the proestrous rat when the circulating E-17β level is highest (Castro-Vasquez and McCann, 1975). Throughout the estrous cycle, the variation in responsiveness to pituitary LRH and E receptor levels varied as well (Park *et al.,* 1976).

As a result of changes at the hypothalamic level, the pars intermedia was less developed in rats and boars treated with E, or with E and P, and the effect was pronounced up to the age of sexual maturity. GRH had a stimulative effect on pars-intermedia cells (Pantić and Šimić, 1978).

Sensitivity of pituitary GTH cells to synthetic mammalian GRH is established in fishes (Kaul and Vollrat, 1974). These workers observed a decreased amount of secretory granules half an hour after a single injection of LRH, and immediately after spawning, these cells were devoid of secretory granules.

Gonadotropic cells of immature female rats pretreated with TP and treated with GRH lost the affinity for staining, so that it was rather difficult to identify them as specific GTH cells, but the effect was reversible (Pantić and Gledić, 1978a,b).

The presence of GTH cells was established in fragments of the anterior pituitary directly or previously cultivated *in vitro* for 28 days before transplantation into the anterior eye chamber or into the subcapsular region of the kidney, indicating that both FSH and LH could be synthesized and released up to 8 months in ectopic sites (Pavić *et al.,* 1977) (Fig. 1).

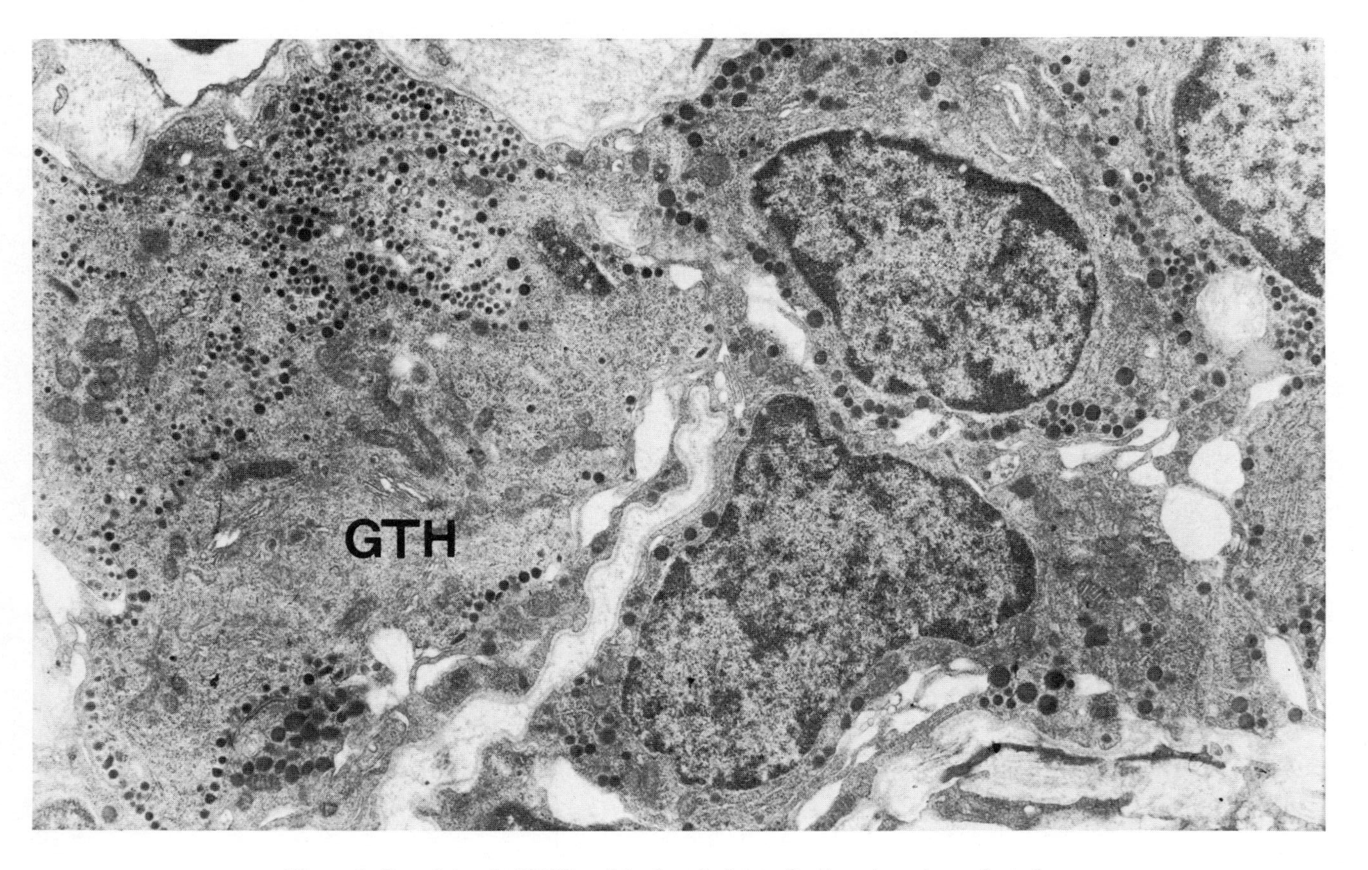

Figure 1. Gonadotropic (GTH) cell in the pituitary after long-term transplantation.

4.2. Sensitivity of GTH Cells to Gonadal Steroids

Hypertrophy and hyperplasia of LH cells were observed after castration in both pars distalis and pars tuberalis of the rat (Gross, 1978) and in pars intermedia of neonatally castrated male piglets (Pantić and Šimić, 1978). These cells are suggested as a primary site of E influence (Ferin *et al.*, 1978, 1979). Long-term treatment (52 weeks) with high doses of 17β-estradiol had a suppressive effect on LH rather than FSH synthesis and release (Etreby and Bab, 1978). Sar and Stumpf (1973) noted that [^{3}H]testosterone was incorporated only in GTH cells, and the labeling index was 15% for T and 60–80% for E. Pituitary cells are sensitive to a challenge of LRH and are, in part, under the control of gonadal steroids (Debeljuk *et al.*, 1972, 1973). A marked hypersensitivity of LH cells to LRH was observed in 7-week-old rats androgenized with 100 μg TP (Fujii *et al.*, 1978).

Hypertrophy of GTH cells during migration and before the spawning period was observed in some teleost fishes. The cells attained a maximal size of 835.3 μm^3 during spawning, their nuclei were polymorphous, and the cytoplasm was degranulated and vacuolated (Pavlović and Pantić, 1975). GTH cells of the central part of the carp mesadeno-phypophysis were predominantly vacuolated, and the degree of vacuolization was stronger in animals treated with E and P; a greater amount of granules filled up the cytoplasm of the peripheral part of this lobe (Pantić and Lovren, 1978).

Gonadotropic cells were sensitive to a single dose of male and female gonadal steroids (TP or E) administered during early stages of fish development, during chick embryogenesis and after hatching, and during the neonatal period of rats and piglets. The degree of sensitivity gradually decreased and in the rat was expressed in the peripubertal stage of development (Figs. 2 and 3) in the adult, and was lowest in the oldest. Similarities in reaction were established in all the animal species we examined, and some of them will be briefly mentioned.

A reciprocal relationship was expressed such that the number of luteotropic hormone (LTH) cells was increased and the gonadotrophs decreased. The cytological feature of GTH cells showed signs of lesser activity and was more expressed when the animals were treated during embryogenesis or up to the 12th day after hatching (Pantić and Škaro, 1974; Škaro-Milić and Pantić, 1976), in the first days of life after partus of rats (Pantić and Genbačev, 1971), and in neonatally treated piglets (Pantić and Gledić, 1978a,b). Fish pituitary cells seemed to be less sensitive than those of mammals (Sekulić and Pantić, 1977). Examination

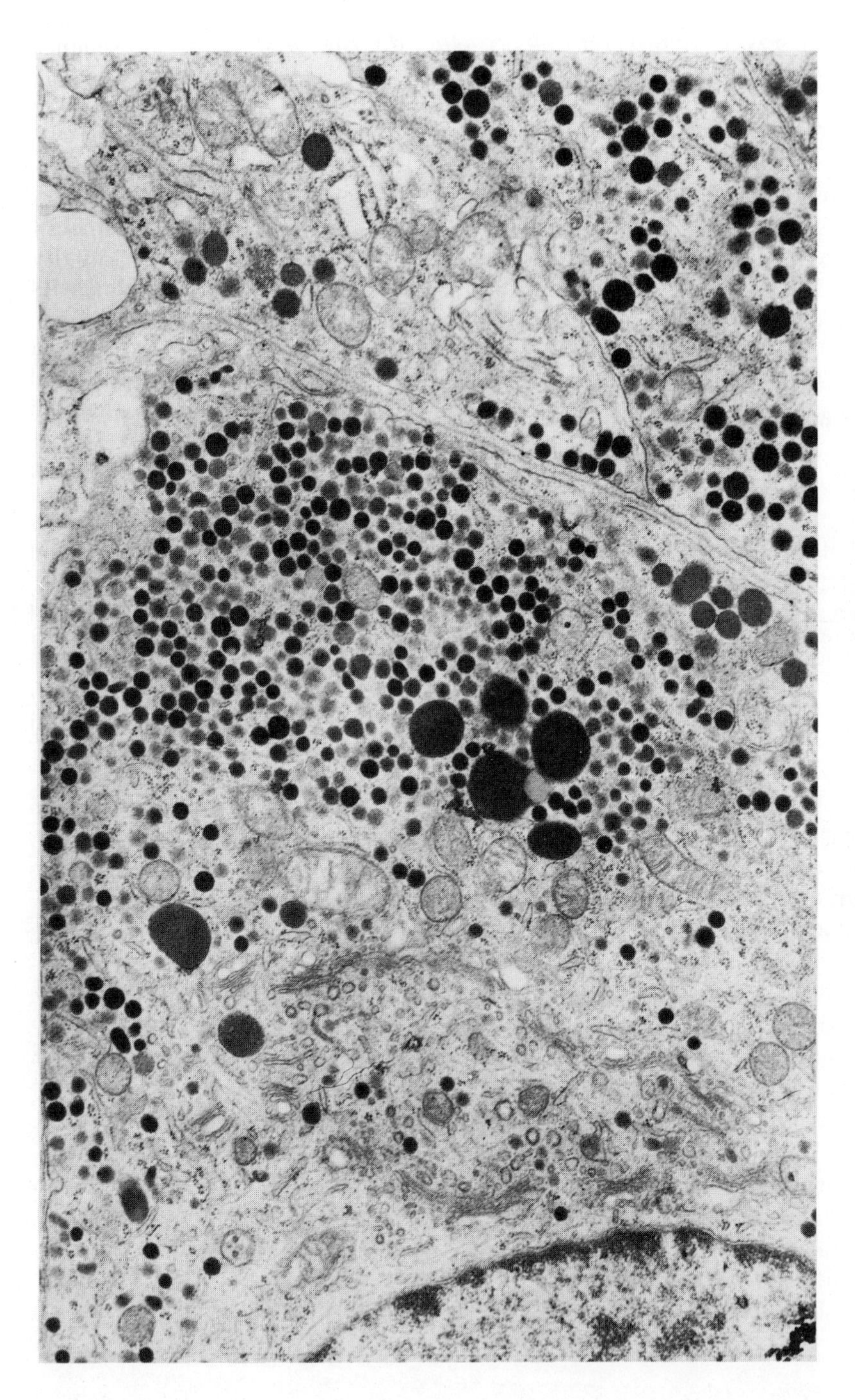

Figure 2. Gonadotropic cell of an intact 80-day-old male rat.

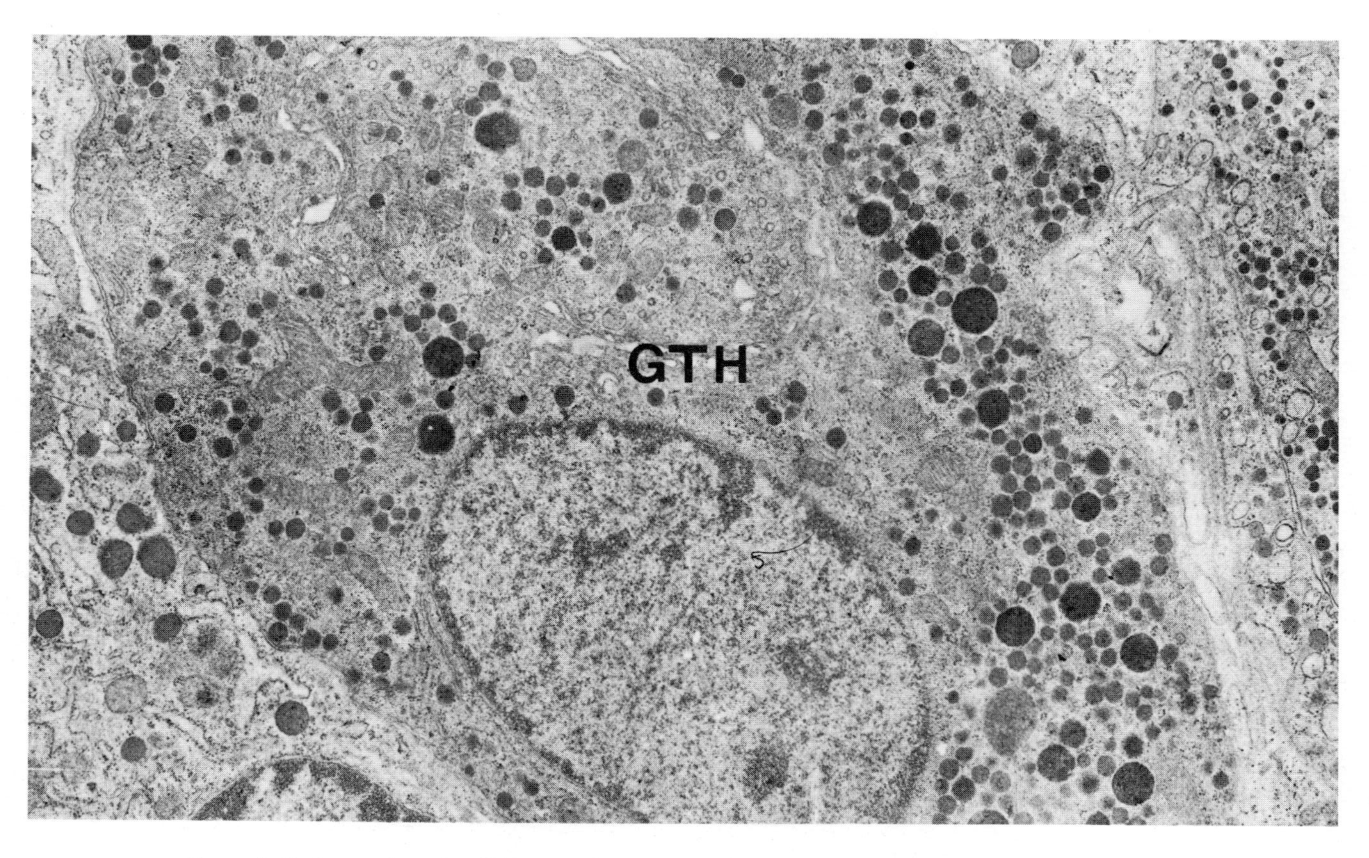

Figure 3. Dark gonadotropic (GTH) cell of an 80-day-old male rat treated neonatally with 1000 μg E.

of GTH cells of the pituitaries of chickens at ages up to 3 months treated with E on day 4, 8, or 13 after hatching showed that small-sized "signet" GTH cells were the predominant GTH cells. A large number of dark cells of gonadotropic cells appeared as a result of the direct and indirect effect of E administered on day 4 after hatching (Škaro-Milić and Pantić, 1976, 1977). The intercellular space between the GTH cells was slightly widened and osmiophobic.

A pronounced reciprocal interrelationship between prolactin and GTH cells was pronounced up to sexual maturation in chickens and rats treated neonatally with a single dose of TP (see Pantić, 1975). GTH cells reacted in a way after treatment with TP similar to that after treatment with E. "Signet" gonadotropic cells present in the pituitaries of 3-month-old chickens treated with a single dose of 2000 μm E on day 4 after hatching and their response to LRH we considered to be a result of indirect and long-term effect through brain neurons.

GTH cells in the pituitaries of rats treated with a single dose of TP on day 4 after birth differentiated slowly, and their number and volume were smaller. The number of [^{3}H]thymidine-labeled cells was 2.2 times smaller in sexually mature animals treated with a single dose or with repeated doses than in controls, showing that DNA synthesis in GTH cells was retarded (Pantić and Gledić, 1978a,b).

As a result of long-term effects on the differentiation of GTH cells, loss of affinity for staining, even using immunocytochemistry, and paucity of both FSH- and LH-producing cells were expressed up to the age of sexual maturity (Vigh *et al.,* 1978).

Gonadotropic cells, revealed with anti-rat LH or anti-hCG as LH cells in the pituitaries of gonadectomized rats treated with E, were enlarged in size, and only some of them retained the characteristics of gonadectomy cells. The appearance and the number of gonadectomy cells were similar in castrated male and female rats (Genbačev and Pantić, 1975).

Clearly reduced synthetic and secretory activities of GTH cells were observed in female rats fed with a diet supplemented with 0.20% of Protamone containing triiodothyronine and l-thyroxine. After 6 months, all the examined female rats were in continuous diestrus, and a normal cycle was restored within 3 months after cessation of treatment (Pantić *et al.,* 1975a,b). These changes in the number, appearance, and subcellular organization of GTH cells we considered to be a severe temporary disorder in the hypothalamic–pituitary–gonadal axis. The prolonged diestrus was a result of a decrease of GTH secretion and altered steroidogenesis in the ovaries.

With the reduction of cell organelles, especially granules, in LH cells and a gradual decrease of LTH band density in the pituitaries of rats bearing MtT tumors, a decreased number of GTH cells and a tendency of their cytoplasm to vacuolize occur (Pantić *et al.*, 1971a,b).

Similarity in the concentration of T in pig and man from the fetal period up to sexual maturity was observed by Meusy-Dessolle (1975) and Raeside and Sigman (1975). Examining the sensitivity of pituitary GTH cells to the deficit of androgen and excess of E, or of E and P, Pantić and Gledić (1978a,b) established that in piglets gonadectomized on the first day of life, the number of GTH cells was increasing, so that in the pituitary of 4-month-old animals, these cells predominated and "signet" cells increased moderately from 4 to 6 months. An increased number of hypertrophic and vacuolated GTH cells was established in pars intermedia of 4- to 6-month-old castrated boars (Pantić and Šimić, 1978).

The degree of retardation of GTH-cell differentiation and synthesizing and release capacity in both rats and pigs was more clearly evident in animals treated with repeated doses than in those treated with a single dose administered during the postnatal period of life.

4.3. Brain as a Site of Gonadal-Steroid Action

The brain has been suggested as a central site of mechanisms of E action (Chader and Villee, 1970). As a key site for the action of androgen in the rat brain, the nucleus intertitialis striae terminalis and the medial nucleus of the amygdala have been suggested (Sheridan *et al.*, 1974). Mapping the "receptor" structure for [^{3}H]-E, Stumpf (1972) concluded that the striae terminalis is the main link between gonadal-hormone target neurons in the mammalian brain, interconnecting the amygdala, preoptic hypothalamic nuclei, and structures in the midline thalamus as well as in the midbrain.

Feedback action of both female gonadal steroids is stimulatory to gonadotropin release, dependent on dosage, age of development, and duration of treatment, and has been demonstrated in prepubertal (Brown-Grant and Greig, 1975; Caligaris *et al.*, 1968, 1971; Ramirez and Sawyer, 1965; Meyer and McCormack, 1967), cyclic (Everett, 1948; Zeilmaker, 1960; Brown-Grant, 1969), pregnant (Everett, 1947), and E-primed ovariectomized rats (Everett, 1947, 1948; Zeilmaker, 1960; Brown-Grant, 1969; Ramirez and Sawyer, 1965; Meyer and McCormack, 1967). This stimulatory effect of E is exerted on the limbic system (especially the bed nucleus of stria terminalis and the lateral septum), while the main sites of P action are expressed through the diagonal band of Broca, the preoptic suprachiasmatic area, and the anterior hypothalamic area (Kawakami *et al.,* 1978a,b). As a feedback site for the action of T or its metabolites in the undifferentiated hypothalamus and adult male rat hypothalamus, Litteria (1973) suggested the arcuate, nucleus paraventricularis, periventricular, and nucleus supraopticus.

The incorporation of LRH into the median eminence from the cerebrospinal fluid was facilitated by estradiol benzoate (Recabarren and Wheaton, 1978). The properties of hypothalamic regulatory hormones, a large number of their structural analogues, and the role of these peptides in regulation of pituitary specific cells have been described by Schally *et al.* (1973) and others. The newborn rat hypothalamus of either sex has the inherent ability to maintain cyclic release of GTH cells. The noncyclic nature in the male is determined by the testes during the first few days of postnatal life. Both male and female rats, under the influence of androgen during 15 days of postnatal life, lost the ability to release GTHs in a cyclic manner (Vigh *et al.,* 1978; Sétáló *et al.,* 1977; Döhler and Haucke, 1978).

Some of our results showing the character of brain-neuron reactions to gonadal steroids administered into the newborn rats will be mentioned. A significantly enlarged nuclear volume in nucleus basolateralis (NBL) and nucleus lateralis posterior (NLP) of corpus amygdaloideum appeared swollen, and we considered it as a sign showing that larger cells lost stimulative properties so that an inhibitory effect of smaller cell types prevailed (Pantić and Drekić, 1980). As a result of structural and functional changes in the other brain neurons, including hypothalamic neurons, the neurohypophysis appeared less developed. The peptidergic and aminergic nerve terminals contain far fewer granules and microvesicles than those in intact animals of the corresponding age (Pantić and Šimić, 1977a,b) (Figs. 4 and 5). As a consequence of such influence of gonadal steroids on brain neurons expressed in the pars neurosa of developing animals, the pars intermedia is less developed. This means that the development of these lobes is closely dependent on the hypothalamo–neuro–pituitary interconnection (Pantić *et al.,* 1978b).

The other pathway of neuronal influence is expressed via portal blood vessels and is dependent on the amount of GRH release. There was a decrease in the number of indentified GTH cells when the animals were neonatally treated with gonadal steroids (Pantić and Gledić, 1977; Vigh *et al.,* 1978). However, during sexual maturation, the number of LTH cells is decreased and GTH cells are identifiable (Pantić *et al.,* 1980b).

5. Sensitivity of Ovarian Tissue

5.1. Growing Follicles

It is known that during initiation of oogenesis, ovarian tissue is very sensitive to FSH (Huhtaniemi *et al.*, 1978) and is hormone-dependent (Döhler, 1978). FSH activates or induces binding sites in granulosa cells of the immature intact rat (Zeleznik *et al.*, 1974). The number of FSH receptors per granulosa cell increases so that follicles grow rapidly in response to E and FSH. An inhibitory effect of intrafollicular E on P accumulation in growing follicles was mentioned by Thanki and Channing (1978).

Primordial follicles consisting of oocytes and follicular (pregranulosa) cells are initiated and then grow into bilaminar and multilaminar follicles. Growth of follicles was impossible without oocytes or follicle cells, which means that interreaction between germ and follicular cells was of significant importance.

Initiation of follicle growth was recognizable when one or more of the flattened follicular cells of primary follicles assumed a cuboid appearance. FSH stimulated the growth of follicles, which will ovulate in response to the LH release in the following proestrus (Hirshfield and Midgley, 1978).

Growth of oocyte nuclei and nucleologenesis are the first signs of oogenesis, and it seems that growth of oocytes and follicular cell hyperplasia are the first stage of folliculogenesis in the fish *Serranus scriba*. Follicular-cell hypertrophy and thickening of the zona pellucida are closely connected with an increase of these cell activities and vitellogenesis. At the stage of follicle maturation, follicular cells are 2 μm in height and the regressive changes in them are more pronounced. A bilayered affinity for staining of zona pellucida was altered and a connection between follicular cells and oocytes was weakened (Pantić and Lovren, 1977).

Antrum formation in ovarian follicles of E-stimulated hypophysectomized immature female rats was stimulated by FSH (Goldenberg *et al.*, 1971). A stimulatory effect of this glycoprotein in activation of GTH receptors was observed by Brown-Grant and Greig (1975).

The fate of growing follicles in mammals is maturation of both follicle and oocyte, followed by ovulation or atretic degeneration. There is no doubt that the initiation of oocyte maturation depends on the stimuli originated by gonadal steroids and gonadotropins. What does initiation of oocyte maturation mean? This is without doubt a complex of events occurring in oocytes arrested in the diplotene stage (G_2 phase), which is

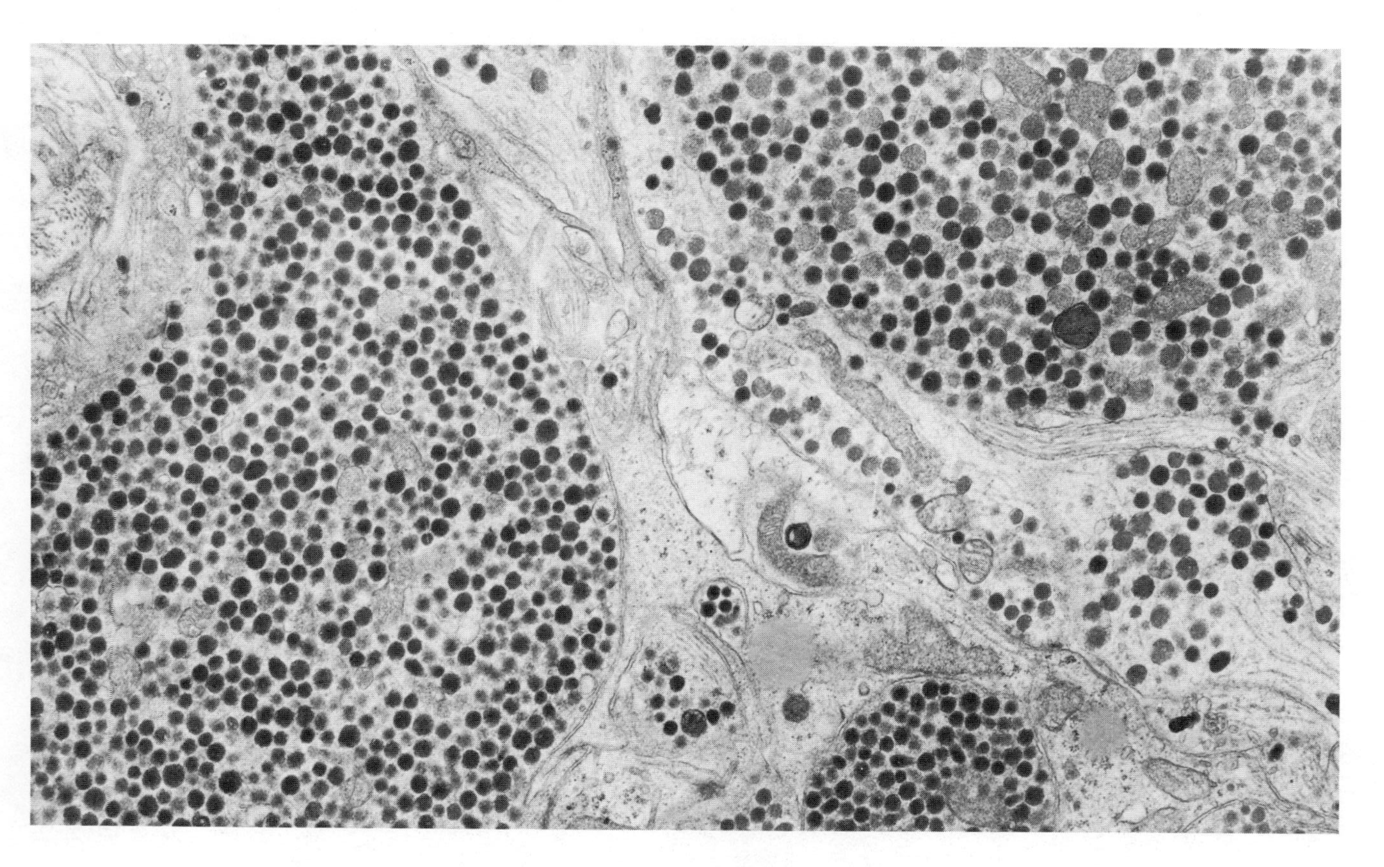

Figure 4. Nerve terminals filled with granules in the neurohypophysis of a 30-day-old intact male rat.

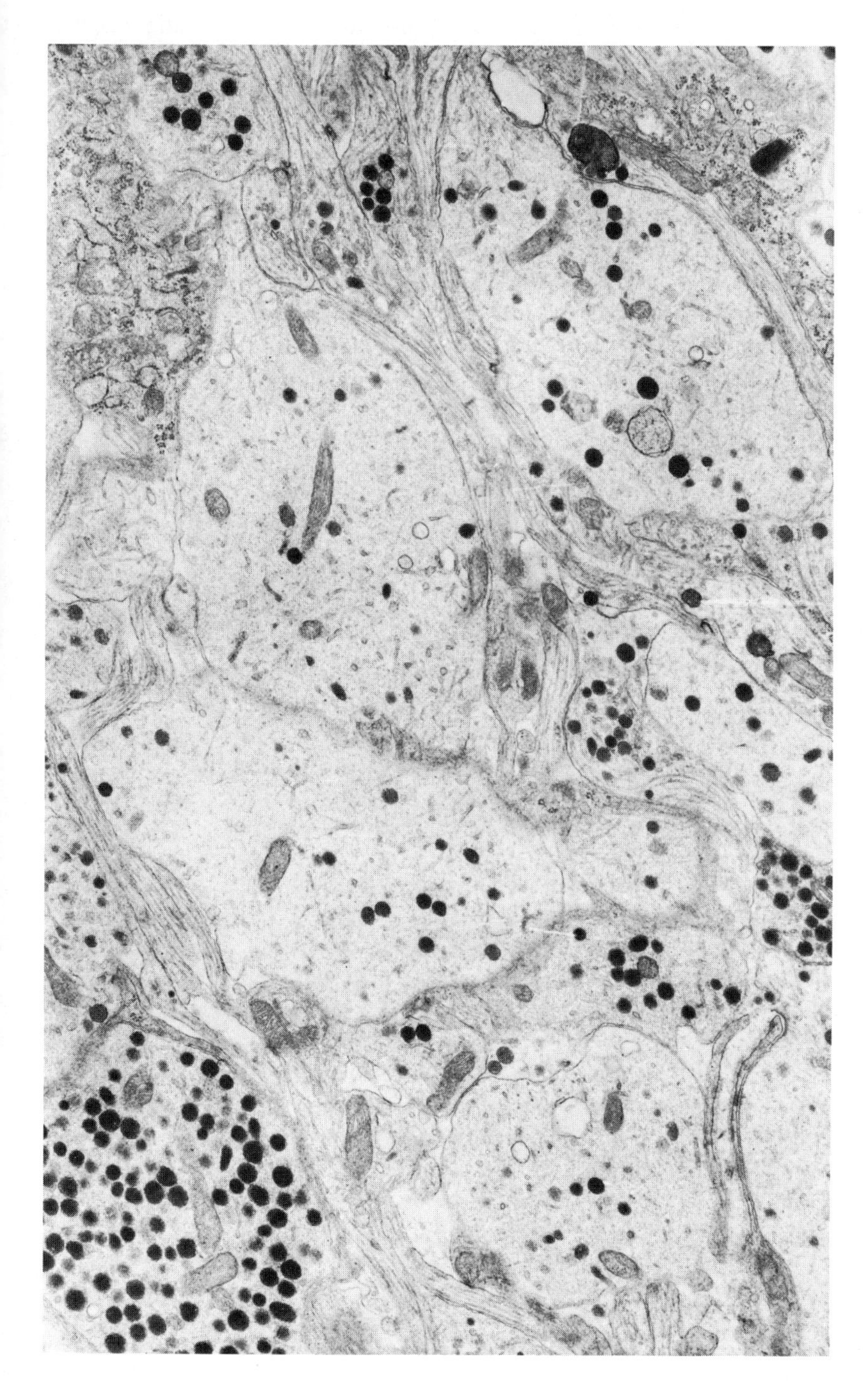

Figure 5. Nerve terminals in the neurohypophysis of a 38-day-old male rat treated neonatally with E.

preceded by the G_1 phase after maturation and activation of the zygote. Various steroids can induce maturation without ovulation in cultured fish *(Salmo)* follicles (Jalabert *et al.*, 1972). Prostaglandin F_2 may induce ovulation of the immature oocytes (Jalabert, 1976). A long-term effect of female or male gonadal steroids (E, E + P, or TP) administered neonatally affects the growth of ovarian follicles in female rats and pigs. Many follicles undergo atresia, and secondary interstitial glands are formed. However, a number of them are cystoid or even cystic in appearance up to puberty in the rat (Pantić and Nikolić, 1980) and in the pig (Jablan-Pantić and Miladinović, 1977). Intercellular spaces between the granulosa cells of the cumulus oophorus, an irregularly developed granulosa layer, and villosity of these cells are the clearly visible first signs of follicle degeneration (Pantić and Lovren, 1979) (Figs. 6 and 7).

In mammalian species, oocyte maturation is highly dependent on the follicle and could be initiated by the direct or indirect action of GTH. However, the sensitivity of the follicular tissue to the effect of GTH inducing oocyte maturation may be increased by FSH, gonadal steroids, or corticosteroids. The reaction is dependent on the manner of plasma-membrane control, polarization, a selective permeability for Ca, Mg, K, and Na ions, and other factors. The activities of ATPase, adenylate cyclase, phosphodiesterase, and other enzymes are controlled by Ca. These enzymes are localized on the surface of oocyte plasma membranes and are involved in the regulation of cAMP levels (Masui and Clarke, 1979).

5.2. *Maturing Follicles*

Theca cells of small follicles have few receptors and could not be target cells for ovulatory action of LH.

Both granulosa and theca cells of preovulatory rat follicles are sensitive to LH, and this gonadotropin has an acute and prolonged stimulatory effect on some early steps of steroidogenesis (Hamberger *et al.*, 1978).

It seems that granulosa cells possess receptors exclusively for FSH, but not for LH or prolactin (Richards and Midgley, 1976). However, C. Y. Lee (1976) showed that FSH binds to granulosa cells of medium-size and large follicles and play a major role in maturation of granulosa cells, to increase the available cell-membrane LH/hCG receptors for these glycoproteins. The receptor molecules were characterized by Channing and Kammerman (1973). They are localized on the surface of plasma membranes of granulosa and theca cells of large follicles and corpus luteum cells.

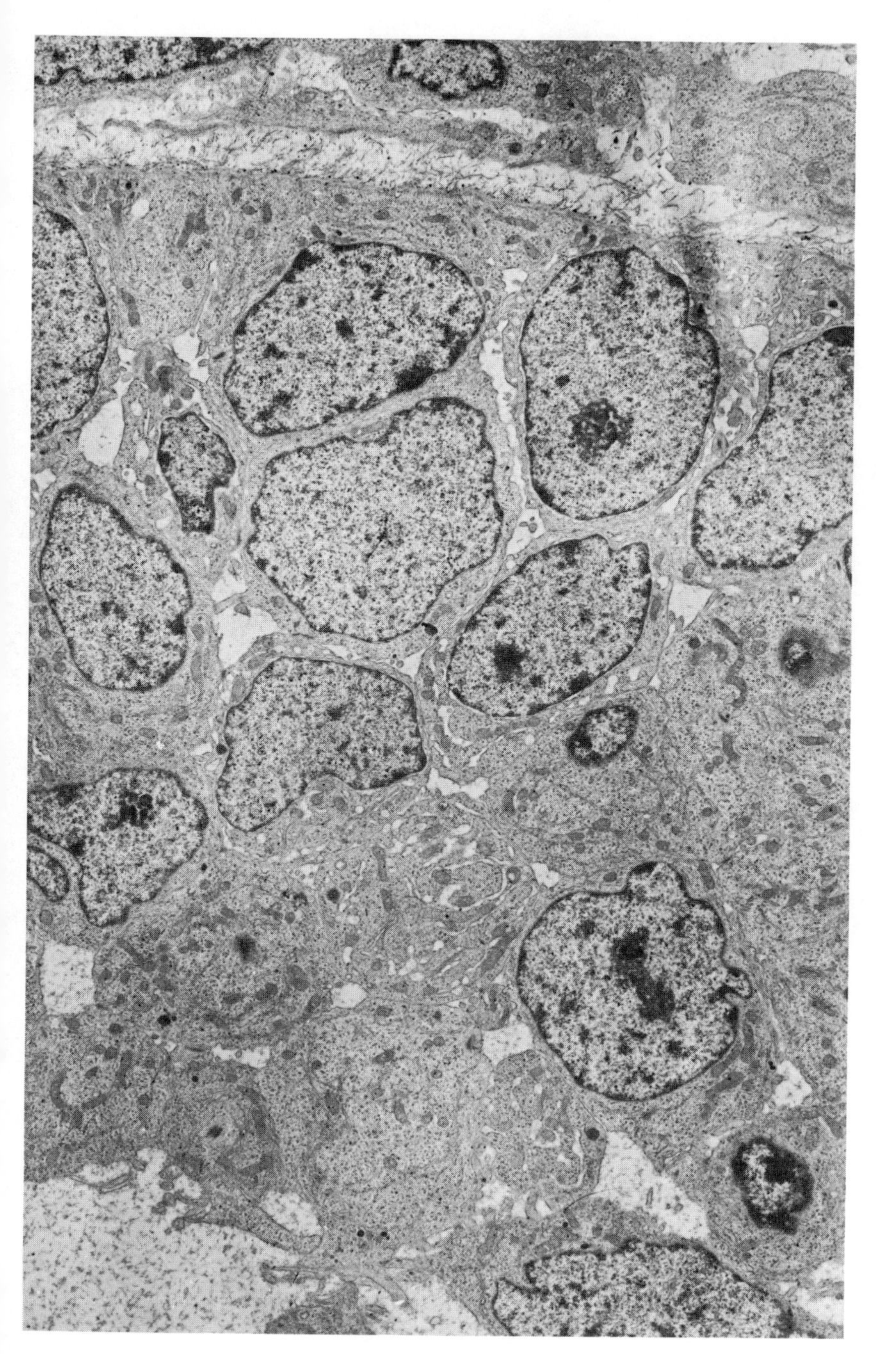

Figure 6. Theca and granulosa cells in the follicle of an 80-day-old intact female rat.

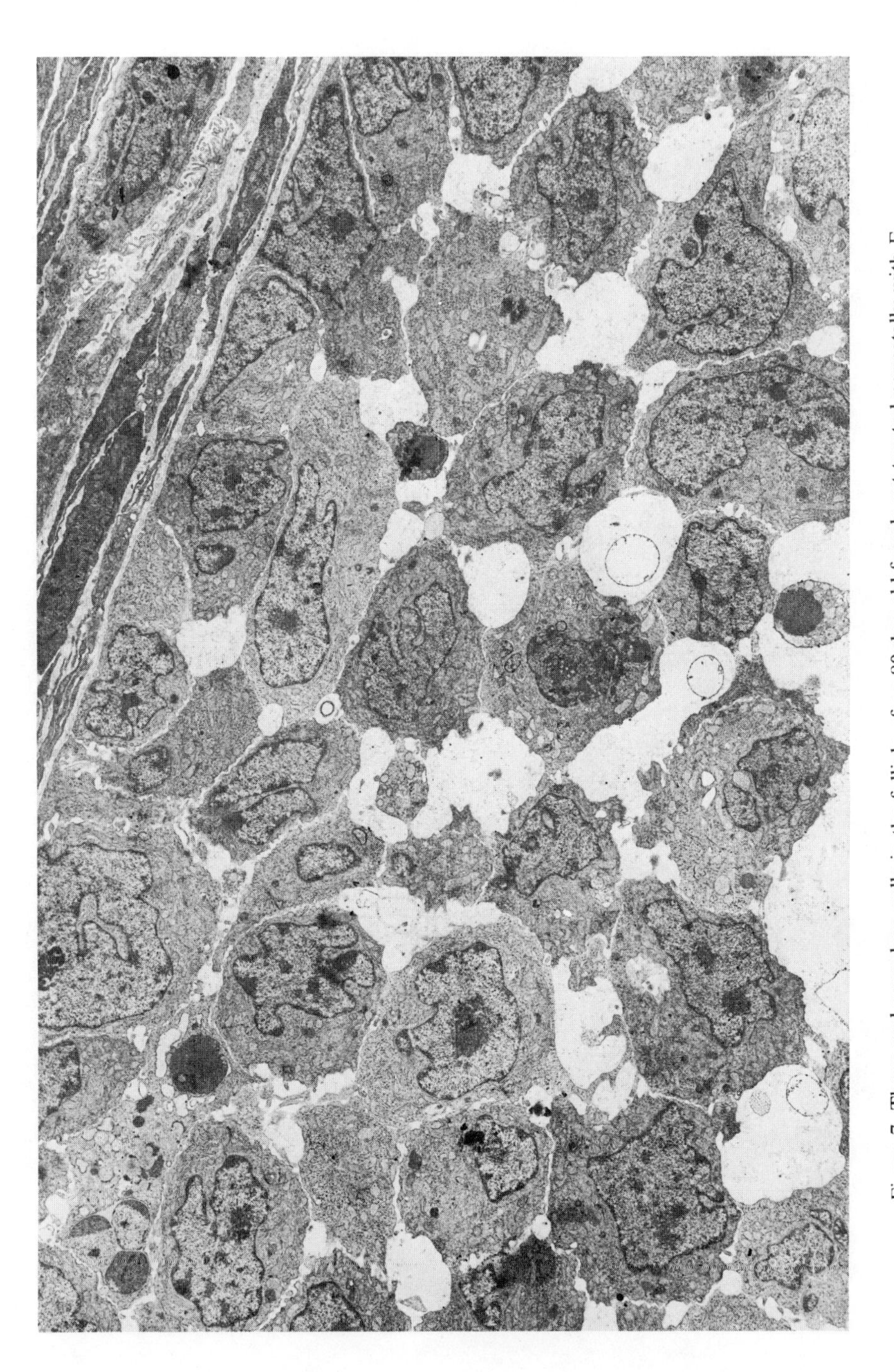

Figure 7. Theca and granulosa cells in the follicle of an 80-day-old female rat treated neonatally with E.

What is the role of LH? This hormone might play a role in the biosynthesis of glycogen in the preovulatory follicles and corpora lutea, in increasing the vascularity of both the preovulatory follicle and the follicle at the time of ovulation. When the basement membrane is broken down, the small blood vessels and erythrocytes penetrate into the stratum granulosum (Reed and Bellinger, 1979). LH acts to produce an increase in prostaglandin synthetase activity (Clark *et al.,* 1978).

No unaltered maturing or ovulating follicles or corpora lutea were observed in the ovaries of pubertal rats treated neonatally with E (Pantić and Nikolić, 1980).

The ovarian weight in immature rats pretreated with T and treated with LRH was clearly increased in relation to the value obtained for intact or only-T-treated rats. The heaviest ovaries, having weights from 1.63 times higher than those in intact animals, were in animals treated with hCG (Pantić *et al.,* 1977). The number of growing follicles and corpora lutea was increased, and the atretic ones were smaller in GRH-treated animals than in corresponding controls. Hyperluteinization was pronounced in all the ovaries in the animals pretreated with TP and was more evident in the animals sacrificed on day 57 than at the age of 43 or 47 days. Signs of luteinization of interstitial cells were more pronounced in TP- than in E-pretreated females.

An extremely high level of hyperluteinization of growing and maturing follicles and interstitial glands was the result of hCG effect. In the animals treated with repeated doses of hCG, the number of corpora lutea is increased and is dependent on the dose and duration of treatment. With an increased number of corpora lutea, the paucity of theca and interstitial cells is more expressed, and these cells are not clearly observable after prolonged treatment with this glycoprotein (Pantić *et al.,* 1978a).

Both LH and hCG are bound mainly on the external surface of corpus luteum and interstitial cell plasma membranes, which contain receptors to which hLH binds with 4 times lower affinity than hCG (Rao *et al.,* 1979). Channing and Kammerman (1974) observed that new corpora lutea bound more LH and hCG than older ones. The corpus luteum took up 7-to 9-fold more hCG than there is in blood, the follicle wall bound hCG at twice blood levels, and corpus albicans only slightly more labeled hCG. A paucity of LH/hCG receptors was observed in porcine and rat ovary (Midgley, 1973) and in monkey ovary (Channing and Kammerman, 1974). Macrophages play a key role in luteolysis of the corpora lutea, and in this regressive pathway, both autophagy and heterophagy are involved (Paavola, 1979).

Estradiol, as a luteolytic agent, has an inhibitory effect on the corpus luteum by blocking the stimulatory action of LH at a step after cAMP. A marked rise in serum LH occurs in both male and female castrated rats. After castration, LH release might be depressed by prolactin (Grandison *et al.*, 1977). This means that elevated prolactin can decrease the serum LH rise after castration, acting via the hypothalamus. FSH release is more sensitive to a dopaminergic stimulus in immature than in adult ovariectomized heifers (Beck and Convey, 1977).

5.3. Role of Ovarian Follicles

Synthesis and release of E are the main roles of granulosa cells. However, for these activities, both theca and granulosa cells are necessary (Markis and Ryan, 1975). These cells, isolated from hypophysectomized rats, are able to synthesize large quantities of E if they are cultured in medium containing highly purified FSH and androgen (Dorrington *et al.*, 1975a,b). The role of theca cells is to supply the granulosa cells with androgen precursor, which is aromatized to estradiol (Fortune and Armstrong, 1978). The biosynthesis of E is greatly increased if T is added to the culture medium (Markis and Ryan, 1975, 1977).

FSH stimulates cAMP and P release by granulosa cells. The magnitude of stimulation decreases with the maturation of follicles and the decrease in receptors for FSH. P secretion by granulosa cells from small follicles is inhibited by 1 μg/ml of 17β-E, and the action of E and T is additive. However, secretion of P in response to FSH is kept inhibited by intrafollicular E and other components of follicular fluid (Thanki and Channing, 1978).

By intercellular modification, FSH enhances the capacity for P release (Channing, 1970a,b; Zeleznik *et al.*, 1974). The properties of granulosa cells, like the acquisition of specific LH/hCG receptors and enhanced steroidogenesis, are manifested during the initiation of antrum formation in ovarian follicles (Hillier *et al.*, 1978). The increase of sensitivity to these hormones is closely related to the maturation of LH receptors (C. Y. Lee, 1976).

Granulosa cells of maturing follicles stimulated by LH release increased amounts of P. These cells isolated from preovulatory follicles are a source of ovarian "inhibin" as a molecule that affects GTH-cell activity (Erickson and Hsueh, 1978) and have a role in prostaglandin synthesis (Clark *et al.*, 1975).

Nonsteroidal factor is found in follicular fluid (Bender *et al.*, 1978).

These authors observed that this factor, acting additively with E, is capable of regulating FSH release.

5.4. Interstitial Cells

Interstitial and theca cells originate from the undifferentiated fibroblastlike stromal cells. Two types of interstitial gland cells differentiate. Primary interstitial gland cells originate from the hypertrophy of undifferentiated fibroblastlike cells during the fetal life of species having a long gestation and in the postnatal period in animals with a short gestation period. Secondary interstitial cells originate from the hypertrophy of the theca interna cells of the atretic follicles. Both types of cells are steroid-hormone-producing cells (Guraya, 1973, 1978).

In ovaries of nonmammalian vertebrates, interstitial cells arise from the hypertrophied theca cells of the large previtellogenic and early vitellogenic follicles that undergo atresia and also from connective cells (Guraya, 1972, 1976b).

The interstitial glandular cells were altered in all the examined animals treated with both female and male gonadal steroids (Pantić and Lovren, 1979) (Fig. 8). Gametogenesis in *Salmo gairdneri* of both sexes was retarded if E was administered to these animals briefly during early development (Pantić *et al.*, 1980a).

The role of interstitial-gland cells in steroidogenesis is confirmed by histochemical examination. Granulosa and theca cells become hypertrophic at the onset of follicular atresia, and they have a role in the phagocytosis and the digestion of vitellogenic granules and other substances from the oocytes.

6. Sensitivity of Testicular Tissue

6.1. General Properties of Testicular Tissue

The seminiferous tubules are surrounded by basement membrane, myoid cells, and connective tissue. At the end of the 3rd week, i.e., shortly after the onset of spermatogenesis, the blood–testes barrier, as a basic and most effective barrier, is developed (Fawcett, 1975; Fawcett *et al.*, 1970; Vitale *et al.*, 1973). The blood–testes barrier is a complex of the intercellular Sertoli cell junction with the role of excluding transport and passage of many substances and preleptotene spermatocytes. This

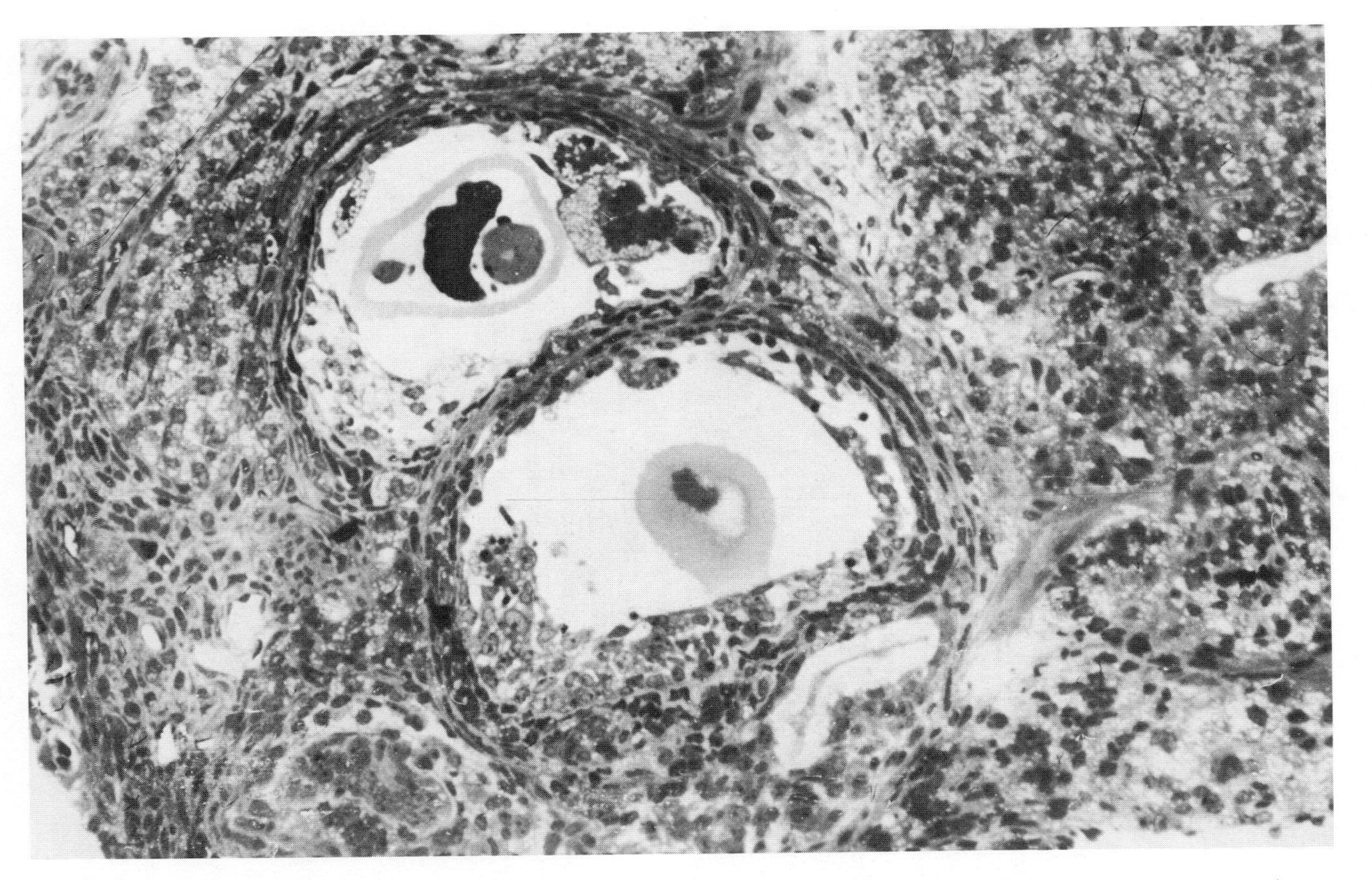

Figure 8. Atretic follicles and interstitium in the ovary of a 38-day-old female rat treated neonatally with E (semithin sections,

complex is not directly dependent on GTH or androgen and seems to be intact for at least 35 days after hypophysectomy (Hagens *et al.*, 1978). In human testes, this barrier is established shortly before or after the proliferation of spermatogonia, giving rise to primary spermatocytes (Furuya *et al.*, 1978).

At the end of the 3rd week, meiosis has begun (Clermont and Perey, 1975) and myoid and Sertoli cells have attained their adult appearance (Bressler, 1978; Bressler and Lustbader, 1978). The Leydig cells in pig occur during the perinatal period (2.5 weeks before and after birth). Both the intertubular and the peritubular cells, which are distinguishable at that time, undergo regression between 3 and 7 weeks after birth (Straaten and Wensing, 1978; Straaten *et al.*, 1978).

6.2. *Steroidogenesis in Testicular Tissue*

Testosterone, as one of the main steroids, is produced in the male embryo at the time of differentiation of enzymatic capacity to convert C_{21} steroids of placental origin to T (George *et al.*, 1978). In the fetal period of the pig, elevation of serum T concentration occurs between 40 and 60 days post coitus and is low when the testes descend (60–100 days post coitus) (Colenbrander *et al.*, 1978).

In very young fetal rat testes, T synthesis occurs, probably autonomously, and although LH receptors are found (Childs, 1978), the testes cannot be stimulated by exogenous gonadotropin (Zaaijer, 1979).

It is known that Leydig cells are the source of T (Moon and Hardy, 1973; Straaten and Wensing, 1978). Considering the importance of pig testes in T production, Colenbrander *et al.* (1978) distinguished three phases of life, the first one during gonadal differentiation occurring in early fetal development, the second in the perinatal period, and the third extending after puberty.

Biosynthesis of T is regulated by a multifactorial control system, such as steroid substrate of maternal origin, as P in fetal testes and fetal pituitary gonadotropins (Pointis *et al.*, 1979). FSH, prolactin, TSH, glucocorticords, and other hormones are involved in testicular steroidogenesis.

Androgen secretion is responsible for the suppression of cyclicity in the neonatal male rat. However, the absence of gonadal activity in the neonatal female allows the hypothalamus to remain in the undifferentiated or in the cyclic state (Gorski, 1966, 1971). During the perinatal period of the rat, the aromatizing system is active, and conversion of androgens to estrogens takes place in the hypothalamus of various species (Ryan and Lee, 1976). The antiestrogen MER-25 prevents the effects of both single and multiple injections of T in the neonatal female rat (McDonald and Doughty, 1974), suggesting that estrogens may play an important role in hypothalamic differentiation.

6.3. Sensitivity during Testicular-Cell Differentiation

It is known that the first signs of mammalian testicular development are the formation of medullary cords containing the germ cells delimited by Sertoli cells and, subsequently, differentiation of Leydig cells (Zaaijer, 1979). Duplication of Sertoli cells *in vivo* and in organ culture was investigated as well.

We established in roosters 3.5 months old a decrease in the weight of the testes, reduction of the seminiferous-tubule diameter, and disturbance or complete inhibition of spermatogenesis if the roosters were treated with single doses of E on days 7 and 11 of embryogenesis, or on days 4, 8, and 12 after hatching. The effect of the hormone was most pronounced and permanent when the treatment was administered earlier during embryogenesis. Dose-dependent changes in animals treated with 500 or 1000 μg E were found on day 4 after hatching and were evident in the testes of roosters at the age of 8 months. Spermatogenesis was recovered and no signs of inhibition effects of E on spermatogenesis were observed in the examined roosters up to the age of 8 months, even if the chicken was treated with a dose of 5000 μg E on day 14 after hatching (Pantić and Kosanović, 1973; Kosanović *et al.*, 1980). In the examined roosters treated with E, the changes were more expressed in the left than in the right gonad and were more extensive and severe with increasing time between the treatment and sacrifice. The results summarized thus far showed that spermatogenesis was inhibited in the first prophase of meiotic division. Teratospermatocytes, destruction of germ epithelium, and degenerative changes were more severe in the left testis. The number, size, and structure of Sertoli cells were clearly altered in all the examined and treated roosters. These cells were smaller in size, their number was decreased, their nuclei were oval, cytoplasm was stained with acridine orange, and the so-called blood–testes barrier was not clearly pronounced (Kosanović *et al.*, 1980).

Cytology and duration of cycle of the germ epithelium of boars was examined by Swierstra (1966). Hyperplasia and hypertrophy of Leydig cells were clearly expressed in boars at the age of 2 and 4 months.

6.4. Sensitivity of Testicular Cells to Gonadotropins

At present, testicular-cell sensitivity to both FSH and LH is considered mainly with respect to the presence of receptors. In an examination of the spermatogenesis of *Carassius* treated in January with hCG or female gonadal steroid, the following results were obtained: hCG injected at this time of year to *Carassius* kept in pools at 18°C has a stimulatory effect. Clear signs of over-stimulation increasing the proliferative rate of germinative cells and the number of spermatozoa were observed in animals treated repeatedly with these hormones (Pantić and Lovren, 1977). Sertoli cells, Leydig cells, and spermatogonia have receptors for FSH. Given that Sertoli cells have 4 times the surface of spermatogonia, it seems that the density of FSH receptors is quite similar in these two cell types (Orth and Christensen, 1978). The Sertoli cell is considered the major FSH target cell in the seminiferous tubules; these cells bind FSH specifically, increasing cAMP synthesis and enhancing incorporation of amino acids into protein (Steinberger *et al.*, 1975). However, it appears that these cells respond directly to FSH and T.

Androgen-binding-protein (ABP) activity in testes and epididymis is increased by highly purified FSH (Tindal and Means, 1976). Testicular binding of highly purified FSH hormone is tissue-specific and dose-dependent (Davies *et al.*, 1975, 1978).

The binding capacity of testicular LH/hCG is significantly increased 6–16 hr after injection of 3–11 IU hCG (Huhtaniemi and Martikaineu, 1978; Huhtaniemi *et al.*, 1978). FSH and prolactin have a potent inhibitory effect on spermatogenesis (Pelletier *et al.*, 1978).

6.5. Sensitivity of Testicular Tissue to Gonadal Steroids

In all the examined species of animals, testicular tissue reaction was clearly pronounced after treatment with a single dose or with repeated doses of female or male gonadal steroids.

Three repeated doses of E alone or in combination with P provoked degenerative changes in Sertoli and germ cells in testes of *Carassius carassius*, rat, and piglet. An inhibitory effect on spermatogenesis was more pronounced if both steroids were administered (Pantić and Gledić, 1977a,b, 1978a,b) (Figs. 9 and 10).

The sensitivity of rooster testes expressed more during embryogenesis; after hatching, it was dose-dependent and decreased from the 12th day, so that recovery of spermatogenesis occurred even if higher single doses were administered after that time. The results obtained thus far

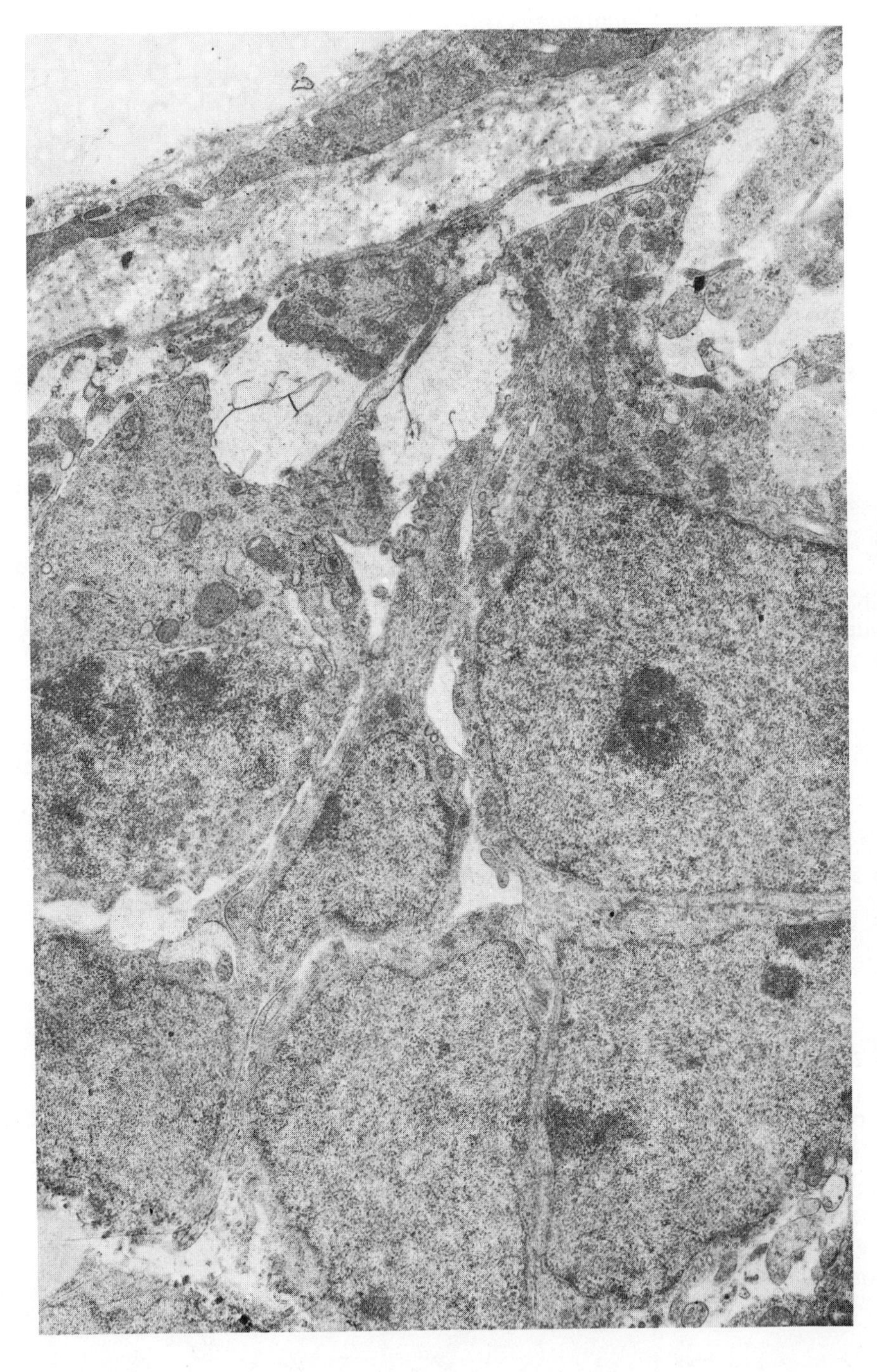

Figure 9. Part of the seminiferous tubule of a 16-day-old intact male rat.

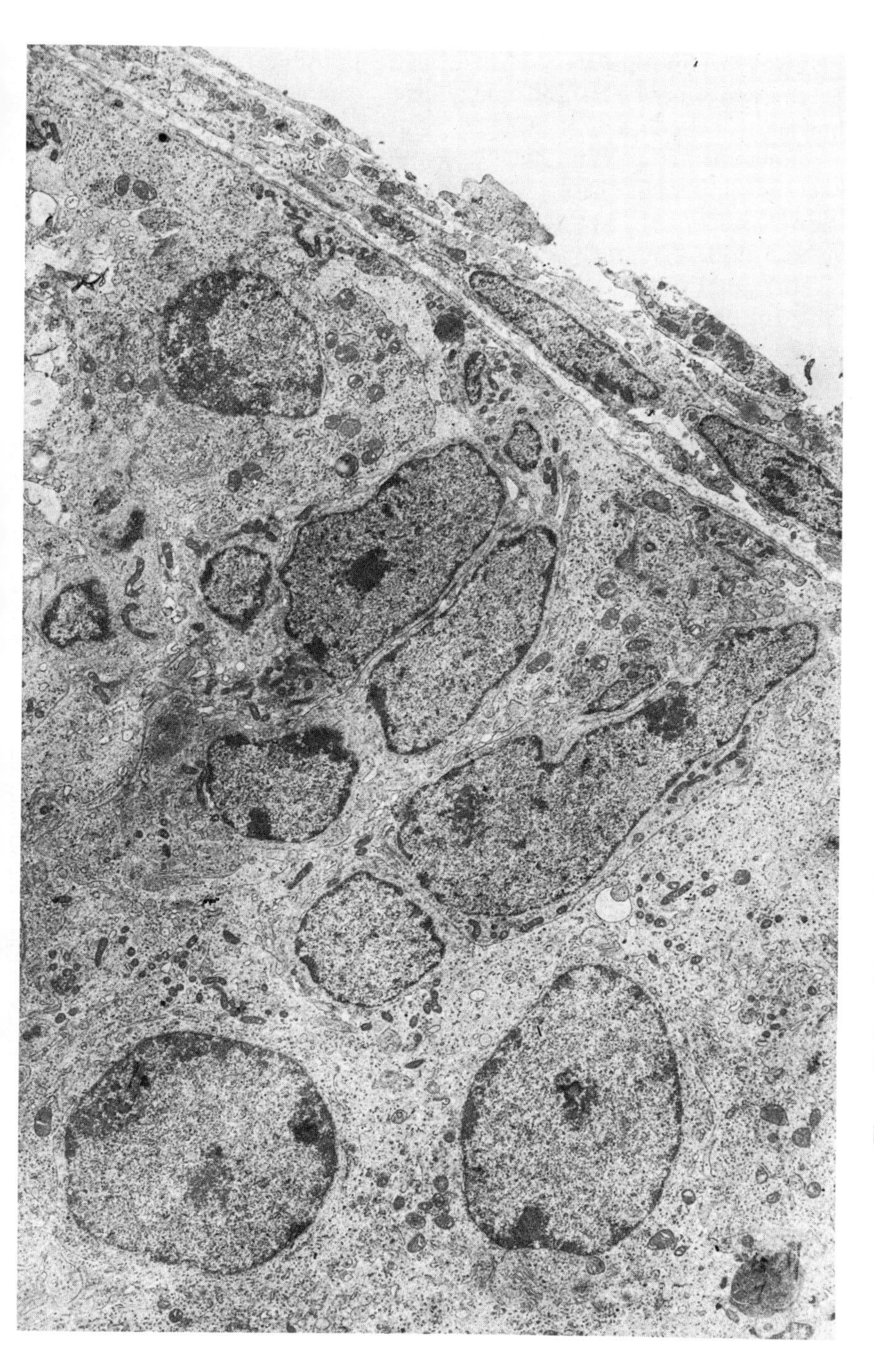

Figure 10. Part of the seminiferous tubule in a 16-day-old male rat treated neonatally with E.

showed that without doubt, a long-term effect was expressed, altering the pathway of differentiation of neurons responsible for production of LRH, changing the LTH- and gonadotropic-cell interrelationship, and diminishing GTH-cell activities; at the same time, however, there was a direct effect on the gonads that was expressed more on the left testis than on the right testis (Pantić and Kosanović, 1973).

Sertoli cells play a key role in the fate of postmitotic germ cells and in the transport of primary spermatocytes from the basal compartment to the adluminal space. Microtubules and microfilaments play an important role in the movement of spermatids in the cytoplasm of these cells during spermiogenesis (Osman and Plöen, 1978).

Assuming that Sertoli cells are target cells for FSH and gonadal steroids, i.e., that in normal developing testes they have receptors for the aforementioned hormones, it seems that if the animals were treated with gonadal steroids (E or TP), they were unable to receive the signals originated from hormones and their abilities to transport T, primary spermatocytes, and various substances were altered. The long-term effect of E or TP was expressed directly or indirectly, or both, on the rat germ-cell epithelium. In a study of the incorporation of [^{3}H]thymidine injected into 70-day-old intact animals and animals treated with a single dose of TP on each of days 19–29 and sacrificed 4 days after injection of 250 μCi [^{3}H]thymidine, this labeled DNA precursor was bound mostly in spermatocytes of intact rats. However, in TP-treated animals, the number of labeled cells was smaller than in controls (Pantić and Gledić, 1978a,b).

At the time of termination of the first wave of spermatogenesis, i.e., at the age of 40–50 days, in rats treated with a single dose of TP on day 4 after birth, spermatogenesis was slowed down so that the cycles of the germ epithelium occurred 10–15 days later than in intact animals, and most spermatocytes degenerated. In rats treated with repeated daily doses of TP from day 19 to day 29, the diameter of the seminiferous tubules was markedly decreased, the spermatocytes in the first meiotic division were the most sensitive cells, and in some tubules only primary and secondary spermatocytes were present. The sensitivity of testicular tissue to a single dose of E, E and P in combination, or T is expressed in pigs as well, and was characterized by undeveloped tubules, altered Sertoli cells, and arrest of spermatogenesis in the preleptotene or the zygotene stage of the first meiotic division. Sertoli cells in boars at the age of 2–4 months are intensively enlarged, and their cytoplasm is vacuolated and expanded to adjacent cells, filling up the tubular lumen. However, these changes are less expressed in pigs than in roosters or rats (Pantić and Gledić, 1977a,b, 1978a,b).

An increased number of Sertoli-cell precursors and spermatogonia, but not of spermatocytes, was observed by Davies (1971) in the testes of 9-day-old mice treated with FSH. He pointed out an increase in the amount of cytoplasm in the seminiferous tubules of immature mice so that the total mass of Sertoli-cell cytoplasm was increased by 62%.

We observed an extensive enlargement in the cytoplasm of these cells in both roosters and pigs treated neonatally with a single dose of E or of E and P in combination. However, in rats treated neonatally with gonadal steroid, these cells appeared more severely involuted than after hypophysectomy (Fig. 11) (Pantić and Gledić, 1978a,b; Kosanović *et al.*, 1980; Pantić and Kosanović, 1973).

FSH is necessary for the restoration and maintenance of spermatogenesis and for the stimulation of ABP secretion by Sertoli cells. Lostroh (1969) pointed out that for complete recovery of spermatogenesis in long-term-hypophysectomized rats, a combined treatment with FSH and LH is necessary.

Recovery of spermatogenesis in about 60–70% of hypophysectomized rats is stimulated by both direct and previously cultivated grafts, transplanted shortly after long-term following a period of 8–28 days after hypophysectomy as a period sufficient to bring about complete involution (Pavić *et al.*, 1977).

A decrease in the weight of the testes, regression of Leydig cells, inhibition of spermatogenesis, and decreased changes in accessory sex glands were the result of a decreased production of FSH and LH in all rats treated with head irradiation and E (Stošić and Pantić, 1970).

6.6. Responsiveness of Testicular Cells to Other Pituitary Hormones and Steroids

Testicular responsiveness to exogenous LH in immature hypophysectomized rats is significantly increased by growth hormone (GH). Prolactin (PRL) has the ability to increase LH-receptor concentration. Maintainance of testicular LH receptors is at least partially dependent on PRL and GH. Rubin *et al.* (1976) pointed out the role of PRL in the regulation of T secretion and testicular development. The mode of PRL action may be an increase in the concentration of esterified cholesterol in the testes (Bartke, 1969, 1971a,b) as a pool of precursors for steroidogenesis (Bartke, 1971a,b). PRL may increase the ability of the testes to produce T in response to LH. In the immature rat (30–35 days old), a parallelism is observed among an increase of plasma PRL level (Bartke

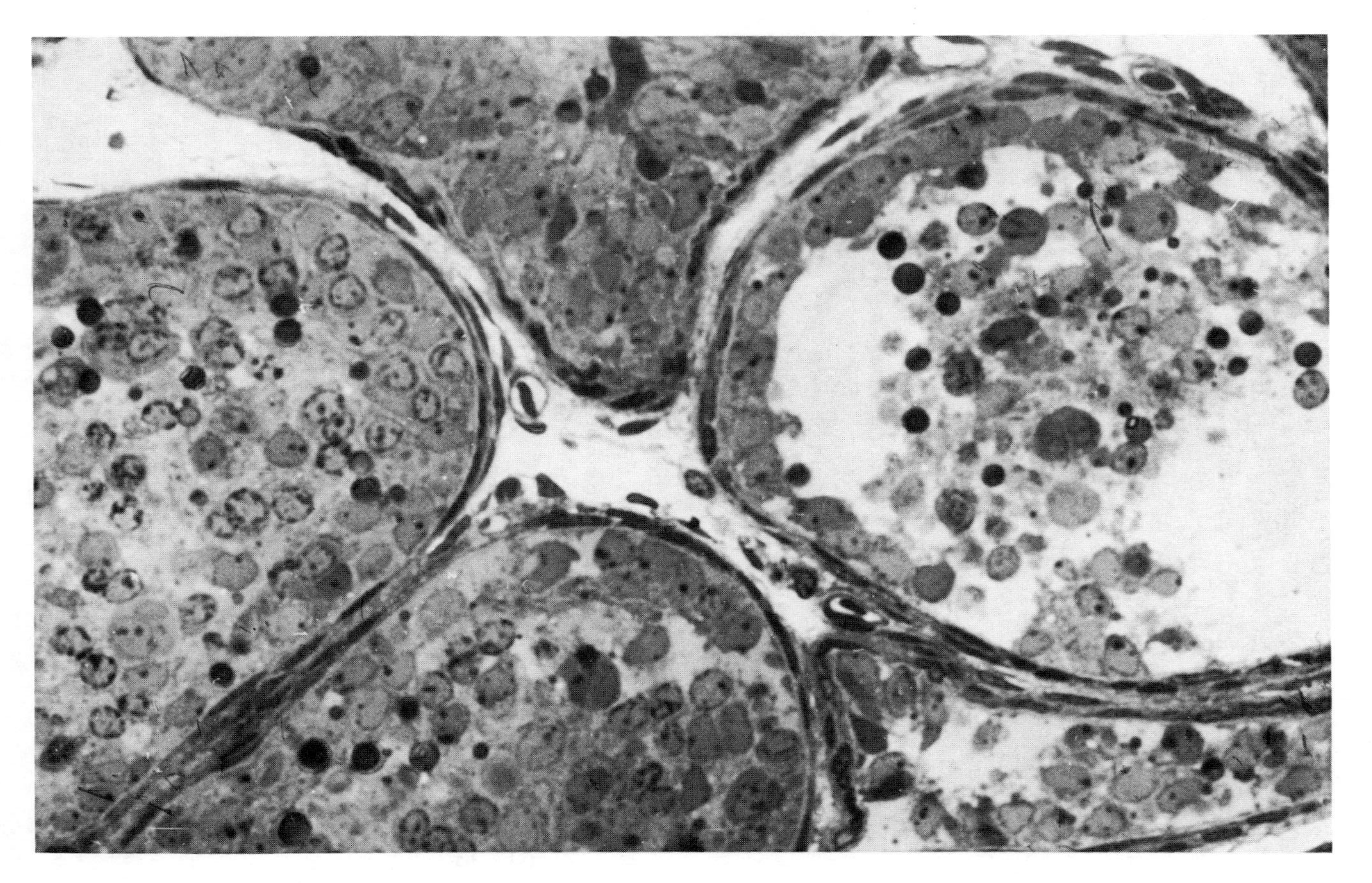

Figure 11. Seminiferous tubules in the inhibited spermatogenesis and cellular degeneration in the testes of a 38-day-old rat treated neonatally with E (semithin section, toluidine blue).

et al., 1977, 1978), endogenous T levels (Knorr *et al.*, 1970), the number of testicular LH cells, and the ability of the testes to respond to LH by T production (Odell and Swerdloff, 1976; Rubin *et al.*, 1976). It has to be mentioned that PRL release might be stimulated by thyrotropin-releasing hormone.

There is a close correlation between thyroid and testicular hyperactivity (Pantić, 1976). The role of hypo- and hyperthyroidism in testicular function has been considered (Leathem, 1972).

In addition to gonadal steroids, other substances are produced in the testicular Sertoli cells. Factors secreted by isolated rat Sertoli cells selectively inhibit FSH secretion by acting directly on the anterior pituitary gonadotrophs (Steinberger and Steinberger, 1976, 1977; Steinberger *et al.*, 1975, 1978). These workers considered that advanced germ cells or spermatogonia or both are a source of inhibin.

T secretion is inhibited by glucocorticoids, and glycocorticoid receptors are present in Leydig cells. Leydig cells in rats atrophied after treatment with ACTH (Asling *et al.*, 1951). In the boar, these cells undergo a strong atrophy from 1 to 3 months after hypophysectomy and can be reactivated by hCG in organ culture, showing the capacity to synthesize T (Morat *et al.*, 1978).

As a result of the effect of a single dose of TP, the first wave of spermatogenesis ended about 10–15 days later than in controls (Pantić and Gledić, 1977a,b). Long-lasting effects of TP in newborn male rats (1–30 days) caused complete inhibition of release and synthesis of GTH and severe atrophy of the testes (Arai and Masuda, 1968).

The effect of gonadal steroids is undoubtedly reflected in the behavior of animals, so that the sexual instinct in male and female pigs was not pronounced during the whole experiment, i.e., up to the adult stage. Such behavior of animals of both sexes is undoubtedly the result of retarded steroidogenesis in testes and ovaries up to the period of sexual maturation (Pantić *et al.*, 1978a,b).

7. *Concluding Remarks*

The sensitivity and character of reactions of pituitary gonadotropic cells to LRH, hCG, and female and male gonadal steroids, and of testicular and ovarian tissue to hCG and gonadal steroids, has been investigated in some species of fishes, chickens, rats, and piglets of different ages. The results described below were obtained.

The size and cellular organization of neurohypophysis and pars intermedia in rats treated neonatally with gonadal steroids might be used as a parameter to show to what extent corresponding brain neurons are altered and under what experimental conditions the reaction is temporary and when permanent. Both lobes are reduced in size, and the cell population is decreased up to puberty. Pituicytes in the neurohypophysis are irregularly distributed and altered, and nerve terminals of aminergic and peptidergic fibers are more devoid of granules and microvesicles. Pars intermedia development depends on these neuron reactions, showing a close inter-relationship with the blood levels of gonadal steroids in all the examined species. Excess of gonadal steroids leads to regression. In the neonatally gonadectomized piglets, which are androgen-deficient from this period onward, GTH cells differentiated even in this lobe.

As a result of the influence of gonadal steroids during embryogenesis and the perinatal period, the paucity of GTH cells is expressed up to puberty. However, the reaction is species-specific and is dependent on the stage of differentiation of both brain-neuron and GTH-cells. The number and the appearance of GTH cells in neonatally treated animals are closely dependent on the degree of hyperplasia of prolactin cells and their properties. GTH cells are less sensitive to estrogen during rat development, and the reaction is far less pronounced in old animals. Recovery of these cells occurs in males treated neonatally or chronically during peripubertal, adult, or old age. Even though the paucity of GTH cells in neonatally treated rats is expressed up to the prepubertal period, they are clearly identifiable during sexual maturation (Pantić *et al.*, 1980b).

Our data obtained thus far showed that the most clearly visible reaction of GTH cells to GRH is expressed in rats. The modulatory role of pretreatment with TP is very important.

An increased number of atretic follicles and formation of cystoid or cystic follicles instead of development of maturing and ovulating follicles in the ovaries of neonatally estrogenized female rats and pigs have to be considered from the point of view of the character and degree of inhibition of GTH-cell differentiation and their synthetic and release capacity, on one hand, and the absence of maturing LH receptors in granulosa, theca, and interstitial cells, on the other. An increase in the number and size of interstitial glands undoubtedly has an influence on the delayed development of maturing and ovulating follicles and their ability for hormone biosynthesis up to puberty. No ovulating follicles or corpora lutea were present in rats up to puberty. In female pigs treated neonatally with T, no pronounced estrus was observed up to the age of 6 months.

The long-term effect of gonadal steroids had an inhibitory effect on the development of the testes in fishes, roosters, and pigs. Leydig and Sertoli cells were atrophic in all roosters and rats, and the inhibition of spermatogenesis was pronounced in the leptothene and zygotene stages of I meiosis. These changes were permanent if the animals were treated during embryogenesis, up to 12 days after hatching of chickens, or after partus of rats. Later on, the inhibitory effect was dose-dependent. Spermatogenesis was retarded if newborn piglets were treated with E and P in combination. However, proliferation and hypertrophy of Leydig cells and appearance of Sertoli cells expressed in piglets from 2 to 4 months old showed signs of anabolic effects. A stimulative effect of hCG on testicular tissue in the examined fishes and rats was observed.

Discussion

Pantić (in reply to a question from Dr. Nolin): GTH cells were identified in groups or as single cells in pars intermedia of different species of animals, using light microscopy, electron microscopy, and immunocytochemistry. With the development of pigs which were gonadectomized the first day of birth, the number of hypotrophied GTH cells increases, not only in pars anterior, but also in pars intermedia and especially in the cranial part of this lobe. My opinion is that the cell differentiation in this lobe is closely associated with the excess or deficit of androgen and if the piglets are deficient in this steroid from the neonatal period, a great number of GTH cells differentiate even in PI.

References

Arai, Y., and Masuda, S., 1968, Long lasting effects of pre-pubertal administration of androgen on male hypothalamic–pituitary–gonadal system, *Endocrinol. Jpn.* **15**(3):375–378.

Asling, C.W., Reinhart, W.O., and Li, C.H., 1951, Effect of andrenocorticotropic hormone on body growth, visceral proportion and white cell counts of normal and hypophysectomized male rats, *Endocrinology* **48:**534–545.

Bartke, A., 1969, Prolactin changes cholesterol stores in the mouse testis, *Nature (London)* **224:**700–701.

Bartke, A., 1971a, Effects of prolactin on spermatogenesis in hypophysectomized mice, *J. Endocrinol.* **49:**311–316.

Bartke, A., 1971b, Effects of prolactin and lutenizing hormone on the cholesterol stores in the mouse testis, *J. Endocrinol.* **49:**317–324.

Bartke, A., 1976, Pituitary–testis relationships, in: *Sperm Action* (P.O. Hubinont, M.L. Hermite, and J. Schwers, eds.), pp. 136–152, S. Karger, Basel.

Bartke, A., Smith, M.S., Michael, S.D., Peron, F.G., and Dalterio, S., 1977, Effects of experimentally induced chronic hyperprolactinemia on testosterone and gonadotropin levels in male rats and mice, *Endocrinology* **100:** 182–186.

Bartke, A., Hafiez, A.A., Bex, J.F., and Dalterio, S., 1978, Hormonal interactions in regulation of androgen secretion, *Biol. Reprod.* **18:**44–45.

Beck, T.W., and Convey, E.M., 1977, Estradiol control of serum luteinizing hormone concentration in the bovine, *J. Anim. Sci.* **45**(5): 1096–1101.

Bender, M.E., Miller, B.J., Possley, M.R., and Keyes, L.P., 1978, Steroidogenic effect of 17β-estradiol in the rabbit: Stimulation of progesterone synthesis in prematurely regressing corpora lutea, *Endocrinology* **103**(5): 1937–1943.

Bern, H.A., Jones, L.A., Mori, T., and Young, P.N., 1975, Exposure of neonatal mice to steroids: Long term effect on the mammary gland and other reproductive structures, *J. Steroid Biochem.* **6:**673–676.

Bern, H.A., Jones, L.A., Mills, K.T., Kohrman, A., and Mori, T., 1976, The use of neonatal mouse in studies of the long term effects of early exposure to hormones and other agents, *J. Toxicol. Environ. Health Suppl.* **1:** 103–116.

Bradham, S.L., 1979, Regulatory subunits of rat brain adenylate cyclase, XIth International Congress of Biochemistry, Toronto, Canada (Abstract 11-7-S134).

Bressler, S.R., 1978, Hormonal control of postnatal maturation of the seminiferous cord, *Ann. Biol. Anim. Biochim. Biophys.* **18**(2B):535–540.

Bressler, S.R., and Lustbader, J.I., 1978, Effect of testosterone on development of the lumen in seminiferous tubules of the rat, *Andrologia* **10**(4):291–298.

Brown-Grant, K., 1969, The induction of ovulation by ovarian steroids in adult rat, *J. Endocrinol.* **43:**553.

Brown-Grant, K., and Greig, F., 1975, A comparison of changes in the peripheral plasma concentration of luteinizing hormone and follicle-stimulating hormone in the rat, *J. Endocrinol.* **65:**389–397.

Caligaris, L., Astrada, J.J., and Taleisnik, S., 1968, Stimulating and inhibiting effects of progesterone on the release of luteinizing hormone, *Acta Endocrinol. (Copenhagen)* **59**(2): 177–185.

Caligaris, L., Astrada, J.J., and Taleisnik, S., 1971, Release of luteinizing hormone induced by estrogen injection into ovariectomized rats, *Endocrinology* **88:**810–815.

Castro-Vasquez, A., and McCann, S.M., 1975, Cyclic variations in the increased responsiveness of the pituitary to luteinizing hormone-release hormone (LHRH) induced by LHRH, *Endocrinology* **97:** 13–18.

Chader, J.G., and Villee, C.A., 1970, Uptake of oestradiol by the rabbit hypothalamus: Specificity of binding by nuclei *in vitro, Biochem. J.* **118:**93–97.

Channing, C.P., 1970a, Effect of stage of estrous cycle and gonadotropin upon luteinization of porcine granulosa cells in culture, *Endocrinology* **87**(1): 156–164.

Channing, C.P., 1970b, Influences of the *in vivo* hormonal environment upon luteinization of granulosa cells in tissue culture, *Recent Prog. Horm. Res.* **26:**589.

Channing, C.P., and Kammerman, S., 1973, Characterisation of gonadotropin receptors of granulosa cells during follicle maturation, *Endocrinology* **92:**531.

Channing, C.P., and Kammerman, S., 1974, Binding of gonadotropins to ovarian cells, *Biol. Reprod.* **10:** 179–198.

Channing, C.P., Sakai, C.N., and Bahl, O.P., 1978, Role of the carbohydrate residues of human chorionic gonadotropin in binding and stimulation of adenosine 3′,5′-monophosphate accumulation by porcine granulosa cells, *Endocrinology* **103**(2):341–348.

Childs, V.G., 1978, Immunocytochemical demonstration of endogenous gonadotropin binding sites in the fetal rat testis, in: *Structure and Function of Gonadotropins* (K.W. McKerns, ed.), Plenum Press, New York.

Clark, M.R., Marsh, J.M., and LeMaire, W.J., 1975, Mechanism of luteinizing hormone regulation of prostaglandin synthesis in rat granulosa cells, *Biophys. Chem.* **253**(21):7757–7761.

Clark, M.R., Marsh, M.J., and LeMaire, J.W., 1978, Stimulation of prostaglandin accumulation in preovulatory rat follicles by adenosine 3′,5′-monophosphate, *Endocrinology* **102**(1):39–44.

Clermont, Y., and Perey, B., 1975, Quantitative study of cell population of the seminiferous tubules in immature rats, *Am. J. Anat.* **100:**241–268.

Colenbrander, B., de Jong, F.H., and Wensing, J.G., 1978, Changes in serum testosterone concentration in the male pig during development, *J. Reprod. Fertil.* **53:**377–380.

Corbier, P., Kerdelhue, B., and Roffi, J., 1979, La crise testiculaire du rat nouveau-né, Congress of European Comparative Endocrinologists, (Abstract 7).

Davies, A.G., 1971, Histological changes in the seminiferous tubules of immature mice following administration of gonadotropins, *J. Reprod. Fertil.* **25:**21–28.

Davies, A.G., Davies, W.E., and Sumner, C., 1975, Stimulation of protein synthesis *in vivo* in immature mouse testis by FSH, *J. Reprod. Fertil.* **42:**415–422.

Davies, A.G., Lawrence, N.R., and Lynch, S.S., 1978, Binding of ^{125}I-labelled FSH in the mouse testis *in vitro, J. Reprod. Fertil.* **53:**249–254.

Debeljuk, L., Arimura, A., and Schally, V.A., 1972, Studies on the pituitary responsiveness to luteinizing hormone-releasing hormone (LH-RH) in intact male rat of different ages, *Endocrinology* **90:**585–588.

Debeljuk, L., Arimura, A., and Schally, V.A., 1973, Stimulation of release of FSH and LH by infusion of LH-RH and some of its analogues, *Neuroendocrinology* **11:**130.

Denef, C., Hautekeete, E., and Dewals, R., 1978a, Monolayer culture of gonadotrophs separated by velocity sedimentation: Heterogeneity in response to luteinizing hormone-releasing hormone, *Endocrinology* **103**(3):736–747.

Denef, C., Hautekeete, E., De Wolf, A., and Vanderschueren, B., 1978b, Pituitary basophils from immature male and female rats: Distribution of gonadotrophs and thyrotrophs as studied by unit gravity sedimentation, *Endocrinology* **103**(3):724–735.

Döhler, K.D., 1978, Is female sexual differentiation hormone-mediated?, *Trends NeuroSci.* **1:**138–140.

Döhler, K.D., and Haucke, J.L., 1978, Thoughts on the mechanisms of sexual brain differentiation, in: *Hormones and Brain Development* (G. Dörner and M. Kawakami, eds.), pp. 153–158, North-Holland, Amsterdam.

Dorrington, J.H., and Armstrong, D.T., 1975a, Follicle-stimulating hormone stimulates estradiol-17 synthesis in cultured Sertoli cells, *Proc. Nat. Acad. Sci. U.S.A.* **72:**2677.

Dorrington, J.H., Moon, Y.S., and Armstrong, D.T., 1975b, Estradiol-17 biosynthesis in cultured granulosa cells from hypophysectomized immature rats; stimulation by follicle-stimulating hormone, *Endocrinology* **97:**1328.

Ellendorff, F., Parvizi, N., Elsaesser, F., and Smidt, D., 1977, The miniature pig as an animal model in endocrine and neuroendocrine studies of reproduction, *Lab. Anim. Sci.* **27**(2) (Part II):822–830.

Elsaesser, F., Parvizi, N., and Ellendorff, F., 1978, Steroid feedback on luteinizing hormone secretion during sexual maturation in the pig, *J. Endocrinol.* **78:**329–342.

Erickson, G.F., and Hsueh, J.W.A., 1978, Secretion of "inhibin" by rat granulosa cells *in vitro*, *Endocrinology* **103**(5):1960–1963.

Etreby, M.F., and Bab, F.M.R., 1978, Effect of 17β-estradiol on cells stained for FSHβ and/or LHβ in the dog pituitary gland, *Cell Tissue Res.* **193**(2):211–218.

Everett, J.W., 1947, Hormonal factors responsible for deposition of cholesterol in the corpus luteum of the rat, *Endocrinology* **41**:364.

Everett, J.W., 1948, Progesterone and estrogen in the experimental control of ovulation time and other features of the estrous cycle in the rat, *Endocrinology* **43**:389.

Fawcett, D.W., 1975, Ultrastructure and function of the Sertoli cell, in: *Handbook of Physiological Endocrinology* (D.W. Hamilton and R.O. Greep, eds.), pp. 21–55, American Physiological Society, Washington, D.C.

Fawcett, D.W., Leak, V.L., and Heidger, N.P., 1970, Electron microscopic observations on the structural components of the blood–testis barrier, *J. Reprod. Fertil. Suppl.* **10**:105–122.

Ferin, M., Bogumil, J., Drewes, J., Dyrenfursh, I., Jewelewicz, R., and Wiele, V.R.L., 1978, Pituitary and ovarian hormonal response to 48h gonadotropin releasing hormone (GnRH) infusions in female rhesus monkeys, *Acta Endocrinol.* **89**(1):48–59.

Ferin, M., Rosenblatt, H., Carmel, P.W., Artunes, J.L., and Wiele, V.R.L., 1979, Estrogen-induced gonadotropin surges in female rhesus monkeys after pituitary stalk section, *Endocrinology* **104**(1):50–52.

Fortune, J.E., and Armstrong, D., 1978, Hormonal control of 17β-estradiol biosynthesis in proestrous rat follicles: Estradiol production by isolated theca versus granulosa, *Endocrinology* **102**(1):227–235.

Fujii, T., Kato, T., and Wakabayashi, K., 1978, Age difference in the response to synthetic luteinizing hormone-releasing hormone (LHRH) in androgenized rats with special reference to the dose of testosterone propionate for androgenization, *Endocrinol. Jpn.* **25**(3):281–287.

Furuya, S., Kumamoto, Y., and Sugiyama, S., 1978, Fine structure and development of Sertoli junctions in human testis, *Arch. Androl.* **1**:211–219.

Genbačev, O., and Pantić, V., 1975, Pituitary cell activities in gonadectomized rats treated with oestrogen, *Cell Tissue Res.* **157**:273–282.

George, F.W., Catt, K.J., Neaves, W.B., and Wilson, W.D., 1978, Studies on the regulation of testosterone synthesis in the fetal rabbit testis, *Endocrinology* **102**:665–673.

Goldenberg, R.L., Viatukaitis, J.L., and Ross, G.T., 1971, Estrogen and follicle stimulating hormone interactions on follicle growth in rats, *Endocrinology* **90**(6):1492–1498.

Gorski, R.A., 1966, Localization and sexual differentiation of the nervous structures which regulate ovulation, *J. Reprod. Fertil. Suppl.* **1**:67–88.

Gorski, R.A., 1971, Gonadal hormones and perinatal development of neuroendocrine functions, in: *Frontiers in Neuroendocrinology* (L. Martini and W.F. Ganong, eds.), pp. 237–290, Oxford University Press, London.

Grandison, L., Hodson, C., Chen, H.T., Advis, J., Simpkins, J., and Meities, J., 1977, Inhibition by prolactin of post-castration rise in LH, *Neuroendocrinology* **23**:312–322.

Grane, L.F., Löw, H., Hall, K., Tally, M., and Grebing, C., 1979, Hormone control of plasma membrane dehydrogenase and its functional implications, XIth International Congress of Biochemistry, Toronto, Canada (Abstract 11-6-S165, p. 625).

Gross, S.D., 1978, Effect of castration and steroid replacement on immunoreactive luteinizing hormone cells in the pars tuberalis of the rat, *Endocrinology* **103**(2):583–588.

Guraya, S.S., 1972, Function of the human ovary during pregnancy as revealed by the histochemical, biochemical and electron microscope techniques, *Acta Endocrinol. (Copenhagen)* **69:** 107–182.

Guraya, S.S., 1973, Interstitial gland tissue of mammalian ovary, *Acta Endocrinol. (Copenhagen)* **72**(Suppl. 171):5–27.

Guraya, S.S., 1976a, Recent advances in the morphology, histochemistry, and biochemistry of steroid-synthesizing cellular sites in the nonmammalian vertebrate ovary, *Int. Rev. Cytol.* **44:**365–409.

Guraya, S.S., 1976b, Recent advances in the morphology, histochemistry and biochemistry of steroid-synthesizing cellular sites in the testes of nonmammalian vertebrates, *Int. Rev. Cytol.* **47:**99–136.

Guraya, S.S., 1978, Recent advances in the morphology, histochemistry, biochemistry and physiology of interstitial gland cells of mammalian ovary, *Int. Rev. Cytol.* **55:** 171–245.

Hagens, L., Plöen, L., and Ekwall, H., 1978, Blood–testis barrier evidence for intact inter-Sertoli cell junctions after hypophysectomy in the adult rat, *J. Endocrinol.* **76:**87–91.

Hamberger, L., Hillensjö, T., and Ahrén, K., 1978. Steroidogenesis in isolated cells of preovulatory rat follicles, *Endocrinology* **103**(3):771–777.

Herbert, D.C., 1975, Localization of antisera to LHβ and FSHβ in the rat pituitary gland, *Am. J. Anat.* **144:**379–385.

Hillier, S.G., Zeleznik, A.J., and Ross, T.G., 1978, Independence of steroidogenic capacity and luteinizing hormone receptor induction in developing granulosa cells, *Endocrinology* **102**(3):937–946.

Hirshfield, A.N., and Midgley, A.R., Jr., 1978, The role of FSH in the selection of large ovarian follicles in the rat, *Biol. Reprod.* **19:**606.

Horikoshi, H., Miyagawa, H., and Baba, Y., 1978, Effect of steroids in combination with LH-RH on the release of LH and FSH in LH-RH-primed immature male rat, *Endocrinol. Jpn.* **25**(4):391–396.

Huhtaniemi, I.T., and Martikainen, H., 1978, Rat testis LH/hCG receptor and testosterone production after treatment with GnRH, *Mol. Cell. Endocrinol.* **11:** 199–204.

Huhtaniemi, I., Martikainen, H., and Tikkala, L., 1978, hCG-induced changes in the number of rat testis LH/hCG receptors, *Mol. Cell. Endocrinol.* **11:**43–50.

Jablan-Pantić, O., and Miladinović, Z., 1977, Arterial supply to the pig ovary and its relationship to ovarian cycle, *Isr. J. Med. Sci.* **14**(8):896 (Abstract).

Jalabert, B., 1976, *In vitro* oocyte maturation and ovulation in rainbow trout *(Salmo gairdneri),* northern pike *(Esox lucius)* and goldfish *(Carrassius auratus), J. Fish. Res. Board Can.* **33:**974–988.

Jalabert, B., Breton, B., and Bry, C., 1972, Maturation et ovulation *in vitro* des ovocytes de la truite arc-en-ciel *Salmo gairdneri, C. R. Acad. Sci.* **275:** 1139–1142.

Jones, A.L., and Bern, H.A., 1977, Long-term effects of neonatal treatment with progesterone, alone and in combination with estrogen, on the mammary gland and reproductive tract of female BALB/Cf3H mice, *Cancer Res.* **37:**67–75.

Kao, L.W.L., and Weisz, J., 1975, Direct effect of testosterone and its 5α-reduced metabolites on pituitary LH and FSH release *in vitro:* Change in pituitary responsiveness to hypothalamic extract, *Endocrinology* **96**(2):253–260.

Kaul, S., and Vollrat, L., 1974, The goldfish pituitary: Cytology, *Cell Tissue Res.* **154:**211–230.

Kawakami, M., Arita, J., Yoshioka, E., Visessuvan, S., and Akema, T., 1978a, Data on the sites of the stimulatory feedback action of gonadal steroids indispensable for follicle-stimulating hormone release in the rat, *Endocrinology* **103:**752–770.

Kawakami, M., Yoshioka, E., Konda, N., Arita, J., and Visessuvan, S., 1978b, Data on the sites of the stimulatory feedback action of gonadal steroids indispensable for luteinizing hormone release in rat, *Endocrinology* **102**(3):791–798.

Kimura, T., and Nandi, S., 1967, Nature of induced persistent vaginal cornification in mice. IV. Changes in the vaginal epithelium of old mice treated neonatally with estradiol or testosterone, *J. Natl. Cancer. Inst.* **39:**75–93.

Knorr, D.W., Vanha-Perttula, T., and Lipsett, M.B., 1970, Structure and function of rat testis through pubescence, *Endocrinology* **86:**1298–1304.

Kosanović, M., Pantić, V., and Šijački, N., 1979, Long term effect of estradiol on the development of rooster testes, *Acta Vet. (Belgrade)* **29**(5):207–221.

Kurosumi, K., 1968, Functional classification of cell types of the anterior pituitary gland accomplished by electron microscopy, *Arch. Histol. Jpn.* **29:**329–362.

Leathem, J.H., 1972, Role of the thyroid, in: *Reproductive Biology* (H. Balin and S. Glasser, eds.), pp. 857–876, Excerpta Medica, Amsterdam.

Lee, C.Y., 1976, The porcine ovarian follicle. III. Development of chorionic gonadotropin receptors associated with increase in adenyl cyclase activity during follicle maturation, *Endocrinology* **99**(1):42–48.

Lee, C., Aloj, S.M., Brady, R.O., and Kohn, L.D., 1976, The structure and function of glycoprotein hormone receptors: Ganglioside interaction with human chorionic gonadotropin, *Biochem. Biophys. Res. Commun.* **69:**852.

Litteria, M., 1973, Inhibitory action of neonatal androgenization on the incorporation of (^{3}H)-lysine in specific hypothalamic nuclei of the adult female rat, *Exp. Neurol.* **41:**395–401.

Lostroh, A.J., 1969, Regulation by FSH and ICSH (LH) of reproductive function in the immature male rat, *Endocrinology* **85:**438–445.

Markis, A., and Ryan, K.J., 1975, Progesterone, androstenedione, testosterone, estrone and estradiol synthesis in hamster ovarian follicle cells, *Endocrinology* **96**(3):694–701.

Markis, A., and Ryan, K.J., 1977, Evidence for interaction between granulosa cells and theca in early progesterone synthesis, *Endocr. Res. Commun.* **4**(3&4):233–246.

Masui, Y., and Clarke, J.H., 1979, Oocyte maturation, *Int. Rev. Cytol.* **57:**186–282.

McDonald, G.P., and Doughty, C., 1974, Effect of neonatal administration of different androgens in the female rat: Correlation between aromatization and the induction of sterilization, *Endocrinology* **61:**95–103.

Meusy-Dessole, N., 1975, Variations quantitatives de la testosterone plasmatique chez la porc male, de la maissance a l'age adulte, *C.R. Acad. Sci. Ser. D* **281:**1875–1878.

Meyer, R.K., and McCormack, C.E., 1967, Ovulation in immature rats treated with ovine follicle stimulating hormone: Facilitation by progesterone and inhibition by continuous light, *J. Endocrinol.* **38:**187.

Midgley, A.R., 1973, Autoradiographic analysis of gonadotropin binding to rat ovarian tissue secretion, in: *Receptors for Reproductive Hormones*, Vol. 36, *Advances in Experimental Biology and Medicine* (B.W. O'Malley and A.R. Means, eds.), pp. 365–378, Plenum Press, New York.

Moon, J.S., and Hardy, M.H., 1973, The early differentiation of the testis and interstitial cells in the fetal pig, and its duplication in organ culture, *Am. J. Anat.* **138:**253–268.

Morat, M., Chevalier, M., and Dufaure, J.P., 1978, Mise en evidence d'une secretion de testosterone par le tissue testiculaire de Verrat hypophysectomise, sous l'action de HCG, en culture organotypique, *C. R. Acad. Sci. Ser. D* **286:**1605.

Mori, T., Bern, H.A., Mills, K.T., and Young, P.N., 1976, Long-term effect of neonatal steroid exposure on mammary gland development and tumorigenesis in mice, *J. Natl. Cancer Inst.* **57:**1057–1062.

Nakane, P.K., 1970, Classification of anterior pituitary cell types with immunoenzyme histochemistry, *J. Histochem. Cytochem.* **18:**9–20.

Odell, W.D., and Swerdloff, R.J., 1976, Etiologies of sexual maturation: A model system based on the sexually maturing rat, *Recent Prog. Horm. Res.* **32:**245–288.

Orth, J., and Christensen, A.K., 1978, Autoradiographic localization of FSH-binding sites on Sertoli cells and spermatogonia in testes from hypophysectomized rats, in: *Structure and Function of the Gonadotropins* (K.W. McKerns, ed.), Plenum Press, New York and London.

Osman, I.D., and Plöen, L., 1978, The terminal segment of the seminiferous tubules and the blood–testis barrier before and after efferent ductule ligation in the rat, *Int. J. Androl.* **1:**235–249.

Paavola, L.G., 1979, The corpus luteum of the guinea pig. IV. Fine structure of macrophage during pregnancy and postpartum luteolysis, and the phagocytosis of luteal cells, *Am. J. Anat.* **154:**337–363.

Pantić, V., 1974, Gonadal steroids and hypothalamo–pituitary–gonadal axis, *INSERM* **32:**97–118.

Pantić, V., 1975, The specificity of pituitary cells and regulation of their activities, *Int. Rev. Cytol.* **40:**153–195.

Pantić, V., 1976, Ultrastructure of deer and roe-buck thyroid, *Z. Zellforsch.* **81:**487–500.

Pantić, V.R., 1980, Adenohypophyseal cell specificities and gonadal steroids, in: *Synthesis and Release of Adenohypophyseal Hormones* (M. Jutisz and K.W. McKerns, eds.), pp. 335–362, Plenum Press, New York.

Pantić, V., and Drekić, D., 1980, Long term effect of estrogen on cells of corpus amygdaloideum nuclei, *Acta Vet. (Belgrade)* (in press).

Pantić, V., and Genbačev, O., 1971, Luteotropic and gonadotropic activities of the pituitaries of rats neonatally treated with oestrogen, VIth Conference of European Comparative Endocrinologists, Montpellier (Abstract 166).

Pantić, V., and Gledić, D., 1977a, Long term effects of gonadal steroids on pig pituitary gonadotropic cells and testes, *Bull. T.LX Acad. Serbian Sci. Arts* **16:**91–109.

Pantić, V., and Gledić, D., 1977b, Reaction of pituitary gonadotropic cells and testes to testosterone propionate (TP), *Bull. T.LVI Acad. Serbian Sci. Arts* **15:**131–146.

Pantić, V., and Gledić, D., 1978a, Gonadotropic and germ cells reaction to oestrogen, in: *Proceedings of the International Symposium of Neuroendocrine Regulatory Mechanisms*, pp. 87–94.

Pantić, V., and Gledić, D., 1978b, Reaction of pituitary gonadotropic and germ cells of male rats and pigs neonatally treated with estrogen, *Gen. Comp. Endocrinol.* **34**(1) (Abstract 54).

Pantić, V., and Kosanović, M., 1973, Testes of roosters treated with a single dose of estradiol dipropionate, *Gen. Comp. Endocrinol.* **21:**108–117.

Pantić, V., and Lovren, M., 1977, Examinations of testes of *Carassius carrassius* treated with choriogonadotropin or female sexual steroids, *Folia Anat. Jugoslav.* **6:**73–84.

Pantić, V., and Lovren, M., 1978, The effect of female gonadal steroids on carp pituitary gonadotropic cells and oogenesis, *Folia Anat. Jugoslav.* **7:**25–34.

Pantić, V., and Lovren, M., 1979, Ultrastructure of the ovary of rats neonatally treated with oestrogen, XIIth Congress of Italian Society of Electron Microscopy (Abstract 131).

Pantić, V., and Nikolić, Z., 1980, Long term effect of estrogen on the rat ovary, *Acta Vet. (Belgrade)* (in press).

Pantić, V., and Šimić, M., 1977a, Effect of gonadal steroids on pituitary pars intermedia cells of some teleostea and rat, *Bull. T.LX Acad. Serbian Sci. Arts* **16:**23–40.

Pantić, V., and Šimić, M., 1977b, Sensitivity of the pituitary pars intermedia to castration or gonadal steroids, *Bull. T.LX Acad. Serbian Sci. Arts* **16:**67–80.

Pantić, V., and Šimić, M., 1978, Effect of gonadal steroids on pars intermedia cells of some teleostei, the rat and the pig, *Gen. Comp. Endocrinol.* **34**(1) (Abstract 27).

Pantić, V., and Škaro, A., 1974, Pituitary cells of roosters and hens treated with a single dose of oestrogen during embryogenesis or after hatching, *Cytobiologie* **9:**72–83.

Pantić, V., Genbačev, O., Milković, S., and Ožegović, B., 1971a, Pituitaries of rats bearing transplantable MtT mammotropic tumor, *J. Microsc.* **3:**405–415.

Pantić, V., Ožegović, B., Genbačev, O., and Milković, S., 1971b, Ultrastructure of transplantable pituitary tumor cells producing luteotropic and adrenocorticotropic hormones, *J. Microsc.* **12:**225–232.

Pantić, V., Genbačev, O., and Jovanović, M., 1975a, Effect of long term Protamone treatment on pituitary cell structure and function in rats, *Acta Vet. (Belgrade)* **25**(4): 179–187.

Pantić, V., Genbačev, O., and Gledić, D., 1975b, The ability of pituitary cells to respond to gonadal steroids, VIIIth Conference of European Comparative Endocrinologists, Bangor (abstracts), p. 145.

Pantić, V., Stolević, E., Gledić, D., and Momčilov, P., 1977, Effect of gonadotropic releasing hormone on pituitary–gonadal axis of female rats and women, *Int. J. Fert.* (in press).

Pantić, V., Šijački, N., and Kolarić, S., 1978a, The role of gonadal steroids in the regulation of behaviour and productive performance of pigs, *Acta Vet. (Belgrade)* **28**(1):31–40.

Pantić, V., Sekulić, M., Lovren, M., and Šimić, M., 1978b, Neurosecretory and pars intermedia cells of fish and mammals, in: *Neurosecretion and Neuroendocrine Activity Evolution, Structure and Function* (W. Bargmann, A. Oksche, A. Polenov, and B. Scharrer, eds.), pp. 257–259, Springer-Verlag, Berlin.

Pantić, V., Šijački, N., and Milinković, R., 1980a, Long term effect of estrogen on gonadal development in fish *Salmo gardnieri, Acta Vet. (Belgrade)* (in press).

Pantić, V.R., Gledić, D., and Martinović, J.V., 1980b, Pituitary gonadotropic cells, serum concentration of gonadotropins and testicles in adult rats neonatally treated with a single dose of estradiol **30:**101–114.

Park, K.R., Saxena, B.B., and Gandy, H.M., 1976, Specific binding of LHRH to the anterior pituitary gland during the estrous cycle in the rat, *Acta Endocrinol.* **82**(1):62–70.

Pavić, D., Živković, N., Pantić, V., and Martinovitch, P., 1977, Glucoprotein synthesizing cells in the anterior pituitaries transplanted into hypophysectomized male rats, *Arch. Sci. Biol.* **27**(1-2):1–8.

Pavlović, M., and Pantić, V., 1975, The adenohypophysis in the teleostea *Alburnus albidus* and *Alosa fallax* in different phases of sexual cycle, *Acta Vet. (Belgrade)* **25**(4): 163–178.

Pelletier, G., Dupont, A., and Puviani, R., 1975, Ultrastructural study of the uptake of peroxidase by the rat median eminence, *Cell Tissue Res.* **156**(4):521–532.

Pelletier, G., Cusan, L., Auclair, C., Kelly, P.A., Désy, L., and Labrie, F., 1978, Inhibition of spermatogenesis in the rat by treatment with (D-Ala^6, des-Gly-$NH_2$10) LHRH ethylamide, *Endocrinology* **103**(2):641–643.

Pointis, G., Latreille, M.T., and Cedard, L., 1979, Regulation of testosterone biosynthesis in the foetal mouse, *Xth Conference of European Comparative Endocrinologists* (Abstract 23).

Ramirez, V.D., and Sawyer, C.H., 1965, Advancement of puberty in the female rat by estrogen, *Endocrinology* **76:**1158–1168.

Rao, V.C., Mitra, S., and Carman, F.R., 1979, Properties of gonadotropin binding sites in rough endoplasmic reticulum (RER) and Golgi fractions (GFs) of bovine corpora lutea, XIth International Congress of Biochemistry, Toronto (Abstract 11-3-S15).

Raeside, J., and Sigman, D.M., 1975, Testosterone levels in early fetal testes of domestic pigs, *Biol. Reprod.* **13:**318–321.

Recabarren, S.E., and Wheaton, J.E., 1978, Estradiol potentiation of hypothalamic uptake of LH-RH from the CSF, *Neuroendocrinology* **27:**1–8.

Reed, M., and Bellinger, J., 1979, Glycogen in guinea-pig ovarian follicles, *Xth Conference of European Comparative Endocrinologists* (Abstract 27).

Richards, J.S., and Midgley, A.R.V., 1976, Protein hormone action: A key to understanding ovarian follicular and luteal cell development, *Biol. Reprod.* **14:**82–94.

Rubin, R.T., Poland, R.E., and Tower, B.B., 1976, Prolactin-related testosterone secretion in normal adult men, *J. Clin. Endocrinol. Metab.* **42:**112–116.

Ryan, R.J., and Lee, C.Y., 1976, The role of membrane bound receptors, *Biol. Reprod.* **14:**16–29.

Sar, M., and Stumpf, E.W., 1973, Pituitary gonadotrophs: Nuclear concentration of radioactivity after injection of (^{3}H)testosterone, *Science* **179:**389–391.

Schally, A.V., Arimura, A., and Kastin, A.J., 1973, Hypothalamic regulatory hormones, *Science* **179:**341–350.

Sekulić, M., and Pantić, V., 1977, Somatotropic and prolactin pituitary cells of Teleostea fish, XVIIth Congress of the Anatomical Association of Yugoslavia (Abstracts), Dubrovnik, *Folia Anat. Jugoslav.* (Addendum).

Sétáló, G., Horváth, J., Schally, A.V., Arimura, A., and Flerkó, B., 1977, Effect of the isolated removal of the median eminence (ME) and pituitary stalk (PS) on the immunohistology and hormone release of the anterior pituitary gland grafted into the hypophysiotrophic area (HTA) and/or of the *in situ* pituitary gland, *Acta Biol. Acad. Hung.* **28**(37):333–349.

Sheridan, J.P., Sar, M., and Stumpf, E.W., 1974, Autoradiographic localization of ^{3}H-testosterone or its metabolites in the neonatal rat brain, *Am. J. Anat.* **140:**589–593.

Škaro-Milić, A., and Pantić, V., 1976, Gonadotropic and luteotropic cells in chicken treated with oestrogen after hatching, *Gen. Comp. Endocrinol.* **28:**283–291.

Škaro-Milić, A., and Pantić, V., 1977, Pituitary cells of roosters treated with a single dose of testosterone propionate (TP) after hatching, *Folia Anat. Jugoslav.* **6:**33–44.

Soji, T., 1978a, Changes of serum and pituitary TSH, LH and FSH concentration following the slow infusion of TRH and LRH, *Endocrinol. Jpn.* **25**(3):231–236.

Soji, T., 1978b, Cytological changes of the pituitary basophils in rats slowly infused with LRH and with LRH and TRH in combination, *Endocrinol. Jpn.* **25**(3):259–274.

Soji, T., 1978c, Cytological changes of the pituitary basophils in rats slowly infused with thyrotropin-releasing hormone (TRH), *Endocrinol. Jpn.* **25**(3):245–258.

Spona, J., 1973, LH-RH stimulated gonadotropin release mediated by two distinct pituitary receptors, *FEBS Lett.* **35:**59–62.

Steinberger, A.L., and Hoffman, E.G., 1978, Immunocytobiology of luteinizing hormone-releasing hormone, *Neuroendocrinology* **25:**111–128.

Steinberger, A., and Steinberger, E., 1976, Secretion of an FSH-inhibiting factor by cultured Sertoli cells, *Endocrinology* **99:**918.

Steinberger, A., and Steinberger, E., 1977, Inhibition of FSH by Sertoli cell-factor (SCF) *in vitro,* in: *The Testis in Normal and Infertile Men* (P. Troen and H.R. Nankin, eds.), pp. 271–279, Raven Press, New York.

Steinberger, A., Elkington, H.S.J., Sanborn, M.B., Steinberger, E., Heindel, J.J., and Lindsey, N.J., 1975, Culture and FSH responses of Sertoli cells isolated from sexually mature rat testis, in: *Hormonal Regulation of Spermatogenesis* (F.S. French, V. Hansson, E.M. Ritzen, and S.N. Nayfeh, eds.), pp. 399–411, Plenum Press, New York.

Steinberger, A., Sanborn, M.B., and Steinberger, E., 1978, FSH and Sertoli cells, in: *Structure and Function of Gonadotropins* (K.W. McKerns, ed.), pp. 517–551, Plenum Press, New York.

Stošić, N., and Pantić, V., 1970, The hypothalamus, pituitary and testes of intact and thyroidectomized rats after head irradiation and treatment with oestrogen, *Arch. Sci. Biol.* **22**(1-4):67–75.

Straaten, H.W.M. van, and Wensing, C.J.G., 1978, Leydig cell development in the testis of pig, *Biol. Reprod.* **17:**86–93.

Straaten, H.W.M. van, Ridder, R.R., and Wensing, C.J.G., 1978, Early deviation of testicular Leydig cells in the naturally unilateral cryptorchid pig, *Biol. Reprod.* **19:**171–176.

Stumpf, E.W., 1972, Estrogen, androgen and glucocorticosteroid concentrating neurons in the amygdala, studied by dry autoradiography, in: *The Neurobiology of the Amygdala, Advances in Behavioral Biology,* Vol. 2 (B.E. Eleftheriou, ed.), pp. 763–774, Plenum Press, New York.

Swierstra, E.E., 1966, Cytology and duration of the cycle of the seminiferous epithelium of the boar: Duration of spermatozoan transit through the epididymis, *Anat. Rec.* **161:**171–186.

Thanki, K.H., and Channing, C.P., 1978, Effects of follicle-stimulating hormone and estradiol upon progesterone secretion by porcine granulosa cells in tissue culture, *Endocrinology* **103**(1):74–80.

Tindal, D., and Means, A.R., 1976, Concerning the hormonal regulation of androgen binding protein in rat testis, *Endocrinology* **99**(3):809–818.

Tougard, C., Picart, R., and Tixier-Vidal, A., 1977, Cytogenesis of immunoreactive gonadotropic cells in the fetal rat pituitary at light and electron microscopic levels, *Dev. Biol.* **58:**148–163.

Vigh, S., Sétálo, G., Török, A., Pantić, V., Flerko, B., and Gledić, D., 1978, Deficiency of FSH and LH cells in rats treated with oestradiol in the early postnatal life, *Bull. T.LXI Acad. Serbian Sci. Arts* **17:**1–7.

Vilchez-Martinez, J.A., Arimura, A., Debeljuk, L., and Schally, A.V., 1974, Biphasic effect of estradiol benzoate on the pituitary responsiveness to LH-RH, *Endocrinology* **94:**1300–1303.

Vitale, R., Fawcett, D.W., and Dym, M., 1973, The normal development of the blood–testis barrier and the effect of clomiphene and estrogen treatment, *Anat. Rec.* **176:**333–344.

Wagner, T.O.F., Adams, T.E., and Nett, T.M., 1979, GNRH interaction with anterior pituitary. I. Determination of the affinity and number of receptors of GNRH in ovine anterior pituitary, *Biol. Reprod.* **20:**140–149.

Zaaijer, J.J.P., 1979, Trends in sex differentiation (vertebrates), *Xth Conference of European Comparative Endocrinologists* (Abstracts).

Zeilmaker, G.H., 1960, The biphasic effect of progesterone on ovulation in the rat, *Acta Endocrinol. (Copenhagen)* **51:**461.

Zeleznik, A.J., Midgley, A.R., and Reichert, L.E., 1974, Granulosa cell maturation in the rat: Increased binding of human chorionic gonadotropin following treatment with follicle stimulating hormone *in vivo, Endocrinology* **95:**818–825.

II

GnRH Analogues as Contraceptive Agents

4

Induction of Menstruation Following Subcutaneous Injection of LH-RH and Intranasal Administration of [D-Ser(TBU)^6,des-Gly-NH_2^{10}]LH-RH Ethylamide after Ovulation in Normal Women

A. Lemay, F. Labrie, and J.P. Raynaud

1. Introduction

Although luteinizing-hormone-releasing hormone (LH-RH) can induce ovulation in the animal (Schally and Arimura, 1977) and in some cases in the human (Nillius, 1976), it has recently been demonstrated that this hypothalamic hormone, as well as potent agonistic analogues, have antifertility effects in the animal. The administration of LH-RH or its agonistic analogue [D-Ala^6,des-Gly-NH_2^{10}] LH-RH ethylamide on diestrus I or II in normal female rats has been shown to delay the expected day of mating and vaginal cornification (Banik and Givner, 1975; Beattie and Corbin, 1977; Ferland *et al.*, 1978; Kledzik *et al.*, 1978a). Gonado-

A. Lemay • Laboratory of Endocrinology of Reproduction, Department of Obstetrics and Gynecology, Hôpital St. François D'Assise, Quebec G1L 3L5, Canada. *F. Labrie* • Department of Molecular Endocrinology, Le Centre Hospitalier de l'Université Laval, Quebec G1V 4G2, Canada. *J. P. Raynaud* • Centre de Recherches Roussel-Uclaf, 93230 Romainville, France.

tropin-releasing peptides also have potent postcoital contraceptive activity in female rats (Corbin *et al.*, 1976; Humphrey *et al.*, 1976; Kledzik *et al.*, 1978b) and rabbits (Hilliard *et al.*, 1976).

Recently, it has been found that these antifertility effects of LH-RH and its agonists are accompanied by a marked loss of ovarian LH, follicle-stimulating hormone (FSH), and prolactin receptors as well as decreased plasma progesterone concentration (Ferland *et al.*, 1978; Kledzik *et al.*, 1978b; Corbin *et al.*, 1976). These data suggested that endogenous gonadotropin release induced by treatment with the neuropeptides leads to inhibition of ovarian receptors and function (Labrie *et al.*, 1979). Despite the luteotropic effect of human chorionic gonadotropin (hCG), luteolytic effects of LH and hCG have also been found in the pseudopregnant rabbit (Spies *et al.*, 1966) and in the pregnant sheep, hamster, and rat (Moor *et al.*, 1968; Yang and Chang, 1968; Banik, 1975). These luteolytic effects were supported by the recent findings that treatment with LH or hCG induces a marked loss of LH receptors in the ovary accompanied by a reduced ability of LH to stimulate ovarian adenylyl cyclase activity (Conti *et al.*, 1976; Hunzicker-Dunn and Birnbaumer, 1976).

Following the observation of such potent antifertility effects of LH-RH and its agonistic analogues in the rat and a better knowledge of the mechanisms involved (down-regulation of ovarian gonadotropin receptor levels and function), we investigated the possibility of a similar luteolytic effect of treatment with gonadotropin-releasing peptides during the luteal phase in normal women. Such an approach could lead to a near-physiological method of control of the luteal-phase length and time of appearance of menses. This chapter summarizes the results obtained so far after subcutaneous treatment with LH-RH (Lemay *et al.*, 1978, 1979a) and intranasal administration of [D-Ser(TBU)^{6},des-Gly-NH_2^{10}]LH-RH ethylamide (Lemay *et al.*, 1979b). Possible clinical applications of such treatment in the control of the menstrual cycle and fertility regulation in normal women are also discussed.

2. Subjects, Treatments, and Methods

2.1. Subjects

All volunteers participating in the studies gave informed consent. They had normal and regular menstrual cycles. There was no endocrinological or gynecological problem as evaluated by medical history, complete physical examination, and routine blood and urine analysis. None had taken birth control pills or any hormonal medication for at least 6 months.

2.2. Subcutaneous LH-RH Treatment

Ten volunteers aged 26–42 were studied for a total of 17 treatment cycles. The study was conducted during at least three consecutive cycles corresponding to pretreatment, treatment, and posttreatment cycles. In some cases, a second treatment was administered during a later cycle. Doses of 250 μg LH-RH (Factrel, Ayerst Laboratories) were given subcutaneously every 4 hr for five injections starting at 08:00 hr on one or two consecutive days. The LH-RH treatment was started 1–9 days after the LH peak (day 0).

2.3. Intranasal [D-Ser(TBU^6,des-Gly-NH_2^{10}]LH-RH Ethylamide Treatment

Six volunteers, aged 26–35, were studied during six consecutive menstrual cycles corresponding to two pretreatment, two treatment, and two posttreatment cycles. Two drops of a solution containing 500 μg [D-Ser(TBU)^6,des-Gly-NH_2^{10}]LH-RH ethylamide (graciously provided by Dr. Van der Ohe and Sandow, Hoechst Pharmaceuticals) were deposited in each nostril at 08:00 and 17:00 hr on one day between days 4 and 9 after the LH peak (day 0).

2.4. Methods

Blood samples were withdrawn daily between 16:00 and 18:00 hr for measurement of serum LH, FSH, progesterone, and estradiol, which were determined by double-antibody radioimmunoassays as described previously (Lemay *et al.*, 1979a). Materials for LH and FSH measurements were supplied through the courtesy of the National Pituitary Agency, National Institutes of Health, Bethesda, Maryland. Radioimmunoassay data were analyzed with a program derived from Model II of Rodbard and Lewald (1970). Statistical significance was measured according to the multiple-range test of Kramer (1956).

3. Results

3.1. Subcutaneous LH-RH Treatment

As illustrated in Fig. 1, the administration of five subcutaneous doses of LH-RH at 4-hr intervals on the fifth day after the LH peak caused a second peak of plasma levels of LH and FSH. A significant increase in serum estradiol level was also noted on the day of LH-RH treatment. At 5 days after the LH-RH treatment, serum progesterone

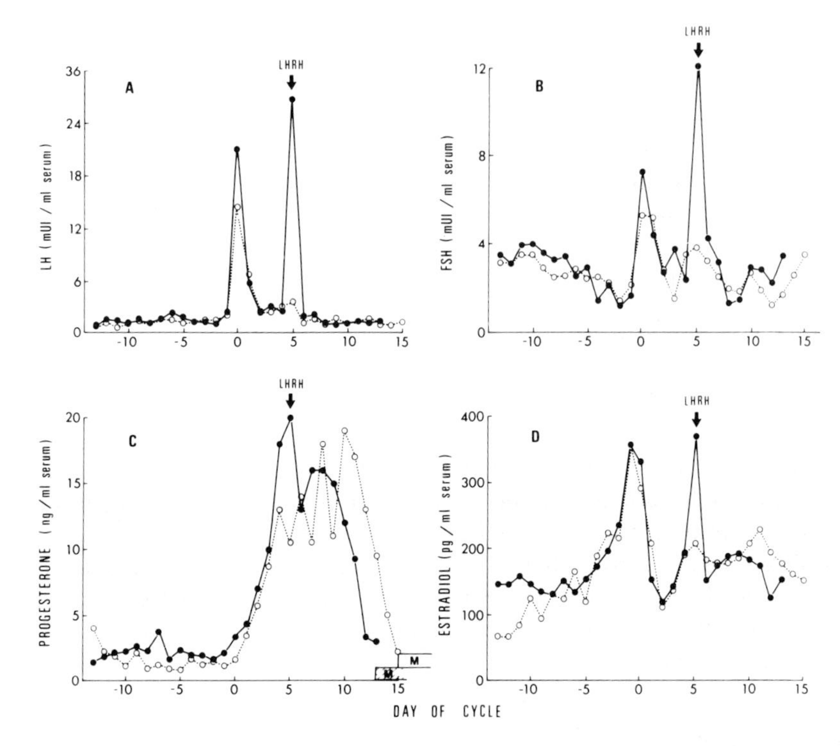

Figure 1. Effect of five subcutaneous 250-μg doses of LH-RH, given every 4 hr on day 5 after the LH surge, on serum LH (A), FSH (B), progesterone (C), and 17β-estradiol (D), and on time of menses in a normal woman. The luteal phase was shortened by 2 days. (M) Menstruation.

levels started to decrease precociously, and the menstruations appeared 2 days earlier than in the control cycle.

This single daily treatment with LH-RH in the luteal phase was given in 13 cycles, while 4 women received the treatment on two successive days in a later cycle (Table I). Results obtained in these 17 treatment cycles showed that LH-RH, when administered between days 1 and 9 after the LH peak, shortened the luteal phase from 1 to 4 days in 16 treatment cycles. Only one woman had no change in the length of her luteal phase after such treatment. Two of three women whose luteal phases were shortened by only 1 day experienced unusual spotting 3 and 5 days before the beginning of their menses. Although the number of subjects is limited, the treatment with LH-RH seemed to be more efficient when applied on two successive days as compared with one day.

Table I. Effect of Treatment with LH-RH on Luteal-Phase Length and Plasma Progesterone Levels in Normal Women

Subject	Age	Day(s) of LH-RH treatment	Usual cycle length (days)	Shortening of luteal phase (days)	Plasma progesterone[a] Control cycle (ng/ml)	Treatment cycle (ng/ml)	Percent of control
H.D.	38	1	26	1	81.2	54.5	67.1
L.Mc.	26	2	30	2	148.0	107.5	72.2
J.D.	37	3	28	0	62.9	65.1	103.4
L.A.	38	3	25	1[b]	103.4	62.6	60.5
L.L.	34	3	28	2	65.0	30.0	52.3
L.L.	34	3,4	28	2	56.0	48.7	75.8
G.S.	42	5	27	1[c]	108.3	96.4	89.0
C.B.	37	5	29	1	75.1	62.5	83.2
U.B.	37	5	29	2	118.7	86.6	72.9
H.D.	38	5	26	2	60.4	17.1	35.9
J.D.	37	6,7	28	4	30.7	16.6	54.0
N.L.	38	7	31	3	62.5	30.7	49.1
Y.L.	40	7	30	4	45.1	14.2	31.4
Y.L.	40	7,8	30	3	—	—	—
L.A.	38	8	25	3	48.6	15.4	31.6
U.B.	37	8	29	3	83.9	26.7	31.8
G.S.	42	8,9	27	4	43.6	28.8	66.0

[a] Data are expressed as the sum of serum progesterone levels measured daily between the day of LH-RH treatment and beginning of menses. Corresponding days were used for the control cycle.
[b] Spotting occurred 5 days before menses.
[c] Spotting occurred 3 days before menses.

Results obtained in the 17 treated cycles suggest that there is a better efficiency of treatment when LH-RH is administered in the middle rather than in the early luteal phase. In fact, as shown in Fig. 2A, in 10 cycles in which LH-RH was administered between days 1 and 5, the mean shortening of the luteal phase was 1.4 ± 0.2 days, as compared with 3.3 ± 0.2 days for the 7 cycles when the neurohormone was injected between days 6 and 9 following the LH peak ($p < 0.01$).

Serum progesterone levels also decreased to a greater extent when LH-RH was administered after day 5 of the luteal phase. This appears to be analogous to the effect of LH-RH treatment on the time of appearance of menses (Fig. 2B). In 10 cycles during which the women received LH-RH injections between days 1 and 5, serum progesterone, on average, decreased to 71.2 ± 6.0% of control values, whereas the mean serum progesterone levels (6 values) were 43.9 ± 5.9% of control values after LH-RH administration between days 6 and 9 after the LH peak ($p < 0.01$).

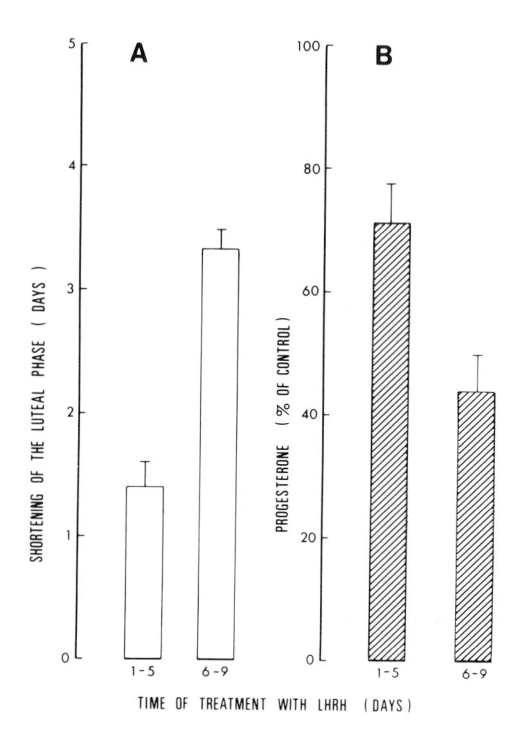

Figure 2. Effect of time of LH-RH treatment on the shortening of the luteal phase (A) and decrease of serum progesterone (B). The day of endogenous LH peak was taken as day 0. Results are expressed as the mean of 10 and 7 treatment cycles, respectively, for the two periods of treatment after the LH peak (days 1–5 and 6–9).

3.2. Intranasal [D-*Ser(TBU)*^{6}*,des-Gly-NH*$_2$^{10}]*LH-RH Ethylamide Treatment*

Since the potent LH-RH agonistic analogue [D-Ser-(TBU)^6,des-Gly-NH_2^{10}]LH-RH ethylamide (Wiegelman *et al.,* 1978) is active by the intranasal route (Sandow *et al.,* 1978), we have studied the effect of this peptide administered intranasally during two successive menstrual cycles in 6 normal women. Figure 3 illustrates the typical pattern of serum LH, FSH, estradiol, and progesterone levels in one woman during six consecutive menstrual cycles: two control pretreatment, two treatment, and two control posttreatment cycles. When the LH-RH analogue was administered on day 8 and 6 of the first and second treatment cycles, respectively, a rise of LH and FSH was observed. On the day of LH-RH analogue treatment, the progesterone and estradiol levels appeared to be increased and declined rapidly during the following 4 days. In this typical case, the luteal phase was shortened by 2 and 4 days when compared to the pretreatment cycles. In the 2 posttreatment cycles, although the levels of progesterone are lower, the length of the luteal phase and the progesterone profile are similar to those in pretreatment cycles.

The luteolytic effect of intranasal administration of [D-

Ser(TBU)^6,des-Gly-NH_2^{10}]LH-RH ethylamide is illustrated in Fig. 4, which shows the serum progesterone profile from the pretreatment, treatment, and posttreatment cycles. There is a clear shortening of the luteal phase after treatment with the LH-RH analogue, as evidenced by the precocious fall of progesterone and time of appearance of the menses. In the combined posttreatment cycles, the progesterone profile and the luteal-phase length return to normal.

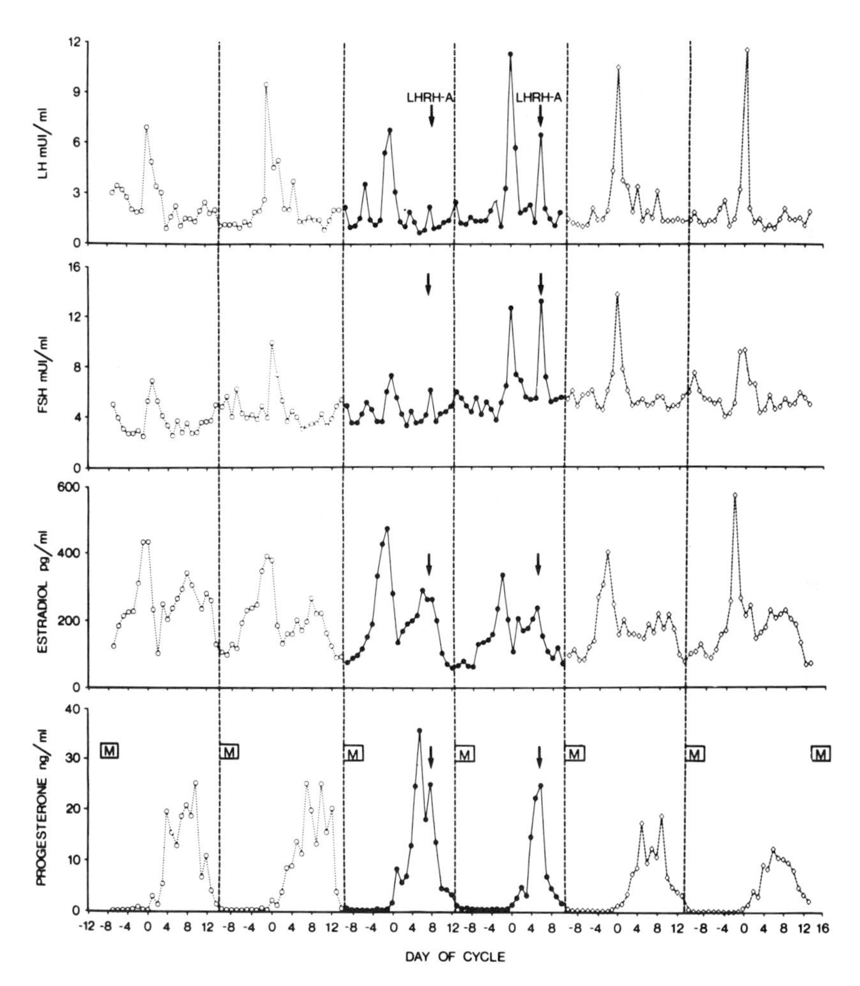

Figure 3. Typical hormonal profile of LH and progesterone and response to the intranasal administration of 500 μg [D-Ser(TBU)^6,des-Gly-NH_2^{10}]LH-RH ethylamide (LHRH-A) at 08:00 and 17:00 hr on days 8 and 6 after the LH peak during the third and fourth menstrual cycles, respectively. The luteal phase was shortened by 2 and 4 days in the two treatment cycles. (M) Menstruation.

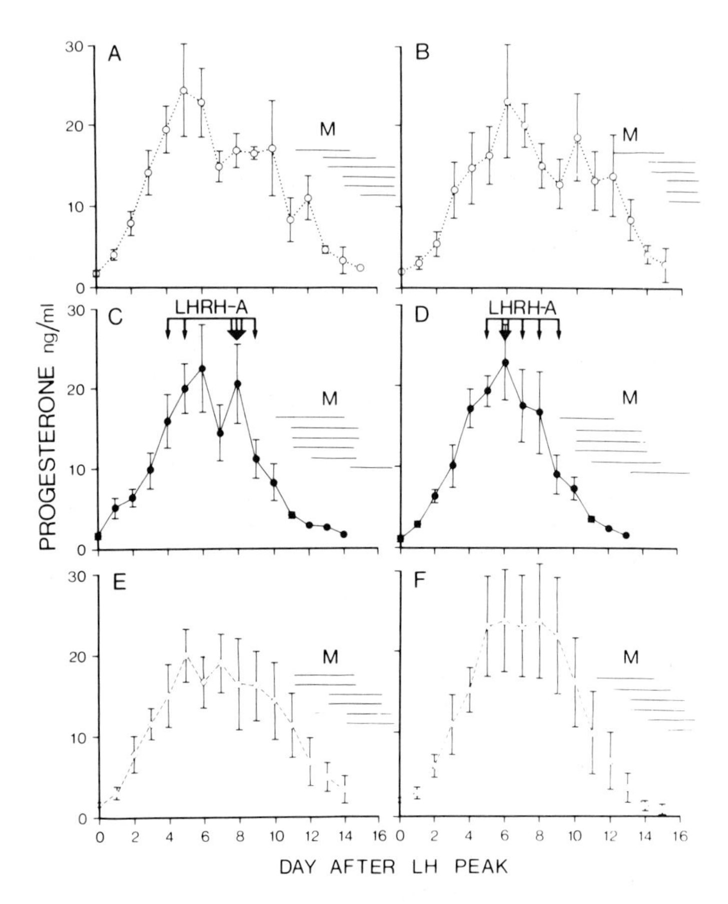

Figure 4. Serum progesterone profile of 6 women studied during successive pretreatment (A, B), treatment (C, D), and post-treatment (E, F) cycles. [D-Ser(TBU)^{6},des-Gly-NH_2^{10}]LH-RH ethylamide (LHRH-A) was administered at 08:00 and 17:00 hr on one day between days 4 and 9 after the LH peak (day 0). (M) Menstruation.

The effects of intranasal administration of the LH-RH analogue on the luteal-phase length and progesterone concentration are summarized in Table II. The luteal phase was shortened from 13.6 ± 0.3 to 10.9 ± 0.3 days after LH-RH analogue treatment ($p < 0.01$), while the length of the luteal phase was 12.8 ± 0.3 days in the posttreatment cycles. The mean shortening was 2.7 ± 0.4 days, with a range of 0.5–4.5 days. Serum progesterone levels were reduced to 61.3 ± 9.2% of control (range 13–128%) after administration of [D-Ser(TBU)^{6},des-Gly-NH_2^{10}]LH-RH ethylamide, and returned to 97.3 ± 13.3% of control in the posttreatment cycles.

Table II. Effect of Intranasal Administration of [D-Ser(TBU)^{6},des-Gly-NH_2^{10}]LH-RH Ethylamide on Luteal-Phase Length and Serum Progesterone Levels

					Plasma progesterone				
						Treatment cycles		Posttreatment cycles	
Subject	Age	Day of LH-RH treatment	Usual cycle length (days)	Shortening of luteal phase (days)	Pretreatment cycles (ng/ml)	(ng/ml)	Percent of control	(ng/ml)	Percent of control
P.H.	24	4	27	0.5	81.5	50.8	62.3	69.8	85.6
S.S.	28	5	26	4.0	83.7	39.7	47.4	55.3	66.0
A.P.	33	5	28	4.5	207.8	68.9	33.1	243.9	117.3
C.B.	30	6	26	4.0	114.3	14.6	12.7	51.7	45.2
A.P.	33	6	28	4.5	173.1	48.1	27.7	206.6	119.3
L.L.	34	7	29	2.0	68.7	52.8	76.8	110.5	160.8
G.R.	35	7	26	2.5	35.4	30.5	86.1	53.3	150.5
G.R.	35	8	26	0.5	22.0	28.1	127.7	23.4	106.3
P.H.	26	8	27	1.5	38.6	34.1	88.3	27.4	70.9
C.B.	30	8	26	2.0	72.0	24.5	34.0	30.4	42.2
L.L.	34	8	29	2.0	52.5	37.3	71.0	87.0	165.7
S.S.	28	9	26	3.0	32.6	22.5	69.0	12.4	38.0
				2.7 ± 0.4			$61.3 \pm 9.2\%$		$97.3 \pm 13.3\%$

4. Discussion

The data presented herein clearly indicate a luteolytic effect of LH-RH and [D-Ser(TBU)^6,des-Gly-NH_2^{10}]LH-RH ethylamide in normal women. When LH-RH was administered subcutaneously, the effective dose was found to be five successive injections at 4-hr intervals. The mean shortening of the luteal phase was 3.3 ± 0.2 days when LH-RH was administered late (days 6–9) as compared with 1.4 ± 0.2 days when LH-RH was given early (days 1–5) after the LH peak (see Fig. 2). This apparently higher efficiency of treatment on days 6–9 (as revealed by daily appearance of menses) is accompanied by a more precocious and marked decrease of serum progesterone levels after LH-RH administration.

A similar luteolytic effect was obtained after intranasal administration of 500 μg [D-Ser(TBU)^6,des-Gly-NH_2^{10}]LH-RH ethylamide at 08:00 and 17:00 hr between days 4 and 9 after the LH peak. In the six women treated during two consecutive cycles, the mean shortening of the luteal phase was 2.7 ± 0.4 days, and serum progesterone levels were reduced to 61.3 ± 9.2% of control (see Table II). Although, in some women, there appeared to be low levels of serum progesterone in the posttreatment cycles, recovery appeared to be complete when the 12 cycles were considered (97.3 ± 13.3% of pretreatment control cycles). This is also supported by the length of the luteal phase, which was within normal limits during the posttreatment cycles (12.8 ± 0.3 vs. 13.6 ± 0.3 days) (see Fig. 4).

A recent study has also shown that five daily injections of 100 μg of a different LH-RH agonist, [des-Gly-NH_2^{10}]LH-RH ethylamide, early in the luteal phase led to a marked decrease in plasma progesterone levels (Koyama *et al.*, 1978). One or two 50-μg does of a similar analogue, [D-Trp^6,des-Gly-NH_2^{10}]LH-RH ethylamide, administered during the middle of the luteal phase were found to induce an early and concomitant decline in circulating levels of estradiol and progesterone and a shortened luteal phase (Casper and Yen, 1979). [D-Leu^6,des-Gly-NH_2^{10}]LH-RH ethylamide and [D-Ser(TBU)^6,des-Gly-NH_2^{10}]LH-RH ethylamide have also recently been found to have a luteolytic activity in the monkey (Raynaud *et al.*, 1978, 1980).

The data obtained in women are strongly supported by more detailed studies performed in rats. As mentioned earlier, previous studies in intact female rats have clearly demonstrated the potent antifertility effects of the acute administration of LH-RH and its agonists as reflected by delayed mating and vaginal cornification, pre- and postcoital contraception, and interruption of pregnancy (Banik and Givner, 1975; Beattie and Corbin, 1977; Ferland *et al.*, 1978; Kledzik *et al.*, 1978a,b; Corbin *et al.*,

1976; Humphrey *et al.*, 1976; Hilliard *et al.*, 1976). Recently, we have found that these antifertility effects are accompanied by a marked loss of ovarian LH, FSH, and prolactin receptors as well as decreased plasma progesterone concentration (Ferland *et al.*, 1978; Kledzik *et al.*, 1978a,b). These data suggest that endogenous gonadotropin release induced by treatment with the gonadotropin-releasing peptides leads to inhibition of ovarian receptors and function.

Recent data indicate that LH-RH and its agonists can also have a direct ovarian site of action. In fact, addition of these peptides to granulosa cells in culture led to an inhibition of the FSH-induced secretion of estrogens and progesterone (Hsueh and Erickson, 1979). Moreover, we have found that treatment of hypophysectomized rats with LH-RH agonists can lead to an inhibition of ovarian LH receptor levels (Séguin, Cusan, and Labrie, unpublished data). It has also been reported that treatment with LH-RH or an LH-RH agonist can terminate pregnancy in hypophysectomized rats (Bex and Corbin, 1979). [D-Trp^6-Pro^9]LH-RH ethylamide has also been found to inhibit follicular development in hypophysectomized rats pretreated with pregnant mare serum gonadotropin (Ying and Guillemin, 1979). The importance of the ovarian site of action of LH-RH agonists remains to be determined. Studies currently in progress in our laboratory and other laboratories should help to clarify further the mechanisms involved in the inhibitory effects of treatment with LH-RH agonists on ovarian function.

Although the mechanism of action of LH-RH or its analogues remains to be clarified, the down-regulation of ovarian receptors and ovarian steroidogenesis is probably the explanation for the luteolytic effect observed in the human. As illustrated in Fig. 5, a second and inappropriate LH surge induced by LH-RH or [D-Ser(TBU)^6,des-Gly-NH_2^{10}]LH-RH ethylamide could desensitize ovarian gonadotropin receptors, causing a premature regression of the corpus luteum and early appearance of menses. This effect of endogenous LH could be additive to the direct effect of LH-RH agonists at the ovarian level. We have also noticed a short-lived rise in serum estradiol levels on the day of LH-RH treatment. Although estrogen itself can cause luteolysis (Gore *et al.*, 1973), the increase in estradiol found after treatment with LH-RH or LH-RH agonists is likely to be too small and too brief to explain the luteolytic effect.

The data presented herein suggest that such a treatment with LH-RH agonists could be used as a near-physiological approach for the control of the menstrual cycle in normal women. This approach is near-physiological since it is based on the regression of corpus luteum function, an event that normally occurs at the end of each menstrual cycle. The aim of the treatment is simply to control in a precise manner

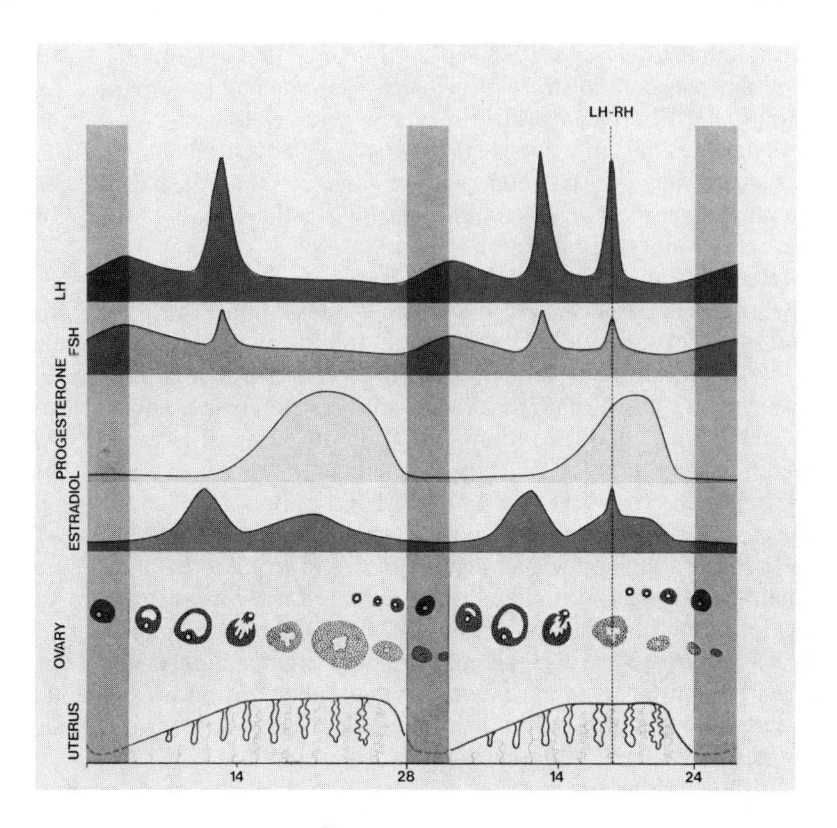

Figure 5. Drawing illustrating the effects of postovulatory administration of LH-RH or an agonistic analogue on the hypophyseal–ovarian axis and endometrium. The normal cycle (28 days) is compared to the treatment cycle (24 days), showing a second peak of LH and FSH, a short-lived rise of estradiol, and the shortening of the luteal phase.

the time of decrease of this ovarian function. Thus, in selective situations, the menstruations could be induced at a more appropriate time or the menstrual cycle could be shortened. During these studies, we have also noticed a decrease in the symptoms of dysmenorrhea in some women. This effect is probably related to the decreased levels of progesterone observed after treatment.

Since progesterone secretion by the corpus luteum is essential for implantation and maintenance of early pregnancy and since luteal-phase insufficiency is associated with infertility, it appears important to investigate the use of such treatment as a possible new contraceptive method in normally ovulating women. However, more detailed studies have to be carried out concerning the efficiency of such a treatment at various doses and time intervals during the luteal phase. The luteolytis also has to be demonstrated in the presence of rising circulating levels of hCG.

As illustrated in this chapter, LH-RH and various LH-RH analogues could be effective luteolytic agents using different routes of administration. However, the intranasal administration of a potent analogue such as [D-Ser(TBU)^6,des-Gly-NH_2^{10}]LH-RH ethylamide would be a more practical and acceptable approach.

5. Summary

Ten normal women were treated with five subcutaneous 250-μg doses of luteinizing-hormone-releasing hormone (LH-RH) at 4-hr intervals on one or two consecutive days between days 1 and 9 following the spontaneous LH surge. In 16 of 17 treatment cycles, the luteal phase was shortened from 1 to 4 days. There appeared to be a better efficiency of treatment when LH-RH was administered on days 6–9 as compared with days 1–5. In fact, the luteal phase was shortened by 3.3 ± 0.2 vs. 1.4 ± 0.2 days ($p < 0.01$) when the neurohormone was injected late as compared with early in the luteal phase. Similarly, the serum progesterone levels were respectively decreased to 44 ± 5 vs. $21 \pm 6\%$ of control levels ($p < 0.01$) in these two periods of treatment. A similar luteolytic effect was obtained after intranasal administration of [D-Ser(TBU)^6,des-Gly-NH_2^{10}]LH-RH ethylamide. Six normal women received 500 μg of the analogue at 08:00 and 17:00 hr on one day between days 4 and 9 following their LH peak during two consecutive cycles. In the 12 treatment cycles, the luteal phase was shortened by 2.7 ± 0.4 days (range 0.5–4.5 days) and plasma progesterone levels were reduced to $61.3 \pm 9.2\%$ of control. As determined by daily blood sampling for LH, FSH, estradiol, and progesterone, normal cycles occurred immediately after treatment. The data presented herein indicate a luteolytic effect of treatment with LH-RH or an LH-RH agonistic analogue in normal women and support the interest in such a new and near-physiological approach for the control of corpus luteum function and menstrual cycle in women. The possible use of this treatment as a contraceptive method is discussed.

ACKNOWLEDGMENTS. A. Lemay is a Scholar at the Medical Research Council of Canada. This research was supported by the Medical Research Council of Canada.

Discussion

NOLIN: Can all these effects of LH-RH be duplicated simply by giving a bolus of LH at the appropriate time?

LEMAY: We have administered 5000 to 10,000 IU of hCG (APL, Ayerst) during a single day after ovulation in three normal volunteers, and no effect was observed on the length of the luteal phase and the serum progesterone levels. The administration of various and repetitive doses of hCG for the treatment of luteal insufficiency is not always effective. Careful studies would have to be done in order to mimic the effect of LH-RH treatment, that is, a second and limited LH surge 1 week after ovulation.

VILLEE: What is the half-life of these LH-RH analogues in the human?

LEMAY: The half-life of LH-RH analogues in the human serum has not been measured so far. However, we know that the LH levels in the blood remain elevated for at least 10 to 12 hr after intranasal administration of 500 μg of D-Ser(TBU)^6,des-Gly-NH_2^{10}]LH-RH ethylamide.

CHANNING: I noticed that after your LH-RH administration, there was a rise in serum estradiol. It has been shown that estrogen is luteolytic in the primate. Could you account the observation of luteolysis as due to the elevated estradiol?

LEMAY: As mentioned in the text, administration of high doses of estrogen can cause luteolysis in the human. The increase in estradiol observed after treatment with LH-RH or the LH-RH analogue is probably too small to explain the luteolytic effect.

RAYNAUD: At variance with the findings in women, we found an increase in progesterone, but we did not observe an increase in estrogen levels in monkeys after the administration of [D-Leu^6,des-Gly-NH_2^{10}]LH-RH ethylamide in the luteal phase.

SPILMAN: In your second series of trials, in which women were treated with the LH-RH analogue for two cycles after two control cycles, my impression from your slides was that menses began during the treated cycles when serum progesterone was at a higher concentration than when menses began during the control cycles. Was this true, and what is your interpretation of this?

LEMAY: It is true that in a few treatment cycles, menstruations appeared while the serum progesterone levels had not yet reached a level as low as in the control cycle. In a few other treatment cycles, some women also experienced unusual spotting 3 to 5 days before the beginning of the menses and while the serum progesterone had not yet reached basal levels. This phenomenon could probably be explained by the observation that the total amount of progesterone secreted in the blood during the luteal phase was less and the withdrawal of progesterone was faster in those treatment cycles than in the control cycles, leading to endometrium instability and premature bleeding.

References

Banik, V.K., and Givner, M.L., 1975, Ovulation, induction and antifertility effects of an LH-RH analog (AY-25,205) in cycling rats, *J. Reprod. Fertil.* **44:**87.

Beattie, C.W., and Corbin, A., 1977, Pre- and post-coital contraceptive activity of LHRH in the rat, *Biol. Reprod.* **16:**333.

Bex, F.J., and Corbin, A., 1979, LHRH or LHRH agonist-induced termination of pregnancy in hypophysectomized rats: Extrapituitary site of action, Endocrinology Society Meeting, Abstract A236.

Casper, R.F., and Yen, S.S.C., 1979, Induction of luteolysis in the human with a long-acting analog of luteinizing hormone-releasing factor, *Science* **205:**408.

Conti, M., Harwood, J.P., Hsueh, A.J.W., Dufau, M.L., and Catt, K.J., 1976, Gonadotropin-induced loss of hormone receptors and desensitization of adenylate cyclase in the ovary, *J. Biol Chem.* **251:**7729.

Corbin, A., Beattie, C.W., Yardley, J., and Toell, T.J., 1976, Postcoital contraceptive effects of an agonistic analogue of luteinizing hormone-releasing hormone, *Endocr. Res. Commun.* **2:**359.

Ferland, L., Auclair, C., Labrie, F., Raynaud, J.P., and Azadian-Boulanger, G., 1978, Effets inhibiteurs de la LH-RH sur les récepteurs ovariens de la LH chez la rate, *C.R. Acad. Sci. Ser D* **286:**1113.

Gore, B.Z., Caldwell, B.V., and Speroff, L., 1973, Estrogen-induced human luteolysis, *J. Clin. Endocrinol. Metab.* **36:**615.

Hilliard, J., Pang, C.N., and Sawyer, C.H., 1976, Effects of luteinizing hormone-releasing hormone on fetal survival in pregnant rabbits, *Fertil. Steril.* **27:**421.

Hsueh, A.J.W., and Erickson, G.F., 1979, Extrapituitary action of gonadotropin-releasing hormone: Direct inhibition of ovarian steroidogenesis, *Science* **204:**854.

Humphrey, R.R., Windsor, B.L., Bousley, F.G., and Edgren, R.A., 1976, Antifertility effects of an analog of luteinizing hormone-releasing hormone (LHRH) in rats, *Contraception* **14:**625.

Hunzicker-Dunn, M., and Birnbaumer, L., 1976, Adenyl cyclase activities in ovarian tissues. IV. Gonadotropin-induced desensitization of the luteal adenylyl cyclase throughout pregnancy and pseudo-pregnancy in the rabbit and the rat, *Endocrinology* **99:**221.

Kledzik, G.S., Cusan, L., Auclair, C., Kelly, P.A., and Labrie, F., 1978a, Inhibition of ovarian LH and FSH receptor levels by treatment with [D-Ala^{6}, des-Gly-NH_2^{10}]LHRH ethylamide during the estrous cycle in the rat, *Fertil. Steril.* **30:**348.

Kledzik, G.S., Cusan, L., Auclair, C., Kelly, P.A., and Labrie, F., 1978b, Inhibitory effect of a luteinizing hormone (LH)-releasing hormone agonist on rat ovarian LH and follicle-stimulating hormone receptor levels during pregnancy, *Fertil. Steril.* **29:**560.

Kramer, C.Y., 1956, Extension of multiple-range test to group means with unequal numbers of replications, *Biometrics* **12:**307.

Koyama, T., Takayoshi, O., Takahino, K., and Motoi, S., 1978, Effect of postovulatory treatment with a luteinizing hormone-releasing hormone analog on the plasma level of progesterone in women, *Fertil. Steril.* **30:**549.

Labrie, F., Auclair, C., Cusan, L., Lemay, A., Bélanger, A., Kelly, P.A., Ferland, L., Kledzik, G., Azadian-Boulanger, G., and Raynaud, J.P., 1979, Inhibitory effects of treatment with LHRH or its agonists on ovarian receptor levels and function, in: *Ovarian Follicular and Corpus Luteum Function, Advances in Experimental Biology and Medicine,* Vol. 112 (C.P. Channing, J. Marsh, and W. Sadler, eds.), pp. 687–693, Plenum Press, New York.

Lemay, A., Labrie, F., Raynaud, J.P., and Azadian-Boulanger, G., 1978, Action lutéolytique de la LHRH chez la femme, *C.R. Acad. Sci.* **286:**527–529.

Lemay, A., Labrie, F., Ferland, L. and Raynaud, J.P., 1979a, Possible luteolytic effects of luteinizing hormone (LH)-releasing hormone in normal women, *Fertil. Steril.* **31:**29–34.

Lemay, A., Labrie, F., Bélanger, A., and Raynaud, J.P., 1979b, Luteolytic effect of intranasal administration of [D-Ser(TBU)^{6},des-Gly-NH_2^{10}]LHRH ethylamide in normal women, *Fertil. Steril.* **32:**646–651.

Moor, R.M., Rowson, L.E., Hay, M.F., and Caldwell, B.V., 1968, The effect of exogenous gonadotrophins on the conceptus and corpus luteum in pregnant sheep, *J. Endocrinol.* **44:**217.

Nillius, S.J., 1976, Therapeutic use of luteinizing hormone-releasing hormone in the human female, in: *Hypothalamus and Endocrine Functions* (F. Labrie, J. Meites, and G. Pelletier, eds.), pp. 93–112, Plenum Press, New York.

Raynaud, J.P., Azadian-Boulanger, G., Mary, I., Mouren, M., Lemay, A., Ferland, L., Auclair, C., and Labrie, F., 1978, Action lutéolytique de la LHRH chez la rate, la guenon et la femme, in: *L'Implantation de l'Oeuf* (F. Du Mesnil du Buisson, A. Psychoyos, and K. Thomas, eds.), pp. 273–283, Masson, Paris.

Raynaud, J.P., Mary, I., Moguilewsky, M., Mouren, M., and Labrie, F., 1980, Inhibition of progesterone secretion in luteal phase by 2 LHRH agonists, *Fertil. Steril.* **34:**593–598.

Rodbard, D., and Lewald, J.E., 1970, Computer analysis of radioligand assay and radioimmunoassay data, in: *Second Karolinska Symposium on Research Methods in Reproductive Endocrinology* (E. Diczfalusy, ed.), p. 79, Bogtrykheriet Forum, Copenhagen.

Sandow, J., Recherberg, W.V., Köning, W., Hahn, M., Jerzabek, G., and Frazer, H., 1978, Physiological studies with highly active analogues of LHR, in: *Hypothalamic Hormones: Chemistry, Physiology and Clinical Applications* (D. Gupta and W. Voelter, eds.), pp. 307–326, Verlag Chemie-Weinheim, New York.

Schally, A.V., and Arimura, A., 1977, Physiology and nature of hypothalamic regulatory hormones, in: *Clinical Neuroendocrinology* (L. Martini and G.M. Besser, eds.), pp. 1–42, Academic Press, New York.

Spies, H.G., Coon, L.L., and Gier, H.T., 1966, Luteolytic effect of LH and hCG on the corpora lutea of pseudopregnant rabbits, *Endocrinology* **78:**67.

Wiegelman, W., Solbach, H.G., Kley, M.K., Nieschlag, E., Rudorff, K.H., and Kruskemper, M.L., 1978, A new LHRH analogue: [D-Ser(TBU)[6]]LHRH-(1–9)nonapeptide ethylamide, in: *Hypothalamic Hormones: Chemistry, Physiology and Clinical Applications* (D. Gupta and Voelter, M. eds.), pp. 327–333, Verlag Chemie-Weinheim, New York.

Yang, W.H., and Chang, M.C., 1968, Interruption of pregnancy in the rat and hamster by administration of PMS or hCG, *Endocrinology* **83:**217.

Ying, S.Y., and Guillemin, R., 1979, [D-Trp[6]-Pro[9]-NET)-luteinizing hormone-releasing factor inhibits follicular developmental in hypophysectomized rats, *Nature (London)* **280:**593.

5

Antifertility Effects of LH-RH and Its Agonists

Frederick J. Bex and Alan Corbin

1. Introduction

The identification, characterization, and synthesis of luteinizing-hormone-releasing hormone (LH-RH) introduced a potential and exciting new approach for the treatment of male and female infertility. This application, which theoretically coincides with the profertility LH-releasing role played by endogenous LH-RH, has been and continues to be exhustively pursued, but use of the chemically and biologically identical synthetic LH-RH in a variety of clinical situations has never achieved its anticipated therapeutic promise. Primary responsibility for the inconsistency of the results has been attributed to the short half-life of the peptide, and it was hoped that this problem might be circumvented by increased exposure (e.g., through more frequent administration or constant infusion) or by employing LH-RH derivatives with prolonged activity. However, despite heroic treatment regimens and the use of a number of LH-RH agonistic analogues with much-increased potency and duration of activity, there has been little improvement in the clinical utility of this approach. In fact, comprehensive reproductive pharmacological evaluation has revealed that the LH-releasing property responsible for the original profertility classification imparts definite correlatable and predictable antireproductive activity. This paradoxical relationship, first

Frederick J. Bex and *Alan Corbin* • Endocrinology Section, Wyeth Laboratories, Inc., Philadelphia, Pennsylvania 19101.

evidenced in laboratory animals by the ability of these agents to terminate pregnancy, disrupt cyclicity, retard puberty, and inhibit spermatogenesis, has been extended to humans, in whom LH-RH and its agonistic analogues have been shown to induce luteolysis and premature menstruation. The identification of the antireproductive nature of LH-RH and agonists, aside from providing an explanation for their disappointing performance in treating various hypogonadal states, has, quite ironically, established the concept of these peptides as a novel class of contraceptive agents.

This chapter will detail the developments regarding the chemistry and reproductive pharmacology of LH-RH and its agonistic analogues that have led to the inception of the antifertility theme, its verification in numerous laboratory models, description of certain components of the inhibitory mechanism of action, and prospects for clinical application in both fertility regulation and anticancer therapy.

2. Antireproductive Properties of the LH-Releasing Peptides in the Female

2.1. Pre- and Postcoital Contraception

The structural determination and synthesis of LH-RH (Matsuo *et al.*, 1971a,b) not only stimulated interest in the therapeutic profertility potential of this compound but also raised the possibility that antagonistic derivatives might be generated that would serve to inhibit or regulate fertility (Schally *et al.*, 1971). D-[Phe]^2-D-[Ala]^6-LH-RH (Wy-18,185) represented one of the first and, until recently (Rivier and Vale, 1978), one of the more potent and predictable LH-RH antagonistic analogues identified as being capable of inhibiting spontaneous ovulation in the rat (Corbin and Beattie, 1975a). This compound suppressed the proestrous ovulatory surge of LH and follicle-stimulating hormone (FSH), presumably through both pituitary and hypothalamic sites of action (Corbin and Beattie, 1976), without affecting basal gonadotropin levels and normal estrous cyclicity. Since an approach requiring precise knowledge of the timing of the ovulatory event would have limited clinical utility, Wy-18,185 was evaluated for its effect on reproductive competence when administered following ovulation and fertile mating (postcoital contraceptive test). Although the absence of antipregnancy activity of Wy-18,185 during the first 7 (preimplantational) days of gestation was not entirely unexpected in view of its inability to depress basal LH levels, the dramatic pregnancy inhibition produced by the parent molecule, LH-RH, originally included as a control, was quite exciting (Fig. 1). More-

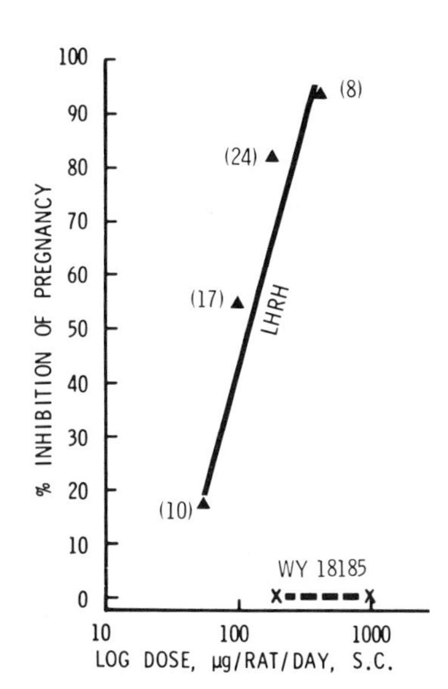

Figure 1. Effect of LH-RH and an LH-RH antagonist, D-Phe²-D-Ala⁶-LH-RH (Wy-18,185), on pregnancy in the rat when administered preimplantationally (days 1–7). The figures in parentheses are the numbers of rats.

over, a similar inhibitory effect on pregnancy was observed when LH-RH was administered during the postimplantational interval, days 7–12 (Fig. 2) (Corbin and Beattie, 1975b). To investigate the specificity of this seemingly paradoxical contragestational effect of the parent molecule, a large number of LH-RH congeners, shown by us or by others (Coy *et*

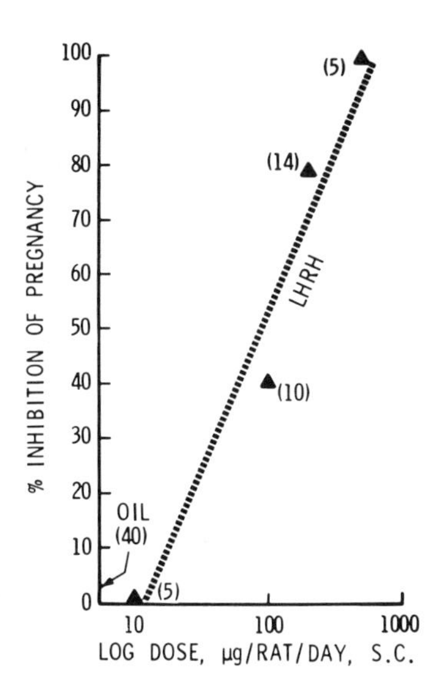

Figure 2. Effect of LH-RH on pregnancy in the rat when administered postimplantationally (days 7–12). The figures in parentheses are the numbers of rats.

al., 1975; Vale *et al.*, 1976) to have agonistic properties (ovulation induction; release of LH or FSH or both in ovariectomized, steroid-blocked rats) were evaluated for their postcoital activity (Corbin *et al.*, 1978, 1979a,b). The potency of these analogues in inducing ovulation (indirect measure of LH release) was found to be directly correlated with their ability to inhibit pregnancy pre- or postimplantationally (Table I). The nature of the antipregnancy effect produced by these peptides observed on autopsy day 14 following day 1–7 treatment was characterized by a lack of uterine implantation sites or resorption of sites that were present (Fig. 3) and, on autopsy day 18 following day 7–12 treatment by resorbing sites or placental scars or both. Although less effective, LH-RH or representative agonists were also quite capable of terminating pregnancy when administered orally (Fig. 4 and 5) or given intramuscularly as a single injection on certain days of gestation (Fig. 6). The contragestational property of these peptides, as has been quite universally established for their LH-releasing capacity, was found not to be species-specific. Either pre- or postimplantational treatment disrupted pregnancy in the rabbit (Corbin and Beattie, 1975b; Corbin *et al.*, 1976; Rippel and Johnson, 1976a; Hilliard *et al.*, 1976), while only postimplantational administration was effective in the hamster (Windsor *et al.*, 1977).

In general, the antifertility potency of LH-RH or its agonistic analogues is a reflection of the amount of LH released, which is in turn modulated by duration of exposure to the LH-releasing peptide (frequency or length of treatment, active half-life of compound), receptor affinity, and mode of administration. In this regard, antipregnancy activity has correlated fairly closely with both an agonist's resistance to degradation (Table II) (Koch *et al.*, 1977) and the ovulation-inducing effectiveness of the route by which it is administered (Table III).

2.1.1. Mechanism of Pre- and Postimplantational Termination of Pregnancy

2.1.1a. Ovarian-LH-Receptor Down-Regulation. The close association between the capacity to release LH and contragestational activity coupled with studies demonstrating the ability of LH to negatively affect ovarian response by reducing its own gonadal-receptor levels (down-regulation) (Richards *et al.*, 1976; Rao *et al.*, 1977) suggested that abnormally elevated endogenous LH levels induced by LH-releasing peptides were responsible for their antireproductive effects. Subsequent investigations revealed that either LH-RH (Bex and Corbin, 1979a) or its agonists (Kledzik *et al.*, 1978a,b) did, in fact, produce a dramatic reduction in ovarian, and specifically luteal, LH-receptor populations.

Table I. Comparison of the Agonist and Postcoital Contraceptive Profile of the D-Ala^{6}-LH-RH and D-Trp^{6}-LH-RH Series in the Rat

Compound	Agonist ovulation induction (MED_{100}, μg/rat, i.v.)[a]	Postcoital contraception (MED_{100}, μg/day, s.c.)[a]	
		Preimplantation[b]	Postimplantation[b]
LH-RH	2.0	200–500	500
D-Ala^{6}-LH-RH (Wy-18,186)	1.0	10	100
D-Ala^{6}-des-Gly^{10}-Pro^{9}-NHEt-LH-RH (Wy-18,481)	1.0	1.0	$>25<50$
D-Ala^{6}-N^{α}-Me-Leu^{7}-des-Gly^{10}-Pro^{9}-NHEt-LH-RH (Wy-40,905)	0.10	1.0	1.0
D-Ala^{6}-N^{α}-Me-Leu^{7}-N^{α}-Me-Arg^{8}-des-Gly^{10}-Pro^{9}-NHEt-LH-RH (Wy-42,310)	10	Inactive at 100	Inactive at 100
D-Trp^{6}-LH-RH (Wy-42,462)	0.10	10	$>1.0<10$
D-Trp^{6}-des-Gly^{10}-Pro^{9}-NHEt-LH-RH (Bachem No. 9022)	0.10	1.0	10
D-Trp^{6}-N^{α}-Me-Leu^{7}-des-Gly^{10}-Pro^{9}-NHEt-LH-RH (Wy-40,972)	0.10	1.0	1.0

[a] Approximate minimal 100% effective dose.
[b] Preimplantational treatment: days 1–7 of pregnancy; postimplantational treatment: days 7–12 of pregnancy.

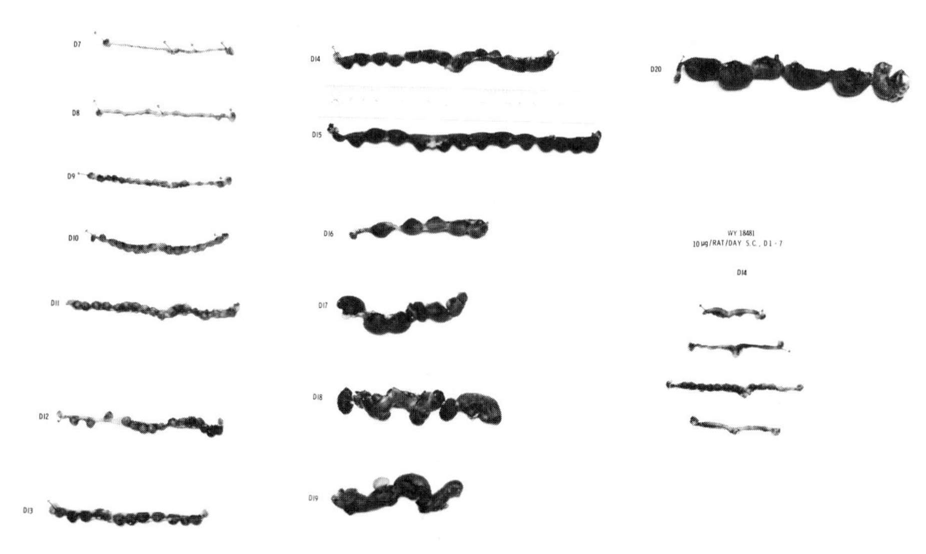

Figure 3. Examples of pregnant rat uterus on sequential days (D) of gestation: effect of preimplantational treatment with an LH-RH agonist, D-Ala[6]-des-Gly[10]-Pro[9]-NHEt-LH-RH (Wy-18,481).

During preimplantational treatment, the immediate (within 24 hr) and sustained loss of LH receptors (Bex and Corbin, 1979a) was associated with increased serum LH, dysphasic serum FSH, decreased serum prolactin (PRL), delay of and reduction in the implantational surge of estradiol, and decreased serum progesterone (Beattie *et al.*, 1977). From the chronology of these effects, the down-regulatory property of LH,

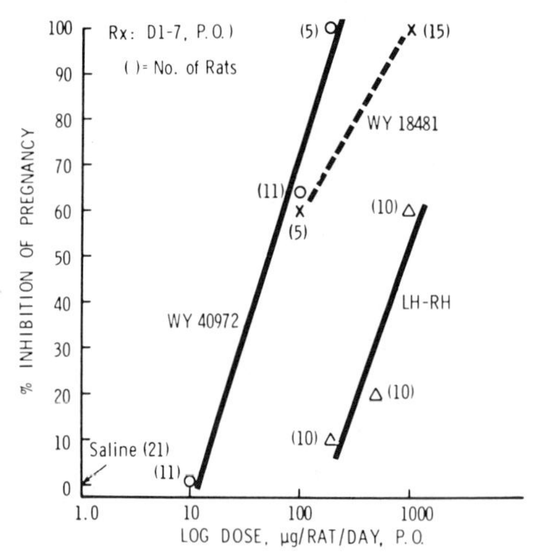

Figure 4. Effect of LH-RH and the LH-RH agonists Wy-18,481 and Wy-40,972 on pregnancy when orally (P.O.) administered preimplantationally (days 1–7). The figures in parentheses are the numbers of rats.

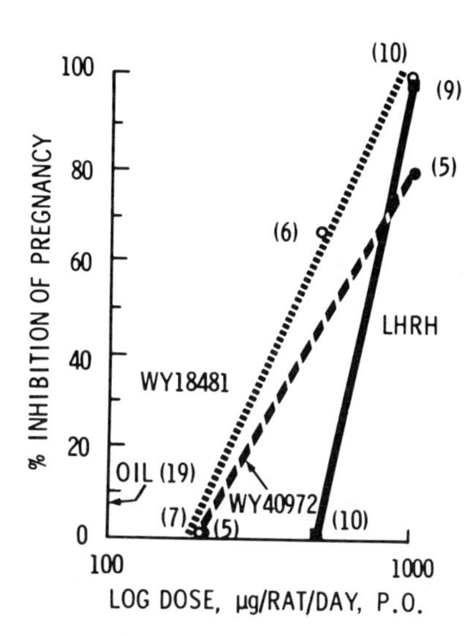

Figure 5. Effect of LH-RH and the LH-RH agonists Wy-18,481 and Wy-40,972 on pregnancy when orally (P.O.) administered postimplantationally (days 7–12). The figures in parentheses are the numbers of rats.

and the supportive role of PRL during this period in maintaining luteal LH receptors and luteal function, it was proposed that the disruption of ovarian steroidogenesis was a result of an LH/PRL imbalance (Yoshinaga, 1978). The delay and attenuation of the preimplantational estradiol surge and the reduced levels of progesterone were, in turn, responsible for the high incidence of implantational failure, and for those blastocysts that did implant, the steroidal milieu was inappropriate for their survival, thereby leading to resorption. Notable, in regard to this proposed preimplantational mechanism of action, is the inability of LH-RH to prevent implantation in the hamster, in which a surge of estradiol, unlike the rat, is not required to initiate this event (Edgren *et al.*, 1977).

2.1.1b. Extrapituitary Antipregnancy Effects. As with the preimplantational LH-RH treatment, postimplantational administration of LH-RH, as well as certain agonists, has been found to produce a rapid and sustained reduction in luteal LH receptors (Bex and Corbin, 1979a). This loss of luteal LH responsiveness is associated with progressively decreasing progesterone levels and failure of the pregnancy (Humphrey *et al.*, 1977). Although a pituitary-mediated refractoriness to LH, in view of its particularly crucial role during this period of pregnancy, would be sufficient to explain the contragestational effects of LH-RH, a direct nonpituitary component that has been identified during the latter half of pregnancy may be involved as well (Macdonald and Beattie, 1979; Bex and Corbin, 1979b). It is noteworthy that administration of LH-RH or an agonist after day 12 of pregnancy in the rat, when the pituitary is no longer necessary, results in fetal resorption independent of the presence

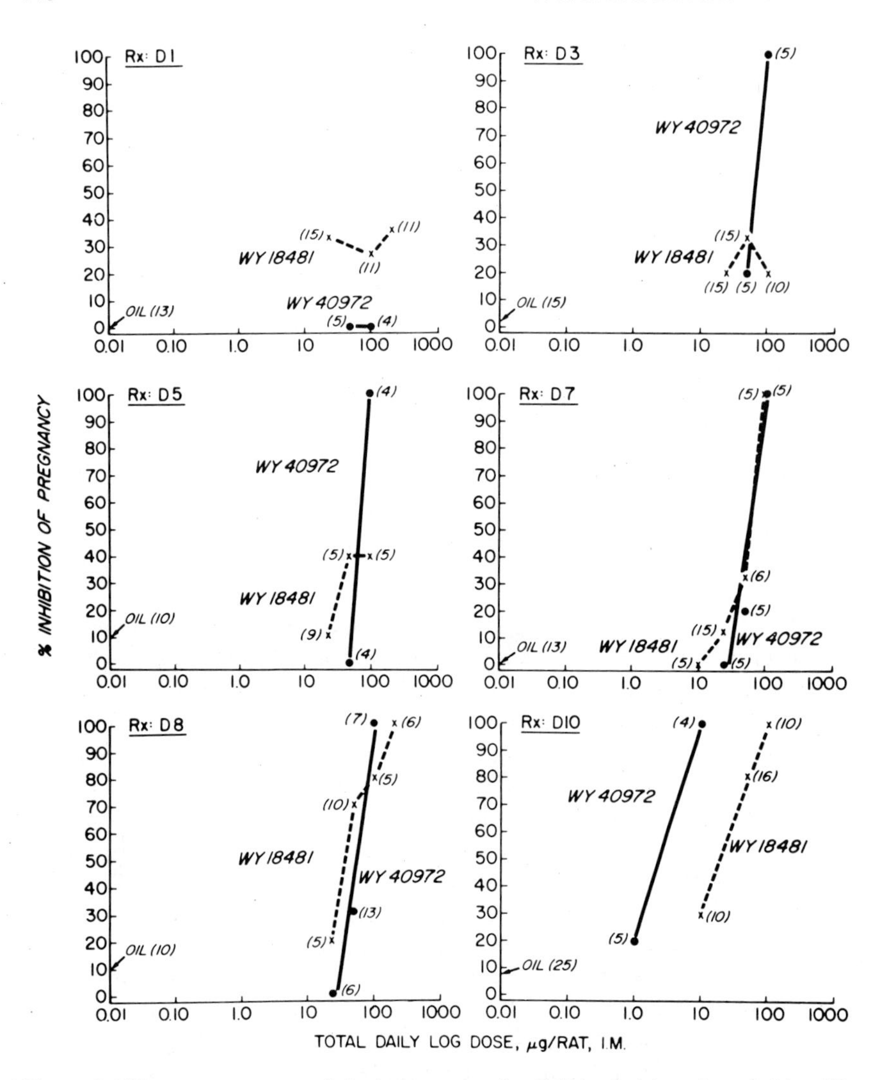

Figure 6. Effect on pregnancy of single intramuscular (I.M.) administration of either Wy-18,481 or Wy-40,972 given on selected days of gestation. The figures in parentheses are the numbers of rats.

of the pituitary (Table IV). This inhibition is associated with a fall in serum progesterone levels that can be reversed by exogenous administration of this hormone. The placenta has been investigated as a likely site for the extrapituitary effect in view of its importance in subserving the pituitary's role in supporting luteal function at this time, further reinforced by the fact that the human placenta synthesizes, stores, and

Table II. Reduced Susceptibility to Enzymatic Cleavage Conferred by Modifications to LH-RH and Relationship to Biological Activity

Compound[a]	Structure												Ovulation induction (MED$_{100}$, μg/rat, i.v.)	Postcoital contraception Rx D1–7 (MED$_{100}$, μg/day, s.c.)
								Enzyme attack						
		1	2	3	4	5	6	7	8	9	10			
LH-RH	Pyro	Glu	His	Trp	Ser	Tyr	Gly	Leu	Arg	Pro	Gly-NH_2		2.0	200–500
Wy-18,186		/	/	/	/	/	Ala	/	/	/	/		1.0	10
Wy-42,462		/	/	/	/	/	Trp	/	/	/	/	Greater resistance to degradation than D-Ala[6]-LH-RH	0.1	10
Wy-18,481		/	/	/	/	/	Ala	/	/	/NHEt	Des		1.0	1.0
Bachem. 9022		/	/	/	/	/	Trp	/	/	/NHEt	Des		0.1	1.0
Wy-40,905		/	/	/	/	/	Ala	N^{α}Me/	/	/NHEt	Des		0.1	1.0
Wy-40,972		/	/	/	/	/	Trp	N^{α}Me/	/	/NHEt	Des		0.1	1.0
									Resistance to degradation					

[a] Listed in approximate increasing order of resistance to degradation from top to bottom.

Table III. Ovulation Induction in Nembutalized Proestrous Rats with D-Trp[6]-N^{α}-Me-Leu[7]-des-Gly[10]-Pro[9]-NHEt-LH-RH (Wy-40,972) via Various Routes of Administration

Route	Ovulation-inducing ED_{100} (μg/rat)
Intravenous	0.10 (ED_{90})
Intramuscular	0.10 (ED_{90})
Intraperitoneal	0.10
Subcutaneous	0.10 (ED_{80})
Aerosol	10
Topical	10
Intranasal	10
Oral	10 (ED_{80})
Rectal	10
Ear canal	50
Vaginal	50 (ED_{80})

Table IV. Termination of Pregnancy Following Administration of LH-RH or D-Ala[6]-des-Gly[10]-Pro[9]-NHEt-LH-RH (Wy-18,481) in Hypophysectomized Rats: Reversal by Progesterone or Ectopic Pituitary Graft

	Termination of pregnancy (%)[b]	
Daily treatment[a]	Intact	Hypx[c]
Oil	0 (9)	0 (25)
500 μg LH-RH	53 (15)	79 (14)
50 μg Wy-18,481	56 (9)	100 (8)
500 μg LH-RH + 5 mg progesterone	—	20 (15)
Oil + ectopic pituitary[d]	—	17 (6)
500 μg LH-RH + ectopic pituitary	—	9 (11)

[a] Treatment was administered on days 13–16 of pregnancy; LH-RH and Wy-18,481 were given as a divided dose at 09:00 and 15:00; progesterone was given as a single injection at 09:00; autopsy was performed on day 18.

[b] Number of animals in parentheses.

[c] Hypophysectomy was performed on day 13 of pregnancy.

[d] Ectopic graft was transplanted to the kidney capsule on day 11.

is responsive to LH-RH (Siler-Khodr and Khodr, 1978; Khodr and Siler-Khodr, 1978a,b). However, measurement of rat placental lactogen (rPL), a component of the placental luteotropic complex, indicates that this hormone reflects advanced stages of fetoplacental resorption presumably initiated by withdrawal of progesterone support (Bex and Corbin, 1979b). While direct ovarian effects of LH-RH and agonists have been identified in the immature female rat (Mayar *et al.*, 1979; Hsueh and Erickson, 1979; Ying and Guillemin, 1979), the possibility that a similar mechanism is responsible for the contragestational activity in the adult is unlikely. In this regard, the pregnancy-terminating effects of LH-RH in rats hypophysectomized on day 12 or later were found to be reversed by an ectopic pituitary graft (Table IV). The nature of the hormones from the ectopic pituitary responsible for protecting the pregnancy, which could quite possibly indicate the site of the extrapituitary effect, has not been identified; however, this graft does release large amounts of PRL and may respond to the LH-RH treatment with significant LH/FSH production. Thus, it might be speculated that effects on placental production of hormones other than rPL—possibly an LH-like chorionic gonadotropin—necessary for luteal function, rather than a direct ovarian action, represent the extrapituitary mechanism for the antipregnancy effects produced by these peptides.

2.1.2. Precoital and Estrous-Cycle Effects

While profertility regimens of LH-RH and its agonists have been employed in the human to induce ovulation with some success, the ensuing luteal phase has generally been deficient and the pregnancy rate poor (Taymor, 1974; Comaru *et al.*, 1976). Relevant to this situation are precoital studies in the rat in which acute administration of LH-RH was found to delay mating and inhibit pregnancy in a dose-related fashion (Banik and Givner, 1975, 1976; Beattie and Corbin, 1977). Chronic precoital administration in the rat, in addition to delaying mating, reduced both the number of rats delivering and the litter size. The inhibitory effects on pregnancy following precoital treatment in rats might be explained in terms of the observed desynchronization of behavioral and biological events resulting in either the shedding of immature ova incapable of viable fertilization or of fertilization occurring after the optimal interval (Banik and Givner, 1976). However, it has been demonstrated that ovulation induced by LH-RH at the appropriate time on proestrus, in rats in which the spontaneous ovulatory LH surge was blocked by pentobarbital, and followed by a normal mating, resulted in only half the animals becoming pregnant (Corbin *et al.*, 1978). Thus, it seems that precoital administration of these peptides can induce luteal incompetency leading to an inappropriate steroidal milieu for pregnancy, much like that described for their postcoital contragestational effects.

2.1.3. Restoration of Fertility Following Pre- or Postcoital Administration

The reversibility of the antipregnancy effects of the LH-releasing peptides represents a crucial aspect for their prospective use as contraceptive agents. The vaginal-smear type of rats receiving either pre- or postimplantational LH-RH/agonist treatment was found to change by midgestation from a leukocytic smear characteristic of pregnancy to one that was predominantly cornified (Corbin *et al.*, 1978) (Fig. 7). Confirmation that this break in smear type represented a resumption of cyclicity was provided by the fact that all animals exposed to males at this time mated and carried to normal, uneventful term (Table V). While 4- or 7-day precoital administration delays the appearance of expected estrus and mating, and reduces the fertility of those that do mate, regular cyclicity was reestablished within 1 week following termination of treatment (Beattie and Corbin, 1977). In fact, administration of an agonist for

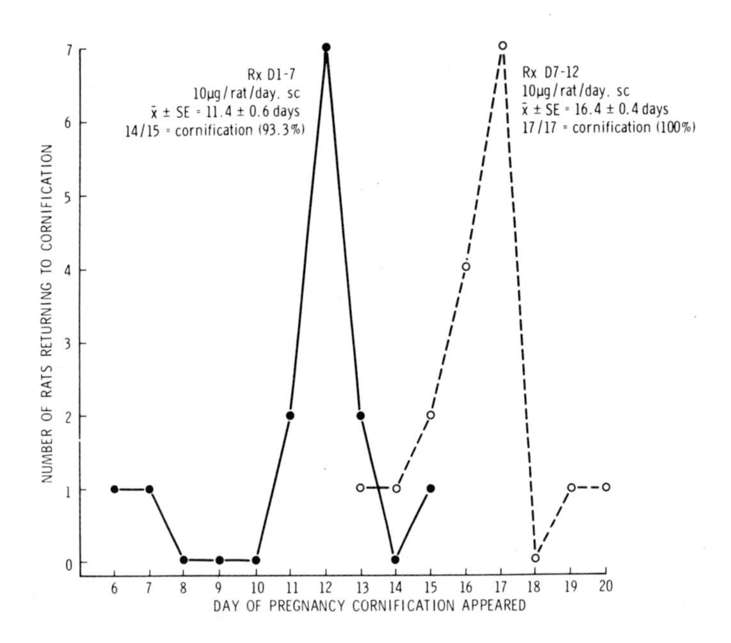

Figure 7. Distribution of appearance of vaginal cornification in pregnant rats treated with Wy-40,972.

Table V. Study of Pregnancy Termination, Vaginal Smear and Return to Fertility in Rats Treated with D-Trp[6]-N^{α}-Me-Leu[7]-des-Gly[10]-Pro[9]-NHEt-LH-RH (Wy-40,972)

		Pregnancy No. 1 + R_x							Pregnancy No. 2, no R_x, return to fertility		
Group	N	Total daily dose (μg/rat, s.c.)	R_x schedule (days of pregnancy)	Number of rats returning to vaginal cornification (total)	Time to first vaginal cornification during pregnancy (days postvaginal sperm)[a]	Inhibition of pregnancy (%)	Day of delivery[a]	Litter size[a]	N	Day of delivery[a]	Litter size[a]
Oil	11	—	1–7	0/11	No cornification	0	22.2 ± 0.2	12.0 ± 2.4	11[b]	23.2 ± 1.2	10.7 ± 0.7
Wy-40,972	15	10	1–7	14/15	11.4 ± 0.60	100	—	0	6	22.7 ± 0.3	12.7 ± 2.3
Oil	6	—	7–12	0/6	No cornification	0	22.8 ± 0.2	13.2 ± 0.5	5[b]	23.2 ± 0.2	12.8 ± 1.0
Wy-40,972	17	10	7–12	17/17	16.4 ± 0.4	100	—	0	6	22.8 ± 0.2	10.5 ± 0.9

[a] Means ± S.E.
[b] Oil controls are not remated because they do not show cornification during pregnancy. Numbers refer to oil controls that were introduced into the second part of the study to coincide with the remating schedule of animals receiving Wy-40,972 and to provide an additional set of term control data.

77 consecutive days, resulting in acyclicity, ovarian atrophy, and uterine regression, has been shown to be followed within 3 weeks by a restoration of normal ovarian weight and function as well as the ability to successfully mate and give birth to a normal complement of healthy, well-formed pups (Johnson *et al.*, 1976a).

2.2. Effects on Puberty

Evidence for the antireproductive effects of the LH-releasing compounds has been extended to prepubertal animals as well. Chronic treatment of immature rats with an agonistic analogue delayed time of vaginal opening and retarded the growth of the ovaries, uteri, and anterior pituitary (Johnson *et al.*, 1976a; Rippel and Johnson, 1976b; Corbin *et al.*, 1978; Vilchez-Martinez *et al.*, 1979). It is therefore apparent that the antifertility activity represents the major characteristic pharmacological property of this category of peptides in the female, a concept that is supported by data from a variety of experimental models and laboratories employing numerous LH-RH congeners. While effective profertility regimens may be designed, success will depend on the ability to circumvent their predominantly antireproductive nature.

2.3. Clinical Contraceptive Application

A number of recent clinical studies designed specifically to assess the contraceptive utility of LH-RH/agonists have yielded very encouraging results. When 250 μg LH-RH was administered subcutaneously (s.c.) at 4-hr intervals for a total of five injections starting 1–9 days following ovulation, the luteal phase was shortened by 1–4 days and progesterone levels were reduced to half that observed in the control subjects (Lemay *et al.*, 1979). Supporting the direct relationship between agonist (LH-releasing activity) and antifertility potency established in laboratory animals, only 100 μg of a superagonist (D-Trp^6-des-Gly^{10}-Pro^9-NHEt-LH-RH) given in divided doses (s.c.) on two days during the midluteal phase (Casper and Yen, 1979) or five daily injections (s.c.) of another agonist (des-Gly^{10}-Pro^9-NHEt-LH-RH) during the early luteal phase (Koyama *et al.*, 1978) were capable of a similar luteolytic and emmenogogic effect. Intranasal dosages of 1 mg D-Ser(TBu)^6-des-Gly^{10}-Pro^9-NHEt-LH-RH (Buserelin®) administered on days 4–9 following the midcycle LH peak also decreased progesterone and shortened the cycle (Lemay *et al.*, 1980). The two cycles following cessation of this treatment were considered to be normal and ovulatory in nature. Daily administration (s.c.) of a 5-μg dose of Buserelin® during an entire menstrual cycle was found to inhibit ovulation as well as the expected luteal progesterone increase (Nillius *et al.*, 1978). Again, the subsequent nontreatment cycle was found to be normal. Similar chronic treatment extended for up to 12 weeks consistently inhibited ovulation and resulted in insufficient luteal phases. It is particularly notable that chronic intranasal administration of this agonist for up to 180 days (600 μg/day) (Bergquist *et al.*, 1979a,b), which produced anovulatory cycles, did not generally affect normal menstrual bleeding. In all these chronic-treatment studies, normal ovulatory cycles returned within an average of 1 month following termination of treatment.

Thus, it is quite clear that the antifertility properties of this class of peptides, first identified in laboratory animals, not only are evident in humans but also present much promise as fertility-regulating agents. Moreover, the human studies verify the direct relationship between LH-releasing activity and luteolysis, which in turn provides a valuable predictive basis for developing this particular contraceptive approach.

3. Antireproductive Properties of the LH-Releasing Peptides in the Male

In the male, exogenous LH and human chorionic gonadotropin (hCG) have been shown to negatively affect testicular-LH-receptor concentrations (Hsueh *et al.*, 1976, 1977; Sharpe, 1976). On the basis of the identification of a similar LH-receptor down-regulatory phenomenon occurring in the female, in response to the abnormally elevated endogenous LH levels induced by LHRH/agonists, and the importance of this process to the antipregnancy effect, it seemed quite possible that these peptides might effectively disrupt male reproduction as well. Recent studies have, in fact, demonstrated that single and repeated short-term administration of various agonists produces a marked and prolonged decline in testicular LH receptors, decreased testosterone levels, inhibition of spermatogenesis, and reduced weights of the testes and seminal vesicles (Auclair *et al.*, 1977a,b; Pelletier *et al.*, 1978; Bex and Corbin, 1978; Cusan *et al.*, 1979; Corbin *et al.*, 1979b; Corbin and Bex, 1979). These results corroborate the predominantly antifertility profile of the LH-releasing peptides evaluated in the female and explain the equivocal results obtained from their clinical application in the attempted treatment of various andrological disorders. Moreover, the antifertility effects observed in the male allude to the possibility that LH-RH/agonists may represent a novel and much-sought-after approach to male-fertility regulation. However, while various short-term regimens with LH-RH treatment have been found to be sufficient to negatively affect the steroidal milieu necessary for the series of discrete events involved in female reproduction (i.e., ovulation, implantation), in the male, where gametogenesis is a continually renewing and overlapping process, chronic suppression of testicular function would presumably be required to inhibit fertility. Considering the likelihood that such treatment would also be associated with adverse and quite possibly irreversible effects on other androgen-dependent organs as well as on sexual behavior, it was necessary to evaluate the impact of long-term administration and the potential for dissociating spermatogenic inhibition from androgen depression and its effects on other male reproductive parameters.

3.1. *Effects of Chronic Administration*

Adult male rats were administered a representative and potent LH-RH analogue [D-Ala[6]-des-Gly[10]-Pro[9]-NHEt-LH-RH (Wy-18,481)] for 35 days at daily dose levels covering a 10^5-fold range (0.01–1000 μg). A significant depression in testicular weight was observed after 7 days with a dose as low as 0.1 μg/day; this effect was maintained throughout the 35-day course of treatment (Fig. 8). The weights of testosterone-dependent accessory sex organs were also significantly reduced, but these changes occurred 7–14 days following the first observation of testicular inhibition (Fig. 9 and 10). An inhibition of testicular steroidogenesis, as

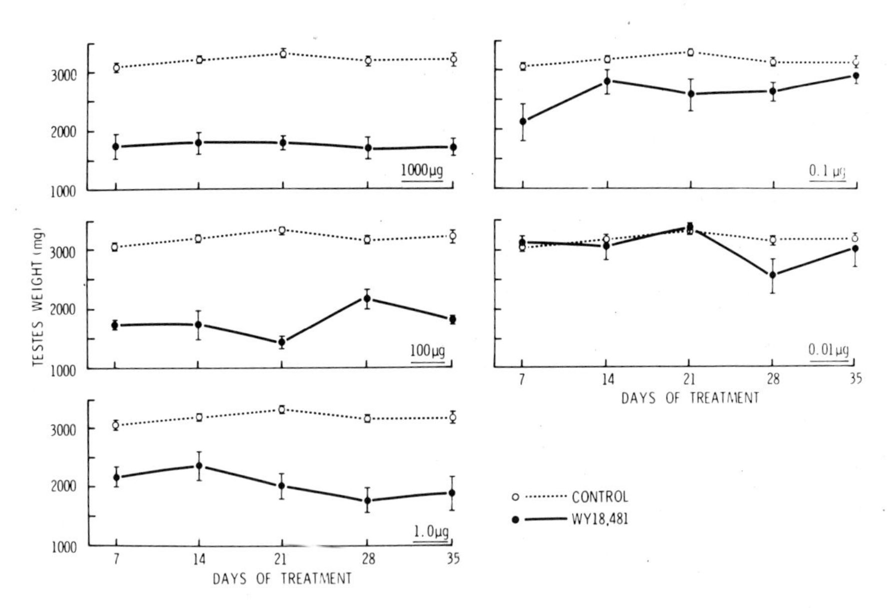

Figure 8. Effect of chronic treatment with Wy-18,481 (0.01–1000 μg/day, s.c.) on testes weight of mature rats.

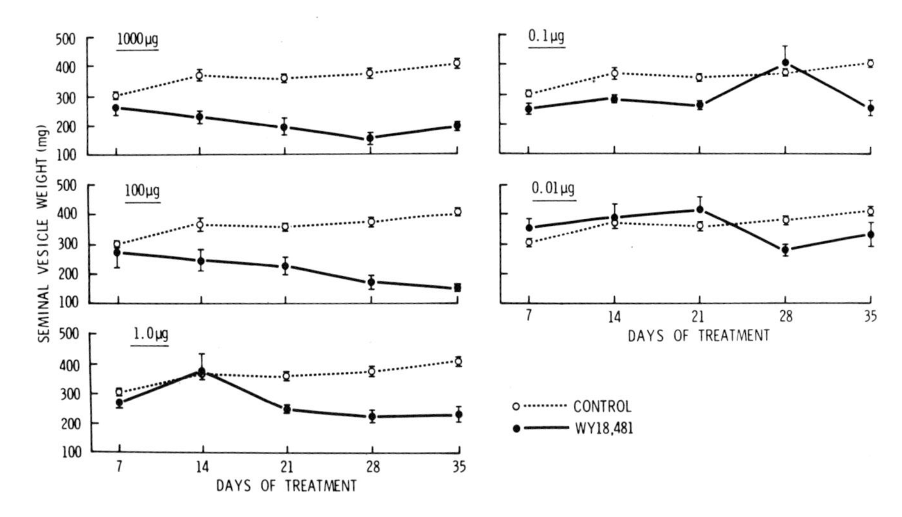

Figure 9. Effect of chronic treatment with Wy-18,481 (0.01–1000 μg/day, s.c.) on seminal-vesicle weight of mature rats.

evidenced by a marked decrease in serum testosterone levels, paralleled the decreased gonadal weight (Fig. 11). Despite the disruptive physiological effects of the 0.1-μg treatment, characterized histologically by progressive disorganization of the seminiferous tubules and loss of sperm in the tubular lumina, there was little influence on ability to mate and a slight depression, if any, in fertility. However, at higher daily dosages

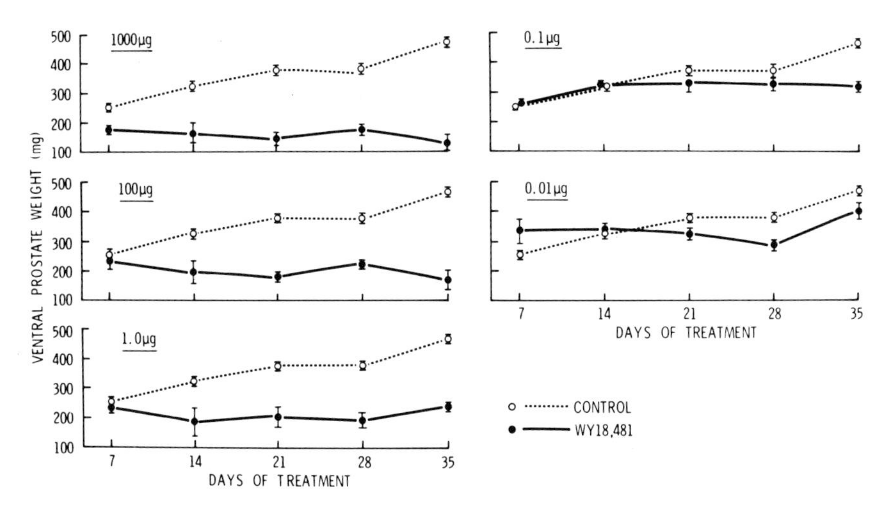

Figure 10. Effect of chronic treatment with Wy-18,481 (0.01–1000 μg/day, s.c.) on ventral-prostate weight of mature rats.

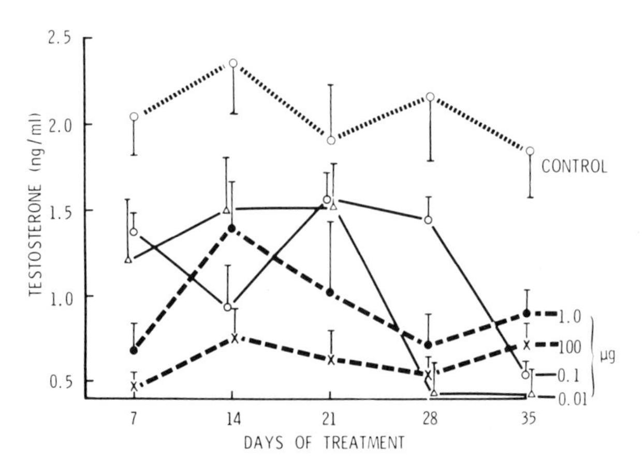

Figure 11. Effect of chronic treatment with Wy-18,481 (0.01–100 μg/day, s.c.) on serum testosterone in mature rats.

(e.g., 100 μg), not only were reproductive-organ weights reduced, but also significant decreases in libido and fertility followed shortly thereafter (Table VI).

3.1.1. Testicular-LH-Receptor Down-Regulation

LH binding was found to be reduced after 7 days of treatment by all dosages of the agonist tested including the 0.01 μg/day dose, which had minimal effects on reproductive-organ weights (Fig. 12). Although it was

Table VI. Effects of D-Ala[6]-des-Gly[10]-Pro[9]-NHEt-LH-RH (Wy-18,481) on Mating and Fertility in Male Rats

Daily Rx	Days Rx	Males mating / Total		Females pregnant / Females inseminated		$\bar{X}$ Normal sites / Total implantation sites
Wy-18,481						
0.1 μg	7	5/5	100%	2/5	40%	6.2/6.5
	14	3/5	60%	1/3	33%	16/16
	21	5/5	80%	4/5	80%	7.5/8.2
	28	5/5	100%	4/5	80%	9.5/10.5
	35	4/5	80%	4/4	100%	14.6/14.8
100 μg	7	4/5	80%	3/4	75%	8.3/9.3
	14	1/5	20%	1/1	100%	13/14
	21	2/5	40%	1/2	50%	10/10
	28	1/5	20%	0/1	0%	0/0
	35	2/5	40%	0/2	0%	0/0

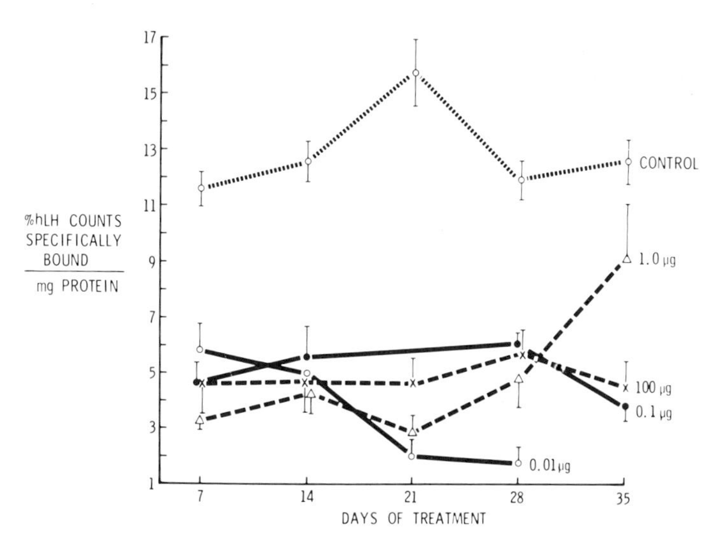

Figure 12. Effect of chronic treatment with Wy-18,481 (0.01–100 μg/day, s.c.) on testicular LH binding in mature rats.

possible that the depressed binding might be due to receptor occupancy, inhibition of this parameter in all instances was associated with lowered LH-dependent steroidogenesis. These observations suggested that:

1. The reduction in reproductive-organ weights produced by higher dosages of the agonist might occur through processes in addition to inhibition of LH binding.

2. Treatment regimens might be designed that result in testosterone levels sufficient to support testosterone-dependent organ weights but not spermatogenesis.

3.1.2. *Prospects for Selective Inhibition of Spermatogenesis*

It had been reported that twice-weekly administration of 0.1 μg Wy-18,481 resulted in a progressive decrease in reproductive-organ weights to 50% of control values by 8 weeks (Pelletier *et al.*, 1978). On this basis, the injection frequency was reduced to once weekly in the hope that transient decreases in testosterone produced by the treatment might desynchronize the spermatogenic process or disrupt epididymal maturation without affecting other testosterone-dependent target tissues. Although this schedule was continued for 11 weeks, representing approximately 1.5 spermatogenic cycles, to maximize the chance of detecting effects at any stage of this continually renewing process, there were no apparent effects on mating performance or fertility (Table VII). There was, in fact, a significant depression in reproductive-organ weights as

Table VII. Effect of Once-Weekly Administration of D-Ala[6]-des-Gly[10]-Pro[9]-NHEt-LH-RH (Wy-18,481) on Mating and Fertility in the Male Rat

Weeks Rx[a]	Males mating[b] / Total	Females pregnant / Females inseminated	$\overline{X}$ Normal sites / Total implantation sites
1	15/21 (71%)	12/15 (80%)	10.9/11.5
2	13/21 (62%)	12/13 (92%)	11.2/11.8
3	8/21 (38%)	6/8 (75%)	9.5/9.8
4	18/21 (86%)	18/18 (100%)	10.1/10.7
5	21/21 (100%)	19/21 (90%)	11.5/11.9
6	16/21 (76%)	13/16 (81%)	10.6/11.2
7	17/21 (81%)	12/17 (71%)	11.7/12.2
8	12/21 (57%)	7/12 (58%)	10.4/11.1
9	19/20 (95%)	16/19 (84%)	10.7/11.4
10	17/20 (85%)	13/17 (76%)	11.5/12.0
11	19/20 (95%)	17/19 (87%)	10.6/11.4

[a] 0.1 μg Wy-18,481, i.m., once weekly.
[b] Males were mated on the evening following each weekly injection.

observed at autopsy following the 11th week of treatment (Table VIII). The results of this experiment and of numerous other experiments employing a variety of agonistic analogues, dosages, and regimens indicate that the inhibitory effects of these peptides on male fertility are consistently associated with and generally preceded by a depression in reproductive-organ weights. Thus, it is quite apparent that it will be difficult to obtain the desired selective spermatogenic inhibition, thereby presenting a situation not unlike that encountered in attempts to develop a steroidal approach to male fertility.

3.2. *Effects of Subacute Administration*

To further differentiate the sequence of hormonal events involved in the male antifertility effects of the agonists and to investigate the

Table VIII. Effect of Once-Weekly Administration of D-Ala[6]-des-Gly[10]-Pro[9]-NHEt-LH-RH on Reproductive-Organ Weights in the Male Rat[a]

Rx (N)	Body weight (g)	Weight of reproductive organs (mg): Adrenal	Ventral prostate	Seminal vesicles	Testes
Oil (5)	510.4 ± 12.9	60.6 ± 3.7	782.8 ± 21.3	680.4 ± 21.3	3449.2 ± 121.3
Wy-18,481 (20)	503.1 ± 8.9	60.1 ± 1.5	625.9 ± 23.1[b]	640.3 ± 17.5	2601.7 ± 133.7[b]

[a] Fertility data are shown in Table VII; 11 weeks of treatment with 0.1 μg Wy-18,481/week. Weights are means ± S.E.M.
[b] $P < 0.01$.

possibility of components other than testicular LH down-regulation, a subacute study was performed in which adult male rats were sacrificed at various intervals following 1, 2, or 3 days daily high-dose (1000 μg) agonist administration (Fig. 13). As expected, the first treatment produced a marked and protracted rise in serum LH, but the magnitude of this response became progressively less on the remaining 2 days. The serum testosterone pattern paralleled that of LH, but was never significantly reduced below that of the controls. However, testes weights of animals autopsied after the third injection were significantly reduced. These findings would indicate that:

1. Depressed LH-dependent steroidogenesis through receptor inhibition becomes a factor after organ-weight depression occurs.

2. Pituitary "exhaustion" in response to at least the higher dosages of the agonists may also contribute to the collapse of reproductive processes; since it is possible that a depression in intratesticular testosterone levels occurs before changes in peripheral concentrations, the LH-receptor down-regulatory phenomenon resulting in reduced steroidogenesis might ultimately be found sufficient to explain testicular inhibition by the agonists. Regarding this point, it has been shown that in addition to LH-RH-induced down-regulation of gonadal receptors for LH, there is also a reduction in PRL receptors (Auclair *et al.*, 1977b). However, the stimulatory effect of PRL at the early stages of the testicular steroidogenic pathway remains intact despite this loss in its receptors (Belanger *et al.*, 1979). Thus, the block in steroidogenesis, identified at the 17-hydroxylase and 17,20-desmolase level, produced by LH-RH, seems likely to be a result of the LH-receptor decrease.

3.3. *Extrapituitary Effects of the LH-Releasing Peptides in the Male*

Studies on the effects of the agonists on reproductive development in the immature male lend support for an additional and presumably direct gonadal component. In this regard, we have found that daily administration of 1 mg Wy-18,481 between days 25 and 35 of age significantly reduces the weight of the testes. A slightly older animal, although still prepubertal, is considerably more sensitive to the inhibitory effects of the agonist. Treatment for 7 days with 100 μg Wy-18,481/day beginning on day 35 of age reduced testes weights by almost 60% and seminal-vesicle weights by nearly 40% (Table IX). This inhibition was maintained throughout the 28-day course of treatment, with a reduction in ventral-prostate weights appearing after 21 days. When this experiment was replicated in males that were hypophysectomized on day 25, the testes weights of the agonist-treated animals were significantly reduced below those of hypophysectomized controls after 14 days' administration, and accelerated depression of sex-accessory-organ weights was observed

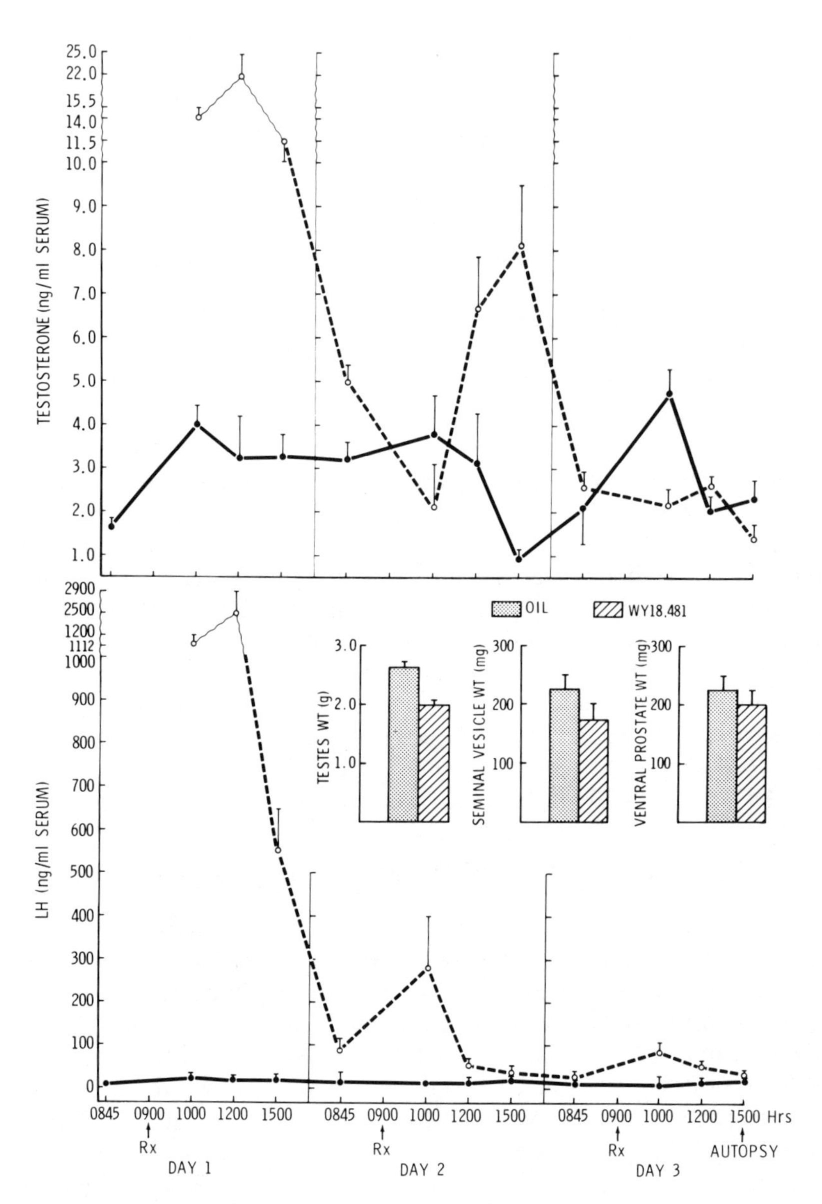

Figure 13. Effect of treatment with 1000 μg Wy-18,481/rat per day, s.c., for 1, 2, or 3 days on serum LH and testosterone in mature male rats. *Insert:* Weights of the testes and sex accessory organs at autopsy following 3 days of treatment.

Table IX. Effect of Chronic Administration of D-Ala[6]-des-Gly[10]-Pro[9]-NHEt-LH-RH (Wy-18,481) on Reproductive Development of Immature Male Rats

Group[a] (*N*)	Weight (mg)[b]			Testosterone (ng/ml)[b]
	Testes	Ventral prostate	Seminal vesicles	
7-Day Rx				
Vehicle (10)	2479 ± 107	148 ± 18	165 ± 25	2.92 ± 0.67
Wy-18,481 (10)	1026 ± 214[c]	137 ± 15	100 ± 11[d]	1.21 ± 0.13[d]
14-Day Rx				
Vehicle (10)	2479 ± 104	171 ± 13	201 ± 15	2.80 ± 0.85
Wy-18,481 (10)	1432 ± 114[c]	164 ± 10	156 ± 11[d]	0.91 ± 0.13[d]
21-Day Rx				
Vehicle (10)	2764 ± 52	315 ± 20	359 ± 14	1.88 ± 0.29
Wy-18,481 (10)	1880 ± 142[c]	184 ± 16[c]	191 ± 19[c]	0.57 ± 0.06[c]
28-Day Rx				
Vehicle (10)	2945 ± 78	387 ± 36	411 ± 22	3.18 ± 0.50
Wy-18,481 (9)	1962 ± 180[c]	224 ± 15[c]	214 ± 21[c]	0.68 ± 0.10[c]

[a] Treatment was initiated for all groups on day 35 of age; 100 μg Wy-18,481/day, s.c. in corn oil, given as a divided dose at 09:00 and 15:00.
[b] Means ± S.E.M.
[c] $P < 0.01$ from vehicle.
[d] $P < 0.05$ from vehicle.

after 28 days (Table X). Histological evaluation of the testes illustrated the significantly greater degenerative effects produced by Wy-18,481 treatment than those occurring as a result of hypophysectomy alone. It has also been reported that administration of an equally potent LH-RH agonist to immature hypophysectomized male rats inhibits FSH-induced increases in testicular LH receptors and depresses testosterone production of the testes *in vitro* (Hsueh, 1979).

Although these results provide compelling support for the involvement of an extrapituitary mechanism in the antireproductive effects of the agonists in the immature male, efforts to identify a similar action in the adult have been less successful. In notable contrast to the results in the immature animal was the observation that the testes and sex-accessory-organ weights of hypophysectomized adult male rats regressed at the same rate whether or not they had been treated with Wy-18,481 (100 μg/day × 35 days) (Table XI). Moreover, in adults, neither Wy-18,481 nor D-Trp[6]-des-Gly[10]-Pro[9]-NHEt-LH-RH has been found to affect *in vitro* testicular steroidogenesis (Badger *et al.*, 1979) (Table XII). However, administration of either peptide initiated *in vivo* prior to the *in vitro* test depressed both basal and stimulated (hCG or human menopausal gonadotropin) testosterone production. To investigate the possibility that the short exposure to the agonist in the *in vitro* studies may not provide

Table X. Effect of Chronic Administration of D-Ala[6]-des-Gly[10]-Pro[9]-NHEt-LH-RH (Wy-18,481) on Immature Hypophysectomized Male Rats

Group[a] (*N*)	Weight (mg)[b]		
	Testes	Ventral prostate	Seminal vesicles
7-Day Rx			
Vehicle (17)	188 ± 23	7.6 ± 0.6	13.9 ± 2.7
Wy-18,481 (17)	188 ± 31	8.1 ± 1.6	18.0 ± 2.7
14-Day Rx			
Vehicle (16)	169 ± 8	5.1 ± 0.5	17.2 ± 1.14
Wy-18,481 (13)	125 ± 10[c]	4.7 ± 0.8	14.6 ± 1.1
21-Day Rx			
Vehicle (12)	238 ± 45	9.6 ± 2.4	18.8 ± 3.3
Wy-18,481 (12)	154 ± 45	5.8 ± 0.6	17.0 ± 1.0
28-Day Rx			
Vehicle (9)	442 ± 108	26.9 ± 16	47.0 ± 22.0
Wy-18,481 (11)	159 ± 5[c]	6.4 ± 0.4	15.7 ± 1.2

[a] All rats were hypophysectomized on day 25 of age and treatment was initiated on day 35 of age; 100 μg Wy-18,481/day, s.c. in corn oil, given as a divided dose at 09:00 and 15:00.
[b] Means ± S.E.M.
[c] $P < 0.01$ from vehicle.

Table XI. Effect of Chronic Administration of D-Ala[6]-des-Gly[10]-Pro[9]-NHEt-LH-RH (Wy-18,481) on Adult Hypophysectomized Male Rats

Group[a] (*N*)	Weight (mg)[b]		
	Testes	Ventral prostate	Seminal vesicles
7-Day Rx			
Vehicle (5)	2599 ± 133	89 ± 25	178 ± 15
Wy-18,481 (3)	2515 ± 164	146 ± 32	162 ± 24
14-Day Rx			
Vehicle (5)	2102 ± 261	87 ± 44	151 ± 23
Wy-18,481 (5)	1675 ± 84	38 ± 10	144 ± 20
21-Day Rx			
Vehicle (5)	967 ± 121	22 ± 3	79 ± 7
Wy-18,481 (5)	981 ± 63	24 ± 4	83 ± 6
28-Day Rx			
Vehicle (5)	655 ± 33	24 ± 4	74 ± 5
Wy-18,481 (5)	541 ± 38	29 ± 3	80 ± 5

[a] Treatment was initiated the day following hypophysectomy in all groups; 100 μg Wy-18,481/day, s.c. in corn oil, given as a divided dose at 09:00 and 15:00.
[b] Means ± S.E.M.

Table XII. Effect of D-Ala[6]-des-Gly[10]-Pro[9]-NHEt-LH-RH (Wy-18,481) on Basal and hCG-Stimulated Testosterone Production by Decapsulated Adult Rat Testes *in Vitro*[a]

In vitro treatment[b]	Testosterone production (ng/2 ml)
	Testis weight (g)[c]
Control (buffer) (33)	156 ± 20
40 mIU hCG (22)	453 ± 52
50 μg Wy-18,481 (11)	198 ± 9
50 μg Wy-18,481 + 40 mIU hCG (11)	414 ± 89

[a] Each testis was randomly assigned to a treatment; incubation was carried out in 2 ml Krebs–Ringer bicarbonate buffer with glucose (1 mg/ml) for 4 hr at 37°C under O_2–CO_2 (95:5).
[b] The number in parentheses is the number of incubation flasks (one testis/flask).
[c] Results represent the mean (± S.E.M.) total testosterone production (ng/2 ml) expressed in terms of decapsulated testis weight in grams.

sufficient time to disclose a direct effect, hypophysectomized adult male rats were treated daily for 4 or 11 days with 100 μg Wy-18,481. An *in vitro* evaluation of testicular steroidogenesis performed the day after the last injection (day 5 or 12) suggests a slight inhibition of basal and hCG-stimulated testosterone production by the agonist (Table XIII). In this regard, it has been reported that adult rat testes will specifically bind D-Trp[6]-LH-RH and that this agonist produces, in hypophysectomized male rats, a greater reduction in LH receptors than that induced by hCG (Arimura *et al.*, 1979). Thus, it is possible that more dramatic reproductive-organ degeneration induced by hypophysectomy in the adult male masked the direct gonadal effect by treatment with the agonist that was identified in the immature animal.

3.4. Reversibility of Antireproductive Effects

Of particular importance to the proposed use of LH-RH agonists as contraceptive agents was the reversibility of the inhibitory effects following cessation of treatment. Male rats chronically administered (35 days) with a daily dose of Wy-18,481 far exceeding that producing maximal inhibitory effects were sacrificed at various intervals up to 70 days posttreatment (Fig. 14). Although at the last time sampled, the weights of the testes and sex accessory organs were still somewhat depressed, they had significantly recrudesced and were pursuing a growth pattern similar to that seen in the control animals. Testicular LH binding and serum testosterone levels had significantly rebounded by day 14 and approximated control values by day 44 (Corbin and Bex, 1979). The restoration of breeding processes became evident between days 14 and

Table XIII. Effect of D-Ala[6]-des-Gly[10]-Pro[9]-NHEt-LH-RH (Wy-18,481) Administered to Hypophysectomized Adult Rats *in Vivo* on Basal and hCG-Stimulated Testosterone Production *in Vitro*[a]

In vivo treatment[b]	*In vitro* treatment	Testosterone production (ng/2 ml) / Testis weight (g)[c]
4-Day Rx		
Vehicle (5)	—	137 ± 36
Vehicle (5)	40 mIU hCG	93 ± 37
Wy-18,481[d] (5)	—	67 ± 12
Wy-18,481[d] (5)	40 mIU hCG	67 ± 16
11-Day Rx		
Vehicle (3)	—	43 ± 22
Vehicle (3)	40 mIU hCG	42 ± 13
Wy-18,481[d] (4)	—	13 ± 7
Wy-18,481[d] (4)	40 mIU hCG	7 ± 1

[a] All rats were hypophysectomized 1 week prior to the initiation of treatment; incubation was carried out in 2 ml Krebs–Ringer bicarbonate buffer with glucose (1 mg/ml) for 4 hr at 37°C under O_2–CO_2 (95:5).
[b] The number in parentheses is the number of incubation flasks (one testis/flask).
[c] Results represent the mean (± S.E.M.) total testosterone production (ng/2 ml) expressed in terms of decapsulated testis weight in grams.
[d] 100 μg Wy-18,481/day, s.c. in corn oil, given as a divided dose at 09:00 and 15:00.

28 and were fully restored by day 44 posttreatment. Again, it is interesting to note that normal fertility was present despite decreased organ weights. Additional return-to-fertility studies employing lower dosages of the agonist over similar or shorter periods indicate that the depression of male reproductive function produced by these compounds is reversible, with the length of time necessary for restoration dependent on the magnitude of the dose and duration of administration.

It is quite clear from a wide variety of studies that in the male, as in the female, the LH-releasing peptides present a predominantly antireproductive pharmacological profile. Since a major component of this effect occurs through pituitary-mediated LH-receptor down-regulation, possibly exacerbated by direct testicular inhibitory effects, present efforts to achieve successful treatments for hypogonadotropic hypogonadism, oligospermia, delayed puberty, or cryptorchidism, through development of more potent and long-acting LH-RH agonists, should be reconsidered. In fact, since the effects of these peptides on male reproduction are reversible, their antifertility (rather than their purported profertility) properties may eventually be exploited for use as male contraceptives. However, the inability of various combinations of doses and dosing schedules utilized thus far to dissociate antispermatogenic effects from numerous attendant side effects (including depression of

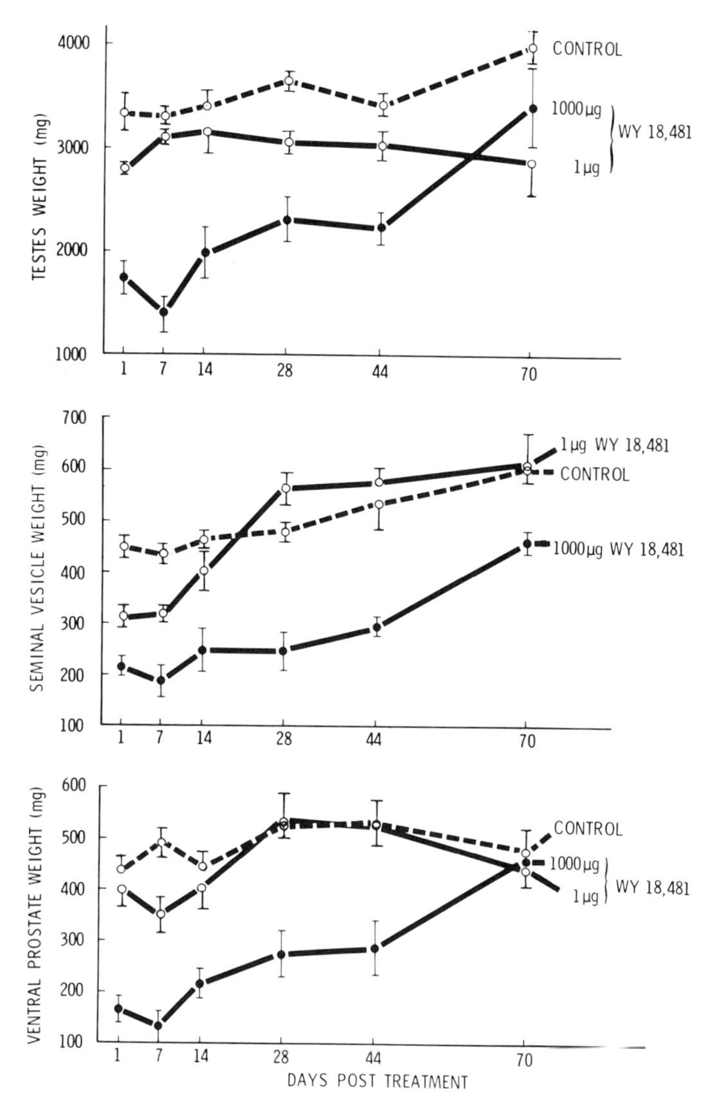

Figure 14. Restoration of the weights of the testes and sex accessory organs in mature male rats following the inhibitory effects produced by chronic treatment with Wy-18,481 (1 or 1000 μg/rat per day, s.c., $\times$ 35 days).

reproductive-organ weights and libidinal effects) presents an obstacle to further development of this approach reminiscent of that encountered during attempts to develop a steroidal method of regulating male fertility.

4. Noncontraceptive Utility of Antireproductive Activity: Antitumor Effects

Inasmuch as LH-RH and its agonists have been shown to exert their antifertility effects through a mechanism involving suppression of gonadal steroidogenesis, there has been increasing interest, aside from the contraceptive ability, in the potential use of these peptides as an alternative or adjunct therapy in the treatment of steroid-dependent tumors. Indeed, various LH-RH agonists have been shown to retard the growth and produce regression of 7,12-dimethylbenz[*a*]anthracene-induced mammary tumors in rodents (Johnson *et al.*, 1976b; DeSombre *et al.*, 1976; Danguy *et al.*, 1977; Dutta *et al.*, 1978). We have recently demonstrated that administration of LH-RH or Wy-18,481 retards the development of tumors in immature hamsters made vulnerable with antilymphocytic serum and inoculated with mouse mammary tumor cells (Corbin *et al.*, 1980). Although the antitumor effect remains evident for a certain amount of time after the cessation of peptide treatment, growth eventually reappears. This tumor recrudescence corresponds with the expected return of gonadal steroidogenesis and can again be inhibited by reinitiating agonist treatment. In general, the antitumor activity of the LH-releasing peptides is very similar to that observed following ovariectomy or treatment with antisteroidal agents. Considering the inhibitory effects of LH-RH agonists on testicular steroidogenesis, this novel application may serve in the treatment of male steroid-dependent tumors as well.

5. Conclusions

It has become firmly established that the LH-releasing class of peptides possesses an essentially antireproductive pharmacological profile; this conclusion is quite different from that originally proposed during the initial stages of the development of these presumptive profertility agents. It has become increasingly apparent that the antifertility property, rather than representing a contradiction of the integral role the parent molecule plays in normal reproductive function, reflects the complex pharmacological nature of these peptides and the processes that they control. Notably, the recent identification of receptor down-regulation by inappropriately elevated LH levels, as well as the extra-pituitary actions of LH-RH and its agonists, have contributed much to resolving the paradoxical pro-/antifertility classification of these peptides. Moreover, extensive reproductive evaluation has consistently supported the potential of these substances for use as contraceptive agents in the female. While it is clear that considerably more development will be necessary before similar optimism can be expressed in terms of their use in regulating male fertility, there definitely exists the possibility that these peptides might be valuable in the treatment of various gonadal steroid-dependent tumors in the male, as well as in the female.

References

Arimura, A., Serafini, P.C., and Sonntag, W., 1979, Does LHRH agonist decrease ovarian and testicular LH/hCG receptors by direct action?, Program of the 61st Annual Meeting of the Endocrine Society, Anaheim, California, Abstract 433.

Auclair, C., Kelly, P.A., Labrie, F., Coy, D.H., and Schally, A.V., 1977a, Inhibition of testicular luteinizing hormone receptor level by treatment with a potent luteinizing hormone releasing hormone agonist or human chorionic gonadotropin, *Biochem. Biophys. Res. Commun.* **76:**855.

Auclair, C., Kelly, P.A., Coy, D.H., Schally, A.V., and Labrie, F., 1977b, Potent inhibitory activity of [D-Leu^6,des-Gly-NH_2^{10}] LHRH ethylamide on LH/hCG and PRL testicular receptor levels in the rat, *Endocrinology* **101:**1890.

Badger, T.M., Beitins, I.Z., Ostrea, T., Crisafulli, J., Little, R., and Saidel, M., 1979, Effects of luteinizing hormone releasing hormone and one of its highly potent analogs (D-Trp^6-Pro^9-NHEt-LRF) on *in vitro* testosterone synthesis, Program of the 61st Annual Meeting of the Endocrine Society, Anaheim, California, Abstract 430.

Banik, U.K., and Givner, M.L., 1975, Ovulation induction and antifertility effects of an LH-RH analogue (Ay-25,205) in cyclic rats, *J. Reprod. Fertil.* **44:**87.

Banik, U.K., and Givner, M.L., 1976, Effects of a luteinizing hormone-releasing hormone analog on mating and fertility in rats, *Fertil. Steril.* **27:**1078.

Beattie, C.W., and Corbin, A., 1977, Pre- and postcoital contraceptive activity of LH-RH in the rat, *Biol. Reprod.* **16:**333.

Beattie, C.W., Corbin, A., Cole, G., Corry, S., Jones, R.C., Koch, K., and Tracy, J., 1977, Mechanism of the post-coital contraceptive effect of LH-RH in the rat. I. Serum hormone levels during chronic LH-RH administration, *Biol. Reprod.* **16:**322.

Belanger, A., Auclair, C., and Caron, S., 1979, Role of prolactin in the LHRH agonist-induced blockage of testicular steroidogenesis, Program of the 61st Annual Meeting of the Endocrine Society, Anaheim, California, Abstract 717.

Bergquist, C., Nillius, S.J., and Wide, L., 1979a, Inhibition of ovulation in women by intranasal treatment with a luteinizing hormone-releasing hormone agonist, *Contraception* **19:**497.

Bergquist, C., Nillius, S.J., and Wide, L., 1979b, Intranasal gonadotropin-releasing hormone agonist as a contraceptive agent, *Lancet* **1979:**215.

Bex, F.J., and Corbin, A., 1978, Inhibition of reproductive processes in the immature and mature male rat with an LHRH agonist, Program of the 3rd Annual Meeting of the American Society of Andrology, Nashville, Tennessee, Abstract 6.

Bex, F.J., and Corbin, A., 1979a, Mechanism of the postcoital contraceptive effect of luteinizing hormone-releasing hormone: Ovarian luteinizing hormone receptor interactions, *Endocrinology* **105:** 139.

Bex, F.J., and Corbin, A., 1979b, LHRH or LHRH agonist-induced termination of pregnancy in hypophysectomized (HYPX) rats: Extrapituitary site of action, Program of the 61st Annual Meeting of the Endocrine Society, Anaheim, California, Abstract 436.

Casper, R.F., and Yen, S.S.C., 1979, Induction of luteolysis in the human with a long-acting analog of luteinizing hormone-releasing factor, *Science* **205:**408.

Comaru, A.M.de M., Rodriques, J., Povoa, L.C., Franco, S., Dimetz, T., Novellino, P., Coy, D.H., and Schally, A.V., 1976, Clinical use of luteinizing hormone-releasing hormone and its long-acting analogues, in: *Recent Advances in Human Reproduction* (A. Campos, D.A. Paz, V.A. Drill, M. Hayashi, W. Rodriques, and A.V. Schally, eds.), p. 140, Excerpta Medica, Amsterdam.

Corbin, A., and Beattie, C.W., 1975a, Inhibition of the pre-ovulatory proestrous gonadotropin surge, ovulation and pregnancy with a peptide analogue of luteinizing hormone releasing hormone, *Endocr. Res. Commun.* **2:** 1.

Corbin, A., and Beattie, C.W., 1975b, Post-coital contraceptive and uterotrophic effects of luteinizing hormone releasing hormone, *Endocr. Res. Commun.* **2:**445.

Corbin, A., and Beattie, C.W., 1976, Effect of luteinizing hormone releasing hormone (LHRH) and an LHRH antagonist on hypothalamic and plasma LHRH of hypophysectomized rats, *Endocrinology* **98:**247.

Corbin, A., and Bex, F., 1979, Inhibition of male reproductive processes with an LHRH agonist, Program of the 1st Pan American Congress of Andrology, Caracas, Venezuela, Abstract 251.

Corbin, A., Beattie, C.W., Yardley, J., and Foell, T.J., 1976, Post-coital contraceptive effects of an agonistic analogue of luteinizing hormone releasing hormone, *Endocr. Res. Commun.* **3:**359.

Corbin, A., Beattie, C.W., Tracy, J., Jones, R., Foell, T.J., Yardley, J., and Rees, R.W.A., 1978, The anti-reproductive pharmacology of LH-RH and agonistic analogues, *Int. J. Fertil.* **23:**81.

Corbin, A., Bex, F.J., Yardley, J.P., Rees, R.W.A., Foell, T.J., and Sarantakis, D., 1979a, Agonist (ovulation induction) and post-coital contraceptive properties of [D-Ala[6]] and [D-Trp[6]]-LHRH series, *Endocr. Res. Commun.* **6:** 1.

Corbin, A., Beattie, C.W., Jones, R., and Bex, F., 1979b, Peptide contraception: Antifertility properties of LH-RH analogues, *Int. J. Gynaecol. Obstet.* **16:**379.

Corbin, A., Rosanoff, E., and Bex, F.J., 1980, LHRH and agonists: Paradoxical antifertility effects, mechanisms of action and therapeutic relevance to steroid dependent tumors, in: *Progress in Cancer Research and Therapy,* Vol. 14 (S. Iacobelli, H.R. Lindner, R.J.B. King, and M.E. Lippman, eds.), p. 533, Raven Press, New York.

Coy, D.H., Labrie, F., Savary, M., Coy, E.J., and Schally, A.V., 1975, LH-releasing activity of potent LH-RH analogs *in vitro, Biochem. Biophys. Res. Commun.* **67:**576.

Cusan, L., Auclair, C., Bélanger, A., Ferland, L., Kelly, P.A., Séguin, C., and Labrie, F., 1979, Inhibitory effects of long-term treatment with a luteinizing hormone-releasing hormone agonist on the pituitary–gonadal axis in male and female rats, *Endocrinology* **104:** 1369.

Danguy, A., Legros, N., Heuson-Stiennon, J.A., Pasteels, J.L., Atassi, G., and Heuson, J.C., 1977, Effects of a gonadotropin-releasing hormone (GnRH) analogue (A-43818) on 7,12-dimethylbenz(a)anthracene-induced rat mammary tumors: Histological and endocrine studies, *Eur. J. Cancer* **13:**1089.

Desombre, E.R., Johnson, E.S., and White, W.F., 1976, Regression of rat mammary tumors effected by a gonadoliberin analog, *Cancer Res.* **36:**3830.

Dutta, A.S., Furr, B.J.A., Giles, M.B., Valaccia, B., and Walpole, A.L., 1978, Potent agonist and antagonist analogues of luliberin containing an azaglycine residue in position 10, *Biochem. Biophys. Res. Commun.* **81:**382.

Edgren, R.A., Windsor, B.L., Bousley, F.G., and Humphrey, R.R., 1977, Absence of a prenidatory effect of luteinizing hormone releasing hormone (LHRH) in hamsters, *Int. J. Fertil.* **22:**40.

Hilliard, J., Pang, C.N., and Sawyer, C.H., 1976, Effects of luteinizing hormone releasing hormone on fetal survival in pregnant rabbits, *Fertil. Steril.* **27:**421.

Hsueh, A.J.W., 1979, Extrapituitary action of gonadotropin releasing hormone (GnRH): Direct inhibition of ovarian and testicular responses, Program of the 61st Annual Meeting of the Endocrine Society, Anaheim, California, Abstract 198.

Hsueh, A.J.W., and Erickson, G.F., 1979, Extrapituitary action of gonadotropin-releasing hormone: Direct inhibition of ovarian steroidogenesis, *Science* **204:**854.

Hsueh, A.J., Dufau, M.L., and Catt, K.J., 1976, Regulation of luteinizing hormone receptors in testicular interstitial cells by gonadotropin, *Biochem. Biophys. Res. Commun.* **72:**1145.

Hsueh, A.J.W., Dufau, M.L., and Catt, K.J., 1977, Gonadotropin-induced regulation of luteinizing hormone receptors and desensitization of testicular 3′,5′-cyclic AMP and testosterone responses, *Proc. Natl. Acad. Sci. U.S.A.* **74:**592.

Humphrey, R.R., Windsor, B.L., Reel, J.R., and Edgren, R.A., 1977, The effects of luteinizing hormone (LHRH) in pregnant rats. I. Postnidatory effects, *Biol. Reprod.* **16:**614.

Johnson, E.S., Gendrich, R.L., and White, W.F., 1976a, Delay of puberty and inhibition of reproductive processes in the rat by a gonadotropin-releasing hormone agonist analog, *Fertil. Steril.* **27:**853.

Johnson, E.S., Seely, J.H., White, W.F., and DeSombre, E.R., 1976b, Endocrine-dependent rat mammary tumor regression: Use of a gonadotropin releasing hormone analog, *Science* **194:**329.

Khodr, G.S., and Siler-Khodr, T., 1978a, Localization of luteinizing hormone-releasing factor in the human placenta, *Fertil. Steril.* **29:**523.

Khodr, G.S., and Siler-Khodr, T., 1978b, The effect of luteinizing hormone-releasing factor on human chorionic gonadotropin secretion, *Fertil. Steril.* **30:**301.

Kledzik, G.S., Cusan, L., Auclair, C., Kelly, P.A., and Labrie, F., 1978a, Inhibitory effect of a luteinizing hormone (LH)-releasing hormone agonist on rat ovarian LH and follicle-stimulating hormone receptor levels during pregnancy, *Fertil. Steril.* **29:**560.

Kledzik, G.S., Cusan, L., Auclair, C., Kelly, P.A., and Labrie, F., 1978b, Inhibition of ovarian luteinizing hormone (LH) and follicle-stimulating hormone receptor levels with an LH-releasing hormone agonist during the estrous cycle in the rat, *Fertil. Steril.* **30:**348.

Koch, Y., Baram, T., Hazum, E., and Fridkin, M., 1977, Resistance to enzymatic degradation of LH-RH analogues possessing increased biological activity, *Biochem. Biophys. Res. Commun.* **74:**488.

Koyama, T., Ohkura, T., Kumasaka, T., and Saito, M., 1978, Effect of postovulatory treatment with a luteinizing hormone-releasing hormone analog on the plasma level of progesterone in women, *Fertil. Steril.* **30:**549.

Lemay, A., Labrie, F., Ferland, L., and Raynaud, J., 1979, Possible luteolytic effects of luteinizing hormone-releasing hormone in normal women, *Fertil. Steril.* **31:**29.

Lemay, A., Labrie, F., and Raynaud, J.P., 1980, Luteolytic activity of LHRH and [D-Ser(TBU)^6,des-Gly-NH_2^{10}]LHRH ethylamide: A new and physiological approach to contraception in women, *Int. J. Fertil.* **5:**203.

Macdonald, G.J., and Beattie, C.W., 1979, Pregnancy failure in hypophysectomized rats following LHRH administration, *Life Sci.* **24:**1103.

Matsuo, H., Baba, Y., Nair, R.M.G., Arimura, A., and Schally, A.V., 1971a, Structure of the porcine LH- and FSH-releasing hormone. I. The proposed animo acid sequence, *Biochem. Biophys. Res. Commun.* **43:**1334.

Matsuo, H., Arimura, A., Nair, R.M.G., and Schally, A.V., 1971b, Synthesis of the porcine LH- and FSH-releasing hormone by the solid-phase method, *Biochem. Biophys. Res. Commun.* **45:**822.

Mayar, M.Q., Tarnavsky, G.K., and Reeves, J.J., 1979, Ovarian growth and uptake of iodinated D-Leu,des-Gly-NH_2^{10}-LHRH ethylamide in HCG treated rats, *Proc. Soc. Exp. Biol. Med.* **161:**216.

Nillius, S.J., Bergquist, C., and Wide, L., 1978, Inhibition of ovulation in women by chronic treatment with a stimulatory LRH analogue—a new approach to birth control?, *Contraception* **17:**537.

Pelletier, G., Cusan, L., Auclair, C., Kelly, P.A., Desy, L., and Labrie, F., 1978, Inhibition of spermatogenesis in the rat by treatment with [D-Ala^6,des-Gly-NH_2^{10}]LHRH ethylamide, *Endocrinology* **103:**641.

Rao, M.C., Richards, J.S., Midgley, A.R., Jr., and Reichert, L.E., Jr., 1977, Regulation of gonadotropin receptors by luteinizing hormone in granulosa cells, *Endocrinology* **101:**512.

Richards, J.S., Ireland, J.J., Rao, M.C., Bernath, G.A., Midgley, A.R., Jr., and Reichert, L.E., Jr., 1976, Ovarian follicular development in the rat: Hormone receptor regulation by estradiol, follicle stimulating hormone and luteinizing hormone, *Endocrinology* **99:**1562.

Rippel, R.H., and Johnson, E.S., 1976a, Regression of corpora lutea in the rabbit after injection of a gonadotropin-releasing peptide, *Proc. Soc. Exp. Biol. Med.* **152:**29.

Rippel, R.H., and Johnson, E.S., 1976b, Inhibition of HCG-induced ovarian and uterine weight augmentation in the immature rat by analogs of GnRH, *Proc. Soc. Exp. Biol. Med.* **152:**432.

Rivier, J.E., and Vale, W.W., 1978, [D-pGlu^1,D-Phe^2,D-Trp^{3,6}]-LRF: A potent luteinizing hormone releasing factor antagonist *in vitro* and inhibitor of ovulation in the rat, *Life Sci.* **23:**869.

Schally, A.V., Kastin, A.J., and Arimura, A., 1971, Hypothalamic follicle-stimulating hormone (FSH) and luteinizing hormone (LH)-regulating hormone: Structure, physiology, and clinical studies, *Fertil. Steril.* **22:**703.

Sharpe, R.M., 1976, hCG-induced decrease in availability of rat testes receptors, *Nature (London)* **264:**644.

Siler-Khodr, T.M., and Khodr, G.S., 1978, Content of luteinizing hormone-releasing factor in the human placenta, *Am. J. Obstet. Gynecol.* **130:**216.

Taymor, M.L., 1974, The use of luteinizing hormone-releasing hormone in gynecologic endocrinology, *Fertil. Steril.* **25:**992.

Vale, W., Rivier, C., Brown, M., Leppaluoto, J., Ling, N., Monahan, M., and Rivier, J., 1976, Pharmacology of hypothalamic regulatory peptides, *Clin. Endocrinol.* **5**(Suppl.):261S.

Vilchez-Martinez, J.A., Pedroza, E., Arimura, A., and Schally, A.V., 1979, Paradoxical

effects of D-Trp^6-leuteinizing hormone-releasing hormone on the hypothalamic pituitary–gonadal axis in immature female rats, *Fertil. Steril.* **31:**677.
Windsor, B.L., Humphrey, R.R., Reel, J.R., and Edgren, R.A., 1977, Postnidatory effects of luteinizing hormone releasing hormone (LHRH) in hamsters, *Int. J. Fertil.* **22:**184.
Ying, S., and Guillemin, R., 1979, (D-Trp^6-Pro^9-NEt)-Luteinizing hormone-releasing factor inhibits follicular development in hypophysectomized rats, *Nature (London)* **280:**593.
Yoshinaga, K., 1978, Effect of an LHRH analog on the pituitary function of pregnant rats, Program of the 60th Annual Meeting of the Endocrine Society, Miami, Florida, Abstract 495.

6

Fertility and Antifertility Effects of LH-RH and Its Agonists

U. K. Banik and M. L. Givner

1. Introduction

Following isolation, purification (Schally *et al.*, 1971; Amoss *et al.*, 1971), and elucidation of the chemical structure of luteinizing-hormone-releasing hormone (LH-RH) (Matsuo *et al.*, 1971a; Burgus *et al.*, 1971) and its synthesis (Matsuo *et al.*, 1971b; Monahan *et al.*, 1971), many investigators (see White *et al.*, 1973; Immer *et al.*, 1974) synthesized a number of analogues that are more potent than LH-RH in releasing LH and follicle-stimulating hormone (FSH) and inducing ovulation in animals and humans (Fugino *et al.*, 1974; Schally *et al.*, 1976; Keye *et al.*, 1976).

During the last few years, we conducted a series of biological experiments using synthetic LH-RH (AY-24,031) (H-Pry-His-Trp-Ser-Tyr-Gly-Leu-Arg-Pro-Gly-NH_2) and its two agonists (Fig. 1): H-Pry-His-Trp-Ser-Tyr [D-Ala]-Leu-Arg-Pro-NH-CH_2-CH_3 (D-Ala^6-LH-RH EA) (AY-25,205) and H-Pry-His-Trp-Ser-Tyr-[D-Trp]-Leu-Arg-Pro-Gly-NH_2 (D-Trp^6-LH-RH) (AY-25,650). The purpose of this chapter is to review some of our basic biological studies on LH-RH and its two agonists with special reference to induction of ovulation, mating behavior, and pregnancy in rats and mice under various experimental conditions. Antifertility effects of these neurohormones in rats are also reviewed. Finally,

U. K. Banik and *M. L. Givner* • Department of Endocrinology and Immunochemistry, Ayerst Research Laboratories, P.O. Box 6115, Montreal H3C 3J1, Canada.

AY-24,031 (LH-RH): H-Pyr-His-Trp-Ser-Tyr-Gly-Leu-Arg-Pro-Gly-NH_2
AY-25,205 (D-Ala^6-LH-RH EA): H-Pyr-His-Trp-Ser-Tyr-[D-Ala]-Leu-Arg-Pro-NH-CH_2-CH_3
AY-25,650 (D-Trp^6-LH-RH): H-Pyr-His-Trp-Ser-Tyr-[D-Trp]-Leu-Arg-Pro-Gly-NH_2

Figure 1. LH-RH and its two agonists.

reference is also made to the possible uses of these substances in improving and controlling fertility in animals and humans.

All experiments were conducted on adult female Sprague–Dawley rats (225 ± 25 g) and Swiss albino mice (30 ± 5 g) obtained from Canadian Breeding Laboratories, St. Constant, Quebec, and Bio Breeding Laboratories, Ottawa, Ontario. The details of the husbandry conditions of animals and other general methodology were described earlier (Banik and Givner, 1975, 1976a,b, 1977). Unless otherwise stated, 10 animals were used in each group of experiments and all injections were given in 0.2 ml saline containing 0.1% gelatin.

2. *Induction of Ovulation by LH-RH and Agonists*

Under normal conditions, the female part of the reproductive process starts with physiologically timed stimuli that reach the higher center of brain (cortex), which, via the hypothalamus, stimulates the pituitary to secrete gonadotropins (FSH and LH). These hormones are involved in the growth of ovarian follicles and rupture of matured follicles to release their eggs in sequence throughout reproductive cycles.

2.1. *Induction of Ovulation in Cycling Rats*

Attempts were made to induce ovulation by a single intravenous (i.v.) dose of 2.5 μg LH-RH in proestrous, estrous, metestrous, and diestrous rats. Synthetic LH-RH was given about 3:30 P.M. in different stages of the cycle, and the evidence for ovulation was obtained by counting freshly shed ova in the oviduct the following respective morning. All (6/6) proestrous animals ovulated with or without LH-RH (Fig. 2). None of the estrous (0/6) or metestrous (0/13) rats ovulated, while 7/7 of the diestrous rats ovulated. As expected, none of the vehicle-treated estrous, metestrous, or diestrous rats ovulated. These results are in general agreement with those of other investigators (Morishita *et al.*, 1974; Yamazaki *et al.*, 1978). These results suggested that the rupture of follicles is dependent on at least two important physiological factors, i.e., nearly or completely mature available follicles and high levels of circu-

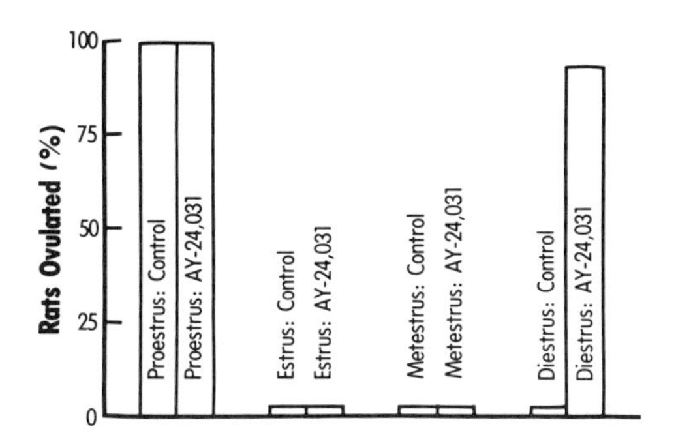

Figure 2. Advancement of ovulation in proestrous, estrous, metestrous, and diestrous rats by a dose of LH-RH (AY-24,031).

lating LH. Both these coincidental factors were inadequate in estrous and metestrous conditions of the cycle.

2.2. *Induction of Ovulation in Fluphenazine-Dihydochloride-Treated Proestrous and in Pregnant, Pseudopregnant, and Diestrous Rats and in Diestrous Mice*

Some CNS depressants and major tranquilizers, administered before the "critical period" of LH release on the day of proestrous, are known to block ovulation in adult rats (Everett and Sawyer, 1950; Barraclough, 1966; Banik and Herr, 1969) due to inhibition of release of LH-RH from the hypothalamus. Therefore, synthetic or natural LH-RH or its agonists administered to such CNS-depressed proestrous rats would induce hypophyseal LH release followed by ovulation (Banik and Givner, 1975, 1977). To block ovulation, fluphenazine dihydochloride (FD), a potent neuroleptic agent, was administered (500 μg/0.2 ml of 0.9% saline/rat) subcutaneously (s.c.) in the dorsal region of the neck between 2:00 and 2:15 P.M. on the day of proestrous. To induce ovulation in FD-treated rats, graded single doses of a lyophilized preparation of LH-RH or of the agonist (D-Ala^6-LH-RH EA or D-Trp^6-LH-RH) were injected in 2% aqueous benzyl alcohol intramuscularly (i.m.) in the same neck region between 3:00 and 3:30 P.M. on the same day. The following morning, evidence for ovulation was obtained following the same method as mentioned earlier (Banik and Givner 1975, 1977).

The relative potency of LH-RH and its two agonists obtained from such experiments is presented in Fig. 3. The ED_{50} values for LH-RH, D-Ala^6-LH-RH EA, and D-Trp^6-LH-RH for induction of ovulation in FD-treated proestrous rats and in pregnant, pseudopregnant, diestrous, and

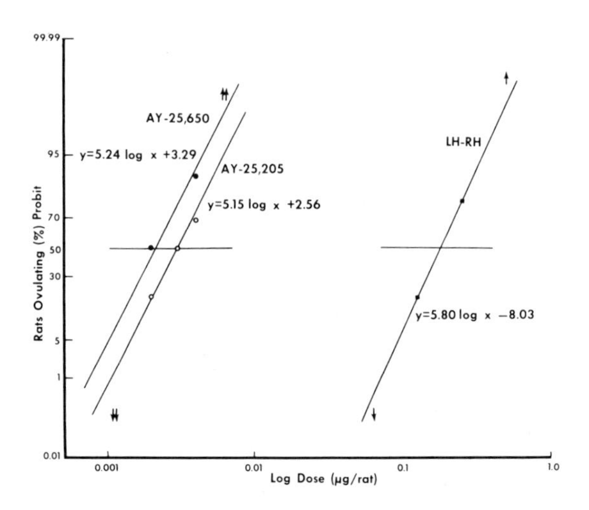

Figure 3. Relative potency of LH-RH and two analogues [D-Ala^{6}-LH-RH EA (AY-25,205) and D-Trp^{6}-LH-RH (AY-25,650)] in inducing ovulation in proestrous rats pretreated with FD.

metestrous rats and diestrous mice were reported earlier (Banik and Givner, 1977) and are summarized in Table I.

2.3. *Induction of Ovulation in Estrogen- and Progestin-Treated Rats*

Suppression of the hypothalamic–hypophyseal–ovarian system by estrogens and progestins led to the development of present-day oral contraceptive pills. The immediate effect of estrogens of oral contraceptives seems to be directed toward the reduction of FSH and that of progestins toward suppression of the midcycle peak of LH (Diczfalusy,

Table I. Relative Potency of LH-RH and Two Agonists

	ED_{50} (ng/animal)		
Test animals	LH-RH	D-Ala^{6}-LH-RH EA (×)[a]	D-Trp^{6}-LH-RH (×)[a]
FD-treated proestrous rats	176	3 (59)	2 (88)
Pregnant rats	14,000	26 (538)	37 (378)
Pseudopregnant rats	11,000	23 (478)	26 (423)
Diestrous rats	1,000	11 (91)	14 (71)
Metestrous rats	>10,000?	128 (>78)	100 (>100)
Diestrous mice	220	5 (44)	4 (55)

[a] (×) Times more potent than LH-RH.

1968). Experiments were conducted to investigate the antiovulatory effects of ethynyl estradiol (EE_2) and medroxyprogesterone acetate (MPA) in rats and their reversal with exogenous administration of pregnant mare serum gonadotropin (PMS) and LH-RH.

Adult 4-day-cycling rats were given a daily oral dose of 100 μg EE_2 for blocking ovulation from the day of estrus to the following proestrus (Banik *et al.*, 1969), whereas a single dose of 1 mg MPA i.m. between 3:00 and 3:30 P.M. on the day before the expected day of proestrus blocked ovulation in all animals (Fig. 4). In EE_2-treated animals, the ovulation was not induced with a daily dose of 5 IU PMS followed by a single dose (0.5 μg) of LH-RH. However, 2.5 μg LH-RH was significantly effective (66.7%) in inducing ovulation in 4/6 estrogen/PMS-treated rats. Likewise, in MPA-treated rats, the same dose of PMS or 0.5 μg LH-RH was not effective in inducing ovulation (Fig. 4). However, a single 0.5-μg dose of LH-RH in the afternoon of the expected day of proestrus was effective in inducing ovulation in MPA-treated rats only when final maturation of follicles was maintained by two doses of PMS.

3. *Induction of Mating Behavior, Ovulation, and Pregnancy by LH-RH and Agonists*

A variety of neurogenic factors are known to trigger endocrine events connected with reproductive functions. The neurogenic stimuli activate the hypothalamic–hypophyseal system, causing release of LH, which in turn leads to ovulation (Wilson *et al.*, 1965). Incidence of pregnancy following coital-induced ovulation in spontaneous ovulators, for example, in rats (Zarrow *et al.*, 1968), other animals, and humans (Clark and Zarrow, 1971; Jöchle, 1973), is now known, even though

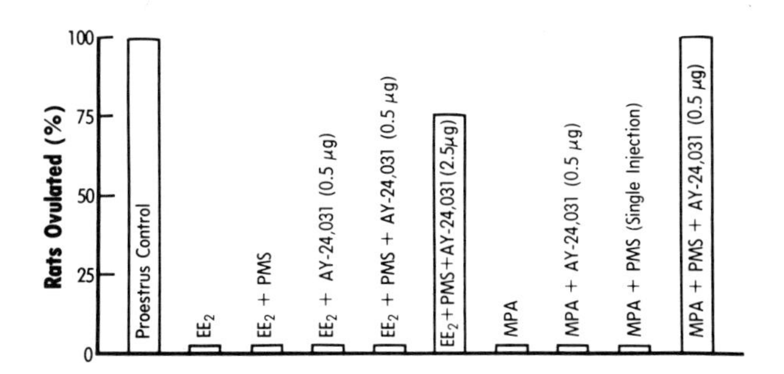

Figure 4. Antiovulatory effect of ethynyl estradiol (EE_2) and medroxyprogesterone acetate (MPA) and its reversal with gonadotropins and synthetic LH-RH (AY-24,031).

elevated levels of serum gonadotropins were not observed in some women after coitus during the mid- to late follicular phase of the menstrual cycle (Stearns *et al.*, 1973; Lee *et al.*, 1974). LH-RH induced mating behavior in ovariectomized/estrogen-treated (Moss and McCann, 1973) and also in hypophysectomized, ovariectomized/estrogen-treated female rats (Pfaff, 1973). Therefore, it seems that the mating-behavioral effect of LH-RH is dependent on circulating estrogen. We have conducted a series of experiments with LH-RH and two agonists on mating, ovulation, and pregnancy in intact rats and mice.

3.1. Advancement of Mating Behavior and Pregnancy in Metestrous, Diestrous, Proestrous, and Estrous Rats by Two Doses of LH-RH

Four-day-cycling adult female rats were given 500 ng LH-RH i.v. at 10:30 A.M. and 1:00 P.M. on different days of the cycle. Cohabitation was allowed in a semidark room between 10:30 A.M. and 4:30 P.M. After 4:30 P.M., each female's mating sign was recorded from microscopic examination of sperm in vaginal smears. All animals were killed on day 9 after the treatment, and the number of implants was recorded. None of the metestrous or diestrous rats with or without LH-RH mated and became pregnant (Fig. 5). Of proestrous rats, 70% mated during cohabitation on the day of proestrus and became pregnant, whereas in the control group, only 10% (1/10) mated and became pregnant. All animals

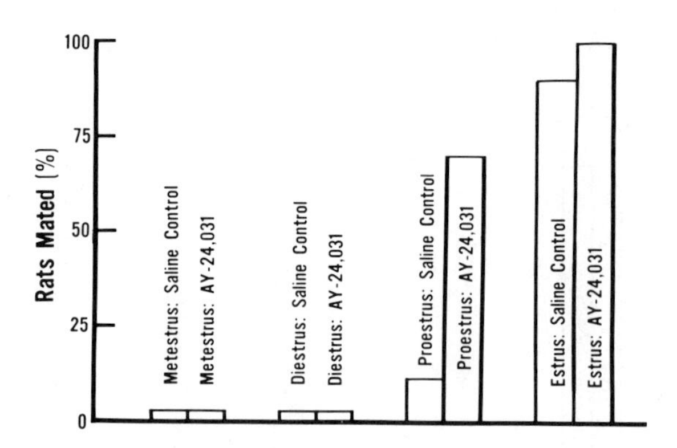

Figure 5. Advancement of mating behavior with synthetic LH-RH (AY-24,031) in metestrous, diestrous, proestrous, and estrous rats.

in the estrous LH-RH-treated group (9/9) mated and became pregnant, and in the control group, 9/10 mated and became pregnant.

3.2. Induction of Ovulation and Mating Behavior in Diestrous Rats by LH-RH and Two Agonists

In a previous study (Banik and Givner, 1977), 1 μg or more of LH-RH and 20 ng or more of D-Ala^6-LH-RH EA and D-Trp^6-LH-RH were significantly effective in inducing ovulation in diestrous rats. In the study reported herein, we wanted to see whether larger doses of these neurohormones, while inducing ovulation, would induce mating behavior as well. Groups of 10 diestrous rats were given i.m. 5000 ng LH-RH or 80 ng D-Ala^6-LH-RH EA (AY-25,205) or D-Trp^6-LH-RH (AY-25,650) at 3:30 P.M. and allowed overnight cohabitation (1♀:1♂) with proven fertile males. Females were inspected on the morning after pairing for mucus plugs or spermatozoa in the vaginal smear. Results presented in Fig. 6 show that LH-RH, and its two agonists, despite inducing ovulation in a

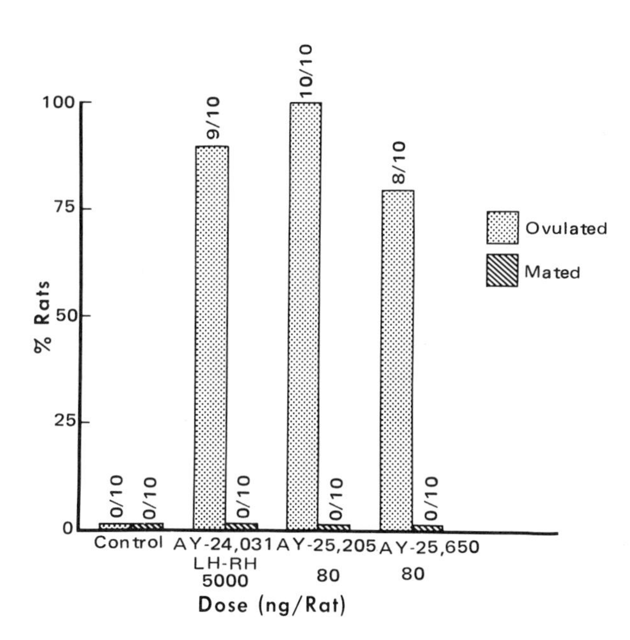

Figure 6. Induction of ovulation and mating behavior in diestrous rats by LH-RH and two agonists.

significant number of rats, were unable to induce mating behavior in any of the rats treated with these neurohormones.

3.3. *Induction of Mating Behavior and Pregnancy in Fluphenazine-Dihydrochloride-Treated Proestrous Rats*

The antiovulatory effect of FD and its reversal with a dose of LH-RH or its agonists in proestrous rats was studied earlier (Banik and Givner, 1975, 1977). The purpose of these experiments was to demonstrate that a single dose of LH-RH or agonist could enhance mating behavior and pregnancy rates in FD-treated proestrous rats.

Groups of proestrous rats were given 500 μg FD s.c. between 2:00 and 2:15 P.M., and LH-RH (250 ng/rat) or 80 ng/rat of one of the two agonists was given s.c. between 3:30 and 4:00 P.M. on the same day. One group of FD-treated rats was given the vehicle. All animals were allowed overnight cohabitation (1♀:1♂) in individual cages, and the following morning, signs of mating were recorded as before. All animals were killed on day 9 postcoitum, and the presence of uterine implantations was recorded. Results presented in Fig. 7 show that LH-RH and agonists significantly enhanced rates of mating behavior and pregnancy in FD-treated proestrous rats. In the vehicle-treated group, 7/20 animals mated, and only 4 of these animals became pregnant.

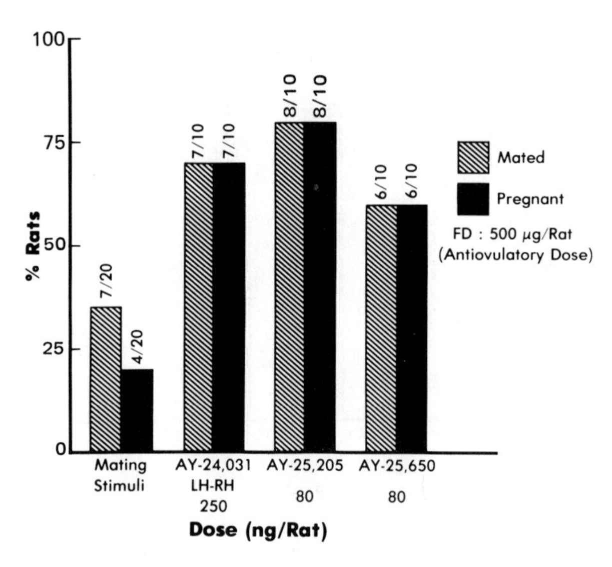

Figure 7. Induction of mating behavior and pregnancy in FD-treated proestrous rats by LH-RH and two agonists.

3.4. Induction of Mating Behavior and Pregnancy in Proestrous Mice

During the last few years, we observed that the estrous cycle in the Swiss albino mice received from our two suppliers was not regular like that of rats. The precise reason for this anomaly is not known to us. Occasionally, we observed that many adult (25–35 g) proestrous mice neither ovulate nor mate, despite showing proestrous or proestrous–estrous vaginal smears. By administering intraperitoneally (i.p.) a single dose of LH-RH, D-Ala[6]-LH-RH EA (AY-25,205), or D-Trp[6]-LH-RH (AY-25,650), we attempted to enhance the rates of ovulation, mating, and pregnancy.

Groups of 10 proestrous mice were given a dose of LH-RH (250 ng/mouse) or 10 ng/mouse of one of the agonists at about 3:30 P.M. and allowed overnight cohabitation with proven fertile males (1♀:1♂). Signs of mating (vaginal plug) and pregnancy (uterine implants) were recorded as in the rat.

Results presented in Fig. 8 show that LH-RH and the two agonists significantly increased the rates of mating behavior and pregnancy in proestrous mice. In the vehicle-treated control group, only 4/10 animals mated, of which 3 became pregnant. In a separate study, we showed previously that LH-RH or D-Ala[6]-LH-RH EA alone was very effective

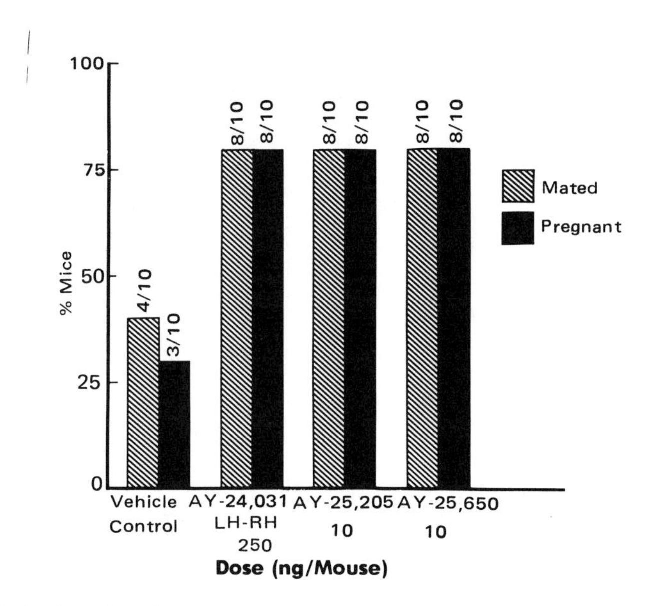

Figure 8. Induction of mating behavior and pregnancy in proestrous mice by LH-RH and two analogues.

in inducing ovulation in diestrous and proestrous mice (Banik and Givner, 1976a).

3.5. *Induction of Ovulation and Mating Behavior in Diestrous Mice*

In previous studies (see above), it was shown that LH-RH or agonists induced ovulation but not mating behavior in diestrous rats. Similar experiments were conducted in diestrous mice. Groups of 10 diestrous mice were given i.p. 5000 ng/mouse of LH-RH or 80 ng/mouse of one of the agonists. Cohabitation was allowed as before, and the signs of mating and the evidence for ovulation were also recorded as before. Results presented in Fig. 9 show that LH-RH and agonists were effective in inducing ovulation in 90% of animals. None of the LH-RH-treated animals mated. In the agonist-treated groups, only 1/10 and 2/10 animals mated. In the vehicle-treated group, none of the animals mated or ovulated (Fig. 9).

4. *Antifertility Effects of LH-RH and Agonists*

Since the isolation and structural elucidation of hypothalamic LH-RH and various agonists, many investigators (Schally *et al.*, 1971; Gay,

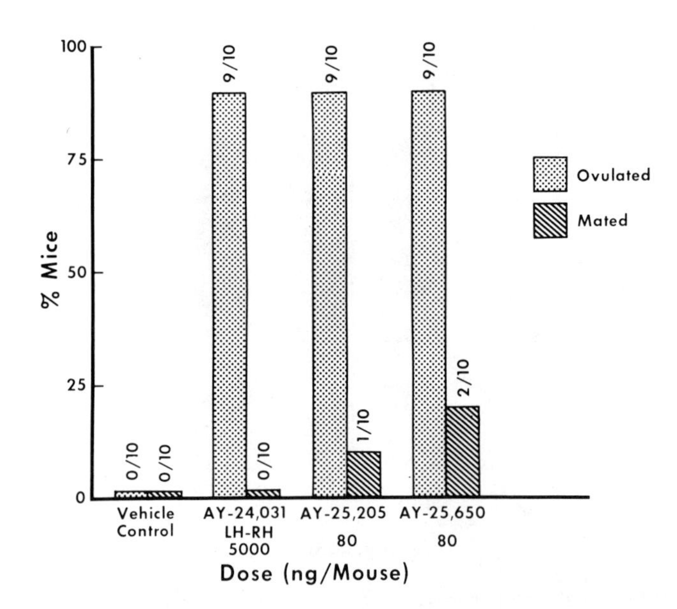

Figure 9. Induction of ovulation and mating behavior in diestrous mice by LH-RH and two agonists.

1972; Guillemin, 1972) have speculated that these neurohormones could be useful for inducing ovulation in women at a desired time to make the rhythm method of contraception more effective. However, convincing experimental evidence to decrease fertility and conception rate with physiological doses of these neurohormones has not been demonstrated as yet. Pharmacological doses of LH-RH and agonistic analogues were attempted to control fertility in experimental animals by various investigators (Corbin *et al.*, 1978). It was reported (Banik and Givner, 1975) that the induction of premature ovulation in diestrous rats by administration of as little as 10 ng D-Ala[6]-LH-RH EA was associated with interference of mating behavior and pregnancy during the ensuing proestrous and estrous periods. The antifertility effects of this analogue disappeared in the subsequent cycle. When the compound was given in 20- to 120-ng doses every 3rd day starting on the afternoon of diestrus, it caused rhythmic antifertility effects on 4-day-cyclic rats that were allowed cohabitation with fertile males except for the first 24 hr after treatment (Fig. 10). However, the use of this compound every 4th day did not interfere with fertility (Fig. 11). This antifertility effect was achieved by inducing ovulation at a "physiologically wrong time" (Banik and Givner, 1976b). Higher doses and chronic treatment with this analogue showed reduced functional capacity of male and female gonads (Johnson *et al.*, 1976; Cusan *et al.*, 1979). Other investigators also speculated about a

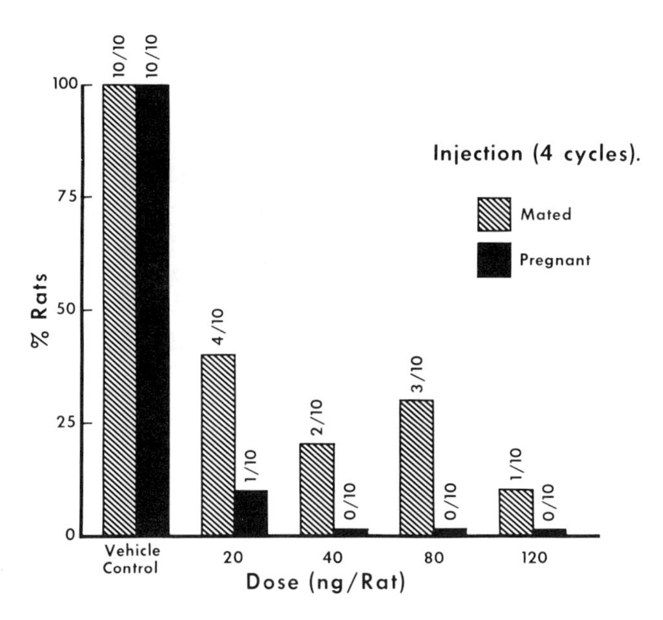

Figure 10. Rhythmic antifertility effects of AY-25,205 injected (i.m.) every 3rd day. Cohabitation was not allowed for 24 hr following injection (4 cycles).

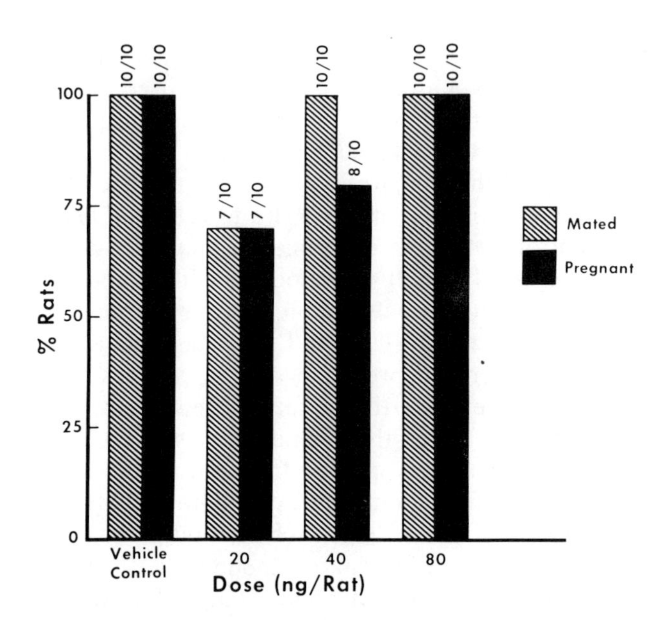

Figure 11. Rhythmic antifertility effects of AY-25,205 injected (i.m.) every 4th day. Uninterrupted cohabitation was allowed (2 cycles).

paradoxical direct inhibitory effect of LH-RH or agonist on the gonad (Mayer *et al.,* 1977; Macdonald and Beattie, 1979).

Equally dramatic effects of large doses of LH-RH (100–1000 μg/day and less) but still pharmacological doses (1–6 μg/day) of some superactive agonists were found to block implantation and terminate gestation when given daily to postcoital rats (Humphrey *et al.,* 1976, 1978; Corbin *et al.,* 1978). Some investigators have already attempted to administer a new LH-RH agonist in large daily doses (87–600 μg) intranasally for blocking ovulation in women of reproductive age (Bergquist *et al.,* 1979). Whether this sort of method of controlling fertility in humans will be acceptable or not, without regular withdrawal bleeding, remains to be seen. Further, on the basis of their recent findings of a luteolytic effect of an LH-RH agonist in nonpregnant women, some investigators (Lemay *et al.,* 1979; Casper and Yen, 1979) (also see Chapter 4) even suggested that these potent neurohormones could be effective in inducing menstruation or early abortion.

If the "rhythm" method or periodic abstinence could be made uniformly effective in preventing pregnancy, it would revolutionize modern family planning. The problems of the rhythm method of contraception have not been solved as yet. The main reason for occasional failure of this method of contraception is perhaps due to induction of

Length of cycle	Days of Cycle																																		
	1	2	3	4	5	6	7	8	9	10	11	12	13	14	15	16	17	18	19	20	21	22	23	24	25	26	27	28	29	30	31	32	33	34	35
25	M	M	M	M							O														m										
26	M	M	M	M								O														m									
27	M	M	M	M									O														m								
28	M	M	M	M										O														m							
29	M	M	M	M											O														m						
30	M	M	M	M												O														m					
31	M	M	M	M													O														m				
32	M	M	M	M														O														m			
33	M	M	M	M															O														m		
34	M	M	M	M																O														m	
35	M	M	M	M																	O														m

For a woman with a 26- to 31-day cycle:

	1	2	3	4	5	6	7	8	9	10	11	12	13	14	15	16	17	18	19	20	21	22	23	24	25	26	27	28	29	30	31	32	33	34	35
26-31	M	M	M	M									(?)	(?)	(?)	(?)	(?)	(?)									?	?	?	?	?				

☐ – infertile day

▨ – fertile day

M – menses

O – theoretical ovulation day

m – theoretical start of next menses

(?) – possible days of ovulation

? – day next menses may begin

Figure 12. Ogino (1930) and Knaus (1930) calendar rhythm method of contraception. (Reproduced with permission from Ross and Piotrow, 1974.)

ovulation after sexual intercourse in some women during the early follicular phase (Jöchle, 1973) and uncertainty of the time of ovulation during the menstrual cycle in many women (Mastroianni, 1974). Many investigators have speculated that the synthetic LH-RH and agonists could be used for inducing ovulation in women at a desired time, and by observing abstinence the failure rate of the rhythm method of contraception could be reduced (Schally *et al.*, 1973; Guillemin, 1972; Banik and Givner, 1976b).

About 50 years ago, Ogino (1930) from Japan and Knaus (1930) from Austria outlined independently the menstruation patterns in women of reproductive age based on basal body temperature and cervical mucus methods (Fig. 12). If theoretical fertile and safe periods were followed for the purpose of the rhythm method of contraception, it is conceivable that the self-determined cervical mucus scoring system used with a dose of 100 μg LH-RH or about 10–20 μg AY-25,205, AY-25,650, or other similar agonist should ensure ovulation in almost all women of reproductive age if given during one of the days of positive cervical-mucus symptom, for example, between days 9 and 14 of the menstrual cycle [or one of the days of estimated time of ovulation as reported by Hilgers *et al.*, (1978)]. Also, the day of ovulation almost dictates the date of menstruation (i.e., 14–15 days after the day of ovulation) if the woman is not pregnant (Fig. 12). It is suspected that the main reason for some failures of the rhythm method of contraception is due to the fact that the date of ovulation is unpredictable in many women for various reasons.

5. Concluding Remarks

Among many potential applications of LH-RH and its agonists, it is conceivable that if an adequate single dose was given at an appropriate time, some of these synthetic neurohormones or similar preparations may enable the following applications:

1. Ensure human ovulation at a definite time, and if women abstain from intercourse for about 4 days before and 2 days after administration, one can reduce the failure rate of the classic rhythm method of contraception. However, this method of contraception may not be useful in women who may have coital-induced ovulation (Jöchle, 1973) during the early follicular stage of the cycle (e.g., days 5–9 of the cycle).

2. Regularize menstrual cycles in women by inducing ovulation during one of the days in the ovulatory period (see Fig. 12) and thereby causing the commencement of menses about 14–15 days (Ross and Piotrow, 1974) following the treatment.

3. Increase the rate of conception in animals and humans by having insemination, artificial or natural, within 12 hr of treatment. It is presumed that such conceptions would occur following fertilization between matured eggs and sperms at optimal time. As Hertig (1967) put it, ". . . if an oocyte lingers longer than Day 14 in the follicle, it has an increasing chance of becoming a *'bad egg'* when fertilized. . . ." He links preovulatory overripeness with abnormal embryonic development. It is expected that a single dose of some of these compounds given at a "physiologically right time" may be useful to avoid such abnormal embryonic development.

ACKNOWLEDGMENTS. Our thanks are due to Drs. H. Immer, K. Sestanj, and T. Dobson of Ayerst Laboratories, Montreal, who prepared LH-RH and analogues (AY-25,205 and AY-25,650). Efficient technical assistance in biological experiments was provided by Messrs. W. Van Leenhoff and P. Gauthier of our department. Statistical analysis of the data was performed by Mr. R. Paquette of our laboratories. Our sincere thanks are due to Ms. V. Knauer for her secretarial assistance.

DISCUSSION

BOGDANOVE: Your approach is very attractive and promising. However, I think I must echo Dr. Haour's concern about possible long-term complications if women are to be given repeated monthly injections or insufflations of these compounds throughout an extended period. As you know, rabbits can be immunized against unconjugated LH-RH. In male rabbits, used to raise antisera to LH-RH, there resulted a total neutralization of their endogenous LH-RH, so that there was a very extensive, hypophysectomy-like, atrophy of the entire reproductive system. Do you know yet whether, or how successfully, such an undesirable complication could be avoided?

BANIK: The dose is too small and only one per cycle, which may not induce antibodies.

PETRUSZ: You can generate anti-LH-RH with a small dose, but this required adjuvant which stimulates the RES.

BOGDANOVE: Ah! But what if the lady gets whooping cough at the same time?

PETRUSZ: Yes, that could serve as a natural adjuvant.

BANIK: We have studied the repeated contraceptive effect of D-Ala^6-LH-RH EA over a period of four cycles. Almost all animals became pregnant within the following two cycles at the end of treatment of four cycles.

REFERENCES

Amoss, M., Burgus, R., Blackwell, R., Vale, W., Fellows, R., and Guillemin, R., 1971, Purification, amino acid composition and N-terminus of the hypothalamic luteinizing hormone releasing factor (LRF) of ovine origin, *Biochem. Biophys. Res. Commun.* **44:**205.

Banik, U.K., and Givner, M.L., 1975, Ovulation induction and antifertility effects of an LH-RH analogue (AY-25,205) in cyclic rats, *J. Reprod. Fertil.* **44:**87.

Banik, U.K., and Givner, M.L., 1976a, Induction of ovulation in adult mice with a synthetic LH-RH (AY-24,031) and an analogue (AY-25,205), *J. Reprod. Fertil.* **47:**95.

Banik, U.K., and Givner, M.L., 1976b, Effects of luteinizing hormone-releasing hormone analogue on mating and fertility in rats, *Fertil. Steril.* **27:**1078.

Banik, U.K., and Givner, M.L., 1977, Comparative ovulation-inducing capacity of a synthetic luteinizing hormone-releasing hormone and two analogs in adult rats and mice, *Fertil. Steril.* **28:**1243.

Banik, U.K., and Herr, F., 1969, Effects of medroxyprogesterone acetate and chlorpromazine on ovulation and vaginal cytology in a strain of 4-day cyclic rats, *Can. J. Physiol. Pharmacol.* **47:**573.

Banik, U.K., Revesz, C., and Herr, F., 1969, Orally active oestrogens and progestins in prevention of pregnancy in rats, *J. Reprod. Fertil.* **18:**509.

Barraclough, C.A., 1966, Modification in the CNS regulation of reproduction after exposure of prepubertal rats to steroid hormones, *Recent Prog. Horm. Res.* **22:**503.

Bergquist, C., Nillius, S.J. and Wide, L., 1979, Inhibition of ovulation in women by intranasal treatment with a luteinizing hormone-releasing hormone agonist, *Contraception* **19:**497.

Burgus, R., Butcher, M., Ling, N., Monahan, M., Rivier, J., Fellows, R., Amoss, M., Blackwell, R., Vale, W., and Guillemin, R., 1971, Structure moléculaire du fracteur hypotrape hypophysaire du luteinisation (LH), *C. R. Acad. Sc. (Paris)* **273:**1611.

Casper, R.F., and Yen, S.C., 1979, Induction of luteolysis in the human with a long-acting analog of luteinizing hormone releasing factor, *Science* **205:**408.

Clark, J.H., and Zarrow, M.X., 1971, The influence of copulation on time of ovulation in women, *Am. J. Obstet. Gynecol.* **109:**1083.

Corbin, A., Beattie, C.W., Tracy, J., Jones, R., Foelie, T.J., Yardley, J., and Rees, R.W.A., 1978, The anti-reproductive pharmacology of LH-RH and agonistic analogues, *Int. J. Fertil.* **23:**81.

Cusan, L., Auclair, C., Bélanger, A., Ferland, L., Kelly, P.A., Séguin, C., and Labrie, F., 1979, Inhibitory effects of long term treatment with a luteinizing hormone-releasing hormone agonist on the pituitary–gonadal axis in male and female rats, *Endocrinology* **104:**1369.

Diczfalusy, E., 1968, Mode of action of contraceptive drugs, *Am. J. Obstet. Gynecol.* **10:**163.

Everett, J.W., and Sawyer, C.H., 1950, A 24-hour periodicity in the "LH-release apparatus" of female rats, disclosed by barbiturate sedation, *Endocrinology* **47:**198.

Fugino, M., Fukuda, T., Shinagawa, S., Kobayashi, S., Yamazaki, I., Nakayama, R., Seely, J.H., White, W.F., and Rippel, F.H., 1974, Synthetic analogs of luteinizing hormone-releasing hormone (LH-RH) substituted in position 6 and 10, *Biochem. Biophys. Res. Commun.* **60:**406.

Gay, V.L., 1972, The hypothalamus: Physiology and clinical use of releasing factors, *Fertil. Steril.* **23:**50.

Guillemin, R., 1972, Physiology and chemistry of the hypothalamic relasing factors for gonadrotropins: A new approach to fertility control, *Contraception* **5:**1.

Hertig, A.H., 1967, The overall problem in man, in: *Comparative Aspects of Reproductive Failure* (K. Benirschke, ed.), pp. 11–41, Springer-Verlag, New York, Heidelberg, and London.

Hilgers, T.W., Abraham, G.E., and Cavanagh, D., 1978, Natural family planning. 1. The peak symptom and estimated time of ovulation, *Obstet. Gynecol.* **52:**575.

Humphrey, R.R., Windsor, B.L., Bausley, F.G., and Edgren, R.A., 1976, Antifertility effects of an analog of luteinizing hormone releasing hormone (LH-RH) in rats, *Contraception* **14:**625.

Humphrey, R.R., Windsor, B.L., Jones, D.L., Reel, J.R., and Edgren, R.A., 1978, The effects of luteinizing hormone releasing hormone (LH-RH) in pregnant rats. 2. Prenidatory effects and delayed parturition, *Biol. Reprod.* **19:**84.

Immer, H., Nelson, V.R., Revesz, C., Sestanj, K., and Gotz, M., 1974, Luteinizing hormone-releasing hormone and analogs: Synthesis and biological activity, *J. Med. Chem.* **17:**1060.

Jöchle, W., 1973, Coitus-induced ovulation, *Contraception* **7:**523.

Johnson, F.S., Seely, J.H., White, W.F., and de Sombre, E.F., 1976, Endocrine dependent rat mammary tumor regression: Use of gonadotropin releasing hormone analog, *Science* **194:**329.

Keye, W.R., Jr., Young, F.R., and Jaffe, R.B., 1976, Hypothalamic gonadotropin releasing hormone: Physiologic and clinical considerations, *Obstet. Gynecol. Surv.* **31:**635.

Knaus, H., 1930, Die periodische Frucht- und Unfruchtbarkeit des Weibes, *Zentralbl. Gynaekol.* **57:**1393.

Lee, P.A., Jaffe, R.B., and Midgley, A.R., 1974, Lack of alteration of serum gonadotropins in men and women following sexual intercourse, *Am. J. Obstet. Gynecol.* **120:**985.

Lemay, A., Labrie, F., Ferland, L., and Raynaud, J.P., 1979, Possible luteolytic effects of luteinizing hormone-releasing hormone in normal women, *Fertil. Steril.* **31:**29.

Macdonald, G.J., and Beattie, C.W., 1979, Pregnancy failure in hypophysectomized rats following LH-RH administration, *Life Sci.* **24:**1103.

Mastroianni, L., 1974, Rhythm: Systematized chance-taking, *Fam. Plann. Perspect.* **6:**209.

Matsuo, H., Baba, Y., Nair, R.M.G., Arimura, A., and Schally, A.V., 1971a, Structure of the porcine LH- and FSH-releasing hormone. 1. The proposed amino acid sequence, *Biochem. Biophys. Res. Commun.* **43:**1334.

Matsuo, H., Arimura, A., Nair, R.M.G., and Schally, A.V., 1971b, Synthesis of the porcine LH- and FSH-releasing hormone by the solid phase method, *Biochem. Biophys. Res. Commun.* **45:**822.

Mayer, M.Q., Tarnavsky, G.K., and Reeves, J.J., 1977, Ovarian growth and uptake of iodinated D-Leu^{6}-des-Gly-$NH_2{}^{10}$-LH-RH ethylamide in HCG treated prepuberal rat, *Proc. West. Sect. Am. Soc. Anim. Sci.* **28:**182.

Monahan, M., Rivier, J., Burgus, R., Amoss, M., Blackwell, R., Vale, W., and Guillemin, R., 1971, Synthesis totale par phase solide d'un decapeptide qui stimule la sécretion des gonadotrophines hypophysaires LH et FSH, *C. R. Acad. Sci. (Paris)* **273:**508.

Morishita, H., Mitani, H., Masuda, Y., Higuchi, K., Tomioka, M., Nagamachi, N., Kawamoto, M., Ozasa, T., and Adachi, H., 1974, Effect of synthetic luteinizing hormone releasing hormone on ovulation during the oestrus cycle in the rat, *Acta Endocrinol. (Copenhagen)* **76:**431.

Moss, R.L., and McCann, S.M., 1973, Induction of mating behavior in rats by luteinizing hormone-releasing factor, *Science* **181:**177.

Ogino, K., 1930, Ovulationstermin und Konzeptionstermin, *Zentrabl. Gynaekol.* **54:**464.

Pfaff, D., 1973, Luteinizing hormone-releasing factor potentiates lordosis behavior in hypophysectomized ovariectomized female rats, *Science* **182:**1148.

Ross, C., and Piotrow, P.T., 1974, Periodic abstinence, Population Reports, Series 1, Number 1, Population Information Program, The John Hopkins University, Baltimore, Maryland.

Schally, A.V., Arimura, A., Baba, Y., Nair, R.M.G., Matsuo, H., Redding, T.W., Debeljuk, L., and White, W.F., 1971, Isolation and properties of the FSH and LH-releasing hormone, *Biochem. Biopys. Res. Commun.* **43:**393.

Schally, A.V., Arimura, A., Kastin, A.J., Matsuo, H., Baba, Y., Redding, T.W., Nair,

R.M.G., Debeljuk, L., and White, W.F., 1973, Gonadotropin-releasing hormone: One polypeptide regulates secretion of luteinizing and follicle stimulating hormones, *Science* **173:** 1036.

Schally, A.V., Kastin, A.J., and Coy, D.H., 1976, LH-releasing hormone and its analogs: Recent basic and clinical investigations, *Int. J. Fertil.* **21:** 1.

Stearns, E.L., Winter, J.S.D., and Faiman, C., 1973, Effects of coitus on gonadotropin, prolactin and sex steroid levels in men, *J. Clin. Endocrinol.* **37:** 687.

White, W.F., Hedlund, M.T., Rippel, R.H., Arnold, W., and Flouret, G.R., 1973, Chemical and biological properties of gonadotropin-releasing hormone synthesized by the solid-phase method, *Endocrinology* **93:** 96.

Wilson, J.R., Alader, N., and Boeuf, B.L., 1965, The effects of intromission frequency on successful pregnancy in the female rat, *Proc. Natl. Acad. Sci. U.S.A.* **53:** 1392.

Yamazaki, I., Nakagawa, H., and Yoshida, K., 1978, Induction of ovulation by luteinizing hormone-releasing hormone (LH-RH) and its analogues in rats under various preovulatory sexual conditions, *Endocrinol. Jpn.* **25:** 503.

Zarrow, M.X., Campbell, P.S., and Clark, J.H., 1968, Pregnancy following coital-induced ovulation in a spontaneous ovulator, *Science* **159:** 329.

Receptors in Cellular Localization of Hormones

7

Stable and Specific Tracers

Jean-Pierre Raynaud, Tiiu Ojasoo, and Viviane Vaché

1. Introduction

Since the introduction of labeled dexamethasone to detect and characterize the glucocorticoid receptor, the use of labeled synthetic hormones as tracers has become increasingly widespread. The progestin receptor is now rarely assayed with labeled progesterone, but with radioligands such as [^{3}H]promegestone (Duffy and Duffy, 1979; McGuire *et al.*, 1977b; Horwitz and McGuire, 1975; Philibert and Raynaud, 1973, 1974a, b) and [^{3}H]-ORG 2058 (Keightley, 1979; Koenders *et al.*, 1979; Jänne *et al.*, 1978; Fleischmann and Beato, 1978); the androgen receptor is assayed with [^{3}H]metribolone rather than labeled dihydrotestosterone (Raynaud *et al.*, 1980a; Ekman *et al.*, 1979b; Ghanadian *et al.*, 1978; Menon *et al.*, 1978; Shain and Boesel, 1978; Cowan *et al.*, 1977; Asselin *et al.*, 1976a; Bonne and Raynaud, 1975, 1976), and the estrogen receptor can, in the opinion of several authors, often be unambiguously detected only with ligands such as [^{3}H]moxestrol (MacLusky *et al.*, 1979a, b; Martin *et al.*, 1978; Okret *et al.*, 1978; Raynaud *et al.*, 1978; Raynaud and Moguilewsky, 1976, 1977; McEwen *et al.*, 1975; Raynaud, 1974). The reasons underlying the ever-increasing use of these synthetic radioligands (tags) are to be found in the distinct advantages they offer over labeled natural hormones as described previously in some detail (Raynaud *et al.*, 1979b) and as briefly summarized herein. Unlike the natural hormones, tags do not

Jean-Pierre Raynaud, Tiiu Ojasoo, and *Viviane Vaché* • Centre de Recherches Roussel-Uclaf, 93230 Romainville, France.

bind with high affinity to the plasma proteins that often contaminate tissue and cytosol preparations (Ballard, 1979; Hähnel and Twaddle, 1979; Krieg *et al.*, 1979; Karr *et al.*, 1978; Wagner, 1978). Unlike estradiol, testosterone, and dihydrotestosterone, neither moxestrol nor metribolone binds extensively to human sex-steroid-binding protein (SBP); unlike progesterone, hydrocortisone, and aldosterone, neither promegestone nor dexamethasone binds with high affinity to human corticosteroid-binding globulin (CBG). Furthermore, tags are resistant to degradation and form more stable (slowly dissociating) receptor complexes than the natural hormones, thus facilitating the detection of receptors and enabling their measurement by exchange assay. For instance, metribolone, unlike dihydrotestosterone, is not degraded on *in vitro* incubation with cytosol (Raynaud *et al.*, 1979b; Krieg *et al.*, 1979; Pousette *et al.*, 1979; Tremblay *et al.*, 1977), and both promegestone and dexamethasone form cytosolic receptor complexes that dissociate considerably more slowly than the corresponding progesterone complex (Raynaud *et al.*, 1979b; Keightley, 1979; Jänne *et al.*, 1978; Moguilewsky and Raynaud, 1977; Philibert and Raynaud, 1977) and hydrocortisone complex (Munck and Leung, 1977; Rousseau *et al.*, 1972). Finally, if possible, tags should be more receptor-specific than the natural hormones. Hydrocortisone and aldosterone bind to at least two different tissue receptors, a mineralocorticoid and a glucocorticoid receptor, that are not easy to distinguish in the absence of ultraspecific radioligands (Raynaud *et al.*, 1979a; Matulich *et al.*, 1976; Rousseau *et al.*, 1972).

Few available tags, however, meet each and every one of these prerequisites and, with the ever-greater need for more sensitive assay techniques (e.g., in the clinic to draw correlations with the success rates of different forms of hormone therapy and in the laboratory to obtain a deeper understanding of steroid hormone action), the disadvantages of existing tags, which are relatively minor when compared to those of the labeled natural hormones, can nevertheless become use-limiting. We therefore propose to point out in this chapter some pitfalls in the current use of these tags, to propose methods for palliating some of their disadvantages in particular as regards hormone-receptor specificity, and to suggest possible compounds for use as "improved tags."

2. Some Pitfalls in the Use of Available Tags

Full awareness of the advantages and limits of available radioligands is necessary to obtain maximum benefit from their use and to avoid misinterpretation of results. Consequently, factors such as nonspecific binding, degradation, and receptor specificity should be checked.

Although the synthetic radioligands have a far lower affinity for specific plasma proteins than the natural hormones, they can nevertheless bind to these proteins. For instance, according to available data, promegestone binds to purified human CBG with an affinity constant of about 8×10^6 M^{-1} compared to 4×10^8 M^{-1} for progesterone. This low promegestone binding is totally dissociated in the presence of dextran-coated charcoal (Asselin *et al.*, 1976b). When using [^{3}H]promegestone, it is therefore necessary to ensure that the chosen assay technique effectively separates high-affinity from low-affinity binding.

For high sensitivity, it is also necessary to have low nonspecific binding in particular in experiments on nuclear binding (Vu Hai and Milgrom, 1978). Published results suggest that, in the cytosol, the nonspecific binding of moxestrol is considerably less than that of estradiol (Okret *et al.*, 1978; Raynaud *et al.*, 1978) and of diethylstilbestrol (Gustafsson, personal communication), but that the nonspecific binding of promegestone is higher than that of progesterone (McGuire *et al.*, 1977a) and of ORG 2058 (Jänne *et al.*, 1976). The intrinsic association constant of promegestone binding to purified human serum albumin is 6×10^5 M^{-1} (Westphal, personal communication) compared to 4×10^4 M^{-1} for progesterone (Westphal, 1971). A compromise is thus required between the specific activity of the radiolabeling necessary for receptor detection and that which leads to too high nonspecific-binding values.

Specific activity is in itself a crucial point to be taken into consideration. It should be not only, as mentioned above, not too high, yet high enough, but also stable. This is one of the major problems that has been encountered with radioligands such as labeled diethylstilbestrol, which is rapidly degraded during storage in the laboratory.

Lack of degradation during cytosol incubation should, if possible, be checked in the tissue under study, since it is known that in certain pathological conditions such as, for instance, benign hypertrophy and cancer of the prostate, enzyme activity can be modified compared to that in the normal tissue (Krieg *et al.*, 1979).

But the major drawback to the use of many radioligands is their lack of receptor specificity. This problem is not inherent to synthetic radioligands since, as shown in Tables II–V, the natural hormones bind, albeit weakly, to receptors other than their own. Nevertheless, by definition, a tag should bind to one receptor only and should form a receptor complex that has a slower dissociation rate than the natural hormone–receptor complex. As a consequence, the endogenous hormone occupying the binding sites may be replaced by the radiolabeled tag in an exchange assay and the total number of binding sites may be measured. In our laboratory, steroids are routinely screened for binding to the cytosol estrogen, androgen, progestin, glucocorticoid, and mineralocorticoid receptors in the target tissues used to evaluate their biological activity (Raynaud *et al.*, 1979a, 1980b); Ojasoo and Raynaud, 1978). The relative binding affinities (RBAs) we have obtained for the compounds listed in Table I (compounds that have been used as tags) are presented in Tables II–V, which yield two kinds of information. On the one hand, the RBAs recorded after short incubation times indicate whether an interaction can occur between the steroid and receptor; on the other hand, the increase or decrease of these RBAs on prolonging incubation time indicates whether the complex thus formed is slow- or fast-dissociating (Raynaud *et al.*, 1980b; Raynaud, 1978; Bouton and Raynaud, 1978; Bouton *et al.*, 1978).

Of the estrogens (Table II), ethynyl estradiol bound to the progestin receptor and minimally to the androgen and glucocorticoid receptors. Its 11-methoxy derivative, moxestrol, did not exhibit any progestin binding,

Table I. Chemical Names of the Test Substances

Estradiol	Estra-1,3,5(10)-triene-3,17β-diol
Ethynyl estradiol	19-Nor-17α-pregna-1,3,5(10)-trien-20-yne-3,17β-diol
RU 2858 (moxestrol)	11β-Methoxy-19-nor-17α-pregna-1,3,5(10)-trien-20-yne-3,17-diol
Diethylstilbestrol	(E)-4,4′-(1,2-Diethyl-1,2-ethenediyl)bis phenol
Progesterone	Pregn-4-ene-3,20-dione
Medroxyprogesterone acetate	17α-Acetyloxy-6α-methyl-pregn-4-ene-3,20-dione
Chlormadinone acetate	17α-Acetyloxy-6-chloro-pregna-4,6-diene-3,20-dione
Norgestrel	13β-Ethyl-17β-hydroxy-18,19-dinor-17α-pregn-4-en-20-yn-3-one
Norethindrone	17β-Hydroxy-19-nor-17α-pregn-4-en-20-yn-3-one
RU 5020 (promegestone)	17α,21-Dimethyl-19-nor-pregna-4,9-diene-3,20-dione
RU 27987	21 S-Hydroxy-17α,21-dimethyl-19-nor-pregna-4,9-diene-3,20-dione
Testosterone	17β-Hydroxy-androst-4-en-3-one
5α-Dihydrotestosterone	17β-Hydroxy-5α-androstan-3-one
RU 1881 (metribolone)	17β-Hydroxy-17α-methyl-estra-4,9,11-trien-3-one
Trenbolone	17β-Hydroxy-estra-4,9,11-trien-3-one
Cortisol (hydrocortisone)	11β,17α,21-Trihydroxy-pregn-4-ene-3,20-dione
Corticosterone	11β,21-Dihydroxy-pregn-4-ene-3,20-dione
Dexamethasone	9α-Fluoro-11β,17α,21-trihydroxy-16α-methyl-pregna-1,4-diene-3,20-dione
Triamcinolone acetonide	9α-Fluoro-11β,21-dihydroxy-16α,17α-[(1-methylethylidene)bis (oxy)]-pregna-1,4-diene-3,20-dione
Prednisolone	11β,17α,21-Trihydroxy-pregna-1,4-diene-3,20-dione
16α-Methylprednisolone	11β,17α,21-Trihydroxy-16α-methyl-pregna-1,4-diene-3,20-dione
Aldosterone	11β,21-Dihydroxy-3,20-dioxo-pregn-4-en-18-al
Fludrocortisone	9α-Fluoro-11β,17α,21-trihydroxy-pregn-4-ene-3,20-dione

Table II. Specificity (Relative Binding Affinities) of Estrogens Used as Tags[a]

Estrogen	Receptors[b] ES Mouse uterus 2 hr, 0°C	ES Mouse uterus 5 hr, 25°C	PG Rabbit uterus 24 hr, 0°C	AND Rat prostate 2 hr, 0°C	MIN Rat kidney 24 hr, 0°C	GLU Rat thymus 24 hr, 0°C
radiol	100	100	0.9 ± 0.2	7.9 ± 1.3	0.5 ± 0.2	1.2 ± 0.1
ynyl estradiol	112 ± 5	245 ± 13	10 ± 2	1.3 ± 0.5	<0.1	2.3 ± 0.0
2858 (noxestrol)	12 ± 1	122 ± 11	0.5 ± 0.1	<0.1	<0.1	2.7 ± 0.5
thylstilbestrol	175 ± 15	100 ± 13	<0.1	<0.1	<0.1	<0.1

[a] he following organs were homogenized in 10 mM Tris-HCl (pH 7.4.), 0.25 M sucrose buffer: immature-mouse uteri (1:25, wt./vol.), stradiol-primed-rabbit uteri (1:50, wt./vol.), castrated-rat prostates (1:5. wt./vol.), and adrenalectomized-rat thymuses (1:10, wt./vol.). erfused kidneys from the adrenalectomized rats were homogenized (1:3, wt./vol.) in Krebs-Ringer phosphate buffer containing glucose. he Tris–sucrose homogenates were centrifuged at 105,000*g* for 60 min at 0–4°C to obtain cytosols, which were incubated as indicated ith [³H]estradiol (58 Ci/mmole), [³H]promegestone (51.4 Ci/mmole), [³H]metribolone (58.2 Ci/mmole), and [³H]dexamethasone (25.9 Ci/mole), respectively, in the presence or absence of 0–2500 nM unlabeled steroid. The rat kidney homogenates were incubated with H]aldosterone (53.2 Ci/mmole) at 0°C with or without unlabeled steroid, and in the presence of an excess of a specific glucocorticoid, U 26988, then centrifuged at 800*g* for 10 min at 0°C. Bound radioactivity was measured by a dextran-coated charcoal adsorption chnique. The RBAs of estradiol, progesterone, testosterone, aldosterone, and corticosterone were respectively taken as equal to 100. esults are the means ± S.E.M., or the means of two determinations differing by less than 15%.

[b] S) Estrogen; (PG) progestin; (AND) androgen; (MIN) mineralocorticoid; (GLU) glucocorticoid.

but had the same glucocorticoid binding. An increase in incubation time and temperature resulted in slightly increased ethynyl estradiol binding (2 times) and markedly increased moxestrol binding (9 times) to the estrogen receptor, implying that these compounds dissociate more slowly than estradiol, an implication that has been confirmed by comparing the dissociation rates of moxestrol and estradiol in different tissues of several species (Raynaud *et al.*, 1978, 1979b; Bouton and Raynaud, 1978; Martin *et al.*, 1978). On the other hand, the lower RBA recorded for diethylstilbestrol after incubation for 5 hr (25°C) compared to 2 hr (0°C) may indicate that, in the absence of *in vitro* degradation, diethylstilbestrol, which is more specific than either estradiol or moxestrol, may dissociate faster than estradiol from the receptor.

Many progestins (Table III) are unsuitable as tags on account of high androgen binding (Raynaud *et al.*, 1979a, 1980b; Labrie *et al.*, 1977). This binding excludes the use of medroxyprogesterone acetate, chlormadinone acetate, norgestrel, and norethindrone, which exhibit markedly higher androgen binding than progesterone. Promegestone competed less than progesterone for androgen, mineralocorticoid, and glucocorticoid binding. The relatively high glucocorticoid binding observed under these experimental conditions for progestins has been shown to reflect antiglucocorticoid activity (Raynaud *et al.*, 1980b).

The chosen synthetic androgen, metribolone, was not as specific for the androgen receptor as dihydrotestosterone (Table IV). Like many

Table III. Specificity (Relative Binding Affinities) of Progestins Used as Tags[a]

Progestin	Receptors[b] ES Mouse uterus 2 hr, 0°C	PG Rabbit uterus 2 hr, 0°C	PG Rabbit uterus 24 hr, 0°C	AND Rat prostate 2 hr, 0°C	MIN Rat kidney 24 hr, 0°C	GLU Rat thyr 24 hr, 0
Progesterone	< 0.1	100	100	5.5 ± 0.6	22 ± 6	77 ±
Medroxyprogesterone acetate	< 0.1	127 ± 4	306 ± 25	51 ± 7	0.9	213
Chlormadinone acetate	< 0.1	175 ± 28	321 ± 35	20 ± 4	<0.1	36
Norgestrel	< 0.1	170 ± 12	907 ± 184	87 ± 23	1.2 ± 0.1	39
Norethindrone	< 0.1	156 ± 18	263 ± 10	43 ± 3	0.2	5.5 ±
RU 5020 (promegestone)	< 0.1	222 ± 7	533 ± 40	1.4 ± 0.4	2.5 ± 0.3	30 ±
RU 27987	< 0.1	194 ± 27	660 ± 40	2.1 ± 0.7	16	50 ± 1

[a] See Table II, footnote *a*.
[b] See Table II, footnote *b*.

other trienes, it bound appreciably to several receptors (Raynaud *et al.*, 1979a; Ojasoo and Raynaud, 1978). The corresponding nonmethylated compound, trenbolone, was more specific, although it did not form an androgen receptor complex that dissociated quite as slowly as the metribolone receptor complex, since the RBA tended to decrease slightly rather than increase with incubation time.

A clear distinction between mineralocorticoid and glucocorticoid binding is difficult (Raynaud *et al.*, 1979a; Matulich *et al.*, 1976; Rousseau *et al.*, 1972), since available radioligands such as [^{3}H]aldosterone and [^{3}H]dexamethasone bind to both receptors (Table V). Furthermore, these receptors are frequently present concurrently. The most specific miner-

Table IV. Specifity (Relative Binding Affinities) of Androgens Used as Tags[a]

Androgen	Receptors[b] ES Mouse uterus 2 hr, 0°C	PG Rabbit uterus 24 hr, 0°C	AND Rat prostate 30 min, 0°C	AND Rat prostate 2 hr, 0°C	MIN Rat kidney 24 hr, 0°C	GLU Rat thymus 24 hr, 0°C
Testosterone	<0.1	1.1 ± 0.3	100	100	3.7 ± 1.0	1.8 ± 0.9
5α-Dihydrotestosterone	<0.1	0.9	94 ± 10	120 ± 3	0.2	0.5 ± 0.2
Metribolone	<0.1	191 ± 28	58 ± 26	203 ± 5	37 ± 7	19 ± 2
Trenbolone	<0.1	16 ± 2	251 ± 55	189 ± 24	2.6	2.3

[a] See Table II, footnote *a*.
[b] See Table II, footnote *b*.

Table V. Specifity (Relative Binding Affinities) of Mineralocorticoids and Glucocorticoids Used as Tags[a]

Mineralocorticoid or glucocorticoid	Receptors[b]						
	ES Mouse uterus	PG Rabbit uterus	AND Rat prostate	MIN Rat kidney		GLU Rat thymus	
	2 hr, 0°C	24 hr, 0°C	2 hr, 0°C	1 hr, 0°C	24 hr, 0°C	1 hr, 0°C	24 hr, 0°C
Cortisol	<0.1	<0.1	<0.1	53 ± 7	17 ± 5	47 ± 10	59 ± 5
Corticosterone	<0.1	2.8 ± 0.3	0.5 ± 0.2	48 ± 6	18 ± 6	100	100
Dexamethasone	<0.1	<0.1	<0.1	137 ± 7	21 ± 3	167 ± 28	455 ± 83
Triamcinolone acetonide	<0.1	12 ± 1	<0.1	106	8.0 ± 2.0	210 ± 22	792 ± 23
Prednisolone	<0.1	<0.1	<0.1	44	34	62 ± 5	118 ± 5
16α-Methylprednisolone	<0.1	<0.1	<0.1	0.6	2.3	137 ± 25	150 ± 9
RU 26988	<0.1	0.15 ± 0.01	<0.1	1.3	<0.1	211 ± 18[c]	268 ± 23
Aldosterone	<0.1	0.7 ± 0.1	<0.1	100	100	25 ± 3	34 ± 5
Fludrocortisone	<0.1	0.4	0.1	293 ± 35	318 ± 30	204	264 ± 27

[a] See Table II, footnote *a*.
[b] See Table II, footnote *b*.
[c] Determined after 4 hr incubation.

alocorticoid tested would appear to be aldosterone; the most specific glucocorticoids, 16α-methylprednisolone and RU 26988 (Table V).

The concurrent presence of several receptors and the lack of total specificity of the radioligands render the interpretation of results difficult in cases of very low receptor concentrations, and it becomes impossible to tell whether a particular receptor is present. For instance, metribolone binds to the androgen receptor in rat prostate cytosol (Bonne and Raynaud, 1975), which does not contain progesterone receptor. It can therefore be used as a specific tag for this receptor in a screening system with rat prostate cytosol. According to the data in Table IV, it also competes for progestin binding, and consequently, in an unknown tissue, radioactivity displacement may be due to binding to the androgen or progestin receptor, or both. It is in this way that the presence of a progestin-like receptor in hyperplastic prostate cytosol was surmised (Asselin *et al.*, 1976a) and found to be present in the stroma rather than in the epithelium (Bashirelahi *et al.*, 1978; Cowan *et al.*, 1977). However, according to certain experiments (Menon *et al.*, 1978), high-affinity metribolone binding in the nuclear extract from hyperplastic prostate has the characteristics of binding to an androgen receptor only; thus, in nuclear experiments, metribolone is a purely androgen-specific tag.

Furthermore, in particular in experiments on cytosol, extreme caution must be exercised in the interpretation of results, since differences in specificity may occur for many reasons including (1) differences in the plasma binding of the competitors; (2) differences in their metabolism, as described for glucocorticoids in the thymus and liver (Disorbo *et al.*, 1977); (3) differences in the experimental conditions used to measure displacement, since the degree of displacement depends on the kinetics of the interaction between steroid and receptor under the chosen incubation conditions; and (4) presence of unsuspected binders. All these points should be checked, if possible, before asserting that the specificity differences are inherent to the tissues or species under study or result from a pathological process.

3. *Methods for Palliating Specificity Disadvantages*

It is possible to eliminate the interference of one (or several) high-affinity binding component(s) by the addition of an excess of a ligand that binds solely and specifically to this (or these) component(s). Thus, the addition of an excess of unlabeled triamcinolone acetonide to the incubation medium has been proposed to measure androgen binding with labeled metribolone, since the triamcinolone acetonide, which does not compete for androgen binding, but competes for binding to the progestin, mineralocorticoid, and glucocorticoid receptors, eliminates any interference by these receptors (compare the specificity profiles of metribolone and triamcinolone acetonide in Tables IV and V). It is in this way that androgen binding has been detected in the human prostate (Asselin *et al.*, 1979; Hicks and Walsh, 1979; Zava *et al.*, 1979; Pertschuk *et al.*, 1978). Similarly, in the assay of the mineralocorticoid receptor with labeled aldosterone, the presence of interfering glucocorticoid binding can be greatly reduced by the addition of the highly specific glucocorticoid RU 26988, which, unlike dexamethasone, does not bind to the mineralocorticoid receptor. All measurements of mineralocorticoid binding in Tables II–V were performed in the presence of RU 26988. By this technique, it has recently been possible to make a more categorical distinction than heretofore between mineralocorticoid and glucocorticoid binding in the brain (Moguilewsky and Raynaud, 1980)

4. *Improvements upon Available Radioligands*

The highly specific glucocorticoid RU 26988 does not bind with high affinity to CBG and, as indicated by the increasing RBA values in Table V, forms a complex with the glucocorticoid receptor that dissociates somewhat more slowly than the corticosterone–receptor complex. It can therefore be used to displace corticosterone in an exchange assay. This observation has been confirmed by direct measurement of the dissociation rates (k_{-1}) of the receptor complexes formed in rat hippocampal cytosol at 0°C [k_{-1} corticosterone $= 2.1 \times 10^{-4}$ sec^{-1} ($t_{\frac{1}{2}} = 0.9$ hr); k_{-1} RU 26988 $= 1.7 \times 10^{-4}$ sec^{-1} ($t_{\frac{1}{2}} = 1.1$ hr)] (Moguilewsky and Raynaud, 1980).

The specificity of the androgen trenbolone is a decided improvement over that of metribolone; trenbolone nevertheless still binds slightly (16%) to the progestin receptor (see Table IV). Furthermore, the complex it forms with the androgen receptor would appear to dissociate fractionally faster than the natural hormone complex and thus would not permit a valid exchange assay.

In the field of progestins, the 21 S-hydroxy derivative of promegestone (RU 27987), its active metabolite, is specific to the progestin receptor and forms a complex with this receptor that dissociates even more slowly than the promegestone complex (see Table III) (Raynaud *et al.*, 1980c). Furthermore, in preliminary experiments, its nonspecific binding would appear to be only about one tenth that of promegestone. Consequently, it is expected that labeled RU 27987 can be used even more effectively than promegestone has been used in the past to assay nuclear binding sites (Kato and Onouchi, 1979; Blaustein and Wade, 1978; Vu Hai and Milgrom, 1978).

5. Current Applications of Synthetic Radioligands

Despite the limitations of available radioligands outlined in this chapter, these compounds nevertheless remain our best tools for the rapid and relatively reliable detection and assay of steroid-hormone receptors. Their contribution to the identification of steroid-hormone receptors in tissues in which this was not possible with natural hormones on account of high metabolic activity and interfering binding is undeniable, and we shall therefore briefly outline in this section some of their major fields of application.

Because metribolone, on account of its trienic structure, is not readily metabolized (Salmon *et al.*, 1971) or degraded on *in vitro* incubation, unlike testosterone and dihydrotestosterone, it has enabled the *in vivo* and *in vitro* study of the androgen receptor in tissues with high metabolic activity (Naftolin *et al.*, 1975; Tremblay *et al.*, 1977; Calandra *et al.*, 1978), e.g., in the brain (Lieberburg *et al.*, 1978), where testosterone is partly aromatized to estrogens, in muscle (Tremblay *et al.*, 1977), and in adrenal gland (Asselin and Melançon, 1977).

Neither moxestrol, promegestone, metribolone, nor dexamethasone binds appreciably to SBP, CBG, or other plasma proteins with high affinity for natural hormones. In particular, unlike estradiol, moxestrol does not bind to estradiol-binding protein (EBP) (Raynaud *et al.*, 1980d; McEwen *et al.*, 1975; Raynaud, 1973), a fetal protein that probably protects the embryo from the deleterious effects of maternal endogenous hormones, and has therefore been successfully used to study the ontogeny of estrogen receptors in the brain and pituitary of the rat (Raynaud *et al.*, 1980d; MacLusky *et al.*, 1979a,b; Raynaud and Moguilewsky, 1976; McEwen *et al.*, 1975) and guinea pig (Plapinger *et al.*, 1977). With metribolone, which does not bind to androgen-binding protein (ABP) (Kirchhoff *et al.*, 1979), the presence of androgen receptors has been revealed in the rat epididymis (Pujol and Bayard, 1979) and human testis

(Stoa *et al.*, 1979) and also in the submandibular gland of the mouse (Verhoeven, 1979), which contains a protein, besides the androgen receptor, with a high affinity for dihydrotestosterone. Dexamethasone, which does not bind to CBG or to many other proteins to which hydrocortisone binds in the liver, kidney, brain, and pituitary (Rousseau and Baxter, 1979), is one of the most widely used tracers to assay glucocorticoid receptors in these organs despite its lack of total receptor specificity.

The ability to detect receptors with tags in tissues in which this was previously difficult or impossible has, however, been most widely exploited and publicized in the clinical field. These were the tools that were to diagnose the possible hormone dependence of a tumor and enable the choice of an appropriate form of therapy. Until the advent of the tag, these assays remained difficult because of low receptor concentrations, high plasma contamination (in particular with needle biopsies), high endogenous hormone concentrations, and sometimes high metabolic activity. Now, any or all of estrogen, progestin, androgen, and glucocorticoid receptors are assayed in human pathology, for instance, in tumors of the breast (Allegra *et al.*, 1979a; Martin *et al.*, 1978; Matsumoto and Sugano, 1978; McGuire, 1978), uterus (Martin *et al.*, 1979; Tamaya *et al.*, 1979; Pollow *et al.*, 1978), prostate (Raynaud *et al.*, 1980a; Ekman *et al.*, 1979a, b; Bashirelahi *et al.*, 1978), kidney (Li *et al.*, 1979; Concolino *et al.*, 1978), testes (Stoa *et al.*, 1979), and pharynx and larynx (Saez *et al.*, 1979). Preliminary correlations with response to hormone treatment have already been drawn, in particular for estrogen and progestin receptors in mammary cancer (Allegra *et al.*, 1979b; Kent-Osborne and McGuire, 1979; McGuire, 1978; Barnes *et al.*, 1977) and androgen receptors in prostate cancer (Walsh *et al.*, 1979; Gustafsson *et al.*, 1978). Although quantitative correlations between receptor concentrations and response to therapy are few, all studies nevertheless conclude that, in the absence of hormone receptors, the chances of successful endocrine treatment are minimal. It is for this reason that it is particularly important to be able to eliminate false-positive results (e.g., due to plasma contamination) with tags such as promegestone and moxestrol.

References

Allegra, J.C., Lippman, M.E., Thompson, E.B., Simon, R., Barlock, A., Green, L., Huff, K.K., Do, H.M.T., and Aitken, S.C., 1979a, Distribution, frequency, and quantitative analysis of estrogen, progesterone, androgen, and glucocorticoid receptors in human breast-cancer, *Cancer Res.* **39:**1447–1454.

Allegra, J.C., Lippman, M.E., Thompson, E.B., Simon, R., Barlock, A., Green, L., Huff, K.K., Do, H.M.T., Aitken, S.C., and Warren, R., 1979b, Relationship between the progesterone, androgen, and glucocorticoid receptor and response rate to endocrine therapy in metastatic breast cancer, *Cancer Res.* **39:**1973–1979.

Asselin, J., and Melançon, R., 1977, Characteristics of androgen binding in rat adrenal tissue using (3H)methyltrienolone (R1881) as tracer, *Steroids* **30:**591–604.

Asselin, J., Labrie, F., Gourdeau, Y., Bonne, C., and Raynaud, J.P., 1976a, Binding of (3H)methyltrienolone (R 1881) in rat prostate and human benign prostatic hypertrophy (BPH), *Steroids* **28:**449–459.

Asselin, J., Labrie, F., Kelly, P.A., Philibert, D., and Raynaud, J.P., 1976b, Specific progesterone receptors in dimethylbenzanthracene (DMBA)-induced mammary tumors, *Steroids* **27:**395–404.

Asselin, J., Melançon, R., Gourdeau, Y., Labrie, F., Bonne, C., and Raynaud, J.P., 1979, Specific binding of 3H-methyltrienolone to both progestin and androgen binding components in human benign prostatic hypertrophy (BPH), *J. Steroid Biochem.* **10:**483–486.

Ballard, P.L., 1979, Delivery and transport of glucocorticoids to target cells, in: *Glucocorticoid Hormone Action* (J.D. Baxter and G.G. Rousseau, eds.), pp. 25–48, Springer-Verlag, Berlin.

Barnes, D.M., Ribeiro, G.G., and Skinner, L.G., 1977, Two methods for measurement of oestradiol-17β and progesterone receptors in human breast cancer and correlation with response to treatment, *Eur. J. Cancer* **13:**1133–1143.

Bashirelahi, N., Young, J.D., and Sanefugi, H., 1978, Androgen and estrogen receptor distribution in epithelial and stroma cells of human prostate, *Fed. Proc. Fed. Am. Soc. Exp. Biol.* **37:**1312.

Blaustein, J.D., and Wade, G.N., 1978, Progestin binding by brain and pituitary cell nuclei and female rat sexual behavior, *Brain Res.* **140:**360–367.

Bonne, C., and Raynaud, J.P., 1975, Methyltrienolone, a specific ligand for cellular androgen receptors, *Steroids* **26:**227–232.

Bonne, C., and Raynaud, J.P., 1976, Assay of androgen binding sites by exchange with methyltrienolone (R 1881), *Steroids* **27:**497–507.

Bouton, M.M., and Raynaud, J.P., 1978, The relevance of kinetic parameters in the determination of specific binding to the estrogen receptor, *J. Steroid Biochem.* **9:**9–15.

Bouton, M.M., Bonne, C., and Raynaud, J.P., 1978, "*In vitro*" screening for antihormones, *J. Steroid Biochem.* **9:**836.

Calandra, R.S., Purvis, K., Naess, O., Attramadal, A., Djoseland, O. and Hansson, V., 1978, Androgen receptors in the rat adrenal gland, *J. Steroid Biochem.* **9:**1009–1015.

Concolino, G., Marocchi, A., Tenaglia, R., DiSilverio, F., and Sparano, F., 1978, Specific progesterone receptor in human renal cancer, *J. Steroid Biochem.* **9:**399–402.

Cowan, R.A., Cowan, S.K., and Grant, J.K., 1977, Binding of methyltrienolone (R 1881) to a progesterone receptor-like component of human prostatic cytosol, *J. Endocrinol.* **74:**281–289.

Disorbo, D.,Rosen, F., McPartland, R.P., and Milholland, R.J., 1977, Glucocorticoid activity of various progesterone analogs: Correlation between specific binding in thymus and liver and biologic activity, *Ann. N. Y. Acad. Sci.* **286:**355–368.

Duffy, M.J., and Duffy, G.J., 1979, Studies on progesterone receptors in human breast carcinomas: Use of natural and synthetic ligands, *Eur. J. Cancer* **15:**1181–1184.

Ekman, P., Snochowski, M., Dahlberg, E., Bression, D. Högberg, B., and Gustafsson,

J.Å., 1979a, Steroid receptor content in cytosol from normal and hyperplastic human prostates, *J. Clin. Endocrinol. Metab.* **49:**205–215.

Ekman, P., Snochowski, M., Dahlberg, E. and Gustafsson, J.Å., 1979b, Steroid receptors in metastatic carcinoma of the human prostate, *Eur. J. Cancer* **15:**257–262.

Fleischmann, G., and Beato, M., 1978, Characterization of the progesterone receptor of rabbit uterus with the synthetic progestin, 16α-ethyl-21-hydroxy-19-norpregn-4-ene-3,20-dione, *Biochim. Biophys. Acta* **540:**500–517.

Ghanadian, R., Auf, G., Chaloner, P.J., and Chisholm, G.D., 1978, The use of methyltrienolone in the measurement of the free and bound cytoplasmic receptors for dihydrotestosterone in benign hypertrophied human prostate, *J. Steroid Biochem.* **9:**325–330.

Gustafsson, J.Å., Ekman, P., Snochowski, M., Zetterberg, A., Pousette, Å., and Högberg, B., 1978, Correlation between clinical response to hormone therapy and steroid receptor content in prostatic cancer, *Cancer Res.* **38:**4345–4348.

Hähnel, R., and Twaddle E., 1979, Factors that may influence the estradiol receptor assay in human tissues: Sex hormone binding globulin and endogenous steroids, *J. Steroid Biochem.* **10:**95–98.

Hicks, L.L., and Walsh, P.C., 1979, A microassay for the measurement of androgen receptors in human prostatic tissue, *Steroids* **33:**389–406.

Horwitz, K.B., and McGuire, W.L., 1975, Specific progesterone receptors in human breast cancer, *Steroids* **25:**497–505.

Jänne, O., Kontula, K., and Vihko, R., 1976, Progestin receptors in human tissues: Concentrations and binding kinetics, *J. Steroid Biochem.* **7:**1061–1068.

Jänne, O., Kontula, K., Vihko, R., Feil, P.D., and Bardin, C.W., 1978, Progesterone receptor and regulation of progestin action in mammalian tissues, *Med. Biol.* **56:**225–248.

Karr, J.P., Kirdani, R.Y., Murphy, G.P., and Sandberg, A.A., 1978, Sex hormone binding globulin and transcortin in human and baboon males, *Arch. Androl.* **1:**123–129.

Kato, J., and Onouchi, T., 1979, Nuclear progesterone receptors and characterization of cytosol receptors in the rat hypothalamus and anterior hypophysis, *J. Steroid Biochem.* **11:**845–854.

Keightley, D.D., 1979, The binding of progesterone, R-5020 and ORG-2058 to progesterone receptor, *Eur. J. Cancer* **15:**785–790.

Kent-Osborne, C., and McGuire, W.L., 1979, The use of steroid hormone receptors in the treatment of human breast cancer, *Bull. Cancer* **66:**203–210.

Kirchoff, J., Soffie, M., and Rousseau, G.G., 1979, Differences in the steroid binding site specificities of rat prostate androgen receptor and epididymal and androgen binding protein (ABP), *J. Steroid Biochem.* **10:**487–497.

Koenders, A.J.M., Geurts-Moespot, J., Beex, L.V.A.M., and Benraad, T.J., 1979, Assay of progesterone receptor binding sites with 17,21-dimethyl-19-norpregna-4,9-diene-3,20-dione and 16α-ethyl-19-nor-4-pregnene-3,20-dione as the radioactive ligands, *J. Endocrinol.* **80:**15P.

Krieg, M., Bartsch, W., Janssen, W., and Voigt, K.D., 1979, A comparative study of binding, metabolism and endogenous levels of androgens in normal, hyperplastic and carcinomatous human prostate, *J. Steroid Biochem.* **11:**615–624.

Labrie, F., Ferland, L., Lagacé, L., Drouin, J., Asselin, J., Azadian-Boulanger, G., and Raynaud, J.P., 1977, High inhibitory activity of R 5020, a pure progestin, at the hypothalamo-adenohypophyseal level on gonadotropin secretion, *Fertil. Steril.* **28:**1104–1112.

Li, J.L., Li, S.A., and Gonzales, R., 1979, Specific steroid hormone binding in human renal carcinoma, *Cancer Res.* **20:**273.

Lieberburg, I., MacLusky, N.J., Roy, E.J., and McEwen, B.S., 1978, Sex steroid receptors in the perinatal rat brain, *Am. Zoo.* **22:**539–544.

MacLusky, N.J., Chaptal, C., and McEwen, B.S., 1979a, The development of estrogen receptor systems in the rat brain and pituitary: Postnatal development, *Brain Res.* **178:**143–160.

MacLusky, N.J., Lieberburg, I., and McEwen, B.S., 1979b, The development of estrogen receptor systems in the rat brain: Perinatal development, *Brain Res.* **178:**129–142.

Martin, P.M., Rolland, P.H., Jacquemier, J., Rolland, A.M., and Toga, M., 1978, Routine analysis of multiple steroid receptors in human breast cancer. I. Technological features, *Biomedicine* **28:**278–287.

Martin, P.M., Rolland, P.H., Gammerre, M., Serment, H., and Toga, M., 1979, Estradiol and progesterone receptors in normal and neoplastic endometrium: Correlations between receptors, histopathological examinations and clinical responses under progestin therapy, *Int. J. Cancer* **23:** 321–329.

Matsumoto, K., and Sugano, H., 1978, Human breast cancer and hormone receptors, in: *Endocrine Control in Neoplasia* (R.K. Sharma and W.E. Criss, eds.), pp. 191–208, Raven Press, New York.

Matulich, D.T., Spindler, B.J., Schambelan, M., and Baxter, J.D., 1976, Mineralocorticoid receptors in human kidney, *J. Clin. Endocrinol. Metab.* **43:**1170–1174.

McEwen, B.S., Plapinger, L., Chaptal, C., Gerlach, J., and Wallach, G., 1975, Role of fetoneonatal estrogen binding proteins in the associations of estrogen with neonatal brain cell nuclear receptors, *Brain Res.* **96:**400-406.

McGuire, W.L. (ed.), 1978, *Progress in Cancer Research and Therapy,* Vol. 10, *Hormones, Receptors, and Breast Cancer,* Raven Press, New York.

McGuire, W.L., Horwitz, K.B., Pearson, O.H., and Segaloff, A., 1977a, Current status of estrogen and progesterone receptors in breast cancer, *Cancer (Philadelphia)* **39:**2934–2947.

McGuire, W.L., Raynaud, J.P., and Baulieu, E.E. (eds), 1977b, *Progress in Cancer Research and Therapy,* Vol. 4, *Progesterone Receptors in Normal and Neoplastic Tissues,* Raven Press, New York.

Menon, M., Tananis, C.E., Hicks, L.L., Hawkins, E.F., McLoughlin, M.G., and Walsh, P.C., 1978, Characterization of the binding of a potent synthetic androgen, methyltrienolone, to human tissues, *J. Clin. Invest.* **61:**150–162.

Moguilewsky, M., and Raynaud, J.P., 1977, Progestin binding sites in the rat hypothalamus, pituitary and uterus, *Steroids* **30:**99–109.

Moguilewsky, M., and Raynaud, J.P., 1980, Evidence for a specific mineralocorticoid receptor in rat pituitary and brain, *J. Steroid Biochem.* **12:**309–314.

Munck, A., and Leung, K., 1977, Glucocorticoid receptors and mechanisms of action, in: *Receptors and Mechanism of Action of Steroid Hormones,* Vol. 8, Part II (J.R. Pasqualini, ed.), pp. 311–397, Marcel Dekker, New York.

Naftolin, F., Ryan, K.J., Davies, I.J., Reddy, V.V., Flores, F., Petro, Z., and Kuhn, M., 1975, The formation of estrogens by central neuroendocrine tissues, *Recent Prog. Horm. Res.* **31:**295–319.

Ojasoo, T., and Raynaud, J.P., 1978, Unique steroid congeners for receptor studies, *Cancer Res.* **38:**4186–4198.

Okret, S., Wrange, Ö., Nordenskjöld, B., Silfverswärd, C., and Gustafsson, J.Å., 1978,

Estrogen receptor assay in human mammary carcinoma with the synthetic estrogen 11β-methoxy-17α-ethynyl-1,3,5(10)-estratriene-3,17β-diol (R2858) *Cancer Res.* **38:**3904–3909.

Pertschuk, L.P., Zava, D.T., Gaetjens, E., Macchia, R.J., Brigati, D.J., and Kim, D.S., 1978, Detection of androgen and estrogen receptors in human prostatic carcinoma and hyperplasia by fluorescence microscopy, *Res. Commun. Chem. Pathol. Pharmacol.* **22:**427–430.

Philibert, D., and Raynaud, J.P., 1973, Progesterone binding in the immature mouse and rat uterus, *Steroids* **22:**89–98.

Philibert, D., and Raynaud, J.P., 1974a, Progesterone binding in the immature rabbit and guinea pig uterus, *Endocrinology* **94:**627–632.

Philibert, D., and Raynaud, J.P., 1974b, Binding of progesterone and R 5020, a highly potent progestin, to human endometrium and myometrium, *Contraception* **10:**457-466.

Philibert, D., and Raynaud, J.P., 1977, Cytoplasmic progestin receptors in mouse uterus, in: *Progesterone Receptors in Normal and Neoplastic Tissues* (W.L. McGuire, J.P. Raynaud, and E.E. Baulieu, eds.), pp. 227–243, Raven Press, New York.

Plapinger, L., Landau, I.T., McEwen, B.S., and Feder, H.H., 1977, Characteristics of estradiol-binding macromolecules in fetal and adult guinea pig brain cytosols, *Biol. Reprod.* **16:**586–599.

Pollow, K., Schmidt-Gollwitzer, M., and Pollow, B., 1978, Characterization of a cytoplasmic receptor for progesterone in normal and neoplastic human endometrial tissues, *J. Mol. Med.* **3:**55–69.

Pousette, Å., Snochowski, M., Bression, D., Högberg, B., and Gustafsson, J.Å., 1979, Partial characterization of (3H)methyltrienolone binding in rat prostate cytosol, *Biochim. Biophys. Acta* **582:**358–367.

Pujol, A., and Bayard, F., 1979, Androgen receptors in the rat epididymis and their hormonal control, *J. Reprod. Fertil.* **56:**217–222.

Raynaud, J.P., 1973, Influence of rat estradiol binding plasma protein (EBP) on uterotrophic activity, *Steroids* **21:**249–258.

Raynaud, J.P., 1974, Estrogen interactions at the hypothalamic subcellular level, in: *Drug Interactions* (P.L. Morselli, S. Garattini, and S.N. Cohen, eds.), pp. 151–162, Raven Press, New York.

Raynaud, J.P., 1978, The mechanism of action of anti-hormones, in: *Advances in Pharmacology and Therapeutics,* Vol. 1, *Receptors* (J. Jacob, ed.), pp. 259–278, Pergamon Press, Oxford.

Raynaud, J.P., and Moguilewsky, M., 1976, Ontogénèse des récepteurs des oestrogènes chez le rat, in: *Systĕme Nerveux, Activité Sexuelle et Reproduction* (A. Soulairac, J.P. Gautray, J.P. Rousseau, and J. Cohen, eds.), pp. 85–92, Masson, Paris.

Raynaud, J.P., and Moguilewsky, M., 1977, Steroid competitition for estrogen receptors in the central nervous system, in: *Progress in Reproductive Biology,* Vol. 2, *Clinical Reproductive Neuroendocrinology* (P.O. Hubinont, M. L'Hermite, and C. Robyn, eds.), pp. 78–87, S. Karger, Basel.

Raynaud, J.P., Martin, P., Bouton, M.M., and Ojasoo, T., 1978, 11β-Methoxy-17-ethynyl-1,3,5(10)-estratriene-3,17β-diol (Moxestrol), a tag for estrogen receptor-binding sites in human tissues, *Cancer Res.* **38:**3044–3050.

Raynaud, J.P., Ojasoo, T., Bouton, M.M., and Philibert, D., 1979a, Receptor binding as a tool in the development of new bioactive steroids, in: *Drug Design,* Vol. VIII (E.J. Ariëns, ed.), pp. 169–214, Academic Press, New York.

Raynaud, J.P., Ojasoo, T. and Vaché, V., 1979b, Unusual steroids in measuring steroid receptors, in: *Steroid Receptors and the Management of Cancer,* Vol. 1 (E.B. Thompson and M.E. Lippman, eds.), pp. 215–232, CRC Press, Boca Raton.

Raynaud, J.P., Bouton, M.M., and Martin, P.M., 1980a, Steroid prostate hyperplasia and adenocarcinoma: Steroid hormone receptor assays and therapy, in: *Steroid Receptors, Metabolism and Prostatic Cancer* (F.H. Schröder and H.J. de Voogt, eds.), pp. 165–181, Excerpta Medica, Amsterdam.

Raynaud, J.P., Bouton, M.M., Moguilewsky, M., Ojasoo, T., Philibert, D., Beck, G., Labrie, F., and Mornon, J.P., 1980b, Steroid hormone receptors and pharmacology, *J. Steroid Biochem.* **12:** 143–158.

Raynaud, J.P., Brown, N., Coussedière, D., Pottier, J., Delettré, J., and Mornon, J.P., 1980c, Role of metabolism and receptor binding in progestin-induced responses in the uterus, in: *Steroid-Induced Uterine Proteins* (M. Beato, ed.), pp. 217–236, Elsevier North-Holland, Biomedical Press, Amsterdam.

Raynaud, J.P., Moguilewsky, M., and Vannier, B., 1980d, Influence of rat estradiol binding plasma protein (EBP) on estrogen binding to its receptor and on induced biological responses, in: *Development of Responsiveness to Steroid Hormones* (A.M. Kaye and M. Kaye, eds.), pp. 59–75, Pergamon Press, Oxford.

Rousseau, G.G., and Baxter, J.D., 1979, Glucocorticoid receptors, in: *Glucocorticoid Hormone Action* (J.D. Baxter and G.G. Rousseau, eds.), pp. 50–77, Springer-Verlag, Berlin.

Rousseau, G., Baxter, J.D., Funder, J.W., Edelman, I.S., and Tomkins, G.M., 1972, Glucocorticoid and mineralocorticoid receptors for aldosterone, *J. Steroid Biochem.* **3:**219–227.

Saez, S., Martin, P.M., and Cignoux, B., 1979, Androgen receptors in the normal mucosa and in epithelioma of human larynx and pharynx, in: *Steroid Receptors and the Management of Cancer,* Vol. I (E.B. Thompson and M.E. Lippman, eds.), pp. 205–213, CRC Press, Boca Raton.

Salmon, J., Raynaud, J.P., and Pottier, J., 1971, Etude métabolique d'un stéroïde triénique: le R 1881, in: *Symposium sur les Progrès des Techniques Nucléaires en Pharmacodynamie,* Saclay, 1970 (G. Valette and Y. Cohen, eds.), pp. 237–247, Masson, Paris.

Shain, S.A., and Boesel, R.W., 1978, Human prostate steroid hormone receptor quantitation: Current methodology and possible utility as a clinical discriminant in carcinoma, *Invest. Urol.* **16:** 169–174.

Stoa, K.F., Hekim, N., Dahl, O., and Horsaeter, P.Å., 1979, Binding of androgens and progestins in the human testis, *J. Steroid Biochem.* **11:**261–265.

Tamaya, T., Motoyama, T., Ohono, Y., Ide, N., Tsurusaki, T., and Okada, H., 1979, Estradiol-17β, progesterone and 5α-dihydrotestosterone receptors of uterine myometrium and myoma in the human subject, *J. Steroid Biochem.* **10:**615–622.

Tremblay, R.R., Dubé, J.Y., Ho-Kim, M.A., and Lesage, R., 1977, Determination of rat muscles androgen-receptor complexes with methyltrienolone, *Steroids* **29:** 185–195.

Verhoeven, G., 1979, Androgen binding proteins in mouse submandibular gland, *J. Steroid Biochem,* **10:** 129–138.

Vu Hai, M.T., and Milgrom, E., 1978, Characterization and assay of the progesterone receptor in rat uterine nuclei, *J. Endocrinol.***76:**33–41.

Wagner, R.K., 1978, Extracellular and intracellular steroid binding proteins: Properties, discrimination, assay and clinical applications, *Acta Endocrinol. (Copenhagen)* **88**(Suppl. 218): 1–73.

Walsh, P.C., Hicks, L.L., Reiner, W.G., and Trachtenberg, J., 1979, The use of androgen receptors to predict the duration of hormonal response in prostatic cancer, in: Abstracts, Meeting of the American Urological Association, New York.

Westphal, U., 1971, *Steroid–Protein Interactions,* Springer-Verlag, Berlin.

Zava, D.T., Landrum, B., Horwitz, K.B., and McGuire, W.L., 1979, Androgen receptor assay with (3H)methyltrienolone (R1881) in the presence of progesterone receptors, *Endocrinology,* **104:** 1007–1012.

8

Specific Inhibitors of Androgen Binding to the Androgen-Binding Protein (ABP)

Jacques I. Quivy, Raymond Devis, Xuan-Hoa Bui, Jean-Pierre Schmit, and Guy G. Rousseau

1. Introduction

Androgen-binding protein (ABP) is a protein produced by the rat testis. It is distinct from the intracellular androgen receptor present in target tissues (Table I). Following binding of testosterone secreted by the Leydig cells on luteinizing hormone stimulation, ABP reaches, via the seminiferous tubules, the epididymis, beyond which it disappears (French and Ritzén, 1973; Hansson *et al.*, 1974). It has been suggested that ABP serves as an androgen-concentrating factor (Hansson *et al.*, 1973; Purvis and Hansson, 1978) that may be essential for sperm maturation in the epididymis. Indirect evidence points to the role of ABP in spermatogenesis and sperm maturation: the supportive effect of follicle-stimulating hormone (FSH) on these processes is mediated by the Sertoli cells (Fakunding *et al.*, 1976), the only known source of ABP is the Sertoli

Jacques I. Quivy • International Institute of Molecular and Cellular Pathology, B-1200 Brussels, Belgium; Unité de Chimie Hormonologique et de Pharmacognosie, Université de Louvain, B-1200 Brussels, Belgium. *Raymond Devis* and *Xuan-Hoa Bui* • Unité de Chimie Hormonologique et de Pharmacognosie, Université de Louvain, B-1200 Brussels, Belgium. *Jean-Pierre Schmit* • Département de Chimie, Université de Sherbrooke, Sherbrooke (Québec) J1K 2R1, Canada. *Guy G. Rousseau* • International Institute of Molecular and Cellular Pathology, B-1200 Brussels, Belgium.

cell (Hagenäs *et al.*, 1975), ABP synthesis is induced by FSH (Hansson *et al.*, 1973; Fakunding *et al.*, 1976), appearance of ABP at puberty coincides with onset of fertility (Hansson *et al.*, 1973), and disappearance of ABP after 100 days of age in H^{re} mutants coincides with onset of sterility (Musto and Bardin, 1976).

The role or even the occurrence of ABP in other species, including man, is far from clear. One reason is the presence of circulating testosterone/estradiol-binding globulin (TeBG), a globulin presumably of liver origin that is very similar, if not identical, to ABP (Hansson *et al.*, 1975) (see also Table I) and that is lacking in rats (Ritzen *et al.*, 1971). Testicular or epididymal tissue becomes easily contaminated with TeBG during experiments, while ABP is also known to circulate in the blood (Gunsalus *et al.*, 1978). The usual method for studying ABP is to label it with [^{3}H]androgen. However, it is then difficult to distinguish it from the androgen receptor and from TeBG because all three proteins have a similar affinity for the radioactive ligand. A radioimmunoassay for ABP has been developed (Gunsalus *et al.*, 1978), but ABP and TeBG may share common antigenic determinants (Weddington *et al.*, 1975).

We (Kirchhoff *et al.*, 1979) and others (Tindall *et al.*, 1978; Cunningham *et al.*, 1979) have investigated whether discrimination based on binding specificity could not be improved by resorting to steroid analogues structurally different from the natural androgens. It was found that indeed some substitutions conferred a higher affinity for ABP or for the androgen receptor (Table II). Yet, no steroid had an absolute specificity for either protein. On the other hand, no steroid is known that allows a clear distinction between TeBG and ABP from the same species (Tindall *et al.*, 1978). We therefore turned to nonsteroidal analogues, since work on estrogens and antiestrogens suggested that such derivatives may display less cross-reactivity with the different steroid-binding proteins than the steroids themselves (Borgna *et al.*, 1979). We report herein

Table I. Comparison of Proteins That Bind Androgens[a]

Property	Rat receptor	Rat ABP	Human TeBG
Molecular weight	276,000	90,000	≈100,000
Stokes radius (nm)	8.4	4.7	4.7
S (low ionic strength)	8	4.6	4.6
Frictional ratio	1.96	1.61	1.87
pI	5.8	4.7	5.5
−SH requirement	+	−	+
Thermal stability	−	+	−

[a] Modified from Mainwaring (1977).

able II. Steroids with Different Affinities for ABP and for the Androgen Receptor (AR)

Trivial name[a]	Systematic name	K_i (nM) for ABP[b]	Affinity ratio (ABP/AR)	Ref. No.[c]
-Methyl-17-methoxy-DHT	2α-Methyl-17β-methoxy-5α-androstan-3-one	0.8	28	1
7-Methoxy-DHT	17β-Methoxy-5α-androstan-3-one	6.8	62	1
1-Oxo-DHT	17β-Hydroxy-5α-androstan-3,11-dione	—	0.02	1
-Oxa-17-methyl-DHT	2-Oxa-17α-methyl-17β-hydroxy-5α-androstan-3-one	—	0.02	1
-Dehydro-DHT acetate	3-Oxo-5α-androst-1-en-17β-yl-acetate	7.7	5	2
'estosterone	17β-Hydroxy-4-androsten-3-one	8.5	0.2	2
7-Methyltestosterone	17α-Methyl-17β-hydroxy-4-androsten-3-one	8.8	0.3	2
-Dehydrotestosterone	17β-Hydroxy-1,4-androstadien-3-one	27	0.2	2
-Methyl-1-dehydro-DHT	1-Methyl-17β-hydroxy-5α-androst-1-en-3-one	36	0.1	2
Methyltrienolone	17β-Hydroxy-17α-methyl-4,9,11-estratrien-3-one	48	0.02	2
-Methyl-DHT	1β-Methyl-17β-hydroxy-5α-androstan-3-one	135	0.07	2
1-Hydroxytestosterone	11α,17β-Dihydroxy-4-androsten-3-one	260	0.07	2
6,17-Dihydroxyprogesterone	16α,17α-Dihydroxy-4-pregnene-3,20-dione	425	32	2
7-Deoxytestosterone	4-Androsten-3-one	450	6	2
Androstanedione	5α-Androstan-3,17-dione	550	0.14	2
Chlormadinone acetate	6-Chloro-4,6-pregnadiene-3,20-dione-17α-yl-acetate	990	0.03	2
Cyproterone acetate	6-Chloro-1α,2α-methylene-4,6-pregnadiene-3,20-dione-17α-yl-acetate	1060	0.03	2
5-Methylprogesterone	6α-Methyl-4-pregnene-3,20-dione	1470	0.01	2

[a] (DHT) Dihydrotestosterone.
[b] The K_i refers to the equilibrium (dissociation) affinity constant for data from reference 2 and to the steroid concentration that inhibits by 50% the binding of 0.2 nM [³H]-DHT for data from reference 1. The affinity ratio is computed from these K_i values.
[c] References: (1) Tindall *et al.* (1978); (2) Kirchhoff *et al.* (1979 and unpublished observations).

on the binding properties of dicycloalkane derivatives and show how some of these compounds may help discriminate among ABP, TeBG, and the androgen receptor.

2. Materials and Methods

2.1. Binding Studies

Binding to androgen receptor, ABP, and TeBG was performed in, respectively, prostate cytosol, epididymis cytosol, and plasma as described elsewhere (Kirchhoff *et al.*, 1979; Rousseau *et al.*, 1980). Duplicate incubations contained [³H]steroid, methyltrienolone for the receptor, and dihydrotestosterone (DHT) for ABP and TeBG. [³H]Steroid was added at various concentrations without or with at least a 500-fold excess of nonradioactive steroid to determine nonspecific binding. Equilibrium conditions at 0°C were 16 hr for the receptor and 3 hr for ABP and TeBG. Separation of bound from free steroid was obtained by adsorbing the latter on activated charcoal during 8 min for the receptor and 2 sec for ABP and TeBG, a time compatible with the dissociation rates. We showed earlier that for ABP, this method gives results identical to steady-state polyacrylamide gel electrophoresis (Kirchhoff *et al.*, 1979).

2.2. Synthesis of Nonsteroidal Compounds

2.2.1. Dicyclohexane Derivatives

The dicyclohexanols PRCL and PRTL are prepared by catalytic hydrogenation of diethylstilbestrol or of *d,l*-hexestrol; PMTL derives from mesohexestrol by the same route (Fig. 1). The dicyclohexanones PRDX and PMDX are synthesized from, respectively, PRCL (or PRTL) and PMTL by chromic oxidation with Jones's reagent. Details are given elsewhere (Devis and Bui, 1979).

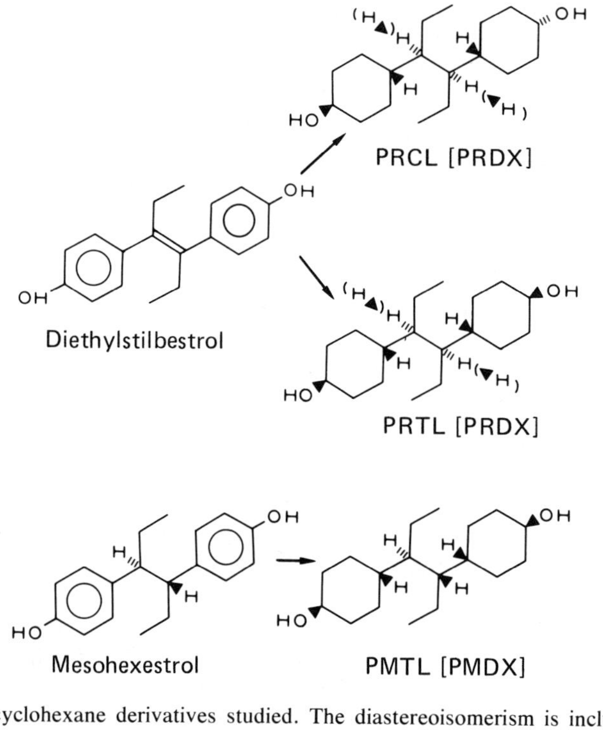

PRCL [PRDX]

Diethylstilbestrol

PRTL [PRDX]

Mesohexestrol

PMTL [PMDX]

Figure 1. Dicyclohexane derivatives studied. The diastereoisomerism is included in the arbitrary nomenclature: the letter M stands for meso; the letter R stands for racemic *(d,l)*, the second enantiomer being shown in parentheses. Configurations of substituents are indicated by convention with rules applied in steroid chemistry; for instance, all hydroxyl groups, except one in PRCL that is axial, are equatorial rather than strictly above the plane of the ring. PRDX and PMDX refer to derivatives in which a ketone group was substituted for each hydroxyl group in the parent compound.

2.2.2. Dicyclopentanone Derivatives

All these compounds (Table III) except B4MH and B4RH are obtained by dialkylation of ethyl-2-oxocyclopentanecarboxylate with α,ω-dibromoalkane chains (Mayer and Alder, 1955) of appropriate methylene number ($n + 2$ in Table III), as modified by Quivy, Devis, and Piraux (in prep.). This reaction yields a mixture of the two diastereoisomers, meso and *d,l* (Table III), one of which (B2MC, B3RC, B4MC, B4MCI, B5RC, B6MC) always crystallizes. The other isomer, in liquid form (B2RC, B4RC), is then about 95% pure by high-pressure liquid chromatography (HPLC). Compounds B4MH (crystalline) and B4RH (liquid form, 97% pure by HPLC) derive from hydrolysis and decarboxylation of B4MC. Structural determination by X-ray crystallography has been achieved (B2RC, B5RC, B6MC, and B4MH) or is in progress. All these compounds have been characterized by their melting point, if any, and by infrared, nuclear magnetic resonance, and mass spectroscopy (Quivy, Devis, and Piraux, in prep.).

Table III. Dicyclopentanone Derivatives

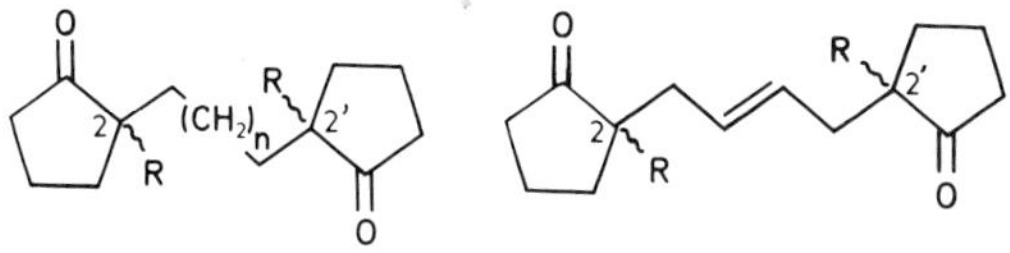

Code	Structure	Stereochemistry relative to 2,2′[a]	*n*	R	ABP binding[b]
B2RC	I	*d,l*	0	COOEt	0
B2MC	I	*meso*	0	COOEt	8
B3RC	I	(*d,l*)	1	COOEt	0
B4MC	I	(*meso*)	2	COOEt	8
B4RC	I	(*d,l*)	2	COOEt	7
B4MH	I	*meso*	2	H	8
B4RH	I	(*d,l*)	2	H	28
B4MCI	II	(*meso*)	—	COOEt	14
B5RC	I	*d,l*	3	COOEt	4
B6MC	I	*meso*	4	COOEt	0

[a] Stereochemistry is in parentheses when the structure has not been completely elucidated.

[b] ABP binding refers to the percentage of inhibition by 10 μM analogue of specific binding of 2.5 nM [^{3}H]-DHT.

3. Results

3.1. Dicyclohexane Derivatives

Three of the five dicyclohexane derivatives clearly inhibit the binding of DHT to rat ABP (Fig. 2). This effect is specific in that slight structural modifications (see Fig. 1) lead to clear differences in the pattern of inhibition. Moreover, in a "cold-chase" experiment, PRDX, the most potent dicyclohexane derivative, behaves exactly like the natural hormone DHT; namely, it does not perturb binding of previously bound hormone (Fig. 3). This suggests that the nonsteroidal analogue occupies on ABP the same site as DHT. This was confirmed in equilibrium binding experiments using a constant concentration of analogue against various concentrations of [^{3}H]-DHT. Lineweaver–Burk plots were compatible with competitive inhibition at the DHT site. Equilibrium dissociation constants for PRDX, PMDX, and PRCL were 80, 210, and 290 nM, respectively (Rousseau *et al.,* 1980).

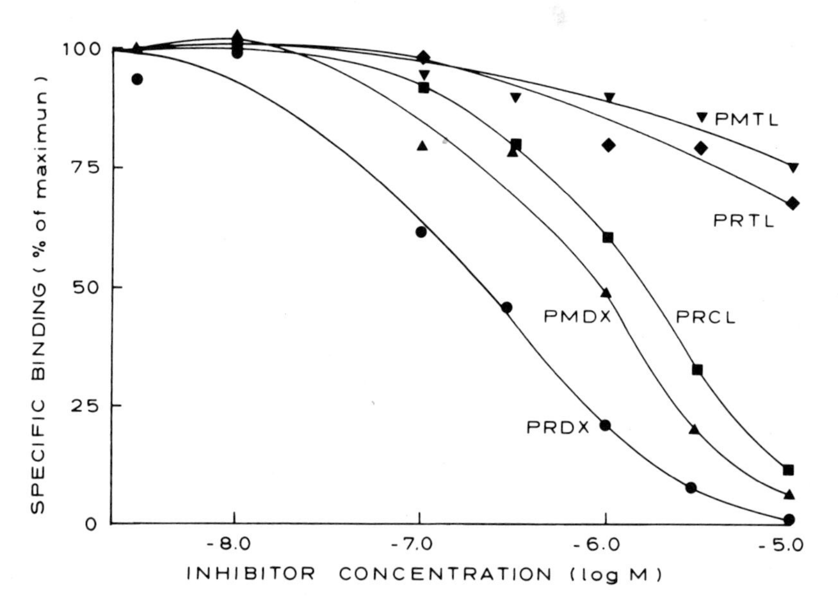

Figure 2. Binding of dicyclohexane derivatives to rat ABP. Epididymis cytosol from castrated rats was incubated at 0°C for 3 hr with 1.5 nM [^{3}H]-DHT in the presence of the indicated concentrations of competitors (see Fig. 1 for nomenclature), and binding to ABP was determined as described in Section 2.1. Nonspecific binding of DHT was 10% of total binding. Specific binding is expressed as a percentage of maximum binding seen in the absence of inhibitor (7100 dpm/assay).

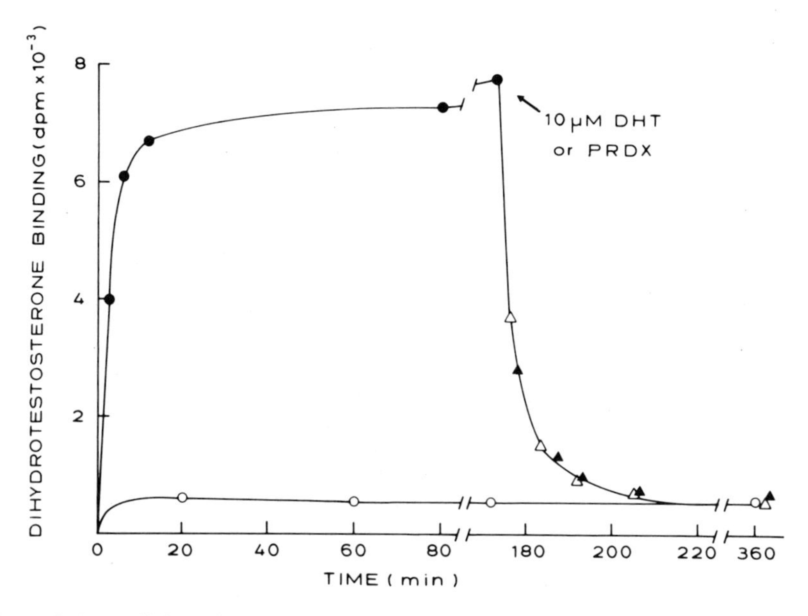

Figure 3. Reversibility of DHT binding to rat ABP in the presence of PRDX. Epididymis cytosol was incubated in the presence of 1.5 nM [^{3}H]-DHT without (●) or with (○) 10 μM nonradioactive DHT, and binding to ABP was determined at the times indicated. After 170 min (arrow), the incubations containing only [^{3}H]-DHT received either 10 μM nonradioactive DHT (△) or PRDX (▲).

While PRDX has the highest affinity for ABP, it exhibits virtually no binding for the androgen receptor. In fact, only PMDX exerts a small inhibition on the specific binding of [^{3}H]methyltrienolone (Fig. 4). Thus, PRDX and PRCL are highly specific for ABP when compared to the receptor. The two enantiomers of PRDX have been separated (Devis and Bui, 1979). Preliminary results suggest that the *d* form has a higher affinity than the *l* form.

Because of the similarity between ABP and TeBG, it was interesting to test the ability of the latter to bind the dicyclohexane derivatives. This was done in species other than the rat, which lacks TeBG. While estradiol exerted the expected inhibition of DHT binding to human TeBG, only one of the five compounds, PRCL, had a weak affinity for that protein (Fig. 5). The steroid specificity of TeBG differs in different species (Mickelson and Petra, 1978). The same is true for ABP (Tindall *et al.*, 1978). We therefore examined the binding of our analogues to both proteins from the same animal. We found that neither testosterone (Fig. 6) nor estradiol could distinguish rabbit ABP from rabbit TeBG, the

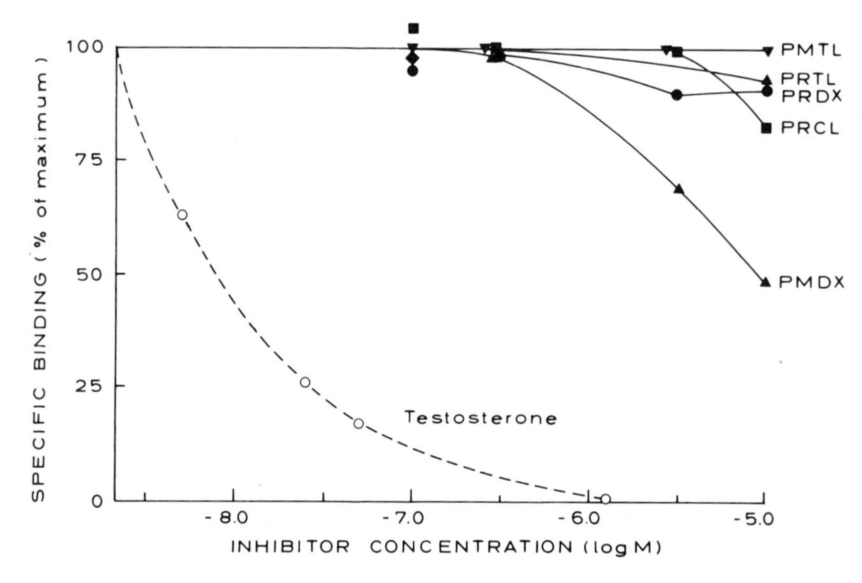

Figure 4. Lack of binding of dicyclohexane derivatives to rat androgen receptor. Prostate cytosol from castrated rats was incubated at 0°C with 2.5 nM [^{3}H]methyltrienolone in the presence of the indicated concentrations of competitors (see Fig. 1 for nomenclature). After 16 hr, binding to the androgen receptor was determined as described in Section 2.1. Nonspecific binding of methyltrienolone was 27% of total binding. Specific binding is expressed as a percentage of maximum binding seen in the absence of inhibitor (7340 dpm/assay).

affinity of each steroid being the same for both proteins. In contrast, the affinity of the nonsteroidal ligands PRCL, PRDX, and PMDX was clearly different for ABP and TeBG. All these results are summarized in Table IV.

3.2. *Dicyclopentanone Derivatives*

We developed this series of compounds following the report that B2MC and its analogues displayed in rats androgenic (Melnikova, 1961) and anabolic (Melnikova and Melnik, 1975; Melnik, 1977) activities. The ketone groups in these analogues might assume a spatial position analogous to that of the functional substituents at C-3 and (or) C-17 in androgenic steroids. Therefore, we synthesized compounds of increasing inter-ring chain length to increase that probability. None of the dicyclopentanone derivatives tested, however, bound to the androgen receptor. Still, one of them, B4RH, clearly bound to ABP, albeit with a low affinity (see Table III).

Table IV. Binding of Dicyclohexane Derivatives[a]

Compound	Inhibition of androgen binding at 1000-fold excess (%)				
	Rat ABP	AR	Rabbit ABP	TeBG	Human TeBG
PRDX	85	9	13	51	7
PMDX	63	27	5	37	0
PRCL	47	0	46	83	27
PRTL	18	4	0	21	0
PMTL	12	0	0	14	0

[a] The [^{3}H]ligand was 1.5 nM DHT for rat ABP, 2.5 nM methyltrienolone for rat androgen receptor (AR), and 5 nM DHT for other experiments.

4. Discussion

We have described dicyclohexane derivatives that bind reversibly to the androgen-binding site of rat ABP but do not interact with the rat androgen receptor. Some of these compounds also bind to rabbit ABP, but with an affinity different from that for rat ABP. This confirms that

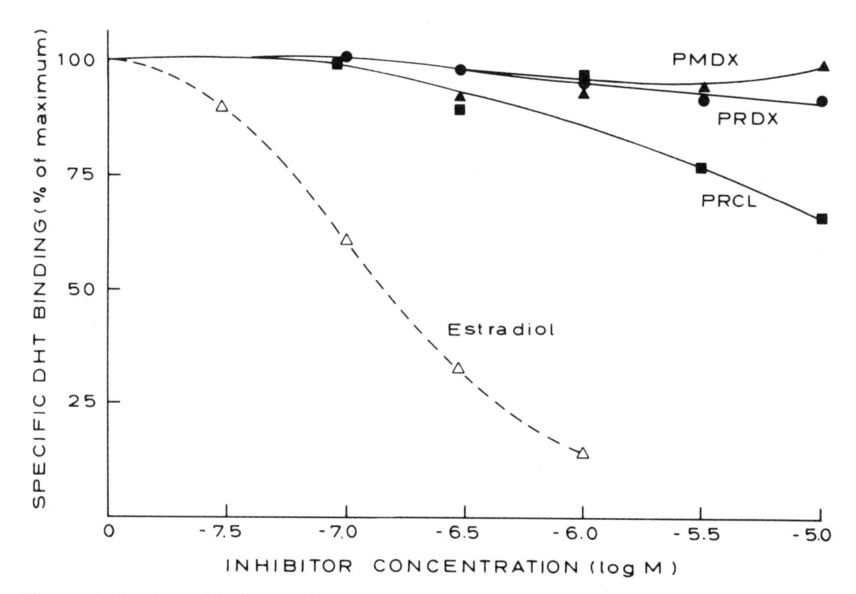

Figure 5. Lack of binding of dicyclohexane derivatives to human TeBG. Human male plasma free from endogenous steroids was incubated for 3 hr at 0°C with 5 nM [^{3}H]-DHT in the presence of the competitors as indicated. Nonspecific binding was 9% of the total. Specific binding was 46,470 dpm/assay.

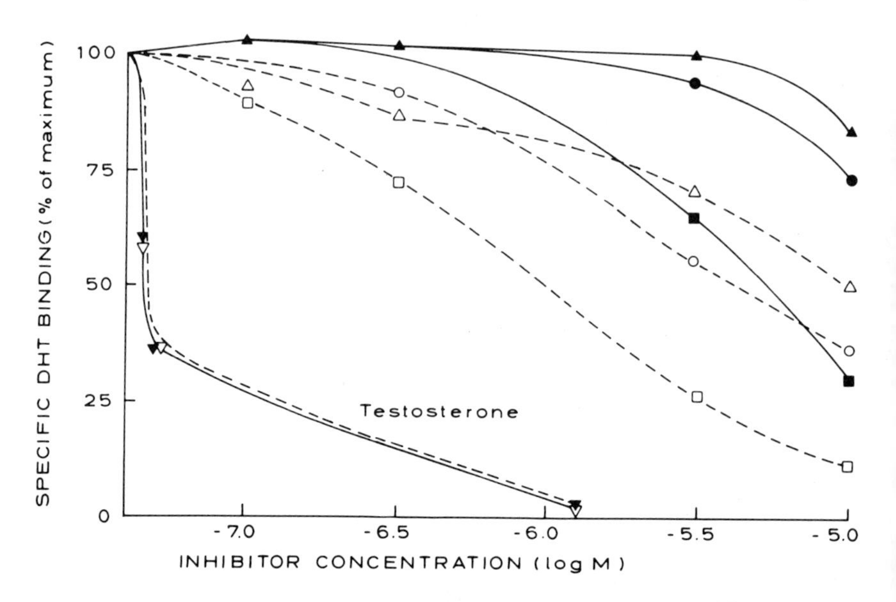

Figure 6. Binding of dicyclohexane derivatives to rabbit ABP and TeBG. Plasma and epididymal tissue were obtained from a mature male New Zealand rabbit. Binding to plasma TeBG (△, ○, □) was determined as described in Fig. 5 in the presence of PRDX (○, ●), PMDX (△, ▲), or PRCL (□, ■). Nonspecific binding was 10% of the total, which was 28,440 dpm/assay. Binding to ABP (▲, ●, ■) was determined as described for rat tissue after charcoal treatment of epididymal cytosol to remove endogenous steroids. Nonspecific binding was 9% of the total, which was 25,070 dpm/assay.

ABP has a different binding specificity in different species. One of the dicyclopentanones tested (B4RH) binds to ABP with low affinity. It is interesting to note that the dicyclopentanone derivative and the dicylohexane derivative that have the highest affinity for ABP (B4RH and PRDX) both have cyclic ketone functions with similar steric environments. Dreiding-model-building, by imposing on PRDX a conformation derived from X-ray data (Devis and Bui, 1979), shows that the ketone groups of B4RH can be easily superimposed onto those of PRDX. We find that none of the dicyclopentanones in the series binds to the androgen receptor; yet this series includes derivatives reportedly endowed with anabolic and androgenic activity. Such effects might therefore result from metabolism to more active compounds or from an indirect stimulation of other organs.

The dicyclohexane derivatives have little or no affinity for human TeBG. However, they bind to rabbit TeBG. While steroids do not allow distinction between TeBG and ABP from the same species (Tindall *et al.*, 1978; Hsu and Troen, 1978), the dicyclohexane derivatives clearly differentiate both proteins in the rabbit. Thus, the binding sites of TeBG and ABP may differ in a given species. Dicyclohexane derivatives might therefore be of interest to investigate the controversial question of whether man possesses an ABP distinct from TeBG. Indeed, evidence for (Hsu and Troen, 1978) and against (Vigersky *et al.*, 1976; Purvis *et al.*, 1978) this possibility has been presented. Other studies remained inconclusive (Lipschultz *et al.*, 1977). Finally, more potent derivatives of the nonsteroidal compounds described here could be developed that would allow a new approach to explore the role of ABP in male reproductive physiology.

ACKNOWLEDGMENTS. We thank Prof. M. De Visscher for continuous support and interest. G.G.R is Maître de Recherches of the F.N.R.S. (Belgium) and recipient of F.R.S.M. Grant 3.4514.75. The assistance of J. Kirchhoff, M.A. Gueuning, and P. Tilkin is gratefully acknowledged. This research was supported in part by the Roussel-UCLAF Company (Romainville, France), by the Ford Foundation (Grant 7900661), and by the "Coopération Belgo-Québecoise" (Project 010501). We thank F. Van de Calseyde for secretarial help.

DISCUSSION

PIETRAS: Do the rabbit ABP and TeBg bind 17α-estradiol in preference to 17β-estradiol?

ROUSSEAU: According to Rosner and Darmstadt (*Endocrinology* **92:**1700, 1973), both 17α- and 17β-estradiol compete poorly with [^{3}H]testosterone for binding to rabbit TeBG. However, the affinity of 17α-estradiol was slightly higher than that of 17β-estradiol. I do not know of any study on the binding of 17α-estradiol to rabbit ABP.

RAYNAUD: Some of these new compounds are derivatives from diethylstilbesterol, which is a potent estrogen. It will be interesting to see if they are able to interfere with the estrogen receptor.

ROUSSEAU: This is a very important point which we hope can be answered in the near future.

REFERENCES

Borgna, J.L., Capony, F., and Rochefort, H., 1979, Mechanisms of action of synthetic antiestrogens: A review, in: *Antihormones* (M.K. Agarwal, ed.), pp. 219–234, Elsevier/North-Holland, Amsterdam.

Cunningham, G.R., Tindall, D.J., and Means, A.R., 1979, Differences in steroid specificity for rat androgen binding protein and the cytoplasmic receptor, *Steroids* **33:**261.

Devis, R., and Bui, X.-H., 1979, Contribution à l'étude des androgènes synthétiques non stéroïdes. Partie I. Perhydrohexestrols et dicétones correspondantes, *Bull. Soc. Chim. France* **2:**1.

Fakunding, J.L., Tindall, D.J., Dedman, J.R., Mena, C.R., and Means, A.R., 1976, Biochemical actions of follicle-stimulating hormone in the Sertoli cell of the rat testis, *Endocrinology* **98:**392.

French, F.S., and Ritzen, E.M., 1973, A high-affinity androgen-binding protein (ABP) in rat testis: Evidence for secretion into efferent duct fluid and absorption by epididymis, *Endocrinology* **93:** 88.

Gunsalus, G.L., Musto, N.A., and Bardin, C.W., 1978, Immunoassay of androgen binding protein in blood: A new approach for study of the seminiferous tubule, *Science* **200:**65.

Hagenäs, L., Ritzen, E.M., Ploën, L., Hansson, V., French, F.S., and Nayfeh, S.N., 1975, Sertoli cell origin of testicular androgen-binding protein (ABP), *Mol. Cell. Endocrinol.* **2:**339.

Hansson, V., Reusch, E., Trygstad, O., Torgersen, O., Ritzen, E.M., and French, F.S., 1973, FSH stimulation of testicular androgen binding protein, *Nature (London) New Biol.* **246:**56.

Hansson, V., Trygstad, O., French, F.S., McLean, W.S., Smith, A.A., Tindall, D.J., Weddington, S.C., Petrusz, P., Nayfeh, S.N., and Ritzen, E.M., 1974, Androgen transport and receptor mechanisms in testis and epididymis, *Nature (London)* **250:**387.

Hansson, V., Ritzen, E.M., French, F.S., Weddington, S.C., and Nayfeh, S.N., 1975, Testicular androgen-binding protein (ABP): Comparison of ABP in rabbit testis and epididymis with a similar androgen-binding protein (TeBG) in rabbit serum, *Mol. Cell. Endocrinol.* **3:**1.

Hsu, A.F., and Troen, P., 1978, An androgen binding protein in the testicular cytosol of human testis, *J. Clin. Invest.* **61:**1611.

Kirchhoff, J., Soffié, M., and Rousseau, G.G., 1979, Differences in the steroid-binding site specificities of rat prostate androgen receptor and epididymal androgen-binding protein (ABP), *J. Steroid Biochem.* **10:**487.

Lipshultz, L.I., Tsai, Y.H., Sanborn, B.M., and Steinberger, E., 1977, Androgen-binding activity in the human testis and epididymis, *Fertil. Steril.* **28:**947.

Mainwaring, W.I.P., 1977, *The Mechanism of Action of Androgens, Monographs on Endocrinology,* Vol. 10, Springer-Verlag, Heidelberg.

Mayer, R., and Alder, E., 1955, Synthesen mit Dicarbonsäuren. XV. Cyclopentanonester-Kondensationen mit Dihalogeniden, *Chem. Ber.* **88:**1866.

Melnik, S.I., 1977, New preparation with a nonsteroidal structure possessing anabolic activity, *Chem. Abstr.* **87:**15822q.

Melnikova, T.A., 1961, Androgen activity of certain derivatives of dicyclopentane, *Tr. Leningr. Khim. Farm. Inst.* **13:**163.

Melnikova, T.A., and Melnik, S.I., 1975, Anabolic activity of some dicyclopentanone derivatives, *Farmakol. Toksikol. (Moscow)* **38:**713.

Mickelson, K.E., and Petra, P.H., 1978, Purification and characterization of the sex steroid-binding protein of rabbit serum: Comparison with the human protein, *J. Biol. Chem.* **253:**5293.

Musto, N.A., and Bardin, C.W., 1976, Decreased levels of androgen binding protein in the reproductive tract of the restricted (H^{re}) rat, *Steroids* **28:**1.

Purvis, K., and Hansson, V., 1978, Androgens and androgen-binding protein in the rat epididymis, *J. Reprod. Fertil.* **52:**59.

Purvis, K., Calandra, R., Sander, S., and Hansson, V., 1978, Androgen binding proteins and androgen levels in the human testis and epididymis, *Int. J. Androl.* **1:**531.

Ritzen, E.M., Nayfeh, S.N., French, F.S., and Dobbins, M.C., 1971, Demonstration of androgen-binding components in rat epididymis cytosol and comparison with binding components in prostate and other tissues, *Endocrinology* **89:**143.

Rousseau, G.G., Quivy, J.I., Kirchhoff, J., Bui, X.H., and Devis, R., 1980 Nonsteroidal compounds which bind epididymal androgen-binding protein but not the androgen receptor, *Nature* **284**:458.

Tindall, D.J., Cunningham, G.R., and Means, A.R., 1978, Structural requirements for steroid binding to androgen binding proteins, *Int. J. Androl. Suppl.* **2**:434.

Vigersky, R.A., Loriaux, D.L., Howards, S.S., Hodgen, G.B., Lipsett, M.B., and Chrambach, A., 1976, Androgen binding proteins of testis, epididymis, and plasma in man and monkey, *J. Clin. Invest.* **58**:1061.

Weddington, S.C., Brandtzaeg, D., Hansson, V., French, F.S., Petrusz, P., Nayfeh, S.N., and Ritzen, E.M., 1975, Immunological cross-reactivity between testicular androgen-binding protein and serum testosterone-binding globulin, *Nature (London)* **258**:257.

9

Profiles of Target-Cell Prolactin and Adrenocorticotropin during Lactational Diestrus

Janet M. Nolin

1. Introduction

Endocrinology can be viewed as a very old science when one considers that ablation of certain of the glands of internal secretion, in particular the gonads, was practiced very early on in our history. (One can even postulate that the bravest among some of our ancestors got hold of a saber-toothed tiger long enough to turn him into a pussycat!) It was not until whole organs, homogenates, or extracts were tested for activity that could repair a deficiency, however, that modern endocrinology began about a century ago. This sort of qualitative bioassay approach continued to be used well past the first third of the present century and involved hormones exclusively in their glands of origin, i.e., hormones in very large amounts. As bioassay sensitivities were improved, detection of a relatively wide range of concentrations became possible in the late 1940's and 1950's, and during this period it even became possible to detect some hormones in transit to their targets, but again, the amounts that could be detected were present at the lower concentrations that could be measured in tissues of origin, and therefore the improved methods could detect

Janet M. Nolin • Department of Physiology, Medical College of Virginia, Richmond, Virginia 23298.

hormones in blood only if they were present in supraphysiological amounts. Then, in the late 1960's and early 1970's, came radioimmunoassays (RIAs) and radioreceptor assays (RRAs) that allowed the study of hormones in transit, not only in physiological, but also in subphysiological, concentrations.

However, measurements of a hormone en route to its target, as sensitive as the assays used to obtain those measurements might be, and although they may give reasonable (?)* estimates of overall hormone availability, do not describe availability to individual cells, and more important, they do not define relationships between actual target cell–hormone interactions and target-cell response. Endocrinology is fundamentally the study of hormonal control. Hormones direct events in their targets, specifically in their target cells. Clearly, the detectability of hormones in their individual target cells, where hormone action actually takes place, might contribute significantly to our study of hormonal control. Predictably, the logical sequence of technical refinements that have occurred over the past years has now led to the development of methodology applicable to that purpose.

Endogenous prolactin (PRL), a protein hormone, was the first to be detected in one of its targets, the milk-secretory cell (Nolin and Witorsch, 1976). Since then, three other hormones of endogenous origin have been studied in their targets: luteinizing-hormone-releasing hormone, a hypothalamic peptide, in pituitary gonadotrophs (Sternberger *et al.*, 1978); glucocorticoid, a steroid, in adrenocorticotrophs (Dornhorst and Gann, 1978); and adrenocorticotropin (ACTH), another peptide, in the cells of the adrenal cortex (Nolin, 1979a, 1980b). An attempt has also been made to study the glycoprotein, follicle-stimulating hormone, in the testis (Hutson *et al.*, 1977).

This approach to endocrinology is of such recent origin that much of what may be learned is yet to come. In this chapter, I would like to summarize some of the work from our laboratory and outline some of the things we have learned and some of the questions that have come into focus.

The method we use to detect target-cell hormone is an immunohistochemical one borrowed, in part, from others who had used a similar approach to detect anterior-pituitary hormones in their cells of origin

* Measurements of hormone availability (in the circulation) by either RRA or RIA are not necessarily measurements of the availability of biologically active hormone. RRA depends on binding of hormone to cells, but binding alone is not always associated with response (Nolin and Witorsch, 1976; Nolin, 1978b,d). RIA depends on antigen–antibody interaction, but immunoactivity is not synonymous with hormone regulatory activity, nor does immunoreactivity necessarily reflect regulatory activity.

(Nakane, 1968; B.L. Baker, 1970). Details of protocols and discussions of useful and not so useful controls, as well as of limitations of the method, can be found in Nolin and Witorsch (1976) and Nolin (1978a,d, 1979b, 1980a).*

We have been using this approach to study lactation: the hormonal regulation of the mammary gland itself as well as both the regulation of, and the roles played in this process by, the maternal adrenal and ovary.

During lactation, domination of the maternal organism by two anterior-pituitary hormones, PRL and ACTH, prevails. The major contribution of ACTH is to stimulate adrenocortical secretion. Glucocorticoids, in particular, are essential to lactation in most species, including the rat, and they act directly on mammary tissue (Topper, 1970). PRL, of course, also acts directly on the mammary gland, but it acts on the rat adrenal cortex as well, to facilitate corticosterone output (Ogle and Kitay, 1979). Like adrenocortical cells, ovarian lutein cells are also direct targets for PRL during lactation (Nolin, 1978a). Lutein cells are very active at this time, but it is not yet certain what direct (or indirect) contribution(s) their secretory products might make to the lactational state. By contrast, the reproductive ovary is quiescent during lactation, at least in the sense that ovulation does not occur. Curiously, ovarian follicles also appear to be PRL targets (see Section 5).

2. *Prolactin in the Normal Rat Mammary Gland*

In mammary glands from normally, continuously suckled mothers, alveoli at various stages of repletion are present, and the entire range of variability can often be seen in a single section (5 mm $\times$ 5 mm $\times$ 5 μm). Alveoli filled to capacity with milk are lined by squamous, nonsecreting

*In addition to the discussion in the references cited, some further points should be made. No molecule can be identified with certainty by immunological criteria alone. Fragments or analogues of the authentic antigen or even completely alien molecules with similar antigenic determinants can bind to antibody generated by the "authentic" antigen. Like RIA, immunohistochemistry (IHC) relies on antigen–antibody interaction. It is necessary to realize that IHC is even more prone than RIA to errors due to cross-reacting molecules, since in IHC the antibody–antigen reaction is noncompetitive. This is why I have insisted that RIA tests for specificity cannot be used to characterize an antiserum for IHC. Testing for immunological specificity must be done within the IHC system itself. In contrast to RIA, however, IHC does not rely solely on immunological criteria. Inherent in this technology is a criterion of biological response, cell morphology, that has so far provided evidence consistent with the interpretation that the entry of immunologically detectable PRL and ACTH into its target cells is accompanied by signs of hormone action, and that exclusion from the cell coincides with lack of response. Also inherent in IHC is the opportunity to compare this correlation among individual cells in the same tissue.

or "resting" cells. Cells lining alveoli with little or no milk are columnar, often tall columnar, about 7–8 μm wide and 20–25 μm from basal to apical pole. Between these two stages, when alveoli are being filled, milk-secretory cells (MSC) are cuboidal. From this morphological picture, it seems apparent that MSC must undergo functional cycles that can be divided into three major phases: lactogenic (onset of milk synthesis and release), galactopoietic (sustained synthesis and release), and resting (cessation of milk secretion) phases (Fig. 1). On the basis of uptake and transcellular transport of tritiated leucine (Saacke and Heald, 1974) and in terms of general morphology, the cells in any given alveolus appear to be in functional synchrony. Resting cells, at the end of an activity cycle, certainly appear to be in synchrony in terms of the absence of both PRL incorporation and secretory activity. Although milk contained in alveoli lined by resting cells holds appreciable amounts of PRL, PRL is undetectable within the cells. By contrast, molecular moeities (receptors?) that can recognize PRL (detectable because they bind exogenous PRL, with what appears to be high affinity) are present in the apical regions of MSC during the resting phase of their functional cycle.

Concomitant with alveolar emptying, while changing shape (becoming columnar) and entering a renewed lactogenic phase, MSC also prepare for and enter a renewed PRL incorporation cycle. This cycle consists of dissociable processes: PRL recognition at the basal plasmalemma [prior movement of anticipant receptors from apical pole of resting cells to basal pole during alveolar emptying? (see Fig. 2C)]; PRL

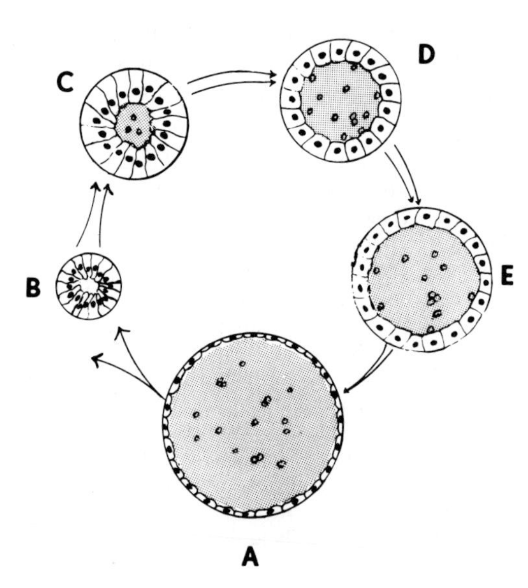

Figure 1. MSC activity cycle. The milk-filled alveolus is lined with flattened, resting cells (A). With emptying, as the alveolus contracts and its capillary bed expands (arrows), the cells become columnar and resume milk synthesis [the lactogenic phase of the cycle (B–C)]. As milk accumulates in the alveolar lumen, the MSC regress in height, becoming cuboidal [the galactopoietic phase (D–E)]. With further accumulation of milk, the cells reenter their resting phase (A). Reprinted from Nolin and Bogdanove (1980) with permission.

internalization (uptake into the basal pole of the cell); polar translocation through the cytoplasm (basal toward apical pole); nuclear uptake and, eventually, transcellular transport and excretion from the cell apex. During the lactogenic phase, these processes are asynchronous cell–cell (PRL is not in the same locus in each of the cells of a single alveolus) and discontinuous within single cells (PRL is present in discretely different areas in individual cells).

In the early lactogenic phase, characterized, in part, by the presence of PRL in MSC nuclei, the milk that begins to accumulate in the alveolar lumen is PRL-free. PRL accumulates in cell apices during this period and does not traverse the cell completely until somewhat later. As milk continues to accumulate and the basal–apical dimension of MSC is reduced and the side–side dimension increased, PRL appears to stream steadily through all the cells in an alveolus, finding its way in increasing amounts into the alveolar space. During this period of galactopoiesis, PRL is not found in nuclei and PRL incorporation appears to be synchronous cell–cell. Transcellular transport of PRL into milk seems a necessary component of this synchrony. Eventually, as MSC again approach the resting phase of their activity cycle, their job done for the moment, PRL transfer from blood to milk ceases altogether. These relationships between MSC activity cycles and PRL incorporation cycles are shown graphically in Fig. 2.

Estimates of the length of time MSC spend in each of the three major phases of their activity cycle and, in particular, how long PRL might be present in nuclei (Fig. 3) revealed that the lactogenic phase lasts about 1½ hr, the galactopoietic phase about 4 hr, and the resting phase about 2½ hr. PRL is present in nuclei for only about 5% of overall cycle length, or about 24 min.

The series of studies that yielded the data and the concepts just outlined provide evidence of PRL–MSC right time–right place relationships that are compatible with an early direct effect of PRL on transcriptional events and with the possibility of a sustained direct effect of PRL on galactopoiesis. We have begun to study this possibility (see Section 3). Documentation for this section, unless otherwise noted, is to be found in Nolin and Witorsch (1976), Nolin (1978a,b, 1979b), and Nolin and Bogdanove (1980).

3. *Experimental Modulation of Milk-Secretory-Cell Prolactin*

In a series of studies aimed at probing whether intracellular PRL is *required* for cell response, we gave graded doses of estrogen to lactating rats and examined the effects of these treatments both on the metabolic

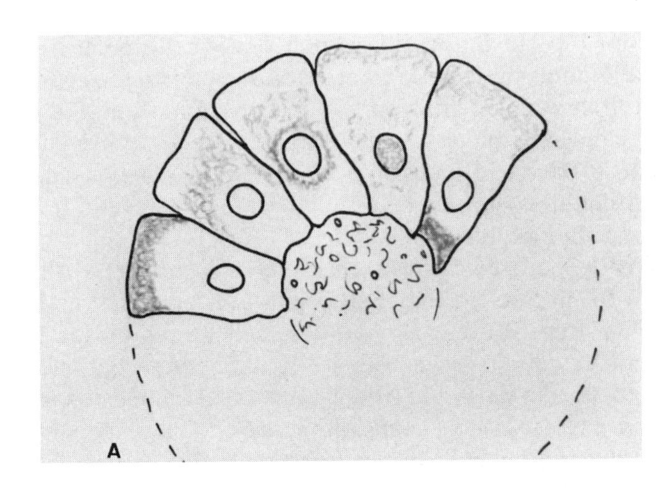
A

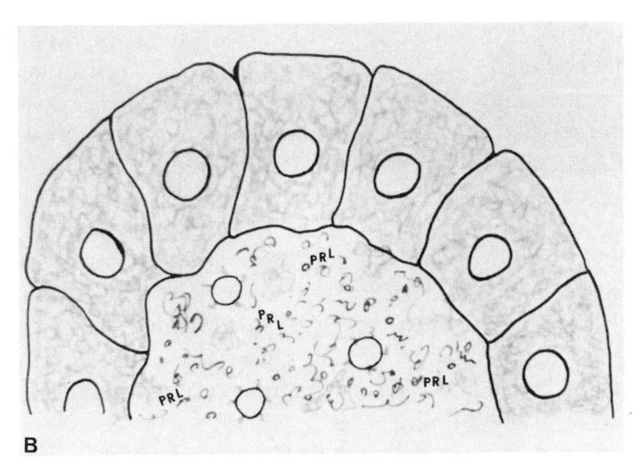
PRL
PRL
PRL
PRL
B

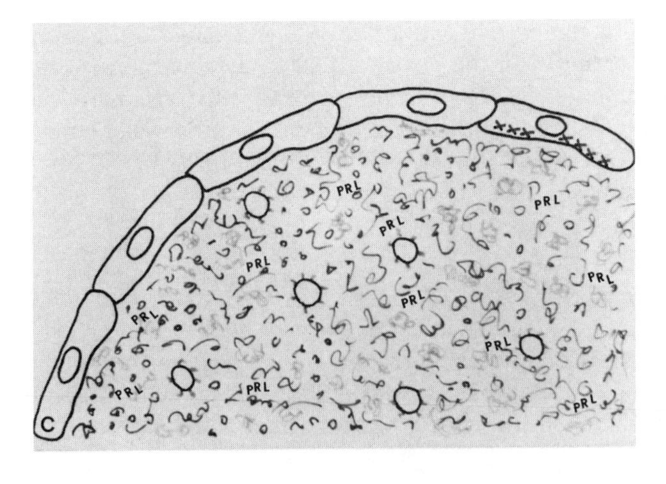
PRL
PRL
PRL
PRL
PRL
PRL
PRL
PRL
PRL
PRL
PRL
C

responses of MSC to PRL and on PRL incorporation (Fig. 4). The effects on both parameters of PRL–MSC interaction were dose-dependent. As the dose of estrogen increased, there was progressive lactational failure and associated alterations in the abilities of MSC to complete the PRL-incorporation process. Gradations of effects bearing an inverse relationship to estrogen dosage included: failure of PRL internalization and even of recognition of the hormone; persistence of recognition and internalization with failure of PRL transport into the central cytoplasm; persistence of recognition, internalization, and polar transport, but failure of nuclear uptake; and with the lowest dose of estrogen tested, persistence of recognition, internalization, and polar transport with limited and disrupted nuclear uptake.* Nuclear uptake was the step most sensitive to the effects of estrogen, and failure of this process occurred in conjunction with the onset of lactational failure (Nolin and Bogdanove, 1980).

This series of experiments demonstrated that PRL recognition, internalization, and transport into the central cytoplasm, components of the PRL-incorporation process that must precede PRL uptake by nuclei,

*These effects were not only qualitative but quantitative; i.e., the extent of the defects in a given cell population varied cell–cell. However, the incidence of cell–cell differences decreased as estrogen dosage increased.

←

Figure 2. PRL incorporation by MSC in relation to their functional cyclicity. (A) Cells in the lactogenic phase of their activity cycle. Note that during this phase, PRL incorporation is asynchronous cell–cell (PRL is not located in the same area of each cell within a given alveolus). Heterogeneity in the presence of PRL within individual cells also exists and suggests intermittent internalization of PRL and discontinuity among the component parts of the PRL-incorporation process that lead to the subsequent redistribution of PRL within the cell. PRL concentrates near the basal pole of nearly half these cells, suggesting that periods of PRL internalization are slightly shorter than the periods in which PRL uptake ceases, or is reduced, while previously internalized hormone is being redistributed. Intranuclear PRL, found in cells only during this lactogenic phase of the cell activity cycle, is found in about a fourth of the cells in a single alveolus at a given time. Notably, a small amount of milk may accumulate in alveolar lumina prior to PRL excretion. (B) Cells in the galactopoietic phase of their activity cycle. Note that during this phase, within individual alveoli, PRL incorporation appears to be synchronous cell–cell, and within individual cells, there appears to be a continous streaming of PRL through the cell cytoplasm from basal to apical pole and out into the alveolar lumen, without nuclear uptake of the hormone. (C) Cells in the resting phase of their activity cycle. PRL is not detectable in resting, inactive cells. Despite abundant PRL in the general circulation, local mechanisms operate to suspend both milk synthesis and PRL incorporation following successful completion of an alveolar quota of milk production. Notably, these cells do contain unoccupied PRL-binding sites (represented by XXX in the upper right cell. Their apical location suggests that they may be anticipant receptors awaiting recycling to the basal pole of the cell at the onset of a new activity cycle when alveolar emptying will have relieved the constriction imposed, by previous alveolar distention, on the local availability of PRL.

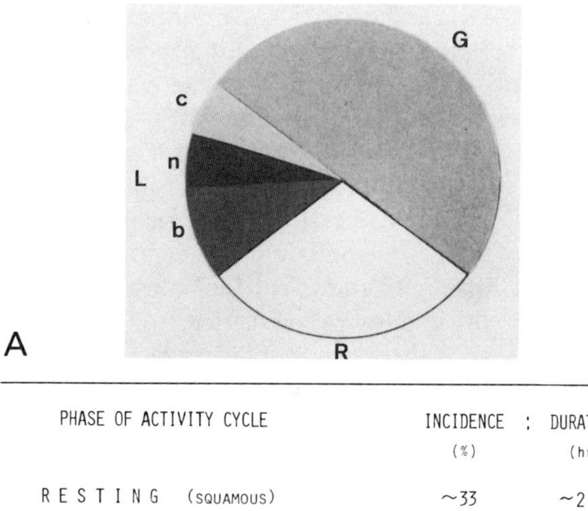

B PHASE OF ACTIVITY CYCLE		INCIDENCE (%)	: DURATION (hr)
RESTING	(SQUAMOUS)	~33	~2.6
INITIAL	(COLUMNAR)	<20	<1.6
	BASAL POLE	~10	~0.8
	NUCLEI	~ 5	~0.4
MIDDLE	(CUBOIDAL)	>50	>4.0

Figure 3. (A) Proportion of MSC in normally lactating rat mammary gland in which PRL incorporation: (1) has ceased [(R) squamous cells in the resting phase of the MSC activity cycle]; (2) occurs as a discontinuous process [(L) columnar cells in the initial or lactogenic phase of the MSC activity cycle]; or (3) occurs as a continous process [(G) galactopoietic, cudoidal cells]. The proportions of cells in which, during the lactogenic phase, PRL is present only in the nuclei (n), in the basal portion (b), or in other cytoplasmic areas (c) of the cell are also shown. Incidence was assessed by examining 25 sections of each of four mammary glands from 24 rats. (B) Relative durations of the three major phases of the MSC activity cycle and of two subdivisions of the lactogenic phase. Calculations of phase durations were based on the assumption that relative durations should be proportional to the frequency with which various stages of the activity cycle were encountered and on the observation that optimal alveolar refilling time in the rat is about 8 hr (Grosvenor *et al.*, 1967).

do not seem adequate, in themselves, to maintain normal alveolar activity cycles, but the question of whether nuclear uptake of PRL by MSC is *required* for cell response needs still further inquiry.

4. Adrenal-Cortical-Cell Prolactin and Adrenocorticotropin

As part of our overall approach to the study of hormone action during lactation, we have studied both PRL and ACTH (Fig. 5) in the adrenal cortex. As noted earlier, both hormones are required by the rat

	PUP WT.	MILK IN GLAND	PRL INCORPORATION PROCESS				
			R	I	T	N	E
>16 μG	NA	0	0	0	0	0	0
16 μG	↓	1	1	1	0	0	0
4 μG	→	2	3	3	1	0	1
1 μG	↗	3	4	4	3	2	3
CONTROL	↗	4	4	4	4	4	4

Figure 4. Differential sensitivity to estrogen of experimentally dissociable aspects of the process of PRL incorporation by MSC from rats exposed to a variety of estrogen treatment regimens. The data are expressed on a scale of 0–4 to assign the extent of deviation from the control value (4). The extent of deviation was directly related to estrogen dosage. Nuclear uptake appears to be the phase of PRL incorporation most sensitive to the effects of estrogen. (R) Recognition; (I) internalization (PRL present in cell base); (T) translocation toward apical pole (polar translocation); (N) nuclear uptake; (E) Excretion (listed as part of the PRL incorporation process, albeit an adjunctive part, because the apparent steady streaming of PRL through galactopoietic MSC seems to be in part a function of PRL excretion). Data taken from Nolin and Bogdanove (1980).

adrenal for maximal output of corticosterone, which, in turn, is essential to the milk biosynthetic process.

In contrast to PRL, ACTH has a long list of "second messengers" or surrogates (Rubin and Laychock, 1978), suggesting that while PRL might be detectable within its adrenal target cells as it is in MSC, ACTH might not need to (or be able to) enter its target cells and might reside

Figure 5. Target-cell PRL has "been around" for a while now (Nolin and Witorsch, 1976). Target-cell ACTH is a relative newcomer (Nolin, 1979a, 1980b). Drawing by Jon Bogdanov.

only in the plasmalemma. This appears not to be the case. Both PRL and ACTH can be incorporated by their adrenal target cells.

During lactation, endogenous PRL is restricted to cells in the two inner zones of the adrenal cortex, the fasciculata and reticularis, the zones responsible for corticosterone production. ACTH, on the other hand, is present in cells of all three zones of the adrenal cortex (not surprisingly, because all three zones are ACTH targets). PRL is present in both cell cytoplasm and nuclei, often reciprocally. ACTH is found in the cytoplasm of cells in the inner two thirds of the cortex and on the nuclear envelope. In glomerulosa cells, it is also present on the nuclear envelope and, in some cells of any given population, inside nuclei. [Glomerulosa cells secrete aldosterone, but the role of this mineralocorticoid in lactation is poorly understood. One might guess that it may play an important role in view of the increased requirements for monitoring of water metabolism imposed by loss of relatively large quantities of sodium and water into milk (Peaker, 1977).]

As in the course of investigations of PRL in the mammary gland, the logical, *first* question brought into focus by these observations (since target cells are in fact where hormone action occurs) is whether entry of these peptidal hormones subserves a requirement for direct interaction of hormone with intracellular constituents as part of their mechanisms of action.

Studies of endogenous PRL and ACTH in individual adrenal cells are even fewer than studies of endogenous hormone in MSC. However, to attempt to put our observations in perspective within the framework of the available information, it may be helpful to consider that in the case of ACTH, its mechanism(s) of action on neither the inner cortical cells (Haynes, 1975) nor glomerulosa cells (Davis, 1975) is as yet understood. This remains true despite the large body of literature that has accumulated since direct binding of ACTH to adrenal cells, in association with stimulation of adenylate cyclase activity leading to the formation of an ACTH surrogate, was first demonstrated in 1970 (Lefkowitz *et al.,* 1970). Nevertheless, there appear to be some guidelines available from the existing literature on the mechanism(s) of action of ACTH. By contrast, how PRL influences the adrenal, in terms of molecular interactions, is completely uncharted territory. In our studies, the presence of both ACTH and PRL in their adrenal targets was associated with an apparent high level of activity and seems to be, as is the case for the PRL–MSC interaction, an integral part of cell response. The idea that entry of ACTH into its adrenal target cells might actually be required, as part of its mechanism of action, does have strong support from biochemical studies demonstrating that it can act directly to stimulate cytosol enzymatic activity crucial to steroidogenesis (for review of this and related work, see McKerns, 1978). ACTH may also need to enter adrenal cells to activate lysosomes that in some way may facilitate its actions (Szego *et al.*, 1974). No analogous evidence is yet available to help explain the presence of PRL in adrenal-cell cytoplasm or of either PRL or ACTH in adrenal-cell nuclei. The very presence of these regulatory hormones in the nucleus, the subcellular organelle that has as its primary function the governing of mitotic and transcriptional activity, may be an important clue.

Uptake of PRL and ACTH by adrenal cells may well be linked to individual cell activity cycles as is uptake of PRL by MSC. The mechanism utilized by MSC to "inactivate" PRL includes its excretion from the cell until, at the end of an activity cycle, PRL is no longer detectable in MSC. Some adrenocortical cells in any given gland are also devoid of regulatory peptide, suggesting that in these cells as well, a "resting" phase, at least in terms of PRL or ACTH incorporation, exists as part of functional cyclicity. The presence of PRL in MSC nuclei precedes excretion during MSC cyclicity. It would seem not unreasonable to suspect that interaction of ACTH or PRL or both with adrenal-cell nuclei might also precede inactivation. How and when in the cell cycle presumably non-PRL(or ACTH-) excreting cells, such as adrenocortical cells, might inactivate their regulatory peptides in preparation for cycle-linked renewed uptake is among the many other questions raised by our findings (and our interpretations so far). In cells in which regulatory peptide resides only in the cytoplasm and is not excreted from the cell and for which no biochemical evidence is available, the possibility remains open that direct regulatory actions on subcellular constituents are not operative and uptake subserves inactivation by cytoplasmic components in anticipation of renewed uptake, exclusively.

Documentation for this section, unless otherwise noted, is to be found in Nolin (1978a, 1979a, 1980b). Also, please see the previous section.

5. *Follicular Prolactin*

Our approach to the study of lactation has also led us to explorations of PRL in the ovary. Dormancy of ovulatory activity is characteristic of most species during most of lactation. After one ovulatory cycle immediately postpartum in the rat, the cloistered rat, at least, settles down to the business at hand. Why ovulation does not occur during lactation is still uncertain, but resolution of this question may be hastened by new evidence that PRL–ovarian interaction may not be restricted to the postovulatory compartment of the ovary, the corpora lutea. Although rat corpus luteum function has been known to be PRL-dependent for nearly 40 years (Astwood, 1941), only recently has evidence begun to accumulate that granulosa cells of the ovarian follicle are also PRL targets (Midgley, 1972; Rolland and Hammond, 1975; Crisp, 1979; Nolin, 1979c).

Endogenous PRL is present in granulosa-cell cytoplasm and either on the nuclear envelope or inside the nucleus in any population at a given time. It is only present in granulosa cells of antral follicles in which it is also present in antral fluid (for related studies of human follicular fluid, see McNatty *et al.,* 1975). But PRL does not limit its follicular migration to follicular fluid; it is present in the developing oocyte as well. During lactation, PRL is present in oocytes of follicles of all sizes. This and its apparent absence from granulosa cells of preantral follicles and its presence in granulosa cells and follicular fluid of rapidly growing follicles suggests long-term storage of PRL in oocytes, with transfollicular transport of PRL serving to replenish supplies required by accelerated oocyte growth. However, transcellular transport by granulosa cells cannot be the only explanation for incorporation of PRL by these cells. Why does PRL gain access to granulosa-cell nuclei? Is this required for its effects on follicular steroidogenesis (see McNatty *et al.,* 1977; Veldhuis, *et al.,* 1979)? Equally curious is the question of why PRL is sequestered in oocyte cytoplasm. In the case of the oocyte, the question is not so much why it is inside this cell (we have already seen that PRL is incorporated by several different kinds of cells), but rather, since this is the first evidence that the oocyte might be regulated directly by PRL, what are its actions there? The lattice of cytoplasmic filaments associated with PRL (PRL is not detectable in the germinal vesicle) have also been associated with mammalian "yolk" substance (Szöllösi, 1972). Perhaps PRL influences oocyte viability by influencing its nutrient support.

There are, however, other possibilities. The fact that the lactating rat is in a prolonged period of normophysiological anovulation prompted the following sequence of ideas: All the PRL-containing oocytes observed so far have been dictyate oocytes and showed no signs of impending reentry into the maturational process. Follicular fluid contains peptidal maturation inhibitor(s) apparently derived from granulosa cells [for review, see Channing (1979) and Chapter 28]. In addition to granulosa cells, some oocytes also appear to contain similarly active factors (Nekola and Nalbandov, 1971; also references and discussion in Schuetz, 1974). A peptide, PRL, is present in all three follicular compartments prior to maturation: granulosa cells, follicular fluid, and oocytes.

Anti-PRL serum induces meiotic maturation in cultured follicles (T.G. Baker and Hunter, 1978). The maturation inhibitor(s) of follicular fluid has (have) (a) molecular weight(s) of 10,000 or less. Rat pituitary PRL has a molecular weight of 23,000. Perhaps PRL influences the production of peptidal maturation inhibitors (T.G. Baker and Hunter, 1978).

However, the molecular size of target-cell PRL is unknown. Although it retains immunoreactivity to antiserum to pituitary PRL, this is not evidence that it is the same molecule (see footnote, page 197). "PRL" has a large variety of seemingly unrelated actions in different cells and even within the same cell. Could there exist in ovarian follicles a family of fragments of the parent PRL molecule, retaining, perhaps, one or more of the several sets of antigenic determinants that might be present in the parent molecule, but differing from each other sufficiently to regulate several different processes, among them oocyte maturation, granulosa-cell luteinization, and follicular steroidogenesis? Could the entry of PRL into nuclei of granulosa cells participate in this process by initiating production of the several "clipping" enzymes, each with highly stringent substrate specificity, which would be required for this potentiation of biological activity through fragmentation of a sufficiently large peptidal hormone into several active smaller peptides?

There might be more comprehensive applicability of this idea. Target-cell "clipzymes" (Fig. 6) might account in general for the pluritrophic actions of PRL. Individual target cells capable of "clipping" circulating PRL (the prohormone for target-cell PRL) could participate in their own regulation by providing themselves with their own essential regulatory peptides. Only one or two particular peptides might be needed in certain cell populations such as adrenal and lutein cells. In MSC, as in granulosa cells, several might be necessary. Hormonal activity can reside in a peptide sequence containing as few as three amino acids. Thyrotropin-releasing hormone is one example. Certainly the size of the PRL molecule (198 amino acids) is compatible with the presence of several single-action small peptide sequences linked together in one multiple-action large peptide. This idea is still a hypothesis. It has only begun to be tested in our laboratory.

Documentation of this section, except where otherwise noted, is to be found in Nolin (1978c, 1979c, 1980a).

6. Conclusion

Endocrinology has been extended from the study of hormones in tissues in which they are synthesized or stored, or both, to the study of hormones in transit to their targets, and now to the study of endogenous

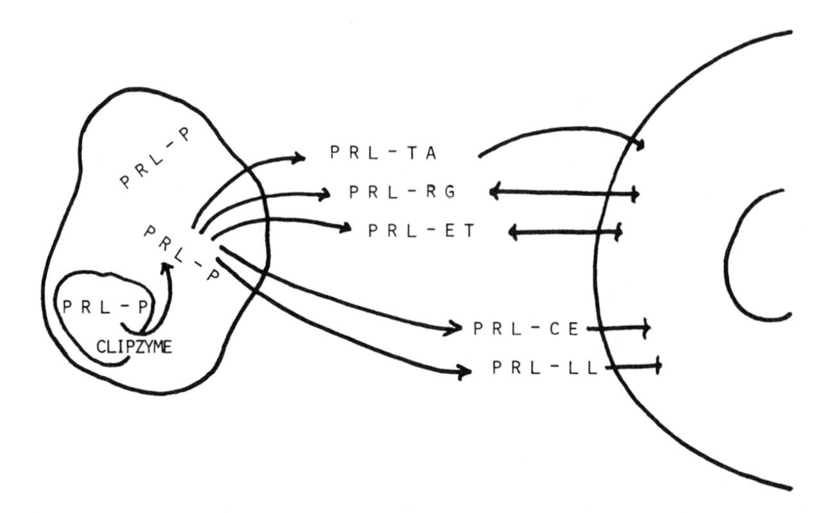

Figure 6. Clipzyme hypothesis. PRL, a protein hormone containing 198 amino acids, is pluritrophic; i.e., it has several targets among which, and even within which, its actions are varied. The clipzyme hypothesis attempts to account for this pluritrophism by proposing that there are inherent in PRL target cells mechanisms for producing various fragments from what we know to be pituitary PRL (or, more specifically, from circulating PRL). In this sense, PRL impinging on the cell from the circulation is the prohormone for target-cell PRL. Different target cells (e.g., ovarian, adrenal, or milk-secretory cells) would "clip" circulating PRL in a highly specific fashion, thus providing essential peptides for their own regulation. Only one particular peptide might be needed in certain cell populations; in others, several might be required. The size of the parent PRL molecule provides the potential for a considerable number of smaller peptides. In this diagram, in which a granulosa cell is shown at left and an oocyte at right, the availability of granulosa-cell clipzymes is stimulated by a direct action of the parentlike PRL molecule on granulosa-cell nuclei. The resultant array of products of clipzyme activity, in turn, becomes available to regulate granulosa-cell luteinization and the follicular steroid profile, oocyte maturation, and, possibly, oocyte nutrition. In the MSC, separate fragments might be required for as many disparate functions as Na^+ transport, lactose synthesis, fatty acid synthesis, triglyceride production, and casein synthesis. Each of these components of the milk-secretory process is known to be PRL-dependent.

hormones in their individual target cells. At present, this approach to endocrinology is characterized by a greater number of heuristic arguments than of scientific observations. Conclusions must be deferred until this ratio changes.

ACKNOWLEDGMENTS. *Sine quibus non* E. M. Bogdanove and our children. A special thanks to Anthony Padua. And Gail Swatt, I thank you. Financial support for this work was provided by the National Science Foundation (PCM 76-23641), for which I would also like to express my appreciation.

Discussion

McKerns: It is most amazing that only a few years ago no one could detect any polypeptide hormone within a target cell and few people believed that peptide hormones had any direct regulatory role in the cytoplasm or nucleus. How did the concept change?

Nolin: As everyone interested in hormone action is well aware, that is a highly provocative question on several counts. Suffice it to say, I think the prevailing theosophy changed!

Channing: Your prolactin antiserum may cross-react with our oocyte-maturation inhibitor.

Nolin: I would like very much to test that possibility.

D. Villee: Is there any possibility of determining the structure of the intracellular immunoreactive material, since it may not be identical to the pituitary hormone as it is secreted?

Nolin: That would, of course, be a necessary test of my hypothesis. Yes, I think it is eminently possible. It would require the talents of a persistent and innovative peptide chemist, but the basic technology is available.

D. Villee: Another question: can you explain the high concentration of prolactin in amniotic fluid?

Nolin: No, but there is some thought that it may participate in the regulation of fluid volume.

Bazer: Decapitated-sheep fetal–placental complex has average to above average amniotic fluid volume. I don't know about prolactin levels in amniotic fluid of these animals, but prolactin, whatever the source, and arginine-vasopressin as well, are believed to be involved in the control of fluid volume in these animals.

D. Villee: Do the adrenal cells which bind prolactin increase their production of glucocorticoids in response to prolactin?

Nolin: Ogle and Kitay (1979) have shown that the answer to your question is basically "yes." We have not yet studied the possible correlations of glucocorticoid secretion and intracellular prolactin (or ACTH).

Bazer: I would like to add that in addition to prolactin, there is a high concentration of receptors for placental lactogens in adrenal tissue, adding to the evidence that the adrenal is a target for lactogenic hormones. Dr. Nolin, have you looked at prolactin in uterine tissue?

Nolin: No, we haven't.

Bazer: In connection with the subject of the lactogenic hormones (and steroids) of the placenta, I would like to add a comment. Placental lactogens and/or steroids are believed to exert an effect on lactational performance of the cow. This so-called "sire of fetus" effect is realized because the genotype of the placenta may affect the hormones which regulate mammary gland development and, eventually, the amount of milk produced (Thatcher *et al.*, 1979). Support for this comes from the early conclusion of the Austrian physician Halban (c. 1865), who stated that the hormones of the corpus luteum prepare the uterus for pregnancy while the hormones of the placenta prepare the mammary gland for lactation.

It is also possible that "uterine milk" or uterine secretion of the sheep and cow may require placental hormones in addition to those of the ovary for full secretory activity. If so, similar hormones may affect both the uterus and the mammary gland to provide a continuing source of nutrients for the prenatal and postnatal development of the conceptus.

S.P. KALRA: Füxe and co-workers localized prolactin in the median eminence of the rat a couple of years ago. I was wondering whether you have similar observations on prolactin localization in the brain?

NOLIN: No, I'm sorry, we have no information on prolactin in the brain.

LEMAY: Since milk is excreted as a hemiocytosis process, have you looked at, or is it possible to look at, prolactin in excreted milk?

NOLIN: Yes, prolactin is detectable in milk, in alveolar milk with immunohistochemistry and in milk excreted from the gland by radioimmunoassay.

CHANNING: I thought I ought to mention that we found that addition of prolactin did not inhibit maturation of isolated cumulus-enclosed porcine oocytes.

NOLIN: Maybe that adds substance to the idea that granulosa cells might turn circulating prolactin into target-cell prolactin. Perhaps without more (or different) granulosa cells than the cumulus cells, this does not occur and any effects "prolactin" might have remain invisible.

CHANNING: We also found, in the pig, that as the follicle grows, there is a decrease in granulosa-cell content of oocyte-maturation inhibitor. Did you observe changes in granulosa-cell content as a function of follicular maturation?

NOLIN: We have not done a careful cell count to assess how many cells in a given follicle contain prolactin at any given time, nor is it possible for us, with immunohistochemistry, to define, except in a relative sense, how much prolactin might be present in single cells. However, we do observe considerable cell–cell variation in the presence (and distribution) of prolactin, particularly among cells of larger follicles. Whether this is simply due to greater functional asynchrony among cells as the follicle matures or to some other factor, I cannot answer at the present time.

BAZER: Dr. Nolin, please review the comments you made on translocation of prolactin as affected by estrogen.

NOLIN: Well, perhaps I can put it in a nutshell. I'll try. In a dose-dependent fashion, estrogen seems to affect the various components of the prolactin-incorporation process in the milk-secretory cell selectively. At the highest dose we tested, the cells could no longer even recognize prolactin, although it was plentiful in the circulation. Without recognition, of course, there was no internalization, i.e., no movement into the basal portion of the cell. Lower doses allowed recognition and internalization, but affected polar translocation, nuclear uptake, and/or transcellular transport. The extent of interruptions in the prolactin-incorporation process was directly related to the degree of lactational failure. The most sensitive step in the prolactin-incorporation process appeared to be nuclear uptake, and failure of this part of the process was accompanied by the onset of failure to respond to prolactin, i.e., by the onset of lactational failure.

BAZER: Is the prolactin-incorporation process influenced by lysosomes?

NOLIN: We do not know yet. Lysosomes may be involved in some target cells but not others. For example, in rat lutein cells, progesterone secretion appears to depend upon an estrogen–prolactin synergism (Nolin and Bogdanove, 1980). Estrogen is a destabilizer of lysosomal membranes. Prolactin has the opposite effect. Perhaps, in lutein cells, some sort of balance is struck as part of the required synergism. However, in milk-secretory cells, with their dependence on glucocorticoid and prolactin, I think it is possible that lysosomes might not be involved in hormone action, since both these hormones are lysosome stabilizers.

PIETRAS: Recent findings by Le Cam *et al.* (1979) indicate that certain primary amines can block the cellular uptake of peptide hormones without inhibiting their early actions in target cells. Could you comment on the significance of the latter findings with respect to your observations?

NOLIN: Most hormones are capable of influencing their targets in several different ways. Very rapid response of a target cell might be induced very soon after initial hormone–cell contact. I think we would not be surprised if that were the case. Uptake and retention may be required for other responses. Our estrogen studies are provocative, I think, in that they demonstrate that disruption of the nuclear uptake of prolactin and the beginning of lactional failure go hand in hand. This would suggest that nuclear uptake of prolactin (and therefore internalization and polar transport into the cell cytoplasm) might be a necessary component of the mechanism of action of prolactin, presumably on the transcriptional events which are required for the secretory response of the milk-secretory cell (MSC) to this hormone. This, of course, does not rule out the possibility of another effect of prolactin on MSC that might occur on initial contact between the two.

REFERENCES

Astwood, E.B., 1941, The regulation of corpus luteum function by hypophysial luteotrophin, *Endocrinology* **28:**309.

Baker, B.L., 1970, Studies on hormone localization with emphasis on the hypophysis, *J. Histochem. Cytochem.* **18:**1.

Baker, T.G., and Hunter, R.H.F., 1978, Interrelationships between the oocyte and somatic cells within the Graafian follicle of mammals. *Ann. Biol. Anim. Biochem. Biophys.* **18:**419.

Channing, C.P., 1979, Intraovarian inhibitors of follicular function, in: *Ovarian Follicular Development and Function* (A.R. Midgley and W.A. Sadler, eds.), pp. 59–64, Raven Press, New York.

Crisp T.M., 1979, Studies on the binding of labelled prolactin (PRL) to rat granulosa cell cultures, *Anat. Rec.* **193:**512 (abstract).

Davis, J.O., 1975, Regulation of aldosterone secretion, in: *Handbook of Physiology,* Section 7, *Endocrinology,* Vol. VI, *Adrenal Gland* (R.O. Greep and E.B. Astwood, section eds.), pp. 77–106, American Physiological Society, Washington, D.C.

Dornhorst, A., and Gann, D.S., 1978, Immunoperoxidase stains cortisol in adrenal and pituitary, *J. Histochem. Cytochem.* **26:**909.

Grosvenor, C.E., Mena, F., Dhariwal, A.P.S., and McCann, S.M., 1967, Reduction of milk secretion by prolactin-inhibiting factor: Further evidence that exteroceptive stimuli can release pituitary prolactin in rats, *Endocrinology* **81:**1021.

Haynes, R.C., Jr., 1975, Theories on the mode of action of ACTH in stimulating secretory activity of the adrenal cortex, in: *Handbook of Physiology,* Section 7, *Endocrinology,* Vol. VI, *Adrenal Gland* (R.O. Greep and E.B. Astwood, section eds.), pp. 69–76, American Physiological Society, Washington, D.C.

Hutson, J.C., Gardner, P.J., and Childs, M.G., 1977, Immunocytochemical localization of a follicle-stimulating hormone-like molecule in the testis, *J. Histochem. Cytochem.* **25:**1119.

Le Cam, A., Maxfield, F., Willingham, M., and Pastan, I., 1979, Insulin stimulation of amino acid transport in isolated rat hepatocytes is independent of hormone internalization, *Biochem. Biophys. Res. Commun.* **88:**873.

Lefkowitz, R.J., Roth, J., Pricer, W., and Pastan, I., 1970, ACTH receptors in the adrenal: Specific binding of ACTH-^{125}I and its relation to adenyl cyclase, *Proc. Natl. Acad. Sci. U.S.A.* **65:**745.

McKerns, K.W., 1978, Regulation of gene expression in the nucleus by gonadotropins, in: *Structure and Function of the Gonadotropins* (K.W. McKerns, ed.), pp. 315–338, Plenum Press, New York.

McNatty, K.P., Hunter, W.M., McNeilly, A.S., and Sawers, R.S., 1975, Changes in the concentration of pituitary and steroid hormones in the follicular fluid of human Graafian follicles throughout the menstrual cycle, *J. Endocrinol.* **64:**555.

McNatty, K.P., McNeilly, A.S., and Sawers, R.S., 1977, Prolactin and progesterone secretion by human granulosa cells *in vitro,* in: *Prolactin and Human Reproduction* (P.G. Crosignani and C. Robyn, eds.), pp. 109–117, Academic Press, New York.

Midgley, A.R., Jr., 1972, Gonadotropin binding to frozen sections of ovarian tissue, in: *Gonadotropins* (B.B. Saxena, C.G. Beling, and H.M. Gandy, eds.), pp. 248–260, John Wiley, New York.

Nakane, P.K., 1968, Simultaneous localization of multiple tissue antigens using the peroxidase-labeled antibody method: A study on pituitary glands of the rat, *J. Histochem. Cytochem.* **16:**557.

Nekola, M.V., and Nalbandov, A.V., 1971, Morphological changes of rat follicular cells as influenced by occytes, *Biol. Reprod.* **4:**154.

Nolin, J.M., 1978a, Intracellular prolactin in rat corpus luteum and adrenal cortex, *Endocrinology* **102:**402.

Nolin, J.M., 1978b, "Anticipant" receptors—a concept based on comparisons between endogenous PRL and free PRL-binding sites in individual milk secretory cells, *Fed. Proc. Fed. Am. Soc. Exp. Biol.* **37:**265 (Abstract 284).

Nolin, J.M., 1978c, Apparent incorporation of endogenous prolactin by dictyate mammalian oocytes, *J. Cell Biol.* **79:**177a (Abstract G1007).

Nolin, J.M., 1978d, Target cell prolactin, in: *Structure and Function of Gonadotropins* (K.W. McKerns, ed.), pp. 151–182, Plenum Press, New York.

Nolin, J.M., 1979a, Intracellular sites of action for endogenous ACTH?, *Proc. Endocr. Soc.,* p. 135 (Abstract 249).

Nolin, J.M., 1979b, The prolactin incorporation cycle of the milk secretory cell: An integral component of the prolactin response cycle, *J. Histochem. Cytochem.* **27:**1203.

Nolin, J.M., 1979c, Does PRL inhibit oocyte maturation? Undergo transcellular transport by granulosa cells (GC)? Act directly on GC nuclei?, *Biol. Reprod.* **20:**66a (Abstract 109).

Nolin, J.M., 1980a, Incorporation of endogenous prolactin by granulosa cells and dictyate oocytes in the postpartum rat: Effects of estrogen, *Biol. Reprod.* **22:**417.

Nolin, J.M., 1980b, Incorporation of regulatory peptide hormones by individual cells of the adrenal cortex: Prolactin-adrenocorticotrophin differences, *Peptides* **1:**249.

Nolin, J.M., and Bogdanove, E.M., 1980, Effects of estrogen on prolactin (PRL) incorporation by lutein and milk secretory cells and on pituitary PRL secretion in the postpartum rat: Correlations with target cell responsiveness to PRL, *Biol. Reprod.* **22:**393.

Nolin, J.M., and Witorsch, R.J., 1976, Detection of endogenous immunoreactive prolactin in rat mammary epithelial cells during lactation, *Endocrinology* **99:**949.

Ogle, T.F., and Kitay, J.I., 1979, Interactions of prolactin and adrenocorticotropin in the regulation of adrenocortical secretions in female rats, *Endocrinology* **104:**40.

Peaker, M., 1977, The aqueous phase of milk: Ion and water transport, in: *Comparative Aspects of Lactation* (M. Peaker, ed.), pp. 113–134, Academic Press, New York.

Rolland, R., and Hammond, J.M., 1975, Demonstration of a specific receptor for prolactin in porcine granulosa cells, *Endocr. Res. Commun.* **2**:281.

Rubin, R.P., and Laychock, S.G., 1978, Prostaglandins and calcium–membrane interactions in secretory glands, *Ann. N.Y. Acad. Sci.* **307**:377.

Saacke, R.G., and Heald, C.W., 1974, Cytological aspects of milk formation and secretion, in: *Lactation II* (B.L. Larson and V.R. Smith, eds.), pp. 147–189, Academic Press, New York.

Schuetz, A.W., 1974, Role of hormones in oocyte maturation, *Biol. Reprod.* **10**:150.

Sternberger, L.A., Petrali, J.P., Joseph, S.A., Meyer, H.G., and Mills, K.R., 1978, Specificity of the immunocytochemical luteinizing hormone-releasing hormone receptor reaction, *Endocrinology* **102**:63.

Szego, C.M., Rakich, D.R., Secler, B.J., and Gross, R.S., 1974, Lysosomal labilization: Rapid, target-specific effect of ACTH, *Endocrinology* **95**:863.

Szöllösi, D., 1972, Changes of some cell organelles during oogenesis in mammals, in: *Oogenesis* (J.D. Biggers and A.W. Schuetz, eds.), pp. 47–64, University Park Press, Baltimore.

Thatcher, W.W., Wilcox, C.J., Bazer, F.W., Collier, R.J., Eley, R.M., Stover, D.G., and Bartol, F.F., 1979, Bovine conceptus effects prepartum and potential carryover effects postpartum, Proceedings of the Beltsville Symposium on Animal Reproduction, May 14–17, 1978, Beltsville, Maryland.

Topper, Y.J., 1970, Multiple hormone interactions in the development of mammary gland *in vitro*, in: *Recent Progress in Hormone Research* (E.B. Astwood, ed.), pp. 287–308, Academic Press, New York.

Veldhuis, J.D., Klase, P.A., and Hammond, J.M., 1979, Prolactin induction or inhibition of steroidogenesis in cultured porcine granulosa cells, *Biol. Reprod.* **20**:70A (Abstract 118).

10

Gonadotropin-like Immunoreactivity in the Rat Ovary

Ultrastructural Localization in the Granulosa, Theca, and Lutein Cells

Paul Ordronneau and Peter Petrusz

1. Introduction

To devise efficient and safe methods of contraception, it is necessary to gain a firm grasp of the biological processes of reproduction. In the female, the gonadotropic hormones have a central role in regulating the reproductive cycles. Thus, it is of prime importance to understand the effect that these hormones have on their target cells. Labeled gonadotropins have been used in autoradiographic studies of the ovary. The majority of these reports have focused on the cells of the corpus luteum (Espeland *et al.,* 1968; Midgley and Beals, 1971; Ryan and Lee, 1976; Abel *et al.,* 1976; Chen *et al.,* 1978; Conn *et al.,* 1978). Fewer electron-microscopic autoradiographic studies have been reported on the cells of the stratum granulosum (Abel *et al.,* 1976; Han, 1979), and none on the theca interna. Since immunocytochemistry (ICC) has been successfully used to demonstrate gonadotropin immunoreactivity in the male reproductive tract (Hutson *et al.,* 1977; Childs *et al.,* 1978), it was decided to employ this technique at the ultrastructural level to localize immuno-

Paul Ordronneau and *Peter Petrusz* • Department of Anatomy, University of North Carolina, Chapel Hill, North Carolina 27514.

reactive gonadotropins in the rat ovary. Three predominant cell types of the ovary, namely, the cells of the corpus luteum, the stratum granulosum, and the theca interna, were studied.

2. Materials and Methods

2.1. Tissue Preparation

Normal cycling female Sprague–Dawley rats (ARS, Madison, Wisconsin) were used. The ovaries were fixed by vascular perfusion of 4% paraformaldehyde–0.5% glutaraldehyde as described elsewhere (Ordronneau and Petrusz, 1980). Pituitary and ovarian tissues were embedded in glycol methacrylate according to the method of Leduc and Bernhard (1968) with modifications as suggested by Spaur and Moriarty (1977). Tissue sections were placed on bare nickel grids.

2.2 Antisera

Rabbit antiserum to the β subunit of rat follicle-stimulating hormone (A-rFSHβ 856-2) was provided by the National Institute of Arthritis, Metabolism and Digestive Diseases (NIAMDD), National Institutes of Health, Bethesda, Maryland. Rabbit antiserum to human chorionic gonadotropin (A-hCG No. 229) was produced in this laboratory against a commercially available human chorionic gonadotropin preparation [(hCG) Organon, Inc., West Orange, New Jersey] by the multiple intradermal injection technique of Vaitukaitis *et al.* (1971). Sheep anti-rabbit gamma-globulin (sARGG) was obtained from Antibodies Incorporated (Davis, California). The rabbit peroxidase–antiperoxidase (PAP) was obtained from Miles Laboratories, Inc. (Elkhart, Indiana).

2.3. Immunoperoxidase Staining

The electron-microscopic ICC procedure was based on the method of Moriarty *et al.* (1973). Amplification was achieved by repeating the sARGG and PAP incubations (see Fig. 1) as suggested by Vacca *et al.* (1975). Briefly, the following incubations are required:

1. The primary antiserum for 46–50 hr at 4°C.
2. sARGG (1:100) for 10 min at room temperature (r.t.).
3. PAP (1:100) for 2–3 min at r.t.
4. sARGG (1:100) for 10 min at r.t.

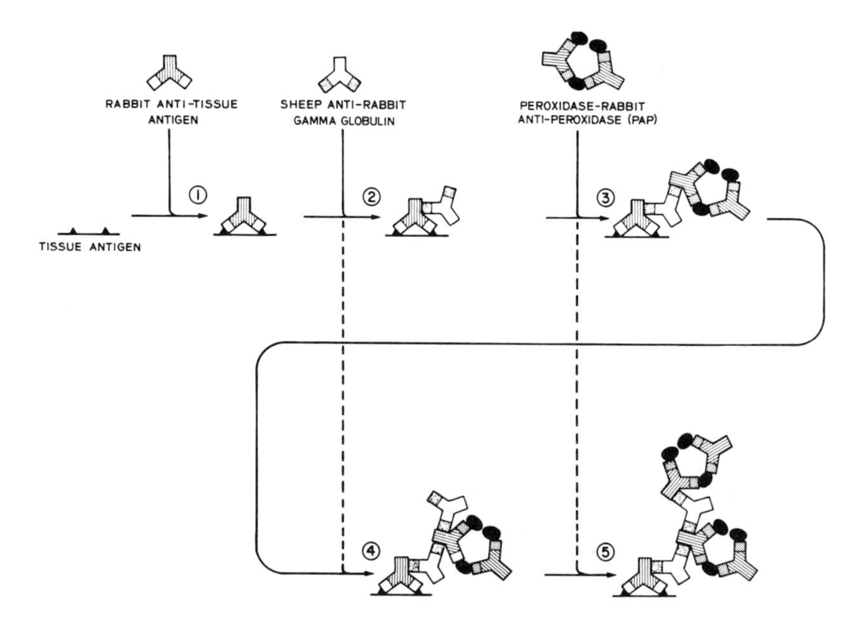

Figure 1. Schematic representation of the immunoperoxidase technique used. See Section 2.3 for details.

5. PAP (1:100) for 2–3 min at r.t.
6. 4-Chloro-1-naphthol (ICN-K & K Laboratories, Plainview, New York), prepared as described by Li *et al.* (1977).
7. 2% Osmium tetroxide for 15 min.

The grids were examined in a JEM 100 B (JEOL Inc., Medford, Massachusetts) electron microscope.

2.4. *Specificity Controls*

The specificity of both the method and the primary antisera was assessed on both pituitaries and ovaries from normal female rats.

The method specificity was checked in two ways:

1. The intensity of staining was evaluated as the dilution of the primary antiserum was increased. At least one dilution was included that gave no staining.
2. The primary antiserum was omitted from the staining sequence.

The hormonal specificity of the primary antisera was assessed by the following criteria:

1. Whether or not consistent staining of morphologically distinct cell types in the rat pituitary was obtained with a given antiserum (for a characterization of the cell types, see Moriarty, 1973).
2. Evaluation of staining obtained after absorption of the primary antisera with increasing amounts of the appropriate antigen. The antigens for absorption were kindly provided by the NIAMDD, and were highly purified (iodination-grade) hormone preparations normally distributed for radioimmunoassay. The absorption procedure involved incubating a given amount of the antigen with the appropriate dilution of the antibody for 7 days at 4°C with occasional shaking. The preparations were then centrifuged at 5000 rpm for 20 min and the supernatants used for immunostaining (Petrusz *et al.,* 1975).

The staining intensity was scored in the following way: 2, heavy staining with high background; 1, well-localized staining; 0, no staining.

3. Results

3.1. Specificity Controls

Tests for method specificity involved: (1) diluting the primary antiserum until staining was abolished and (2) omitting the primary antiserum. Both these assessments were performed on sections from the pituitary and the ovary.

For the pituitary, staining was abolished with both antisera by increasing the dilution. However, the endpoint for each was different. For the A-rFSHβ serum, staining was abolished at a dilution of 1:10,000. For the A-hCG serum, abolition of the staining occurred at 1:1,000,000 (Fig. 2).

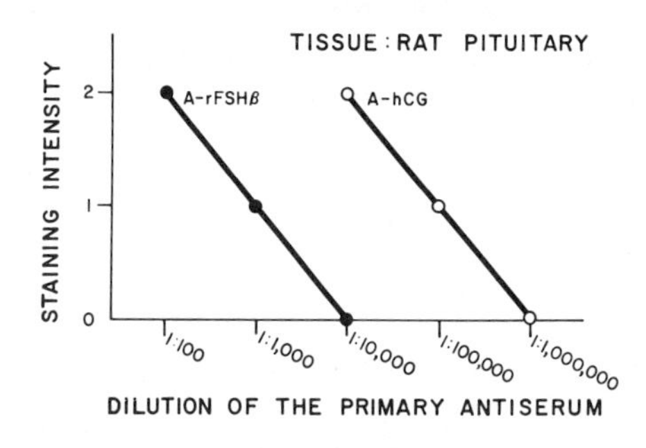

Figure 2. Control for method specificity: Results of immunoperoxidase staining in rat pituitary gonadotrophs as the dilutions of the primary antisera are increased. For scoring of staining intensity, see Section 2.4.

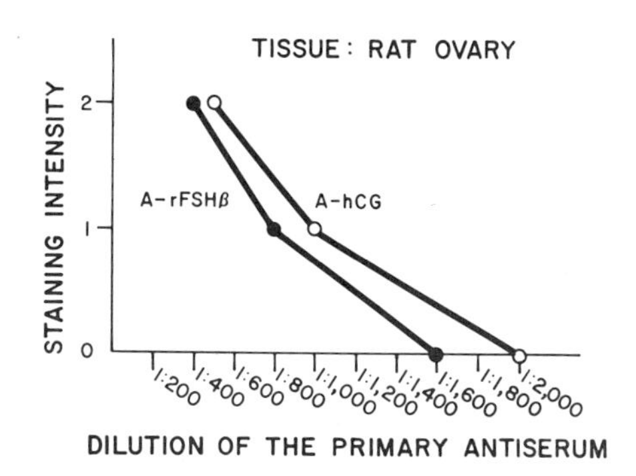

Figure 3. Control for method specificity: Results of immunoperoxidase staining in ovarian cells as the dilutions of the primary antisera are increased. For scoring of staining intensity, see Section 2.4.

In the ovary, with the A-rFSHβ serum, staining was abolished at a dilution of 1:1600. With the A-hCG serum, staining disappeared at 1:2000 (Fig. 3).

Routinely, as part of the staining procedure, the primary antiserum was omitted on one grid. If any positive results were obtained on the control grid, the staining was discarded. With very few exceptions, the control grids were negative.

It can be concluded from these method-specificity tests that any staining seen is due to the primary antisera and not to any of the other reagents in the staining procedure. Endogenous peroxidase was never encountered in either the pituitary or the ovary. Apparently, the procedures employed in the preparation of the tissues destroy these activities.

The specificity of the primary antisera was reflected in: (1) the staining of distinct cell types in the pituitary and (2) the absorption of each antiserum with its homologous and heterologous antigens.

There are three known types of gonadotrophs in the female rat pituitary (for review, see Moriarty, 1976). However, two of these cell types [Type I and Type II (Kurosumi and Oota, 1968)] comprise the bulk of the gonadotroph population. Since both luteinizing hormone (LH) and FSH are contained in both these cell types (Moriarty, 1976), it was sufficient to establish that the antisera stained only one population of cells known as gonadotrophs. Indeed, both antisera stained only gonadotrophs (Fig. 4).

Results of the absorption of the A-rFSHβ serum in both pituitary and ovary are shown in Fig. 5. In both tissues, absorption of the antiserum with increasing amounts of the β subunit of FSH resulted in abolishment of the staining. However, when heterologous antigens were used [LHβ or thyroid-stimulating hormone (TSH)β], the staining was unaffected. Similarly, if the A-hCG serum was absorbed with increasing

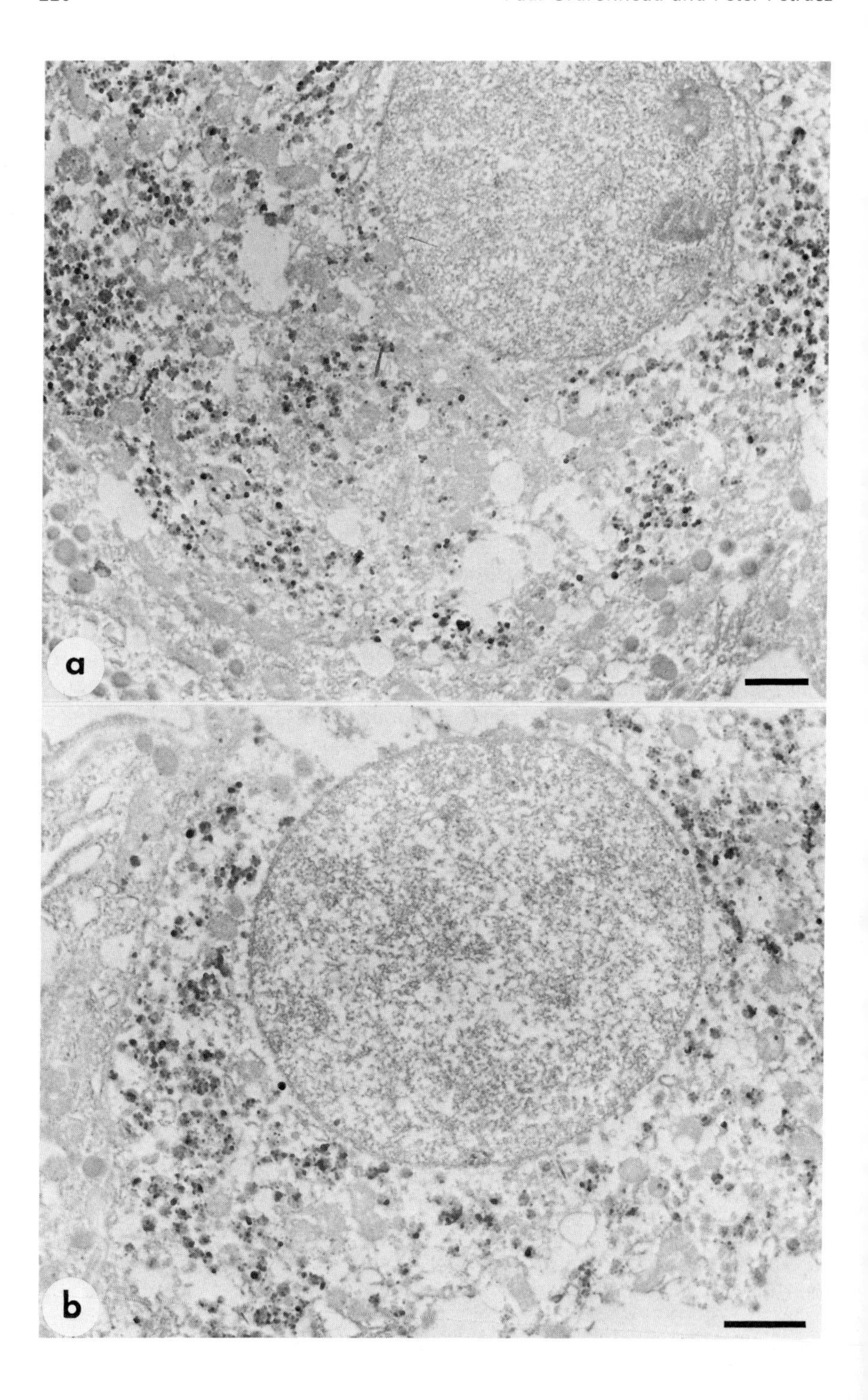
a
b

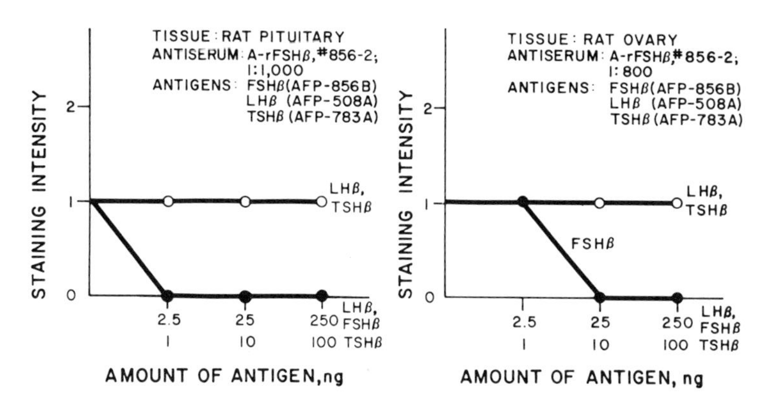

Figure 5. Control for antibody specificity: Results of immunoperoxidase staining in pituitary and ovary after absorption of the A-rFSHβ serum with various antigens. Only absorption of the antiserum with FSHβ reduces staining intensity. For details of the absorption procedure and scoring of staining intensity, see Section 2.4.

amounts of LHβ, the staining was abolished. Absorption of this antiserum with either FSHβ or TSHβ had no effect on immunostaining (Fig. 6).

These results indicate that there is a degree of specificity of each antiserum for its hormone. However, no attempt is made to distinguish between LH and FSH. There are several reasons for this. First, there is a good deal of homology in the primary sequence between the gonadotropins (Pierce, 1976). Since the binding site for which a given antibody is specific represents only a small portion of the amino acid sequence [6–7 amino acids (Atassi, 1975)], antibodies raised against one of the gonadotropins might be expected to cross-react with the other. Second, the staining in the pituitary demonstrates that the antisera are specific for the gonadotropins and no other adenohypophyseal hormone. However, such staining does not allow for the discrimination between the two hormones, since both are present within the same cell (Moriarty, 1976). Finally, although hCG has primarily a luteinizing effect on the ovary, it is known to contain intrinsic FSH activity in both biological and immunological terms (Robyn *et al.*, 1969; Louvet *et al.*, 1976; Siris *et al.*, 1977). For these reasons, it is impossible, with our present tools, to make a distinction between LH and FSH. However, the antisera do appear to be specific for the gonadotropins.

←

Figure 4. Immunoperoxidase staining in pituitary gonadotrophs with (a) A-rFSHβ and (b) A-hCG as the primary antiserum. No other cell type in the pituitary was stained with either antiserum. Bars: 1 μm.

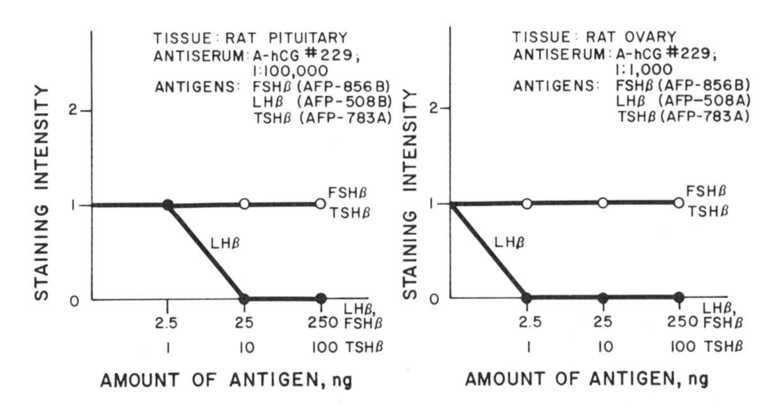

Figure 6. Control for antibody specificity: Results of immunoperoxidase staining in pituitary and ovary after absorption of the A-hCG serum with various antigens. Only absorption of the antiserum with LHβ reduces staining intensity. For details of the absorption procedure and scoring of staining intensity, see Section 2.4.

3.2. Gonadotropin-like Immunoreactivity in the Normal Adult Ovary

Two compartments of the ovary were investigated in this study: the follicle and the corpus luteum. The A-rFSHβ serum was employed at working dilutions of 1:800 and 1:1000, while the A-hCG serum was used at a dilution of 1:1000. The localizations described below were the same with both antisera. Preantral follicles [Type 5b and Type 6 (Pedersen and Peters, 1968)] were investigated. Two cell types of the follicle were examined, the granulosa cell and the cells of the theca interna. Staining was observed to some degree in all the granulosa cells. In general, the same subcellular structures were stained in all cells. These included the plasma membrane, the cytoplasm, presumptive mitochondria, the chromatin, and the nucleolus (Figs. 7 and 8).

In the theca interna, all the cells contained immunoreactive sites. These sites were found over the cytoplasm, presumptive mitochondria, the chromatin, and the nucleolus (Figs. 9 and 10).

In the second compartment, the corpus luteum, most of the cells were positively stained. However, an occasional unstained cell was seen. The reactive sites included the plasma membrane, the cytoplasm, presumptive lysosomes, presumptive mitochondria, the chromatin, and the nucleolus (Figs. 11 and 12). No immunoreactive sites were seen over the numerous lipid droplets of the cytoplasm.

All the staining sites mentioned above could be abolished by increasing the dilution of the primary antisera or by absorbing them with their homologous antigens.

4. Discussion

A similar general pattern of gonadotropin-like immunoreactivity is seen in all three cell types studied. The granulosa, theca, and lutein cells all have immunoreactive sites over their cytoplasm, mitochondria, chromatin, and nucleolus. Occasionally, sites are seen over the plasma membrane. Since the granulosa cells were more intensively studied than the other two cell types, the chance of finding staining over the plasma membrane was greater. The most obvious examples of lysosomal staining are seen in the lutein cells, where secondary lysosomes are more abundant.

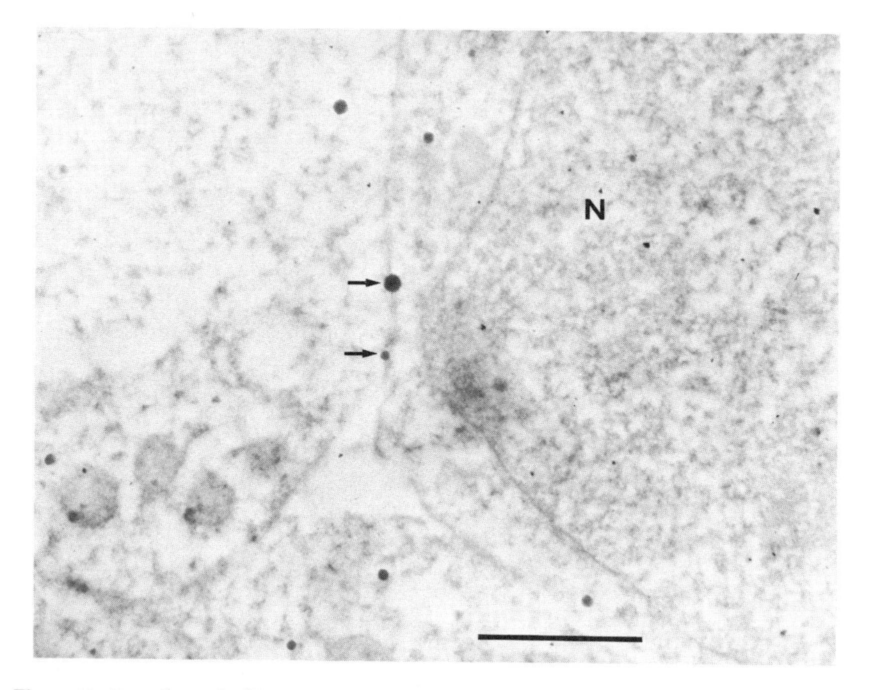

Figure 7. Gonadotropin-like immunoreactivity in granulosa cells of the adult rat. Immunoreactive sites are seen over the plasma membrane (arrows). (N) Nucleus of one granulosa cell. Bar: 1 μm.

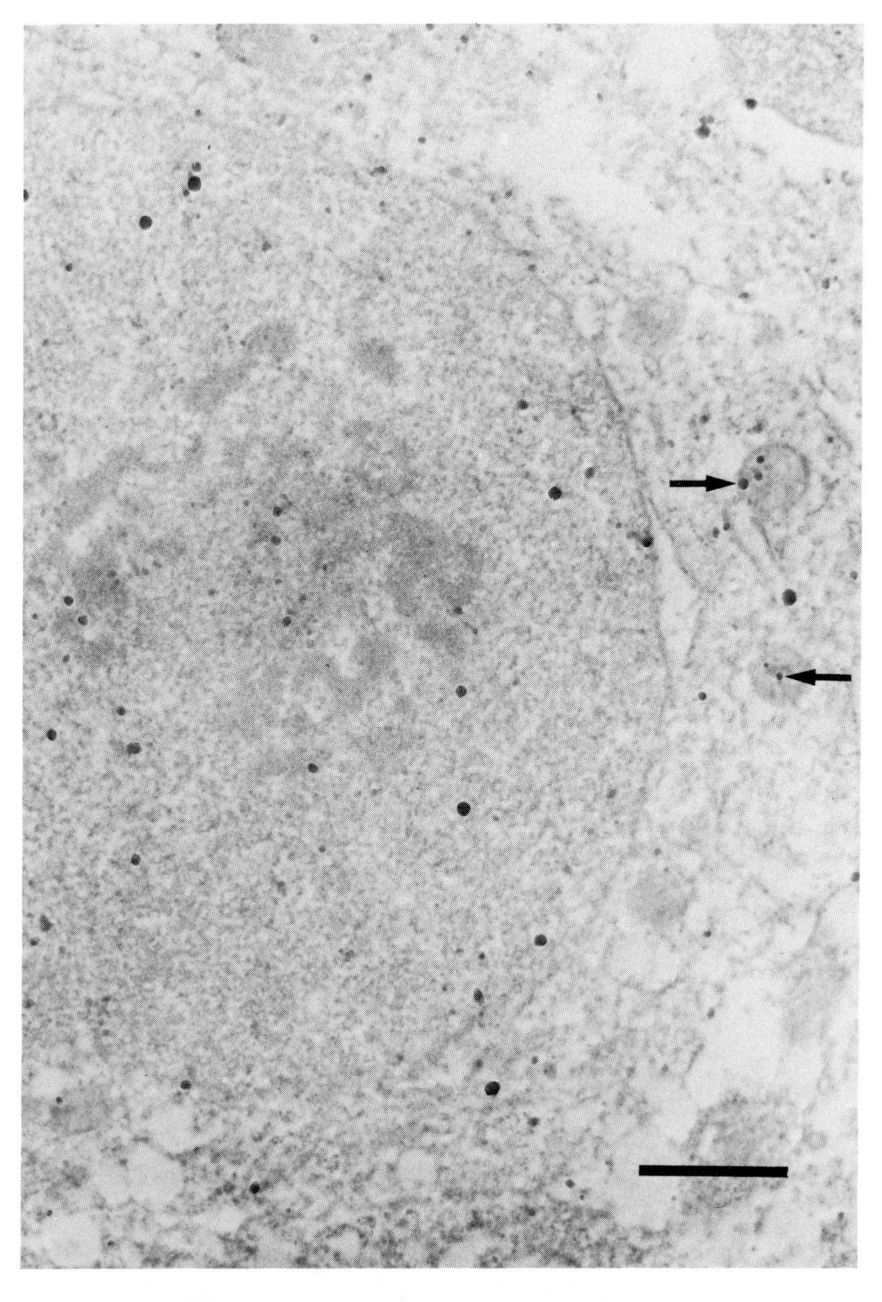

Figure 8. Gonadotropin-like immunoreactivity in granulosa cells of the adult rat. Immunoreactive sites are seen over the cytoplasm, mitochondria (arrows), chromatin, and nucleolus. Bar: 1 μm.

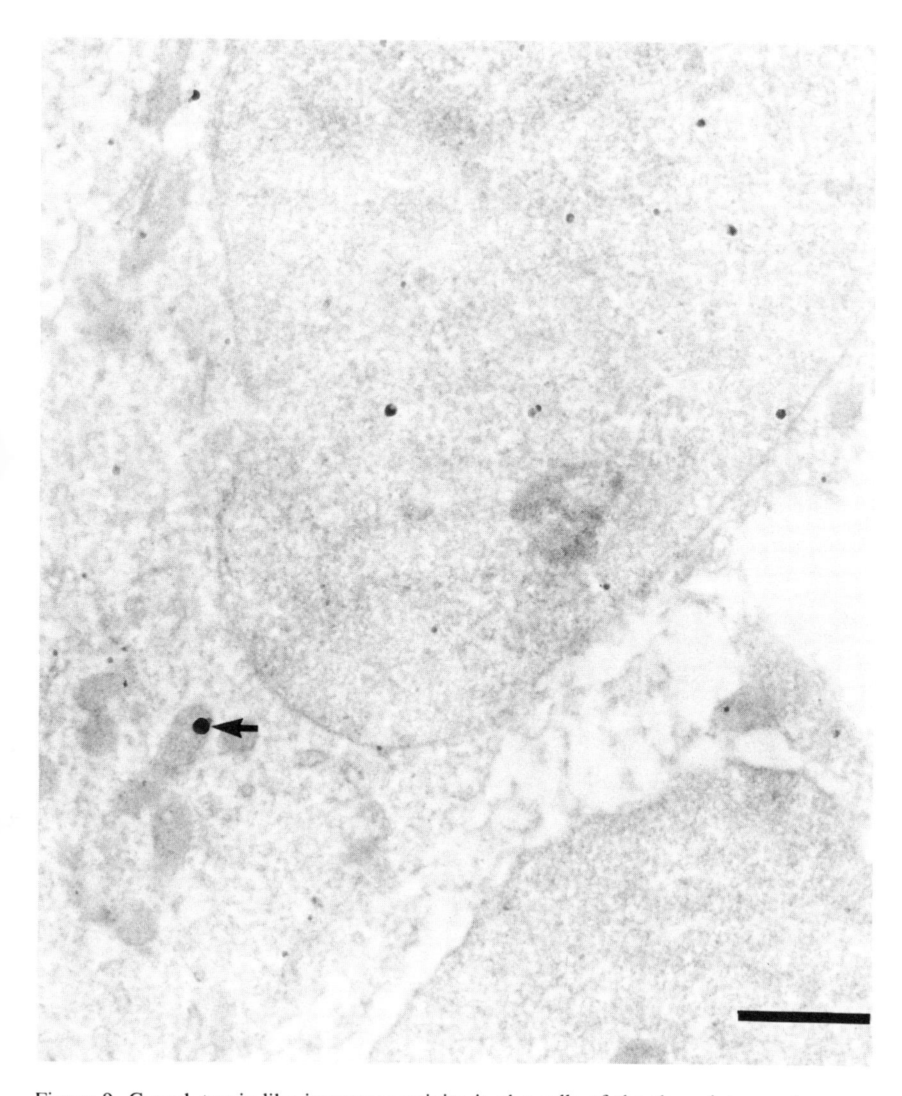

Figure 9. Gonadotropin-like immunoreactivity in the cells of the theca interna. Immunoreactive sites are seen over the cytoplasm, mitochondria (arrow), and chromatin. Bar: 1 μm.

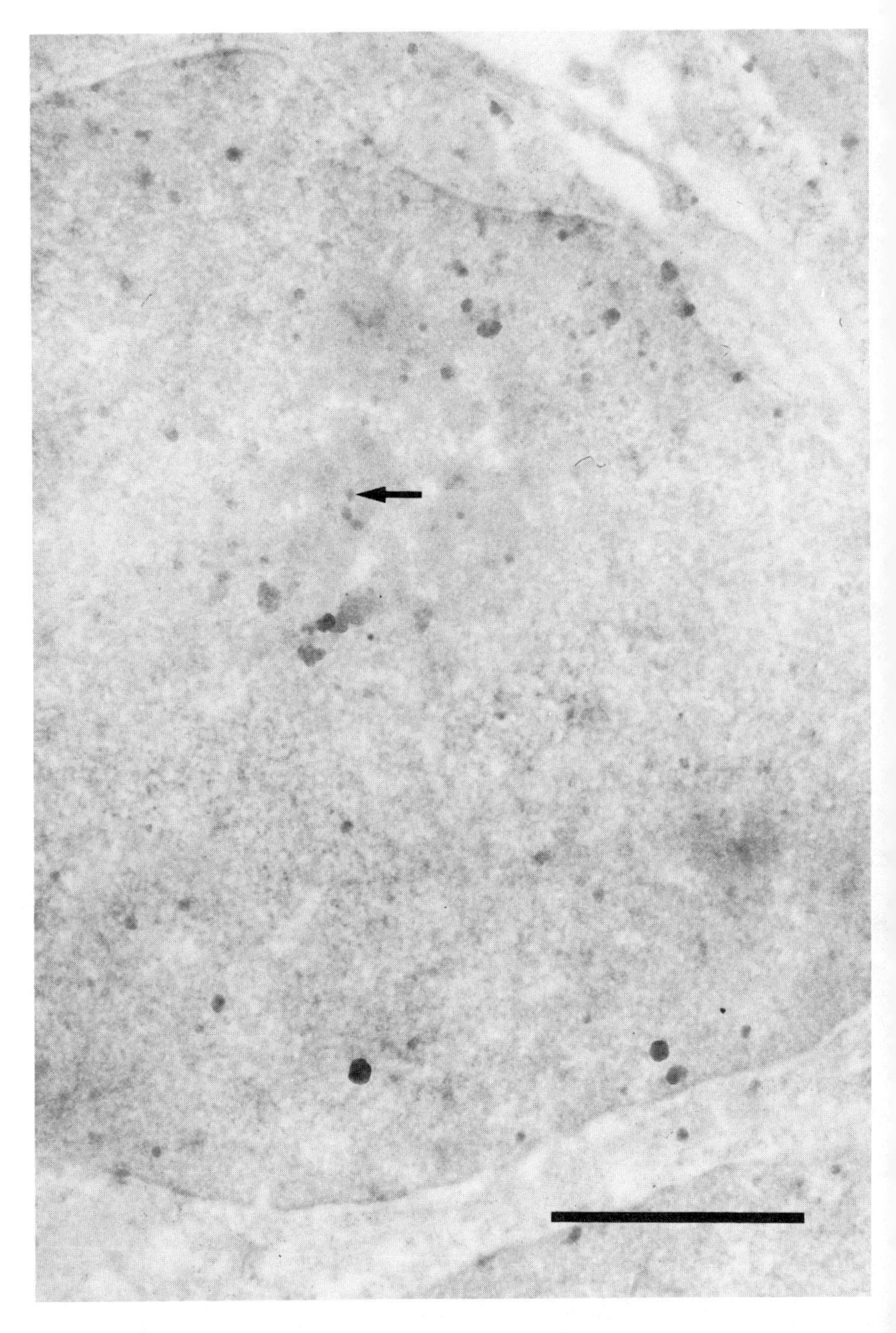

Figure 10. Gonadotropin-like immunoreactivity in the cells of the theca interna. Immunoreactive sites are seen over the chromatin and nucleolus (arrow). Bar: 1 μm.

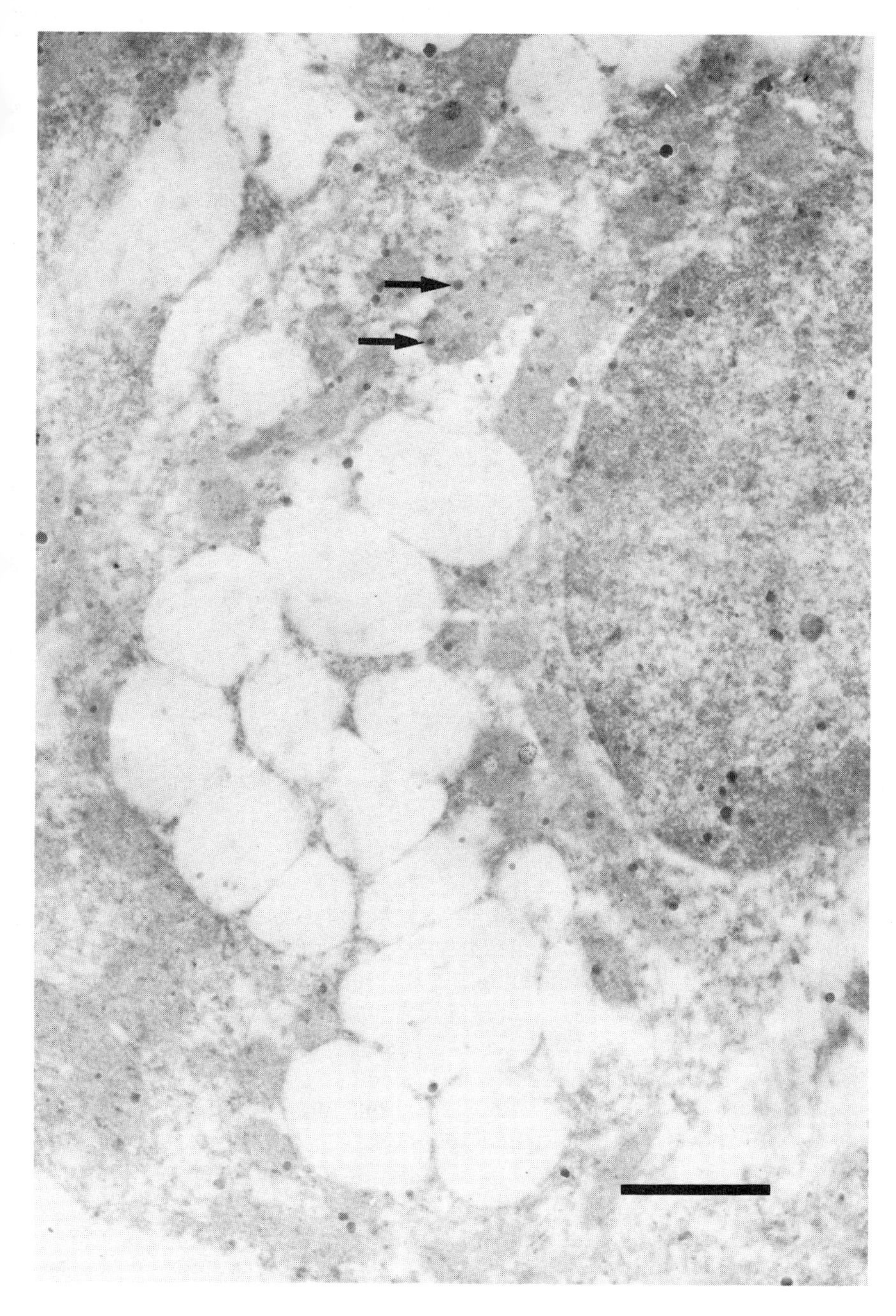

Figure 11. Gonadotropin-like immunoreactivity in lutein cells of the adult rat. Immunoreactive sites are seen over the cytoplasm, mitochondria (arrows), chromatin, and nucleolus. Bar: 1 μm.

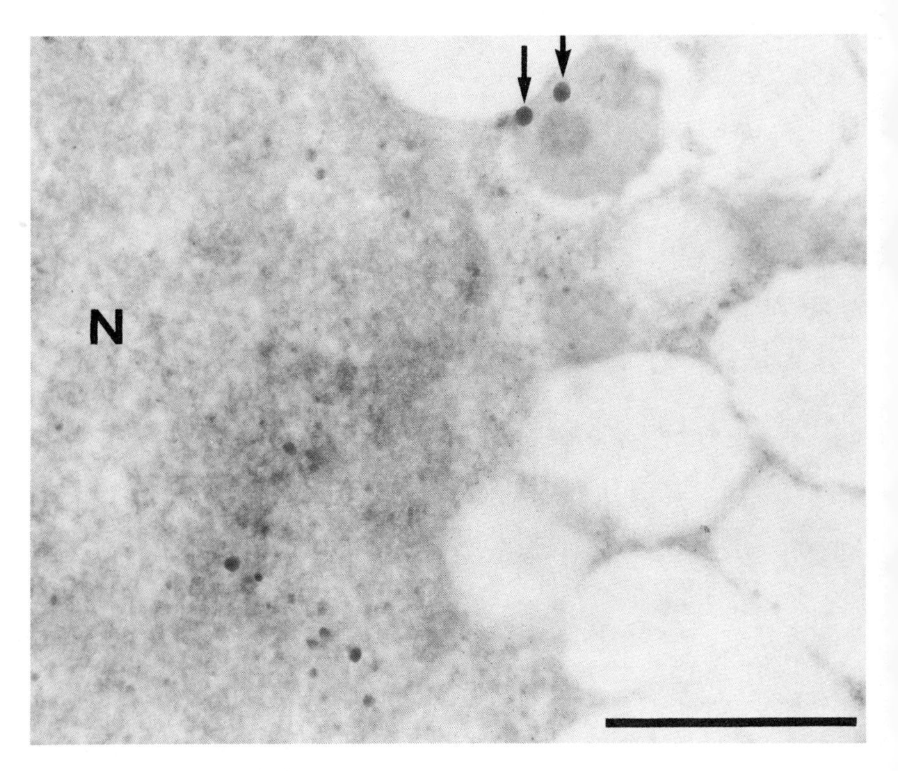

Figure 12. Gonadotropin-like immunoreactivity in lutein cells of the adult rat. Immunoreactive sites are seen over a lysosome (arrows). (N) Nucleus of the lutein cell. Bar: 1 μm.

In 1978, Petrusz and Sar proposed a model for the fate of gonadotropins within their target cells. It was suggested at that time that the gonadotropins bind to the membrane receptors and are internalized via encytosis. The endocytic vesicle would then fuse with primary lysosomes to form secondary lysosomes. The whole hormone or a fragment of it would then be released from the lysosome to interact with various intracellular compartments. The localizations described here support this model. Furthermore, immunoreactive sites are seen over metabolically active structures (mitochrondria, chromatin, and nucleolus). This suggests that the lysosomal pathway is not simply one of degradation.

There are mechanisms, well founded in cellular biology, that can account for the internalization of protein hormones. Substances that enter the cell by such mechanisms may do so to be inactivated or to participate in intracellular processes. This study defines some of the subcellular sites that should be investigated as possible locations of inactivation or action of gonadotropins.

ACKNOWLEDGMENTS. The authors wish to thank Ms. Marion Blackburn for the artwork. This research was supported in part by grants from the U.S.P.H.S. (HD 10306) and the Population Council (M74-116).

REFERENCES

Abel, J.H., McClellan, M.C., Chen, T.T., Sawyer, H.R., and Niswender, G.D., 1976, Subcellular compartmentalization of ovarian granulosa–luteal cell function, *J. Cell Biol.* **70:**367a.

Atassi, M.Z., 1975, Antigenic structure of myoglobin: The complete immunochemical anatomy of a protein and conclusions relating to antigenic structures of proteins, *Immunochemistry* **6:**43.

Chen, T.T., McClellan, M.C., Diekman, M.A., Abel, J.H., Jr., and Niswender, G.D., 1978, Localization of human chorionic gonadotropin in lysosomes of ovine luteal cells, in: *Structure and Function of the Gonadotropins* (K.W. McKerns, ed.), pp. 591–612, Plenum Press, New York.

Childs, G.V., Hon, C., Russell, L.R., and Gardner, P.J., 1978, Subcellular localization of gonadotropins and testosterone in the developing fetal rat testis, *J. Histochem. Cytochem.* **26:**545.

Conn, P.M., Conti, M., Harwood, J.P., Dufau, M.L., and Catt, K.J., 1978, Internalization of gonadotropin-receptor complex in ovarian luteal cells, *Nature (London)* **274:**598.

Espeland, D.H., Naftolin, F., and Paulsen, C.A., 1968, Metabolism of labeled ^{125}I-HCG by the rat ovary, in: *Gonadrotropins 1968* (E. Rosemberg, ed.), pp. 177–184, Geron-X, Los Altos, California.

Han, S.S., 1979, Preliminary observations on time-dependent intrafollicular distribution of ^{125}I-HCG, in: *Ovarian Follicular Development and Function* (A.R. Midgley and W.A. Sadler, eds.), pp. 107–111, Raven Press, New York.

Hutson, J.C., Gardner, P.L., and Moriarty, G.C., 1977, Immunocytochemical localization of FSH-like molecule in the testis, *J. Histochem. Cytochem.* **25:**119.

Kurosumi, K., and Oota, Y., 1968, Electron microscopy of the two types of gonadotrophs in the anterior pituitary glands of persistent estrous and diestrous rats, *Z. Zellforsch. Mirosk. Anat.* **85:**34.

Leduc, E.H., and Bernhard, W., 1968, Recent modification of the glycol methacrylate embedding procedure, *J. Ultrastruct. Res.* **19:**196.

Li, J.Y., Dubois, M.P., and Dubois, P.M., 1977, Somatotrophs in the human fetal anterior pituitary: An electron microscopic–immunocytochemical study, *Cell Tissue Res.* **181:**545.

Louvet, J.-P., Harman, S.M., Nisula, B.C., Ross, G.T., Birken, S., and Canfield, T., 1976, Follicle stimulating activity of human chorionic gonadotropin: Effect of dissociation and recombination of subunits, *Endocrinology* **99:**1126.

Midgley, A.R., Jr., and Beals, T.F., 1971, Analysis of hormones in tissues, in: *Principles of Competitive Protein Binding Assays* (W.D. Odell and W.H. Daughaday, eds.), pp. 339–350, Lippincott, Philadelphia.

Moriarty, G.C., 1973, Adenohypophysis: Ultrastructural cytochemistry, a review, *J. Histochem. Cytochem.* **21:**855.

Moriarty, G.C., 1976, Ultrastructural–immunocytochemical studies of rat pituitary gonadotrops in cycling female rats, *Gunma Symp. Endocrinol.* **13:**207.

Moriarty, G.C., Moriarty, C.M., and Sternberger, L.A., 1973, Ultrastructural immunocytochemistry with unlabeled antibodies and the peroxidase–antiperoxidase complex: A technique more sensitive than radioimmunoassay, *J. Histochem. Cytochem.* **21:**825.

Ordronneau, P., and Petrusz, P., 1980, Immunocytochemical demonstration of anterior pituitary hormones in the pars tuberalis of long-term hypophysectomized rats, *Amer. J. Anat.* **158:**491.

Pedersen, T., and Peters, H., 1968, Proposal for a classification of oocytes and follicles in the mouse ovary, *J. Reprod. Fertil.* **13:**555.

Petrusz, P., and Sar, M., 1978, Light microscopic localization of gonadotropin binding sites in ovarian target cells, in: *Cell Membrane Receptors for Drugs and Hormones* (R.W. Straub and L. Bolis, eds.), pp. 167–182, Raven Press, New York.

Petrusz, P., DiMeo, P., Ordronneau, P., Weaver, C., and Keefer, D.A., 1975, Improved immunoglobulin–enzyme bridge method for light microscopic demonstration of hormone-containing cells of the rat adenohypophysis, *Histochemistry* **46:**9.

Pierce, J.G., 1976, Structural homologies of glycoprotein hormones, in: *Endocrinology* (V.H.T. James, ed.), pp. 99–103, Exerpta Medica, Amsterdam.

Robyn, C., Petrusz, P. and Diczfalusy, E., 1969, Follicle stimulating hormone-like activity in human chorionic gonadotropin preparations, *Acta Endocrinol.* **60:**137.

Ryan, R.J., and Lee, C.Y., 1976, The role of membrane bound receptor, *Biol. Reprod.* **14:**16.

Siris, E.S., Nisula, B.C., Catt, K.J., Horner, K., Birken, S., Canfield, R.E., and Ross, G.T., 1977, New evidence for intrinsic follicle stimulating hormone-like activity in human chorionic gonadotropin and luteinizing hormone, *Endocrinology* **102:**1356.

Spaur, R.G., and Moriarty, G.C., 1977, Improvements of glycol methacrylate. I. Its use as an embedding medium for electron microscopical studies. *J. Histochem. Cytochem.* **25:**163.

Vacca, L.L., Rosario, S.L., Zimmerman, E.A., Tomashefsky, P., Ng, P.-Y., and Hsu, K.C., 1975, Application of immunoperoxidase techniques to localize horseradish peroxidase-tracer in the central nervous system, *J. Histochem. Cytochem.* **23:**208.

Vaitukaitis, J.L., Robbins, J.B., Nieschlag, E., and Ross, G.T., 1971, A method for producing specific antisera with small doses of immunogen, *J. Clin. Endocrinol. Metab.* **33:**988.

IV

Uterine and Mammary Receptors

11

Hormonal Modulation of Progesterone Receptors

Judith Saffran and Bonnie K. Loeser

1. Introduction

According to current theories of hormone action, all the steroid hormones, including vitamin D, exert their effects in a similar way (Buller and O'Malley, 1976; Chan and O'Malley, 1976; DeLuca, 1979; Gorski and Gannon, 1976; Yamamoto and Alberts, 1976). There is general, but not universal, agreement that the hormone diffuses passively through the cell membrane (Baulieu, 1978). After entering the target cell, the hormone binds to a cytoplasmic receptor protein that is present only in target cells and that is fairly specific for the hormone. The binding is of high affinity, with equilibrium dissociation constants in the neighborhood of 10^{-9} M. Because there is a limited number of receptor molecules per cell, the binding of the hormone is saturable. The complex of hormone and receptor then undergoes a process of "activation" or "transformation" that enables it to enter the nucleus and bind to chromatin. This binding initiates the biochemical events that are characteristic of the hormone. There is increased transcription, resulting eventually in an increase in protein synthesis.

Specific proteins are synthesized under the influence of the hormone. Examples of specific proteins induced in the uterus by progesterone are uteroglobin in the rabbit (Beier, 1974; Rahman *et al.*, 1975), purple

Judith Saffran and *Bonnie K. Loeser* • Department of Obstetrics and Gynecology and Department of Biochemistry, Medical College of Ohio, Toledo, Ohio 43699.

protein in the pig (Basha *et al.*, 1979; Roberts and Bazer, 1976), and 17β-estradiol dehydrogenase in the human (Pollow *et al.*, 1978; Tseng, 1978). In addition, there is increased synthesis of nonspecific tissue proteins, so that the reproductive tract grows under the influence of sex hormones.

Because the effects of the hormone depend on the presence of receptors for the hormone, to understand how hormones exert their effects, it is necessary to know how the concentration of receptors is regulated. For example, evidence is accumulating that some endocrine disorders may be the result of receptor defects, rather than of hormonal insufficiency or excess. Disorders of peptide hormones have been studied more extensively than those of steroid hormones [insulin receptors (Flier *et al.*, 1975; Kahn *et al.*, 1976; Olefsky, 1976)], but some steroid-receptor defects are already known [androgen receptors (Bardin *et al.*, 1975; Kaufman *et al.*, 1976)] and others may exist (Oberfield *et al.*, 1979).

The regulation of ovarian steroid receptors can be divided into long-term and short-term regulation, and both are influenced by hormones (Brenner and West, 1975). Long-term regulation refers to changes in the concentration of receptors over the course of days, while short-term regulation is restricted to changes that take place within hours and minutes.

An example of long-term receptor regulation is the fluctuation of estrogen and progesterone receptors in the uterus and hypothalamus in response to changes in the secretion of these hormones by the ovary (Bayard *et al.*, 1978; Booth and Colas, 1977; Chen and Leavitt, 1979; Koligian and Stormshak, 1977a; Giannopoulos and Tulchinsky, 1979; Kielhorn and Hughes, 1977; Milgrom *et al.*, 1972; Rodriquez *et al.*, 1979; VuHai *et al.*, 1978; West *et al.*, 1977, 1978). Estrogens and progesterone act both synergistically with and antagonistically to one another. Progesterone usually acts on the uterus after it has been stimulated by estrogen, but progesterone also suppresses estrogenic effects on the uterus.

It is now known that estrogens increase the concentration of estrogen receptors in the uterus (Hsueh *et al.*, 1976) and also induce the synthesis of progesterone receptors (Evans *et al.*, 1978; Faber *et al.*, 1972; Feil *et al.*, 1972; Freifeld *et al.*, 1974; Illingworth *et al.*, 1977; Jänne *et al.*, 1975; Leavitt *et al.*, 1977; Luu Thi *et al.*, 1975; Milgrom *et al.*, 1973; Moguilewsky and Raynaud, 1977; Reel and Shih, 1975; Spona *et al.*, 1979). Progesterone has also been reported to induce cytoplasmic estrogen receptors of rat uterine endometrium (Martel and Psychoyos, 1978). This may account for the synergism.

Progesterone, on the other hand, has a generally suppressive action on receptors. It decreases the concentration of estrogen receptors in the reproductive tract, and thus it modulates estrogen action in the estrous and menstrual cycles (Coulson and Pavlik, 1977; Elsner *et al.*, 1977;

Hsueh *et al.*, 1975; Koligian and Stormshak, 1977b; West *et al.*, 1976, 1977, 1978). Progesterone also decreases the concentration of its own receptor in the uterus, sometimes to undetectable levels (Freifeld *et al.*, 1974; Isomaa *et al.*, 1979; Koseki *et al.*, 1977a; Leavitt *et al.*, 1974; Luu Thi *et al.*, 1975; Milgrom *et al.*, 1973; Vu Hai *et al.*, 1977; Walters and Clark, 1978).

Over short time spans, steroid hormones change the distribution of receptors between the cytoplasm and nucleus of target cells. The estrogen-receptor system of the uterus has been extensively studied and will be used as an example, but the results seem to be fairly generally applicable to other steroid receptors.

After 17β-estradiol is injected into a rat, it binds to the cytoplasmic estrogen receptor in the uterus, and the complex is translocated to the nucleus (Baudendistel *et al.*, 1978; Cidlowski and Muldoon, 1974; Hsueh *et al.*, 1976; Juliano and Stancel, 1976; Myatt *et al.*, 1978; Zava *et al.*, 1976). As a result, the concentration of cytoplasmic receptors decreases while the concentration of nuclear receptors increases. The decrease in cytoplasmic receptors causes the uterus to become, for a time, insensitive to the action of estrogen (Anderson *et al.*, 1974; Katzenellenbogen, 1975). But the concentration of estrogen receptors in the cytoplasm is gradually restored. From a minimal concentration 1–4 hr after the injection, the receptor level increases to normal after about 12 hr, and after 24 hr, it may be 50–100% greater than it was originally. Thus, sensitivity to estrogen returns. At the same time, the concentration of nuclear receptors returns to its original low level.

The replenishment of the cytoplasmic receptors may be caused by the release of "used" molecules from the nucleus, or by the synthesis of new receptor protein, or both (Baudendistel *et al.*, 1978; Cidlowski and Muldoon, 1974; Hsueh *et al.*, 1976; Měster and Baulieu, 1975; Sarff and Gorski, 1971; Zava *et al.*, 1976).

The fate of the hormone–receptor complex after it has acted in the nucleus is not known. It is not certain whether the hormone and receptor then part company, and if they do, whether this takes place in the nucleus or the cytoplasm. But it is becoming evident that what happens after the receptor has acted in the nucleus is important for hormone action, and if the receptor is not processed in the correct way, hormone action may slow down or cease (Horwitz and McGuire, 1978a,b).

This becomes more obvious when we see what happens after an antiestrogen is administered. Antiestrogens inhibit the actions of estrogens on target tissues, but some antiestrogens are weak estrogens themselves (Baudendistel *et al.*, 1978; Horwitz *et al.*, 1978). They bind to the cytoplasmic estrogen receptor and translocate it to the nucleus (Capony and Rochefort, 1978; Koseki *et al.*, 1977b). However, the processing of the receptor complex in the nucleus is defective and the receptor is retained in the nucleus for an abnormally long time—for days rather than for the normal time span required for estrogen action. It has been suggested that this prevents the replenishment of the cytoplasmic receptor and results in a depression of estrogen action (Baudendistel *et al.,* 1978; Clark *et al.,* 1973; Katzenellenbogen *et al.,* 1978, 1979; Koseki *et al.,* 1977b; Ruh and Baudendistel, 1978; Ruh *et al.,* 1979; Whalen *et al.,* 1975).

We must conclude that although it is not certain what happens to steroid receptors after they have exerted their effects, the normal flux of receptors between the cytoplasm and nucleus of target cells is important for hormone action.

Much less is known of the regulation of progesterone-receptor concentration, especially short-term regulation. In some respects, the progesterone receptor has unique properties: (1) Progesterone action requires prior stimulation by estrogen, probably because the progesterone receptor is induced by estrogen. In the absence of estrogen, the concentration of progesterone receptors is very low (Philibert and Raynaud, 1973, 1974). (2) Progesterone is rather rapidly metabolized in its target organs. The pathway of metabolism is very similar to that of testosterone in its target organs. The first step is the reduction of progesterone to dihydroprogesterone (5α-pregnane-3,20-dione) (Saffran *et al.,* 1974a,b; Tabei *et al.,* 1974; Verma and Laumas, 1976), and this is exactly analogous to the reduction of testosterone to dihydrotestosterone (17β-hydroxy-5α-androstane-3-one) in the male reproductive tract. Both reactions are controlled by hormones. The 5α-reductase that converts progesterone to dihydroprogesterone is stimulated by estrogen (Armstrong and King, 1971; Rahman *et al.,* 1978; Saffran *et al.,* 1974b), and that which reduces testosterone to dihydrotestosterone is stimulated by androgens (Moore and Wilson, 1973). However, it is not certain whether metabolism is important for progesterone action. It seems not to be important in the uterus (Rahman *et al.,* 1975; Sanyal and Villee, 1973), but it may have a role in brain, hypothalamus, and pituitary (Cheng and Karavolas, 1973; Karavolas *et al.,* 1976, 1979; Nuti and Karavolas, 1977; Tabei *et al.,* 1974). (3) Progesterone has a suppressive action on both estrogen and progesterone receptors. It suppresses its own receptor to very low levels, and this has made it difficult to study progesterone-receptor levels after progesterone administration.

Other technical problems have increased the difficulty of measuring progesterone receptors. The receptor is fairly unstable, and the progesterone–receptor complex dissociates rapidly. Progesterone binds nonspecifically to corticosteroid-binding globulin (CBG) and other proteins, and natural glucocorticoids (cortisol, corticosterone) compete with progesterone for binding. The problems have been partially overcome. Buffers have been devised that stablize the progesterone receptor. Progesterone receptors are more stable in phosphate buffer than in the commonly used tris buffer (Faber *et al.,* 1978). Glycerol and SH-protective agents also stabilize the receptor. In addition, there are available synthetic progestins that bind to the receptor with high affinity and that exhibit little binding to CBG. Several are available in tritium-labeled form for binding studies (Feil *et al.,* 1979; Isomaa *et al.,* 1979; Raynaud, 1977; Srivastava, 1978).

The objective of the research reported in this chapter was to study the short-term regulation of progesterone receptors in the uterus, i.e., the subcellular distribution of progesterone receptors in the uterus after the *in vivo* administration of progesterone to guinea pigs.

2. *Experimental Methods and Results*

The experimental animal was the ovariectomized guinea pig injected with estrogen to induce the synthesis of progesterone receptors. Any estrogen is satisfactory. It is convenient to give a single injection of a long-acting estrogen and to use the animals after 5–7 days. In our experiments,. 125 μg estradiol cypionate® (Upjohn) was administered subcutaneously. Batra *et al.* (1978) used one injection of poly-estradiol-phosphate to induce progesterone-receptor synthesis in rabbits.

Progesterone was injected subcutaneously at doses of 1 or 10 mg/kg. In some experiments, a dose of 200 μg/kg was given.

The progesterone was dissolved in saline containing enough ethanol to give a clear solution. This required 70% ethanol for the largest dose of progesterone. Such an aqueous–ethanol solvent has been found to allow more rapid absorption of progesterone than oil (Lafaye and Frigot, 1978). Control animals received only the solvent.

2.1. *Tritium-Labeled Steroids*

[1,2-^{3}H]Progesterone (specific activity ≈ 50 Ci/mmol) and [17α-^{3}H]-R5020 (17,21-dimethyl-19-nor-4,9-pregnadiene-3,20-dione) were obtained from New England Nuclear. All unlabeled steroids were purchased from Sigma Chemical Company.

2.2. *Measurement of Cytoplasmic and Nuclear Progesterone Receptors*

The uterus was removed and homogenized in 5 mM phosphate buffer containing 10% glycerol and 10 mM monothioglycerol (PGT). Cytosol and nuclear fractions were prepared as described by us (Saffran *et al.,* 1976).

We wished to measure both unbound and total (bound and unbound) progesterone receptors in cytosol. To validate the measurements, we measured the rate of binding of [^{3}H]-R5020 at 0 and 20°C in cytosol containing predominantly unbound or bound receptors. Uterine cytosol of ovariectomized, estrogen-treated guinea pigs should contain mainly unbound progesterone receptors. Progesterone of adrenal cortical origin may contribute to a small amount of bound receptor, but care was taken not to stress the animals. To obtain cytosol with bound receptor, the same cytosol was incubated with an approximately saturating concentration (10^{-8} M) of unlabeled progesterone.

Cytosol containing free receptor bound [^{3}H]-R5020 rapidly at 0°C (Fig. 1). Equilibrium was reached in 30–60 min and was stable for at least 24 hr. At 20°C, the rate of binding was more rapid (not shown), and equilibrium was reached in 10–15 min.

The progesterone–receptor complex dissociates and exchanges its progesterone with [^{3}H]-R5020. At 0°C the rate of exchange was slow. After 30 min, exchange was about 20% complete, and after 2 hr was 40–50% complete. Equilibrium was reached within 24 hr. At 20°C, the rate of exhange was much more rapid. It was complete within 1–2 hr and remained stable for 4 hr. Similar values were obtained at equilibrium at

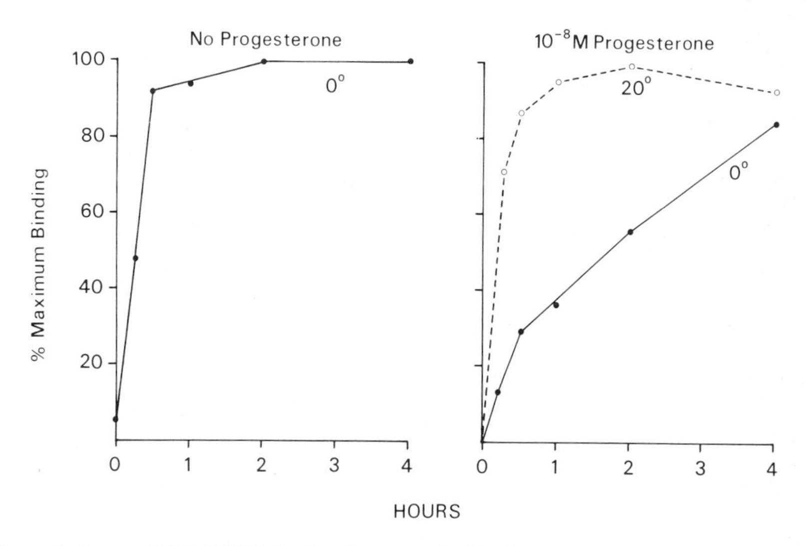

Figure 1. Rate of [^{3}H]-R5020 binding by cytosol. Uterine cytosol from estrogen-treated, ovariectomized guinea pigs was incubated at 0 and at 20°C with 10^{-8} M [^{3}H]-R5020 for varying periods of time. Protein-bound and free [^{3}H]-R5020 were then separated with dextran-coated charcoal. For each incubation, binding in the presence of a 1000-fold molar excess of unlabeled progesterone was subtracted to correct for nonsaturable binding and give specific binding.

0 and 20°C, and were the same in any one cytosol, whether the receptor was free or bound.

We estimated unbound receptor after an incubation of 30 min at 0°C. We measured the total receptor concentration by incubating at 20°C for 60 min. The progesterone–receptor complex dissociates fairly rapidly even at 0°C, so that it is not possible to use binding at 0°C as an accurate measure of unbound receptor, as is possible for the estrogen receptor.

We compared the measurement of total receptor concentration after incubating with a single saturating concentration of [^{3}H]progesterone with receptor measurement by Scatchard analysis. For the Scatchard assay, we incubated cytosol at 20°C for 60 min with increasing concentrations of [^{3}H]progesterone (10^{-9} to 5×10^{-8} M), using dextran-coated charcoal to separate bound and unbound progesterone. We plotted the ratio of bound/free progesterone vs. the molarity bound as described by Scatchard (1949). The equilibrium dissociation constants and concentrations of binding sites were calculated from the curve. The two methods gave similar results, and the single-point assay was routinely used to measure progesterone receptors. This method does tend to underestimate the receptor concentration by 10–15%.

Nuclear progesterone receptors were also measured by exchange with [^{3}H]-R5020 or [^{3}H]progesterone. This was done in one of two ways: (1) Nuclei were suspended in 5 mM PGT buffer and incubated with [^{3}H]progestin (20°C, 60 min). Binding in the presence of a 1000-fold molar excess of progesterone was subtracted to correct for nonspecific binding. At the end of the incubation, the nuclei were centrifuged at 1000g, and the sedimented nuclei were washed and extracted with ethanol (Saffran *et al.*, 1976). [^{3}H]Progestin–receptor complex in the supernatant was measured after the solution was treated with dextran-coated charcoal to absorb free progestin (Saffran *et al.*, 1976). (2) Progesterone–receptor was extracted from the nuclei with PGT buffer containing 0.3–0.5 M KC1, and specific exchange with [^{3}H]progestin was carried out on the extract (20°C, 60 min).

3. *In Vivo Experiment*

Progesterone was injected subcutaneously, and at zero time and at intervals after the injection, cytoplasmic and nuclear receptors were measured. Most experiments were carried out after the injection of 1 and 10 mg/kg. In a small number of experiments, a dose of 200 μg/kg was used.

The choice of a physiological dose of progesterone presents a problem. Progesterone has many biological effects, and the dose depends on the endpoint chosen. One of the most important actions of progesterone is the maintenance of pregnancy, and work, mainly with rats, indicated that this requires a fairly high dose of progesterone (Csapo and Wiest, 1969; Lerner *et al.*, 1962; Nutting and Meyer, 1964; Wu, 1972). The doses ranged from 5 to 40 mg/kg body weight, and we chose a dose of 10 mg/kg to represent this aspect of progesterone action.

Progesterone has many other effects. Among them, it converts the estrogen-dominated uterine endometrium to a secretory endometrium. This is the basis of the Clauberg and McPhail bioassays of progesterone in rabbits (Giannina and Meli, 1976). It induces the synthesis of uteroglobin in rabbits (Rahman *et al.*, 1975) and is required for deciduoma formation in rat uterus (Madjerek, 1972). These effects all have maximal responses with lower doses of progesterone (0.5–2 mg/kg). We used a dose of 1 mg/kg as representative of this type of progesterone action.

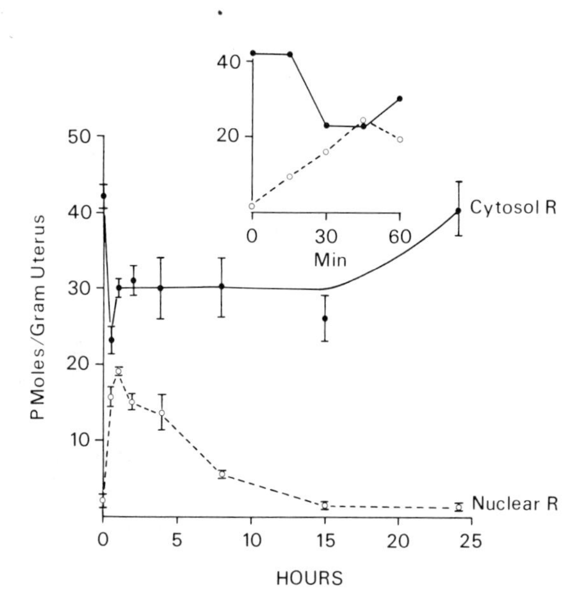

Figure 2. Concentration of progesterone receptors in uterine cytosol and nuclei after the injection of progesterone (1 mg/kg). Doses of 500 μg progesterone, dissolved in 0.5 ml 30% ethanol in 0.9% NaC1 (vol./vol.) were injected subcutaneously into 500-g guinea pigs (1 mg/kg), and at intervals after the injection, cytosol and nuclear fractions were prepared (Saffran *et al.*, 1976). Uteri were homogenized in 20 ml PGT buffer per gram uterus, yielding a cytosol having approximately 3 mg protein/ml. Total cytoplasmic receptors were measured by incubating with [^{3}H]-R5020 for 60 min at 20°C, in the presence and absence of a 1000-fold molar excess of unlabeled progesterone. Nuclear receptors were measured in the 0.3 M KC1 extract of nuclei by exchange with [^{3}H]-R5020 (20°C, 60 min) with or without unlabeled progesterone. Each point is the mean ± S.E. of 6 experiments (2 guinea pigs per experiment).

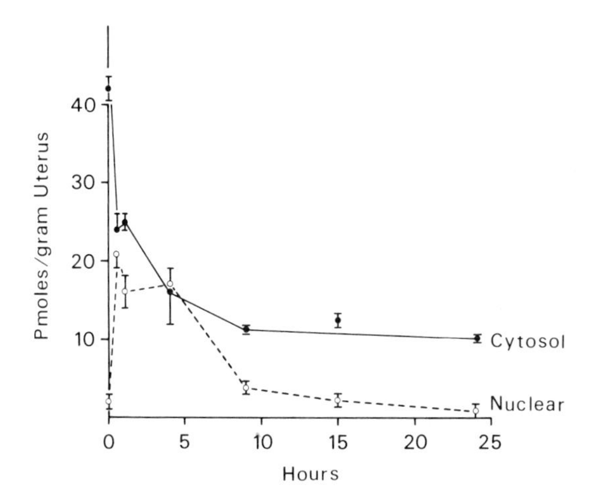

Figure 3. Concentration of progesterone receptors in uterine cytosol and nuclei after the injection of 10 mg/kg progesterone. Doses of 5 mg progesterone dissolved in 0.5 ml 70% ethanol in 0.9% NaC1 (vol./vol.) were injected subcutaneously into 500-g guinea pigs (10 mg/kg). At intervals thereafter, receptors were measured in uterine cytosol and nuclei as described in the Fig. 2 caption, except that the cytosol was treated with dextran-coated charcoal before incubation with [^{3}H]-R5020.

After the injection of progesterone, the concentration of cytoplasmic receptors fell, while the concentration of nuclear receptors increased. The exact amounts depended on the dose of progesterone and on the time after the injection. After the injection of 1 mg/kg, the cytoplasmic receptors decreased to approximately two thirds of the initial value within 30–60 min (Fig. 2). Thereafter, the concentration gradually increased again and after 24 hr had almost returned to the original value. At the same time, the concentration of receptors in the nucleus (measured in the KC1 extract) increased to a maximum within 30–60 min. The receptors were retained in the nucleus at a somewhat lower concentration for 4 hr and then decreased to the control level by 8 hr.

After the injection of 10 mg/kg, the cytoplasmic receptors fell rapidly during the first 30 min after the injection and then continued to decrease gradually (Fig. 3). The nuclear-receptor concentration increased correspondingly and reached a maximum value after 30 min. It was retained in the nucleus at less than the maximum concentration for 4 hr and then decreased to the preinjection level after 8 hr.

Thus, the cytoplasmic receptors responded differently to a low and a high dose of progesterone, but the nuclear receptors responded in the same way.

After a dose of 200 μg/kg (not shown), there was no apparent change in the concentration of receptors in the cytosol. Despite this, there was a gradual increase in the level of nuclear receptors, although to a smaller extent than after 1 and 10 mg/kg.

4. Does Progesterone Affect the Properties of the Progesterone Receptor?

4.1. Unbound vs. Bound Receptors in Cytosol

In the absence of progesterone, the receptor in uterine cytosol had properties characteristic of free receptor and bound [^{3}H]-R5020 rapidly at 0°C (Fig. 4). At 60 min after the injection of solvent alone, only 85% of the receptor seemed to be free, possibly because the injection caused the release of progesterone from the adrenal cortex. When binding measurements were made 60 min after the injection of progesterone, the cytosol contained both bound and unbound receptor. The degree of receptor occupancy depended on the dose injected, and was approximately 40, 80, and 90% after doses of 200 μg/kg, 1 mg/kg, and 10 mg/kg, respectively.

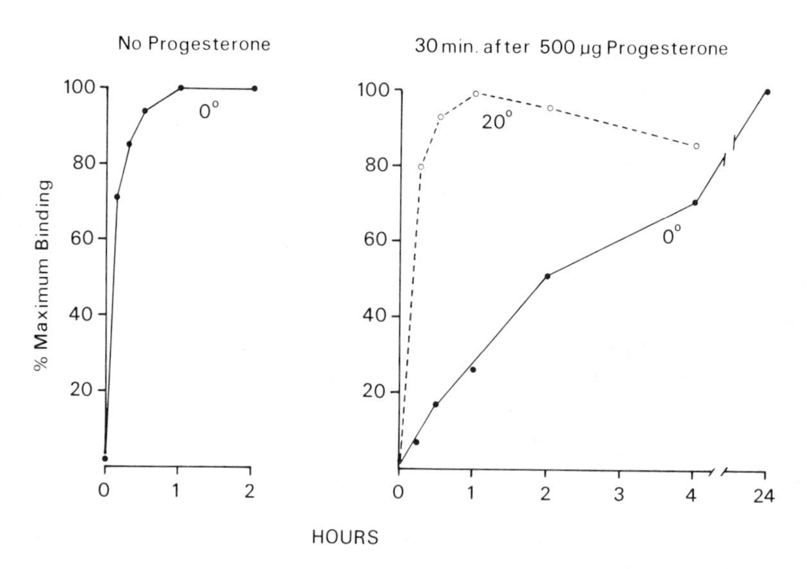

Figure 4. Rate of binding of [^{3}H]-R5020 in uterine cytosol after the injection of progesterone. At 30 min after the injection of 500 μg progesterone (1 mg/kg), or of solvent only (30% ethanol in 0.9% NaCl), uterine cytosol was prepared, and the rate of specific [^{3}H]-R5020 binding at 0°C was measured. [^{3}H]-R5020 bound after 30 min at 0°C was used as an estimate of unbound receptor. Each point is the mean of 2 incubations in a typical experiment.

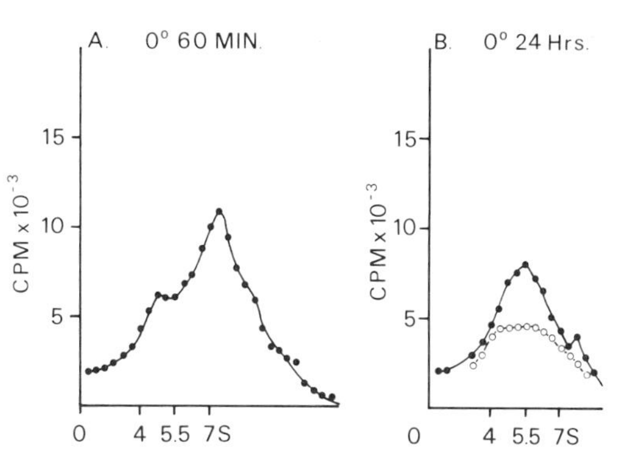

Figure 5. Sucrose-density-gradient centrifugation of uterine cytosol after the injection of progesterone. At 1 hr after the injection of 1 mg/kg progesterone, uterine cytosol was prepared exactly as described (Saffran *et al.*, 1976) and incubated for 60 min at 0°C (A) or for 24 hr at 0°C (B) with 2×10^{-8} M [^{3}H]-R5020. A volume of 0.2 ml was then layered onto a 5–30% sucrose gradient in 5 mM PGT buffer and centrifuged for 24 hr. ^{14}C-labeled standards of ovalbumin (3.6 S), transferrin (5.5 S), and gamma-globulin (7 S) were added to each gradient. (●) [^{3}H]-R5020 only; (○) [^{3}H]-R5020 and a 100-fold excess of unlabeled progesterone.

4.2. Sucrose-Gradient Analysis

At 60 min after the injection of 1 mg progesterone/kg, the cytosol was incubated with [^{3}H]-R5020 at 0°C for 1 hr or for 24 hr (Fig. 5). The incubation mixture was then layered onto a sucrose gradient and centrifuged for 24 hr. After incubation for 1 hr, a binding peak with a sedimentation coefficient of 7 S and a small amount of 4 S binding were seen. This is typical of "unactivated" progesterone receptor. After the 24-hr incubation (conditions that activated the progesterone receptor) (Saffran *et al.*, 1976), the sedimentation coefficient of the binding peak decreased to 5.5 S.

4.3. Temperature Stability

Uterine cytosols from uninjected guinea pigs and from guinea pigs after the injection of 1 or 10 mg/kg progesterone were incubated with [^{3}H]-R5020 at 30°C, a temperature at which the receptor is unstable (Fig. 6). Specific binding was maximal after 15 min in all cytosols and then gradually decayed. The rate of decrease of binding was slightly slower in the cytosol of animals that had received 10 mg/kg progesterone.

4.4. Scatchard Analysis

When the cytosols were incubated with increasing concentrations of [^{3}H]progesterone, saturation of binding was seen at 7.5×10^{-9} M progesterone in all cases (control and after the administration of progesterone). When the results were plotted by the method of Scatchard, the

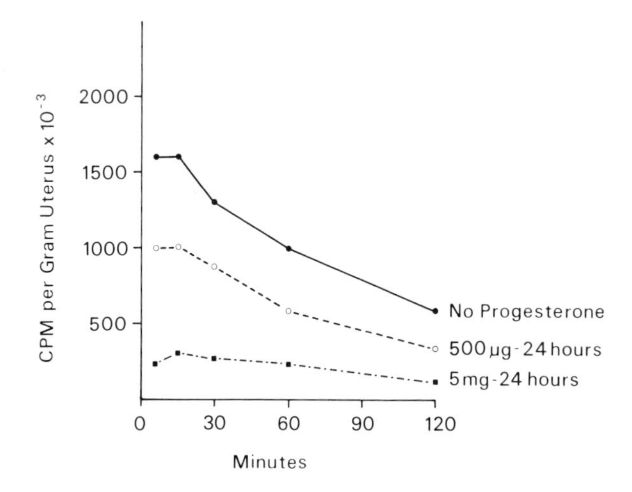

Figure 6. Stability of [^{3}H]-R5020 binding at 30°C. Uterine cytosol of uninjected guinea pigs and of guinea pigs 24 hr after the injection of 1 and 10 mg/kg progesterone was incubated for varying periods of time at 30°C with [^{3}H]-R5020 with or without a 1000-fold excess of unlabeled progesterone. Each point is the mean of 4 incubations.

equilibrium dissociation constant (K_d) was 6–9 mM in the control cytosols, and was not changed at any time after the injection of 1 mg/kg progesterone. After 10 mg/kg, the K_d varied from 10 to 15 nM.

5. Nuclear Receptors

Although the properties of the cytoplasmic progesterone receptors were not changed after the administration of progesterone, the properties of the nuclear receptors were changed in some respects. In particular, methods that were adequate for the measurement of nuclear receptors in *in vitro* experiments did not give a good measure of nuclear receptors after the administration of progesterone. When receptors were measured by suspending the nuclei in buffer containing [^{3}H]progestin (Saffran and Loeser, 1979), the specifically bound tritium remained associated with the nuclei after cell-free incubations of uterine cytosol and nuclei. Approximately 60% of the receptor was found in the nuclear pellet and 40% in the supernatant, after treatment with dextran-coated charcoal. After the injection of progesterone, at least 80% of the specific, protein-bound tritium was found in the supernatant, and only about 20% remained in the nuclear pellet. The results were the same whatever the dose of progesterone injected, and at all times after the injection. The tritium that leaked into the supernatant was not absorbed by dextran-coated charcoal and was therefore bound to a macromolecule. The binding was temperature-sensitive. No binding was seen at 37°C. Unlabeled progestins inhibited binding in a concentration-dependent manner, in the following order: R5020 > progesterone > 5;α-pregnanedione > testosterone > cortisol = 5β-pregnane-3α,20α-diol. Cortisol and pregnanediol caused no suppression of binding. Testosterone was slightly inhibitory.

On sucrose-gradient centrifugation, a binding peak with a sedimentation coefficient of 3.5 S was seen (Fig. 7).

The nuclear receptor can be extracted with buffer containing KC1. The optimum concentration of KC1 was 0.5 M (Chen and Leavitt, 1979). Approximately 60% of the receptor was extracted into 0.15 M KC1 and 80–85% into 0.3 M KC1. On sucrose-gradient analysis on gradients made with 5 mM PGT buffer, the binding peak had a sedimentation coefficient of 3.5 S (Fig. 7). The peak was abolished by incubation in the presence of a 100-fold molar excess of unlabeled progesterone.

The sum of receptors measured in the nuclear pellet and supernatant was only about 75% of that measured in the 0.5 M KC1 extract of nuclei. The pellet may contain progesterone–receptor complex that does not readily dissociate and exchange its progesterone with an [^{3}H]progestin.

The binding properties of the nuclear-receptor fractions (soluble in

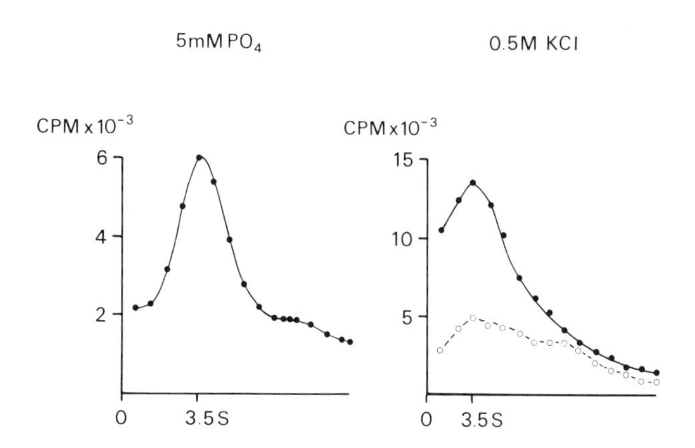

Figure 7. Sucrose-density-gradient centrifugation of progesterone receptor extracted from nuclei after the injection of progesterone. At 1 hr after the injection of 1 mg/kg progesterone, uterine nuclei were suspended in 5 mM PGT buffer or 5 mM PGT containing 0.5 M KC1 (protein concentration of 40 mg/ml) for 10 min at 0°C. The nuclei were centrifuged at 1000 g, and the extracts were incubated with 1.2×10^{-8} M [^{3}H]-R5020 with (○) or without (●) 10^{-6} M unlabeled progesterone for 24 hr at 0°C. Aliquots were layered onto 5–30% sucrose gradients prepared in 5 mM PGT and centrifuged.

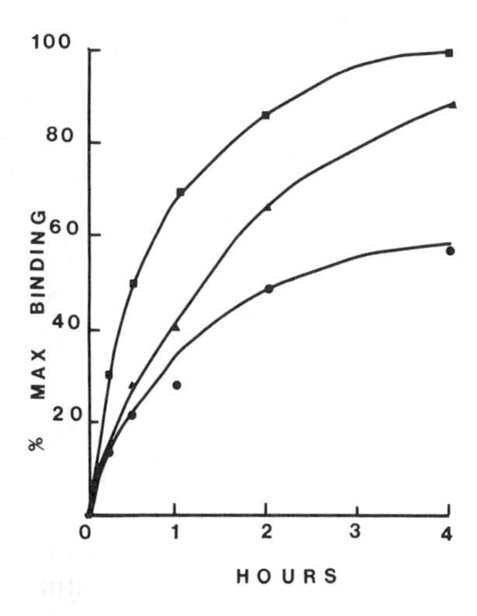

Figure 8. Rate of [^{3}H]-R5020 binding by nuclear fractions. At 30 min after the injection of 1 mg/kg progesterone, uterine nuclei were suspended in 5 mM PGT buffer or PGT containing 0.3 M KC1 (protein concentration 4 mg/ml) for 10 min at 0°C. The nuclei were centrifuged at 1000 *g*, and the supernatant fractions and the pellet from the 5 mM PGT extraction were incubated with 10^{-8} M [^{3}H]-R5020 at 0°C. Incubation in the presence of a 1000-fold excess of unlabeled progesterone was subtracted. (■) 5 mM PGT; (▲) 0.3 M KCl; (●) pellet.

0.5 M KC1, soluble in 5 mM PGT, and remaining in the pellet) were compared by measuring the rate of [^{3}H]-R5020 binding at 0°C. The initial binding rate (0–30 min) was greatest for the 5 mM PGT extract, followed by the 0.5 M KC1 extract and the nuclear pellet, in decreasing order (Fig. 8). The association rate constants were 8×10^6 and 3×10^6 $M^{-1} \cdot min^{-1}$ for the 5 mM PGT- and 0.5 M KC1-soluble fractions, respectively (Groyer *et al.,* 1979). The PGT-soluble nuclear extract probably contains a larger proportion of the receptor in unbound form than does the KC1 solution. The association rate constant was 4×10^7 $M^{-1} \cdot min^{-1}$ for cytosols having mainly unbound receptors, indicating that the nuclear fractions probably contain both bound and unbound receptors in different proportions.

6. Metabolism of Progesterone in the Uterus

To determine whether progesterone was metabolized in guinea pig uterus, 200 μCi [1,2-^{3}H]progesterone, mixed with 1 mg/kg unlabeled progesterone, was injected subcutaneously, and after 60 min uterine cytosol and nuclear fractions (pellet, 5 mM PGT-soluble, 0.5 M KC1-soluble) were prepared. The fractions were extracted with ether and analyzed on thin-layer plates of Silica G_f in chloroform–acetone 9:1 or chloroform–ethyl acetate 13:1. All the recovered tritium ran with the mobility of progesterone. This is characteristic of guinea pig uterus, which seems to lack the enzyme 5α-reductase, which converts progesterone to 5α-pregnanedione.

7. Discussion

After the administration of progesterone to guinea pigs, there was a rapid translocation of progesterone receptors in the uterus from the cytoplasm to the nucleus. The decrease in cytoplasmic receptors and the increase in nuclear receptors agreed fairly well, both quantitatively and temporally, during the first hour after the injection of progesterone. The amounts and the time course of cytoplasmic depletion and nuclear accumulation of receptor were similar. However, after the first hour, the fate of the cytoplasmic receptor seemed to depend on the dose of progesterone that had been injected. After a moderate dose (1 mg/kg), the receptor was gradually and almost completely replenished over the course of 24 hr. But after a larger dose of progesterone (10 mg/kg), there was no replenishment of the cytoplasmic receptor; instead, the receptor concentration continued to drop further.

After both doses of progesterone, receptors were retained in the nucleus for about 4 hr and then decreased to the preinjection level by 8 hr.

The movement of progesterone receptors after the lower dose of progesterone qualitatively resembled the flux of estrogen receptors in the uterus after the injection of 17β-estradiol. However, there was a quantitative difference. At 24 hr after progesterone, the progesterone receptor was only partially replenished, whereas the estrogen receptor after 17β-estradiol increased to 150–200% of its original concentration, probably because of synthesis of new receptor protein.

The decrease in progesterone receptors after the higher dose of progesterone is an old observation (Freifeld *et al.,* 1974; Leavitt *et al.,* 1974; Milgrom *et al.,* 1973) and is very puzzling. There is as yet no explanation for it. Late in pregnancy when progesterone levels are elevated in the human, the concentration of cytoplasmic progesterone receptors in the uterus is also low (Giannopoulos and Tulchinsky, 1979; Kreitmann *et al.,* 1978). We have also observed this in pregnant guinea pigs (unpublished).

The properties of the cytoplasmic receptor were not changed by the administration of progesterone, but there was a change in the solubility properties of the nuclear receptor. A large fraction (approximately 80%) became soluble in hypotonic phosphate buffer, whereas in *in vitro* experiments, the presence of KC1 or NaC1 in the buffer (at least 0.15 M) was required to extract receptor from the nuclei. The leakage of nuclear receptor into hypotonic buffer was also noted by Walters and Clark (1978). Its significance is not known. However, it raises doubts about the accurate separation cytoplasmic and nuclear progesterone receptors after the uterus is homogenized in hypotonic buffers.

ACKNOWLEDGMENTS. This work was supported by Grant HD-09359 from the National Institutes of Health (National Institute of Child Health and Human Development) and by a Biomedical Research Support Grant (SO7-RR-05700-09) to the Medical College of Ohio.

DISCUSSION

SPILMAN: Would you comment on the apparent discrepancy that long-term progesterone treatment causes a decrease in uterine progesterone receptors, but yet you can still see the usual biological effects of progesterone. Does this suggest that the number of progesterone receptors is in great excess of what is needed for a normal biological response?

SAFFRAN: Spare receptors may exist, but I don't know whether this would account for a response to progesterone concomitant with a reduction in the concentration of receptors. The decreased receptor level may have a regulatory significance which we don't yet understand.

D. VILLE: (1) Do you have any information regarding the circulating levels of progesterone in your animals given 1 mg/kg and 10 mg/kg of progesterone? (2) Does dihydroprogesterone have any biological effect on target cells of progesterone? (3) Are there any other metabolities of progesterone found in target cells?

SAFFRAN: (1) We did not measure the progesterone concentration in blood after injecting progesterone. (2) Dihydroprogesterone seems not to be active in the uterus. It has very little ability to induce uteroglobin, form a secretory endometrium, or maintain pregnancy. It does increase blastocyst viability. It may be active in brain, hypothalamus, and pituitary. It has activity in the sexual behavior of some animals. (3) Metabolites more highly reduced than dihydroprogesterone are found in the uterus and in neuroendocrine tissue (e.g., 3-hydroxy-5α-pregnane-20-one and 20-hydroxy-5α-pregnane-3-one). However, there is evidence that these metabolites are not bound to the progesterone receptor and are not retained as is dihydroprogesterone in some tissues.

MCKERNS: Are we still to think of the mechanism of action of progesterone as being similar to that proposed for estrogen?

SAFFRAN: Yes. The mechanisms of action of progesterone and of other steroid hormones, including estrogens, have many characteristics in common. There are more similarities than differences. *In vivo* and *in vitro* experiments indicate that a progesterone receptor that is specific for target tissue and for steroids with progestational activity is present in uterus, vagina, mammary gland, and anterior pituitary along with estrogen receptor. Both receptors are saturable and bind hormone with an affinity that correlates with physiological activity. The complex of hormone and receptor undergoes activation and nuclear translocation. The sedimentation properties of receptors in cytosol and nuclei are similar, but not identical, for progesterone and estradiol.

The regulation of estrogen and that of progesterone receptors during the reproductive cycle are interdependent. For example, prior estrogen action is necessary to induce the synthesis of the progesterone receptor, and progesterone decreases the concentration of both receptors. Both progesterone and estradiol undergo metabolism in the uterus, but the role of metabolism is not understood. Our knowledge is still very incomplete in understanding receptor regulation.

PIETRAS: Edwards *et al.* (1979) have presented evidence which indicates that the relatively long residence time of antiestrogens in target-cell nuclei may be attributable to a receptor–antiestrogen flux rather than a steady state. Have you investigated this hypothesis in your experimental system?

SAFFRAN: We have not done experiments with inhibitors of protein synthesis to suppress receptor synthesis, as was reported by Chamness and co-workers.

BANIK: Your work on the need of uterine receptors for progesterone action is very important for finding a compound that would act at the endometrial level to interfere with implantation of fertilized eggs. I am very optimistic that this approach would be the most physiological method for controlling fertility in humans and animals. From your studies, are you suggesting that progesterone or a similar compound could decrease cytoplasmic or nuclear receptors of endometrial tissue and interfere with the process of implantation?

SAFFRAN: Your question is difficult to answer because although progesterone is needed for implantation and the maintenance of pregnancy, progesterone decreases the concentration of its own receptor. At the end of pregnancy, when progesterone levels are high in humans, the receptor level is very low. Perhaps progesterone receptors are present in a form that does not bind the [^{3}H]progestins that are used to measure them. When other methods for measurement of receptors, such as immunoassays, are available, a careful study of receptor concentrations during pregnancy and especially at the time of implantation should be made.

REFERENCES

Anderson, J.N., Peck, E.J., and Clark, J.H., 1974, Nuclear receptor estradiol complex: A requirement for uterotrophic responses, *Endocrinology* **95:**174–178.

Armstrong, D.T., and King, E.R., 1971, Uterine progesterone metabolism and progestational response: Effects of estrogens and prolactin, *Endocrinology* **89:**191–197.

Bardin, C.W., Bullock, L.P., Jänne, O., and Jacob, S.T., 1975, Genetic regulation of the androgen receptor—a study of testicular feminization in the mouse, *J. Steroid Biochem.* **6:**515–520.

Basha, S.M.M., Bazer, F.W., and Roberts, R.M., 1979, The secretion of uterines purple phosphatase by cultured explants of porcine endometrium dependency upon the state of pregnancy of the donor animal, *Biol. Reprod.* **20:**431–441.

Batra, S., Sjoberg, N.O., and Horbert, G.T., 1978, Estrogen and progesterone interactions in the rabbit uterus *in vivo* after steroid administration, *Endocrinology* **102:**268–272.

Baudendistel, L.J., Ruh, M.F., Nadel, E.M., and Ruh, T.S 1978, Cytoplasmic oestrogen receptor replenishment: Oestrogens vs. anti-oestrogens, *Acta Endocrinol.* **89:**599–611.

Baulieu, E.-E., 1978, Cell membrane, a target for steroid hormones, *Mol. Cell Endocrinol.* **12:**247–254.

Bayard, F., Damilano, S., Robel, P., and Baulieu, E.-E., 1978, Cytoplasmic and nuclear estradiol and progesterone receptors in human endometrium, *J. Clin. Endocrinol. Metab.* **46:**635–648.

Beir, H.M., 1974, A hormone-sensitive endometrial protein in blastocyst development, *J. Reprod. Fertil.* **37:**221–237.

Booth, B.A., and Colas, A.E., 1977, Properties of two progesterone-binding proteins of the rat uterus, *J. Steroid Biochem.* **8:**1171–1177.

Brenner, R.M., and West, N.B., 1975, Hormonal regulation of the reproductive tract in female mammals, *Annu. Rev. Physiol.* **37:**273–302.

Buller, R.E., and O'Malley, B.W., 1976, The biology and mechanism of steroid hormone receptor interaction with the eukaryotic nucleus, *Biochem. Pharmacol. 25* 1612.

Capony, F., and Rochefort, H., 1978, High-affinity binding of the antiestrogen [^{3}H]tamoxifen to the 8S estradiol receptor, *Mol. Cell Endocrnol.* **11:**181–198.

Chan, L., and O'Malley, B.W., 1976, Mechanism of action of the sex steroid hormones, *N. Engl. J. Med.* **294:**1322–1328, 1372–1381, 1430–1437.

Chen, T.J., and Leavitt, W.W., 1979, Nuclear progesterone receptor in hamster uterus: Measurement by [^{3}H]-progesterone exchange during the estrous cycle, *Endocrinology* **104:**1588–1597.

Cheng, Y.-J., and Karavolas, H.J., 1973, Conversion of progesterone to 5α-pregnane-3,20-dione and 3α-hydroxy-5α-pregnane-20-one by rat medial basal hypothalami and the effects of estradiol and stage of the estrous cycle on the conversion, *Endocrinology* **93:**1157–1162.

Cidlowski, J.A., and Muldoon, T.G., 1974, Estrogenic regulation of cytoplasmic receptor populations in estrogen-responsive tissues of the rat, *Endocrinology* **95:**1621–1629.

Clark, J.H., Anderson, J.N., and Peck, E.J., 1973, Estrogen receptor–antiestrogen complex: Atypical binding by uterine nuclei and effects on uterine growth, *Steroids* **2:**707–718.

Coulson, P.B., and Pavlik, E.J., 1977, Effects of estrogen and progesterone on cytoplasmic estrogen receptor and rates of protein synthesis, *J. Steroid Biochem.* **8:**205–212.

Csapo, A.J., and Wiest, W.G., 1969, An examination of the quantitative relationship between progesterone and the maintenance of pregnancy, *Endocrinology* **85:**735–746.

DeLuca, H., 1979, Recent advances in our understanding of the vitamin D endocrine system, *J. Steroid Biochem.* **11:**35–52.

Edwards, D.P., Chamness, G.C., and McGuire, W.L., 1979, Estrogen and progesterone receptor proteins in breast cancer, *Biochim. Biophys. Acta* **560:**457–486.

Elsner, C.W., Illingworth, D.V., dela Cruz, K., Flickinger, G.L., and Mikhail, G., 1977, Cytosol and nuclear estrogen receptor in the genital tract of the rhesus monkey, *J. Steroid Biochem.* **8:**151–155.

Evans, P.W., Sholiton, L.J., and Leavitt, W.W., 1978, Progesterone receptor in the rat anterior pituitary: Effect of estrogen priming and adrenalectomy, *Steroids* **31:**69–81.

Faber, L.E., Sandmann, M.L., and Stavely, H.E., 1972, Progesterone-binding proteins of the rat and rabbit uterus, *J. Biol. Chem.* **247:**5648–5649.

Faber, L.E., Sandmann, M.L., and Stavely, H.E., 1978, Effect of buffers and electrolytes on the stability of uterine progesterone receptors, *Biochem. Med.* **19:**78–89.

Feil, P.D., Glasser, S.R., Toft, D.O., and O'Malley, B.W., 1972, Progesterone binding in the mouse and rat uterus, *Endocrinology* **91:**738–746.

Feil, P.D., Marni, W.J., Mortel, R., and Bardin, C.W., 1979, Nuclear progesterone receptors in normal and malignant human endometrium, *J. Clin. Endocrinol. Metab.* **48:**327–334.

Flier, J.S., Kahn, C.R., Roth, J., and Bar, R.S., 1975, Antibodies that impair insulin receptor binding in an unusual diabetic syndrome with severe insulin resistance, *Science* **190:**63–65.

Freifeld, M.L., Feil, P.D., and Bardin, C.W., 1974, The *in vivo* regulation of the progesterone receptor in guinea pig uterus: Dependence on estrogen and progesterone, *Steroids* **23:**93–103.

Giannina, T., and Meli, A., Estrogen–progesterone interaction: Factors influencing progesterone-induced endometrial changes of the uterus of the immature rabbit, *Proc. Soc., Exp. Biol. Med.* **151**:591–593.

Giannopoulos, G., and Tulchinsky, D., 1979, Cytoplasmic and nuclear progestin receptors in human myometrium during the menstrual cycle and in pregnancy at term, *J. Clin. Endocrinol. Metab.* **49**:100–106.

Gorski, J., and Gannon, F., 1976, Current models of steroid hormone action: A critique, *Annu. Rev. Physiol.* **38**:425–450.

Groyer, A., Picard-Groyer, M.T., and Robel, P., 1979, Glucocorticoid receptor in human embryo fibroblasts. I. Kinetic and Physicochemical properties, *Mol. Cell Endocrinol.* **13**:255–267.

Horwitz, K.B., and McGuire, W.L., 1978a, Estrogen control of progesterone receptor in human breast cancer correlation with nuclear processing of estrogen receptor, *J. Biol. Chem.* **253**:2223–2228.

Horwitz, K.B., and McGuire, W.L., 1978b, Actinomycin D prevents nuclear processing of estrogen receptor, *J. Biol. Chem.***253**:6319–6322.

Horwitz, K.B., Koseki, Y., and McGuire, W.L., 1978, Estrogen control of progestero receptor in human breast cancer: Role of estradiol and antiestrogen, *Endocrinology* **103**:1742–1751.

Hsueh, A.J.W., Peck, E.J., and Clark, J.H., 1975, Progesterone antagonism of the oestrogen receptor and oestrogen-induced uterine growth, *Nature (London)* **254**:337–339.

Hsueh, A.J.W., Peck, E.J., and Clark, J.H., 1976, Control of uterine estrogen receptor levels by progesterone, *Endocrinology* **98**:438–444.

Illingworth, D.V., Elsner, C., de Groot dela Cruz, K., Flickinger, G.L., and Mikhail, G., 1977, A specific progesterone receptor of myometrial cytosol from the rhesus monkey, *J. Steroid Biochem.* **8**:157–160.

Isomaa, V., Isotala, H., Orava, M., and Jänne, O., 1979, Regulation of cytosol and nuclear progesterone receptors in rabbit uterus by estrogen, antiestrogen and progesterone administration, *Biochim. Biophys. Acta* **585**:24–33.

Jänne, O., Kontula, K., Luukkainen, T., and Vihko, R., 1975, Oestrogen-induced progesterone receptor in human uterus, *J. Steroid Biochem.* **6**:501–509.

Juliano, J.V., and Stancel, G.M., 1976, Estrogen receptors in the rat uterus retention of hormone–receptor complexes, *Biochemistry* **15**:916–920.

Kahn, C.R., Flier, J.S., Bar, R.S., Archer, J.A., Gorden, P., Martin, M.M., and Roth, J., 1976, The syndromes of insulin resistance and acanthosis nigricans: Insulin receptor disorders in man, *N. Engl. J. Med.* **294**:739–745.

Karavolas, H.J., Hodges, D.R., and O'Brien, D.J., 1976, Uptake of [^{3}H]-progesterone and [^{3}H]-5α-dihydroprogesterone by rat tissues *in vivo* and analysis of accumulated radioactivity: Accumulation of 5α-dihydroprogesterone by pituitary and hypothalamic tissues, *Endocrinology* **98**:164–175.

Karavolas, H.J., Hodges, D.R., and O'Brien, D.J., 1979, The *in vivo* uptake and metabolism of [^{3}H]progesterone and [^{3}H]-5α-dihydroprogesterone by rat CNS and anterior pituitary: Tissue concentration of progesterone itself or metabolites?, *J. Steroid Biochem.* **11**:863–872.

Katzenellenbogen, B.S., 1975, Synthesis and inducibility of the uterine estrogen-induced protein IP during the rat estrous cycle: Clues to uterine estrogen sensitivity, *Endocrinology* **96**:289–297.

Katzenellenbogen, B.S., Katzenellenbogen, J.A., Ferguson, E.R., and Krauthammer, N., 1978, Anti-estrogen interaction with uterine estrogen receptors studies with radiolabeled antiestrogen (C1-628), *J. Biol. Chem.* **253:**697–707.

Katzenellenbogen, B.S., Tsai, T., Tatee, T., and Katzenellenbogen, J.A., 1979, Estrogen and antiestrogenic action: Studies in reproductive target tissues and tumors, in: *Steroid Hormone Receptor Systems* (W.W. Leavitt and J.H. Clark, eds.), pp. 111–132, Plenum Press, New York.

Kaufman, M., Straisfeld, C., and Pinsky, L., 1976, Male pseudohermaphroditism presumably due to target organ unresponsiveness to androgens: Deficient 5α-dihydrotestosterone binding in cultured skin fibroblasts, *J. Clin. Invest.* **58:**345–350.

Kielhorn, J., and Hughes, A., 1977, Variations in uterine cytosolic oestrogen receptor levels during the rat oestrous cycle, *Acta Endocr–850.*

Koligian, K.B., and Stormshak, F., 1977a, Nuclear and cytoplasmic estrogen receptors in ovine endometrium during the estrous cycle, *Endocrinology* **101:**524–533.

Koligian, K.B., and Stormshak, F., 1977b, Progesterone inhibition of estrogen receptor replenishment in ovine endometrium, *Biol. Reprod.* **17:**412–416.

Koseki, Y., Zava, D.T., and Chamness, G.C., 1977a, Progesterone interaction with estrogen and antiestrogen in the rat uterus—receptor effects, *Steroids* **30:**169–177.

Koseki, Y., Zava, D.T., Chamness, G.C., and McGuire, W.L., 1977b, Estrogen receptor translocation and replenishment by the antiestrogen tamoxifen, *Endocrinology* **101:**1104–1100.

Kreitman, B., Derache, B., and Bayard, F., 1978, Measurement of corticosteroid binding globulin, progesterone and progesterone "receptor" content in human endometrium, *J. Clin. Endocrinol. Metab.* **47:**350–353.

Lafaye, A., and Frigot, P., 1978, Influence du véhicule sue la vitesse d'action de la progesterone au niveau de la motilite utérine chez la ratte gestante recemment biovariectomisée, *Ann. Pharm. Fr.* **36:**391–396.

Leavitt, W.W., Toft, D.O., Strott, C.A., and O'Malley, B.W., 1974, A specific progesterone receptor in the hamster uterus: Physiologic properties and regulation during the estrous cycle, *Endocrinology* **94:**1041–1053.

Leavitt, W.W., Chen, T.J., and Allen, T.C., 1977, Regulation of progesterone receptor formation by estrogen action, *Ann. N. Y. Acad. Sci.* **286:**210–225.

Lerner, L.J., Brennan, D.M., Yiacas, E., de Phillipo, M., and Borman, A., 1962, Pregnancy maintenance in ovariectomized rats with 16α17α-dihydroxyprogesterone derivatives and other progestogens, *Endocrinology* **70:**283–287.

Luu Thi, M.T., Baulieu, E.-E., and Milgrom, E., 1975, Comparison of the characteristics and the hormonal control of endometrial and myometrial progesterone receptors, *J. Endocrinol.* **66:**349–356.

Madjerek, Z.S., 1972, A new bioassay of progestational activity, *Acta Morphol. Neerl.-Scand.* **10:**259–268.

Martel, D., and Psychoyos, A., 1978, Progesterone-induced oestrogen receptors in the rat uterus, *J. Endocrinol.* **76:**145–154.

Měster, J., and Baulieu, E.-E., 1975, Dynamics of oestrogen-receptor distribution between the cytosol and nuclear fractions of immature rat uterus after oestradiol administration, *Biochem. J.* **146:**617–623.

Milgrom, E., Atger, M., Perrot, M., and Baulieu, E.-E., 1972, Progesterone in uterus and plasma. IV. Uterine progesterone receptors during the estrous cycle and implantation in the guinea pig, *Endocrinology* **90:**1071–1078.

Milgrom, E., Thi, L., Atger, M., and Baulieu, E.-E., 1973, Mechanisms regulating the concentration and conformation of progesterone receptors in the uterus, *J. Biol. Chem.* **248:**6366–6374.

Moguilewsky, M., and Raynaud, J.P., 1977, Progestin binding sites in the rat hypothalamus, pituitary and uterus, *Steroids* **30:**99–109.

Moore, R.J., and Wilson, J.D., 1973, The effect of androgenic hormones on the reduced nicotinamide adenine dinucleotide phosphate: △4-3-Ketosteroid 5α-oxidoreductase of rat ventral phosphate, *Endocrinology* **93:**581–592.

Myatt, L., Chaudhuri, G., Elder, M.G., and Limm, L., 1978, The oestrogen receptor in the rat uterus in relation to intrauterine devices and the oestrous cycle, *Biochem. J.* **176:**523–529.

Nuti, K.M., and Karavolas, H.J., 1977, Effect of progesterone and its 5α-reduced metabolites on gonadotrophin levels in estrogen-primed ovariectomized rats, *Endocrinology* **100:**777–781.

Nutting, E.F., and Meyer, R.K., 1964, Effect of various steroids on nidation and fetal survival in ovariectomized rats, *Endocrinology* **74:**573–578.

Oberfield, S.E., Levine, L.S., Carey, R.M., Bejar, R., and New, M.I., 1979, Pseudohypoaldosteronism: Multiple target organ unresponsiveness to mineralocorticoid hormones, *J. Clin. Endocrinol. Metab.* **48:**228–234.

Olefsky, J.M., 1976, The insulin receptor: Its role in insulin resistance of obesity and diabetes, *Diabetes* **25:**1154–1162.

Philibert, D., and Raynaud, J.P., 1973, Progesterone binding in immature mouse and rat uterus, *Steroids* **22:**89–98.

Philibert, D., and Raynaud, J.P., 1974, Progesterone binding in immature rabbit and guinea pig uterus, *Endocrinology* **94:**627–632.

Pollow, K., Schmidt-Gollwitzer, M., Borqoi, E., and Pollow, P., 1978, Influence of estrogens and gestagens on 17β-hydroxysteroid dehydrogenase in human endometrium and endometrial carcinoma, *J. Mol. Med.* **3:**81–89.

Rahman, S.S.U., Billiar, R.B., and Little, B., 1975, Induction of uteroglobin in rabbits by progestogens, estradiol-17β and ACTH, *Biol. Reprod.* **12:**305–314.

Rahman, S.S., Billiar, R.B., and Little, B., 1978, *In vivo* uterine metabolism of progesterone and 5α-pregnane-3,20-dione during the synthesis and release uteroglobin in ovariectomized steroid-treated rabbits, *Endocrinology* **102:**1901–1908.

Raynaud, J.P., 1977, R5020, a tag for the progestin receptor, in: *Progesterone Receptors in Normal and Neoplastic Tissues* (W.L. McGuire, J.P. Raynaud, and E.-E. Baulieu, eds.), pp. 9–21, Raven Press, New York.

Reel, J.R., and Shih, Y., 1975, Oestrogen-inducible uterine progesterone receptors: Characteristics in the ovariectomized immature and adult hamsters, *Acta Endocrinol.* **80:**344–354.

Roberts, R.M., and Bazer, F.W., 1976, Phosphoprotein phosphatase activity of progesterone-induced purple glycoprotein of procine uterus, *Biochem. Biophys. Res. Commun.* **68:**450–455.

Rodriquez, J., Sen, K.K., Seski, J.C., Memon, M., Johnson, T.R., and Memon, K.M.J., 1979, Progesterone binding by human endometrial tissue during the proliferative and secretory phases of the menstrual cycle and by hyperplastic and carcinomatous endometrium, *Am. J. Obstet. Gynecol.* **133:**660–665.

Ruh, T.S., and Baudendistel, L.J., 1978, Antiestrogen modulation of the salt resistant nuclear estrogen receptor, *Endocrinology* **102:**1838–1846.

Ruh, T.S., Baudendistel, L.J., Nicholson, W.F., and Ruh, M.F., 1979, The effects of antioestrogens on the oestrogen receptor, *J. Steroid Biochem.* **11:**315–322.

Saffran, J., and Loeser, B.K., 1979, Nuclear binding of guinea pig uterine progesterone receptor in cell-free preparations, *J. Steroid Biochem.* **10:**43–51.

Saffran, J., and Loeser, B., 1980, The effect of progesterone *in vivo* on the subcellular distribution and properties of progesterone receptors in guinea pig uterus, *J. Steroid Biochem.* **13:**589–598.

Saffran, J., Loeser, B.K., Haas, B.M., and Stavely, H.E., 1974a, Metabolism of progesterone in rat uterus, *Steroids* **23:**117–132.

Saffran, J., Loeser, B.K., Haas, B.M., and Stavely H.E., 1974b, Metabolism of progesterone in subcellular fractions of rat uterus, *Steroids* **24:**839–847.

Saffran, J., Loeser, B.K., Bohnett, S.A., and Faber, L.E., 1976, Binding of progesterone receptor by nuclear preparations of rabbit and guinea pig uterus, *J. Biol. Chem.* **251:**5607–5613.

Sanyal, M.K., and Villee, C.A., 1973, Production and physiologic effects of progesterone metabolites in pregnant and pseudopregnant rats, *Endocrinology* **92:**83–93.

Sarff, M., and Gorski, J., 1971, Control of estrogen-binding protein concentration under basal conditions and after estrogen administration, *Biochemistry* **10:**2557–2563.

Scatchard, G., 1949, The attraction of proteins for small molecules and ions, *Ann. N. Y. Acad. Sci.* **51:**660–672.

Shapiro, S., and Forbes, S.H., 1978, Alterations in human endometrial protein synthesis during the menstrual cycle and in progesterone-stimulated organ culture, *Fertil. Steril.* **30:**175–180.

Spona, J., Bieglmayer, C., and Pirker, R., 1979, Progesterone receptor in the rat anterior pituitary: Transformation and nuclear translocation, *FEBS Lett.* **97:**269–274.

Srivastava, A.K., 1978, Differentiation of binding proteins for D-norgestral in human uterine tissue and plasma, *J. Steroid Biochem.* **9:**1241–1244.

Tabei, T., Haga, H., Heinrichs, W.L., Hermann, W.L., 1974, Metabolism of progesterone by rat brain, pituitary gland and other tissues, *Steroids* **23:**651–666.

Tseng, L., 1978, Steroid specificity in the stimulation of human endometrial estradiol dehydrogenase, *Endocrinology* **102:**1398–1403.

Verma, U., and Laumas, K., 1976, Cellular and subcellular metabolism of progesterone by human proliferative and secretory phase endometrium and myometrium, *J. Steroid Biochem.* **7:**275–282.

Vu Hai, M.T., Logeat, F., Warembourg, M., and Milgrom, E., 1977, Hormonal control of progesterone receptors, *Ann. N.Y. Acad. Sci.* **286:**199–209.

Vu Hai, M.T., Logeat, F., and Milgrom, E., 1978, Progesterone receptors in rat uterus: Variations in cytosol and nuclei during the estrous cycle and pregnancy, *J. Endocrinol.* **76:**43–48.

Walters, M.R., and Clark, J.H., 1978, Cytosol and nuclear compartmentalization of progesterone receptors of the rat uterus, *Endocrinology* **103:**601–609.

West, N.B., Verhage, H.G., and Brenner, R.M., 1976, Suppression of the estradiol receptor system by progesterone in the oviduct and uterus of the cat, *Endocrinology* **99:**1010–1016.

West, N.B., Verhage, H.G., and Brenner, R.M., 1977, Changes in nuclear estradiol receptor and cell structure during estrous cycles and pregnancy in the oviduct and uterus of cats, *Biol. Reprod.* **17:**138–143.

West, N.B., Norman, R.L., Sandow, B.A., and Brenner, R.M., 1978, Hormonal control of nuclear estradiol receptor content in luminal epithelium in the uterus of the golden hamster, *Endocrinology* **103:**1732–1741.

Whalen, R.E., Martin, J.V., and Olsen, K.L., 1975, Effect of oestrogen antagonist on hypothalamic oestrogen receptors, *Nature (London)* **258:**742–743.
Wu, J.T., 1972, Effect of progesterone on rat blastocyst viability during prolonged delay in implantation, *Endocrinology* **91:**1386–1388.
Yamamoto, K.R., and Alberts, B.M., 1976, Steroid receptors: Elements for modulation of eukaryotic transcription, *Annu. Rev. Biochem.* **45:**721–746.

12

Hormonal Control of Steroid Receptors in Human Endometrium during the Menstrual Cycle

Paul Robel, Rodrigue Mortel, and Etienne-Emile Baulieu

1. Introduction

In the past two decades, the development of saturation-analysis methods has permitted specific and accurate measurements of circulating sex-steroid hormones, and their cyclic changes during the menstrual cycle are well documented. For the past ten years, a large body of evidence has been accumulated indicating that hormones interact with a receptor system before triggering cellular responses. Hormone receptors were first identified and physicochemically studied for steroids, particularly estradiol in the rat uterus. Once reliable measurements of hormone receptors became available in animal models, correlations were attempted between receptor concentrations and circulating levels of various hormones. It was discovered that the concentration of receptor molecules is not fixed, but varies with the physiological state of the animals. In particular, variations of the concentrations and subcellular distributions of receptors were observed during the estrous cycle and mimicked by

Paul Robel, Rodrigue Mortel, and *Etienne-Emile Baulieu* • Unité de Recherches sur le Métabolisme Moléculaire et la Physio-Pathologie des Stéroides de l'Institut National de la Santé et de la Recherche Médicale (U 33 INSERM) and ER 125 CNRS, 94270 Bicêtre, France.

injecting estradiol and progesterone into hormone-deprived animals (Baulieu *et al.*, 1975).

In the immature rat uterus, estradiol receptor is found almost exclusively in the soluble fraction of the cytoplasm, which is commonly called the cytosol. After injection of estradiol *in vivo*, estradiol–receptor complexes are translocated into the nuclear compartment of the cell. A considerable amount of effort has been devoted to the search for nuclear "acceptors," which may be the sites where characteristic changes of gene expression occur. However, it should be stressed that the presence of receptor in the nucleus does not necessarily mean triggering of hormone action. For example, antiestrogens are capable of transferring the receptor into the nucleus; nevertheless, no response may occur (Sutherland *et al.*, 1977). Another example is the finding in some target tissues, including human endometrium, of readily available (apparently unoccupied) nuclear receptor sites, which estradiol can reach directly. Therefore, it appears that on occasion, the formation of hormone–receptor complexes in the cytoplasm is not a mandatory step for hormone action. In practical terms, the cytosol/nuclear receptor ratio is worth considering but should be interpreted with caution.

In addition, the presence of free and occupied receptor sites in both cytosol and nuclei must be kept in mind. Because of the slow dissociation rate of estradiol from receptor sites, techniques have been designed to measure separately available sites at low temperature or both available and occupied sites by "exchange" at higher temperature. Actually, the latter measurements are fraught with difficulties because of the easy inactivation of receptors. Available receptor sites may be either unoccupied or occupied by low-affinity natural or synthetic ligands. Finally, it should be emphasized that techniques worked out for the measurement of receptor sites occupied by natural hormones (estradiol or progesterone) may have to be adjusted in certain physiological or pharmacological conditions when receptor sites are occupied by other ligands. For example, estradiol-receptor sites may be occupied by (1) natural estrogens other than estradiol (estrone, estriol, catechol-estrogens), (2) synthetic steroidal estrogens (ethinyl estradiol), (3) nonsteroidal estrogen (diethylstilbestrol), (4) antiestrogen, (5) androgen with significant affinity for estrogen receptor (androst-5-ene-3β,17β-diol-5α-androstane-3β,17β-diol), or (6) synthetic progestagens of the 19-nor-testosterone series. Likewise, progesterone-receptor sites may be occupied by synthetic progestagens derived from either progesterone or 19-nor-testosterone.

For pathophysiological purposes, we have considered as absolutely necessary the measurement of both cytoplasmic and nuclear receptor concentrations, whether occupied or unoccupied by any ligand.

The presence of estradiol receptor in the human female reproductive tract was suggested by selective retention of radioactive estradiol in the normal human uterus (Davis *et al.*, 1963) and endometrium (Brush *et al.*, 1967; Evans and Hähnel, 1971). The first quantitative approach to total receptor concentration in human endometrium was that of Tseng and Gurpide (1972a), who applied the technique of endometrium-slice superfusion. Wiest and Rao (1971) should be credited for the first unambiguous demonstration of progesterone receptor in human endometrium. A systematic study of estradiol and progesterone receptors in both the cytoplasmic and the nuclear fraction was conducted by our group in normal women (Bayard *et al.*, 1978).

2. *Methodological Problems*

2.1. *General*

It is generally impossible to obtain endometrial tissue from women deprived of estrogens. However, when the endometrium is exposed to endogenous hormones, part or all of the receptor sites may be occupied by hormones, and a large fraction of the hormone–receptor complexes is found in the nuclear compartment. Therefore, interpretation of published results must take into account which of the following fractions has been measured: (1) free cytoplasmic receptor sites, (2) free and occupied cytoplasmic receptor sites, (3) free nuclear receptor sites, (4) free and occupied nuclear receptor sites, and (5) total cellular receptor sites. Indeed, the physiological significance of the results will be different according to the category of receptor site assayed. Unfortunately, the many published reports are difficult to interpret because the assay conditions used were mostly empirical. For instance, the assay will measure part of available sites when less than saturating concentrations of hormones are used. Similarly, it will determine only part of occupied sites when exchange reactions are incomplete. In cases in which receptor concentrations are determined following incubation of tissue slices with radioactive hormones, it is clear that such a technique does not measure endogenous nuclear receptor but translocated cytosolic hormone–receptor complexes. Such improper terminology, associated or not with improper methodology, explains most discrepancies found in the literature.

The labeling of receptor sites, particularly the occupied ones, with tritiated hormones of high specific activity is based on the knowledge of equilibrium and kinetic association and dissociation rate constants. They provide ground for the so-called "exchange" techniques, in which nonradioactive hormone filling the receptor sites is replaced by radioactive hormone. Besides the receptor, crude endometrial extracts contain other hormone-binding components that may be divided into low-affinity nonsaturable binding and high-affinity saturable binding components. The latter are similar to plasma proteins that specifically bind natural steroid hormones. Consequently, the binding of hormone to the receptor must be differentiated from its binding to both kinds of nonreceptor binding components. A rational approach to such discrimination is at present based on knowledge of the physicochemical properties of estradiol and progesterone receptors, namely, affinity and specificity of the binding, heat stability, and other characteristics related to size, electrophoretic mobility, and sensitivity to various reagents.

2.2. Physicochemical Properties of the Estradiol Receptor

2.2.1. Estradiol-Binding Parameters

Most data related to estradiol binding constants in human endometrium have been obtained with the dextran-coated charcoal adsorption technique. The reported K_{Deq} varied from 0.1 to 0.3 nM at 0°C (Krishnan *et al.*, 1973; Crocker *et al.*, 1974; Makler and Eisenfeld, 1974; Tsibris *et al.*, 1978) and from 0.3 to 0.6 nM at 30°C (Robertson *et al.*, 1971; Bayard *et al.*, 1978). These values are in keeping with those reported by Baulieu *et al.* (1975) in other mammalian species. At 4°C, the association rate constant is 1×10^{-7} M^{-1} min^{-1}, and the dissociation rate constant is very small, on the order of 1×10^{-3} min^{-1} (Makler and Eisenfeld, 1974).

2.2.2. Binding Specificity

The affinity displayed by estradiol for the receptor is about 2–10 times greater than that of its metabolites, estrone and estriol. Other compounds such as 17α-ethinyl estradiol and 11β-methoxy-17α-ethinyl estradiol (moxestrol) also bind well to the receptor. Among nonsteroidal compounds, there are conflicting reports concerning the relative affinity of diethylstilbestrol as compared to estradiol. If the important binding to nonspecific components is taken into consideration, it seems established that diethylstilbestrol has higher affinity for the receptor than estradiol. Antiestrogens of the nonsteroidal series, such as *cis*-clomifen, C1-628, or U 11,100 A (nafoxidin) have weak affinity for the receptor (Notides *et al.*, 1972; Krishnan *et al.*, 1973). Androgens have very low affinity for the receptor, although androst-5-ene-3β,17β-diol and 5α-andro-stane-3β,17β-diol are better competitors for estradiol binding ($K_i \approx 10^{-7}$ M) than testosterone or dihydrotestosterone. Corticosteroids, progesterone, and C21 progestagens used in oral contraceptives do not bind the receptor, although chlormadinone acetate has been reported to compete for estradiol binding (Notides *et al.*, 1972; Krishnan *et al.*, 1973; Makler and Eisenfeld, 1974; Pollow *et al.*, 1975a; Raynaud *et al.*, 1978).

Interestingly, nonsteroidal compounds such as diethylstilbestrol and tamoxifen or steroidal derivative (moxestrol), while they bind well to the receptor, do not bind significantly to the sex-hormone-binding plasma protein (SBP).

2.2.3. Gradient Ultracentrifugation

The use of gradient-sedimentation analysis in a medium of low ionic strength has shown a predominantly 8 S variety (Notides *et al.*, 1972; Krishnan *et al.*, 1973; Pollow *et al.*, 1975a; Raynaud *et al.*, 1978), whereas Bayard *et al.* (1978) reported a 4–5 S form. The reason for such discrepancy has not been established, but differences due to the various physiological states or instability of the receptor cannot be ruled out. For instance, diisopropylfluorophosphate, a proteolytic enzyme inhibitor, favors the appearance of an 8 S component (Notides *et al.*, 1972). Because of these considerations, it appears that gradient ultracentrifugation is not a reliable procedure to distinguish the estradiol–receptor complex from other specific proteins sedimenting in the 4–5 S region.

2.3. Physicochemical Properties of the Progesterone Receptor

2.3.1. Progesterone-Binding Parameters

The K_{Deq} has been measured generally between 0 and 4°C, and most authors report values of 1–4 nM for endometrial receptor (Haukkamaa and Luukkainen, 1974; Bayard *et al.*, 1978). However, Rao *et al.* (1974) indicated a slightly higher affinity (0.3–0.4 nM). In any case, the affinity of progesterone for the endometrial receptor is not very different from that of progesterone for corticosteroid-binding globulin (CBG) [3 nM at 0–4°C (Seal and Doe, 1966)]. The dissociation rate of progesterone–receptor complexes is very rapid, but is reduced by the addition of glycerol to assay buffer (Feil *et al.*, 1972; Bayard *et al.*, 1978). Thus, exchange of occupied receptor sites with radioactive hormone is possible at 0–4°C, but the measurement must take into account the rapid dissociation of progesterone–receptor complexes even at low temperature. This is accomplished either by adding glycerol to assay buffer or by the use of synthetic ligands with slower dissociation rates [R 5020 (Philibert and Raynaud, 1974) ORG 2058 (Jänne *et al.*, 1976)].

2.3.2. Binding Specificity

Extensive reports of structural requirements of progesterone receptors have been published (H.E. Smith *et al.*, 1974; Kontula *et al.*, 1975). The results demonstrated a good correlation of relative binding affinity with progestational activity as measured by the subcutaneous Clauberg assay. Removal of the 19-methyl group is the only modification of the A ring that does not decrease or eliminate receptor binding. The α,β-unsaturated carbonyl system is essential. The C20 carbonyl group of the 17β side chain is also essential, although this CH_3CO 17β side chain can be replaced by a 17α-ethinyl-17β-hydroxyl group. Removal of the angular methyl group at C10 increases binding of progesterone, and two radioactive derivatives of 19 nor-progesterone, R 5020 and ORG 2058, have been proposed for the assay of progesterone recptor. They are characterized by a slow dissociation rate from the receptor and minimal binding to CBG.

Other steroid hormones, i.e., androgens, estrogens, and glucocorticosteroids, have negligible affinity for the progesterone receptor in human endometrium (Wiest and Rao, 1971; Haukkamaa and Luukkainen, 1974).

The binding affinities of synthetic progestagens used as contraceptive steroids appear to fall roughly into three groups. Compounds that bind strongly are *d*-norgestrel, norethisterone, medroxyprogesterone acetate, and megestrol acetate. Moderately strong binding is shown by *dl*-norgestrel and chlormadinone acetate, but a large number of estrane progestagens show only insignificant binding affinity (norethisterone acetate, norethynodrel, ethynodiol diacetate, lynestrenol). There is, however, good evidence that the latter compounds are metabolized to norethisterone (Briggs, 1975).

2.3.3. Gradient Ultracentrifugation

Gradient-sedimentation analysis in a medium of low ionic strength has shown a predominant 4 S variety, even in the presence of proteolytic enzyme inhibitors (Wiest and Rao, 1971; Bayard *et al.*, 1978). However, when patients had been pretreated with estrogens prior to hysterectomy, a predominant 7.5 S variety was observed (Philibert and Raynaud, 1974). In addition, Pollow *et al.* (1975a) reported an 8 S receptor variety in normal premenopausal endometrium. Such discrepancies might be related to the influence of receptor concentration on its sedimentation rate, already observed in guinea pig uterus (Milgrom *et al.*, 1973).

2.4. General Properties of Estradiol and Progesterone Receptors

Estradiol and progesterone receptors are of proteinaceous nature, as shown by their sensitivity to proteolytic enzymes, particularly pronase and papain (Makler and Eisenfeld, 1974; Jänne *et al.*, 1975). They are very unstable at temperatures of 37°C or more (Hähnel *et al.*, 1974), and the binding of steroids is inhibited by sulfhydryl reagents (Rao *et al.*, 1974; Makler and Eisenfeld, 1974; Jänne *et al.*, 1975). Furthermore, progesterone binding to the progesterone receptor is inhibited by metal ions (Hg^{2+}, Cu^{2+}, Ag^{+}, and Zn^{2+}) (Kontula *et al.*, 1974), probably by interference with the SH groups. Consequently, addition of EDTA and of −SH reducing agents is commonly used to protect the receptor sites. Conversely, heat or sulfhydryl reagents have been proposed to inactivate receptor binding in order to measure residual binding of hormone to nonspecific components. However, such treatments often result in a significant increase of nonsaturable binding, and consequently an underestimation of receptor concentration.

A pH optimum for binding of close to 8 has been reported for both receptors (Hähnel, 1971; Rao *et al.*, 1974; Jänne *et al.*, 1975), with a greater sensitivity to acidic environment. The isoelectric point of estradiol receptor is close to 5 (Pollow *et al.*, 1975a), that of progesterone receptor is 4.8 (Jänne *et al.*, 1975), and both can be selectively precipitated by basic proteins such as protamine (Steggles and King, 1970). This last property is yet to be used for measurement of endometrial receptors. Precipitation of receptors can equally be achieved by ammonium sulfate to a fractional saturation of 35% (Kontula *et al.*, 1975). This procedure results in a purification of about 10-fold, eliminating most nonsaturable binding proteins and allowing a more accurate appraisal of binding specificity and affinity.

Few attempts have been made to obtain highly purified receptors from human uterus. One report, in which both endometrium and myometrium were included, claims a 40,000-fold purification of progesterone receptor by ammonium sulfate fractionation, affinity chromatography, and ion-exchange chromatography (R.G. Smith *et al.*, 1975). The purified receptor sedimented at 3.7 S on sucrose-gradient ultracentrifugation. Recently, a 20,000-fold purification of uterine estradiol receptor has been described (Coffer *et al.*, 1977).

2.5. Endogenous Hormones in Human Endometrium

Proliferative endometrium contains about 1 ng and secretory endometrium 0.5 ng estradiol/g tissue. Such amounts (Guerrero *et al.*, 1975;

Batra *et al.*, 1977) produce concentrations of endogenous estradiol in cytosol of 0.5 nM or less, which can be neglected in most receptor assay procedures utilized.

The concentration of progesterone in the endometrium has been determined by several authors (Bayard *et al.*, 1975; Kreitmann *et al.*, 1978; Batra *et al.*, 1977; Guerrero *et al.*, 1975; Haukkamaa and Luukkainen, 1974). When the reported concentrations were recalculated in nanograms per gram of tissue, rather widespread values were obtained. In the follicular phase, progesterone levels ranged between 2 and 9 ng/g tissue, and values for the mean secretory phase varied between 8.5 and 27 ng/g. Therefore, at least in secretory endometrium, the concentration of endogenous progesterone will bring about a very significant isotopic dilution of added tracer. In fact, the concentration of progesterone-receptor sites doubled when cytosol was stripped of endogenous progesterone using the dextran-coated-charcoal procedure (Haukkamaa, 1974).

2.6. *Plasma Proteins in Endometrial Cytosol*

As initially demonstrated in the rat uterus (Milgrom and Baulieu, 1970), a major difficulty encountered in developing an assay to measure progesterone receptor is the presence of a CBG-like component in uterine cytosol despite thorough washing of tissues before homogenization. In the report of Verma and Laumas (1973), the only observed binding component showed the physicochemical properties and binding specificity of CBG. In other reports, a combination of physical methods such as gradient ultracentrifugation, polyacrylamide gel electrophoresis, and appropriate binding competition experiments (Young and Cleary, 1974; Philibert and Raynaud, 1974; Bayard *et al.*, 1978; Kreitmann *et al.*, 1978) indicated the simultaneous presence of progesterone receptor and CBG. An absolute requirement in setting up the assay is the demonstration that progesterone can be displaced from binding sites only by progestagens and not by cortisol. A convenient approach is to determine the binding of [^{3}H]progesterone in the presence of competitors and of a 100 to 1,000-fold excess of nonradioactive cortisol. This technique prevents the binding of progesterone to CBG (Bayard *et al.*, 1978) by saturation of CBG binding sites with the nonradioactive cortisol. Kreitmann *et al.* (1978) have measured the concentration of CBG in human endometrium and observed wide variations from sample to sample that were apparently unrelated to the phase of the menstrual cycle. The reported values averaged 25 pmol/g, which represents approximately 5% of the plasma values (about 500 nM). In endometrial cytosol, the concentration of CBG is generally severalfold greater than that of progesterone receptor (Kreitmann *et al.*, 1978). Since, as previously indicated, the affinity of progesterone is similar for CBG and receptor, it is quite conceivable that most saturable binding sites in endometrial samples with low receptor content may be due to contamination by plasma CBG. No enrichment of endometrial cytosol in CBG vs. other plasma proteins has been observed (Kreitmann *et al.*, 1978). The SBP is one of the plasma proteins that contaminate human endometrium. However, its concentration in non-pregnant-female plasma is 10 times lower than that of CBG, approximately 60 nM (Heyns and De Moor, 1971). Likewise, its affinity for estradiol (K_{Deq} = 1.3 nM at 4°C) (Mercier-Bodard *et al.*, 1970) is much lower than that of estradiol for the receptor. Consequently, the binding of estradiol to SBP in cytosol preparations can be generally considered negligible (Bayard *et al.*, 1978).

2.7. Metabolism of Ligands

Following the work of Tseng and Gurpide (1972a,b), estradiol dehydrogenase activity has been well demonstrated in human endometrium and its changes correlated with hormonal status. This enzyme converts estradiol into estrone and is located mainly in the glandular epithelium of secretory endometrium (Scublinsky *et al.*, 1976), particularly in the endoplasmic reticulum and mitochondrial subcellular fractions (Pollow *et al.*, 1975b). NAD is the preferred cofactor, and the K_m for estradiol is 3.3 μM at 40°C and pH 9.5. The oxidation of estradiol to estrone by estradiol dehydrogenase is greatly enhanced in secretory endometrium, as shown by Tseng and Gurpide (1972b, 1974) as well as Pollow *et al.* (1975b). In addition, the activity of this enzyme can be increased by progestagens either *in vivo* or *in vitro* (Tseng and Gurpide, 1975a; Gurpide *et al.*, 1977).

Progesterone is also metabolized by human endometrium, as reported by Sweat and Bryson (1970). According to these authors, the major metabolites were 5α-pregnane-3,20-dione (pregnanedione), an undefined dihydroxy compound, and 6β-hydroxy progesterone. However, when NADPH was added, the major products were pregnanedione and 20α-OH-pregn-4-ene-3-one (20α-OH progesterone) (Collins and Jewkes, 1974). Pollow *et al.* (1975c) demonstrated 5α-reductase, 5β-reductase, and 20α-hydroxysteroid dehydrogenase activity. The latter has been reported to be more active during the secretory phase of the menstrual cycle. However, these enzyme activities do not preclude the use of the corresponding natural hormones in receptor measurements because no significant metabolism of ligand was observed under the assay conditions (Hähnel, 1971; Bayard *et al.*, 1978). This lack of ligand metabolism is due mainly to the localization of most enzyme activities in the particulate fractions of the cytoplasm and to the relatively large K_m values of enzyme compared to the hormone concentrations needed for receptor assays. However, the *in vivo* metabolism of estradiol and progesterone in human endometrium partly explains the dissimilarities observed between plasma and tissue concentrations of both hormones.

2.8. Main Characteristics of Reliable Assays for Estradiol and Progesterone Receptors in Human Endometrium

2.8.1. General

Pathophysiological changes in receptor concentration cannot be evaluated without a thorough validation of the specificity of the technique adopted. It is imperative that the binding assay be specific for estradiol or progesterone receptor or both and that it give an accurate evaluation of binding constants. It must be realized that no evidence exists for significant affinity changes of receptors with physiological states, and therefore the only parameter that is subjected to variations is the concentration of receptor sites. It is equally important to be mindful that human endometrium is generally exposed to endogenous ovarian hormones and that receptor sites can be expected to be either filled or unfilled with hormones and located both in the cytoplasm and in the nuclear fractions. It should also be recalled that since estradiol dissociates very slowly from receptor sites at low temperature, only unfilled sites will be measured in these conditions, whereas the sum of filled and unfilled sites will be determined when an exchange technique is used (Anderson *et al.*, 1972). Conversely, the dissociation of progesterone–receptor complexes being very rapid, some exchange will necessarily occur regardless of the assay conditions. Therefore, in the case of progesterone receptors, only the sum of unfilled and occupied receptor sites can be rigorously measured.

2.8.2. Choice of Ligand

The criteria for selection of the best radioactive ligand are: high affinity, slow dissociation rate, strict binding specificity, and absence of metabolism. In the case of estrogen receptors, [^{3}H]estradiol has been generally utilized, since it fulfills, roughly, all these conditions. Assay conditions have been described under which no metabolism occurs and under which binding to SBP and to androgen receptor [present in very minute amounts (unpublished)] has been prevented by adding 20 nM dihydrotestosterone to the incubation buffers (Bayard *et al.*, 1978). Recently, the use of [^{3}H]moxestrol has been proposed by Raynaud *et al.* (1978), because it binds minimally to SBP, androgen receptor, and nonsaturable low-affinity proteins.

In the case of progesterone receptors, the use of [^{3}H]progesterone is acceptable providing the rapid dissociation of progesterone–receptor complexes is alleviated by the addition of glycerol and that binding of the hormone to CBG is prevented by an excess of nonradioactive cortisol added to the incubation buffers (Feil *et al.*, 1972; Milgrom *et al.*, 1972; Bayard *et al.*, 1978). As an alternative, a synthetic progestagen, such as [^{3}H]-R 5020, which does not bind to CBG, can be used (Raynaud, 1977), but this compound binds rather strongly to nonsaturable components (Seematter *et al.*, 1978) and in addition displays relatively high affinity for glucocorticosteroid receptor. The latter difficulty can be circumvented by the addition of cortisol to assay buffers. Another synthetic progestin, [^{3}H]-ORG-2058, seems to have properties similar to those of R 5020 (Jänne *et al.*, 1976).

2.8.3. Choice of Binding Assay

Few systematic efforts have been made to define the most efficient and accurate techniques for measuring receptors. In general, low-ionic-strength buffers containing EDTA, SH-reducing agents, and eventually glycerol are used at a slightly alkaline pH (7.8–8), because they are thought to protect cytosol receptors. The dextran-coated-charcoal adsorption technique has been utilized for binding assay in most reports (Wiest and Rao, 1971; Krishnan *et al.*, 1973; Crocker *et al.*, 1974; Evans *et al.*, 1974; Rao *et al.*, 1974; Young and Cleary, 1974; MacLaughlin and Richardson, 1976; Bayard *et al.*, 1978), and likewise gel filtration (Makler and Eisenfeld, 1974; Trams *et al.*, 1973) has been employed. Several authors (Verma and Laumas, 1973; Philibert and Raynaud, 1974) have reported on the use of equilibrium dialysis, which apparently results in large losses of binding sites. More selective procedures such as ammonium sulfate or protamine sulfate precipitation have not been currently adopted.

The measurement of binding-site concentration can be performed either by constructing a Scatchard plot or by performing a single-point analysis at a saturating concentration of ligand. The former approach allows in addition the K_{Deq} determination, thus confirming the specificity of the measured binding. However, in the presence of endogenous hormone, once isotopic equilibrium has been reached, the plot provides an acceptable estimate only of receptor-site concentration, with an underestimation of the affinity constant. The single-point analysis is obtained by labeling the sites with a saturating concentration of radioactive hormone. It requires smaller amounts of material and, under carefully controlled conditions, gives results similar to the multipoint analysis necessary for constructing a Scatchard plot (Haukkamaa and Luukkainen, 1974).

Regardless of the method employed, binding to nonsaturable proteins must be subtracted by parallel incubation(s) with the same concentration(s) of radioactive hormone plus a large excess of nonradioactive hormone. On occasion, nonsaturable binding has been evaluated following inactivation of the receptor by heat or SH reagents. The binding to specific plasma proteins or possibly to other receptors must be prevented, as previously indicated, by the use of appropriate radioactive ligand or by the addition of appropriate nonradioactive competitors, or by both measures.

In the case of progesterone receptor, endogenous progesterone, depending on its concentration, may interfere with the binding assay. Such difficulty has been avoided by exposing the cytosol to dextran-coated charcoal at 4°C for less than 30 min, a procedure that removes most endogenous progesterone prior to incubation (Young and Cleary, 1974). Likewise, the concentration of endogenous progesterone can be measured by radioimmunoassay and the result used to correct the specific activity of the added tracer (Bayard *et al.,* 1978). If such correction is neglected, the concentration of progesterone-receptor sites can be underestimated to as much as one half or one third the actual values. This is particularly true for progesterone-receptor measurement in gestational endometrium.

2.8.4. Practical Constraints

For clinical-investigation purposes, measurements of receptor concentration are most often performed on endometrial samples obtained by biopsy. The weight of such samples rarely exceeds 200 mg, part of which must be kept for histological examination. Therefore, it becomes advantageous to use assay techniques that allow measurement of both estradiol-receptor and progesterone-receptor concentration in cytosol prepared from samples of 50 mg or more. Recently, Levy and co-workers have introduced and proposed a glass fiber filter exchange assay, which permits the measurement of nuclear receptors on the same amount of tissue (C. Levy *et al.,* 1980b).

2.8.4a. Distribution of Estrogen and Progesterone Receptors in Human Endometrium. The location of the biopsy inside the uterine cavity is often unknown, and therefore the interpretation of receptor assay would be greatly facilitated if receptors were evenly distributed throughout the endometrial lining. Conflicting results have been published on this matter. For Robertson *et al.* (1971) and Tsibris *et al.* (1978), the concentration of receptor in endometrium showed a progressive decrease throughout the length of the organ from the fundus to the cervix, whereas Lunan and Green (1975) reported more receptors in the body than in the fundus. Brush *et al.* (1967) indicated that in some cases the uptake of estradiol by different regions of the endometrium varies considerably. Bayard *et al.* (1978) also reported large differences in receptor concentration between biopsies of the fundus and of the body of the uterus, but they concluded that no systematic trend was observed, and the ratio of progesterone to estradiol receptors remained practically constant. In any case, a compromise can be reached by taking biopsies systematically from the midregion of the endometrium.

2.8.4b. Preservation of Endometrial Samples. Endometrial samples, kept in isotonic saline or homogenization buffer at 0–4°C, must be processed relatively rapidly (within 1 hr) to prevent inactivation of receptor sites. Receptors are also inactivated when endometrial curettings are kept in liquid nitrogen. However, cytosol and nuclear estrogen and progesterone receptors remained stable when endometrial samples were immersed in a preservation medium and then frozen in liquid nitrogen (Bayard *et al.*, 1978). Likewise, Koenders *et al.* (1978) reported no loss of cytosol receptor sites when samples were frozen in liquid nitrogen, lyophylized in glass vials stoppered under vacuum, and stored at 4°C. Both techniques allow the shipment of endometrial biopsies from clinical departments to distant biochemical laboratories and permit convenient and simultaneous receptor measurements in small series of samples.

3. *Estradiol and Progesterone Receptors in the Normal Menstrual Cycle*

The dating of normal endometrium is usually based on a combination of several criteria, namely: day of cycle (when the regularity of cycles is known), basal body temperature, histological evaluation, and serial measurements of plasma luteinizing hormone, estradiol, and progesterone. At best, the dating of the biopsy can be made with ± 1-day precision. In most reports dealing with receptor measurement, only broad terms are used, such as "proliferative" or "secretory" endometrium or both. In the few publications in which receptors had been measured on a daily basis, the criteria for dating were usually not well defined. In our opinion, valuable information can be obtained by separate investigations of the four following periods of the menstrual cycle: (1) the early proliferative phase; (2) the late proliferative phase, when the preovulatory plasma estradiol surge occurs; (3) the early secretory phase extending from the day of ovulation to the day of implantation; and (4) the late secretory phase.

3.1. Estradiol Receptor

3.1.1. Retention of [^{3}H]Estradiol in Whole Tissue

The first indication of variable retention of [^{3}H]estradiol throughout the menstrual cycle came from *in vivo* experiments. At 2 hr following intravenous injection of radioactive hormone, the concentration of retained radioactivity was higher in proliferative than in secretory endometrium (Brush *et al.,* 1967). This was confirmed by *in vitro* incubation of tissue slices (Evans and Hähnel, 1971; Trams *et al.,* 1973). Tseng and Gurpide (1972a) measured the amount of estradiol tightly bound to nuclei after *in vitro* superfusion or incubation of endometrial slices. Under carefully controlled conditions, the amount of nuclear estradiol–receptor complexes was found to be 3.1 pmol/mg DNA in proliferative, 1.6 in early secretory, 0.6 in midsecretory, and 0.5 in late secretory endometrium (Gurpide *et al.,* 1976; Tseng *et al.,* 1977). A similar approach was utilized by Crocker *et al.* (1974), who observed that the nuclear uptake of estradiol reached a peak in the late proliferative phase.

3.1.2. Cytosol Receptor

In most reports, the adopted methodology has allowed the measurement of unfilled receptor sites, and failed to account for the portion occupied by hormone. Under these conditions, receptor concentration was found to be higher in proliferative than in secretory endometrium (Trams *et al.,* 1973; Crocker *et al.,* 1974; Evans *et al.,* 1974; Pollow *et al.,* 1975a; Schmidt-Gollwitzer *et al.,* 1978). In one published article, a maximum of 2.3 pmol/mg cytosol protein was measured in the late proliferative phase (Crocker *et al.,* 1974). All other publications indicate a continuous decrease of mainly unfilled receptor sites from the early proliferative until the late secretory phase. This trend is undoubtedly due to a progressive increase of occupied receptor sites, because when the incubation temperature was raised to 25°C, the concentration of receptors increased more than 2-fold (Evans *et al.,* 1974). In addition, an inverse relationship was observed between the apparent K_D and the apparent concentration of receptor, a situation typical of measurements performed in the presence of endogenous hormone and non-steady-state conditions (Pollow *et al.,* 1976).

Several authors have determined the sum of unfilled and occupied receptor sites after exchange (Robertson *et al.,* 1971; Bayard *et al.,* 1978; Sanborn *et al.,* 1978; Levy *et al.,* 1980c). All of them agree that an increased amount of estradiol receptor appears at midcycle, with mean values reported between 1.2 and 3.5 pmol/mg DNA.

3.1.3. Nuclear Receptor

Several investigators have measured specific nuclear binding following incubation of tissue slices with estradiol. However, it should be recognized that such a technique does not measure endogenous nuclear receptors, but rather nuclearly translocated hormone–receptor complexes. An exchange assay of nuclear receptors in human endometrium has been reported (Bayard *et al.*, 1978). The concentration of nuclear receptor sites was found to be doubled between the early and late proliferative phases, becoming almost equal to cytosolic sites. Then, during the secretory phase, the nuclear-receptor level decreased but at a slower rate than the cytoplasmic receptor. Recently, it has been shown that endometrial nuclei may contain a significant proportion of receptor sites that can be labeled with [^{3}H]estradiol at 0°C and presumably are unoccupied by endogenous hormones (Levy *et al.*, 1980a).

3.2. Progesterone Receptor

3.2.1. Cytosol Receptor

Due to the rapid dissociation of endogenous progesterone–receptor complexes in the assay conditions used, both unfilled and occupied receptor sites have been measured by most authors, but often without consideration of isotopic dilution brought in by endogenous progesterone. Binding to CBG has been eliminated either by the addition of cortisol to incubations with progesterone or by the use of synthetic progestagens. In some early publications, very high values were reported (Rao *et al.*, 1974; Haukkamaa and Luukkainen, 1974), without significant changes throughout the cycle. In general, however, there is consensus that a remarkable increase of receptor concentration occurs in the late proliferative or midcycle period (MacLaughlin and Richardson, 1976; Syrjälä *et al.*, 1978; Sanborn *et al.*, 1978; Bayard *et al.*, 1978; Levy *et al.*, 1980c). Although early papers published by Pollow's group (Pollow *et al.*, 1975a, 1976) reported very low amounts of progesterone receptor in the proliferative phase, their recent reports showed agreement that a midcycle rise does occur (Schmidt-Gollwitzer *et al.*, 1978).

There is some divergence of opinion concerning the absolute concentration of progesterone receptor. In the late proliferative phase, the mean values reported and expressed were (in pmol/mg DNA): 0.7 (MacLaughlin and Richardson, 1976), 2.8 (Levy *et al.*, 1980c), 12 (Syrjälä *et al.*, 1978), 23 (Sanborn *et al.*, 1978), and 30 (Schmidt-Gollwitzer *et al.*, 1978). However, the levels of progesterone receptor recorded are always severalfold higher than those of estradiol receptor at the same phase of the menstrual cycle. The highest values were obtained with the use of synthetic progestins. The reasons for this discrepancy are not quite clear. However, it has been shown that in addition to binding to progesterone receptor, R 5020 (the most used synthetic progestagen) has considerable affinity for glucocorticoid receptors and appears to bind to serum and tissue proteins with greater avidity than does progesterone (Lippman *et al.*, 1977; Seematter *et al.*, 1978; Powell *et al.*, 1979).

3.2.2. Nuclear Receptor

Our group is the only one that has published results concerning nuclear receptor (Bayard *et al.*, 1978; Levy *et al.*, 1980c). The values reported were consistently lower than those observed for the cytosol receptor. A definite increase occurred in the early secretory phase, the value reaching 0.6 pmol/mg DNA (2500 sites/cell).

3.3. Hormonal Correlations of Estradiol and Progesterone Receptors

The simultaneous increases of estradiol and progesterone receptors at midcycle follow the plasma estradiol surge in the late proliferative phase and strongly suggest a positive relationship between estradiol and endometrial receptors. Indeed, during the proliferative phase, estradiol blood levels are positively correlated with total progesterone- and estradiol-receptor sites (Levy *et al.*, 1980c) and with cytoplasmic progesterone-receptor sites as well (Schmidt-Gollwitzer *et al.*, 1978).

A similar correlation was also reported between progesterone receptor and cytosol estradiol in human myometrium (Kontula, 1975). However, the concentration of unoccupied estradiol-receptor sites correlated inversely with the concentration of estradiol in the blood (Trams *et al.*, 1973; Schmidt-Gollwitzer *et al.*, 1978). The inductive action of estradiol was confirmed in estrogen-treated postmenopausal women by the appearance of progesterone receptors in endometrial cytosol (Jänne *et al.*, 1975) and the elevation of myometrial progesterone receptor to values reaching those reported for the proliferative phase (Illingworth *et al.*, 1975).

The large decrease of both estradiol and progesterone receptors in the secretory phase occurs when plasma and endometrial progesterone concentrations are increased. Indeed, a negative correlation has been reported between plasma progesterone and cytosol estradiol and progesterone receptors (Schmidt-Gollwitzer *et al.*, 1978; Jänne *et al.*, 1975). However, in one report in which statistical evaluation was performed, this inverse relationship was found not to be significant (Levy *et al.*,

1980c). Tseng and Gurpide (1975b) have demonstrated that progesterone and synthetic progestins reduce the level of estradiol receptor in human endometrium.

Another important feature of progesterone action is the large increase of microsomal 17β-hydroxysteroid dehydrogenase activity reported in secretory endometrium by Tseng and Gurpide (1974) as well as by Pollow *et al.* (1975b). This effect explains the relatively low concentration of estradiol in endometrium during the secretory phase (Schmidt-Gollwitzer *et al.,* 1978) and may partly account for the antiestrogenic characteristics of progestins (Gurpide *et al.,* 1977). Human endometrial estradiol dehydrogenase showed a severalfold increase under the influence of progestins, either *in vivo* or following *in vitro* incubation (Tseng and Gurpide, 1975a; Pollow *et al.,* 1978). Since progesterone increases estradiol dehydrogenase activity and decreases the progesterone-receptor level, an inverse relationship between the two parameters could be predicted and was indeed recently reported by Levy *et al.* (1980c), contrary to an earlier article by Schmidt-Gollwitzer *et al.* (1978).

4. General Conclusions

The changes of estradiol and progesterone receptors in human endometrium throughout the normal menstrual cycle are in keeping with the findings previously described in experimental animals. The increase of estradiol receptor in the preovulatory period is related to the plasma estradiol surge, leading to synthesis of more receptor in the cytoplasm and to nuclear translocation of the hormone–receptor complexes. At the same time, cytoplasmic progesterone receptor increases equally as a result of the inductive effect of estradiol (Milgrom *et al.,* 1972, 1973; Brenner *et al.,* 1974; Vu Hai *et al.,* 1978).

The postovulatory decrease of the estradiol receptor is possibly related to progesterone effects on both the conversion of estradiol to estrone and the level of the estradiol receptor (Mester *et al.,* 1974; Hsueh *et al.,* 1975). Consequently, there is a decrease of the estrogen-dependent synthesis of progesterone receptor. In addition, progesterone has been observed to directly "inactivate" its own receptor (Milgrom *et al.,* 1973; Brenner *et al.,* 1974; Tseng and Gurpide, 1975b). However, only cytoplasmic receptor sites decrease immediately following ovulation, whereas the nuclear receptor sites of both receptors do not. Rather, the nuclear progesterone-receptor concentration increases, probably as a consequence of sustained nuclear transfer of hormone–receptor complexes (Bayard *et al.,* 1978; Levy *et al.,* 1980c). This observation stresses the critical importance of measuring both filled and unfilled cytosolic and nuclear receptor sites whenever receptor physiology is being investigated.

During the menstrual cycle, the concentration of nuclear progesterone and estradiol receptors does not exceed that of cytoplasmic receptors because the levels of both hormones in the endometrium are far below the values needed to saturate the receptor sites. Consequently, a large proportion of receptor sites remains unoccupied and located in the cytoplasm.

However, this is not the case for gestational endometrium, in which very low cytosol (MacLaughlin and Richardson, 1976; Kreitmann *et al.*, 1978; Levy *et al.*, 1980c) and high nuclear progesterone-receptor concentrations were reported (Kreitmann *et al.*, 1978; Levy *et al.*, 1980c). The concentration of endogenous progesterone in gestational endometrium is markedly elevated and might explain the predominantly nuclear location of the progesterone receptors in the decidua of 8–12 weeks pregnancy.

The interest in measurement of sex-steroid-hormone receptors for the clinical investigation of gynecological disorders is illustrated by the results obtained in cases of endometrial hyperplasia (Haukkamaa and Luukkainen, 1974; Gurpide *et al.*, 1976; Syrjälä *et al.*, 1978), anovulatory cycles, and luteal insufficiency (Levy *et al.*, 1980c). The high risk of endometrial hyperplasia and adenocarcinoma in patients with anovulatory cycles (Gusberg, 1976) may prove to be related to unopposed prolonged estradiol secretion. It creates sustained high concentrations of nuclear estradiol–receptor complexes and a state of estrogen hyperreceptivity.

Recent reviews have appeared dealing with the application of progesterone- and estradiol-receptor measurements in human endometrial adenocarcinoma (McGuire *et al.*, 1977; Brush *et al.*, 1978; Richardson and MacLaughlin, 1978). The knowledge of the hormonal control of these receptors that has been acquired has permitted investigators to evaluate biochemical changes in endometrial adenocarcinoma samples following *in vivo* challenge with progestagen, estrogen, or antiestrogen (Robel *et al.*, 1978). Preliminary results are encouraging, and if they are confirmed, a more rational program could be designed for the treatment of patients with advanced or metastatic endometrial cancer.

References

Anderson, J., Clark, J.H., and Peck, E.J., 1972, Oestrogen and nuclear binding sites: Determination of specific sites by ^{3}H-oestradiol exchanges, *Biochem. J.* **126:**561–567.

Batra, S., Grundsell, H., and Sjöberg, N.O., 1977, Estradiol-17β and progesterone concentrations in human endometrium during the menstrual cycle, *Contraception* **16:**217–224.

Baulieu, E.-E., Atger, M., Best-Belpomme, M., Corvol, P., Courvalin, J.C., Mester, J., Milgrom, E., Robel, P., Rochefort, H., and De Catalogne, D., 1975, Steroid hormones receptors, *Vitam. Horm.* **33:**649–736.

Bayard, F., Louvet, J.P., Monrozies, M., Boulard, A., and Pontonnier, G., 1975, Endometrial progesterone concentrations during the menstrual cycle, *J. Clin. Endocrinol. Metab.* **41:**412–414.

Bayard, F., Damilano, S., Robel, P., and Baulieu, E.-E., 1978, Cytoplasmic and nuclear estradiol and progesterone receptors in human endometrium, *J. Clin. Endocrinol. Metab.* **46:**635–648.

Brenner, R.M., Resko, J.A., and West, N.B., 1974, Cyclic changes in oviductal morphology and residual cytoplasmic estradiol binding capacity induced by sequential estradiol-progesterone treatment of spayed rhesus monkeys, *Endocrinology* **95:**1094–1104.

Briggs, M.H., 1975, Comparative steroid binding to the human uterine progesterone-receptor, *Curr. Med. Res. Opinion* **3:**95–98.

Brush, M.G., Taylor, R.W., and King, R.J.B., 1967, The uptake of (6,7-^{3}H) oestradiol by the normal human female reproductive tract, *J. Endocrinol.* **39:**599–607.

Brush, M.G., King, R.J.B., and Taylor, R.W., 1978, *Endometrial Cancer,* Baillière Tindall, London.

Coffer, A.I., Milton, P.J.D., Pryse-Davies, J., and King, R.J.B., 1977, Purification of oestradiol receptor from human uterus by affinity chromatography, *Mol. Cell. Endocrinol.* **6:**231–246.

Collins, J.A., and Jewkes, D.M., 1974, Progesterone metabolism by proliferative and secretory human endometrium, *Am. J. Obstet. Gynecol.* **118:**179–185.

Crocker, S.G., Milton, P.J.D., and King, R.J.B., 1974, Uptake of 6,7 ^{3}H-estradiol 17β by normal and abnormal human endometrium, *J. Endocrinol.* **62:**145–152.

Davis, M.E., Wiener, M., Jacobson, H.I., and Jensen, E.V., 1963, Estradiol metabolism in pregnant and non-pregnant women, *Am. J. Obstet. Gynecol.* **87:**979–990.

Evans, L.H., and Hähnel, R., 1971, Oestrogen receptors in human uterine tissue, *J. Endocrinol.* **50:**209–229.

Evans, L.H., Martin, J.D., and Hähnel, R., 1974, Oestrogen receptor concentration in normal and pathological human uterine tissues, *J. Clin. Endocrinol. Metab.* **38:**23–32.

Feil, P.D., Glasser, S.R., Toft, D.O., and O'Malley, B.W., 1972, Progesterone binding in the mouse and rat uterus, *Endocrinology* **91:**738–746.

Guerrero, R., Landgren, B.M., Montiel, R., Cekan, Z., and Diczfalusy, E., 1975, Unconjugated steroids in the human endometrium, *Contraception* **11:**169–177.

Gurpide, E., Gusberg, S.B., and Tseng, L., 1976, Estradiol binding and metabolism in human endometrial hyperplasia and adenocarcinoma, *J. Steroid Biochem.* **7:**891–896.

Gurpide, E., Tseng, L., and Gusberg, S.B., 1977, Estrogen metabolism in normal and neoplastic endometrium, *Am. J. Obstet. Gynecol.* **129:**809–816.

Gusberg, S.B., 1976, The individual at high risk for endometrial carcinoma, *Am. J. Obstet. Gynecol.* **126:**535–542.

Hähnel, R., 1971, Properties of the estrogen receptor in the soluble fraction of human uterus, *Steroids* **17:**105–132.

Hähnel, R., Twaddle, E., and Bundle, L., 1974, Influence of enzymes on the estrogen receptors of human uterus and breast carcinoma, *Steroids* **24:**489–506.

Haukkamaa, M., 1974, Binding of progesterone by rat myometrium during pregnancy and by human myometrium in late pregnancy, *J. Steroid Biochem.* **5:**73–79.

Haukkamaa, M., and Luukkainen, T., 1974, The cytoplasmic progesterone receptor of human endometrium during the menstrual cycle, *J. Steroid Biochem.* **5**:447–452.

Heyns, W., and De Moor, P., 1971, The binding of 17β-hydroxy-5α-androstan-3-one to the steroid binding β globulin in human plasma, as studied by means of ammonium sulfate precipitation, *Steroids* **18**:709–730.

Hsueh, A.J.W., Peck, E.J., and Clark, J.H., 1975, Progesterone antagonism of the oestrogen receptor and oestrogen induced uterine growth, *Nature (London)* **254**:337–339.

Illingworth, D.V., Wood, G.P., Flickinger, G.L., and Mikhail, B., 1975, Progesterone receptor of the human myometrium, *J. Clin. Endocrinol. Metab.* **40**:1001–1008.

Jänne, O., Kontula, K., Luukkainen, T., and Vihko, R., 1975, Oestrogen induced progesterone receptor in human uterus, *J. Steroid Biochem.* **6**:501–509.

Jänne, O., Kontula, K., and Vihko, R., 1976, Progestin receptors in human tissues: Concentrations and binding kinetics, *J. Steroid Biochem.* **7**:1061–1068.

Koenders, A.J., Geurts-Moespot, H., Kho, K.H., and Benraad, T.J., 1978, Estradiol and progesterone receptor activities in stored lyophilised target tissue, *J. Steroid Biochem.* **9**:947–950.

Kontula, K., 1975, Progesterone-binding protein in human myometrium: Binding site correlation in relation to endogenous progesterone and estradiol-17β levels, *J. Steroid Biochem.* **6**:1555–1561.

Kontula, K., Jänne, O., Luukkainen, T., and Vihko, R., 1974, Progesterone-binding protein in human myometrium: Influence of metal ions on binding, *J. Clin. Endocrinol. Metab.* **38**:500–503.

Kontula, K., Jänne, O., Vihko, R., De Jaer, E., De Visser, J., and Zeelen, F., 1975, Progesterone-binding proteins: *In vitro* binding and biological activity of different steroidal ligands, *Acta Endocrinol.* **78**:574–592.

Kreitmann, B., Derache, B., and Bayard, F., 1978, Measurement of the corticosteroid binding globulin, progesterone and progesterone "receptor" content in human endometrium, *J. Clin. Endocrinol. Metab.* **47**:350–353.

Krishnan, A.R., Hingorani, V., and Laumas, K.R., 1973, Binding of ^{3}H-oestradiol with receptors in the human endometrium and myometrium, *Acta Endocrinol.* **74**:756–758.

Levy, C., Mortel, R., Eychenne, B., Robel, P., and Baulieu, E.-E., 1980a, Unoccupied nuclear oestradiol-receptor sites in normal human endometrium, *Biochem. J.* **185**:733–738.

Levy, C., Eychenne, B., and Robel, P., 1980b, Assay of nuclear estradiol receptor by exchange on glass fiber filters, *Biochim. Biophys. Acta* **630**:301–305.

Levy, C., Robel, P., Gautray, J.P., DeBrux, J., Verma, U., Descomps, B., Baulieu, E.-E., and Eychenne, B., 1980c, Estradiol and progesterone receptors in human endometrium: Normal and abnormal menstrual cycles and early pregnancy, *Am. J. Obstet. Gynecol.* **136**:646–651.

Lippman, M., Huff, K., Bolman, G., and Neifeld, J.P., 1977, Interactions of R5020 with progesterone and glucocorticoid receptors in human breast cancer and peripheral blood lymphocytes *in vitro*, in: *Progesterone Receptors in Normal and Neoplastic Tissues* (W.L. McGuire, J.P. Raynaud, and E.-E. Baulieu, eds.), pp. 193–210, Raven Press, New York.

Lunan, C.B., and Green, B., 1975, Oestradiol-17β uptake *in vitro* into the nuclei of endometrium from different regions of the human uterus, *Acta Endocrinol.* **78**:353–363.

MacLaughlin, D.T., and Richardson, G.S., 1976, Progesterone binding by normal and abnormal human endometrium, *J. Clin. Endocrinol. Metab.* **42**:667–678.

Makler, A., and Eisenfeld, A.J., 1974, *In vitro* binding of ^{3}H-estradiol to macromolecules from the human endometrium, *J. Clin. Endocrinol. Metab.* **38:**628–633.

Mercier-Bodard, C., Alfsen, A., and Baulieu, E.-E., 1970, Sex steroid binding plasma protein (SBP), *Acta Endocrinol.* **64** (Suppl. 147):204–224.

McGuire, W.L., Raynaud, J.P., and Baulieu, E.-E. (eds.), 1977, *Progesterone Receptors in Normal and Neoplastic Tissues,* Raven Press, New York.

Mester, J., Martel, D., Psychoyos, A., and Baulieu, E.-E., 1974, Hormonal control of oestrogen receptor in uterus and receptivity for ovoimplantation in the rat, *Nature (London)* **250:**776–778.

Milgrom, E., and Baulieu, E.-E., 1970, Progesterone in uterus and plasma. I. Binding in rat uterus 105,000 g supernatant, *Endocrinology* **87:**276–287.

Milgrom, E., Atger, M., Perrot, M., and Baulieu, E.-E., 1972, Progesterone in uterus and plasma. VI. Uterine progesterone *receptors* during the estrus cycle and implantation in the guinea pig, *Endocrinology* **90:**1071–1078.

Milgrom, E., Luu Thi, M., Atger, M., and Baulieu, E.-E., 1973, Mechanisms regulating the concentration and the conformation of progesterone *receptor*(s) in the uterus, *J. Biol. Chem.* **248:**6366–6374.

Notides, A.C., Hamilton, D.E., and Rudolph, J.H., 1972, Estrogen binding proteins of the human uterus, *Biochim. Biophys. Acta* **271:**214–224.

Philibert, D., and Raynaud, J.P., 1974, Binding of progesterone and R 5020, a highly potent progestin, to human endometrium and myometrium, *Contraception* **10:**457–466.

Pollow, K., Lübbert, H., Boquoi, E., Kreuzer, G., and Pollow, B., 1975a, Characterization and comparison of receptors for 17β-estradiol and progesterone in human proliferative endometrium and endometrial carcinoma, *Endocrinology* **96:**319–328.

Pollow, K., Lübbert, H., Boquoi, E., Kreuzer, G., Jeske, R., and Pollow, B., 1975b, Studies on 17β-hydroxysteroid dehydrogenase in human endometrium and endometrial carcinoma, *Acta Endocrinol.* **79:**134–145.

Pollow, K., Lübbert, H., Boquoi, E., and Pollow, B., 1975c, Progesterone metabolism in normal endometrium during the menstrual cycle and in endometrial carcinoma, *J. Clin. Endocrinol. Metab.* **41:**729–737.

Pollow, K., Boquoi, E., Schmidt-Gollwitzer, M., and Pollow, B., 1976, The nuclear estradiol and progesterone receptors of human endometrium and endometrial carcinoma, *J. Mol. Med.* **1:**325–342.

Pollow, K., Schmidt-Gollwitzer, M., Boquoi, E., and Pollow, B., 1978, Influence of estrogens and gestagens on 17β-hydroxysteroid dehydrogenase in human endometrium and endometrial carcinoma, *J. Mol. Med.* **3:**81–89.

Powell, B., Garola, R.E., Chamness, G.C., and McGuire, W.L., 1979, Measurement of progesterone receptor in human breast cancer biopsies, *Cancer Res.* **39:**1678–1682.

Rao, B.R., Wiest, W.G., and Allen, W.M., 1974, Progesterone "receptor" in human endometrium, *Endocrinology* **95:**1275–1281.

Raynaud, J.P., 1977, R 2050, a tag for the progestin receptor, in: *Progesterone Receptors in Normal and Neoplastic Tissues* (W.L. McGuire, J.P. Raynaud, and E.-E. Baulieu, eds.), pp. 9–21, Raven Press, New York.

Raynaud, J.P., Martin, P.M., Bouton, M.M., and Ojasoo, T., 1978, 11β-Methoxy-17-ethynyl-1,3,5(10)-estratriene-3,17β-diol (Mexestrol, a tag for estrogen receptor binding sites in human tissues, *Cancer Res.* **38:**3044–3050.

Richardson, G.S., and MacLaughlin, D.T. (eds.), 1978, *Hormonal Biology of Endometrial Cancer,* UICC Technical Reports Series, Vol. 42, UICC, Geneva.

Robel, P., Levy, C., Wolff, J.P., Nicolas, J.C., and Baulieu, E.-E., 1978, Réponse à un antioestrogen comme critère d'hormonosensibilité du cancer de l'endomètre, *C. R. Acad. Sci. (Paris)* **287:**1353–1356.

Robertson, D.M., Mester, J., Beilby, J., Steele, S.J., and Kellie, A.E., 1971, The measurement of high affinity oestradiol receptors in human uterine endometrium and myometrium, *Acta Endocrinol.* **68:**534–542.

Sanborn, B.M., Kuo, H.S., and Held, B., 1978, Estrogen and progestagen binding site concentrations in human endometrium and cervix throughout the menstrual cycle and in tissue from women taking oral contraceptives, *J. Steroid Biochem.* **9:**951–955.

Schmidt-Gollwitzer, M., Genz, T., Schmidt-Gollwitzer, K., Pollow, B., and Pollow, K., 1978, Correlation between estradiol and progesterone receptor levels, 17β-HSD activity and endometrial tissue levels of estradiol, estrone and progesterone, in: *Endometrial Cancer* (M.G. Brush, R.J.B. King, and R.W. Taylor, eds.), Baillière Tindall, London.

Scublinsky, A., Marin, C. and Gurpide, E., 1976, Localization of estradiol 17β-dehydrogenase in human endometrium, *J. Steroid Biochem.* **7:**745–747.

Seal, U.S., and Doe, R.P., 1966, Corticosteroid-binding globulin: Biochemistry, physiology, and phylogeny, in: *Steroid Dynamics* (G. Pincus, T. Nakao, and J.F. Tait, eds.), pp. 63–90, Academic Press, New York.

Seematter, R.J., Hoffman, P.G., Kuhn, R.W., Lockwood, L.C., and Siiteri, P.K., 1978, Comparison of ^{3}H-progesterone and (6.7-^{3}H)-17,21-dimethyl-19-norpregna-4,9-diene-3,20-dione for the measurement of progesterone receptors in human malignant tissue, *Cancer Res.* **38:**2800–2806.

Smith, H.E., Smith, R.G., Toft, D.O., Neergard, J.R., Burrows, E.P., and O'Malley, B.W., 1974, Binding of steroids to progesterone receptor proteins in chick oviduct and human uterus, *J. Biol. Chem.* **249:**5924–5932.

Smith, R.G., Iramain, C.A., Buttram, V.J., Jr, and O'Malley, B.W., 1975, Purification of human uterine progesterone receptor, *Nature (London)* **253:**271–272.

Steggles, A.W., and King, R.J.B., 1970, Use of protamine to study 6,7-^{3}H-estradiol-17β binding in rat uterus, *Biochem. J.* **118:**695–701.

Sutherland, R., Mester, J., and Baulieu, E.-E., 1977, Tamoxifen is a potent "pure" antioestrogen in chick oviduct, *Nature (London)* **267:**434–435.

Sweat, M.L., and Bryson, M.J., 1970, Metabolism of progesterone in proliferative human endometrium and myometrium, *Am. J. Obstet. Gynecol.* **106:**193–201.

Syrjälä, P., Kontula, K., Jänne, O., Kauppila, A., and Vihko, R., 1978, Steroid receptors in normal and neoplastic uterine tissue, in: *Endometrial Cancer* (M.G. Brush, R.J.B. King, and R.W. Taylor, eds.), pp. 242–251, Baillière-Tindall, London.

Trams, G., Engel, B., Lehmann, F., and Maass, H., 1973, Specific binding of oestradiol in human uterine tissue, *Acta Endocrinol.* **72:**351–360.

Tseng, L., and Gurpide, E., 1972a, Nuclear concentration of estradiol in superfused slices of human endometrium, *Am. J. Obstet. Gynecol.* **114:**995–1001.

Tseng, L., and Gurpide, E., 1972b, Changes in the *in vitro* metabolism of estradiol by human endometrium during the menstrual cycle, *Am. J. Obstet. Gynecol.* **114:**1002–1008.

Tseng, L., and Gurpide, E., 1974, Estradiol and 20α-dihydroprogesterone dehydrogenase activities in human endometrium during the menstrual cycle, *Endocrinology* **94:**419–423.

Tseng, L., and Gurpide, E., 1975a, Induction of human endometrial estradiol dehydrogenase by progestins, *Endocrinology* **97:**825–833.

Tseng, L., and Gurpide, E., 1975b, Effects of progestins on estradiol receptor levels in human endometrium, *J. Clin. Endocrinol. Metab.* **41:**402–404.

Tseng, L., Gusberg, S.B., and Gurpide, E., 1977, Estradiol receptor and 17β-dehydrogenase in normal and abnormal human endometrium, *Ann. N. Y. Acad. Sci.* **286:**190–198.

Tsibris, J.C.M., Cazenave, C.R., Cantor, B., Notelovitz, M., Kalra, P.S., and Spellacy, W.N., 1978, Distribution of cytoplasmic estrogen and progesterone receptors in human endometrium, *Am. J. Obstet. Gynecol.* **132:**449–454.

Verma, U., and Laumas, K.R., 1973, *In vitro* binding of progesterone to receptors in the human endometrium and the myometrium, *Biochim. Biophys. Acta* **317:**403–419.

Vu Hai, M.T., Logeat, F., and Milgrom, E., 1978, Progesterone receptors in the rat uterus: Variations in cytosol and nuclei during the oestrus cycle and pregnancy, *J. Endocrinol.* **76:**43–49.

Wiest, W.G., and Rao, B.R., 1971, Progesterone binding proteins in rabbit uterus and human endometrium, in: *Advances in the Biosciences,* Vol. 7, p. 251, Pergamon Press, Vieweg.

Young, P.C.M., and Cleary, R.E., 1974, Characterization and properties of progesterone-binding components in human endometrium, *J. Clin. Endocrinol. Metab.* **39:**425–439.

13

Parturition, Lactation, and the Regulation of Oxytocin Receptors

Melvyn S. Soloff and Maria Alexandrova

1. Introduction

Modulation of hormone action can result from changes in the concentration of hormone receptors, as well as from changes in blood levels of the hormone. A number of studies have shown that target cells exposed to elevated concentrations of hormones develop reduced numbers of receptor sites for the hormones. In most systems studied, however, the changes in receptor concentration are not reflected physiologically by an altered response to the hormone. The oxytocin receptor is one of the exceptions. In this chapter, we will cite experimental evidence that supports the view that the actions of oxytocin on two of its targets—uterine smooth muscle and mammary myoepithelial cells—are regulated by the concentration of oxytocin receptors. This regulation appears to be important for the initiation of parturition and lactation in the rat. Factors that may play a role in regulating oxytocin-receptor concentration will be considered.

Melvyn S. Soloff • Department of Biochemistry, Medical College of Ohio, Toledo, Ohio 43699. *Maria Alexandrova* • Institute of Experimental Endocrinology, Slovak Academy of Sciences, Bratislava, Czechoslovakia.

2. Oxytocin Receptors

Specific receptor sites for oxytocin in the uterus and mammary gland have been defined on the basis of tissue and ligand specificity, and by a correspondence between the dose of oxytocin giving a half-maximal uterine response and the apparent K_d for oxytocin binding (for a review, see Soloff and Pearlmutter, 1979). Oxytocin-receptor sites from rat, human, sow, and ewe myometrium, ewe endometrium, rat mammary gland, and isolated cells from the rat mammary gland have apparent K_d values in the range of 0.5–5 nM. Specific oxytocin-binding sites are localized in the plasma-membrane fractions from the uterus and mammary gland.

With the possible exception of the involuting mammary gland (Schroeder *et al.,* 1977), there is no evidence of receptor regulation by positive or negative cooperativity. These conclusions are based on linear Scatchard plots, which indicate a single class of binding sites, and on the similarity in dissociation rate constants obtained by chemical and isotopic dilution of oxytocin (Schroeder *et al.,* 1977; Pearlmutter and Soloff, 1979). Several studies have shown, however, that the concentration of oxytocin-receptor sites is subject to regulation.

3. Changes in the Concentration of Oxytocin Receptors in the Myometrium in the Estrous Cycle

The sensitivity of rat uterine smooth muscle is greatest in proestrus and estrus (for references, see Soloff and Pearlmutter, 1979). This greater sensitivity appears to be the result of an increase in the concentration of myometrial oxytocin receptors near the time of estrus. In the ewe, which has an estrous cycle of about 17 days, the concentration of oxytocin receptors is maximal during estrus (Fig. 1). The concentration of receptor

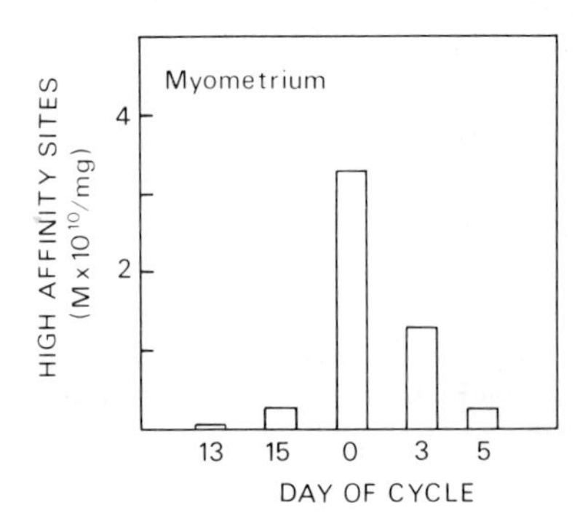

Figure 1. Changes in the concentration of oxytocin receptors per milligram of protein in the myometrium of the ewe during the estrous cycle. Adapted from Roberts *et al.* (1976).

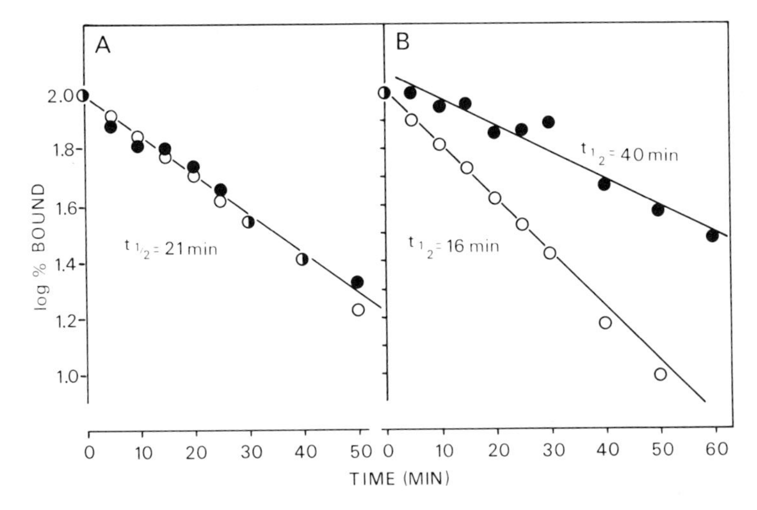

Figure 2. Comparison of the rate of dissociation of oxytocin from isolated mammary cells from lactating rats (A) and from rats with involuted mammary glands (B). The oxytocin-receptor complexes were formed by incubating 5.6 × 10^7 cells/ml with 4 nM [^{3}H]oxytocin for 30 min. Dissociation was initiated with a 1:50 dilution of the cell suspension containing either no (•) or 1 μM nonradioactive oxytocin (○). From Schroeder *et al.* (1977).

sites is relatively low 4 days prior to estrus, rises to a peak on the day of estrus, and falls to near baseline levels 5 days later. There is no change in the affinity of ewe myometrial membranes for oxytocin, the apparent K_d being in the range of 0.5–0.7 nM (Roberts *et al.*, 1976).

4. Oxytocin-Receptor Regulation in the Rat Mammary Gland

In isolated cells of the involuting mammary gland of the rat, the $t_{\frac{1}{2}}$ for dissociation of oxytocin is about 16 min in the presence of 1 μM oxytocin but about 40 min in the absence of the nonradioactive hormone (Fig. 2). The $t_{\frac{1}{2}}$ for oxytocin dissociation from mammary cells prepared from lactating rats is about 20 min, regardless of whether the dissociation rate is measured in the presence or absence of 1 μM oxytocin (Fig. 2). The 2-fold decrease in the dissociation rate constant seen with the involuting mammary gland suggests that the myoepithelial cell may be sensitive to lower concentrations of oxytocin by retaining the hormone longer. Presumably, after lactation has ceased, the production and release of oxytocin are diminished. When oxytocin concentrations are high, as in the case of the administration of exogenous hormone, oxytocin dissociates from its receptor site at the same rate as in the lactating mammary gland.

Apart from the case of the involuting gland, myoepithelial-cell action appears to be modulated by changes in the concentration of oxytocin receptors. The number of binding sites for oxytocin in a crude-membrane fraction from the rat mammary gland increases linearly throughout pregnancy and is greatest during lactation (Fig. 3). This increase in concentration of receptor sites corresponds to increases in the sensitivity of myoepithelial cells to oxytocin. Sala and Freire (1974) found that the sensitivity of mouse mammary strips to oxytocin increases more than 4-fold from day 9 to day 18 of pregnancy, and increases a further 2.5-fold during lactation.

5. *Changes in the Concentration of Myometrial Oxytocin Receptors Preceding Parturition*

In contrast to the pattern of oxytocin binding to mammary membranes during pregnancy, the concentration of oxytocin receptors in the rat myometrium remains low throughout most of pregnancy, rises abruptly several hours before labor, reaches a peak during labor, and then declines significantly by 1 day after parturition (Fig. 4). Scatchard analyses indicate that oxytocin is bound to myometrial membranes with an apparent K_d of 1–2 nM throughout pregnancy (Alexandrova and Soloff, 1980a). The degradation of [^{3}H]oxytocin by myometrial membranes is generally uniform throughout pregnancy, never exceeding 30% after the 1-hr incubation period (Alexandrova and Soloff, 1980a). Because the sensitivity of the rat myometrium to oxytocin also increases abruptly

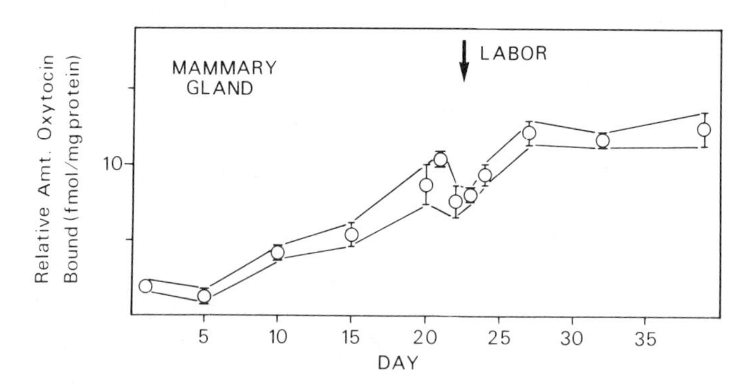

Figure 3. Changes in the relative amount of oxytocin bound specifically per milligram of protein by rat mammary membranes during pregnancy and lactation. Rats were maintained lactating with 8 pups. Each point is the mean ± S.E. of samples from 6 rats. From Soloff *et al.* (1979). © 1979 by the American Association for the Advancement of Science.

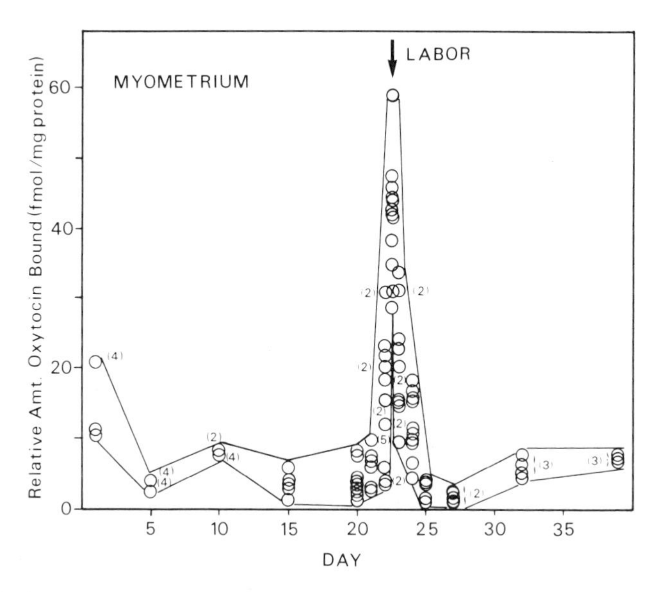

Figure 4. Relative amount of [^{3}H]oxytocin bound specifically by particulate fractions from the rat myometrium during pregnancy and parturition. Unless indicated by a number in parentheses, each point is a myometrial sample from one rat. Parturition occurred between the afternoon of day 22 and the morning of day 23. From Soloff *et al.* (1979). © 1979 by the American Association for the Advancement of Science.

several hours before term (Fuchs and Poblete, 1970), we postulated that the initiation of labor is related to the increased number of oxytocin receptors (see Section 7).

6. Possible Factors That Increase the Concentration of Myometrial Oxytocin Receptors

6.1. Estrogens and Progesterone

The sensitivity of the rat myometrium to oxytocin can be raised 13-fold by daily injections of diethylstilbestrol for 3 days (Follett and Bentley, 1964). Progesterone has the opposite effect. Progesterone injected into ovariectomized rats diminishes spontaneous contractions of the uterus *in vitro* and markedly reduces the response to oxytocin (Berger and Marshall, 1961).

Although estrogens may act at a number of loci in causing an increased sensitivity to oxytocin, the main effect seems to be on oxytocin binding to myometrial receptors. The injection of a single dose of diethylstilbestrol into ovariectomized rats increases the affinity of uterine membranes for oxytocin by more than 4-fold by 24 hr (Soloff, 1975). The concentration of oxytocin-binding sites per uterus is doubled at the same time. The administration of estradiol-17β for 4 days to young rabbits with intact ovaries causes a 3-fold increase in the concentration of oxytocin receptors per milligram of particulate protein (Nissenson *et al.,* 1978). Pharmacological doses of progesterone following estrogen administration cause a reduction in oxytocin binding to barely detectable levels and the abolition of an oxytocin effect on the contraction of rabbit myometrial strips (Nissenson *et al.,* 1978).

In view of effects of exogenous estrogen on myometrial receptors for oxytocin, the increase in oxytocin-receptor concentration occurring near estrus (see Fig. 1) appears to be the result of endogenous estrogen action. The antagonistic effects of progesterone may account for the sudden rise in oxytocin-receptor concentration in the rat myometrium near term because plasma progesterone levels fall markedly at this time (Fig. 5). Indeed, the surge in estradiol/progesterone ratios in rat plasma occurring near the time of parturition (Fig. 5) resembles the spike in oxytocin-receptor concentrations in the myometrium (see Fig. 4). Fur-

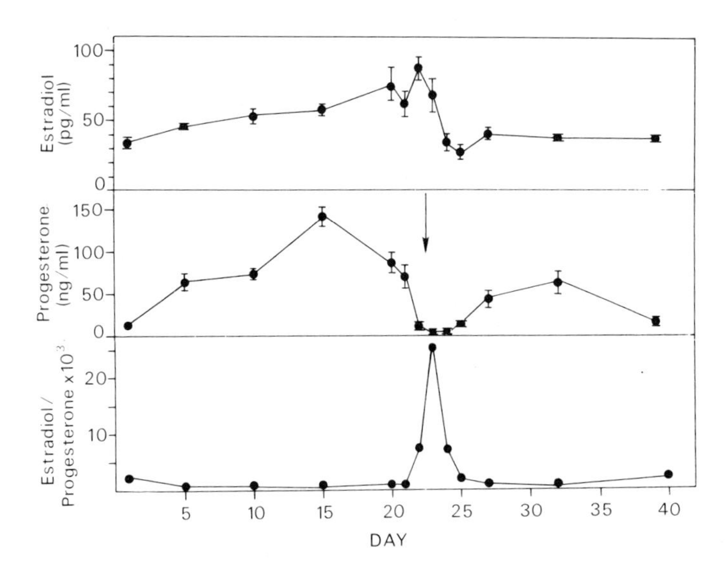

Figure 5. Concentrations of estradiol-17β and progesterone and the estradiol/progesterone ratio in peripheral plasma of rats during pregnancy. Each point is the mean ± S.E. of at least 8 rats. Blood was drawn from the abdominal aorta between 10 A.M. and noon. The concentrations of both steroids were determined by radioimmunoassay. From Soloff *et al.* (1979). © 1979 by the American Association for the Advancement of Science.

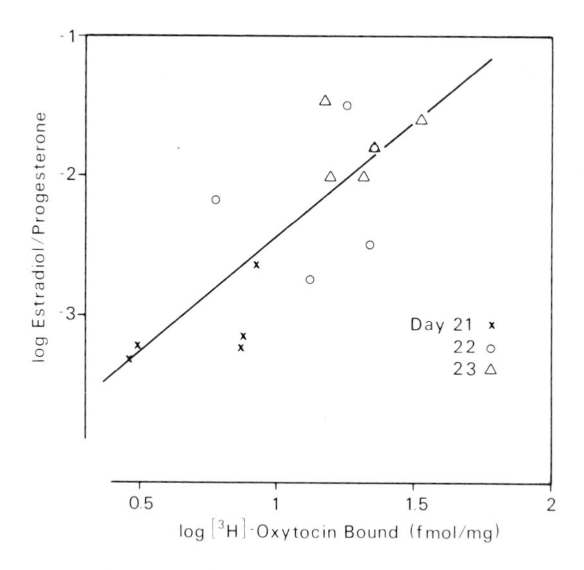

Figure 6. Comparison of the log ratio of plasma estradiol/progesterone concentrations and the log relative concentration of oxytocin receptors in the myometrium of individual rats preceding (days 21 and 22) and following (day 23) parturition. The correlation coefficient is 0.8, $n = 15$, and $p < 0.001$. From Alexandrova and Soloff (1980a). © 1980 by The Endocrine Society.

thermore, the concentration of myometrial oxytocin receptors is proportional to the plasma estradiol/progesterone ratio in individual rats near the time of parturition (Fig. 6). The concentration of receptors is also inversely proportional to the concentration of plasma progesterone, but there is no apparent relationship with plasma estradiol (Alexandrova and Soloff, 1980a).

6.2. *Myometrial Estrogen and Progestin Receptors*

Because estrogen promotes the synthesis of its own receptor and progestins antagonize estrogen action (Sarff and Gorski, 1971; Mester and Baulieu, 1975; Pavlik and Coulson, 1976; West *et al.*, 1976; Hseuh *et al.*, 1976; Clark *et al.*, 1977), we determined whether the marked increase in plasma estradiol/progesterone ratios is associated with an increase in myometrial estrogen-receptor concentration near term. The concentrations of estradiol receptors per milligram of DNA in both the nuclear fraction and the cytosol rise abruptly and are maximal near the time of parturition (Fig. 7). Changes in the concentration of progestin receptors during pregnancy are not remarkable when expressed in terms of protein concentration (Fig. 7). In the cytosol, the amount of progestin bound per milligram of DNA is maximal on day 20, whereas in the nuclear fraction, the greatest amount is bound on day 22. The rise in progestin binding in the nuclear fraction is not abrupt, however, as in the case of the nuclear estradiol receptor.

6.3. Relationship between Oxytocin-Receptor and Estrogen-Receptor Concentrations in the Myometrium

The relatively large error in estrogen-receptor levels shown in Fig. 7 may be due to differences in the length of gestation among individual rats. To minimize the error, we took more frequent samplings starting on the morning of day 22. The concentration of oxytocin receptors is significantly elevated ($p < 0.05$) by the afternoon of day 22 and rises to a maximum during labor on day 23 (Fig. 8). Maximal or near-maximal values for the concentration of myometrial estrogen receptors in both

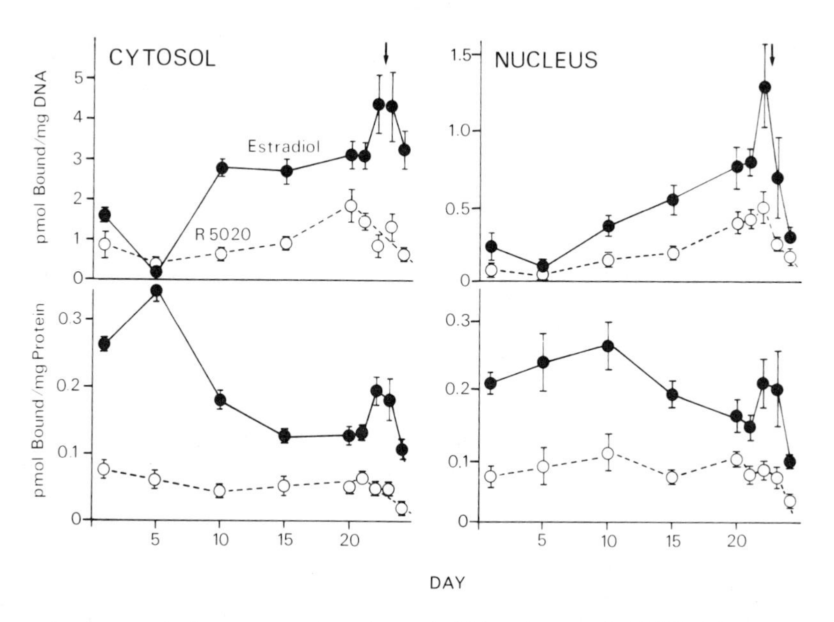

Figure 7. Changes in the concentration of estradiol (•) and progestin [R5020 (○)] receptors in the cytosol and nuclear fractions of the rat myometrium during gestation. The amount of steroid bound is expressed per milligram of DNA (top panels) and per milligram of protein (bottom panels). Parturition occurred between days 22 and 23 (arrows). Each point is the mean ± S.E. of samples from 8 rats. From Alexandrova and Soloff (1980a). © 1980 by The Endocrine Society.

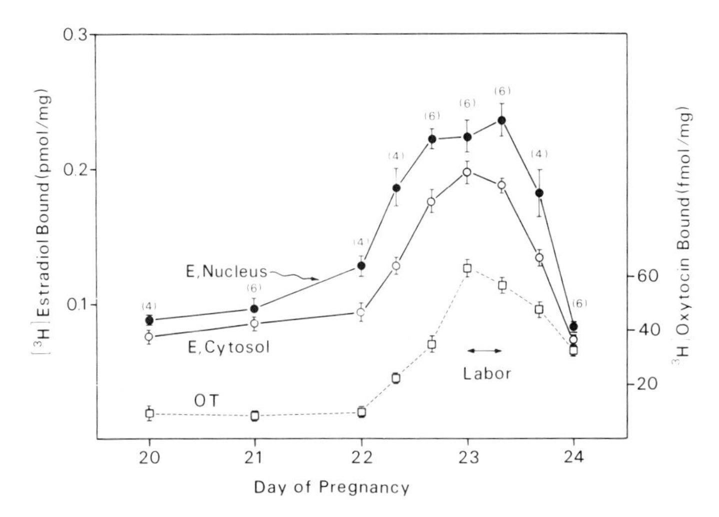

Figure 8. Increases in the concentration of oxytocin receptors (□) and of estrogen receptors in the nuclear (•) and cytosolic (○) fractions of rat myometrium in the perinatal period. Each point is the mean ± S.E. of the number of samples indicated in parentheses. The time of labor is indicated by the double-headed horizontal arrow. From Alexandrova and Soloff (1980a). © 1980 by The Endocrine Society.

cytosol and nuclear fractions are reached by the evening of day 22, more than 12 hr before the maximum in oxytocin-receptor concentrations (Fig. 8). The beginning of the rise in estrogen-receptor concentration in the nuclear fraction precedes the elevation in the cytosol receptor by several hours. The rise in the estrogen-receptor concentration in the nucleus on the morning of day 22 is significantly ($p < 0.05$) elevated over levels found on day 21, whereas the concentration of cytosol receptor on the morning of day 22 is unchanged from the previous day. The concentrations of estrogen receptors in both the cytosol and nucleus fall abruptly after parturition, whereas oxytocin-receptor levels decline more gradually. An examination of the relationship between the concentrations of nuclear estrogen receptors and oxytocin receptors near the time of parturition reveals that the two are directly proportional to each other (Fig. 9). Similar results are obtained when the concentrations of cytosolic estrogen receptors and oxytocin receptors are compared.

6.4. *Prostaglandins*

Because uterine prostaglandin synthesis increases prior to parturition (Vane and Williams, 1973; Harney *et al.,* 1974; Labhsetwar and Watson,

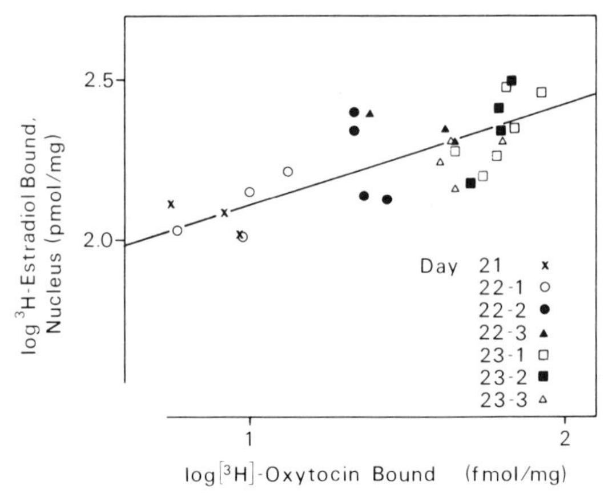

Figure 9. Comparison of the log concentration of nuclear estrogen receptors and oxytocin receptors in the myometrium of individual rats in the perinatal period. The rats are divided into the following groups: 21 (day 21); 22-1, 22-2, and 22-3 are morning, afternoon, and evening of day 22, respectively; 23-1, uterine contractions occurring but no pups born; 23-2, one or more pups born with some pups still *in utero;* 23-3, 2–3 hr after the birth of the last pup. The correlation coefficient is 0.75, $n = 28$, and $p < 0.001$. From Alexandrova and Soloff (1980a). © 1980 by The Endocrine Society.

1974; Carminati *et al.,* 1975), and inhibitors of prostaglandin synthesis prolong the onset of labor (Aiken, 1972; Chester *et al.,* 1972), endogenous prostaglandins are thought to be involved in the initiation of parturition. To determine the possible effects of prostaglandins on oxytocin receptors, we measured the concentration of myometrial estrogen and oxytocin receptors in the rat myometrium after a single injection of prostaglandin $F_{2\alpha}$ ($PGF_{2\alpha}$). The administration of $PGF_{2\alpha}$ on day 18 of pregnancy results in premature parturition on day 20, whereas control rats give birth on day 22 or 23 (Fuchs, 1972; Chatterjee, 1976; Alexandrova and Soloff, 1980c). As in the case of spontaneous labor, $PGF_{2\alpha}$-induced labor on day 20 is associated with a rise in the concentrations of myometrial estrogen and oxytocin receptors (Fig. 10). The administration of progesterone to $PGF_{2\alpha}$-treated rats inhibits the increases in concentration of receptors for estrogen and oxytocin (Fig. 10) and prevents premature parturition. $PGF_{2\alpha}$ treatment also depresses plasma progesterone to less than 20% of control levels, but does not significantly affect the plasma concentration of estradiol (Alexandrova and Soloff, 1980c). Thus, in induced abortion, there are changes in the plasma estradiol/progesterone ratio and in the concentrations of myometrial receptors for estrogen and oxytocin that are similar to the changes seen in spontaneous labor. Furthermore, the abortifacient activity of $PGF_{2\alpha}$ appears to be mediated by a rise in the concentration of oxytocin receptors. When $PGF_{2\alpha}$ administration is ineffective in interrupting gestation, as on day 15 of pregnancy, there is a marginal effect (if any) on the concentrations of estrogen and oxytocin receptors in the myometrium (Alexandrova and Soloff, 1980c).

6.5. Uterine Stretch

The gravid horn of unilaterally pregnant rabbits contracts with greater force and frequency than the nongravid horn in response to oxytocin (Fuchs and Fuchs, 1960). Csapo and Lloyd-Jacob (1962) postulated that the intrauterine volume plays a role in the timing of parturition. This idea suggested to us that distension of the uterus by the growing fetuses may contribute to an increase in oxytocin-receptor concentration and, accordingly, an increase in myometrial sensitivity. We compared the concentration of oxytocin receptors and estrogen receptors in gravid and nongravid uterine horns in unilaterally pregnant rats at or near term (Alexandrova and Soloff, 1980b). The gravid myometrium undergoes hyperplasia, as reflected by an approximate 6-fold increase in the amount of DNA (Fig. 11). Increases in the amount of protein in the cytosol, nuclear-extract, and crude-membrane fractions are greater than 6-fold, indicating that the presence of fetuses also contributes to hypertrophy of the myometrial cells (Fig. 11). Despite the large difference in weight between the gravid and nongravid horns of

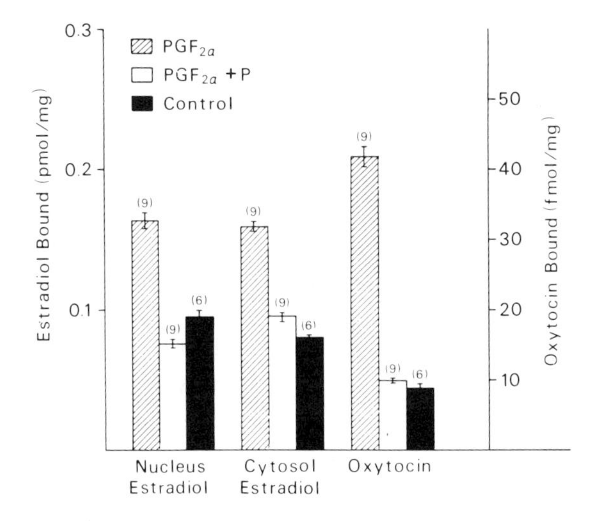

Figure 10. Effect of $PGF_{2\alpha}$ on the myometrial concentration of estradiol receptors in the nuclear and cytosol fractions and on oxytocin receptors. $PGF_{2\alpha}$, 0.5 mg, was administered subcutaneously on day 18 of pregnancy, and myometrial samples were taken during labor on day 20. Progesterone, 2 mg, was administered to one group of $PGF_{2\alpha}$-treated animals twice daily on days 18 and 19. Each value is the mean ± S.E. of myometrial samples from the number of rats indicated in parentheses. From Alexandrova and Soloff (1980c). © 1980 by The Endocrine Society.

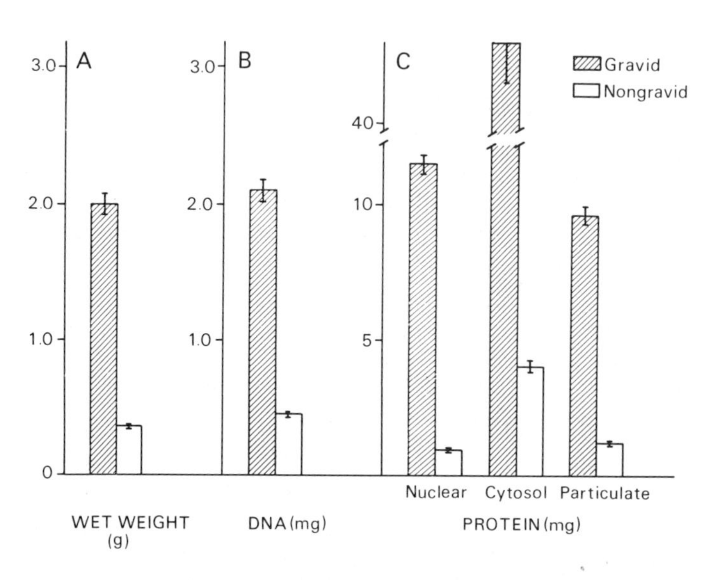

Figure 11. Comparison of the wet weight (A), the content of DNA (B) and the amount of protein in the nuclear, cytosol, and crude-membrane fractions (C) of the myometrium from the gravid and contralateral nongravid horns of unilaterally pregnant rats during or near the time of parturition. Each value is the mean ± S.E. of 10 replicates. From Alexandrova and Soloff (1980b). © 1980 by The Endocrine Society.

individual rats, however, the concentrations of estrogen and oxytocin receptors per cell are the same in both horns (Fig. 12). These data indicate that uterine stretch does not play a role in stimulating the marked rise in concentration of receptors for oxytocin and estrogen at term. It would appear, however, that an increased uterine volume contributes to a greater response to oxytocin by virtue of an expanded contractile apparatus (Csapo, 1950; Michael and Schofield, 1969), hyperplasia, and hypertrophy.

6.6. *Metal Ions*

Magnesium ions and other divalent cations are important for optimal contractile activities of oxytocin and its analogues on the myometrium and isolated mammary strips (for references, see Pearlmutter and Soloff, 1979). The addition of Mg^{2+} to the medium bathing mammary strips, for example, causes a parallel displacement to the left of log dose–response curves for a series of oxytocin analogues, while the maximal response is unchanged (Polacek and Krejci, 1969). These results suggest that the addition of Mg^{2+} causes an increase in the affinity of oxytocin receptors for the analogues. Comparable results have been obtained with the isolated uterus and a number of oxytocin analogues (see Permutter and Soloff, 1979). Studies with other analogues, however, have shown that both the maximal response of the uterus and the apparent affinity for the peptides are diminished in the absence of Mg^{2+} (Walter et al., 1968). The reduction in response has been assumed to be due to a decrease in the intrinsic activity of the peptides. However, it is also possible that the diminished response is the result of a lower concentration of receptor sites for oxytocic peptides. In support of this possibility, Perlmutter and Soloff (1979) found that certain divalent cations are essential for the binding of oxytocin to its receptor sites. Increasing concentrations of divalent nickel, magnesium, and manganese cause an increase in the concentration of binding sites for [^{3}H]oxytocin in a crude-membrane fraction from the mammary gland of the lactating rat. The apparent affinity for the hormone is unchanged by these metal ions. On the other hand, increasing concentrations of cobalt increase the affinity of receptor sites for oxytocin, but do not affect the concentration of sites available for oxytocin binding. This increase in affinity for oxytocin appears to be due to both a faster rate of association and a slower rate of dissociation of oxytocin. Calcium inhibits the effect of Mg^{2+}, but has no effect on oxytocin binding in itself. Divalent copper and iron have no effect.

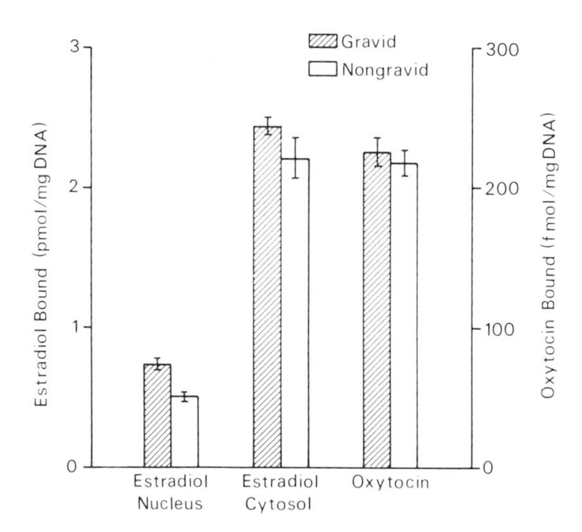

Figure 12. Comparison of the gravid and nongravid uterine horns from unilaterally pregnant rats at or near the time of parturition with respect to the amount of estrogen receptors per cell (nuclear and cytosol fractions). The amount of oxytocin receptor per cell is shown on the right. Each value is the mean ± S.E. of 10 samples. From Alexandrova and Soloff (1980b). © 1980 by The Endocrine Society.

An active metal ion actually appears to be capable of affecting both the affinity and the concentration of receptor sites for oxytocin. Pearlmutter and Soloff (1979) postulated that the oxytocin receptor contains two distinct regions that interact with metal ions (Fig. 13). The binding of metal ion to region A (availability) results in a receptor with the maximum number of oxytocin-binding sites available, but these sites have a low affinity for oxytocin. The binding of metal ion to region B (binding) results in a receptor site that is inaccessible to oxytocin, but that has a potentially high affinity for the peptide. The active metal ions are capable of reacting with both A and B sites, but when the metal-ion concentration is limiting, one or the other site is filled first, depending on the metal. In the case of nickel, magnesium, and manganese, the B site is filled with submillimolar concentrations of metal ion so that the affinity of the receptor for oxytocin is maximal. As the concentration of Ni^{2+}, Mg^{2+}, or Mn^{2+} is increased, filling a greater number of A sites, more high-affinity receptor sites become available for binding oxytocin. In the case of Co^{2+}, the A site is filled with the lowest concentrations of Co^{2+} studied, so that the maximal number of oxytocin-binding sites are

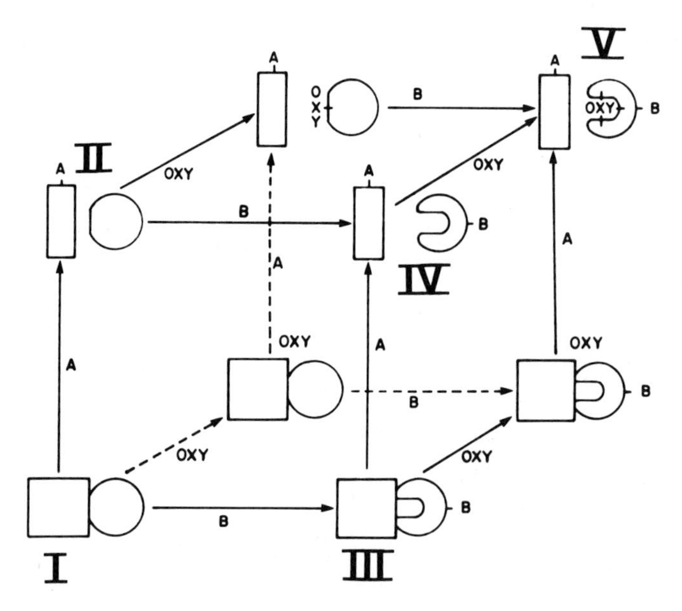

Figure 13. A model describing the mechanisms by which metal ions promote the specific binding of oxytocin to mammary receptor sites. The receptor (I) is composed of two metal-binding regions, A (availability) and B (binding). Oxytocin can be bound to species II with low affinity but high capacity and to species III with potentially high affinity but low capacity. Oxytocin is bound to species IV with both high affinity and high capacity. From Pearlmutter and Soloff (1979).

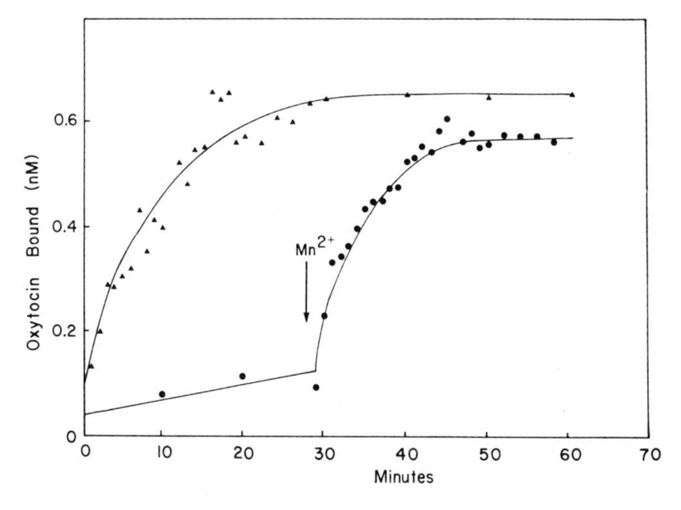

Figure 14. Effect of 5 mM Mn^{2+} on the specific binding of [^{3}H]oxytocin to mammary receptor. The metal ion was added either simultaneously with the hormone (▲) or after a delay of 30 min (●). The initial oxytocin concentration was 3.5 nM; 6 mg membrane protein/ml. From Pearlmutter and Soloff (1979).

available. These receptor sites have a low affinity for oxytocin, however, because the B region is not occupied by Co^{2+}. As the concentration of Co^{2+} is increased, more B sites are occupied and the affinity of the receptor for oxytocin increases.

The effects of the metal ion are very rapid. The delayed addition of Mn^{2+} to a mixture of oxytocin and receptor results in an oxytocin association rate identical to that when the oxytocin and metal ion are added simultaneously (Fig. 14). The dissociation of metal ion from a preformed metal ion–receptor complex with EDTA results in the dissociation of about 70% of the oxytocin bound within 1 min (Fig. 15).

Comparable detailed studies on metal ions have not been carried out with myometrial oxytocin receptors. We assume from initial studies, however (Soloff and Swartz, 1973, 1974), that the mechanisms of metal-ion action are similar in both myometrial and myoepithelial systems.

7. Mechanisms of Parturition

7.1. Oxytocin Receptors

On the basis of the observations already discussed, we have postulated that labor is initiated when the concentration of oxytocin receptors exceeds a threshold level. This postulate is consistent with the data

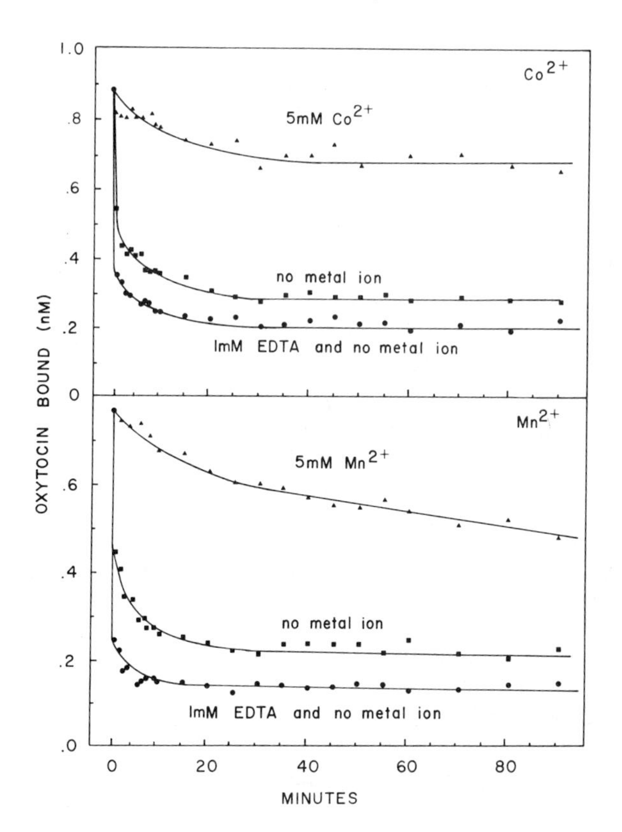

Figure 15. Requirement of metal ion for an intact oxytocin–receptor complex. Oxytocin (4 nM) was preequilibrated with either 5 mM Co^{2+} (top panel) or Mn^{2+} (bottom panel) and mammary membranes. At $t = 0$, the dissociation of the oxytocin–receptor complex was initiated by a 1:50 dilution in buffer containing 5 mM metal ion (▲), buffer without metal ion (■), or buffer containing no metal ion and 1 mM EDTA (•). From Pearlmutter and Soloff (1979).

shown in Fig. 4 and with the observations that the administration of oxytocin to pregnant rats does not induce parturition except when given 6–8 hr or less before actual term (Fuchs and Poblete, 1970). Similar results have been found with the rabbit (Fuchs, 1964). Our hypothesis is also compatible with a number of observations showing that oxytocin concentrations in the blood do not increase prior to parturition but only during expulsion of the fetuses (for references, see Soloff *et al.*, 1979). Therefore, parturition may not be triggered by an increase in oxytocin secretion, but by an increase in the concentration of myometrial oxytocin receptors. This increase would cause an increase in the total quantity of oxytocin bound, resulting in the ability of the uterus to react to the concentrations of oxytocin existing in the circulation. In other words, the response of a target organ to oxytocin is regulated by the target organ itself, rather than by an increase in the blood concentration of the hormone. The increased binding of oxytocin may be a reflection of a greater number of myometrial cells capable of reacting to oxytocin in a coordinated fashion. During lactation, the concentration of oxytocin receptors in the mammary gland is already maximal (see Fig. 2). Milk ejection appears to be preceded by or to occur concomitantly with a spike rise in the concentration of oxytocin in the systemic circulation (for references, see Bisset, 1968; Gorewit, 1979). The contraction of myoepithelial cells in the lactating mammary gland therefore requires increases in the concentration of both oxytocin receptors and circulating oxytocin.

7.2. Progesterone Withdrawal and $PGF_{2\alpha}$

The rise in myometrial oxytocin-receptor concentrations preceding parturition in the rat seems to be a reflection of progesterone withdrawal. The removal of this estrogen antagonist appears to allow the expression of estrogen activity in causing an increase in oxytocin-receptor concentration. It is possible that the actions of estrogen on oxytocin receptors are not direct. Since estrogens stimulate the synthesis of prostaglandins by the rat uterus (Castracane and Jordan, 1975; Ham *et al.*, 1975), estrogen-induced increases in endogenous $PGF_{2\alpha}$ levels might stimulate the increase in myometrial oxytocin receptors. However, treatment of ovariectomized rats with indomethacin, an inhibitor of prostaglandin synthesis, does not affect the activity of estrogen in inducing increases in the concentration of oxytocin receptors in the rat myometrium (Table I). Other lines of evidence also suggest that the initiation of labor does not involve a direct uterotonic effect of prostaglandins on the myometrium. First, there is a lag period of about 48 hr between the administration of $PGF_{2\alpha}$ and labor in the rat (Fuchs, 1972; Chatterjee, 1976.) Second, $PGF_{2\alpha}$ does not induce labor when given after day 21 of gestation, a time when oxytocin is effective (Fuchs, 1972). Third, PGE_1, PGE_2, and $PGF_{2\alpha}$ are almost equipotent uterotonic agents, yet only $PGF_{2\alpha}$ terminates gestation when administered on day 18 (Fuchs *et al.*, 1974). Liggens *et al.* (1972) have found that $PGF_{2\alpha}$ is not oxytocic in pregnant sheep, but it enhances the response of the uterus to oxytocin. The observation that exogenous progesterone negates the effects of $PGF_{2\alpha}$ on the length of gestation and on the concentration of myometrial estrogen and oxytocin receptors (see Fig. 10) is evidence for the luteolytic role of $PGF_{2\alpha}$. Thus endogenous prostaglandins likely play a role in causing a decline in plasma progesterone (Strauss and Stambaugh, 1974; Shaikh *et al.*, 1977),

Table I. Lack of Effect of $PGF_{2\alpha}$ and Indomethacin on the Concentration of Oxytocin Receptors in the Myometrium of Ovariectomized Rats[a]

Treatment	Number of rats	Relative concentration of oxytocin bound (fmol/mg protein)
Oil	4	3.33 ± 0.42
Estrogen	5	5.95 ± 0.33
Estrogen + indomethacin	5	6.40 ± 0.30
$PGF_{2\alpha}$	4	3.52 ± 0.48
$PGF_{2\alpha}$ + progesterone	4	1.78 + 0.50

[a] Ovariectomized rats, 200–225 g, were injected subcutaneously with estrogen (diethylstilbestrol, 5 μg in oil), indomethacin (1 mg/kg body weight in oil, twice, 12 hr apart), $PGF_{2\alpha}$ (0.5 mg of the tromethamine salt in 0.2 M sodium phosphate), progesterone (2 mg in oil, twice, 12 hr apart), or oil, and were killed 24 hr later. The amount of specific oxytocin binding by myometrial membranes was determined as described by Alexandrova and Soloff (1980a). Results are means ± S.E.

which in turn affects the concentration of oxytocin receptors in rat myometrium.

Estrogens stimulate the release of prolactin into the circulation of the pregnant rat, and progesterone antagonizes this action (Ajika *et al.*, 1972; Amenomori *et al.*, 1970; Morishige *et al.*, 1973; Vermouth and Deis, 1974). Indeed, the concentration of serum prolactin rises markedly at or just before parturition in the rat (Linkie and Niswender, 1972; Morishige *et al.*, 1973). We have considered the possibility that prolactin or another anterior-pituitary hormone mediates the effect of estrogen on myometrial oxytocin receptors. Hypophysectomy, however, does not influence the ability of estrogen to cause increases in the concentration of estrogen and oxytocin receptors in the myometrium of ovariectomized rat (Table II). These effects of estrogens therefore appear to be directly on the myometrial cells.

7.3. *Myometrial Estrogen Receptors*

Because of the relationship between the concentrations of estrogen and oxytocin receptors in the rat myometrium, we speculated that the increase in estrogen-receptor concentration mediates the actions of estrogen in stimulating increases in oxytocin-receptor concentration (Alexandrova and Soloff, 1980a). It is possible, however, that the increases in concentration of both receptors are separate manifestations of estrogen action and are not causally related to each other. The occupancy of existing estrogen-receptor sites by estradiol may be sufficient to cause the increase in oxytocin-receptor concentration. But because changes in the concentration of myometrial estrogen receptors precede changes in the concentration of oxytocin receptors, both before and after parturition (see Fig. 8), we propose as a working hypothesis that the levels of myometrial oxytocin receptors are regulated by the levels of estrogen receptors. This model is supported by our findings that in the pregnant guinea pig the increase in oxytocin-receptor concentration per myometrial cell is proportional to rise in nuclear estrogen-receptor concentration near term, while plasma progesterone levels are unchanged (Alexandrova and Soloff, 1980d). Factors other than plasma steroid ratios appear to be involved in causing the increases in estrogen/and oxytocin-receptor concentrations in the guinea pig.

7.4. A Model for Parturition in the Rat

Csapo (1975) postulated that withdrawal of progesterone secretion converts the myometrium from a suppressed to a spontaneously active state, capable of reacting to uterotonic agents. Our findings in the rat are in accord with this postulate. Berger and Marshall (1961) and Torok and Csapo (1976) suggested that progesterone may inhibit the activity of the myometrium by causing activator calcium to be sequestered. Our findings, however, indicate that the effects of progesterone on the pregnant-rat uterus may be due to factors that result in the suppression of oxytocin-receptor concentrations. We postulate that the fall in plasma

Table II. Lack of Effect of Hypophysectomy on Estrogen-Induced Increases in Oxytocin Binding by the Myometrium of Ovariectomized Rats[a]

Treatment	Relative concentration of oxytocin bound (fmol/mg protein)
Hypox–ovex, oil	3.85[b]
Ovex, oil	3.33 ± 0.42[c]
Hypox–ovex, estrogen	7.72[b]
Ovex, estrogen	5.95 ± 0.33[c]

[a] Hypophysectomized–ovariectomized and ovariectomized rats were treated with 5 μg estrogen (diethylstilbestrol) subcutaneously and were killed 24 hr later.
[b] Mean of duplicate determinations.
[c] Values from Table I.

progesterone levels relieves the suppression of myometrial estrogen-receptor concentrations. The resulting rise in concentration of estrogen receptors in the cytosolic and nuclear fractions of the myometrium, accompanied by increasing plasma estradiol concentrations, causes an increase in the concentration of myometrial oxytocin receptors. Parturition is then initiated when the concentration of oxytocin receptors exceeds a threshold and the sites become occupied by oxytocin from the bloodstream, irrespective of a rise in circulating oxytocin levels. The abrupt change in myometrial oxytocin-receptor levels near term explains the sudden change in sensitivity of the rat uterus to oxytocin near term. In species, such as the guinea pig, in which the sensitivity of the myometrium to oxytocin increases steadily toward term (Bell, 1941), oxytocin-receptor concentrations also increase in a steady manner and reach a peak during labor (Alexandrova and Soloff, 1980d).

We do not know whether the increase in oxytocin-receptor concentration at term is due to the unmasking of existing sites or to the *de novo* synthesis of receptor. In the former case, the interaction of certain divalent cations with oxytocin receptors may result in the rapid appearance of oxytocin-binding sites. In view of the activation of muscle contraction by increases in the intracellular concentration of Ca^{2+}, it is possible that local increases in the concentration of other divalent cations are important in the regulation of cellular activity. The phosphorylation or dephosphorylation of oxytocin receptors or of regulatory proteins also could theoretically account for a rapid change in the concentration of receptor sites. Alternatively, because estrogens stimulate the synthesis of specific uterine proteins, such as induced protein (Notides and Gorski, 1966) it is possible that the synthesis of oxytocin receptors is under estrogen control.

Because changes in the concentration of progesterone receptors in the myometrium of the pregnant rat are generally unremarkable (see Fig. 7), we have concluded that progestin receptors are not involved in the onset of parturition. Uterine stretch also does not contribute to the rise in myometrial oxytocin receptors near term. Increases in endogenous levels of $PGF_{2\alpha}$ may initiate the decline in progesterone synthesis, causing the sequence of events culminating in parturition. It is not clear, however, why in the apparent absence of progesterone withdrawal in species such as the guinea pig (Heap and Deanesly, 1966; Challis *et al.*, 1971), oxytocin-receptor levels in the myometrium are still maximal near term (Alexandrova and Soloff, 1980d). The reasons that oxytocin-receptor concentrations in the uterus and mammary gland appear to be regulated differently are also not understood.

The questions posed in this chapter are among the many intriguing problems concerning oxytocin action. Needless to say, regulation of the concentration of oxytocin receptors is fundamental to the processes of both parturition and lactation. Although we have only scratched the surface in our attempts to understand the mechanisms of parturition, the insights gained from our studies on oxytocin-receptor regulation provide a model for further testing.

ACKNOWLEDGMENTS. Studies from our laboratories were supported in part by Grant HD08406 and Contract N01-CB-63983 from the National Institutes of Health and by Grant 1-693 from the National Foundation March of Dimes.

REFERENCES

Aiken, J.W., 1972, Aspirin and indomethacin prolong parturition in rats: Evidence that prostaglandins contribute to expulsion of foetus, *Nature (London)* **240:**21.

Ajika, K., Krulich, L., Fawcett, C.P., and McCann, S.M., 1972, Effect of estrogen on plasma and pituitary gonadotropins and prolactin and on hypothalamic releasing and inhibiting factors, *Neuroendocrinology* **9:**304.

Alexandrova, M., and Soloff, M.S., 1980a, Oxytocin receptors and parturition. I. Control of oxytocin receptor concentration in the rat myometrium at term, *Endocrinology* **106:**730.

Alexandrova, M., and Soloff, M.S., 1980b, Oxytocin receptors and parturition. II. Concentrations of receptors for oxytocin and estrogen in the gravid and nongravid uterus at term, *Endocrinology* **106:**736.

Alexandrova, M., and Soloff, M.S., 1980c, Oxytocin receptors and parturition. III. Increases in estrogen receptor and oxytocin receptor concentrations in the rat myometrium during $PGF_{2\alpha}$-induced abortion, *Endocrinology* **106:**739.

Alexandrova, M., and Soloff, M.S., 1980d, Oxytocin receptors and parturition in the guinea pig, *Biol. Reprod.* **22:**1106.

Amenomori, Y., Chen, C.L., and Meites, J., 1970, Serum prolactin levels in rats during different reproductive states, *Endocrinology* **86:**506.

Bell, G.H., 1941, The behaviour of the pregnant uterus of the guinea-pig. *J. Physiol.* **100:**263.

Berger, E., and Marshall, J.M., 1961, Interactions of oxytocin, potassium, and calcium in the rat uterus, *Am. J. Physiol.* **201:**931.

Bisset, G.W., 1968, The milk-ejection reflex and the actions of oxytocin, vasopressin and synthetic analogues on the mammary gland, in: *Handbook of Experimental Pharmacology,* Vol. 23, *Neurohypophysial Hormones and Similar Polypeptides* (B. Berde, ed.) pp. 475–544, Springer-Verlag, New York.

Carminati, P., Luzzani, F., Soffientini, A., and Lerner, L.J., 1975, Influence of day of pregnancy on rat placental, uterine, and ovarian prostaglandin synthesis and metabolism, *Endocrinology* **97:**1071.

Castracane, V.D., and Jordan, V.C., 1975, The effect of estrogen and progesterone on uterine prostaglandin biosynthesis in the ovariectomized rat, *Biol. Reprod.* **13:**587.

Challis, J.R.G., Heap, R.B., and Illingworth, D.V., 1971, Concentrations of oestrogen and progesterone in the plasma of nonpregnant, pregnant and lactating guinea-pigs, *J. Endocrinol.* **30:**347.

Chatterjee, A., 1976, The possible mode of action of prostaglandins, XII. Differential effects of prostaglandin $F_{2\alpha}$ in inducing premature evacuation of conceptus in the intact and castrated pregnant rat, *Prostaglandins* **12:**1053.

Chester, R., Dukes, M., Slater, S.R., and Walpole, A.L., 1972, Delay of parturition in the rat by anti-inflammatory agents which inhibit the biosynthesis of prostaglandins, *Nature (London)* **240:**37.

Clark, J.H., Hsueh, A.J.W., and Peck, E.J., Jr., 1977, Regulation of estrogen receptor replenishment by progesterone, *Ann. N. Y. Acad. Sci.* **286:**161.

Csapo, A.I., 1950, Actomyosin of the uterus, *Am. J. Physiol.* **160:**46.

Csapo, A.I., 1975, The "seesaw" theory of the regulatory mechanism of pregnancy, *Am. J. Obstet. Gynecol.* **121:**578.

Csapo, A.I., and Lloyd-Jacob, M.A., 1962, Placenta, uterine volume, and the control of the pregnant uterus in rabbits, *Am. J. Obstet. Gynecol.* **83:**1073.

Follett, B.K., and Bentley, P.J., 1964, Bioassay of oxytocin: Increased sensitivity of the rat uterus in response to serial injections of stilboestrol, *J. Endocrinol.* **29:**277.

Fuchs, A.-R., 1964, Oxytocin and the onset of labour in rabbits, *J. Endocrinol.* **30:**217.

Fuchs, A.-R., 1972, Prostaglandin effects on rat pregnancy. I. Failure of induction of labor, *Fertil. Steril.* **23:**410.

Fuchs, A.-R., and Fuchs, F., 1960, The effect of oxytocic substances upon the rabbit uterus *in situ*, *Acta Physiol. Scand.* **49:**103.

Fuchs, A.-R., and Poblete, V.F., Jr., 1970, Oxytocin and uterine function in pregnant and parturient rats, *Biol. Reprod.* **2:**387.

Fuchs, A.-R., Mok, E., and Sundaram, K., 1974, Luteolytic effects of prostaglandins in rat pregnancy, and reversal by luteinizing hormone, *Acta Endocrinol.* **76:**583.

Gorewit, R.C., 1979, Method for determining oxytocin concentrations in unextracted sera; characterization in lactating cattle, *Proc. Soc. Exp. Biol. Med.* **160:**80.

Ham, E.A., Cirillo, V.J., Zanetti, M.E., and Kuehl, F.A., Jr., 1975, Estrogen-directed synthesis of specific prostaglandins in uterus, *Proc. Natl. Acad. Sci. U.S.A.* **72:**1420.

Harney, P.J., Sneddon, J.M., and Williams, K.I., 1974, The influence of ovarian hormones upon the motility and prostaglandin production of the pregnant rat uterus *in vitro*, *J. Endocrinol.* **60:**343.

Heap, R.B., and Deanesly, R., 1966, Progesterone in systemic blood and placentae of intact and ovariectomized pregnant guinea-pigs, *J. Endocrinol.* **34:**417.

Hsueh, A.J.W., Peck, E.J., Jr., and Clark, J.H., 1976, Control of uterine estrogen receptor levels by progesterone, *Endocrinology* **98:**438.

Labhsetwar, A.P., and Watson, D.J., 1974, Temporal relationship between secretory patterns of gonadotropins, estrogens, progestins, and prostaglandin-F in periparturient rats, *Biol. Reprod.* **10:**103.

Liggens, G.C., Grieves, S.A., Kendall, J.Z., and Knox, B.S., 1972, The physiological roles of progesterone, oestradiol-17β and prostaglandin $F_{2\alpha}$ in the control of ovine parturition, *J. Reprod. Fertil. Suppl.* **16:**85.

Linkie, D.M., and Niswender, G.D., 1972, Serum levels of prolactin, luteinizing hormone, and follicle stimulating hormone during pregnancy in the rat, *Endocrinology* **90:**632.

Mester, J., and Baulieu, E.-E., 1975, Dynamics of oestrogen-receptor distribution between cytosol and nuclear fractions of immature rat uterus after oestradiol administration, *Biochem. J.* **146:**617.

Michael, C.A., and Schofield, B.M., 1969, The influence of ovarian hormones on the actomyosin content and the development of tension in uterine muscle, *J. Endocrinol.* **44:**501.

Morishige, W.K., Pepe, G.J., and Rothchild, I., 1973, Serum luteinizing hormone, prolactin and progesterone levels during pregnancy in the rat, *Endocrinology* **92:**1527.

Nissenson, R., Flouret, G., and Hechter, O., 1978, Opposing effects of estradiol and progesterone on oxytocin receptors in rabbit uterus, *Proc. Natl. Acad. Sci. U.S.A.* **75:**2044.

Notides, A., and Gorski, J., 1966, Estrogen-induced synthesis of a specific uterine protein, *Proc. Natl. Acad. Sci. U.S.A.* **56:**230.

Pavlik, E.J., and Coulson, P.B., 1976, Modulation of estrogen receptors in four different target tissues: Differential effects of estrogen vs. progesterone, *J. Steroid Biochem.* **7:**369.

Pearlmutter, A.F., and Soloff, M.S., 1979, Characterization of the metal ion requirement for oxytocin receptor interaction in rat mammary gland membranes, *J. Biol. Chem.* **254:**3899.

Polacek, I., and Krejci, I., 1969, Effect of magnesium on the response of the rat mammary gland strip to oxytocin analogues, *Eur. J. Pharmacol.* **7:**85.

Roberts, J.S., McCracken, J.A., Gavagan, J.E., and Soloff, M.S., 1976, Oxytocin-stimulated release of prostaglandin $F_{2\alpha}$ from ovine endometrium *in vitro:* Correlation with estrous cycle and oxytocin-receptor binding, *Endocrinology* **99:**1107.

Sala, N., and Freire, F., 1974, Relationship between ultrastructure and response to oxytocin of the mammary myoepithelium throughout pregnancy and lactation: Effect of estrogen and progesterone, *Biol. Reprod.* **11:**7.

Sarff, M., and Gorski, J., 1971, Control of estrogen binding protein concentration under basal conditions and after estrogen administration, *Biochemistry* **10:**2557.

Schroeder, B.T., Chakraborty, J., and Soloff, M.S., 1977, Binding of [^{3}H]oxytocin to cells isolated from the mammary gland of the lactating rat, *J. Cell Biol.* **74:**428.

Shaikh, A.A., Naqvi, R.H., and Saksena, S.K., 1977, Prostaglandins E and F in uterine venous plasma in relation to peripheral plasma levels of progesterone and 20α-hydroxyprogesterone in the rat throughout pregnancy and parturition, *Prostaglandins* **13:**311.

Soloff, M.S., 1975, Uterine receptors for oxytocin: Effects of estrogen, *Biochem. Biophys. Res. Commun.* **65:**205.

Soloff, M.S. and Pearlmutter, A.F., 1979, Biochemical actions of neurohypophysial hormones and neurophysin, in: *Biochemical Actions of Hormones,* Vol. 6 (G. Litwack, ed.), pp. 265–333, Academic Press, New York.

Soloff, M.S., and Swartz, T.L., 1973, Characterization of a proposed oxytocin receptor in rat mammary gland, *J. Biol. Chem.* **248:**6471.

Soloff, M.S., and Swartz, T.L., 1974, Characterization of a proposed oxytocin receptor in the uterus of the rat and sow, *J. Biol. Chem.* **249:**1376.

Soloff, M.S., Alexandrova, M., and Fernstrom, M.J., 1979, Oxytocin receptors: Triggers for parturition and lactation?, *Science* **204:**1313.

Strauss, J.F., III, and Stambaugh, R.L., 1974, Induction of 20α-hydroxysteroid dehydrogenase in rat corpora lutea of pregnancy by prostaglandin $F_{2\alpha}$, *Prostaglandins* **5:**73.

Torok, I., and Csapo, A.I., 1976, The effects of progesterone, prostaglandin $F_{2\alpha}$ and oxytocin on the calcium-activation of the uterus, *Prostaglandins* **12:**253.

Vane, J.R., and Williams, K.I., 1973, The contribution of prostaglandin production to contractions of the isolated uterus of the rat, *Br. J. Pharmacol.* **48:**629.

Vermouth, N.T., and Deis, R.O., 1974, Prolactin release and lactogenesis after ovariectomy in pregnant rats: Effect of ovarian hormones, *J. Endocrinol.* **63:**13.

Walter, R., Dubois, B.M., and Schwartz, I.L., 1968, Biological significance of the amino acid residue in position 3 of neurohypophyseal hormones and the effect of magnesium on their uterotonic action, *Endocrinology* **83:**979.

West, N.B., Verhage, H.G., and Brenner, R.M., 1976, Supression of the estradiol receptor system by progesterone in the oviduct and uterus of the cat, *Endocrinology* **99:**1010.

V

Germ-Cell Regulation and Secretory Proteins

14

Control of Spermatogonial Multiplication

M. T. Hochereau-de-Reviers

1. Introduction

In the seminiferous tubules of the testis of adult mammals, spermatogonia are enclosed in the basal compartment of the seminiferous epithelium, with the nuclei and the basal cytoplasm of Sertoli cells, close to the basal membrane (Dym and Fawcett, 1970).

Two types of spermatogonia have been described. Dusty, or type A, spermatogonia are the least-differentiated germ cells, while crusty, or type B, spermatogonia are more differentiated; the last generation of type B spermatogonia form primary spermatocytes (Regaud, 1901), which enter the adluminal compartment of the tubules (Dym and Fawcett, 1970).

Spermatozoa are formed by continuous division and renewal of the stem spermatogonia. The quantitative production of spermatocytes and spermatozoa depends on the total number of stem spermatogonia per testis, the number of spermatogonial generations between stem cells and primary spermatocytes, the scheme of stem-cell renewal, and the yield of spermatogonial divisions.

This chapter reviews the variations in the different generations of spermatogonia, in stem-cell numbers, and in the yield of spermatogonial

M. T. Hochereau-de-Reviers • INRA, Station de Physiologie de la Reproduction, 37380 Nouzilly, France.

divisions. The pharmaceutical control of this phase of spermatogenesis will not be analyzed herein.

2. Material and Methods

Experimental control of spermatogonial multiplication was analyzed in rats, rams, and bulls. Seasonal variations were observed in rams and red deer stags.

After the animals were killed or castrated, the testes were fixed in Bouin–Hollande solution and treated as previously described (Hochereau, 1967).

The cycle of the seminiferous epithelium was classified into eight stages as described by Roosen-Runge and Giesel (1950). In the bull, ram, and red deer stag, Sertoli-cell nuclei and A_0 and A_1 spermatogonia were counted separately at stage 8 and leptotene primary spermatocytes at stage 2. In the rat, A spermatogonia and preleptotene primary spermatocytes were counted at stage 8 and A spermatogonia and zygotene primary spermatocytes at stage 3.

Adult rats received one injection of busulfan (10 mg/kg body weight) and 10 daily injections of ram rete testis fluid (RTF) and pregnant sow ovarian extract (PSOE). RTF was collected and treated as described by Blanc *et al.* (1978) and PSOE as described by Picaper *et al.* (1979). The rats were killed 10 days later.

In the nonbreeding season, adult rams, normal or hemicastrated for a month, were bled serially every 20 min for 10 hr. They were castrated a few days later. Correlations between follicle-stimulating hormone (FSH), luteinizing hormone (LH), and testosterone mean plasma levels and histological testicular parameters were calculated as described by Hochereau-de-Reviers *et al.* (1980a).

Adult rams were hypophysectomized during the sexual season and treated with human chorionic gonadotropin (hCG) (500 IU/day), pregnant mare serum gonadotropin (PMSG) (600 IU/day), or testosterone (0.5 mg/day) during 20 or 40 days after hypophysectomy as described by Courot *et al.* (1979).

3. Definitions of Parameters

3.1. Numbers of Stem Spermatogonia

Different nomenclature and schemes of stem spermatogonia renewal have been utilized during the past ten years. We prefer to use the term *AO reserve stem spermatogonia* (Clermont and Bustos-Obregon, 1968) to signify an ensemble of cell types which Huckins (1971) subdivides into single, paired, or aligned spermatogonia, and the term A_1 stem spermatogonia for cells that are in the G_1 phase during Stages 6–8 of the seminiferous epithelium cycle. They enter the S phase and encounter mitosis during Stage 1. These cells are at the origin of the cyclic activity of the seminiferous epithelium.

Their total number per testis was calculated from their number per cross section at stage 8 of the cycle, and the total length of the seminiferous tubules was calculated according to the method described by Attal and Courot (1963).

3.2. *Yield of Spermatogonial Multiplication*

Two types of spermatogonia, A and B can be identified in mammalian testes; between them, one or more generations of intermediate spermatogonia have been identified.

The theoretical number of spermatocytes and spermatozoa produced can be calculated according to the formula

$$\text{Total number of spermatocytes produced by a stem cell} = (2^K - 1)(2^{n-K})$$

where K is the order number of the spermatogonial division after which the stem cell is isolated and n is the number of spermatogonial generations between stem cells and primary spermatocytes.

The cyclic release of spermatozoa from a given portion of the mammalian seminiferous epithelium is related to the constant delay occurring between two successive mitoses of A_1 spermatogonia, i.e., the duration of the seminiferous epithelium cycle. This mitosis takes place during stage 1 of the cycle after the release of the spermatozoa in the lumen.

The number of spermatogonial multiplications between A_1 spermatogonia and primary spermatocytes has been determined by different methods (for reviews, see Courot *et al.*, 1970; Hochereau-de-Reviers, 1971): morphological and morphometrical analysis of spermatogonia; numbering of spermatogonia of each type at each stage of the seminiferous epithelium cycle; analysis of mitotic index after, or without, blockage by colchicine; analysis of labeling index just after labeling; index of labeled mitoses.

The number of spermatogonial generations between A_1 spermatogonia and primary spermatocytes varies among species (4–6), but is fixed for a given species: 6 in the bull (Hochereau, 1967) (Fig. 1), the ram (Hochereau-de-Reviers *et al.*, 1976b), the red deer stag (Hochereau-de-Reviers and Lincoln, 1978), the rat (Clermont and Bustos-Obregon, 1968), and the mouse (Monesi, 1962; Oakberg, 1971).

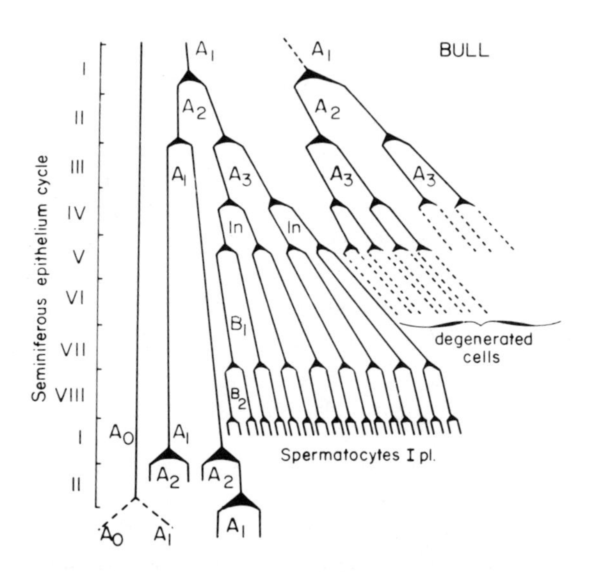

Figure 1. Scheme of spermatogonial multiplications and stem-cell renewal in the bull. A_0 reserve stem cells divide occasionally to give rise to new A_0 and A_1 spermatogonia. Most new A_1 spermatogonia arise from A_2 spermatogonial divisions. Only one third of the theoretical production of primary spermatocytes is obtained, since cells degenerate.

The respective number of type A or B spermatogonial generations varies among species. In the *Cercopithaecus* monkey (Clermont and Antar, 1973), there are 4 generations of type B spermatogonia. In the rat (Clermont and Bustos-Obregon, 1968), there are 4 generations of type A spermatogonia, 1 of intermediate, and 1 of type B spermatogonia.

The yield of spermatogonial multiplications is calculated from the ratio of the number of primary spermatocytes to that of A_1 spermatogonia per tubular cross section. This yield is lower than the theoretical one, since degeneration of cells occurs throughout the multiplication processes. In the rat, only half (Clermont and Bustos-Obregon, 1968) or less (Huckins, 1978) of the theoretical number of gametes is obtained; in the bull, only a third (Hochereau-de-Reviers, 1976).

4. Factors That Affect Variation: Results and Discussion

4.1. Age

A_0 spermatogonia are present in the impuberal calf, and their total number per testis does not vary significantly [$(206 \pm 26) \times 10^6$ vs. (325

$\pm$ 50) $\times$ 10^6] between 4 and 48 months of age (Hochereau-de-Reviers, 1976) (Fig. 2).

The total number of A_1 spermatogonia in the bull increases with age long after puberty: in the 4-month-old calf, few A_1 spermatogonia are observed (2–3 $\times$ 10^6); their number increases until 18 months of age (670 $\times$ 10^6/testis) and does not vary thereafter. The yield of spermatogonial divisions increases with age until adulthood in the calf (Attal and Courot, 1963) and the ram (Courot *et al.*, 1970). This arises from numerous degenerations that affect the first cellular generations and prevent them from developing to primary spermatocytes; they are less important in the adult.

4.2. Sertoli-Cell Numbers

Positive correlations (P = 0.01) between the numbers of Sertoli cells and A_1 spermatogonia per testis have been observed in adult males of different species (rat, ram, bull) (Hochereau-de-Reviers and Courot, 1978). The number of Sertoli cells is established before puberty, since Sertoli-cell mitosis ceases early in life (A. Steinberger and E. Steinberger,

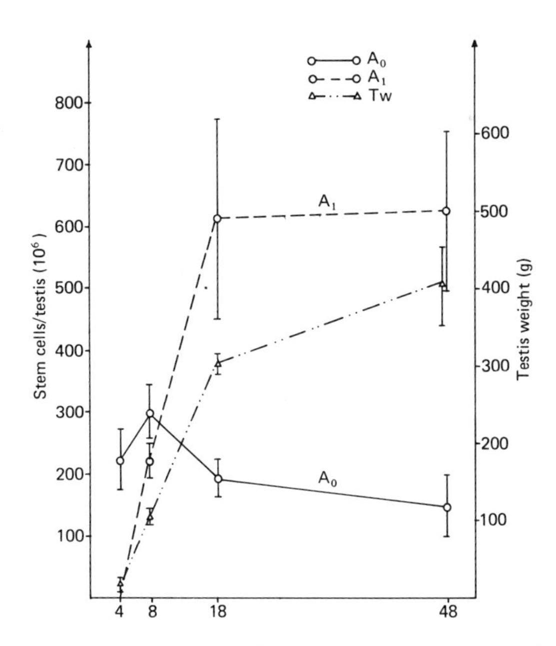

Figure 2. Evolution of the total numbers of type A_0 and A_1 spermatogonia per testis and of testis weight during the life of the bull (4–48 months of age). From Hochereau-de-Reviers (1971).

Table I. Interbreed Variations of Number of Stem Spermatogonia and Yield of Spermatogonial Multiplications in the Ram[a]

Rams	Spermatogonia/testis ($\times 10^6$)		
	A_0	A_1	Leptotene/A_1
Postpubertal (6 months old)			
Ile-de-France, spring-born	88 ± 23	322 ± 31	42 ± 5
Finn × Dorset, winter-born	66 ± 20	174 ± 29	45 ± 7
Finn × Dorset, summer-born	102 ± 21	261 ± 26	23 ± 3
Adult (1 year old)			
Ile-de-France	355 ± 54	441 ± 5	30 ± 3
Romanov	181 ± 19	222 ± 25	44 ± 5

[a] Results are means ± S.E.M.

1971; Nagy, 1972). This implies that Sertoli cells exert control on the A_1 spermatogonial populations. This is probably via hormonal control of the seminiferous tubules. The numbers of A_1 spermatogonia in hemicastrate adult rams was increased compared to control animals (558 ± 60 vs. 781 ± 82) without any augmentation of Sertoli-cell numbers (Hochereau-de-Reviers *et al.*, 1976a).

4.3. Genetics

In postpubertal lambs of the same age, born and killed at the same period of the year, there are genetic variations in the stem-cell number per testis and in the yield of spermatogonial divisions.

Yields of spermatogonial divisions are higher, but stocks of A_1 spermatogonia per testis are lower, in the Romanov than in the Ile-de-France breed (de Reviers *et al.*, 1980) (Table I). In 6-month-old postpub-

Table II. Intrabreed Variations of Number of Stem Spermatogonia in the Ile-de-France Ram[a]

Sire	Number of half brothers	Spermatogonia/testis ($\times 10^6$)		
		A_0	A_1	Leptotene/A_1
A	4	255 ± 48	328 ± 27	22.6 ± 3.7
B	3	173 ± 32	233 ± 49	25.4 ± 6.9
C	3	721 ± 551	410 ± 88	25.4 ± 2.5
D	5	296 ± 10	438 ± 15	27.8 ± 3.2

[a] Results are means ± S.E.M.

ertal lambs of the Ile-de-France or Finn × Dorset crossbreed, yields of spermatogonial multiplications are similar; however, total numbers of A_1 spermatogonia are lower in the Finn × Dorset than in the Ile-de-France (Land and Hochereau-de-Reviers, 1980).

Within the Ile-de-France breed, half brothers fathered by the same sire were similar, while there were significant variations among those fathered by different sires. No statistical differences are observed in total number of A_0 spermatogonia per testis and yield of spermatogonial divisions, while differences were observed in the number of A_1 spermatogonia per testis. The number of A_1 cells was correlated with that of Sertoli cells per testis (Table II).

4.4. Season

4.4.1. Season of Slaughter

In seasonal breeders, for example, rams and red deer stags, there are seasonal variations in the yield of spermatogonial divisions (Ortavant, 1958) as well as in the number of stem cells (Hochereau-de-Reviers *et al.*, 1976a; Hochereau-de-Reviers and Lincoln, 1978) (Table III) and in nuclear size of the Sertoli cells (Hochereau-de-Reviers *et al.*, 1976a), which could reflect variation in their synthetic activity.

4.4.2. Season of Birth

In Ile-de-France rams, born in winter or summer and killed when adult, during the sexual season, the numbers of Sertoli cells and total A_1 spermatogonia were different (de Reviers *et al.*, 1980). However, yields of spermatogonial multiplications as well as total numbers of A_0 stem cells per testis were similar in both groups of rams, being independent of the season of birth (Table IV).

Table III. Spermatogonial Variations According to Season of Slaughter in the Ram and the Red Deer Stag[a]

	Spring			Autumn		
	Spermatogonia/testis ($\times 10^6$)			Spermatogonia/testis ($\times 10^6$)		
Species	A_0	A_1	Leptotene/A_1	A_0	A_1	Leptotene/A_1
Ram	252 ± 50	343 ± 34	19.5 ± 2.2	530 ± 161	558 ± 60	26 ± 2
Red deer	58 ± 5	37 ± 6	5 ± 2	32 ± 4	49 ± 7	14 ± 2

[a] Results are means ± S.E.M.

Table IV. Spermatogonial Variations According to Season of Birth in the Ile-de-France Ram[a]

Season of birth	Spermatogonia/testis ($\times 10^6$)		
	A_0	A_1	Leptotene/A_1
Winter	310 ± 30	281 ± 21	32 ± 4
Summer	355 ± 54	441 ± 54	31 ± 3

[a] Results are means ± S.E.M.

4.5. Effect of Temperature

4.5.1. Elevation of Scrotal Temperature

In the ram, type B spermatogonia are much more sensitive to heat stress than intermediate or type A spermatogonia (Waites and Ortavant, 1968), while in the rat, type B spermatogonia have been claimed to be heat-resistant (Chowdhury and Steinberger, 1964). With repeated applications of heat stress in the rat (Bowler, 1972), the total number of damaged tubules increases and the proposed sensitive periods of the spermatogenetic cycle are mitoses of type A_1 to A_4 spermatogonia. Similarly, Chiu and Irvin (1978) observed that the synthesis of basic nuclear protein of type A spermatogonia is more sensitive to elevated temperature than that of late differentiated spermatogonia or primary spermatocytes.

In the testis of cryptorchid adult rams, most A_1 spermatogonia were destroyed while A_0 spermatogonia were still present (Hochereau-de-Reviers *et al.*, 1979). In the cryptorchid rat, only resting spermatogonia were observed, while dividing spermatogonia disappeared. Cooling of the cryptorchid testis in the pig resulted in a development of spermatogenesis (Frankenhuis and Wensing. 1979), showing that the blocking effect of cryptorchidism on spermatogenesis was mediated through elevation of temperature and was reversible.

4.5.2. Ambient Temperature

In the pig, elevation of the ambient temperature does not affect type A_1 spermatogonia, but does decrease the efficiency of spermatogonial divisions, as shown by the reduced number of leptotene primary spermatocytes (Wettemann and Desjardins, 1979).

4.6. Hormonal Plasma Levels

In prepubertal 120-day-old normal or hemicastrated lambs, the total number of A_0 spermatogonia per testis was negatively correlated with mean plasma FSH between 120 and 180 days of age (Hochereau-de-Reviers *et al.*, 1980) and the yield of spermatogonial divisions was positively correlated with mean plasma LH concentrations between 120 and 180 days of age as previously observed in adult rams (Hochereau-de-Reviers *et al.*, 1976a). In adult normal or hemicastrated rams, the total number of A_0 spermatogonia per testis was not significantly correlated with other parameters. The daily production of A_1 spermatogonia per testis was negatively correlated with the yield of spermatogonial multiplications, which was positively correlated with mean plasma LH levels, but not with testosterone levels (Table V). This suggests that the action of LH on this phase of spermatogenesis is not being mediated through testosterone in the ram (Hochereau-de-Reviers *et al.*, unpublished data).

In the human male, negative correlations between levels of FSH and number of spermatogonia or of primary spermatocytes per tubular cross section have been observed (de Kretser *et al.*, 1973), but no correlation has been observed between serum LH and either the sperm count or the stage of development of spermatogenesis as evaluated by testicular biopsy (Franchimont *et al.*, 1972). However, [^{125}I]-hCG receptors have been located in Ap and B spermatogonia (Fabbrini *et al.*, 1974). In the rat testis, FSH receptors have been shown with electron microscopy on cellular membrane of Sertoli cells and spermatogonia (Orth and Christensen, 1978).

In the rat, after vitamin A deficiency, no variations of FSH have been observed when only Sertoli cells and spermatogonia were present in the tubules (Krueger *et al.*, 1974), while after X-irradiation or busulfan treatment, which selectively depleted spermatogonia, FSH plasma levels were modified (Debeljuk and Mancini, 1971; Gomes *et al.*, 1973).

4.7. Hypophysectomy and Hormonal Supplementation

Variations of spermatogonial multiplications and of the stem-cell population after hypophysectomy in the rat have been disputed (Courot *et al.*, 1971) In hypophysectomized rats, primary spermatocytes were produced in such a way that it was often claimed that this step of spermatogenesis was independent of hormonal control (E. Steinberger, 1971; Matsuyama *et al.*, 1971). However, Clermont and Morgentaler (1955) and Courot *et al.* (1971) observed a decrease in type A spermatogonia and in the yield of spermatogonial multiplications after hypophy-

Table V. Correlations[d] between Histological Parameters and Hormonal Values in Normal or Hemicastrated Adult Rams[a]

Parameters	T.W.	Tot. vol. Leydig	Sertoli	A_0	D.P.A_1	D.P. leptotene	Lept./A_1[b]	FSH[c]	LH[c]	T[c]
Testis weight	1	0.83	0.33	−0.01	0.31	0.78	−0.21	−0.13	0.44	0.63
TV Leydig		1	0.12	0.19	0.02	0.73	0.53	0.11	0.44	0.67
Total number Sertoli			1	0.19	0.25	0.32	−0.20	−0.25	0.06	0.12
A_0 spermatogonia				1	−0.07	0.02	0.05	−0.20	0.12	0.04
Daily production A_1 spermatogonia					1	0.45	−0.70[c]	−0.27	−0.27	−0.08
Leptotene						1	0.22	−0.09	0.17	0.39
Lept./A_1[b]							1	0.13	0.57	0.32
Plasma FSH[c]								1	0.07	−0.31
Plasma LH[c]									1	0.29
Plasma testosterone[c]										1

[a] From Hochereau-de-Reviers *et al.* (unpublished data).
[b] (Lept./A_1) Yield of spermatogonial divisions.
[c] Mean of 31 bleedings.
[d] Statistical significance: $r \geqslant 0.576 : P \leqslant 0.05$
$0.497 \leqslant r \leqslant 0.576 : 0.1 < P < 0.05$.

sectomy. In the rat, some gonadotropic hypophyseal cells of the pars tuberalis still remain along the median eminence after hypophysectomy (Dubois *et al.*, 1971), resulting in low levels of gonadotropin in the blood plasma (Crumeyrolle *et al.*, 1979, personnal communication).

In hypophysectomized rams, after a long delay (40–80 days), 70–80% of A_1 spermatogonia had degenerated (Hochereau-de-Reviers *et al.*, 1976b). Moreover, new A_1 were not formed from divisions of A_0 spermatogonia, the total numbers of which decreased 40 days after hypophysectomy (Table VI) to the level observed in impuberal lambs (Hochereau-de-Reviers and Courot, 1978). In hypophysectomized lambs, divisions of gonocytes and, later, of A_0 spermatogonia still occurred, since their total number per testis increased (Courot, 1967), but no A_1 spermatogonia appeared (Hochereau-de-Reviers and Courot, 1978). In hypophysectomized human males, the degree of regression depends on the delay after hypophysectomy (Mancini, 1971).

In prepubertal hypophysectomized rats, FSH maintained testicular growth and spermatogenesis (Courot *et al.*, 1971) and stimulated [^{3}H]thymidine-labeling intensity of type A spermatogonia and of leptotene primary spermatocytes (Ortavant *et al.*, 1972). In a normal rat, this hormone promoted the mitotic activity of spermatogonia (Mills and Means, 1972) and decreased the degeneration of type A spermatogonia, mostly at the type A_4 generation level (Means, 1977).

In prepubertal clomiphen-treated rats, type A spermatogonia decreased and FSH plus testosterone or testosterone alone promoted their increase, but never restored the normal values (Kalra and Prasad, 1967). In impuberal estrogenized male rats, PMSG increased spermatogonial multiplications and primary spermatocytes (E. Steinberger and Duckett, 1967).

In adult hypophysectomized rats, FSH was shown to be efficient in maintaining spermatogonial multiplications and differentiation (Courot *et al.*, 1971; Vernon *et al.*, 1975). However, the role of FSH in spermatogenesis of adult hypophysectomized rats has been disputed (Clermont and Harvey, 1967), since LH or testosterone was able to maintain or restore spermatogenesis (Boccabella, 1963; Clermont and Harvey, 1967; Matsuyama *et al.*, 1971; Ahmad *et al.*, 1975; Chemes *et al.*, 1976; Harris et *et al.*, 1977; Rivarola *et al.*, 1977).

Precise quantitative analysis of spermatogonia in hypophysectomized, testosterone-treated adult rats showed that undifferentiated type $A(A_0–A_2)$ spermatogonia are sustained by testosterone or pituitary hormones, but that testosterone only partially restored A_3 to intermediate spermatogonial multiplication (Chowdhury, 1979). However, normal spermatogenesis was maintained despite low levels of intratesticular testosterone (Cunningham and Huckins, 1979b).

Table VI. Endocrine Control of Spermatogonial Multiplications in Hypophysectomized Adult Rams[a]

Duration of hypophysectomy	Treatment	A_0 spermatogonia/testis ($\times 10^6$)	Daily production ($\times 10^7$)			Yield of spermatogonial multiplications	
			A_1	Intermediate	Leptotene	Int./A_1	Lept./A_1
Intact controls	—	206 ± 42	2.6 ± 0.1	11.8 ± 1.8	94.5 ± 9.0	4.61	36.9
20 days	Hypox	279 ± 42	0.9 ± 0.3	2.2 ± 0.8	3.4 ± 1.6	2.44	3.78
	Hypox + testosterone	221 ± 58	2.1 ± 0.3	7.0 ± 1.8	34.4 ± 4.4	3.33	16.38
	Hypox + PMSG	258 ± 36	3.9 ± 0.8	11.5 ± 2.6	104 ± 16	2.95	26.67
	Hypox + hCG	257 ± 71	3.9 ± 0.8	10.8 ± 2.4	26 ± 4.8	2.77	6.67
40 days	Hypox	89 ± 27	0.8 ± 0.2	2.15 ± 0.60	5.4 ± 1.1	2.62	6.59
	Hypox + testosterone	140 ± 29	1.1 ± 0.5	3.3 ± 0.6	16.0 ± 2.9	2.76	13.45
	Hypox + PMSG	218 ± 46	2.0 ± 0.4	3.5 ± 1.0	42.7 ± 9.7	1.75	21.35
	Hypox + hCG	92 ± 19	1.1 ± 0.1	2.1 ± 0.4	1.03 ± 0.60	1.89	0.93

[a] Results are means ± S.E.M.

In adult hypophysectomized rams, testosterone was unable to maintain normal spermatogenesis (Monet-Kuntz *et al.,* 1976). After supplementation for 20 or 40 days with gonadotropin-like hormones, neither hCG nor PMSG was able to maintain a normal yield of spermatogonial multiplications (Table VI). However, PMSG supplementation allowed a better daily production of leptotene primary spermatocytes in increasing both A_1 spermatogonia and the yield of spermatogonial multiplication. Testosterone supplementation supported these parameters better than hCG, for 40 days (Table VI). We postulate that synergy between FSH and LH for a definite ratio of these hormones is necessary for maintenance of Sertoli-cell function and spermatogonia. However, testosterone secretion was at first greatly increased but not maintained more than 15 days in hCG-treated animals, and longer but at a variable level among animals in PMSG-treated rams (Monet-Kuntz *et al.,* unpublished); this could be due to down-regulation of LH receptors (Chasalow *et al.,* 1979).

4.8. Antihormones

4.8.1. Antiandrogens

In the growing prepubertal rat, the total number of type A spermatogonia per testis increased significantly after cyproterone or cyproterone acetate treatment. This could result either from a stimulation in the formation of stem-cell stock or from a blocking of the differentiation of stem cells into more differentiated spermatogonia (Viguier-Martinez and Hochereau-de-Reviers, 1977) (Fig. 3) as the yield of spermatogonial divisions is depressed (Neumann and Von Berswordt-Wallrabe, 1966; Heinert and Taubert, 1973; Flickinger and Loving, 1976; Viguier-Martinez and Hochereau-de-Reviers, 1977). Since we observed the same effects with both antiandrogens [i.e., pure antiandrogen (cyproterone), which increases gonadotropin, or antiandrogen and progestagen (without any increase of plasma gonadotropins)], androgens could be implicated in the differentiation of A_1 spermatogonia in the prepubertal rat. Androgens may have acted directly or through Sertoli cells, since the Sertoli-cell nuclear area was depressed after antiandrogens (Aumüller *et al.,* 1975; Viguier-Martinez and Hochereau-de-Reviers, 1977).

4.8.2. Antisera against Gonadotropins

Newborn rats have been deprived of endogenous gonadotropins by administration of antisera neutralizing these hormones from birth. Proliferation of spermatogonia and their transformation to spermatocytes were reduced in the absence of gonadotropins (Eshkol and Lunenfeld, 1971–1972).

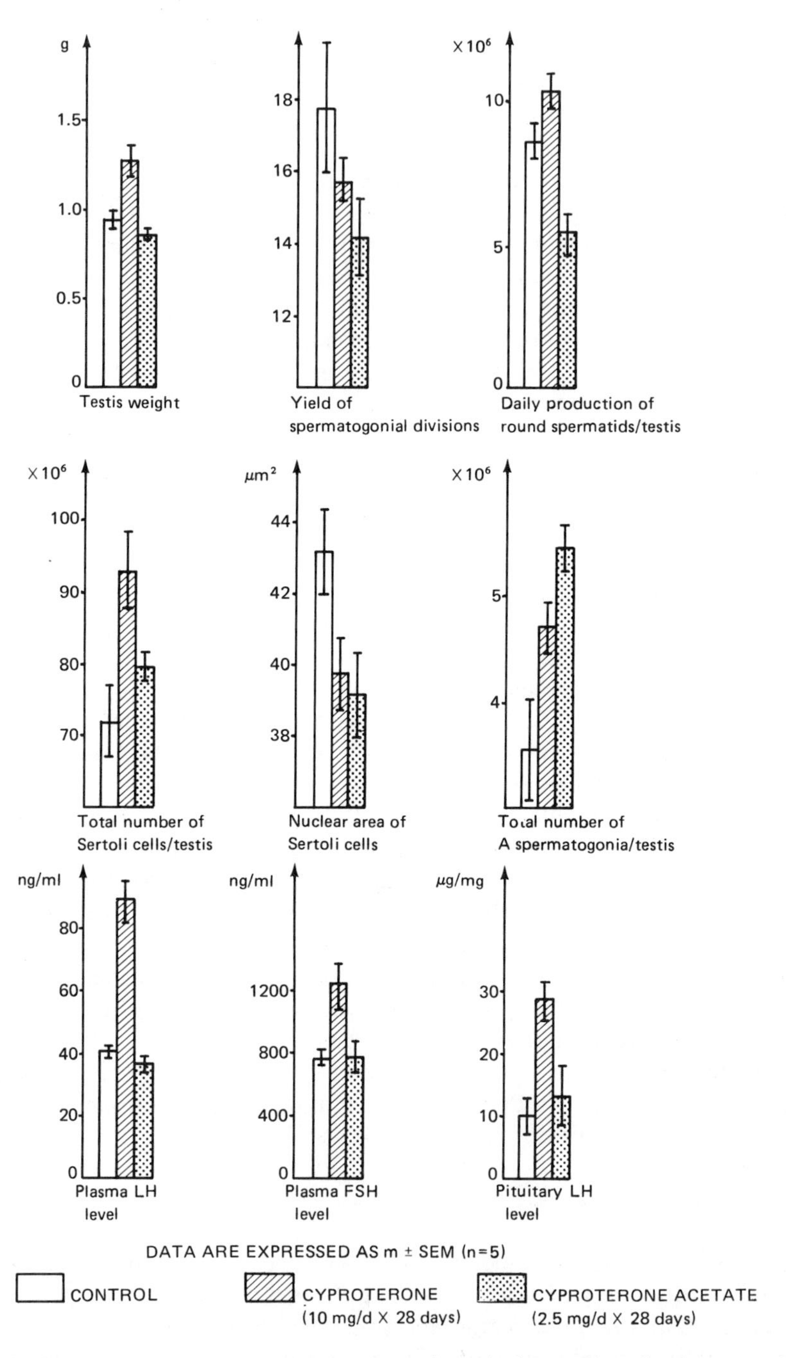

Figure 3. Evolution of male genital tract and gonadotropins in prepubertal male rats treated for 28 days with cyproterone (10 mg/day) or cyproterone acetate (2.5 mg/day).

In the prepubertal rat, FSH antisera decreased testis growth (Raj *et al.*, 1977a), while in the adult rat they were ineffective (Dym and Moudgal, in the discussion of Raj *et al.*, 1978). However, a decrease in fertility was observed after immunization against FSH in adult rats (Turner and Johnson, 1971).

LH antisera inhibited Leydig-cell function and thus blocked spermiogenesis (Raj *et al.*, 1978b).

In growing rabbits, immunization against LH inhibited the maturation of the gonads; spermatogenesis was inhibited at the spermatogonial level. FSH immunization induced a decrease in testis weight, but spermatogenesis was apparently normal (Monastirky *et al.*, 1971).

In the adult Bonney monkey after treatment with anti-FSH serum, a significant reduction in total sperm count and in the percentage of live spermatozoa and their motility was observed (Murty *et al.*, 1979).

4.9. *In Vitro Systems*

Tissue culture provided experimental models in which the environmental conditions were controlled. *In vitro* culture of impuberal 4-day-old rat testes resulted in the onset of spermatogenesis up to the pachytene stage without gonadotropins or testosterone (E. Steinberger *et al.*, 1963; A. Steinberger and E. Steinberger, 1966). In the culture of cryptorchid mouse testis—nearly comparable to impuberal testis—development of type A spermatogonia could be obtained only if serum was added (Aizawa and Nishimune, 1979).

With testis of mature animals, rapid degeneration of differentiated germ cells was observed, and only gonocyte-like cells remained in *in vitro* culture (E. Steinberger *et al.*, 1963). In ram testes cultured *in vitro* for 10 days, few differentiated spermatogonia divided and gave rise to primary spermatocytes evolving to the pachytene stage, but no renewal of A_1 spermatogonia had taken place and only A_0 spermatogonia were observed at the end of the culture (Hochereau-de-Reviers and Ghatnekar, 1974, unpublished data).

The difference between the responses of adult and impuberal testes to *in vitro* culture conditions could result from the differentiation of Sertoli cells. When differentiated, they could synthesize a factor—normally excreted into the lumen of the tubule or into the intertubular tissue—that remains present in the culture medium and blocks spermatogonial renewal.

4.10. *Chalones and Inhibin*

In many somatic tissues, local factors have been claimed to arrest the multiplication or differentiation, or both, of cells. These factors, named "chalones" by Bullough (1962), appeared to be water-soluble proteins. Such a factor has been identified in the testes of adult (Clermont and Mauger, 1974; Irons and Clermont, 1979) or growing (Clermont and Mauger, 1976) rats, but denied by Cunningham and Huckins (1979a). Adult rats receiving a single injection of busulfan (10 mg/kg) and daily injections of RTF or PSOE for 10 days were killed at the end of the treatment or after a delay of 10 days (Table VII). The preleptotene primary spermatocytes counted at the end of the treatment were depressed after RTF and PSOE, as compared to busulfan control animals. These cells were derived from cells that were type A_2 spermatogonia 10 days before or from cells that were type A_1 spermatogonia at stage 6, 20 days before. Type A spermatogonia deriving from type A_0 or A_1 spermatogonia, or both, present 10 or 20 days before were slightly decreased after all treatments, but significantly only after RTF treatment (Table VII). Thus, proteinaceous factors, present in both ovarian and testicular fluid, acted on spermatogonial multiplications in the adult rat testis. Is the factor(s) a chalone? Thumann and Bustos-Obregon (1978) observed an action of rat testis extract on proliferation of type A spermatogonia in culture *in vitro*. But the fact that it acted directly on seminiferous epithelium does not eliminate a central action of the same factors on the hypothalamic–hypophyseal axis.

Inhibin, known to inhibit FSH (Baker *et al.*, 1976; Setchell, 1977) and LH (Blanc *et al.*, 1978), is secreted by Sertoli cells in RTF (Blanc and Dacheux, 1976) and in sperm (Franchimont *et al.*, 1977) or in culture medium (de Jong *et al.*, 1978; A. Steinberger, 1979). Inhibin treatment of prepubertal rats resulted in a decrease of pachytene primary spermato-

Table VII. Corrected Numbers[a] of Type A Spermatogonia and Preleptotene Spermatocytes at Stage 8, 10 or 20 Days after Treatment with Busulfan, at the End of Daily Injections or 10 Days Later[b]

Days after treatment	Treatment[c]	Type A spermatogonia (stage 8)[a]	Preleptotene spermatocytes (stage 8)
	Control	2.09 ± 0.22 a	43.53 ± 1.92 a
10 days	Busulfan	0.48 ± 0.66 b	35.2 ± 2.0 b
	Busulfan + RTF b	0.29 ± 0.04 c	26.0 ± 4.3 c
	Control	2.13 ± 0.19 a	50.33 ± 2.28 a
20 days	Busulfan + BSA b	0.73 ± 0.17 b	18.3 ± 7.55 b
	RTF b	0.32 ± 0.12 c	4.3 ± 2.29 c
	PSOE b	0.57 ± 0.13 bc	2.63 ± 2.45 c

[a] a, b, and c are significantly different (P = 0.05).
[b] Results are means ± S.E.M.
[c] (RTF) ram rete testis fluid: daily injection, 2 ml/rat, 0.6 mg prot./ml; (BSA) bovine serum albumin: 1 ml/rat, 1 mg prot./ml; (PSOE) pregnant sow ovarian extract: daily injection, 1 ml/rat, 0.1 mg prot./ml.

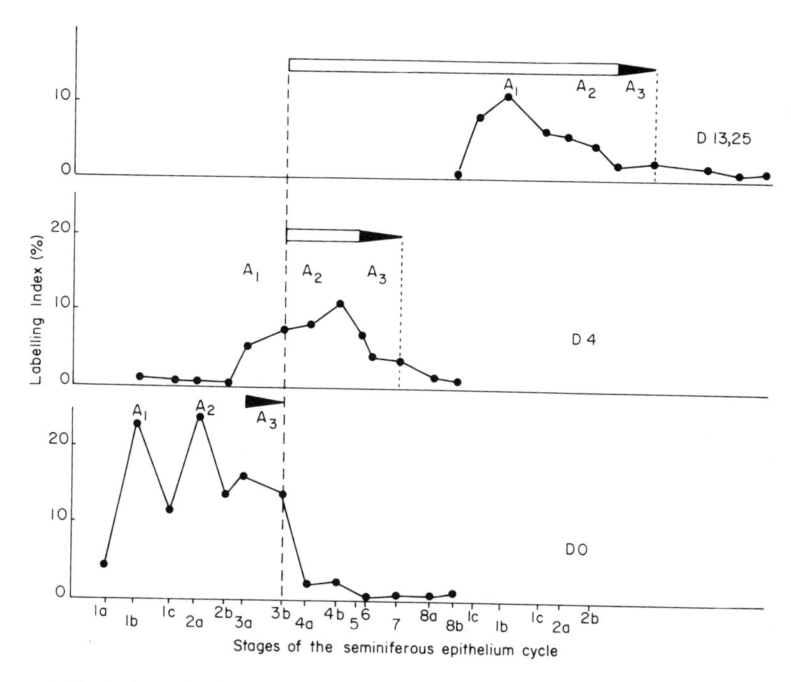

Figure 4. Evolution of type A spermatogonia labeling after [^{3}H]thymidine injection in the bull. (DO) 1 hr after injection (D_4) 4 days after injection; (D 13,25) 13 and 25 days after injection. A_1 stem cells arise mostly from A_1 and A_2 labeled spermatogonia.

cytes (de Jong *et al.*, 1978) and of DNA synthesis in the testis (Franchimont *et al.*, 1979). Is inhibin the active substance in our preparations?

On the other hand, testicular or ovarian extracts also exhibit an inhibitory action on Sertoli-cell multiplication (Hochereau-de-Reviers and Courot, 1978; Picaper *et al.*, 1979). Thus, this action is probably not restricted to type A spermatogonia. No clear-cut conclusion as to inhibin or chalone could actually be drawn in the absence of a highly purified factor.

5. *Stem-Cell Renewal: A Critical Point of View*

Evidence of variations of both compartments of A_0 reserve and A_1 cyclic renewing spermatogonia has been presented. Analysis of labeling index has been done in ram and bull testes after [^{3}H]thymidine injections (Fig. 4), leading to the conclusion that type A_0 spermatogonia could give rise to new A_1 spermatogonia, but that most of the resting type A_1 spermatogonia are issued from peaks of labeling corresponding to type

A_1 and A_2 spermatogonia (Hochereau-de-Reviers, 1971, 1976; Hochereau-de-Reviers *et al.*, 1976b). Morphological analysis of those cells in different species seemed to indicate that stem-cell renewal occurs before the first sign of differentiation leading to meiosis in the spermatogonial population, for example, nuclear morphological differentiation in A_3 spermatogonia of bull and ram testes. Morphological analysis of type A_0, A_1, and A_2 spermatogonia in G_1, S, and G_2 phases in bull or ram testes indicated that all type A spermatogonia in the beginning of the G_1 phase resemble type A_0 spermatogonia and that all of them, at the end of the S phase and during the G_2 phase, were identical to A_2 spermatogonia just prior to mitosis (Fig. 5) (Hochereau-de-Reviers, 1970).

On the other hand, type A_0 spermatogonia were said to be radioresistant but could enter mitosis to repopulate the testis in stem cells after X-irradiation (Dym and Clermont, 1970; Huckins, 1971, 1978; Oakberg, 1971 De Rooij, 1973; Van Keulen and De Rooij, 1974).

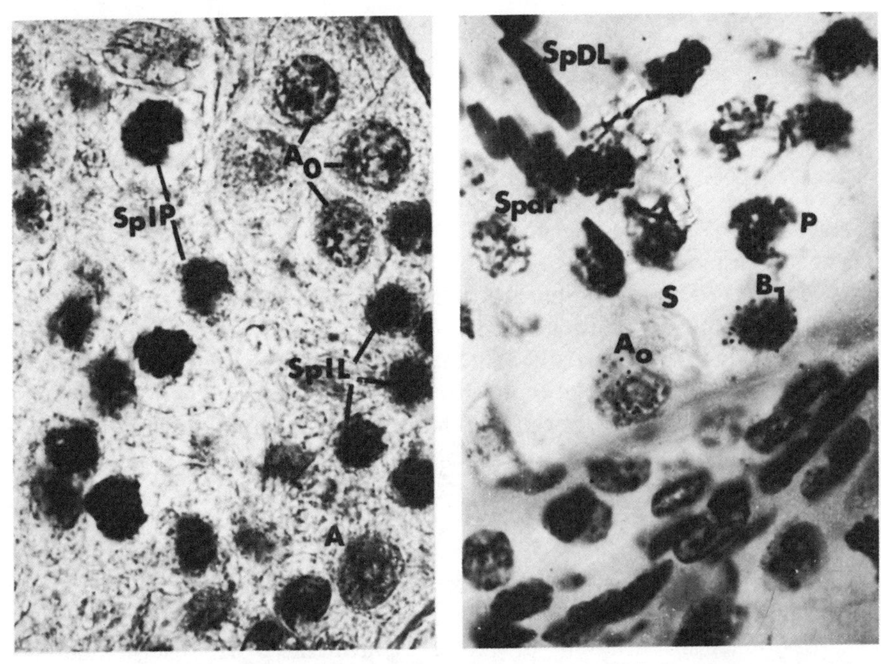

Figure 5. (A) Stage 1 of the seminiferous epithelium cycle in the bull. (A_1) A_1 spermatogonia; (A_0) A_0 spermatogonia; (SpIP) pachytene spermatocytes; (SpIL) leptotene spermatocytes. (B) Stage 6: Autoradiography of testis castrated 1 hr after [^{3}H]thymidine injection. A_0 spermatogonia with a single central nucleolus and B_1 spermatogonia are labeled. (P) Pachytene spermatocytes; (Spdr) round spermatids; (SpDL) elongated spermatids; (S) Sertoli cell nucleus.

Conclusions could be drawn that type A_0 spermatogonia are G_0-phase-blocked stem cells and that other type A spermatogonia are in the G_1, S, and G_2 phases and enter into cyclic activity. The passage from the G_0 phase to cyclic activity is probably under endocrinological or local control, or both, either directly or via Sertoli-cell interaction.

6. Conclusions

1. Two major parameters govern the production of primary spermatocytes and finally that of spermatozoa per testis: (a) the total number of type A_1 spermatogonia that are at the origin of cyclic development of the seminiferous epithelium and (b) the yield of spermatogonial multiplications.
2. The total number of type A1 spermatogonia per testis increases during puberty and in adulthood is highly correlated with the total number of Sertoli cells per testis, which is established before puberty. Nevertheless, this could be modified by the endocrinological environment.
3. Elevation of scrotal temperature or cryptorchidism results in variable effects on spermatogonia; there are species variations.
4. Very high plasma levels of gonadotropins are correlated with the pathological arrest of spermatogenesis. However, in normal subjects, correlations between gonadotropin levels and the efficiency of spermatogenesis have been observed. This could be mediated only partially by testosterone. However, a low level of free intratesticular testosterone does not result in absence of spermatogenesis.
5. After hypophysectomy in prepubertal or adult animals, hormonal supplementation supporting normal spermatogenesis varies with age and species. This is related, probably, to variations in hormonal requirements for each step, i.e., formation and renewal of A_1 spermatogonia, differentiation of spermatogonia with a normal yield of spermatocytes, their passage into the adluminal compartment, and the maturation and support of Sertoli-cell metabolism.
6. Proteinaceous factors secreted by Sertoli cells controlling these phenomena are probably implicated, but their mechanism of action (local or central) remains to be elucidated. They could possibly act at both local and central sites.

ACKNOWLEDGMENTS. Personal results presented in this chapter were obtained in collaboration with M. R. Blanc, C. Cahoreau, A. Caraty, C. Cornu, M. Courot, J. L. Dacheux, M. P. Dubois, D. H. Garnier, A. Lanoiselée-Perrin-Houdon, A. Locatelli, N. Martinat, C. Monet-Kuntz, R. Ortavant, J. Pelletier, C. Perreau, G. Picaper, C. Pisselet, J. C.

Poirier, J. P. Ravault, M. M. de Reviers, and M. Terqui, and M. C. Viguier-Martinez. This work was done with the help of Grants 78.7.2752 DGRST and 25-75-48 INSERM and of C.N.R.S. and I.N.R.A. funds. The English manuscript was reviewed by L. and C. Oldham.

Discussion

BANIK: Your results with PMSG and hCG on spermatogenesis in rams are in general agreement with those of Dr. John McCloud of New York, who maintained spermatogenesis and sperm production with hMG and hCG in hypophysectomized men. In fact, if I remember correctly, two of three or four spouses of the hypophysectomized men became pregnant.

HOCHEREAU-DE-REVIERS: Yes, similar results have been mentioned by Mancini (1971).

BANIK: I understand that one of the important physiological functions of Sertoli cells is to hold spermatids for final maturation of sperms.

HOCHEREAU-DE-REVIERS: The relationship between Sertoli and spermatids is part of the entire role of Sertoli cells. It begins very early in spermatogenesis, as we observed high correlation between Sertoli cells and A_1 spermatogonia total numbers; even the latter cells are in the basal compartment of the seminiferous epithelium.

CHANNING: Did your inhibin inhibit FSH levels? Did you size the molecule? Is inhibin found in spermatic-vein blood? Ovarian inhibin is about 18,000 daltons, but binds to other protein, making molecular-weight determinations difficult. I would suggest obtaining pig follicular fluid as a source of inhibin F. Large follicular fluid is a poor source of inhibin F. There is more in small (1–2 mm) and medium (3–5 mm) follicular fluid.

HOCHEREAU-DE-REVIERS: We have not measured FSH in our treated rats killed 10 days after the end of treatment. In those killed at the end of the 10-day treatments, we have observed lower but not significantly different plasma levels of FSH. In cryptorchid rams, RTF injections induced a decrease in both gonadotropins, FSH and LH(Blanc *et al.,* 1978; Cahoreau *et al.,* 1978). The same effect was observed with the inhibin preparation of Dr. Franchimont [Lincoln and Franchimont, 1978 (personal communication)]. We have not sized the active inhibin in RTF and ram testis extract. In pregnant sow ovary extract, the active material is restricted to molecular weights of about 40,000 to 60,000 daltons (Picaper *et al.,* 1979). In the ram, an Australian group (Baker *et al.,* 1978) have analyzed inhibin content in RTF, lymph, and vein blood. RTF is 10-fold richer in inhibin than lymph.

References

Ahmad, N., Haltmayer, G.C., and Eik-Nes, K., 1975, Maintenance of spermatogenesis with testosterone or dihydrotestosterone in hypophysectomized rats, *J. Reprod. Fertil.* **44:** 103–107.

Aizawa, S., and Nishimune, Y., 1979, *In vitro* differentiation of type A spermatogonia in mouse cryptorchid testis. *J. Reprod. Fertil.* **56:**99–104.

Attal, J., and Courot, M., 1963. Développement testiculaire et établissement de la spermatogenèse chez le taureau, *Ann. Biol. Anim. Biochim. Biophys.* **3:**219–242.

Aumüller, G., Schenk, B., and Neumann, F., 1975, Fine structure of monkey *(Macaca mulatta)* Sertoli cells after treatment with cyproterone, *Andrologia* **7**:317–328.
Baker, H.W.G., Bremner, W.J., Burger, H.G., De Kretser, D.M., Dulmanis, A., Eddie, L.W., Hudson, B., Keogh, E.J., Lee, V.W.K., Rennie, G.C., 1976, Testicular control of FSH secretion, *Recent Prog. Horm. Res.* **32**:429–469.
Baker, H.W.G., Burger, H.G., De Kretser, D.M., Eddie, L.W., Higginson, R.E., Hudson, B., Lee, V.W.K., 1978, Studies on purification of inhibin from ovine testicular secretions using an *in vitro* bioassay, *Int. J. Andr. Suppl.* **2**:115–124.
Blanc, M.R., and Dacheux, J.L., 1976, Existence of inhibin activity in the ram rete testis fluid: Effect of physiological doses on plasma LH and FSH, *IRCS Med. Sci.* **4**:460.
Blanc, M.R., Cahoreau, C., Courot, M., Dacheux, J.L., Hochereau-de-Reviers, M.T., and Pisselet, C., 1978, Plasma follicle stimulating hormone (FSH) and luteinizing hormone (LH) suppression in the cryptorchid ram by a nonsteroid factor (inhibin) from ram rete testis fluid, *Int. J. Androl. Suppl.* **2**:139–146.
Boccabella, A.V., 1963, Reinitiation and restoration of spermatogenesis with testosterone propionate and other hormones after a long term post hypophysectomy regression period, *Endocrinology* **72**:787–798.
Bowler, K., 1972, The Effect of repeated applications of heat on spermatogenesis in the rat: A histological study, *J. Reprod. Fertil.* **28**:325–334.
Bullough, W.S., 1962, The control of mitotic activity in the adult mammalian tissue, *Biol. Rev.* **37**:307.
Cahoreau, C., Blanc, M.R., Dacheux, J.L., Pisselet, C., Courot, M., 1978, Inhibin activity in ram rete testis fluid: Depression of plasma FSH and LH in the castrated and cryptorchid ram, *J. Reprod. Fert. Suppl.* **26**:97–116.
Chasalow, F., Marr, H., Haour, F., and Saez, J.M., 1979, Testicular steroidogenesis after human chorionic gonadotropin desensitization in rats, *J. Biol. Chem.* **254**:5613–5617.
Chemes, H.E., Podesta, E., and Rivarola, M.A., 1976, Action of testosterone, dihydrotestosterone and 5α-androstan-3α-17β diol on spermatogenesis of immature rat testis, *Endocrinology* **72**:787–798.
Chiu, M.L., and Irvin, J.L., 1978, Basic chromosomal proteins in evaluation of the arrest and restoration of spermatogenesis, *Biol. Reprod.* **19**:984–993.
Chowdhury, A.K., 1979, Dependence of testicular germ cell on hormones: A quantitative study in hypophysectomized testosterone treated rats, *J. Endocrinol.* **82**:331–340.
Chowdhury, A.K., and Steinberger, E., 1964, A quantitative study of the effect of heat on germinal epithelium of rat testis, *Am. J. Anat.* **115**:509–524.
Clermont, Y., and Antar, M., 1973, Duration of the cycle of the seminiferous epithelium and the spermatogonial renewal in the monkey *Macaca arctoides, Am. J. Anat.* **136**:153–166.
Clermont, Y., and Bustos-Obregon, E., 1968, Reexamination of spermatogonial renewal in the rat by means of seminiferous tubules mounted "in toto," *Am. J. Anat.* **122**:237–248.
Clermont, Y., and Harvey, S.C., 1967, Effects of hormones on spermatogenesis in the rat, in: *Endocrinology of the Testis,* Ciba Foundation Colloquium No. 16 (G.E.W. Wolstenholme and M. O'Connor, eds.), pp. 173–189, Churchill, London.
Clermont, Y., and Mauger, A., 1974, Existence of a spermatogonial chalone in the rat testis, *Cell Tissue Kinet.* **7**:165–172.
Clermont, Y., and Mauger, A., 1976, Effect of a spermatogonial chalone in the growing rat testis. *Cell Tissue Kinet.* **9**:99–104
Clermont, Y., and Morgentaler, H., 1955, Quantitative study of spermatogenesis in the hypophysectomized rat, *Endocrinology* **57**:369–382.
Courot, M., 1967, Endocrine control of the supporting and germ cells of the impuberal testis, *J. Reprod. Fertil. Suppl.* **2**:89–101.

Courot, M., Hochereau-de-Reviers, M.T., and Ortavant, R., 1970, Spermatogenesis, in: *The Testis,* Vol. I (A.D. Johnson, W.R. Gomes, and N.L. Vandemark, eds.), pp. 339–432, Academic Press, New York.

Courot, M., Ortavant, R., and de Reviers, M.M., 1971, Variations du contrôle gonadotrope du testicule selon l'âge des animaux, *Exp. Anim.* **4:**201–211.

Courot, M., Hochereau-de-Reviers, M.T., Monet-Kuntz, C., Locatelli, A., Pisselet, C., Blanc, M.R., and Dacheux, J.L., 1979, Endocrinology of spermatogenesis in the hypophysectomized ram, *J. Reprod. Fertil. Suppl.* **26:**165–173.

Cunningham, G.R., and Huckins, C., 1979a, Failure to identify a spermatogonial chalone in adult irradiated testes, *Cell Tissue Kinet.* **12:**81–89.

Cunningham, G.R., and Huckins, C., 1979b, Persistence of complete spermatogenesis in the presence of low intratesticular concentrations of testosterone, *Endocrinology* **105:**177–186.

Debeljuk, L., and Mancini, R.E., 1971, Pituitary gonadotrophin activity in male rats with damage of the germinal epithelium induced by busulfan, *J. Reprod. Fertil.* **26:**247–250.

de Jong, F.H., Welschen, R., Hermans, W.P., Smith, S.D., and Van der Molen, H.J., 1978, Effects of testicular and ovarian inhibin-like activity, using *in vitro* and *in vivo* systems, *Int. J. Androl. Suppl.* **2:**125–138.

de Kretser, D.M., Burger, H.G., and Hudson, B., 1973, The relationship between germinal cells and serum FSH levels in males with infertility, *J. Clin. Endocrinol. Metab.* **38:**787–793.

de Reviers, M. Hochereau-de-Reviers, M.T., Blanc, M.R., Brillard, J.P., Courot, M., and Pelletier, J., 1980, Control of Sertoli and germ cells populations in cock and sheep testis, *Reprod. Nutr. Develop.* **20:**241–249.

De Rooij, D.G., 1973, Spermatogonial stem cell renewal in the mouse. I. Normal situation, *Cell Tissue Kinet.* **6:**281–287.

Dubois, M.P., de Reviers, M.M., and Courot, M., 1971, Activité gonadotrope de l'éminence médiane après hypophysectomie chez le rat: Etude en immunofluorescence, *Exp. Anim.* **4:**213–226.

Dym, M., and Clermont, Y., 1970, Role of spermatogonia in the repair of the seminiferous epithelium following X irradiation of the rat testis, *Am. J. Anat.* **128:**265–281.

Dym, M., and Fawcett, D.W., 1970, The blood testis barrier in the rat and the physiological compartmentation of the seminiferous epithelium, *Biol. Reprod.* **3:**308–326.

Eshkol, A., and Lunenfeld, B., 1971–1972, Biological effects of antibodies to gonadotropins, *Gynecol. Invest.* **2:**23–56.

Fabbrini, A., Spera, G., Santiemma, V., and Fraioli, F., 1974, Autoradiographic and spectrometric study of human gonadotrophins; Localization in target organs, in: *Male Fertility and Sterility* (R.E. Mancini and L. Martini, eds.), pp. 389–404, Academic Press, New York.

Flickinger, C.J., and Loving, L.K., 1976, Fine structure of the testis and epididymis of rats treated with cyproterone acetate, *Am. J. Anat.* **146:**359–384.

Franchimont, P., Millet, D., Vendrely, E., Letame, J., Legros, J.J., and Netter, A., 1972, Relationship between spermatogenesis and serum gonadotrophin levels in azoospermia and oligospermia, *J. Clin. Endocrinol. Metab.* **34:**1003–1008.

Franchimont, P., Chari, S., Hazee-Hagelstein, M.T., Debruche, M.L., and Duraiswami, S., 1977, Evidence for existence of inhibin, in: *The Testis in Normal and Infertile Men* (P. Troen and H.R. Nankin, eds.), pp. 253–270, Raven Press, New York.

Franchimont, P., Demoulin, A., Hazee-Hagelstein, M.T., Verstraelen-Proyard, J., and Bourguignon, J.P., 1979, Control of gonadotrophin secretion by inhibin, *Acta Endocrinol. Suppl.* **225:**469.

Frankenhuis, M.T., and Wensing, C.J.G., 1979, Induction of spermatogenesis in the naturally cryptorchid pig. *Fertil. Steril.* **31:**428–433.
Gomes, W.R., Hall, R.W., Jain, S.K., and Boots, L.R., 1973, Serum gonadotropin and testosterone levels during loss and recovery of spermatogenesis in rats, *Endocrinology* **93:**800–809.
Harris, M.E., Bartke, A., Weisz, J., and Watson, D., 1977, Effects of testosterone and dihydrotestosterone on spermatogenesis, rete-testis fluid and peripheral androgen levels in hypophysectomized rats. *Fertil. Steril.* **28:**1113–1117.
Heinert, G., and Taubert, H.D., 1973, Effect of cyproterone and cyproterone acetate on testicular function in the rat: A karyometric study, *Endokrinologie* **61:**168–178.
Hochereau, M.T., 1967, Synthèse de l'ADN au cours des multiplications et du renouvellement des spermatogonies chez le taureau, *Arch. Anat. Microsc. Morphol. Exp.* **56**(Suppl. 3-4):85–96.
Hochereau-de-Reviers, M.T., 1970, Etude des divisions spermatogoniales et du renouvellement de la spermatogonie souche chez le taureau, D. Sci. thesis, Paris, C.N.R.S. A03976.
Hochereau-de-Reviers, M.T., 1971, Etude cinétique des spermatogonies chez les mammifères—revue, in: *La Cinétique de Prolifération cellulaire,* Vol. 18, INSERM Symposia Series Seminar, pp. 189–216.
Hochereau-de-Reviers, M.T., 1976, Variation in the stock of testicular stem cells and in the yield of spermatogonial divisions in ram and bull testes, *Andrologia* **8:**137–146.
Hochereau-de-Reviers, M.T., and Courot, M., 1978, Sertoli cells and development of seminiferous epithelium, *Ann. Biol. Anim. Biochim. Biophys.* **18:**573–583.
Hochereau-de-Reviers, M.T., and Lincoln, G.A., 1978, Seasonal variation in the histology of the testis of the red deer, *Cervus elaphus, J. Reprod. Fertil.* **54:**209–213.
Hochereau-de-Reviers, M.T., Loir, M., and Pelletier, J., 1976a, Seasonal variations in the response of the testis and LH levels to hemicastration of adult rams, *J. Reprod. Fertil.* **46:**203–209.
Hochereau-de Reviers, M.T., Ortavant, R., and Courot, M., 1976b, Type A spermatogonia in the ram, in: *Progress in Reproductive Biology,* Vol. 1, *Sperm Action* (Hubinont, ed.), pp. 13–19, S. Karger Basel.
Hochereau-de-Reviers, M.T., Blanc, M.R., Cahoreau, C., Courot, M., Dacheux, J.L., and Pisselet, C., 1979, Histological testicular parameters in bilateral cryptorchid adult rams. *Ann Biol. Anim. Biochim. Biophys.* **19:**1141–1146.
Hochereau-de-Reviers, M.T., Blanc, M.R., Courot, M., Garnier, D.H., Pelletier, J., and Poirier, J.C., 1980, Hormonal profiles and testicular parameters in the lamb, in: *Testicular Development, Structure and Function* (E. Steinberger and A. Steinberger, eds.), pp. 237–247, Raven Press, New York.
Huckins, C., 1971, The spermatogonial stem cell population in adult rats. I. Their morphology, proliferation and maturation, *Anat. Rec.* **169:**533–558.
Huckins, C., 1978, The morphology and kinetics of spermatogonial degeneration in normal adult rats: An analysis using a simplified classification of the germinal epithelium, *Anat. Rec.* **190:**905–926.
Irons, M.J., and Clermont, Y., 1979, Spermatogonial chalone. Effect on the phases of the cell cycle of type A spermatogonia in the rat, *Cell Tissue Kinet.* **12:**425–433.
Kalra, S.P., and Prasad, M.R.N., 1967, Effect of FSH and testosterone propionate on spermatogenesis in immature rats treated with clomiphene, *Endocrinology* **81:**965–975.
Krueger, P.M., Hogden, G.D., and Sherins, R.J., 1974, New evidence for the role of the Sertoli cell and spermatogonia in feed-back control of FSH secretion in male rats, *Endocrinology* **95:**355–362.

Land, R.B., and Hochereau-de-Reviers, M.T., 1980, Variations in testis growth and histology in 6 month old rams, normal or hemicastrated, according to season of birth, *J. Reprod. Fertil.*, to be published.

Mancini, R.E., 1971, Effect of gonadotropin preparations and of urinary FSH and LH on human spermatogenesis, in: *The Regulation of Mammalian Reproduction* (S.J. Segal, R. Crozier, P.A. Corfman, and P.G. Condliffe, eds.), pp. 151–162, Charles C. Thomas, Springfield, Illinois.

Matsuyama, S., Ogasa, Q., and Yokoki, H., 1971, Quantitative study of spermatogenesis in androgen treated hypophysectomized rats, *Natl. Inst. Anim. Health Q. (Tokyo)* **11:**46–52.

Means, A.R., 1977, Mechanism of action of follicle stimulating hormone (FSH), in: *The Testis*, Vol. IV (A.D. Johnson and W.R. Gomes, eds.), pp. 168–188, Academic Press, New York.

Mills, N.C., and Means, A.R., 1972, Sorbitol dehydrogenase of rat testis: Change of activity during development, after hypophysectomy and following gonadotrophic hormone administration, *Endocrinology* **91:**147–156.

Monastirky, R., Laurence, K.A., and Tovar, E., 1971, The effects of gonadotropin immunization of prepubertal rabbits on gonadal development, *Fertil. Steril.* **22:**318–324.

Monesi, V., 1962, Autoradiographic study of DNA synthesis and the cell cycle in spermatogonia and spermatocytes of mouse testis using tritiated thymidine, *J. Cell Biol.* **14:**1–18.

Monet-Kuntz, C., Terqui, M., Locatelli, A., Hochereau-de-Reviers, M.T., and Courot, M., 1976, Effets de la supplémentation en testosterone sur la spermatogenèse de béliers hypophysectomisés, *C. R. Acad. Sci. Ser. D* **283:**1763–1766.

Murty, G.S.R.C., Rani, C.S., Mougdal, N.R., 1979, Effects of passive immunization with specific antiserum to FSH on the spermatogenic process and fertility of adult male Bonnet monkeys *(Maccaca radiata)*, *J. Reprod. Fert.* **26:**147–163.

Nagy, F., 1972, Cell division kinetics and DNA synthesis in the immature Sertoli cells of the rat testis, *J. Reprod. Fertil.* **28:**389–395.

Neumann, F., and Von Berswordt-Wallrabe, R., 1966, Effects of the androgen antagonist cyproterone acetate on the testicular structure, spermatogenesis and accessory sexual glands of testosterone treated adult hypophysectomized rats, *J. Endocrinol.* **35:**363–371.

Oakberg, E.F., 1971, Spermatogonial stem cell renewal in the mouse, *Anat. Rec.* **169:**515–532.

Ortavant, R., 1958, Le cycle spermatogénétique chez le bélier, D. Sci. thesis, Paris, 127 pp, CNRS A-3118.

Ortavant, R., Courot, M., and Hochereau-de-Reviers, M.T., 1972, Gonadotrophic control of tritiated thymidine incorporation in the germinal cells of the rat testis, *J. Reprod. Fertil.* **31:**451–453.

Orth, J., and Christensen, A.K., 1978, Autoradiographic localization of specifically bound ^{125}I labelled follicle stimulating hormone on spermatogonia of the rat testis, *Endocrinology* **103:**1944–1951.

Picaper, G., Hochereau-de-Reviers, M.T., Martinat, N., and Dubois, M.P., 1979, Inhibition of post hemicastration rise in Sertoli cell stocks in prepubertal rats by proteinaceous pregnant sow corpora lutea and pubertal rat testis extracts, *IRCS Med. Sci.* **7:**132.

Raj, H.G.M., Dym, M., Chemes, H.E., Kotite, N.J., Nayfeh, S.N., and French, F.S., 1978a, Effects of passive immunity to FSH on male reproduction in the immature and adult rat, *Proc. 3rd Ann. Meet. Am. Soc. Androl.* (abstract).

Raj, H.G.M., Dym, M. Sairam, M.R., and Chendy, R., 1978b, Effects of selective depletion of luteinizing hormone and testosterone on the pituitary gonadal axis, *Int. J. Androl. Suppl.* **2:**184–189.

Regaud, C., 1901, Etude sur la structure des tubes séminifères et sur la spermatogenèse chez les mammifères, *Arch. Anat. Microsc.* **101:**231–380.

Rivarola, M.A. Podesta, F.J., Chemes, H.E., and Calandra, R.S., 1977, Hormonal influences on the initiation of spermatogenesis, in: *Endocrinology,* Vol. I (V.H.T. James, ed.), pp. 307–313, Excerpta Medica, Amsterdam.

Roosen-Runge, E.C., and Giesel, L.O., 1950, Quantitative studies on spermatogenesis in the albino rat, *Am. J. Anat.* **87:**1–30.

Setchell, B.P., Davies, R.V., and Main, S.J., 1977, Inhibin, in: *The Testis,* Vol. IV (A.D. Johnson and W.R. Gomes, eds.), pp. 189–238, Academic Press, New York.

Steinberger, A., 1979, Inhibin production by Sertoli cells in culture, *J. Reprod. Fertil. Suppl.* **26:**31–45.

Steinberger, A., and Steinberger, E., 1966, *In vitro* culture of rat testicular cells, *Exp. Cell Res.* **44:**443–452.

Steinberger, A., and Steinberger, E., 1971, Replication pattern of Sertoli cells in maturing rat testis *in vivo* and in organ cultures, *Biol. Reprod.* **4:**84–87.

Steinberger, E., 1971, Hormonal control of mammalian spermatogenesis, *Physiol. Rev.* **51:**1–22.

Steinberger, E., and Duckett, G.E., 1967, Hormonal control of spermatogenesis, *J. Reprod. Fertil. Suppl.* **2:**75–87.

Steinberger, E., Steinberger, A., and Perloff, H., 1963, *In vitro* growth of testicular fragments from rats of various ages, *Anat. Rec.* **145:**288.

Thumann, A., and Bustos-Obregon, E., 1978, An "in vitro" system for the study of rat spermatogonial proliferation control, *Andrologia* **10:**22–25.

Turner, P.C., and Johnson, A.D., 1971, The effect of anti FSH serum on the reproductive organs of the male rat, *Int. J. Fertil.* **16:**169–176.

Van Keulen, C.J.G., and de Rooij, D.G., 1974, The recovery from various gradations of cell loss in the mouse seminiferous epithelium and its implication for the spermatogonial stem cell renewal theory. *Cell Tissue Kinet.* **7:**549–558.

Vernon, R.G., Go, V.L.W., and Fritz, I.B., 1975, Hormonal requirements of the different cycles of the seminiferous epithelium during reinitation of spermatogenesis in long term hypophysectomized rats, *J. Reprod. Fertil.* **42:**77–94.

Viguier-Martinez, M.C., and Hochereau-de-Reviers, M.T., 1977, Comparative action of cyproterone and cyproterone acetate on pituitary and plasma gonadotropin levels, the male genital tract and spermatogenesis in the growing rat, *Ann. Biol. Anim. Biochim. Biophys.* **17:**1069–1076.

Waites, G.W., and Ortavant, R., 1968, Effets précoces d'une bréve élévation de la température testiculaire sur la spermatogenèse du bélier, *Ann. Biol. Anim. Biochim. Biophys.* **8:**323–331.

Wettemann, R.P., and Desjardins, C., 1979, Testicular function in boars exposed to elevated temperature, *Biol. Reprod.* **20:**235–241.

15

Secretory Proteins in the Male Reproductive System

Peter Petrusz, Oscar A. Lea, Mark Feldman, and Frank S. French

1. Introduction

The male reproductive system in mammals consists of two anatomically and functionally interconnected parts: the testis, where both spermatozoa and male hormones (androgens) are formed, and the duct system with the attached glands, where final maturation and storage of spermatozoa occur. The seminiferous tubules of the testis, the duct of the epidydimis, and the subsequent segments of the duct system draining the accessory sex glands are highly specialized to provide a suitable environment in which the final maturational changes of spermatozoa can take place and to assure optimal composition of the semen.

Although the concept of the ''blood–testis barrier'' (or, more correctly, the blood–seminiferous tubule barrier) was introduced as early as around the turn of the century (cf. Setchell and Waites, 1975), it was only in 1971 that Kormano *et al.* (1971) first stated that the seminiferous tubules (and the fluid collected from the efferent ducts) contain specific proteins different from plasma proteins in both quality and quantity. In contrast, the idea that sperm maturation in the epididymis depends on

Peter Petrusz • Department of Anatomy and Laboratories for Reproductive Biology, University of North Carolina, Chapel Hill, North Carolina 27514. *Oscar A. Lea* • Department of Pharmacology, University of Bergen, Bergen, Norway. *Mark Feldman* and *Frank S. French* • Department of Pediatrics and Laboratories for Reproductive Biology, University of North Carolina, Chapel Hill, North Carolina 27514.

specific secretory proteins can be traced back to the 1880's (for a review, see Orgebin-Crist *et al.,* 1975). The prostate gland has also long been recognized to secrete a great number of different substances and contribute significantly to the volume of semen (Price and Williams-Ashman, 1961; Mann, 1964), but little has been known about its secretory proteins other than lysosomal-type enzymes (Mann, 1963; Helminen and Ericsson, 1970).

The development as well as the maintenance of the normal structure and secretory function of the entire male reproductive tract are largely dependent on androgenic hormones (for a review, see Tveter *et al.,* 1975). In addition, there is abundant evidence to indicate that androgens are of vital importance for the processes of both spermiogenesis and sperm maturation (see Steinberger, 1971; Blaquier *et al.,* 1972; Dyson and Orgebin-Crist, 1973). The required high concentration of androgens within the seminiferous tubules and the lumens of the duct system is apparently assured through the secretion of specific androgen-binding components by the epithelial cells lining the various segments of the system. The first such specific component, androgen-binding protein (ABP), was described by Ritzén *et al.* (1973), and is produced in the testes by the Sertoli cells. More recently, two additional secretory proteins, specific for the rat epididymis and prostate, respectively, have been recognized, isolated, and characterized: acidic epididymal glycoprotein (AEG) (Lea *et al.,* 1978) and prostatein (Lea *et al.,* 1977a,b, 1979). The production of these proteins, according to preliminary evidence, is also androgen-dependent. Thus, it appears that throughout the male reproductive system, the same general mechanism is utilized to ensure an optimal environment for the production and maturation of spermatozoa: androgens (alone or in synergism with other hormones) maintain the secretory activity of the various epithelial cells, while some of the secretory proteins, in turn, serve as androgen concentrators in the environment surrounding the spermatozoa. In addition, there is increasing evidence that such proteins might also interact directly with spermatozoa, either as membrane-coating proteins or through other mechanisms, and contribute to the attainment of full motility (Acott and Hoskins, 1978). The most important chemical and biological characteristics of ABP, AEG, and prostatein are summarized in Table I.

Our general knowledge of morphological aspects of protein synthesis and secretion has been drived from biochemical and electron-microscopic studies on model systems such as the pancreas, salivary glands, pituitary, or liver. It has been established that exportable (secretory) proteins are first collected or segregated in the cisternal space of the endoplasmic reticulum (Redman *et al.,* 1966; Blobel and Sabatini, 1970), then transported to the Golgi apparatus either directly or in small "shuttle" vesicles

Table I. Some Physicochemical and Biological Properties of ABP, AEG, and Prostatein

Properties	ABP[a]	AEG[b]	Prostatein[c]
Source	Sertoli cell	Epididymis	Ventral prostate
Steroid binding?	Yes	No	Yes
Interaction with sperm?	Not known	Yes	Yes
Present in serum?	Yes	Yes	Not known
Molecular weight	100,000	33,000	40,000
Stokes radius (Å)	47	26	25
Sedimentation coefficient (S)	4.6	3.0	3.2
Isoelectric point (pH)	4.6–4.7	4.7	4.8
Carbohydrate content (%)	25	6–7	Not known

[a] From Hansson *et al.* (1975a) and Feldman *et al.* (1981).
[b] From Lea *et al.* (1978 and unpublished results).
[c] From Lea *et al.* (1979 and unpublished results).

(Jamieson and Palade, 1967). In the Golgi region, the proteins might undergo further glycosylation and are packaged in secretory granules or vacuoles (Castle *et al.*, 1972; Hopkins and Farquhar, 1973). The granules then migrate toward the apical surface, where they release their contents by the process termed exocytosis. Experimental evidence establishing this "classic" secretory pathway has been reviewed by Palade (1975) and, more recently, by Jamieson and Palade (1977). This general mode of protein secretion has been observed by many investigators in a great variety of cell types, including the epithelium of the rat ventral prostate (Flickinger, 1974). However, the mechanisms of protein secretion from Sertoli cells and from epididymal epithelial cells are still not entirely clear (cf. Fawcett, 1975; Hamilton, 1975).

2. *Testicular Androgen-Binding Protein (ABP)*

It is generally accepted that testicular fluid is produced by the Sertoli cells by active secretion into the seminiferous tubules (Setchell, 1970; Fawcett, 1975). Serum proteins are found at only very low concentrations in this compartment, and both physiological and morphological evidence indicates that macromolecules present in the tubular fluid are likely to be produced in the tubular epithelium or transported there through the Sertoli-cell cytoplasm (Setchell and Waites, 1975; Fawcett, 1972; Dym and Fawcett, 1970).

The finding of relatively high concentrations of ABP in testis and epididymis of the rat (French and Ritzén, 1973a,b; Hansson *et al.*, 1973a), rabbit (Ritzén and French, 1974), and other species (Ritzén *et al.*, 1980) represented a significant turning point in our understanding of the regulation of testicular function, especially of spermatogenesis and its hormonal control. ABP is specifically produced in the testis in response to follicle-stimulating hormone (FSH) and androgens (Hansson *et al.*, 1973b, 1975b; Sandborn *et al.*, 1975; Elkington *et al.*, 1975). It is then transported with testicular efferent-duct fluid to the caput epididymidis, where it is concentrated and later absorbed or degraded (French and Ritzén, 1973a,b). Evidence for the Sertoli-cell origin of ABP has been summarized by Hagenäs *et al.* (1975).

Early attempts to utilize antibodies for radioimmunoassay or immunocytochemical (ICC) detection of ABP have met with difficulties because antibodies to rabbit ABP cross-reacted with a circulating testosterone-binding globulin present in that species (Weddington *et al.*, 1975). More recently, purification of rat ABP has been achieved (Gunsalus *et al.*, 1978a,b; Feldman *et al.*, 1981), and antibodies to ABP became available. Thus, we are now able to report the first results on ICC localization of ABP in cultured rat Sertoli cells (see also White *et al.*, 1979; Kierszenbaum *et al.*, 1979, 1980), in testis, and in epididymis.

The double-bridge immunoperoxidase technique was used essentially as described by Ordronneau and Petrusz in Chapter 10, except that paraffin was used as an embedding medium (see Petrusz *et al.*, 1975). ABP-containing secretory granules were found in the cytoplasm of cultured Sertoli cells (Fig. 1). After treatment with FSH and certain pharmacological agents such as 1-methyl-3-isobutyl xanthine (MIX) or dibutyryl cyclic AMP, a large number of Sertoli cells underwent morphological changes associated with centrifugal migration of the granules and eventually with a decrease in cytoplasmic ABP immunoreactivity and an increase of ABP in the tissue-culture medium as measured by radioimmunoassay.

In the testes of immature, hypophysectomized rats stimulated with FSH and testosterone, ABP was localized in distinct clusters or patches consisting of rather small granules at the basal and apical regions of Sertoli cells of the cytoplasm (Fig. 2) and also in the peritubular spaces (not shown). Different tubules, and even different segments of the same tubule, varied greatly in their content of immunoreactive ABP. This localization pattern seems to indicate that the amount of ABP produced by a given Sertoli cell may vary, perhaps depending on the functional state of the cell and perhaps also on that of the associated germinal elements. Furthermore, the striking basal accumulations (see Fig. 2) are consistent with the possibility that ABP, under certain conditions, may be secreted into the peritubular space rather than into the tubular lumen

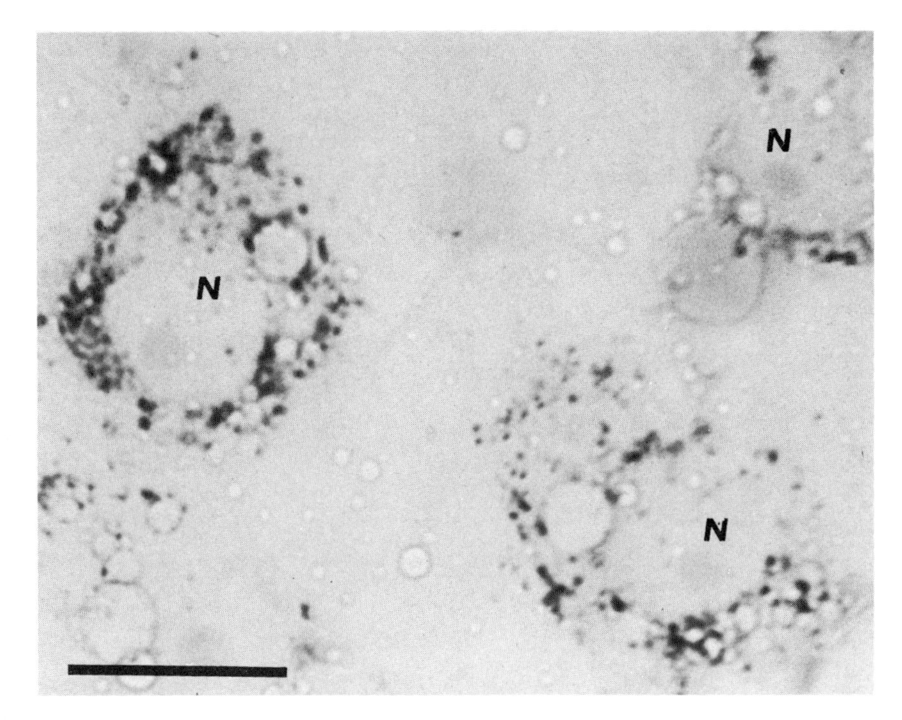

Figure 1. Immunoperoxidase localization of ABP in cultured rat Sertoli cells. Darkly stained secretory granules are widely distributed throughout the cytoplasm. (N) Nucleus. Scale bar: 25 μm (Kierszenbaum *et al.*, 1979).

only. This could provide an explanation for the presence of radioimmunoassayable ABP in the circulating blood in intact immature rats (Musto *et al.*, 1978; Gunsalus *et al.*, 1978b).

ABP is transported with efferent-duct fluid to the epididymis, where it "disappears"—i.e., becomes inactivated or absorbed, mostly in the caput region (French and Ritzén, 1973a,b). ICC localization has shown (Figs. 3 and 4) that ABP in the caput epididymidis is first bound to the microvilli of principal cells, then enters the cells and is concentrated in a well-defined supranuclear region that is well known in this tissue to correspond to the Golgi complex closely associated with lysosomes (cf. Hamilton, 1975).

Electron-microscopic ICC studies will be necessary to determine the precise route of the absorbed protein from the cell surface to the Golgi–lysosomal region. ABP must be at least partially degraded at this site, since it is no longer present in the basal regions of the cells in a form recognizable by the antibody. This finding strongly indicates that ABP does not reach the circulation in immunoreactive form from the

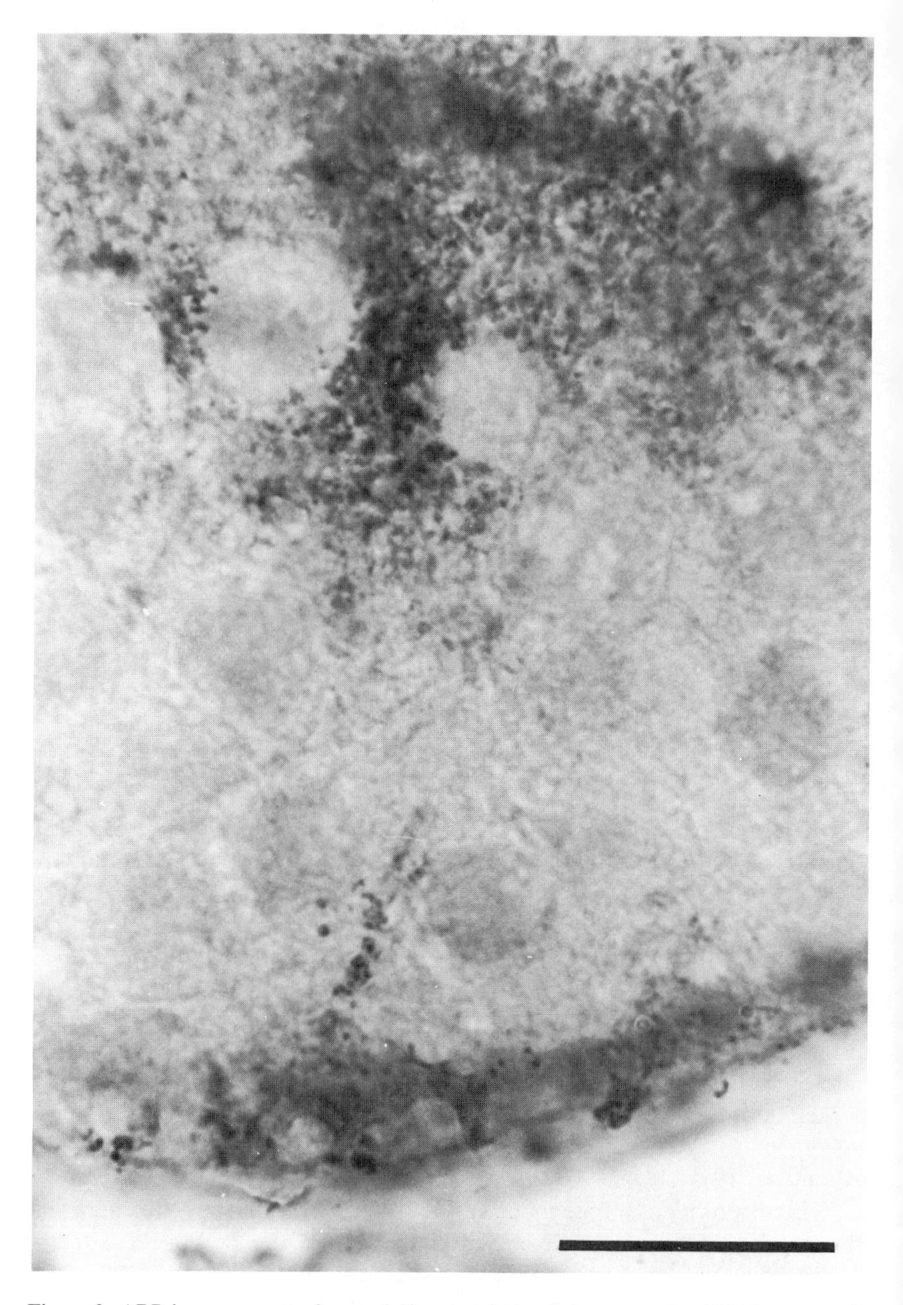

Figure 2. ABP in a segment of a seminiferous tubule of the rat testis. ABP is present in small granules arranged in clusters in the peripheral *(bottom)* and in the central *(top)* regions of the seminiferous epithelium, presumably in the cytoplasm of Sertoli cells. Scale bar: 25 μm.

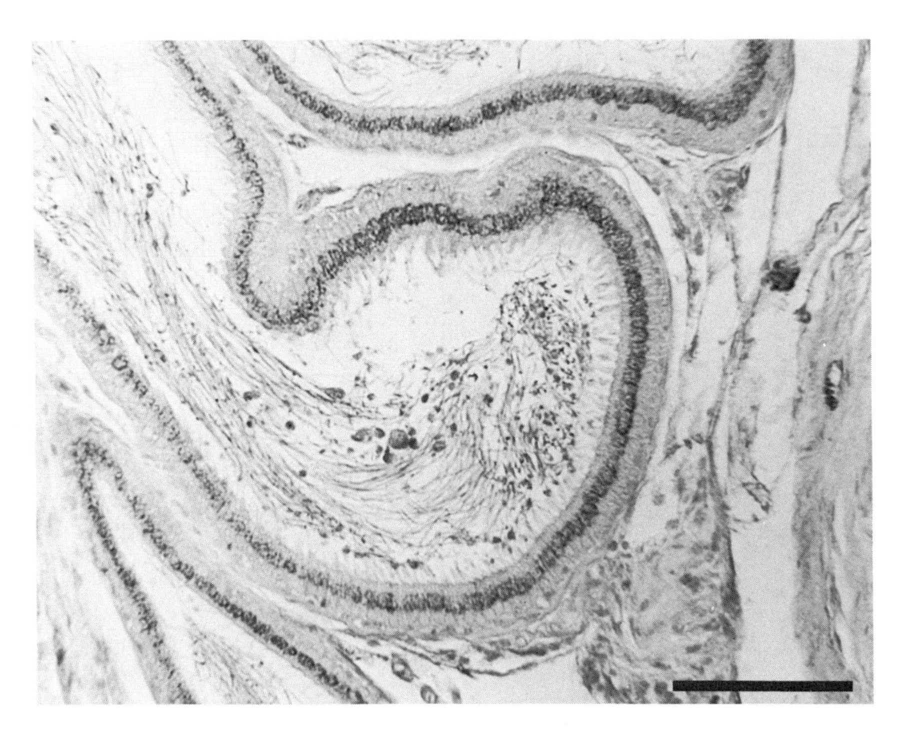

Figure 3. ABP in the rat epididymis. Epididymal epithelial cells in the caput uniformly take up ABP and concentrate it in the supranuclear cytoplasm. Scale bar: 100 μm.

epididymis, and further emphasizes the possibility of secretion into the blood from the perinuclear (basal) region of Sertoli cells. ICC localization of ABP in epididymis also confirms earlier results obtained by techniques based on androgen binding (French and Ritzén, 1973a; Hansson *et al.*, 1974b) indicating that the ABP content of epididymis is highest in the initial segment, decreases gradually along the corpus, and is practically absent from the cauda. Thus, ICC localization studies confirm and extend earlier biochemical data on the cellular origin, secretion, and absorption of ABP in the rat. In addition, they offer an excellent model system for studying the mechanisms of secretion, transport, absorption, and hormonal regulation of this important protein.

3. Acidic Epididymal Glycoprotein (AEG)

Evidence from several mammalian species indicates that specific proteins are secreted by the epididymis (Amann *et al.*, 1973; Koskimies and Kormano, 1975; Cameo and Blaquier, 1975; Huang and Johnson,

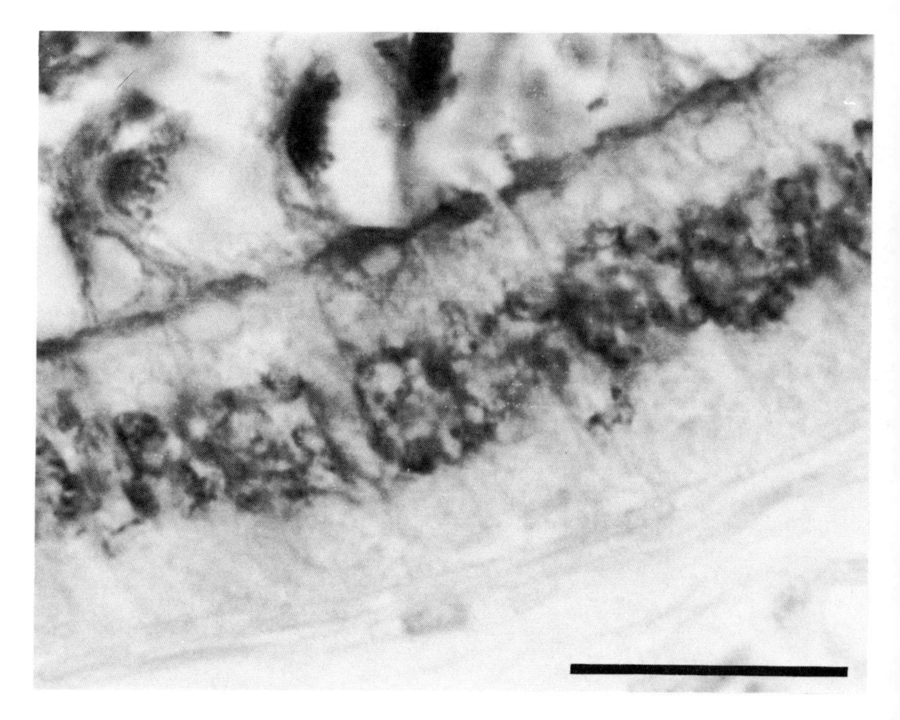

Figure 4. ABP in caput epididymidis. High-power view of a portion of the wall of the epididymal duct, showing masses of ABP attached to large microvilli as well as accumulations of ABP in the Golgi–lysosomal region just superficial to the nuclei. Note the absence of immunoreactive material in the basal portion of the epithelial cells. Scale bar: 25 μm.

1975) under the control of androgenic hormones (Cameo and Blaquier, 1975; Gigon-Depeiges and DuFaure, 1977). However, conclusive proof of glycoprotein synthesis and secretion in the epididymis was lacking, although polysaccharides (Maneely, 1955; Rambourg *et al.*, 1969; Hamilton, 1975) including sialic acid had been identified (Fournier, 1966; Rajalakshmi and Prasad, 1969; Prasad and Rajalakshmi, 1977) and androgen had been shown to stimulate protein synthesis (Cameo and Blaquier, 1975). Epididymal cells had also been shown to contain some of the machinery for the synthesis of glycoprotein: they have an extremely well-developed Golgi apparatus, although no specific secretory granule has been positively identified in the rat (Hoffer *et al.*, 1973) or in most of the other species studied (Hamilton, 1975; Flickinger *et al.*, 1978).

Evidence has also accumulated to suggest that epididymal protein secretion might be instrumental in sperm maturation. Spermatozoa, as they pass through the epididymis, are known to undergo maturational changes in morphology, metabolism, and functions (Bedford, 1975). If this transit is prevented, sperm maturation, as determined by fertilizing ability, is greatly diminished (Orgebin-Crist *et al.*, 1975). One specific change noted is the progressive acquisition of a negative surface charge on the plasma membrane (Cooper and Bedford, 1971). Caudal sperm exhibit greater electrophoretic mobility to the anode than do caput sperm (Nevo *et al.*, 1961; Bangham, 1961). It has been proposed that extrinsic glycoproteins are responsible for this negative surface charge on sperm (Bedford, 1963; Fournier-Delpech *et al.*, 1977) and that these glycoproteins are acquired during passage through the epididymis (Hunter, 1969; Barker and Amann, 1970; Kopecny and Pech, 1977). Indeed, forward motility of immature sperm is stimulated by a glycoprotein found in epididymal plasma (Brandt *et al.*, 1977; Acott and Hoskins, 1978). Thus, it appears that sperm maturation is vitally dependent on a functional relationship with epididymal epithelium (Orgebin-Crist *et al.*, 1975) and that this functional dependence is due in part to the secretion of specific epididymal glycoproteins.

Recently, a possible "missing link" of the hypothesis discussed above was discovered when an acidic glycoprotein, AEG, was purified and localized in the rat epididymis (Lea *et al.*, 1978). It has been shown immunoelectrophoretically and immunocytochemically that AEG is a major secretory protein of the rat epididymis, representing about 2–3% of total soluble protein. The carbohydrate moiety comprises about 7% of the molecule. When the midcorpus is ligated, AEG levels decrease in the cauda; when the efferent ducts are ligated, AEG levels remain unaffected in the epididymis. Localization of AEG by the immunoperoxidase technique revealed that it is secreted largely in the caput and corpus beginning in the region distal to the initial segment (Fig. 5). It appears to be secreted by a specific cell type, the principal cell. Specific staining of AEG was also noted in the clear cells of the cauda, which have been reported to be absorptive in nature (Moore and Bedford, 1977). In this region, AEG heavily coats the microvillous border of the principal cells, but does not appear inside these cells (see Fig. 6). In addition, the surface membrane of the luminal sperm appears coated with AEG. According to results yet to be published in detail, the secretion of AEG is also androgen-dependent: the AEG content in epididymis decreases dramatically after castration or hypophysectomy, and can be restored by testosterone treatment. According to preliminary radioimmunoassay results, AEG is normally present in the serum of adult male rats. Its mode of entry into the general circulation is not known. In conclusion, it appears that AEG, an androgen-dependent secretory glycoprotein that attaches itself to sperm surface membrane, is a likely candidate for a "sperm-maturation factor" in the rat epididymis.

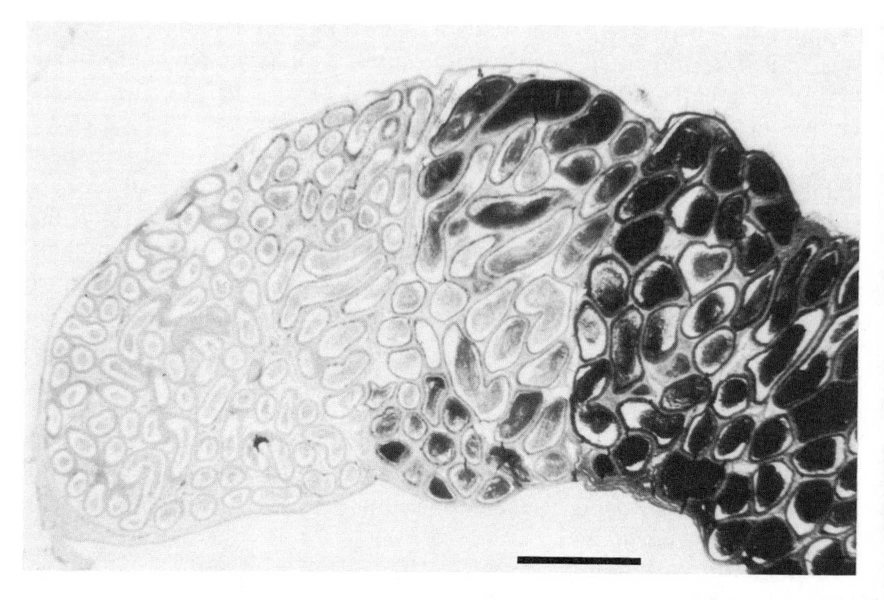

Figure 5. ICC localization of AEG in the rat epididymis. In this longitudinal section, the initial segment, the caput, and the proximal part of the corpus are seen. AEG is totally absent from the initial segment, appears in the caput, and is present at high concentrations in the corpus, where it is found both inside epithelial cells and in the lumen. Scale bar: 1 mm.

4. Prostatein

The ventral prostrate in the rat and many other species has long been recognized as an organ primarily concerned with secretion (Price and Williams-Ashman, 1961; Brandes and Groth, 1963). The epithelial cells lining the acini contain an abundance of organelles associated with protein synthesis and export (Brandes, 1974; Helminen and Ericsson, 1971). The secretory process here takes the well-established form (Flickinger, 1974). Analysis of prostatic secretions from a large number of species has demonstrated a great variety of proteins, including metalloproteins, flavoproteins, mucoproteins, and several enzymes, the most prominent being prostatic acid phosphatase (Mann, 1963, 1964). Growth and development of the prostate as well as its secretory functions are highly dependent on androgenic hormones (Grayhack and Wendel, 1974; Määtälä and Korhonen, 1977; Parker *et al.*, 1978; Heyns *et al.*, 1978) .

Much work has been carried out on the prostatic intracellular (cytosol) androgen-receptor protein and its role in the regulation of prostate function (Stumpf *et al.*, 1971; King and Mainwaring, 1974;

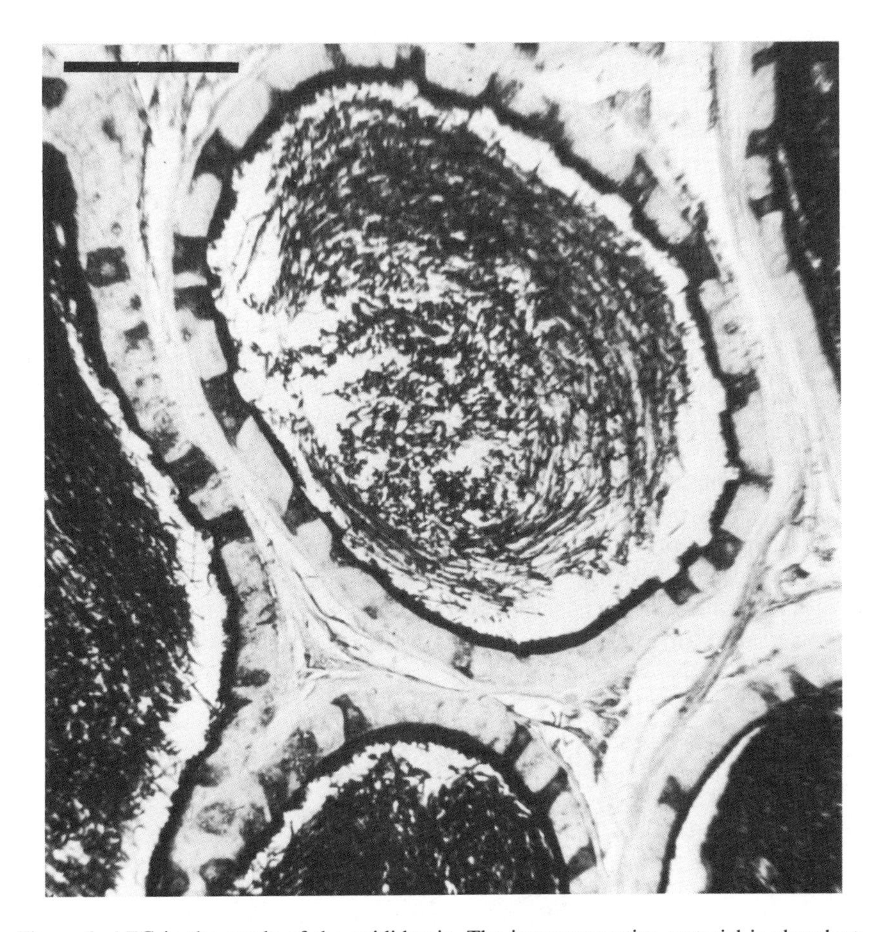

Figure 6. AEG in the cauda of the epididymis. The immunoreactive material is abundant in the lumen, where it appears to be attached to sperm. It forms a heavy coat on the microvillous border, and is also detectable in the cytoplasm of certain epithelial cells that lack microvilli (clear cells). Scale bar: 100 μm.

Tymoczko *et al.,* 1977; Wilson *et al.,* 1977). However, other androgen-binding proteins of this gland have received attention only recently. Lea *et al.* (1977a,b, 1979) isolated and purified from the ventral prostate of the rat a steroid-binding protein and determined some of its physical and chemical properties. Antisera to this protein, prostatein, were raised in rabbits and were used initially for quantitation of prostatein in tissue extracts and purified fractions by rocket immunoelectrophoresis, and also for ICC localization. Prostatein has been found exclusively in the ventral prostate, where it constitutes as much as 25–30% of total soluble protein. ICC localization showed prostatein to be present in the epithelial cells and in the lumen of acinar glands (Fig. 7) in the rat ventral prostate, but not in adjacent glandular structures such as the dorsal prostate or seminal vesicles (Fig. 8). Prostatein is present in the seminal plasma (Lea *et al.,* 1979), and according to ICC evidence, it coats the membrane of spermatozoa. Although prostatein has a low binding affinity for androgens (see Lea *et al.,* 1979), a high concentration of the protein within the lumen of the acinar glands could maintain a relatively high concentration of androgens in the vicinity of the epithelial cells.

A group of Belgian investigators has recently described a steroid-binding protein from the prostate that appears similar to prostatein in several respects, although their reports indicate substantial differences in both steroid specificity and subunit structure (Heyns and DeMoor, 1977; Heyns *et al.,* 1977, 1978). Prostatein also has characteristics similar to the estramustine-binding protein (Forsgren and Högberg, 1978). Estramustine, a nitrogen mustard derivative of estradiol-17β, is beneficial in the treatment of advanced prostatic carcinoma in the human. The therapeutic nature of this drug depends in part on the prostate's ability to concentrate estrogens. When this compound was tritiated, and its location after injection into rats determined using whole-body autora-

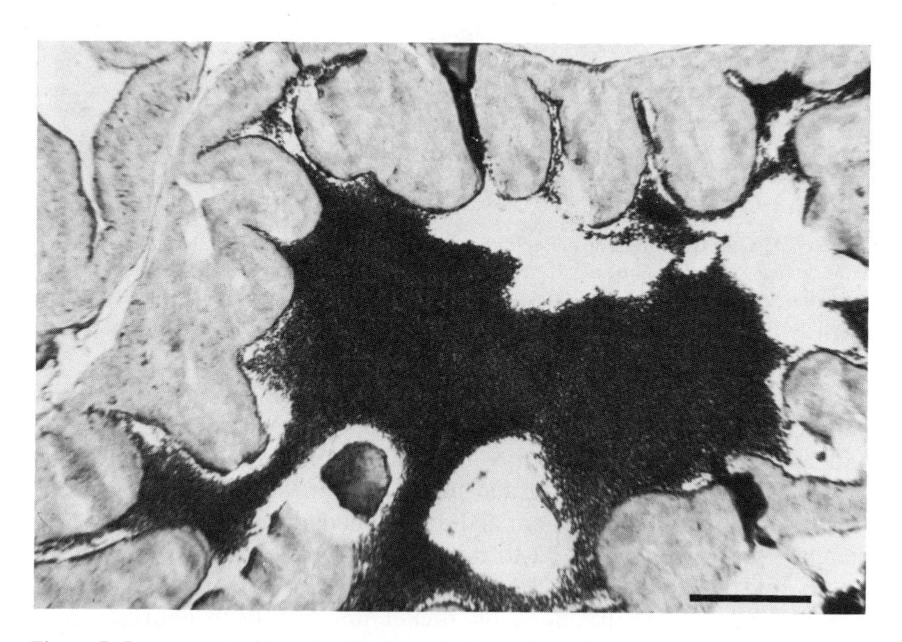

Figure 7. Immunoperoxidase localization of prostatein in the ventral prostate of the rat. Note heavy staining of the secretory product filling the acinar lumen. Scale bar: 100 μm.

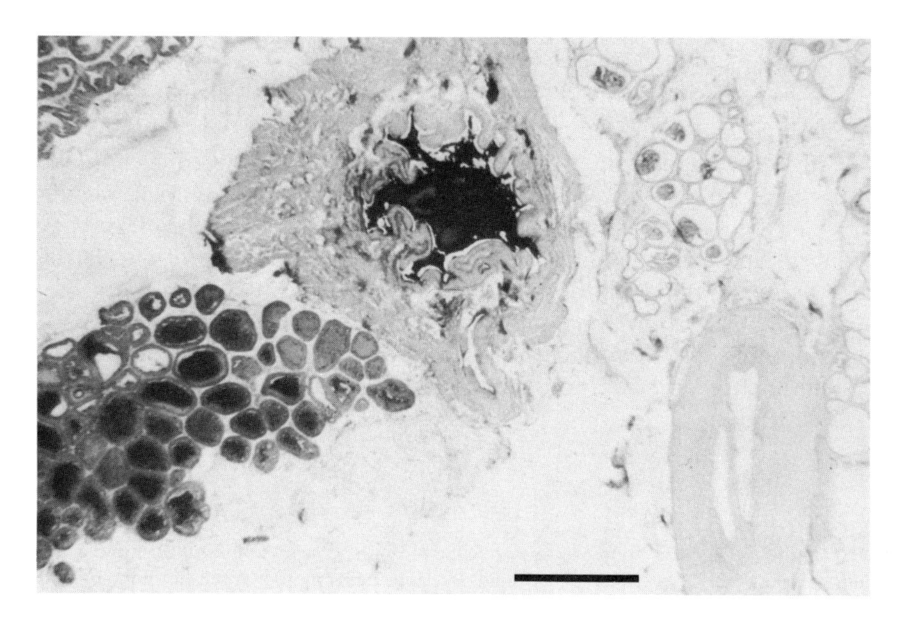

Figure 8. Low-power view of a section from the prostate complex of the rat, stained for prostatein with the immunoperoxidase technique. The dark reaction product is present only in the acini of the ventral prostate *(left)* and in the lumen of the urethra *(center)*, but is absent from the adjacent portions of the dorsal prostate and seminal vesicles *(right)*. Scale bar: 1 mm.

diography, it was found to be sequestered in the ventral prostate, specifically in the epithelium and acinar lumen (Appelgren *et al.*, 1977). The molecule responsible for its uptake was isolated from the prostate and characterized as very similar to prostatein (Forsgren *et al.*, 1979).

The precise secretory mechanism, hormonal regulation, and role in seminal fluid of prostatein remain to be established. Because of its abundant production in the ventral prostate, however, prostatein should be a useful probe for studies on androgen regulation of messenger RNA and protein synthesis and for the investigation of secretory processes in prostate epithelial cells.

5. Summary and Conclusions

Three androgen-dependent secretory proteins, produced in specific regions of the reproductive system of the male rat, have been described: ABP, AEG, and prostatein. ABP and prostatein are androgen-binding proteins and thus contribute to the maintenance of the high androgen concentration necessary in the environment where spermatozoa mature and are transported. AEG and prostatein appear to bind to the surface membrane of spermatozoa and may be contributing to their acquisition of motility and full fertilizing ability. Both these latter proteins are detectable in the ejaculate. ABP and AEG have been detected in blood by radioimmunoassays, but the mechanism and the functional significance of their entry into the circulation are not clear. Nevertheless, they might prove valuable indices reflecting the functional state of the Sertoli cells and epididymal epithelial cells, respectively. Studies on the secretory mechanisms, hormonal regulation, interactions with sperm, and role in fertility of these and other secretory proteins of the male reproductive system might lead to entirely new approaches to the artificial control of male fertility.

Discussion

D. Villee: (1) Is AEG bound to any specific site on sperm or simply diffusely over the entire surface? (2) Does ABP bind estradiol too, thus being more of a sex-hormone-binding protein rather than an androgen-binding protein?

Petrusz: AEG seems to bind diffusely over the entire surface. (2) The steroid specificity of ABP has been studied by direct binding and competitive inhibition of labeled-dihydrotestosterone (DHT) binding by unlabeled steroids (Wilson *et al.*, 1977; Tindall *et al.*, 1978). ABP has the greatest affinity for DHT, followed by testosterone; the binding of estradiol or progesterone is very low or undetectable. These data indicate that ABP should not be regarded as a general sex-steroid-binding protein, but rather as a specific androgen-binding protein. The latter interpretation is in good agreement with the postulated functions of ABP, among them the concentration and transport of biologically active androgens in the seminiferous tubules and epididymis.

Hocherau-de-Reviers: (1) Are there variations of ABP according to the seminiferous epithelium cycle? (2) In a given species, have you evidenced individual variations in contact of ABP, AEG, or prostatein? (3) Could they be related with sperm quantity?

Petrusz: To all three questions, the answer is that since immunocytochemical localization of ABP has been achieved only recently, no systematic studies have yet been conducted. However, our impression is that there are definite variations in ABP content and distribution in different segments of the seminiferous tubules, and these might well be related to the cycle of the seminiferous epithelium.*

Bazer: (1) Can dihydrotestosterone stimulate and/or maintain spermatogenesis? (2) Is it possible that androgen-binding protein protects testosterone from the 5α-reductase activity?

I would like to comment that the possible secretion of ABP into the lumen of the seminiferous tubule (exocrine secretion) or toward the lymphatics and venous system of the testis (endocrine secretion) is reminiscent of movement of the porcine progesterone-

* Such relationship has recently been demonstrated by Parvinen *et al.* (1980), in *Testicular Development, Structure, and Function* (A. Steinberger and E. Steinberger, eds.), pp. 425–432, Raven Press, New York.

induced uteroferrin, which is always secreted into the lumen of the uterine glands during pregnancy, but at the end of the estrous cycle moves into the endometrial stroma. The ABP dilemma is another reason to determine whether certain cells can secrete from the apical and basal portion of the cell and, if so, the mechanism involved.

PETRUSZ: (1) Dihydrotestosterone (DHT) can maintain spermatogenesis in hypophysectomized rats, provided treatment starts immediately after hypophysectomy. The degree of maintenance seems variable, depending on the dose, route of administration, and other factors. Penetration into the seminiferous tubules seems to be important, and it appears more difficult to maintain an effective androgen concentraion in rete testis fluid with DHT than with testosterone (Ahmad *et al.*, 1975; Harris *et al.*, 1977). (2) ABP is being regarded as a specific extracellular (secretory) androgen-binding and -transporting protein. As far as I know, androgen binding to intracellular ABP has not been demonstrated conclusively. In addition, it is known that ABP binds DHT with higher affinity than testosterone. Nevertheless, it is not impossible that ABP, by concentrating biologically active androgens in the extracellular compartment, has a role in "protecting" them from intracellular metabolism.

REFERENCES

Acott, T.S., and Hoskins, D.D., 1978, Bovine sperm forward motility protein, *J. Biol. Chem.* **253:**6744.

Ahmad, N., Haltmeyer, G.C., and Eik-Nes, K.B., 1975, Maintenance of spermatogenesis with testosterone or dihydrotestosterone in hypophycectomized rats, *J. Reprod. Fertil.* **44:**103.

Amann, R.P., Killian, G.J., and Benton, A.W., 1973, Differences in the electrophoretic characteristics of bovine rete testis fluid and plasma from cauda epididymis, *J. Reprod. Fertil.* **35:**321.

Appelgren, L.E., Forsgren, B., Gustafsson, J..-Å., Pousette, Å., and Högberg, B., 1977, Autoradiography of tritiated estramustine in castrated rats, *Acta Pharmacol. Toxicol.* **41**(Suppl. 1):106.

Bangham, A.D., 1961, Electrophoretic characteristics of ram and rabbit spermatozoa, *Proc. R. Soc. London Ser. B* **155:**292.

Barker, L.D.S., and Amann, R.P., 1970, Epididymal physiology. I. Specificity of antisera against bull spermatozoa and reproductive fluids, *J. Reprod. Fertil.* **22:**441.

Bedford, J.M., 1963, Changes in the electrophoretic properties of rabbit spermatozoa during passage through the epididymis, *Nature (London)* **200:**1178.

Bedford, J.M., 1975, Maturation, transport, and fate of spermatozoa in the epididymis, in: *Handbook of Physiology,* Section 7, Vol. V (D.W. Hamilton and R.O. Greep, eds), pp. 303–317, American Physiological Society, Washington, D.C.

Blaquier, J.A., Cameo, M.S., and Burgos, M.H., 1972, The role of androgens in the maturation of epididymal spermatozoa in the guinea pig, *Endocrinology* **90:**839.

Blobel, G., and Sabatini, D.D., 1970, Controlled proteolysis of nascent polypeptides in rat liver cell fractions. I. Location of the polypeptides within ribosomes, *J. Cell Biol.* **45:**130.

Brandes, D., 1974, Fine structure and cytochemistry of male sex accessory organs, in: *Male Accessory Sex Organs* (D. Brandes, ed.), pp. 18–44, Academic Press, New York.

Brandes, D., and Groth, D.P., 1963, Functional ultrastructure of rat prostatic epithelium, in: *Biology of the Prostate and Related Tissues* (E.P. Vollmer, ed.), pp. 47–62, National Cancer Institute Monograph No. 12.

Brandt, H., Acott, T.S., Johnson, D.J., and Hoskins, D.D., 1977, The initiation of forward motility in bovine caput spermatozoa by rete testicular and epididymal fluid proteins,

Proceedings of the 10th Annual Meeting, Society for the Study of Reproduction, pp. 29–30.

Cameo, M.S., and Blaquier, J.A., 1975, Androgen-controlled specific proteins in rat epididymis, *J. Endocrinol.* **69:**47.

Castle, J.D., Jamieson, J.D., and Palade, G.E., 1972, Radioautographic analysis of the secretory process in the parotid acinar cell of the rabbit, *J. Cell Biol.* **53:**290.

Cooper, G.W., and Bedford, J.M., 1971, Acquisition of surface charge by plasma membrane of mammalian spermatozoa during epididymal maturation, *Anat. Rec.* **169:**300.

Dym, M., and Fawcett, D.W., 1970, The blood–testis barrier in the rat and the physiological compartmentation of the seminiferous epithelium, *Biol. Reprod.* **3:**308.

Dyson, A.L.M.B., and Orgebin-Crist, M.C., 1973, Effect of hypophysectomy, castration and androgen replacement upon the fertilizing ability of rat epididymal spermatozoa, *Endocrinology* **93:**391.

Elkington, J.S.H., Sanborn, B.M., and Steinberger, E., 1975, The effect of testosterone propionate on the concentration of testicular and epididymal androgen binding activity in hypophysectomized rat, *Mol. Cell. Endocrinol.* **2:**157.

Fawcett, D.W., 1972, Observations on the organization of the interstitial tissue of the testis and on the occluding cell functions in the seminiferous epithelium, *Adv. Biosci.* **7:**83.

Fawcett, D.W., 1975, Ultrastructure and function of the Sertoli cell, in: *Handbook of Physiology,* Section 7, Vol. V (D.W. Hamilton and R.O. Greep, eds.), pp. 21–56, American Physiological Society, Washington, D.C.

Feldman, M., Lea, O.A., Petrusz, P., Tres, L.L., Kierszenbaum, A.L., and French, F.S., 1981, Androgen binding protein (ABP): Purification from rat epididymis, characterization and immunocytochemical localization, *J. Biol. Chem.* (in press).

Flickinger, C.J., 1974, Protein secretion in the rat ventral prostate and the relation of Golgi vesicles, cisternae and vacuoles, as studied by electron microscope radioautography, *Anat. Rec.* **180:**427.

Flickinger, C.J., Howards, S.S., and English, H.F., 1978, Ultrastructural differences in efferent ducts and several regions of the epididymis of the hamster, *Am. J. Anat.* **152:**557.

Forsgren, B., and Högberg, B., 1978, Binding of estramustine, a nitrogen mustard derivative of estradiol-17β, in cytosol from rat ventral prostate, *Acta Pharm. Suec.* **15:**23.

Forsgren, B., Björk, P., Carlström, K., Gustafsson, J.-Å., Pousette, Å., and Högberg, B., 1979, Purification and distribution of a major protein in rat prostate that binds estramustine, a nitrogen mustard derivative of estradiol-17β, *Proc. Natl. Acad. Sci. U.S.A.* **76:**3149.

Fournier, S., 1966, Distribution of sialic acid in the genital system of adult normal and castrated Wistar rats, *C. R. Soc. Biol.* **160:**1087.

Fournier-Delpech, S., Danzo, B.J., and Orgebin-Crist, M.C., 1977, Extraction of concanavalin A affinity material from rat testicular and epididymal spermatozoa, *Ann. Biol. Anim. Biochim. Biophys.* **17:**207.

French, F.S., and Ritzén, E.M., 1973a, A high affinity androgen-binding protein (ABP) in rat testis: Evidence for secretion into efferent duct fluid and absorption by epididymis, *Endocrinology* **93:**88.

French, F.S., and Ritzén, E.M., 1973b, Androgen binding protein in efferent duct fluid of rat testis, *J. Reprod. Fertil.* **32:**479.

Gigon-Depeiges, A., and DuFaure, J.P., 1977, Secretory activity of the lizard epididymis and its control by testosterone, *Gen. Comp. Endocrinol.* **33:**473.

Grayhack, J.T., and Wendel, E.F., 1974, Hormone dependence of carcinoma of the prostate, in *Male Accessory Sex Organs* (D. Brandes, ed.), pp. 425–432, Academic Press, New York.

Gunsalus, G.L., Musto, N.A., and Bardin, C.W., 1978a, Immunoassay of androgen binding protein in blood: A new approach for study of the seminiferous tubule, *Science* **200:**65.

Gunsalus, G.L., Musto, N.A., and Bardin, C.W., 1978b, Factors affecting blood levels of androgen binding protein in the rat, *Int. J. Androl. Suppl.* **2:**482.

Hagenäs, L., Ritzén, E.M., Plöen, L., Hasson, V., French, F.S., and Nayfeh, S., 1975, Sertoli cell origin of testicular androgen-binding protein (ABP), *Mol. Cell. Endocrinol.* **2:**339.

Hamilton, D.W., 1975, Structure and function of the epithelium lining the ductuli efferentes, ductus epididymidis, and ductus deferens in the rat, in: *Handbook of Physiology,* Section 7, Vol. V (D.W. Hamilton and R.O. Greep, eds.), pp. 259–301, American Physiological Society, Washington, D.C.

Hansson, V., Djöseland, O., Reusch, E., Attramadal, A., and Torgensen, O., 1973a, An androgen binding protein in the testis cytosol fraction of adult rats: Comparison with the androgen binding protein in the epididymis, *Steroids* **21:**457.

Hansson, V., Reusch, E., Trygstad, O., Torgensen, O., Ritzén, E.M., and French, F.S., 1973b, FSH stimulation of testicular androgen binding protein, *Nature (London) New Biol.* **246:**56.

Hansson, V., French, F.S., Weddington, S.C., and Nayfeh, S.N., 1974a, FSH stimulation of testicular androgen binding protein (ABP): Androgen "priming" increases the response to FSH, in: *Hormone Binding and Target Cell Activation in the Testis* (M.L. Dufau and A.R. Means, eds.), pp. 287–290, Plenum Press, New York.

Hansson, V., Trygstad, O., French, F.S., McLean, W.S., Smith, A.A., Tindall, D.J., Weddington, S.C., Petrusz, P., Nayfeh, S.N., and Ritzén, E.M., 1974b, Androgen transport and receptor mechanisms in testis and epididymis, *Nature (London)* **250:**387.

Hansson, V., Ritzén, E.M., French, F.S., and Nayfeh, S.N., 1975a, Androgen transport and receptor mechanisms in testis and epididymis, in: *Handbook of Physiology,* Section 7, Vol. V (D.W. Hamilton and R.O. Greep, eds.), pp. 173–202, American Physiological Society, Washington, D.C.

Hansson, V., Weddington, S.C., Petrusz, P., Ritzén, M.E., Nayfeh, S.N., and French, F.S., 1975b, FSH stimulation of testicular androgen binding protein (ABP): Comparison of ABP response and ovarian augmentation, *Endocrinology* **97:**469.

Harris, M.E., Bartke, A., Weisz, J., and Watson, D., 1977, Effects of testosterone and dihydrotestosterone on spermatogenesis fluid and peripheral androgen levels in hypophysectomized rats, *Fertil. Steril.* **28:**1113.

Helminen, H.J., and Ericsson, J.L.E., 1970, On the mechanism of lysosomal enzyme secretion: Electron microscopic and histochemical studies on the epithelial cells of the rat's ventral prostate lobe, *J. Ultrastruct. Res.* **33:**528.

Helminen, H.J., and Ericsson, J.L.E., 1971, Evidence for a Golgi-mediated merocrine type of secretion of acid phosphatase in prostatic epithelium, *J. Ultrastruct. Res.* **36:**532.

Heyns, W., and DeMoor, P., 1977, Prostatic binding protein: A steroid binding protein secreted by the rat prostate, *Eur. J. Biochem.* **78:**221.

Heyns, W., Peeters, B., and Mous, J., 1977, Influence of androgens on the concentration of prostatic binding protein (PBP) and its mRNA in rat prostate, *Biochem. Biophys. Res. Comm.* **77:**1492.

Heyns, W., Vandamme, B., and DeMoor, P., 1978, Secretion of prostatic binding protein by rat ventral prostate—influence of age and androgen, *Endocrinology* **103:**1090.

Hoffer, A.P., Hamilton, D.W., and Fawcett, D.W., 1973, The ultrastructure of the principal cells and intraepithelial leukocytes in the initial segment of the rat epididymis, *Anat. Rec.* **175:**169.

Hopkins, C.R., and Farquhar, M.G., 1973, Hormone secretion by cells dissociated from rat anterior pituitaries, *J. Cell Biol.* **59:**276.

Huang, H.F.S., and Johnson, A.D., 1975, Amino acid composition of epididymal plasma of mouse, rat, rabbit and sheep, *Comp. Biochem. Physiol.* **50B:**359.

Hunter, A.G., 1969, Differentiation of rabbit sperm antigens from those of seminal plasma, *J. Reprod. Fertil.* **20:**413.

Jamieson, J.D., and Palade, G.E., 1967, Intracellular transport of secretory proteins in the pancreatic exocrine cell. I. Role of the peripheral elements of the Golgi complex, *J. Cell Biol.* **34:**577.

Jamieson, J.D., and Palade, G.E., 1977, Production of secretory proteins in animal cells, in: *International Cell Biology* (B.R. Brinkley and K.R. Porter, eds.), pp. 308–317, Rockefeller University Press, New York.

Kierszenbaum, A.L., Feldman, M., Lea, O., Spruill, W.A., Tres, L.L., Petrusz, P., and French, F.S., 1979, Secretion and uptake of androgen binding protein (ABP) in testis, epididymis and primary cultures of Sertoli cells (SC) and epididymal epithelial cells (EEC), *J. Cell Biol.* **83:**228a.

Kierszenbaum, A.L., Feldman, M., Lea, O., Spruill, W.A., Tres, L.L., Petrusz, P., and French, F.S., 1980, Localization of androgen-binding protein in proliferating Sertoli cells in culture, *Proc. Natl. Acad. Sci. U.S.A.* **77:**5322.

King, R.J.B., and Mainwaring, W.I.P., 1974, *Steroid–Cell Interactions*, Butterworth, London.

Kopecny, V., and Pech, V., 1977, An autoradiographic study of macromolecular synthesis in the epithelium of the ductus epididymidis in the mouse, *Histochemistry* **50:**229.

Kormano, M., Koskimies, A.I., and Hunter, R.L., 1971, The presence of specific proteins, in the absence of many serum proteins, in the rat seminiferous tubule fluid, *Experientia* **27:**1461.

Koskimies, A.I., and Kormano, M., 1975, Proteins in fluids from different segments of the rat epididymis, *J. Reprod. Fertil.* **43:**345.

Lea, O.A., Petrusz, P., and French, F.S., 1977a, Isolation and characterization of prostatein, a major secretory protein of the rat ventral prostate, *Fed. Proc. Fed. Am. Soc. Exp. Biol.* **36:**780.

Lea, O.A., Petrusz, P., and French, F.S., 1977b, Prostatein: A dihydrotestosterone binding protein secreted by the rat ventral prostate, Proceedings of the 59th Annual Meeting Endocrinology Society, p. 165.

Lea, O.A., Petrusz, P., and French, F.S., 1978, Purification and localization of acidic epididymal glycoprotein (AEG): A sperm coating protein secreted by the rat epididymis, *Int. J. Androl. Suppl.* **2:**592.

Lea, O.A., Petrusz, P., and French, F.S., 1979, Purification and characterization of prostatein, a major secretory protein of the rat ventral prostate, *J. Biol. Chem.* **254:**6196.

Määttälä, P., and Korhonen, L.K., 1977, Early ultrastructural changes in the rat ventral prostate induced by castration, *Acta Endocrinol.* **85**(Suppl. 212):46.

Maneely, R.B., 1955, The distribution of polysaccharide complexes and of alkaline glycerophosphatase in the epididymis of rat, *Acta Anat. (Basel)* **24:**314.

Mann, T., 1963, Biochemistry of the prostate gland and its secretion, in: *Biology of the Prostate and Related Tissues* (E.P. Vollmer, ed.), pp. 235–246, National Cancer Institute Monograph No. 12.

Mann, T., 1964, *The Biochemistry of the Semen and of the Male Reproductive Tract*, Methuen, London.

Moore, H.D.M., and Bedford, J.M., 1977, Androgen-dependence of rat epididymal cells: Changes in ultrastructure and protein uptake following short-term castration, *Anat. Rec.* **187:**659.

Musto, N.A., Gunsalus, G.L., and Bardin, C.W., 1978, Further characterization of androgen binding protein in epididymis and blood, *Int. J. Androl. Suppl.* **2:**424.

Nevo, A.C., Michaeli, I., and Schindler, H., 1961, Electrophoretic properties of bull and rabbit spermatozoa, *Exp. Cell Res.* **23**:69.

Orgebin-Crist, M.C., Danzo, B., and Davies, J., 1975, Endocrine control of the development and maintenance of sperm fertilizing ability in the epididymis, in: *Handbook of Physiology,* Section 7, Vol. V (D.W. Hamilton and R.O. Greep, eds.), pp. 319–338, American Physiological Society Washington, D.C.

Palade, G.E., 1975, Intracellular aspects of the process of protein synthesis, *Science* **189**:347.

Parker, M.G., Scrace, G.T., and Mainwaring, W.I.P., 1978, Testosterone regulates the synthesis of major proteins in rat ventral prostate, *Biochem. J.* **170**:115.

Petrusz, P., DiMeo, P., Ordronneau, P., Weaver, C., and Keefer, D.A., 1975, Improved immunoglobulin–enzyme bridge method for light microscopic demonstration of hormone-containing cells of the rat adenohypophysis, *Histochemistry* **46**:9.

Prasad, M.R.N., and Rajalakshmi, M., 1977, Recent advances in the control of male reproductive functions, in: *Reproductive Physiology II* (R.O. Greep, ed.), *Int. Rev. Physiol.* **13**:153–199, University Park Press, Baltimore.

Price, D., and Williams-Ashman, H.G., 1961, The accessory reproductive glands of mammals, in: *Sex and Internal Secretions,* Vol. 1 (W.C. Young and G.W. Corner, eds.), pp. 366–441, Williams and Wilkins, Baltimore.

Rajalakshmi, M., and Prasad, M.R.N., 1969, Changes in the sialic acid content of the accessory glands of the male rat, *J. Endocrinol.* **41**:471.

Rambourg, A., Hernandez, W., and Leblond, C.P., 1969, Detection of complex carbohydrates in the Golgi apparatus of rat cells, *J. Cell Biol.* **40**:395.

Redman, C.M., Siekevitz, P., and Palade, G.E., 1966, Synthesis and transfer of amylase in pigeon pancreatic microsomes, *J. Biol. Chem.* **241**:1150.

Ritzén, E.M., and French, F.S., 1974, Demonstration of an androgen binding protein (ABP) in rabbit testis: Secretion in efferent duct fluid and passage into epididymis, *J. Steroid Biochem.* **5**:151.

Ritzén, E.M., Dobbins, M.C., Tindall, D.J., French, F.S., and Nayfeh, S.N., 1973, Characterization of an androgen binding protein (ABP) in rat testis and epididymis, *Steroids* **21**:593.

Ritzén, E.M., Hansson, V., and French, F.S., 1980, The Sertoli cell, in: *The Testis* (H. Burger and D. de Kretser, eds.), Raven Press, New York (in press).

Sanborn, B.M., Elkington, J.S.H., Chowdhury, M., Tcholakian, R.K., and Steinberger, E., 1975, Hormonal influences on the level of testicular androgen binding activity: Effect of FSH following hypophysectomy, *Endocrinology* **96**:304.

Setchell, B.P., 1970, Testicular blood supply, lymphatic drainage and secretion of fluid, in: *The Testis,* Vol. 1 (A.D. Johnson, W.R. Gomes, and N.L. VanDemark, eds.), pp. 101–239, Academic Press, New York.

Setchell, B.P., and Waites, G.M.H., 1975, The blood–testis barrier, in: *Handbook of Physiology,* Section 7, Vol. V (D.W. Hamilton and R.O. Greep, eds.), pp. 143–172, American Physiological Society, Washington, D.C.

Steinberger, E., 1971, Hormonal control of mammalian spermatogenesis, *Physiol. Rev.* **51**:1.

Stumpf, W.E., Baerwaldt, C., and Sar, M., 1971, Autoradiographic cellular and subcellular localization of sexual steroids, in: *Basic Actions of Sex Steroids on Target Organs* (P.O. Hubinont, F. Leroy, and P. Galand, eds.), pp. 3–20, S. Karger, Basel.

Tindall, D.J., Cunningham, G.R., and Means, A.R., 1978, Structural requirements for steroid binding to androgen binding proteins, *Int. J. Androl. Suppl.* **2**:434.

Tveter, K.J., Hansson, V., and Unhjem, O., 1975, Androgen binding and metabolism, in: *Molecular Mechanisms of Gonadal Hormone Action,* Vol. 1 (J.A. Thomas and R.L. Singhal, eds.), pp. 17–76, University Park Press, Baltimore.

Tymoczko, J.L., Liang, T., and Liao, S., 1977, Androgen receptor interactions in target cells: Biochemical evaluation, in: *Receptors and Hormone Action,* Vol. 2 (B.W. O'Malley and L. Birnbaumer, eds.), pp. 121–156, Academic Press, New York.

Weddington, S.C., Brandtzaeg, P., Hansson, V., French, F.S., Petrusz, P., Nayfeh, S.N., and Ritzén, E.M., 1975, Immunological cross reactivity between testicular androgen-binding protein and serum testosterone-binding globulin, *Nature (London)* **258:**257.

White, M.G., Tres, L.L., and Kierszenbaum, A.L., 1979, Electron microscopic study of rat primary Sertoli and epididymal epithelial cell cultures, *Anat. Rec.* **193:**719.

Wilson, E.M., Lea, O.A., and French, F.S., 1977, Androgen-binding proteins of the male rat reproductive tract, in: *Receptors and Hormone Action,* Vol. 2 (B.W. O'Malley and L. Birnbaumer, eds.), pp. 491–531, Academic Press, New York.

16

Monitoring the Metabolic Rate of Germ Cells and Sperm

Roy H. Hammerstedt

1. Overview of the Metabolic Needs of Germ Cells, Sperm, and Ova: Changes during Differentiation and Specialization

The development of mature, fertile sperm from the germinal epithelium occurs within two organs. Spermatogenesis within the testis requires 6–10 weeks, depending on the species, and results in immotile, infertile testicular sperm. During this phase, germ-cell metabolism emphasizes biosynthesis as the cells divide, differentiate, synthesize macromolecules, and develop unique organelles. After spermiation, testicular sperm are transported rapidly to the epididymis. Transport through the epididymis requires 4–16 days, and although important intracellular modifications occur as sperm undergo maturation, the sperm are quiescent with regard to bioenergetics. After ejaculation, sperm metabolism is largely biodegradative to support the function of the highly motile and fertile sperm. Interaction of sperm with components of fluids secreted by the female reproductive tract results in a process termed capacitation. The velocity of the motile sperm, and presumably the rate of biodegradation, increase as a consequence of capacitation.

After fertilization, a dramatic change in metabolism occurs in the ovum. The complex degradative pathways required to utilize carbohy-

Roy H. Hammerstedt • Paul M. Althouse Laboratory, Biochemistry Graduate Program, Department of Microbiology, Cell Biology, Biochemistry and Biophysics, The Pennsylvania State University, University Park, Pennsylvania 16802.

drates are nonfunctional as the initial cleavage divisions of the fertilized ovum occur. Ova utilize only simple compounds such as lactate or pyruvate. As cleavage and embryogenesis continue, the zygote reacquires the ability to use complex nutrients to support the biosynthetic processes associated with cell division and growth. Thus, different aspects of metabolism are featured during this developmental sequence (Fig. 1). The remainder of this chapter will be restricted to germ cells during spermatogenesis and sperm.

One goal of biochemical endocrinology is to describe modulations of biological systems in terms of known chemical reactions. A variety of experimental models have been used to attain this goal, e.g., chemical systems, purified enzymes, multienzyme systems, isolated organelles, cells in culture, perfused organs, and intact organisms. At each stage, the experimenter learns more about the whole system at the cost of uncertainty about details of individual reactions. Suspensions of whole cells, e.g., sperm, provide an experimental system of moderate complexity. First, basic metabolic features of sperm can be established in a chemically defined medium, and then the effects of components of reproductive tract fluids can be evaluated.

To quantify the changing metabolic events in cells, several requirements must be met. An adequate supply of intact cells of a single type

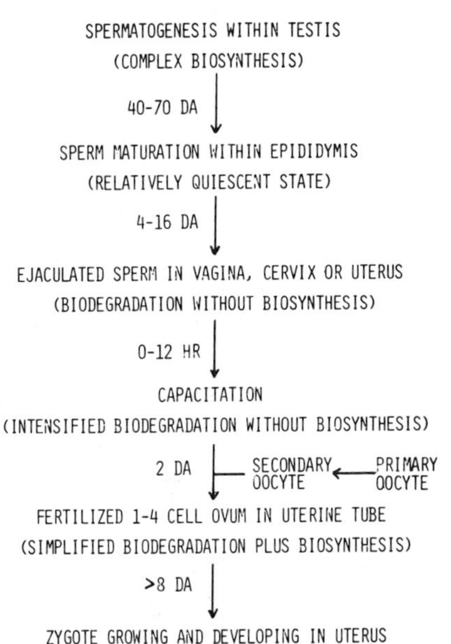

Figure 1. General features of energy metabolism during development of sperm and embryo.

must be obtained. Specific and sensitive analytical methods must be validated for use. A hypothesis, which can be evaluated unequivocally by experimentation, must be formulated and clearly stated regarding the metabolic needs of each cell type. These factors are interdependent, and each may limit progress. As the sensitivity of the methods chosen is increased, fewer cells are required for analysis. This permits use of a more refined cell-isolation procedure to obtain a better-defined cell population for use in the experiment.

2. Selection of an Animal Model

2.1. Methods for Obtaining Cells

An ideal model would allow isolation of all types of germ cells and sperm from a single species. Since this may not be feasible, interspecies comparisons must suffice while new techniques of cell separation are being developed.

Available separation techniques are summarized below and in Table I.

Fractionation of dissociated testicular cells. Unit-gravity sedimentation has been used to separate six to nine classes of ram (Loir and Lanneau, 1974, 1977) and rodent (Lam *et al.*, 1970; Lee and Dixon, 1972; Meistrich, 1972; Bruce *et al.*, 1973; Galena and Terner, 1974; Davis and Schuetz, 1975; Romrell *et al.*, 1976; Barcellona and Meistrich, 1977; Bellvé *et al.*, 1977; Meistrich, 1977; Salhanick and Terner, 1979) testicular germ cells. Centrifugal elutriation (Grabske *et al.*, 1975; Nakamura and Hall, 1976, 1977) and density-gradient centrifugation (Meistrich and Trostle, 1975; Alemán *et al.*, 1978) have also been used.

Chronic cannulation of rete testis. This surgical technique has been successfully used with bulls, rams, and boars to study testicular sperm (reviewed by Voglmayr, 1975; Setchell, 1978). A procedure for collection of rete testis fluid and testicular sperm from rats was described (Free and Jaffe, 1979), but may be of limited use, since too few testicular sperm are obtained for metabolic or biochemical analyses.

Ligation of efferent ducts. Adequate quantities of rete testicular sperm have been obtained from rats, rabbits, and monkeys (Cooper and Waites, 1974; Harris and Bartke, 1974; Waites and Einer-Jensen, 1974; Cooper *et al.*, 1976). This technique could also be used to isolate sperm from other species. The effects of increased hydrostatic pressure on testicular function or the entrapped sperm have not been examined in detail (see Free and Jaffe, 1979).

Expression of sperm from epididymis. Mincing of tissue in buffer

Table I. Methods for Isolation of Germ Cells and Sperm

Stage of development	Cell-isolation technique	Applicability of technique to:			
		Rat	Bull, ram	Monkey	Man
I. Testicular[a]					
Spermatogonia Spermatocytes Round spermatids	Unit-gravity fractionation or elutriation	Successful	Successful	Untested; reasonable possibility	Untested; reasonable possibility
Testicular sperm	Indwelling rete testis cannula	No	Yes	Unlikely	Testes of living man unavailable
	Efferent-duct ligation	Yes	Yes	Yes	
II. Epididymal[b]					
Caput	Expression from tissues	No	Yes	Yes	Amounts limiting
Corpus	Expression from tissues	No	Yes	Yes	Amounts limiting
Cauda	Expression from tissues	Yes	Yes	Yes	Yes
	Indwelling vas deferens cannula	No	Yes	Yes	Technically possible
III. Ejaculated					
	Artificial vagina, masturbation, or electrostimulation	Yes	Yes	Yes	Yes

[a] Objective evaluation based on published success.
[b] Subjective evaluation based on published values for sperm numbers per epididymis.

permits the isolation of caput, corpus, or cauda epididymal sperm. Sufficient quantities of sperm from the caput or corpus may not be obtained from a single rodent, but more than 5×10^9 sperm can be obtained from rams.

Retrograde flushing of cauda epididymis. This technique can be used for most animals. With care, an undiluted "plug" of sperm can be obtained.

Cannulation of vas deferens. Cauda epididymal sperm from the bull (Amann *et al.*, 1963, 1974; Sexton *et al.*, 1971), ram (Tadmor and Schindler, 1966; Tischner, 1967; Voglmayr *et al.*, 1977), and boar (Johnson and Pursel, 1975) can be obtained by cannulation. Success seems possible with rabbits (Bech and Koefoed-Johnsen, 1973; Holtz and Foote, 1974) and rats (Back *et al.*, 1974).

Collection of ejaculated sperm by artificial vagina, electrostimulation, or masturbation. One or more of these techniques can be used with all animals.

2.2. Factors That Influence Selection of an Animal Model

2.2.1. Ease of Isolating Cells in Adequate Numbers

The first consideration in selecting an animal model is the potential of the system to provide a sufficient quantity of each individual class of germ cells and epididymal sperm (Table I). On the basis of this criterion, the bull or ram appears to offer the greatest potential for obtaining adequate numbers of each cell type.

The first step in isolating cells from the germinal epithelium is dispersion of testicular tissue. Fawcett *et al.* (1973) have classified the testes of 14 mammalian species on the basis of abundance of Leydig cells and loose connective tissue, the degree of development and location of the interstitial lymphatics, and the topographical relationship of these components to the seminiferous tubules. Rodents (rat, mouse, and hamster) have large lymphatic sinusoids and a minimum amount of interstitial connective tissue. Seminiferous tubules from these species are easily isolated and disrupted. Another group of animals (man, monkey, ram, and bull) has conspicuous lymph vessels and a moderate amount of loose connective tissue. Disruption is more difficult, but is possible. The boar, oppossum, and others constitute a third class with extensive tissue and small, widely scattered lymphatic vessels. Isolation of germ cells from seminiferous tubules of these species might be difficult. Simple mincing of rodent tissue, with or without treatment with trypsin or DNAase to remove damaged cells, releases germ cells (Galena and Terner, 1974; Davis and Schuetz, 1975; Bellvé *et al.*, 1977; Loir and

Lanneau, 1977; Meistrich, 1977; Alemán *et al.,* 1978; Salhanick and Terner, 1979). This approach has been used to isolate germ cells from ram testes (Loir and Lanneau, 1974). Purified collagenase has also been used (Alemán *et al.,* 1978) to isolate germ cells from testes.

Testes from adult animals contain all types of germ cells. The relative sedimentation rates of these cells at unit gravity have been determined for the ram (Loir and Lanneau, 1974), mouse (Lee and Dixon, 1972; Bellvé *et al.,* 1977), and rat (Lam *et al.,* 1970). Loir and Lanneau (1974) found the following relative sedimentation rates for ram germ cells: spermatozoa, 1.0; late elongated spermatid, 1.35; early elongated spermatid, 2.37; with round spermatids, secondary spermatocytes, and diplotene, zygotene, and pachytene spermatocytes and spermatogonia sedimenting much faster. Rat and mouse germ cells sediment in a similar manner.

Germinal cells appear sequentially in the prepuberal testis. By use of prepuberal males of carefully selected age or body weight (testis development), preparation of a cell suspension rich in a selected germ-cell type is possible (Meistrich, 1972; Davis and Schuetz, 1975; Bellvé *et al.,* 1977; Loir and Lanneau, 1977; Alemán *et al.,* 1978). This simplifies cell isolation, but since a large proportion of the germinal cells degenerate during the first wave of spermatogenesis (see Gondos, 1977) and fertility of bulls is low immediately after puberty (Martig and Almquist, 1969; Rottensten, 1972; Almquist, unpublished data), germ cells and sperm obtained by this approach may not be representative of the cells found in the mature male. Procedures to isolate nonflagellate germ cells from sexually mature rats have been described (Galena and Terner, 1974; Salhanick and Terner, 1979).

Epididymal sperm can be obtained by mincing epididymal tissue. Total sperm content of the epididymis differs greatly among species (Table II). With the mild mincing required to release intact viable sperm, all the sperm cannot be obtained. Between 50 and 90% of the sperm in segments of the ram epididymis were recovered when the tissue was suspended in buffer, minced with a scalpel and forceps, and the resulting suspension filtered through a 37-μm nylon mesh (Hammerstedt, unpublished data).

2.2.2. Similarities in the Kinetics of Germ-Cell Development among Species

Similarity of germ-cell development in the animal model to human spermatogenesis and epididymal maturation is important. No biochemical studies have compared human germ cells with those of other animals. If the enzyme content of a given cell type (e.g., round spermatids or cauda

Table II. Species Differences in Extragonadal Reserves and Transit Time of Sperm through the Epididymis[a]

Reserve and time	Man	Bull (dairy)	Rat (Wistar)	Monkey (rhesus)	Rabbit (New Zealand White)	Ram (Ile-de-France)	Hamster (Syrian)
Sperm content (× 10^9)							
Caput	0.04	19	0.25	1.2	0.36	23	0.15
Corpus	0.04	4.7	—	4.2	0.12	11	—
Cauda	0.11	38	0.45	5.7	1.6	126	1.02
Total excurrent ducts	0.19	69	>0.7	>13	2.2	>165	1.2
Transit time through epididymis (days)							
Caput	0.72	2.5	3.0	1.1	2.2	2.4	2.0
Corpus	0.71	0.6	—	3.8	0.8	1.2	—
Cauda	1.76	5.2	5.1	5.6	9.7	12.8	13.6

[a] Adapted from Amann *et al.* (1976) and Amann and Howards (1980). Data are for animals at sexual rest, except for humans, for which the interval from the last ejaculation is unknown.

sperm) in all species is identical, any animal model could be chosen. However, there are species differences in the life span of each cell type and in the time required for epididymal transit (Tables III and IV), and these factors must be considered.

Comparison of seven potential animal models with man (Table III) reveals a 2-fold difference in the duration of the cycle of the seminiferous epithelium. On this basis, the bull appears to present the closest parallel to the human system. An equally important consideration is the relative rate of progression through germ-cell development (Table IV). Comparison of the relative life spans of each germ-cell type leads to the conclusion that the ram or rabbit presents the closest parallel to the human system.

A third consideration is the time involved in epididymal maturation. Epididymal transit time for sperm differs among species (see Table II). Most mammals have a transit time of 5–10 days, with values ranging from 4 days for humans to 16 days for sexually inactive rams and boars. For most species, sperm isolated from the distal corpus epididymis have good fertility, but only sperm isolated from the cauda epididymis have both good fertility and low embryonic mortality (Orgebin-Christ *et al.*, 1975; Fournier-Delpech *et al.*, 1979). Transit time of sperm through the caput plus corpus epididymis is less than 5 days (Table II). Sperm are held in the cauda epididymis for 2–13 days until removed by ejaculation or voiding in the urine. Since the time required for epididymal maturation does not differ greatly among species, this factor is of little consequence in selecting an animal model.

Table III. Time Required for Germ-Cell Development (Days)[a]

Life span	Man	Bull (dairy)	Rat (Sprague–Dawley)	Monkey (rhesus)	Rabbit (New Zealand White)	Ram (Ile-de-France)	Mouse (Swiss)	Hamster (Syrian)
Duration of cycle of seminiferous epithelium	16.0	13.5	12.9	9.5	10.7	10.4	8.9	8.8
B spermatogonia	6.3	4.3	2.0	2.9	1.3	2.2	1.5	1.6
Preleptotene, leptotene, and zygotene spermatocytes	9.2	9.1	7.8	6.0	7.3	6.6	4.8	4.6
Pachytene and diplotene spermatocytes	15.6	11.4	12.2	9.5	9.5	9.0	8.5	8.1
Secondary spermatocytes	0.8	1.6	0.7	0.6	1.2	1.0	0.8	0.7
Golgi-phase spermatids	7.9	0.7	2.9	1.8	2.1	1.8	1.7	2.3
Cap-phase spermatids	1.6	4.5	5.0	3.7	5.2	4.4	3.6	3.5
Acrosome-phase spermatids	6.5	8.3	5.0	4.0	3.4	4.1	4.5	2.9
Maturing spermatids	7.9	4.2	9.2	4.4	4.3	4.0	5.1	5.8
Total								
Primary spermatocytes	24.8	20.6	20.0	15.5	16.8	15.6	13.3	12.7
Round spermatids	9.5	5.2	7.9	5.5	7.3	6.2	5.4	5.8
Elongating spermatids	6.5	8.3	5.0	4.0	3.4	4.1	4.5	2.9
Maturing spermatids	7.9	4.2	9.2	4.4	4.3	4.0	5.1	5.8

[a] Adapted by R. P. Amann from the literature.

Table IV. Rate of Progression through Germ-Cell Development[a]

Fraction of life span as:	Man	Bull (dairy)	Rat (Sprague–Dawley)	Monkey (rhesus)	Rabbit (New Zealand White)	Ram (Ile-de-France)	Mouse (Swiss)	Hamster (Syrian)
B spermatogonia	0.25	0.21	0.10	0.19	0.08	0.14	0.11	0.12
Total								
Spermatocytes	1.00	1.00	1.00	1.00	1.00	1.00	1.00	1.00
Round spermatids	0.38	0.25	0.40	0.35	0.43	0.40	0.41	0.46
Elongating spermatids	0.26	0.40	0.25	0.26	0.20	0.26	0.34	0.23
Maturing spermatids	0.32	0.20	0.46	0.28	0.26	0.26	0.38	0.46

[a] Calculated by R. P. Amann from the data in Table III, normalized to the number of days of germ-cell development spent as spermatocytes.

2.2.3. Is There a "Best" Model?

Factors that influence the choice of an animal model are summarized in Table V. No species appears to provide an optimal model. Whole-cell incubations require $0.1–0.5 \times 10^9$ cells, while-enzyme characterization studies require more than 10^{10} cells. Available physical facilities and costs for animal maintenance must be considered, since many investigators can house small animals or rams, but not bulls.

Although sheep are seasonal breeders, rams have a long breeding season and provide adequate sperm numbers for more than 6 months of the year. Sperm can be collected during the breeding season, frozen, and saved for enzyme analyses during the late spring and summer. Hemicastration (aseptic surgery) of a ram could provide ample numbers of each type of germ cell or sperm for any type of experiment. The second testicle can be removed 2–3 days later.

3. General Aspects of Metabolism

3.1. The ATP Cycle

A simple representation of energy metabolism is depicted in Fig. 2. A cell in balance with its environment has regulated its rates of ATP synthesis and consumption; the result is a unique ATP/ADP ratio for a specific metabolic state. Many metabolic pathways are available to generate ATP, and the exact choice of pathway depends on the complement of enzymes within the cell and the types of substrates available. The rate of ATP consumption is a function of factors such as amount of biosynthesis occurring in the cell, ion balance across cell membranes, and rate and extent of motility of cell.

If a cell is subjected to a new set of energy requirements, the rates of ATP generation and consumption *must* change. As a result, an ATP turnover rate is established that is characteristic of the cell under the new conditions. If the rates of both synthesis and degradation change equally, the ATP level will not be altered and the changes in energy metabolism will not be detected if the concentrations of ATP, ADP, and AMP are the only parameters monitored. Therefore, to detect all types of changes in metabolic state of the cell, the kinetics of ATP synthesis and degradation must be determined.

3.2. Interaction between Bioenergetics and Other Cell Functions

In his text, Atkinson (1977) stressed the coupling of metabolism (Fig. 3) through an examination of three hypothetical functional blocks (or

Table V. Basis for Selection of Animal Model[a]

Factors to be considered	Bull	Rat	Monkey	Rabbit	Ram	Mouse	Hamster
A. Ease of isolating cells in adequate numbers							
1. Testicular phase							
a. Elutriate germ cells from testicular suspension							
(1) Testicular tissue disruption	3	1	3	3	3	1	1
(2) Sedimentation rates of germ cells	1	1	1	1	1	1	1
b. Cannulate rete testis	1	X	4	X	1	X	X
2. Epididymal phase							
a. Mince caput tissue	1	3	1	2	1	4	3
b. Mince corpus tissue	1	X	1	3	1	X	X
c. Mince cauda tissue	1	2	1	1	1	3	2
d. Cannulate vas deferens	1	X	3	X	1	X	X
B. Comparability of cells obtained to those of man							
1. Duration of cycle of seminiferous epithelium (Table III)	1	2	5	3	4	5	5
2. Relative rate of progression through germ-cell development (Table IV)	5	4	1	2	1	3	4
3. Interval for epididymal maturation (Table II)	1	1	1	1	1	1	1

[a] Arranged in decreasing order of utility within parameters by considering the ease of experimental manipulation, number of sperm obtained per testis or epididymis, and similarity to man. (1) Species of choice; (X) method not useful. The relative importance of various parameters is subjective.

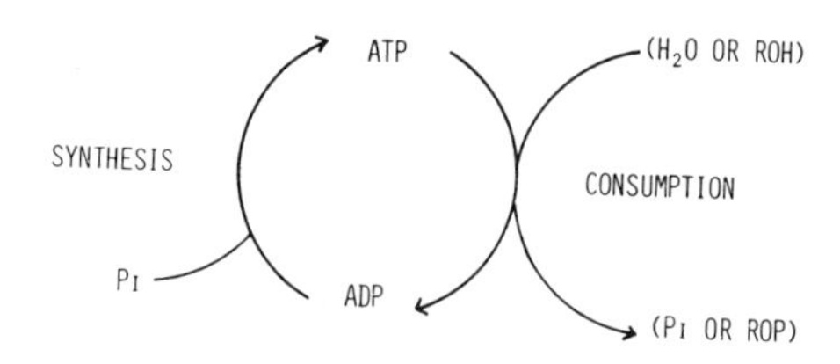

Figure 2. The ATP cycle.

types) of reactions and the relationships among the blocks. During degradative metabolism (catabolism), nutrients can be oxidized to carbon dioxide. Most of the electrons liberated during these oxidations are transferred to oxygen with concomitant production of ATP (via electron-transport phosphorylation). Other electrons are used to generate the NADPH that serves as the reducing agent for biosynthesis. The major metabolic pathways in this block are the glycolytic sequence and the tricarboxylic acid cycle. These same sequences, and the pentose phosphate pathway, also supply the starting materials for all the cell's biosynthetic processes. Biosynthesis (anabolism) is much more complex. The starting materials produced by glycolysis and the tricarboxylic acid cycle are converted into hundreds of cellular components. ATP serves as the universal coupling agent or energy-transducing compound, and NADPH serves as the necessary reducing agent. The requirements for growth are dominated by the synthesis of complex macromolecules, e.g., proteins, nucleic acids, components of membranes. Intracellular metab-

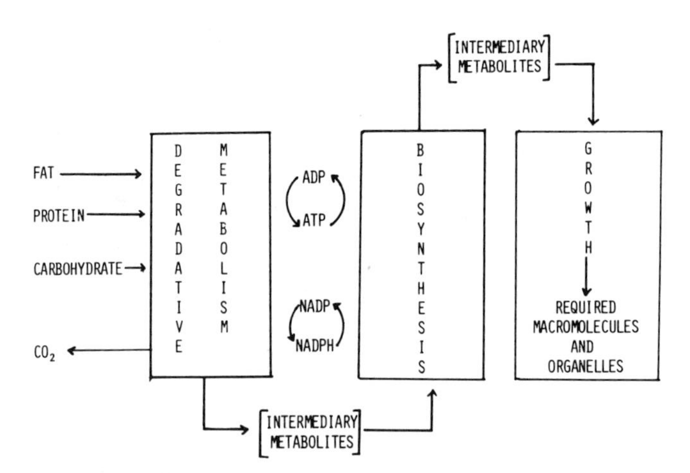

Figure 3. Schematic representation of metabolic processes in a complex cell type. Modified from Atkinson (1977).

olites are consumed by the reactions of biosynthesis and growth and are continuously replenished by catabolic processes.

3.3. Metabolism of Endogenous Reserves

All organisms require nutrients to maintain cellular function. Exogenous nutrients are taken into the cell and either used immediately or converted to other forms [potentially glycogen, lipid (triglyceride or phospholipid), protein, or polymers of simple acids] and stored for later use as needed by the organism (becoming endogenous reserves). The nature of these endogenous reserves and the mode of regulating their utilization vary among cells.

Deposition of endogenous reserves and mobilization of these reserves are highly regulated. Use of endogenous reserves provides a means of satisfying the constant ATP requirement of basal cellular function. The advantage of this process to continued cell function is obvious, but its regulation and quantitative importance are not understood. A symposium on these topics was held 17 years ago (Lamanna, 1963). A computer search revealed fewer than ten papers concerned with this topic since then, and only two dealt with the interaction of exogenous and endogenous metabolism. Quantitative data are absent.

A simple estimate of the contribution of endogenous substrates to energy metabolism can be made by monitoring O_2 consumption in the presence and absence of exogenous substrate. Subtraction of the respiration value in the absence of substrate from that observed for incubations containing exogenous substrate provides an estimate of the relative contributions of each to cellular metabolism. The validity of this approach is uncertain, since only a few studies have tested the underlying assumptions. Addition of exogenous substrate to suspensions of bacteria and yeast has been reported to enhance, have no effect on, or reduce the consumption of endogenous energy reserves (Dietrich and Burris, 1967; Rambeck *et al.*, 1971). In these studies, endogenous reserves were prelabeled by incubation of cell suspensions with ^{14}C-labeled compounds, after which the extracellular substrates were removed and the cells were incubated while the effects of nonradioactive exogenous substrates on $^{14}CO_2$ production were monitored. Comparable studies have not been done with mammalian cells.

3.4. Characteristics of Sperm Metabolism

The metabolic properties of sperm have been studied in my laboratory for nine years. From our data and data from other laboratories, I conclude that the following features of bull and ram spermatozoal metabolism are pertinent to a general understanding of sperm metabolism*:

1. Once motility has been initiated, sperm require an exogenous substrate for maintaining ATP levels (Cascieri *et al.,* 1976); endogenous reserves are not adequate. Simple compounds such as glucose, fructose, lactate, and pyruvate allow maintenance of ATP level.

2. Limited metabolism of amino acids, acetate, and other monomers occurs (Mann, 1964). Sperm can degrade exogenous fatty acids (Hartree and Mann, 1961; Mills and Scott, 1969; Niel and Masters, 1972).

3. Glucose is transported into the cell (Hiipakka and Hammerstedt, 1978a,b), phosphorylated by hexokinase, and metabolized exclusively by the Embden–Meyerhof pathway to pyruvate (Hammerstedt, 1975a,b). No glycogen metabolism or degradation via the pentose phosphate pathway is detectable (Hammerstedt, 1975a; Hiipakka and Hammerstedt, 1978a,b).

4. Pyruvate is reduced to lactate, which is excreted or diverted to the tricarboxylic acid cycle (Hammerstedt, 1975b). The distribution of lactate is variable (Hammerstedt, 1975b). Lactate dehydrogenase is located in both the cytoplasm and the mitochondria (Hutson *et al.,* 1977; Milkowski and Lardy, 1977; Storey and Kane, 1977; Van Dop *et al.,* 1977; Dawson, 1979). Sperm appear unique (Dawson, 1979) in that the reducing equivalents generated during cytoplasmic metabolism enter the mitochondria via lactate rather than by an alternate carrier system. Under some incubation conditions, acetate rather than lactate is excreted (Hutson *et al.,* 1977; Van Dop *et al.,* 1977).

5. Substrates for endogenous metabolism are unknown. Most (Lardy and Phillips, 1941a,b) but not all (Darin-Bennett *et al.,* 1973) data support the suggestion that phospholipids are the most important reservoir. Few triglycerides are found in the sperm (Quinn and White, 1967; Scott *et al.,* 1967).

6. Little or no biosynthesis occurs during an incubation lasting a few hours, although numerous investigators have reported that a small fraction of the radioactivity from glucose is incorporated into other cellular components (see Voglmayr, 1975). Addition of lactate to a sperm suspension metabolizing glucose has no effect on the glucose degradation rate (Hammerstedt, unpublished observation). No gluconeogenesis from lactate has been detected, although ^{14}C from glycerol does appear in carbohydrate isolated from the incubation medium (White, 1957).

A summary outline of sperm metabolism is presented in Fig. 4.

* Data for sperm from other species have been excluded, since many differences among species have been reported.

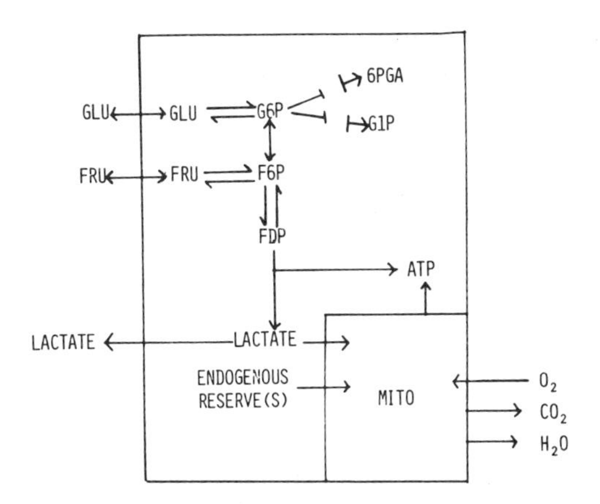

Figure 4. Schematic diagram of metabolism in sperm. Limited catabolic processes yield the ATP used for functions such as motility, ion balance, and preliminary steps of catabolism. The quantitative importance of each metabolic process varies with species, physiological state, and temperature of incubation. After spermiation, the male germ cell is essentially incapable of biosynthesis and growth.

4. Methods for Analysis of Metabolic Rate

4.1. General Features

To successfully detect effects of any compound or cell treatment on energy metabolism, the assay procedures must meet rigid requirements. The analytical method chosen must be (1) sufficiently sensitive so that postulated metabolic regulators can be tested at levels corresponding to their physiological concentration; (2) specific in being responsive only to changes in energy metabolism and not to changes in the transport properties of the cell membrane; and (3) general enough to allow the detection of all changes in energy metabolism regardless of the metabolic pathway.

The methods considered herein to evaluate metabolism of intact cells are restricted to measurements of substrate consumption and product accumulation or the characteristics of a few key intracellular intermediates (e.g., ATP). Numerous other methods are available. A clear description of their general applicability to the study of metabolic regulation was provided by Denton and Pogson (1976).

4.2. Methods Validated for Use with Sperm Suspensions

4.2.1. Adenine Nucleotide Level

Methods to inactivate cells and release the nucleotides were tested to assure that amounts released reflected actual intracellular levels. Either 5% trichloroacetic acid or boiling water inactivation was satisfactory when the adenine nucleotide content of ram or bull sperm was being measured (Cascieri *et al.*, 1976; Hiipakka and Hammerstedt, 1978b). However, only trichloroacetic acid treatment was satisfactory for rabbit sperm (Hammerstedt *et al.*, 1979a). The amount of ATP in the cell extracts is measured by an automated luciferase assay (Hammerstedt, 1973); ADP and AMP levels are determined after their conversion to ATP in prior enzyme incubations.

4.2.2. Glucose Consumption

Glucose consumption is often determined by measuring the amount of glucose remaining in solution after incubation. This approach is insensitive, since, in many incubations, excess substrate is provided and only a small portion (<10%) of the total glucose is consumed.

An alternate method of measuring glucose consumption is based on the release of ^{3}HOH from [2-^{3}H]glucose. After transport across the cell membrane and phosphorylation, [2-^{3}H]glucose-6-phosphate is converted to [1-^{3}H]fructose-6-phosphate by phosphoglucoisomerase. The transfer of hydrogen from C_2 of glucose to C_1 of fructose involves an intermediate in which the hydrogen resides on the enzyme; thus, hydrogen (or tritium) may exchange with solvent (^{3}HOH release) or be transferred from the enzyme back to the carbohydrate (^{3}H retention). This method is valid only if glucose is not metabolized via glycogen storage reactions or the pentose phosphate pathway and if the precise ratio between moles of ^{3}HOH released and moles of glucose consumed is determined for the cells being studied. These conditions were met by bull, ram, and rabbit sperm (Hammerstedt, 1975a; Hiipakka and Hammerstedt, 1978b; Hammerstedt *et al.*, 1979a).

4.2.3. Simultaneous Measurement of Exogenous and Endogenous Metabolic Rate

A set of parallel incubations has been used to simultaneously evaluate metabolic parameters of sperm. One flask, placed in a shaking incubator bath, contains [2-^{3}H]glucose. Repetitive sampling of this cell suspension permits estimation of glucose consumption rate (Section 4.2.2) and adenine nucleotide levels (Section 4.2.1). Another flask, placed in a respirometer, contains [3,4-^{14}C]glucose. The $^{14}CO_2$ released from glucose is trapped in KOH contained in the center well of the respirometer flask, and the rate of oxygen consumption is measured by the change in volume of the air over the cell suspension. Rationale, details, and assumptions implicit in this approach have been discussed (Hammerstedt, 1975b). Pertinent features are summarized below.

The total rate of glucose consumption is estimated by the rate of ^{3}HOH release from [2-^{3}H]glucose. For each mole of glucose degraded via the Embden–Myerhof pathway, 2 mol ATP and 2 mol lactate are produced. If the lactate is excreted, no further ATP is synthesized. However, if the lactate is transported into the mitochondrion, oxidized by mitochondrial lactate dehydrogenase, oxidatively decarboxylated to acetyl-CoA + CO_2 by the pyruvate dehydrogenase complex, and completely degraded via the tricarboxylic acid cycle and the electron-transport system, much more ATP is made.

The complete oxidation of glucose is measured via the $^{14}CO_2$ collected from the incubation using [3,4-^{14}C]glucose. The ^{14}C of [3,4-^{14}C]glucose that is transformed to [1-^{14}C]lactate and excreted will not be collected in the KOH within the center well of the respirometer. However, for every mole of [1-^{14}C]lactate that is metabolized in the mitochondrion, one mole of $^{14}CO_2$ will be found in the KOH of the center well. The ATP yield during mitochondrial degradation of lactate can be estimated as: 5 mol NADH generated during the oxidation of lactate to CO_2 yields 15 mol ATP, 1 mol $FADH_2$ from tricarboxylic acid metabolism yields 2 mol ATP, and one substrate-level phosphorylation in the tricarboxylic acid cycle yields 1 mol ATP. Thus, 18 mol ATP theoretically are recovered for every mole of lactate completely oxidized. Furthermore, the oxygen consumed during the mitochondrial metabolism of the lactate derived from extracellular glucose can be estimated from the balanced equation for lactate oxidation ($C_3H_6O_3 + 3O_2 \rightarrow 3CO_2 + 3H_2O$).

Finally, any oxidation of endogenous reserves occurring during metabolism of exogenous glucose is estimated by subtracting the O_2 consumption related to glucose metabolism (3 micromoles $^{14}CO_2$ produced from [3,4-^{14}C]glucose) from the total observed O_2 consumption. Morton and Lardy (1967a–c) calculated that 5 μmol ATP are formed per micromole O_2 consumed when sperm utilize endogenous substrate(s).

4.2.4. Rate of ATP Turnover

Under steady-state conditions, the rates of synthesis and degradation of ATP are balanced, resulting in characteristic levels of adenine nucleotides that are easily measured (Section 4.2.1). For example, although cauda epididymal and ejaculated bull sperm have equivalent nucleotide levels (Cascieri *et al.*, 1976), they could differ in their rates of ATP synthesis and degradation. To detect this difference, the rates of ATP synthesis and degradation would have to be compared (see Fig. 2).

One method to estimate the rate of ATP synthesis is to establish the exact carbon balance over the period of incubation and, if the ATP yield for each catabolic pathway is known, calculate the rate of ATP synthesis.

Applications of these calculations to the study of bacterial suspensions are given in Decker *et al.* (1970), Thauer *et al.* (1977), Gottschalk and Andreesen (1979), and Jones (1979). The method outlined in Section 4.2.3 can be used for calculations for sperm suspensions. Ejaculated bovine sperm incubated at 35°C in the presence of glucose synthesize 4–6 μmol ATP/hr per 10^8 sperm (Hammerstedt, 1975b) and have an ATP content of approximately 30 nmol ATP/10^8 sperm. Under these incubation conditions, the ATP pool must be replenished every 20–30 sec. If the temperature of incubation is lowered to 27°C, the rate of glucose degradation decreases about 5-fold, but no change in ATP concentration occurs (Hammerstedt and Hay, 1980). The ATP pool would have to be replenished every 3–5 min under these new incubation conditions if all intracellular conditions remained similar to those existing at 35°C. Since many assumptions are necessary when this method of calculation is used, a direct method would be useful.

Boyer and co-workers utilize ^{18}O-exchange reactions to evaluate the characteristics of ATP synthesis and hydrolysis (Boyer, 1967). Their general methods can be used with cell suspensions if it is assumed that after $H_2{}^{18}O$ addition all intracellular H_2O pools rapidly equilibrate with extracellular water (a process limited only by diffusion of H_2O across membranes). Under steady-state conditions, the rate of ATP synthesis equals the rate of incorporation of ^{18}O into orthophosphate by direct hydrolysis ($ATP + H_2{}^{18}O \rightarrow ADP + PO_3{}^{18}O$) or phosphoryl transfer plus hydrolysis ($ATP + ROH \rightarrow ADP + ROPO_3$; $ROPO_3 + H_2{}^{18}O \rightarrow ROH + PO_3{}^{18}O$). Previously used methods for $PO_3{}^{18}O$ analysis were tedious and have undoubtedly hindered application of this method. Recent methods (Midelfort and Rose, 1976) are simpler and more rapid and may stimulate interest in this approach.

An alternative method to determine the rate of ATP turnover is use of a single pulse of $^{32}PO_4$ to label cells. At steady state and under conditions where phosphate transport is not limiting, the rate of incorporation of ^{32}P into Y-[^{32}P]-ATP is equal to the rate of ATP synthesis. However, when we studied phosphate metabolism in erythrocytes and HeLa cells, several modes of phosphate transport were detected and phosphate transport was the limiting step under some incubation conditions (Niehaus and Hammerstedt, 1976).

Further experiments were designed to determine the rate of ATP turnover in bovine sperm cells (Hammerstedt and Niehaus, unpublished data). Phosphate equilibration and extent of ATP labeling were low (Table VI), and the rates of incorporation of extracellular orthophosphate into spermatozoal ATP changed during epididymal maturation (Fig. 5). Testicular sperm had a $t_{\frac{1}{2}}$ of more than 5 min for maximal ATP labeling,

Table VI. Incorporation of Extracellular [^{32}P]Orthophosphate into Intracellular Orthophosphate and ATP of Ejaculated Bovine Spermatozoa[a]

Incubation (min)	Relative specific radioactivity of intracellular		
	PO_4	γ-ATP	β-ATP
10	0.14	0.044	—
30	0.11	0.049	0.049
60	0.10	0.053	0.052
180	0.10	0.071	0.073

[a] Washed ejaculated bovine spermatozoa (1×10^8 cells/ml) were incubated in buffer (Hammerstedt, 1975b) plus 5 mM glucose at 27°C in the presence of [^{32}P]orthophosphate (250 μCi/μmol). Aliquots were removed and processed to determine the specific radioactivity of intracellular orthophosphate and the β- and γ-phosphates of ATP and ADP (Niehaus and Hammerstedt, 1976). Values are expressed relative to the specific radioactivity of extracellular orthophosphate. PO_4 refers to total orthophosphate without regard to ionic species.

while those for cauda epididymal and ejaculated sperm were approximately 2 min. These data are consistent with the rates calculated from substrate balance studies (see above) and establish that the rates of ATP synthesis for cauda epididymal and ejaculated sperm are equivalent. When the temperature of incubation was increased to 37°C, maximum labeling of γ-[^{32}P]-ATP was achieved in less than 30 sec. This is consistent with the higher metabolic rate at higher temperatures (Hammerstedt and Hay, 1980). If ATP turnover is to be used to monitor overall metabolic rate, the rate of phosphate transport must exceed the rate of ATP synthesis. This condition is apparently satisfied for sperm cells, but the lack of equilibration of internal and external $^{32}PO_4$ is of concern. Babcock *et al.* (1975) have described in greater detail the impermeability of the sperm plasma membrane to extracellular orthophosphate.

4.2.5. *Calorimetry*

All methods for generating or utilizing ATP are inefficient, since a perfect transfer of the energy released during the degradation of organic compounds to the synthesis of ATP cannot be achieved. The energy not conserved as ATP is released to the surroundings as heat. The heat produced is a state function and therefore depends only on the chemical identity of starting substrates and end products and not on the nature of the enzymes used in the conversion.

Heat production reflects the sum of all metabolism, so that any

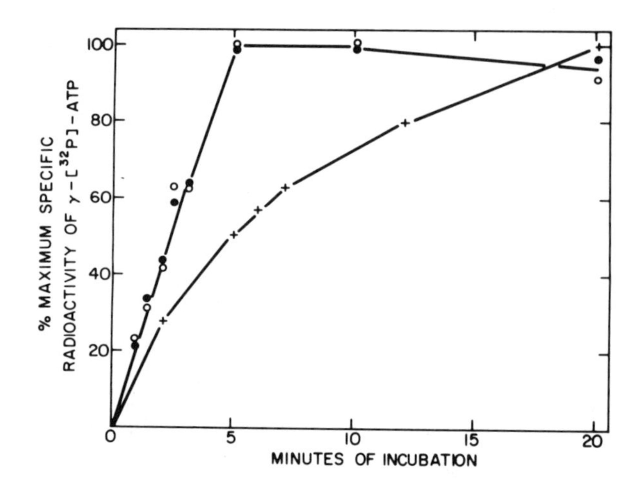

Figure 5. Rate of incorporation of extracellular orthophosphate into the Y-phosphate of bovine sperm. Data for single incubations of nonmotile testicular (+), motile cauda epididymal (○), and motile ejaculated (●) spermatozoa. Incubation conditions and analyses are described in the Table VI footnote. Three to five other incubations, using samples from 2–5 bulls, gave similar profiles.

alteration in energy balance will result in the readjustment of all heat-producing reactions to a new steady-state rate. This summation of small changes should make the measurement of heat production more responsive to changes in energy metabolism than any other method available. The method is responsive only to changes in energy metabolism, since problems associated with influx of labeled substrates and equilibration of metabolite pools are not encountered. Since almost all reactions release heat, a unique assay for each metabolite (essentially impossible) is not required to determine the energy balance within the cell.

Calorimetry was tested as a method to evaluate sperm viability nearly 20 years ago (Rothschild, 1962). The usefulness of the technique was apparent, but the need for a large sample and the insensitivity of the equipment limited its application. Sensitive microcalorimeters that allow measurement of the heat produced by approximately 1×10^8 sperm cells are now available. Thus, microcalorimetry may be a feasible method to evaluate general aspects of cellular metabolism. Recent reviews (Forrest, 1969; Kemp, 1975; Spink and Wadsö, 1975; Lamprecht, 1976; Schaarschmidt and Lamprecht, 1976; Wadsö, 1976) consider the application of calorimetry to the study of energy metabolism by suspensions of microorganisms and mammalian cells. The effects of hormone treatment on the heat produced by hepatocytes (Jarrett *et al.,* 1979) and plant tissues

(Anderson, Lovrien, and Brenner, unpublished observations) have been described. Heat measurements quickly provide reliable information on the metabolic status of cells and should be especially useful when studying the consumption of secondary metabolites or endogenous reserves.

A calorimeter for the study of cell suspensions has been built for my laboratory by Dr. Rex Lovrien of the Department of Biochemistry of the University of Minnesota. This instrument is a differential enthalpy-measuring calorimeter in which the reference heat flow is automatically subtracted (operationally analogous to a twin-beam spectrophotometer). The instrument has one reference and four experimental chambers that can be maintained at any temperature between 5 and 40°C within ± 0.1°C by Peltier thermoelectric pumps, and the desired atmosphere (e.g., air, O^2–CO^2, N^2) can be equilibrated with the cell suspension. Samples (1–2 ml) are placed into each chamber, the chambers sealed, and the isolation boxes closed. Temperature equilibration occurs in about 20 min. Flow of heat from each chamber to the heat sink is detected by thermopiles, and differences in heat flow between reference and sample chambers are displayed on a strip-chart recorder.

Experiments in progress (Inskeep and Hammerstedt, unpublished data) involve determination of the absolute rate of heat production by ejaculated sperm as a function of temperature and comparison of these values with those for adenine nucleotide level(s) and rates of glucose utilization, oxygen consumption, and lactate production. The limited metabolic capabilities of sperm (see Fig. 4) simplify the calculations. The simultaneous incubations (see Section 4.2.3) permit calculation of the moles of exogenous glucose that are converted to lactate or CO_2. Heat produced in the conversion of glucose to lactate is calculated as: (rate of lactate production) × (26 mcal/micromol glucose converted to lactate). Heat produced in the oxidation of glucose to $CO_2 + H_2O$ is calculated as: (rate of $^{14}CO_2$ production) × (672 mcal/micromole glucose converted to CO_2). The heat associated with metabolism of endogenous substrates (exact chemical identity unknown) must be equal to the total measured heat minus heat associated with metabolism of exogenous substrate. Thus, a direct quantitation of the relative importance of endogenous metabolism to the cell can be made without specific assays for each potential endogenous metabolite.

4.2.6. Computer-Based Method to Evaluate Sperm Motility

A dominant feature in the energetics of ejaculated sperm is the ATP demand of cellular motility; this process accounts for approximately 70% of the total ATP consumed by motile sperm (Rikmenspoel, 1965). Careful quantitation of the motility of sperm concomitant with other metabolic parameters would be useful. Visual observations of sperm motility are both inaccurate and imprecise, since the 95% confidence interval for a single subjective estimation of the percentage of motile sperm is equal to ±30–60 percentage units (Deibel *et al.*, 1976). Other objective measurements such as track motility or hand cartography of high-speed movies are tedious. Details of a computer-processing procedure developed at the Pennsylvania State University have been published (Liu and Warme, 1977). Briefly, sperm in a dilute suspension ($3–6 \times 10^7$ cells/ml) are viewed under phase-contrast optics and photographed on movie film at 25 frames/sec. The image is viewed frame by frame by a television camera, digitized, and fed into a computer. After images from a sequence of frames are entered into the computer, the presence or absense of motility and individual cellular velocities are determined on a cell-by-cell basis.

The main restrictions in the procedure are that the sample must be free of particulate matter similar in size to a sperm head and the sample should not contain more than 50 cells per field of view. Characteristics of the sperm population in the sample can be predicted within statistically defined limits. If five scenes are recorded for each of two slides (2-min photography time) and evaluated by computer (10-min analysis time), the predicted standard deviation will be less than 5.5 percentage units for motility and less than 9.8 μm/sec for velocity (Amann, 1979). A manuscript describing validation of the system has been published (Amann and Hammerstedt, 1980).

5. Application of These Methods to the Evaluation of Sperm Metabolism

5.1. Effect of Sample Preparation on Metabolic Characteristics

Seminal plasma contains a variety of particles, sperm fragments, and high- and low-molecular-weight molecules. A variety of sample-preparation procedures have been described (e.g., dilution, low-speed centrifugation and washing, dialysis). Each method yields a sperm suspension containing a different amount of all or selected components of seminal plasma. Both the composition of the suspending medium and the method of sample preparation alter the metabolic characteristics of the sperm (Salisbury and Lodge, 1962; Mann, 1964) and must be considered when evaluating results. Ejaculated bovine sperm were prepared in a standard buffer using four different methods of cell preparation (Hammerstedt, 1975b). The metabolic parameters outlined in Section 4.2.3 were measured and the relative importance of exogenous glucose and endogenous substrate for total ATP synthesis was determined (Table VII). Spermatozoal ATP, ADP, and AMP concentrations were similar regardless of how the sample was prepared; therefore, the rates of ATP synthesis and degradation were in balance in all preparations. However, reliance of cellular metabolism on exogenous glucose differed. Washed sperm derived a greater percentage of their ATP by degradation of exogenous glucose than did sperm that were prepared by dialysis or simple dilution. Since the metabolic needs of sperm were met through different processes, several parameters must be studied concurrently to minimize the likelihood of Type II (Snedecor and Cochran, 1969) experimental error (failure to detect a difference when one really exists).

Table VII. Effects of Preparation Method on Metabolic Parameters of Ejaculated Bovine Sperm[a]

	Preparation method		
	Dilution	Centrifugation and washing	Dialysis
A. Observed metabolite level or substrate consumption rate			
ATP (nmol/10^8 sperm)	30	34	28
ADP (nmol/10^8 sperm)	5	8	7
AMP (nmol/10^8 sperm)	7	8	6
Energy charge[b]	0.78	0.72	0.78
Glucose consumed (μmol/hr/10^8 sperm)	0.86	1.08	1.12
Total O_2 consumed (μmol/hr/10^8 sperm)	0.97	0.72	0.78
Lactate oxidized (μmol/hr/10^8 sperm)	0.08	0.10	0.06
O_2 consumed by oxidation of endogenous substrates[c]	0.73	0.42	0.60
B. Relative importance of exogenous and endogenous substrates to total ATP synthesis			
Metabolic pathway used to generate ATP			
Glucose → lactate[d]	1.7	2.2	2.2
Lactate → CO_2 + H_2O[d]	1.4	1.8	1.1
Total exogenous glucose[d]	3.1	4.0	3.3
Endogenous substrate(s)[d]	3.7	2.1	3.0
Total exogenous glucose plus endogenous substrates[d]	6.8	6.1	6.3
Total ATP derived from exogenous glucose (%)	45	66	52

[a] Data taken from Hammerstedt (1975b). The values in Section B differ slightly from the original presentation because corrected conversions were used for ATP yield during complete oxidation of glucose. Error analyses are presented in the original publication.
[b] Energy charge = (ATP + ½ ADP) ÷ (ATP + ADP + AMP) (Atkinson, 1977).
[c] Calculated as described in Section 4.2.3.
[d] Total ATP obtained during conversion of listed substrate to product. Calculations are described in Section 4.2.3 and data are expressed as μmol ATP generated/hr per 10^8 sperm.

5.2. Comparison of Ejaculated Sperm from Several Species

A standardized cell-preparation method (centrifugation and washing) and buffer were used to prepare ejaculated sperm from bulls, rams, and rabbits. The relative importance of exogenous and endogenous substrates to overall ATP synthesis differed among sperm from the three species (Table VIII). Bull and rabbit sperm rely on endogenous reserves for ATP synthesis even if exogenous glucose is available. However, washed ram

Table VIII. Comparison of Metabolic Characteristics of Ejaculated Bull, Ram, and Rabbit Sperm[a]

		Bull[e]	Ram[f]	Rabbit[g]
A.	Observed metabolite level or substrate consumption rate			
	ATP (nmol/10^8 sperm)	34	22	22
	ADP (nmol/10^8 sperm)	8	7	9
	AMP (nmol/10^8 sperm)	8	4	9
	Energy charge[b]	0.72	0.78	0.65
	Glucose consumed (μmol/hr/10^8 sperm)	1.08	0.48	1.00
	Total O_2 consumed (μmol/hr/10^8 sperm)	0.72	0.69	0.81
	Lactate oxidized (μmol/hr/10^8 sperm)	0.10	0.23	0.07
	O_2 consumed by oxidation of endogenous substrates[c]	0.42	0	0.60
B.	Relative importance of exogenous and endogenous substrates to total ATP synthesis			
	Metabolic pathway used to generate ATP			
	Glucose → lactate[d]	2.2	1.0	2.0
	Lactate → CO_2 + H_2O[d]	1.8	4.1	1.3
	Total exogenous glucose[d]	4.0	5.1	3.3
	Endogenous substrate(s)[d]	2.1	0	3.0
	Total exogenous glucose plus endogenous substrates[d]	6.1	5.1	6.3
	Total ATP derived from exogenous glucose (%)	66	100	52

[a] Methods of calculation and units are defined in Table VII. Error analyses are presented in the original publications.
[b-d] See Table VII, footnotes *b-d*.
[e] From Hammerstedt (1975b).
[f] Unpublished data (Hammerstedt and Hay). The conversion factor for relation of glucose consumed to rate of ^{3}HOH release from [2-^{3}H]glucose is 1.25 (Hiipakka and Hammerstedt, 1978b). Mean of 65 incubations using sperm from 17 animals.
[g] From Hammerstedt *et al.* (1979a).

sperm do not metabolize endogenous reserves if exogenous glucose is available.

5.3. Changes in Sperm Metabolism Concomitant with Epididymal Maturation

An extensive review (Voglmayr, 1975) emphasizes the metabolic fate of exogenously supplied carbon and discusses the effect of incubation conditions on cellular metabolism. We have taken a different approach and quantified the relative importance of several metabolic pathways to sperm before and after epididymal maturation (unpublished data presented in Table IX).

The calculated rate of ATP synthesis increased 2- to 3-fold during epididymal transit (from 2.2 to 7.6 μmol ATP/hr per 10^8 cells). This

Table IX. Comparison of Metabolic Characteristics of Testicular, Cauda Epididymal, and Ejaculated Ram Sperm[a]

	Testicular[e]	Cauda epididymal[f]	Ejaculated[f]
A. Observed metabolite level or substrate consumption rate			
ATP (nmol/10^8 sperm)	19	33	22
ADP (nmol/10^8 sperm)	6	6	7
AMP (nmol/10^8 sperm)	2	0.3	4
Energy charge[b]	0.8	0.9	0.8
Glucose consumed (μmol/hr/10^8 sperm)	0.1	0.7	0.5
Total O_2 consumed (μmol/hr/10^8 sperm)	0.4	1.1	0.7
Lactate oxidized (μmol/hr/10^8 sperm)	0.1	0.3	0.2
O_2 consumed by oxidation of endogenous substrates[c]	0.20	0.3	0
B. Calculated relative importance of exogenous and endogenous substrates to total ATP synthesis			
Metabolic pathway used to generate ATP			
Glucose → lactate[d]	0.3	1.5	1.0
Lactate → CO_2 + H_2O[d]	0.9	4.7	4.1
Total exogenous glucose[d]	1.2	6.2	5.1
Endogenous substrate(s)[d]	1.0	1.4	0
Total exogenous glucose plus endogenous substrates[d]	2.2	7.6	5.1
Total ATP derived from exogenous glucose (%)	55	82	100

[a] Methods of calculation and units are defined in Table VII and represent unpublished data (Hammerstedt and Hay) using sperm collected and prepared as described by Hammerstedt *et al.* (1979b).
[b–d] See Table VII, footnotes *b-d*.
[e] Mean of 12 incubations using sperm from 6 animals.
[f] Mean of 2–8 incubations using sperm from 2 to 6 animals.

change is consistent with the change in rate of ATP labeling from $^{32}PO_4$ by bull sperm as described in Section 4.2.4. In addition, the relative importance of endogenous metabolites to overall ATP synthesis changes dramatically. Endogenous metabolism contributes 45% of the total ATP supply in testicular sperm but only 18% in cauda epididymal sperm, and is absent in ejaculated sperm. Potential endogenous reserves must be present in all three types of sperm, since O_2 consumption is observed when they are incubated without exogenous substrate (Voglmayr, 1975).

5.4. *Importance of These Observations*

Obvious differences in the relative importance of several pathways involved in sperm metabolism are evident if the cells are prepared and incubated under standardized conditions. Well-designed experiments permit evaluation of the metabolic potential of sperm and test their capacity to survive in an artificial medium. However, cellular function *in vivo* reflects the interaction of sperm with testicular, epididymal, uterine, and oviductal fluids. Since each fluid has a unique composition (see Brackett and Mastrianni, 1974; Crabo, 1965; Waites, 1977; Setchell, 1978; Brooks, 1979), the sperm are presented with changes in the type of exogenous substrate(s) available to the cell. As a result, the balance between choice of exogenous vs. endogenous substrate to satisfy ATP needs probably changes. Therefore, our data obtained using glucose as the only exogenous source in standardized incubation conditions address only a small facet of the total problem.

Sperm from different species differ greatly in metabolic characteristics. This precludes extrapolation of data for sperm from animal models to human sperm. Our studies on sperm membranes (reviewed in Hammerstedt, 1979) also emphasize similar and potentially related differences. The great diversity of means used by sperm to satisfy common bioenergetic needs (maintenance of cell viability and motility) and plasma-membrane structure (necessary to permit fusion with the oocyte) provides a fascinating, yet frustrating, field of future experimentation.

6. *Need for Multiple Methods to Estimate Metabolic Rate*

6.1. *Introduction*

Interaction of hormones with target cells may result in one of two effects on cellular ATP metabolism. The first, a direct effect on ATP-yielding pathway(s), will force readjustment of metabolism to alternate pathways. The second effect, interaction with ATP-consuming reactions (e.g., motility, protein synthesis, substrate cycling, ion balance), requires a change in the flux through ATP generating system(s). The preceding sections outline simple methods useful to establish basal conditions within a metabolically simple cell. Sperm are flexible and adapt their metabolic rate to meet new requirements. The following sections present a model system for metabolism (based on ram testicular sperm) and then describe the effects of noxious chemicals on flux through ATP-generating systems. These calculations using sperm illustrate a maximum cellular response to extracellular agents and may aid in the selection of the most appropriate metabolic parameter to detect alterations in metabolic rate of any cell type.

6.2. Description of the Model

Consider a simple metabolic system resembling testicular ram sperm (Table X). Under steady-state conditions, ATP synthesis and degradation are balanced at a rate of 2 μmol ATP/hr per 10^8 cells. About 50% of the total ATP is supplied from endogenous reserves. The balance of the ATP is supplied by degradation of extracellular glucose, with about one third of the total glucose consumed entering the tricarboxylic acid cycle and the balance being excreted as lactate. Maximum ATP demand may approach 4–6 μmol/hr per 10^8 cells, values observed for cauda epididymal or ejaculated sperm.

Experiments with bull sperm (Cascieri *et al.*, 1976) established that maximum flux (V_{max}) through the endogenous pathways yields enough ATP to satisfy requirements of immotile testicular sperm (2 μmol/hr per 10^8 cells), but not enough for motile cauda epididymal or ejaculated sperm (6 μmol/hr per 10^8 cells). The V_{max} of the Embden–Meyerhof pathway must yield more than 6 μmol ATP/hr 10^8 cells, since cauda epididymal and ejaculated sperm remain motile (presumably maintaining ATP levels) under anaerobic conditions if glucose, fructose, or mannose is present (Mann, 1964). ATP levels are maintained in the presence of dinitrophenol (DNP) plus glucose but not DNP and endogenous reserves alone (Cascieri *et al.*, 1976). These maximum values for flux through the pathways, the known ATP yield of metabolic pathways, and thermodynamic constants permit simulation of cellular metabolism.

Assume that the endogenous substrate is palmitate. Complete oxidation of palmitate via β-oxidation and the tricarboxylic acid cycle plus the electron-transport chain is described by the equation $C_{16}H_{32}O_2 + 23O_2 \rightarrow 16CO_2 + 16H_2O$. Degradation of 1 μmol palmitate yields 130 μmol ATP and releases 2384 kcal of heat. Therefore, a V_{max} of 0.016 μmol palmitate oxidized/hr per 10^8 cells would then yield the 2 μmol ATP/hr per 10^8 cells supplied by endogenous metabolism. Addition of

Table X. Theoretical Effect of Metabolic Inhibitors or Factors on ATP Yield by Testicular Ram Sperm[a]

	Agent added	Exogenous substrate	Glucose → lactate: Glucose consumed (μmol)	Glucose → lactate: ATP formed (μmol)	Glucose → CO_2 + H_2O: Glucose consumed (μmol)	Glucose → CO_2 + H_2O: ATP formed (μmol)	Palmitate → CO_2 + H_2O: Palmitate consumed (μmol)	Palmitate → CO_2 + H_2O: ATP formed (μmol)
A. If ATP requirement remains at 2 μmole/hr per 10^8 cells and inhibitor has an effect on ATP-generating system								
1.	None	Glucose	0.057	0.12	0.025	0.88	0.0077	1.0
2.	None	None	0	0	0	0	0.0154	2.0
3.	DNP[b,j]	None	0	0	0	0	0.0154	0.13
4.	CN[c,j]	None	0	0	0	0	0	0
5.	DNP[b]	Glucose	0.50	1.0	0.22	0.88	0.0154	0.12
6.	CN[c]	Glucose	1.0	2.0	0	0	0	0
7.	AsO_4[d]	Glucose	0.064	0.064	0.0275	0.94	0.0077	1.0
8.	Cytochalasin b[e]	Glucose	0	0	0	0	0.0154	2.0
9.	Oxamate[f]	Glucose	0	0	0.028	1.0	0.0077	1.0
10.	AsO_2[g]	Glucose	0.50	1.0	0	0	0.0077	1.0
B. If ATP requirement is 6 μmol/hr per 10^8 cells.								
11.	None[h]	Glucose	0.23	0.46	0.098	3.54	0.0164	2.0
12.	None[i]	Glucose	0.29	0.58	0.12	4.42	0.0077	1.0
13.	None[j]	None	0	0	0	0	0.154	2.0

[a] Details of the model are presented in the text.
[b] Dinitrophenol, an uncoupler of oxidative phosphorylation.
[c] Cyanide, an inhibitor of the electron-transport chain.
[d] Arsenate, an inhibitor of glyceraldehyde-3-phosphate dehydrogenase. The yield between glucose and pyruvate is decreased to one ATP.
[e] A noncompetitive inhibitor of glucose transport (Hiipakka and Hammerstedt, 1978a,b).
[f] An inhibitor of lactate dehydrogenase. In this model, it is assumed to interact only with cytoplasmic lactate dehydrogenase.
[g] An inhibitor of the pyruvate dehydrogenase complex. In this model, AsO_2 is assumed not to interact with the α-ketoglutarate complex.
[h] Endogenous metabolism raised to maximum rate and glucose metabolism increased to satisfy ATP requirement.
[i] Endogenous metabolism held at basal level and glucose metabolism increased to provide required ATP.
[j] ATP demand is greater than ATP supply and cell death will follow.

glycolytic inhibitors would have no effect on the ATP yield of this pathway. However, compounds blocking electron flow would stop degradation because the tricarboxylic acid cycle would not function; ATP yield would be zero. Compounds uncoupling oxidative phosphorylation permit degradation to occur but decrease ATP yield to 8 μmol/micromole of palmitate because the only ATP-yielding reaction would be the substrate-level phosphorylation of the tricarboxylic acid cycle.

Degradation of 1 μmol exogenous glucose to lactate ($C_6H_{12}O_6 \rightarrow 2CH_3CHOHCOOH$) yields 2 μmol ATP and releases 28 kcal heat. The V_{max} for this pathway must be equal to or greater than 4 μmol/hr per 10^8 cells (yielding ≥8 μmol ATP/hr per 10^8 cells) because the ATP needs of motile sperm are satisfied under anaerobic conditions if glucose is present. Inhibitors of glucose transport of the Embden–Meyerhof pathway would prevent glucose consumption, lactate production, and ATP synthesis. An inhibitor of mitochondrial lactate oxidation would have no effect. The molar ATP yield during the conversion of glucose to lactate would be unaffected by inhibitors or uncouplers of the electron-transport chain.

Degradation of 1 μmol exogenous glucose to $CO_2 + H_2O$ ($C_6H_{12}O_6 + 6O_2 \rightarrow 6CO_2 + 6H_2O$) yields 38 μmol ATP and releases 672 kcal heat. A V_{max} for this process is set at 2 μmol/hr per 10^8 cells (yielding 76 μmol ATP when all ATP-yielding reactions are functioning but only 8 μmol ATP when only substrate-level phosphorylation is possible) because ATP levels are maintained when both DNP and either glucose (Cascieri *et al.*, 1976) or lactate (Hammerstedt, unpublished data) are present as exogenous substrate. Inhibitors of glucose transport, the Embden–Meyerhof pathway, and electron transport prevent this metabolism, and no ATP will be produced, while uncouplers of oxidative phosphorylation decrease ATP yield.

It is convenient to assume that endogenous metabolism proceeds at a constant rate unless forced to change to meet overall cellular ATP requirement. Furthermore, the ratio between excretion and mitochondrial oxidation of lactate is constant unless one of the enzymes involved is inhibited.

6.3. *Predicted Effects of Metabolic Inhibitors*

The theoretical effects of metabolic inhibitors on metabolic characteristics of ram testicular sperm are presented in Table XI. If the ATP requirement remains constant at 2 μmol/hr per 10^8 cells, addition of metabolic inhibitors alters the ATP yield of one pathway and forces a shift of metabolism to maintain ATP production (Table XI, Section A). Inhibitors represent classes of compounds that serve as uncouplers of oxidative phosphorylation (DNP), block electron transport (CN); decrease ATP yield of the Embden–Myerhof pathway (AsO_4); completely block glucose metabolism (cytochalasin b); inhibit lactate dehydrogenase, forcing all pyruvate to the tricarboxylic acid cycle (oxamate); and inhibit the pyruvate dehydrogenase complex, forcing all pyruvate to lactate (AsO_2). If cellular demand for ATP is increased by an undefined agent or condition (Table XI, Section B), a shift in the relative contribution of ATP-yielding pathways occurs. When ATP supply is less than ATP demand (examples 3, 4, and 13), ATP concentration will decrease until the sperm die or ATP demand is reduced (see Fig. 1 of Cascieri *et al.*, 1976).

Table XI. Calculated Metabolic Parameters for Cell Suspensions Treated with Metabolic Inhibitors[a]

	Agent added	Exogenous	Percentage of control incubation						
			ATP concentration[b]	Glucose consumption	Lactate production	Total O_2 consumption	Total CO_2 production	$^{14}CO_2$ production[c]	Heat production
	A. ATP requirement remains at 2 μmol/hr per 10^8 cells								
1.	None[d]	Glucose	100	100	100	100	100	100	100
2.	None	None	100	N.O.	N.O.	108	90	N.O.	100
3.	DNP	None	*6*	N.O.	N.O.	108	90	N.O.	100
4.	CN	None	*0*	N.O.	N.O.	*0*	*0*	N.O.	*0*
5.	DNP	Glucose	100	*878*	*877*	*512*	*573*	*885*	*540*
6.	CN	Glucose	100	*1200*	*1754*	*0*	*0*	*0*	*76*
7.	AsO_4	Glucose	100	111	112	*74*	105	110	105
8.	Cytochalasin b	Glucose	100	*0*	*0*	108	90	*0*	100
9.	Oxamate	Glucose	100	*34*	*0*	105	107	112	100
10.	AsO_2	Glucose	100	*610*	*877*	*54*	*45*	*0*	*88*
	B. ATP requirement increased to 6 μmol/hr per 10^8 cells								
11.	None	Glucose	100	*400*	*404*	*288*	*305*	*392*	*296*
12.	None	Glucose	100	*500*	*509*	*274*	*309*	*480*	*291*
13.	None	None	*33*	N.O.	N.O.	108	90	N.O.	100

[a] Abbreviations and effects of inhibitors are presented in Table X. Numbers in *italics* represent differences from the control incubation that would be easily detected during routine incubations. (N.O.) Data not obtainable because exogenous substrate was not included.
[b] Calculated as: (Σ ATP yield from all appropriate ATP-yielding pathways $\div$ ATP demand) $\times$ 100. Values less than 100 represent unbalanced cases in which the observed ATP level will decrease to zero and cell death will occur if the demand for ATP is not diminished.
[c] From [3,4-^{14}C]glucose as outlined in Section 4.2.3.
[d] Calculated values for the control incubation are taken from Table X and are: substrate consumption (μmol/hr per 10^8 cells) for glucose = 0.082 and oxygen = 0.327; product production (μmol/hr per 10^8 cells) for lactate = 0.114; total CO_2 = 0.273 and $^{14}CO_2$ = 0.150; heat production = 36.75 kcal/hr per 10^8 cells.

Further calculations illustrate the difficulty in detecting the effects of the inhibitors when metabolic parameters are measured (Table XI). When the ATP-yielding reactions are directly altered by the inhibitor, changes in ATP level are diagnostic only when ATP supply is less than ATP demand (examples 3 and 4). Glucose consumption and lactate production rates detect total blockage of important steps in glucose metabolism (examples 5, 6, and 8–10) but not decreased ATP yield in the Embden–Myerhof pathway (example 7). Oxygen consumption reflects alteration in electron flow (examples 4–6 and 10) as well as decreased ATP yield in the Embden–Myerhof pathway (example 7). Total CO_2 production responds to alteration in tricarboxylic acid cycle activity (examples 4–6 and 10), while $^{14}CO_2$ production responds to flux through the pyruvate dehydrogenase complex (examples 5, 6, 8, and 10). Heat production is dominated by the reactions involving the reduction of oxygen (examples 4–6). When overall ATP demand is altered (examples 11–13), any parameter may be used.

Many experimenters utilize only the simplest metabolic measurements (e.g., total oxygen consumption and total CO_2 production) and therefore would not detect the perturbations presented as examples 2, 3, 8, and 9. If the exact effect of a metabolic effector is not known, many metabolic parameters must be measured.

No single metabolic characteristic is diagnostic of all changes in metabolic rate. The compensatory effects of integrated metabolism, similarity of metabolic end products, and differing ATP yields preclude any common method of detection. Parameters directly related to all branches of the metabolic pathway must be measured to detect changes in energy state.

6.4. *Use of the Model*

Under actual conditions, where the compound tested may cause only a partial change in flux through a pathway, detection would be even more difficult than illustrated in Table XI. In cells more complex than sperm for which other modes of glucose degradation exist (pentose cycle) or carbon can be accumulated as glycogen, the ability to detect these changes becomes even less likely.

Detection of secondary effects is possible. If a hormone has the effect of stimulating ATP demand, any parameter could be used. Under these circumstances, calorimetry has an advantage in that specific assays for each metabolite are not required and incubations can be done in the physiological fluids.

The considerations outlined above provide insight for further evaluation of our published data (Hammerstedt and Amann, 1976) on the effects of physiological levels of androgens on spermatozoal metabolism. Neither glucose consumption nor ATP level was altered when testicular, cauda epididymal, or ejaculated bovine sperm were incubated with a variety of androgens. Since even a moderate increase in ATP demand would have been detected, the androgens probably had no effect on ATP-requiring processes such as motility, ion balance, or substrate cycling. Direct effects on ATP-yielding reactions would not have been detected using only these two descriptors. A detailed study of ejaculated sperm using the approach of multiple metabolic probes (Section 4) detected no change in any of eight parameters. Thus, as originally concluded (Hammerstedt and Amann, 1976), androgens probably had no direct effect on ATP-yielding reactions of bovine sperm.

7. Conclusions and Projections

The important feature of cellular metabolism is the relative contribution of individual pathways to overall ATP synthesis, rather than the number of micromoles of substrate consumed. Metabolism is a complex blend of many pathways, with the major one(s) dictated by availability of substrates, cofactors, enzymes, and other reactants. Thus, to evaluate the relative importance of several metabolic pathways, measurements must be chosen to permit an exact quantification of each component part. A number of parallel (simultaneous) analytical determinations are required if detection of a relatively subtle change (i.e., satisfying ATP requirements by change in flux through alternate ATP-yielding pathways) is required. The limited metabolic capacity of sperm permits evaluation of the blend of endogenous and exogenous metabolism and their relative importance to the cell.

A bewildering array of metabolic effectors of sperm have been described (see Salisbury and Lodge, 1962; Mann, 1964). In most studies, only one metabolic characteristic was studied. It is apparent, from data presented herein, that factors such as buffer composition, method of sperm preparation, and species providing the sperm preclude a simple interpretation. Comparisons can be made only when these factors are held constant.

A logical path of experimentation is from cells with limited metabolic capacity (sperm) to more complex cells with multiple metabolic pathways and requirements (e.g., spermatocytes, spermatids, or ova under basal metabolic conditions) to more complex cells with altered cell functions (hormonally stimulated cells). The introductory portion of this chapter (Section 1) listed the requirements for the successful quantification of the metabolic needs of cells. Adequate numbers of germ cells and sperm at several stages of development can be obtained (Section 2). Methods to monitor several potential pathways for ATP generation have been proposed (Section 4) and tested on cells that have a limited metabolic capacity (Sections 5 and 6). The methods have not been validated for use with the more complex spermatocytes, spermatids, or ova. Such experiments are necessary before testable hypotheses can be formulated regarding the changes in metabolic capacity and ATP requirements associated with spermatogenesis, spermiation, capacitation, and fertilization.

ACKNOWLEDGMENTS. This chapter was approved on February 12, 1980, as Paper No. 5912 in the Journal Series of the Pennsylvania Agricultural Experiment Station. Financial support was provided by National Institutes of Health Grant HD-05859.

Discussion

D. VILLEE: (1) Do sperm cells utilize other sugars than glucose and fructose and amino acids as sources of energy? (2) Does thyroxine alter consumption and heat production in sperm cells?

HAMMERSTEDT: (1) Mannose is also degraded by sperm; both pyruvate and lactate can also serve as substrates. (2) Hoskins and Casillas (1975) have reviewed the literature regarding the effects of hormones on sperm metabolism. I do not recall the details, but I believe that high levels of thyroxine did alter metabolic parameters. I do not remember the parameters examined.

MCKERNS: Are you planning to do studies at each point of known hormone action in the production of sperm?

HAMMERSTEDT: These experiments have not been directly funded. I plan in the near future to seek funding for research directed toward the characterization of ATP metabolism in germ cells.

PETRUSZ: Exogenous androgen had no effect on sperm metabolism—how about endogenous androgen? Had the cells been exposed to normal levels of androgen before being collected for the experiments?

HAMMERSTEDT: The sperm had been exposed to the normal complement of androgens. The cells were washed and resuspended in buffer for the experiments. No effect of any exogenous androgen on metabolism was noted. In separate experiments, about two thirds of the endogenous steroids were removed by charcoal treatment. Exogenously supplied steroids still had no effect.

SZÖLLÖSI: (1) Reports exist in the literature that some RNA is synthesized by sperm mitochondria. Could this change significantly the results on total ATP production? (2) Have you studied the changes of ATP production during mitochondrial differentiation during spermatogenesis?

HAMMERSTEDT: Periodic reports of both nucleic acid and/or protein synthesis by sperm have appeared. The conclusions are not universally accepted. With regard to my calculations, the reported rates of synthesis are very low and would not significantly change my conclusions.

References

Alemán, V., Trejo, R., Morales, E., Hernández-Jáuregui, P., and Delhumeau-Ongay, G., 1978, A simple and rapid technique to isolate enriched populations of spermatocytes and spermatids from the immature rat testis, *J. Reprod. Fertil.* **54:**67.

Amann, R., 1979, Computerized measurements of sperm velocity and percentage of motile sperm, in: *The Spermatozoon* (D. Fawcett and J.M. Bedford, eds.), pp. 431–435, Urban and Schwartzenberg, Baltimore.

Amann, R.P., and Hammerstedt, R.H., 1980, Validation of a system for computerized measurement of spermatozoal velocity and percentage of motile sperm, *Biol. Reprod.* **23:**647.

Amann, R.P., and Howards, S.S., 1980, Daily spermatozoal production and epididymal spermatozoal reserves of the human male, *J. Urol.* **124:**211.

Amann, R.P., Hokanson, J.F., and Almquist, J.O., 1963, Cannulation of the bovine ductus deferens for quantitative recovery of epididymal spermatozoa, *J. Reprod. Fertil.* **6:**65.

Amann, R.P., Kavanaugh, J.F., Griel, L.C., Jr., and Voglmayr, J.K., 1974. Sperm production of Holstein bulls determined from testicular spermatid reserves, after cannulation of rete testis or vas deferens, and by daily ejaculation, *J. Dairy Sci.* **57:**93.

Amann, R.P., Johnson, L., Thompson, D.L., Jr., and Pickett, B.W., 1976, Daily spermatozoal production, epididymal spermatozoal reserves and transit time of spermatozoa through the epididymis of the rhesus monkey, *Biol. Reprod.* **15:**586.

Atkinson, D.A., 1977, *Cellular Energy Metabolism and Its Regulation,* Academic Press, New York.

Babcock, D.F., First, N.L., and Lardy, H.A., 1975, Transport mechanism for succinate and phosphate localized in the plasma membrane of bovine spermatozoa, *J. Biol. Chem.* **250:**6488.

Back, D.J., Shenton, J.C., and Glover, T.D., 1974, The composition of epididymal plasma from the cauda epididymidis of the rat. *J. Reprod. Fert.* **40:**211.

Barcellona, W.J., and Meistrich, M.L., 1977, Ultrastructural integrity of mouse testicular cells separated by velocity sedimentation, *J. Reprod. Fertil.* **50:**61.

Bech, J., and Koefoed-Johnsen, H.H., 1973, Spermiemorfologi og plasmas sammensaetning i bitestikler fra normale tyre samt to tyre med abnormt saedbillede, The Royal Veterinary and Agricultural University Sterility Research Institute Annual Report, p. 9.

Bellvé, A.R., Cavicchia, J.C., Millette, C.F., O'Brien, D.A., Bhatnagar, Y.M., and Dym, M., 1977, Spermatogenic cells of the prepuberal mouse: Isolation and morphological characterization, *J. Cell Biol.* **74:**68.

Boyer, P.D., 1967, ^{18}O and related exchanges in enzymic formation and utilization of nucleoside triphosphates, in: *Current Topics in Bioenergetics,* Vol. 2 (D.R. Sanadi, ed.), pp. 99–149, Academic Press, New York.

Brackett, B.G., and Mastrianni, L., Jr., 1974, Composition of oviductal fluid, in: *The Oviduct and Its Functions* (A.D. Johnson and C.W. Foley, eds.), pp. 133–159, Academic Press, New York.

Brooks, D.E., 1979, Biochemical environment of sperm maturation, in: *The Spermatozoon* (D. Fawcett and J.M. Bedford, eds.), pp. 23–24, Urban and Schwartzenberg, Baltimore.

Bruce, W.R., Furrer, R., Goldberg, R.B., Meistrich, M.L., and Mintz, B., 1973, Genetic control of the kinetics of mouse spermatogenesis, *Genet. Res.* **22:**155.

Cascieri, M., Amann, R.P., and Hammerstedt, R.H., 1976, Adenine nucleotide changes at initiation of bull sperm motility, *J. Biol. Chem.* **251:**787.

Cooper, T.G., and Waites, G.M.H., 1974, Testosterone in rete testis fluid and blood of rams and rats, *J. Endocrinol.* **62:**619.

Cooper, T.G., Danzo, B.J., Dipietro, D.L., McKenna, T.J., and Orgebin-Crist, M.C., 1976, Some characteristics of rete testis fluid from rabbits, *Andrologia* **8:**87.

Crabo, B., 1965, Studies on the composition of epididymal content in bulls and boars, *Acta Vet. Scand.* **6**(5):1.

Darin-Bennett, A., Polos, A., and White, I.G., 1973, A re-examination of the role of phospholipids as energy substrates during incubation of ram sperm, *J. Reprod. Fertil.* **34:**543.

Davis, J.C., and Schuetz, A.W., 1975, Separation of germinal cells from immature rat testes by sedimentation at unit gravity, *Exp. Cell Res.* **91:**79.

Dawson, A.G., 1979, Oxidation of cytosolic NADH formed during aerobic metabolism in mammalian cells, *Trends Biochem. Sci.* **4:**171.

Decker, K., Jungermann, K., and Thauer, R.K., 1970, Energy production in anaerobic organisms, *Angew. Chem. Int. Ed. Engl.* **9:**138.

Deibel, F.C., Jr., Smith, J.E., Crabo, B.G., and Graham, E.F., 1976, Evaluation of six assays of sperm quality by means of their accuracy, precision, and sensitivity in separating known induced levels of damage, VIIIth International Congress on Animal Reproduction and Artificial Insemination, Cracow, p. 888.

Denton, R.M., and Pogson, C.I., 1976, *Outline Studies in Biology: Metabolic Regulation,* John Wiley (Halsted Press), New York.

Dietrich, S.M.C., and Burris, R.H., 1967, Effect of exogenous substrates on the endogenous respiration of bacteria, *J. Bacteriol.* **93:**1467.

Fawcett, D.W., Neaves, W.B., and Flores, M.N., 1973, Comparative observations on intertubular lymphatics and the organization of the interstitial tissue of the mammalian testis, *Biol. Reprod.* **9:**500.

Forrest, W.W., 1969, Bacterial calorimetry, in: *Biochemical Calorimetry* (H.D. Brown, ed.), Chapter VIII, pp. 165–198, Academic Press, New York.

Fournier-Delpech, S., Colas, G., Courot, M., Ortavant, R., and Brice, G., 1979, Epididymal sperm maturation in the ram: Motility, fertilizing ability and embryonic survival after uterine artificial insemination in the ewe, *Ann. Biol. Anim. Bioch. Biophys.* **19:**597.

Free, M.J., and Jaffe, R.A., 1979, Collection of rete testis fluid from rats without previous efferent duct ligation, *Biol. Reprod.* **20:**269.

Galena, H.J., and Terner, C., 1974, Conversion of progesterone to androgens by non-flagellate germinal cells isolated from seminiferous tubules of rat testis, *J. Endocrinol.* **93:**269.

Gondos, B., 1977, Testicular development, in: *The Testis,* Vol. IV (A.D. Johnson and W.R. Gomes, eds.), p. 9–37, Academic Press, New York.

Gottschalk, G., and Andreesen, J.R., 1979, Energy metabolism in anaerobes, in: *International Review of Biochemistry: Microbial Biochemistry,* Vol. 21 (J.R. Quayle, ed.), pp. 86–115, University Park Press, Baltimore.

Grabske, R.J., Lake, S., Gledhill, B.L., and Meistrich, M.L., 1975, Centrifugal elutriation: Separation of spermatogenic cells on the basis of sedimentation velocity, *Cell. Physiol.* **86:**177.

Hammerstedt, R.H., 1973, An automated method for ATP analysis utilizing the luciferin–luciferase reaction, *Anal. Biochem.* **52:**449.

Hammerstedt, R.H., 1975a, Tritium release from [2-^{3}H]-D-glucose as a monitor of glucose consumption by bovine sperm, *Biol. Reprod.* **12:**545.

Hammerstedt, R.H., 1975b, Use of high speed dialysis to prepare bovine sperm for metabolic studies, *Biol. Reprod.* **13:**389.

Hammerstedt, R.H., 1979, Characterization of sperm surfaces using physical techniques, in: *The Spermatozoon* (D. Fawcett and J.M. Bedford, eds.), pp. 205–216, Urban and Schwarzenberg, Baltimore.

Hammerstedt, R.H., and Amann, R.P., 1976, Effects of physiological levels of exogenous steroids on metabolism of testicular, cauda epididymal and ejaculated bovine sperm, *Biol. Reprod.* **15:**678.

Hammerstedt, R.H., and Hay, S.R., 1980, Effect of incubation temperature on motility and cAMP content of bovine sperm, *Arch. Biochem. Biophys.* **199:**427.

Hammerstedt, R.H., Keith, A.D., Boltz, R.C., Jr., and Todd, P.W., 1979a, Use of amphiphilic spin labels and whole cell isoelectric focusing to assay charge characteristics of sperm surfaces, *Arch. Biochem. Biophys.* **194:**565.

Hammerstedt, R.H., Keith, A.D., Hay, S., DeLuca, N., and Amann, R.P., 1979b, Changes in ram sperm membranes during epididymal transit, *Arch. Biochem. Biophys.* **196:**7.

Harris, M.E., and Bartke, A., 1974, Concentration of testosterone in testis fluid of the rat, *Endocrinology* **95:**701.

Hartree, E.F., and Mann, T., 1961, Phospholipids in ram semen: Metabolism of plasmalogen and fatty acids, *Biochem. J.* **80:**464.

Hiipakka, R.A., and Hammerstedt, R.H., 1978a, 2-Deoxyglucose transport and phosphorylation by bovine sperm, *Biol. Reprod.* **19:**368.

Hiipakka, R.A., and Hammerstedt, R.H., 1978b, Changes in 2-deoxyglucose transport during epididymal maturation of ram sperm, *Biol. Reprod.* **19:**1030.

Holtz, W., and Foote, R.W., 1974, Cannulation and recovery of spermatozoa from the rabbit ductus deferens, *J. Reprod. Fertil.* **39:**89.

Hoskins, D.D., and Casillas, E.R., 1975, Hormones, second messengers, and the mammalian spermatazoan, *Adv. Sex Horm. Res.* **1:**283.

Hutson, S.M., Van Dop, C., and Lardy, H.A., 1977, Mitochondrial metabolism of pyruvate in bovine spermatozoa, *J. Biol. Chem.* **252:**1309.

Jarrett, I.G., Clark, D.G., Filsell, O.H., Harvey, J.W., and Clark, M.G., 1979, The application of microcalorimetry to the assessment of metabolic efficiency in isolated rat hepatocytes, *Biochem. J.* **180:**631.

Johnson, L.A., and Pursel, V., 1975, Cannulation of the ductus deferens of the boar: A surgical technique, *Am. J. Vet. Res.* **36:**315.

Jones, C.W., 1979, Energy metabolism in aerobes, in: *International Review of Biochemistry: Microbial Biochemistry,* Vol. 21 (J.R. Quayle, ed.), pp. 49–84, University Park Press, Baltimore.

Kemp, R.B., 1975, Microcalorimetric studies of tissue cells and bacteria, *Pestic. Sci.* **6:**311.

Lam, D.M.K., Furrer, R., and Bruce, W.R., 1970, The separation, physical characterization, and differentiation kinetics of spermatogonial cells of the mouse, *Proc. Natl. Acad. Sci. U.S.A.* **65:**192.

Lamanna, C., 1963, Endogenous metabolism with special reference to bacteria, *Ann. N. Y. Acad. Sci.* **102:**515.

Lamprecht, A., 1976, Application of calorimetry to the evaluation of metabolic data for whole organisms, *Biochem. Soc. Trans.* **4:**565.

Lardy, H.A., and Phillips, P.H., 1941a, The interrelation of oxidative and glycolytic processes as sources of energy for bull spermatozoa, *Am. J. Physiol.* **133:**602.

Lardy, H.A., and Phillips, P.H., 1941b, Phospholipids as a source of energy for motility of bull spermatozoa, *Am. J. Physiol.* **134:**542.
Lee, I.P., and Dixon, R.L., 1972, Antineoplastic drug effects on spermatogenesis studied by velocity sedimentation cell separation, *Toxicol. Appl. Pharmacol.* **23:**20.
Liu, Y.T., and Warme, P.K., 1977, Computerized evaluation of sperm cell motility, *Comput. Biomed. Res.* **10:**1465.
Loir, M., and Lanneau, M., 1974, Separation of ram spermatids by sedimentation at unit gravity, *Exp. Cell Res.* **83:**319.
Loir, M., and Lanneau, M., 1977, Separation of mammalian spermatids, in: *Methods in Cell Biology,* Vol. XV (D.M. Prescott, ed.), Chapter 3, pp. 55–77, Academic Press, New York.
Mann, T., 1964, *Biochemistry of Semen and the Male Reproductive Tract,* Methuen, London.
Martig, R.C., and Almquist, J.O., 1969, Reproductive capacity of beef bulls. III. Postpubertal changes in fertility and sperm morphology at different ejaculation frequencies, *J. Anim. Sci.* **28:**375.
Meistrich, M.L., 1972, Separation of mouse spermatogenic cells by velocity sedimentation, *J. Cell. Physiol.* **80:**299.
Meistrich, M.L., 1977, Separation of spermatogenic cells from rodent testes, in: *Methods in Cell Biology,* Vol. XV (D.M. Prescott, ed.), Chapter 2, pp. 16–54, Academic Press, New York.
Meistrich, M.L., and Trostle, P.K., 1975, Separation of mouse testis cells by equilibrium density centrifugation in renografin gradients, *Exp. Cell Res.* **92:**231.
Midelfort, C.F., and Rose, I.A., 1976, A stereochemical method for detection of ATP terminal phosphate transfer in enzymatic reactions, *J. Biol. Chem.* **251:**5881.
Milkowski, A.L., and Lardy, H.A., 1977, Factors affecting the redox state of bovine epididymal spermatozoa, *Arch. Biochem. Biophys.* **181:**270.
Mills, S.C., and Scott, T.W., 1969, Metabolism of fatty acids by testicular and ejaculated ram spermatozoa, *J. Reprod. Fertil.* **18:**367.
Morton, B.E., and Lardy, H.A., 1967a, Cellular oxidative phosphorylation. I. Measurement in intact spermatozoa and other cells, *Biochemistry* **6:**43.
Morton, B.E., and Lardy, H.A., 1967b, Cellular oxidative phosphorylation. II. Measurement in physically modified spermatozoa, *Biochemistry* **6:**50.
Morton, B.E., and Lardy, H.A., 1967c, Cellular oxidative phosphorylation. III. Measurement in chemically modified cells, *Biochemistry* **6:**57.
Nakamura, M., and Hall, P.F., 1976, Inhibition by 5-thio-D-glucopyranose of protein biosynthesis *in vitro* in spermatids from rat testis, *Biochem. Biophys. Acta* **447:**474.
Nakamura, M., and Hall, P.F., 1977, Effect of 5-thio-D-glucose on protein synthesis *in vitro* by various types of cells from rat testes, *J. Reprod. Fertil.* **49:**395.
Niehaus, W.G., Jr., and Hammerstedt, R.H., 1976, Mode of orthophosphate uptake and ATP labeling by mammalian cells, *Biochim. Biophys. Acta* **443:**515.
Niel, A.R., and Masters, C.J., 1972, Metabolism of fatty acids by bovine spermatozoa, *Biochem. J.* **127:**375.
Orgebin-Crist, M.-C., Danzo, B.J., and Davies, J., 1975, Endocrine control of the development and maintenance of sperm fertilizing ability in the epididymis, in: *Handbook of Physiology,* Section 7, *Endocrinology,* Vol. V, *Male Reproductive System* (D.W. Hamilton and R.O. Greep, eds.), pp. 319–338, American Physiological Society, Washington, D.C.
Quinn, P.J., and White, I.G., 1967, Phospholipid and cholesterol content of epididymal and

ejaculated ram spermatozoa and seminal plasma in relation to cold shock, *Aust. J. Biol. Sci.* **20:**1205.

Rambeck, W., Wacker, H., and Simon, H., 1971, Ausmass des glykolytischen Abbaus zelleigenir Reservestoffe in verarmter Hefe bei Glucose-gabe, *Z. Physiol. Chem.* **352:**59.

Rikmenspoel, R., 1965, The tail movement of bull spermatozoa: Observations and model calculations, *Biophys. J.* **5:**365.

Romrell, L.J., Bellvé, A.R., and Fawcett, D.W., 1976, Separation of mouse spermatogenic cells by sedimentation velocity, *Dev. Biol.* **49:**119.

Rothschild, L., 1962, Anaerobic heat production and fructolysis of bull spermatozoa at different temperatures, *J. Exp. Biol.* **39:**387.

Rottensten, K., 1972, Alderens indflydelse på tyrenes frugtbarhed, The Royal Veterinary and Agricultural University Sterility Research Institute Annual Report, p. 219.

Salhanick, A.I., and Terner, C., 1979, Androgen synthesis in absence of Leydig and Sertoli cells in a germ cell fraction from rat seminiferous tubules, *Biol. Reprod.* **21:**293.

Salisbury, G.W., and Lodge, J.R., 1962, Metabolism of spermatozoa, *Adv. Enzymol.* **24:**35.

Schaarschmidt, B., and Lamprecht, I., 1976, Calorimetric characterization of microorganisms, *Experientia* **32:**1230.

Scott, T.W., Voglmayr, J.K., and Setchell, B.P., 1967, Lipid composition and metabolism in testicular and ejaculated ram spermatozoa, *Biochem. J.* **102:**456.

Setchell, B.P., 1978, *The Mammalian Testis,* Cornell University Press, Ithaca, New York.

Sexton, T.J., Amann, R.P., and Flipse, R.J., 1971, Free amino acids and protein in rete testis fluid, vas deferens plasma, accessory sex gland fluid, and seminal plasma of the conscious bull, *J. Dairy Sci.* **54:**412.

Snedecor, G.W., and Cochran, W.G., 1969, *Statistical Methods,* 6th Ed., p. 27, Iowa State University Press, Ames.

Spink, C., and Wadsö, I., 1975, Calorimetry as an analytical tool in biochemistry and biology, in: *Methods of Biochemical Analysis,* Vol. 23 (D. Glick, ed.), pp. 2–159, John Wiley, New York.

Storey, B.T., and Kayne, F.J., 1977, Energy metabolism of spermatozoa. VI. Direct intramitochondrial lactate oxidation by rabbit sperm mitochondria, *Biol. Reprod.* **16:**549.

Tadmor, A., and Schindler, H., 1966, Establishment of a fistula in the vas deferens of rams, *Isr. J. Agric. Res.,* p. 157.

Thauer, R.K., Jungermann, K., and Decker, K., 1977, Energy conservation in chemotrophic anaerobic bacteria, *Bacteriol. Rev.* **41:**100.

Tischner, M., 1967, Cyklicznosc przesuwania sie plemnikow przez przetoke nasieniowodu tryka (Cyclicity in spermatozoa transport through a fistulated vas deferens in the ram), *Acta Biol. Cracov. Ser. Zool.* **10:**283.

Van Dop, C., Hutson, S.M., and Lardy, H.A., 1977, Pyruvate metabolism in bovine epididymal spermatozoa, *J. Biol. Chem.* **252:**1303.

Voglmayr, J.K., 1975, Metabolic changes in spermatozoa during epididymal transit, in: *Handbook of Physiology,* Section 7, *Endocrinology,* Vol. V, *Male Reproductive System* (D.W. Hamilton and R.O. Greep, eds.), pp. 437–452, American Physiological Society, Washington, D.C.

Voglmayr, J.K., Musto, N.A., Saksena, S.K., Brown-Woodman, P.D.C., Marley, P.B., and White, I.G., 1977, Characteristics of semen collected from the cauda epididymidis of conscious rams, *J. Reprod. Fertil.* **49:**245.

Wadsö, I., 1976, A system of micro-calorimeters and its use in biochemistry and biology, *Biochem. Soc. Trans.* **4:**561.

Waites, G.M.H., 1977, Fluid secretion, in: *The Testis,* Vol. IV (A.D. Johnson and W.R. Gomes, eds.), pp. 91–123, Academic Press, New York.

Waites, G.M.H., and Einer-Jensen, N., 1974, Collection and analysis of rete testis fluid from macaque monkeys, *J. Reprod. Fertil.* **41:**505.

White, I.G., 1957, Metabolism of glycerol and similar compounds by bull spermatozoa, *Amer. J. Physiol.* **189:**307.

17

Study of a Glycoprotein Produced by the Rat Epididymis

F. Bayard, L. Duguet, M. Mazzuca, and J.C. Faye

1. Introduction

In the mammalian species, spermatozoa are neither motile nor fertile immediately after their release from the germinal epithelium. By the time they reach the cauda epididymidis, they have acquired capacity for both motility and fertility. This maturation process appears to be mediated by epididymal factors that are androgen-dependent (Bedford, 1975; Hamilton, 1975; Orgebin-Crist *et al.*, 1975).

It has been well demonstrated that an epididymal protein secretion exists, since the protein composition of the epididymal fluid differs from that of both the blood plasma and the rete testis fluid (Barker and Amann, 1970; Amann *et al.*, 1973; Koskimies and Kormano, 1975; Huang and Johnson, 1975; Lea *et al.*, 1978; Garberi *et al.*, 1979; Kohane *et al.*, 1979). The fact that typical secretory vacuoles are absent in the different types of epithelial cells lining the lumen of the canal suggests a mechanism of secretion different from the mechanism described in gland cells (Hoffer *et al.*, 1973; Flickinger, 1979). The androgenic control of this protein secretion has also been described (Cameo and Blaquier, 1976). Among the proteins secreted, those that may have the greater functional importance are the glycoproteins. By coating the cell membrane, these glycoproteins could enable the acquisition of ionized negative groups by the surface of the spermatozoa during its transit from the caput to the

F. Bayard, L. Duguet, M. Mazzuca, and *J. C. Faye* • Laboratoire d'Endocrinologie Expérimentale, INSERM U 168, CHU Toulouse Rangueil, 31054 Toulouse Cedex, France.

epididymal cauda (Bedford, 1975; Edelman and Millette, 1974; Gwatkin, 1976; Gordon *et al.*, 1975; Nicholson *et al.*, 1977; Fournier-Delpech *et al.*, 1977; Millette, 1977; Koehler, 1978; Flechon, 1979). In this way, they may play an important role in fertilization by initiating the forward mobility of the spermatozoa (Acott and Hoskins, 1977; Acott *et al.*, 1979) or at a later step of fertilization by allowing successful contact with the egg investments and fusion with the egg membranes (Edelman and Millette, 1974; Gwatkin, 1976). One of the most acidic proteins in epididymal homogenate was described previously by Fournier-Delpech (Fournier, 1968; Fournier-Delpech *et al.*, 1973). This protein was shown to be associated with sialic acid and to be androgen-dependent. In this chapter, we describe studies on the purification, immunohistological localization, and variations of concentration of this sialo-protein (SP) in the epididymis of normal, castrated, and androgen-treated rats.

2. Purification of the Sialoprotein

Adult male Wistar rats were used for purification of the SP. The epididymides were removed, trimmed of fat, and homogenized in an ice-cooled 50 mM Tris-HCl buffer, pH 7.4, containing 1 mM $MgCl_2$, 1 mM $CaCl_2$, and 1 mM $MnCl_2$ (Buffer A). The homogenate was centrifuged at 12,000*g* for 15 min at 4°C, and the supernatant was mixed with an equal volume of a resin solution of concanavalin A (Con A) covalently bound to Sepharose 4B suspended in Buffer A. Gentle shaking was maintained for 4 hr at 4°C. After the resin was loaded into a plastic column, washing was continued with Buffer A until the optical density at 280 nm reached 0 for at least 100 ml. Elution was then carried out with the first 15 ml of a 50 mM Tris-HCl buffer, pH 7.4, containing 2 M α-dimethylglycopyranoside and no divalent ions. This eluate was concentrated using a vacuum dialyzer. The resultant concentrate was layered on a Sephadex G-75 column (100-ml bed volume) equilibrated at 4°C with a 50 mM acetate buffer, pH 6. The SP was eluted between fractions 60 and 70 of 1 ml each. These fractions were pooled and concentrated using the vacuum dialyzer. The concentrate was further purified at 4°C by preparative flat-bed isoelectric focusing on a Sephadex G-75 gel for 12 hr at constant power (3 W) using a 3.5–5 pH range of ampholines and an LKB 2117 "Multiphor" apparatus. Samples of pH ranging from 4.5 to 4.8 (at 23°C) were collected. At each step, the purity of the SP was monitored by polyacrylamide gel disk electrophoresis as described by Davis (1964).

Figure 1 presents the data of the electrophoretic analysis at the first two steps of purification. The final fraction (after isoelectric focusing) was analyzed (Fig. 2) on 4–12% polyacrylamide gels in the presence of sodium dodecyl sulfate (SDS)–mercaptoethanol according to Weber and

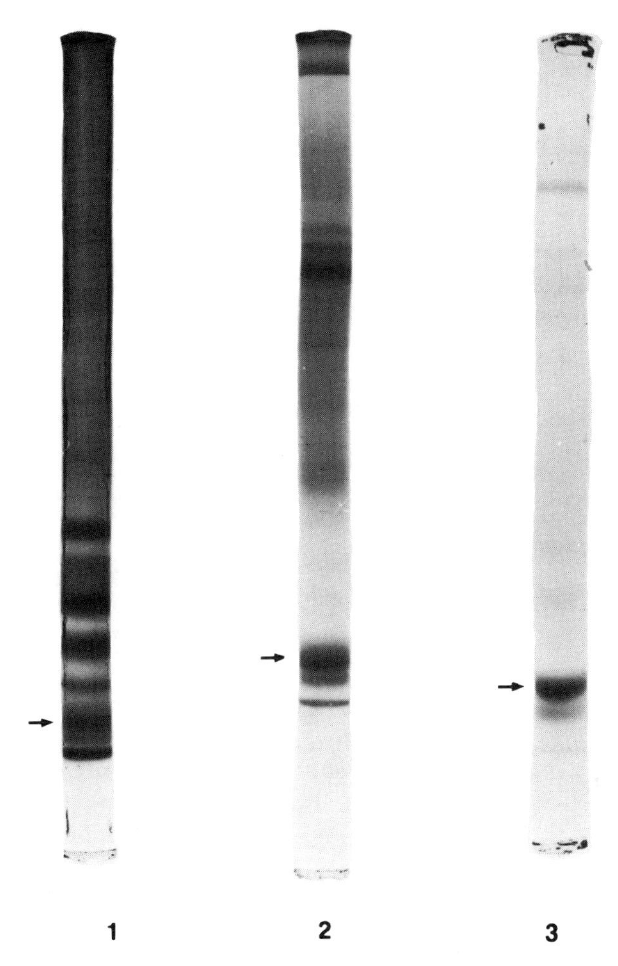

Figure 1. Purification of the SP: analysis by polyacrylamide gel disk electrophoresis. (1) Supernatant of the homogenate (12,000g × 15 min); (2) Con A–Sepharose 4B eluate; (3) Sephadex G-75 eluate. From Davis (1964).

Osborn (1969). Comparison of the mobility of the SP with the mobility of several standard proteins (bovine serum albumin, ovalbumin, alcohol dehydrogenase, α-chymotrypsinogen, lysozyme) showed a molecular weight of 37,500 ± 1225 daltons. When the apparent molecular weight, extrapolated from the appropriate standard curve of Fig. 2, was plotted against gel concentration, a linear relationship was observed, suggesting a low percentage of carbohydrate in the molecule (Segrest and Jackson, 1972).

The isoelectric point was found to be 4.7 ± 0.05.

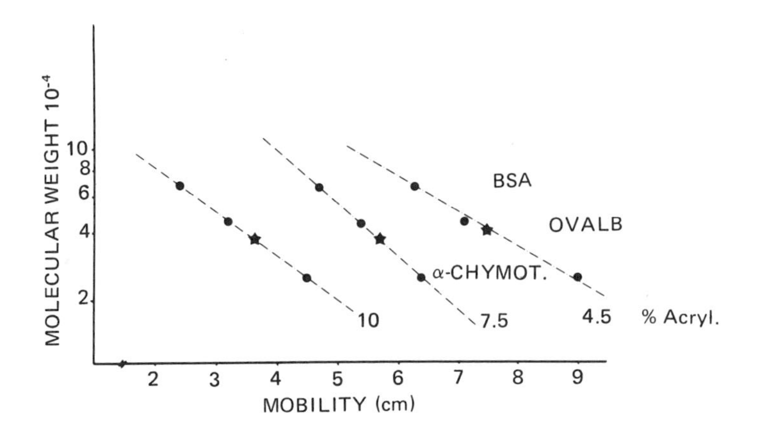

Figure 2. Analysis of the purified SP (★) on 4–12% polyacrylamide gel electrophoresis in the presence of SDS–mercaptoethanol. From Weber and Osborn (1969).

3. *Radioimmunoassay and Concentration of the Sialoprotein in the Epididymis*

Antibodies against the SP were raised in female New Zealand white rabbits and in female goats using the method of Vaitukaitis *et al.* (1971). As shown in Fig. 3, the antisera were specific for the SP, since only one immunoprecipitation band was observed using disk-immunoelectrophoretic analysis of the supernatant (12,000*g* for 15 min) of the epididymal homogenate of a normal adult rat. Using these antibodies, a radioimmunoassay procedure similar to the classic double-antibody method of Midgley (1966) for human chorionic gonadotropin and luteinizing hormone was used to measure the SP in different organs and conditions.

Using this radioimmunoassay, the SP could not be detected in rat serum or in the supernatant fraction of rat testis, seminal vesicle, or prostate homogenates. The SP was detected only in the epididymis and the deferent duct of all strains of adult rats tested. No cross-reaction was observed with dog, rabbit, guinea pig, or human epididymal homogenates.

As shown in Fig. 4, the SP is not detectable in the epididymidis before day 20 and becomes detectable as soon as testosterone is measurable in the testis and in the testicular venous blood (Knorr *et al.*, 1970). The production of the SP then increases progressively and in parallel with the testosterone concentration in the testes (Knorr *et al.*, 1970) to reach a plateau between 3 and 4 months of age. In 3-month-old animals (Fig. 5), the SP decreases significantly 8 days after castration and becomes undetectable between 1 and 2 months after castration. When Silastic capsules (I.D. 1.89 mm, O.D. 3.18 mm, length 2 cm) containing

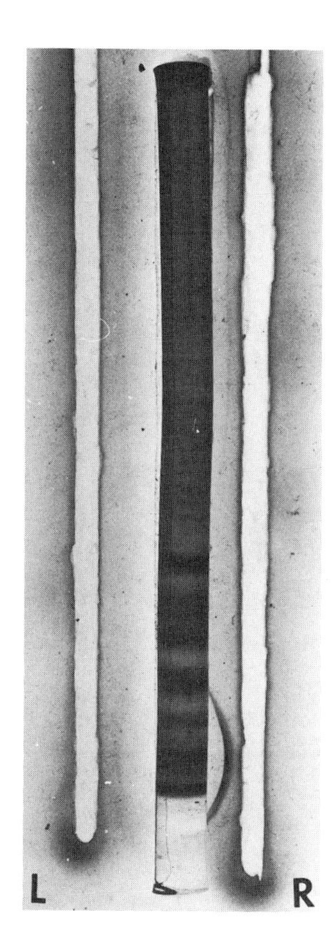

Figure 3. Diskimmunoelectrophoretic analysis of the supernatant of a rat epididymal homogenate (12,000*g* × 15 min). A rabbit antiserum diluted 1:5 was used in the right well (R); the preimmune serum of the same rabbit was used in the left well.

testosterone are implanted under the skin of the castrated animals, an increase of SP concentration is reached. Two (but not more) testosterone capsules are necessary to restore the normal SP concentration as well as to restore weight, DNA, protein, and androgen-receptor content of the organ (Pujol and Bayard, 1979), although one testosterone capsule is sufficient to restore the same parameters of the other male accessory organs, e.g., the prostate (Pujol and Bayard, 1979). These observations support the proposition that a high testosterone supply to the epididymis is needed to sustain the normal production of the SP. A route via venous connections between the testis and the epididymis (Einer-Jensen, 1974) may be of importance, since the hormone transfer demonstrated from spermatic vein to artery in the pampiniform plexus does not appear to supply the epididymis (Free and Jaffe, 1978). Implantation of capsules containing estradiol or diethylstilbestrol with or without the testosterone treatment were ineffective on SP concentration.

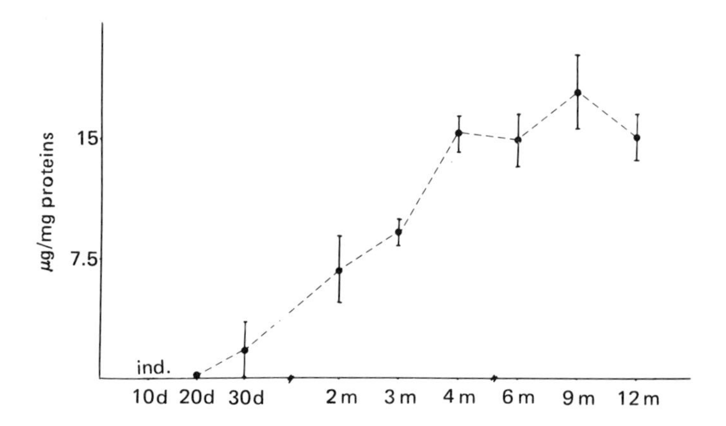

Figure 4. SP concentration in the total epididymis of rats of different ages (mean ± S.D., N = 5 for each group).

4. Localization of the Sialoprotein in the Epididymis and on Spermatozoa

The specific activity of the SP was then measured by radioimmunoassay in the homogenates of different epididymal portions and of the deferent duct. The SP is detected in the caput (Fig. 6), and its specific

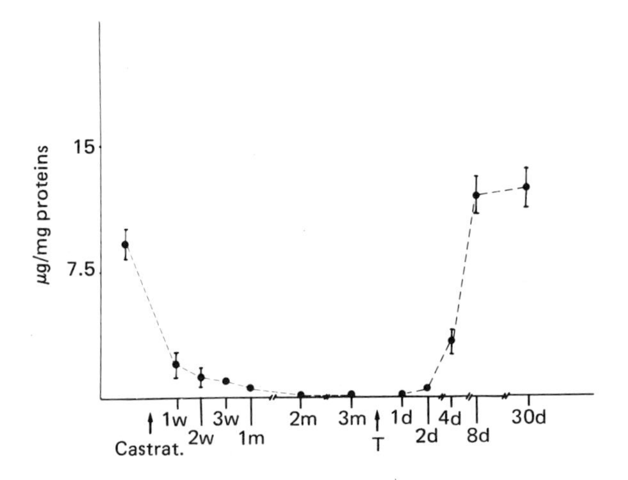

Figure 5. Effect of castration and testosterone treatment on SP concentration in the total epididymis of 3- to 6-month-old rats (mean ± S.D., N = 5 for each group).

activity increases abruptly in this part of the organ, plateaus in the corpus, increases again in the cauda, and finally decreases in the deferent duct.

Immunohistochemical studies were carried out in the same epididymal portion and the deferent duct. After being fixed in Bouin's fluid, the tissues were embedded in paraffin and cut at 5 μm. Immunohistochemical reactions were performed according to Coon's indirect method using a specific rabbit antibody in the first stage and a solution of anti-rabbit immunoglobulin sheep serum conjugated with peroxidase. Then the horseradish peroxidase was localized with 3,3′-diaminobenzidine tetrachloride and fresh hydrogen peroxide according to the method of Graham and Karnovsky (1966). No staining of the epithelial cells or of the spermatozoa was observed on control sections using preimmune rabbit serum in place of the specific primary antiserum.

No labeling is observed in the initial segment of the epididymal duct. All the other parts are labeled, and the labeling is located mainly in the apical part of the "principal cells" (Fig. 7). In the caput and corpus, some epithelial cells are also immunoreactive, and the sterocilia of these cells appeared less developed than those of adjacent cells (Fig. 7B and C). In the cauda, no such cells appear, but images comparable to images

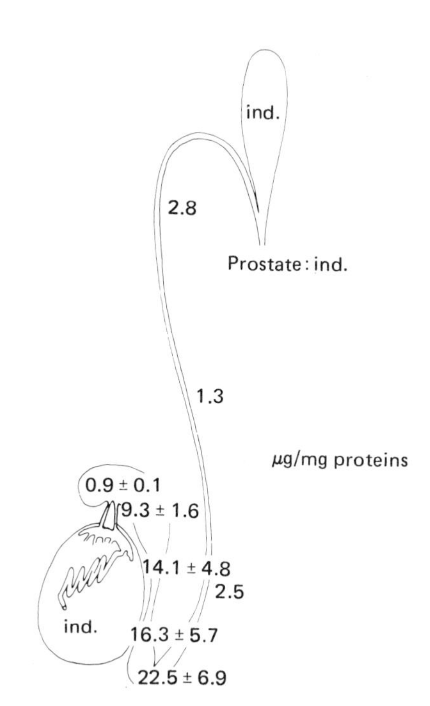

Figure 6. SP concentration at different levels of the rat sexual tract (mean ± S.D., N = 5 measurements for the epididymis; for the deferent duct, the mean of 2 determinations is reported).

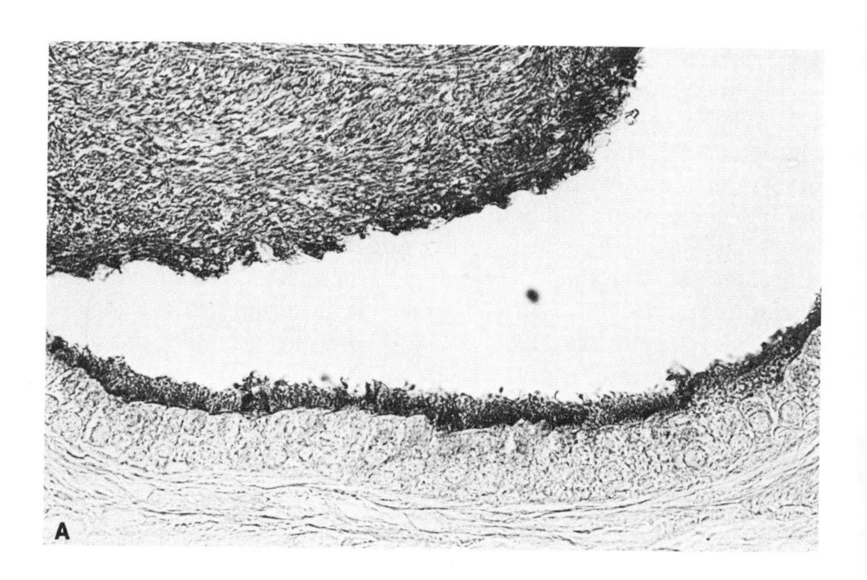

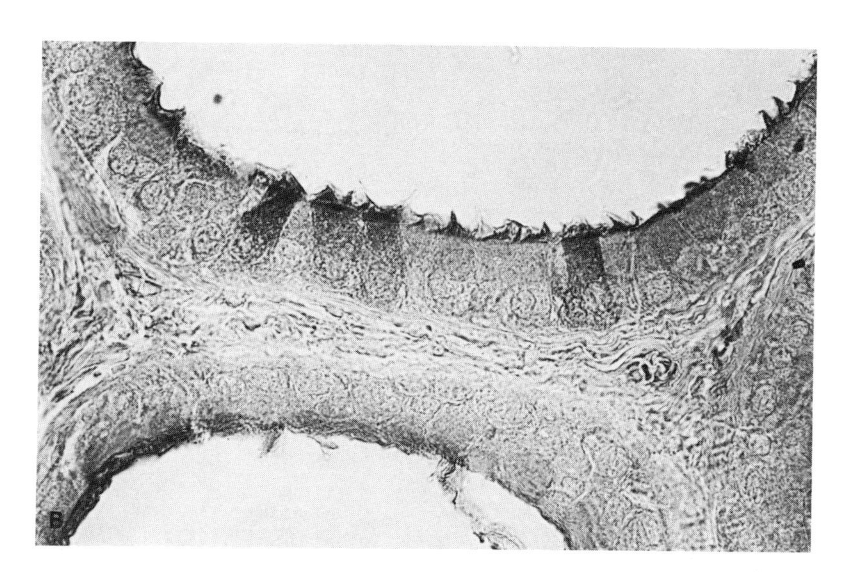

Figure 7. Immunoperoxidase localization of the SP at different levels of the rat epididymis and in the deferent duct. (A–C) Caput, distal from the initial segment, and corpus. The labeling is located in the apical part of the "principal cells." Some epithelial cells are also immunoreactive, and the stereocilia of these cells appeared less developed than those of adjacent cells. (D) Cauda. In the same preparation, adjacent tubules are labeled or unlabeled. (E) Deferent duct, median portion.

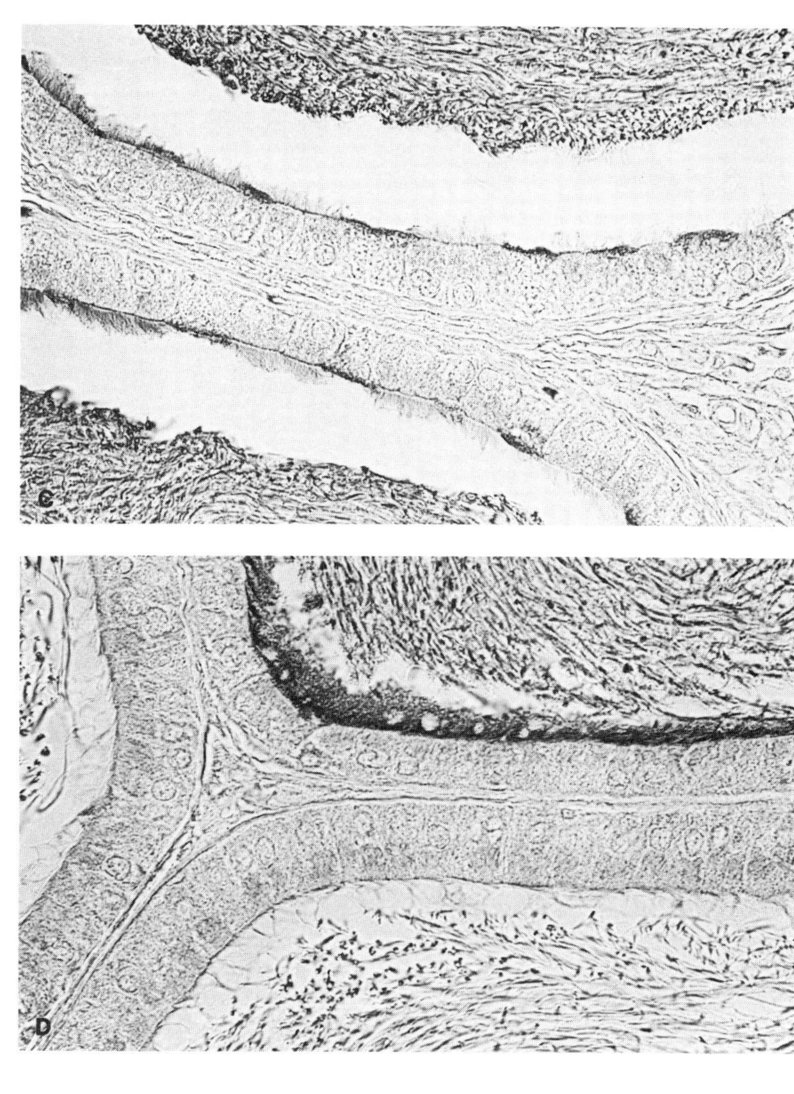
C
D

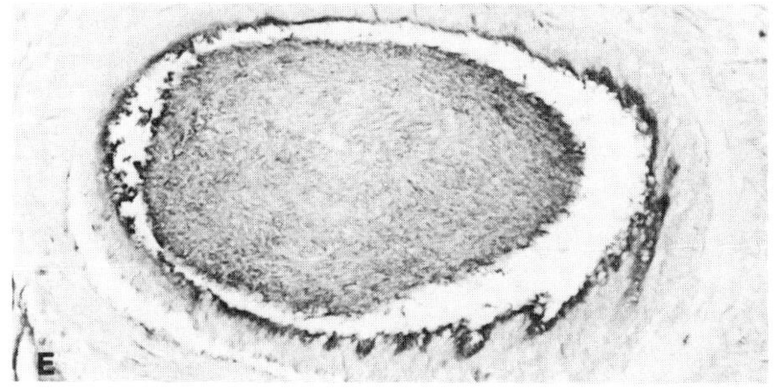
E

observed on the cell luminal surface of the thyroid vesicles (Wetzel *et al.,* 1965), suggesting a protein resorption, are observed (Fig. 7D). Moreover, in this part of the epididymis, the labeling is not uniform, some tubules being labeled and others unlabeled in the same preparation. Finally, in the deferent duct (Fig. 7E), the microvilli of the "principal cells" lining the lumen are also labeled, with, again, images suggesting protein resorption. In all these portions, spermatozoa are also coated with the label. A complementary study of ejaculated spermatozoa by a membrane-immunofluorescence technique (Loosfelt *et al.,* 1970) showed a positive reaction (Fig. 8), which was also observed on spermatozoa collected from different portions of the epididymis and the deferent duct, but was not observed on spermatozoa collected from the rete testis.

5. Discussion

This work specifies and extends the previous observation of Fournier (1968) and Fournier-Delpech *et al.* (1973). From other data reported in the literature, it appears that the SP is very similar to the acidic

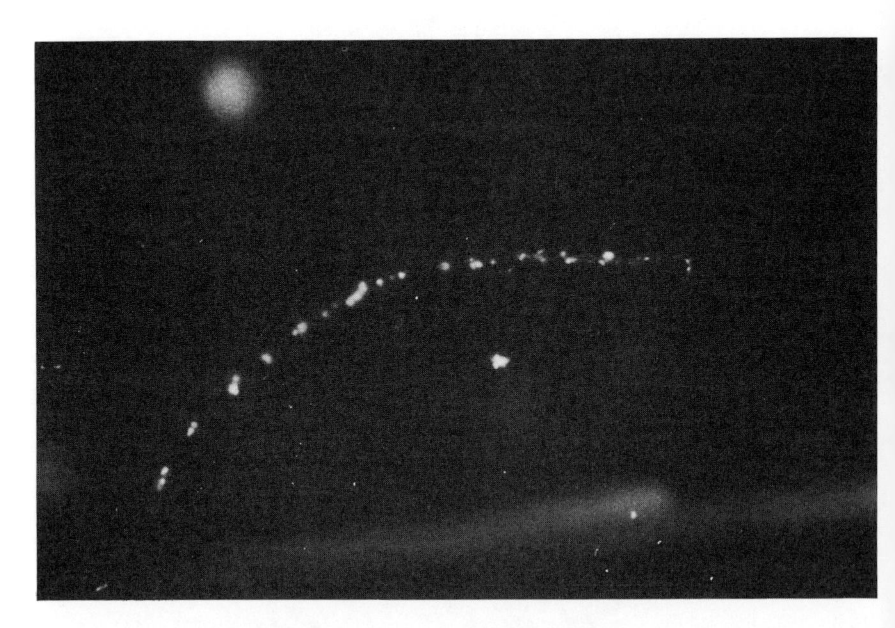

Figure 8. Membrane immunofluorescence of ejaculated spermatozoa. Similar pictures can be obtained from spermatozoa collected from the epididymis or the deferent duct, but not from the rete testis.

epididymal glycoprotein described by Lea *et al.* (1978) or to the protein D-E described by Cameo and Blaquier (1976), Garberi *et al.* (1979), and Kohane *et al.* (1979). Our findings clearly indicate that the SP is produced by the epithelial cells, most probably the "principal cells," of the caput (distal to the initial segment) and of the corpus epididymidis under the influence of testosterone. The high concentration measured in the cauda epididymidis contrasting with the very low concentration in the deferent duct concurs with the heterogeneity of labeling of the caudal epididymal duct and the peculiar images of protein resorption to suggest a process of SP degradation in this portion of the organ—a process that may continue along the length of the deferent duct. Such a process, if confirmed, would appear to be very active, since no immunoreactivity could be detected in the body of the epithelial cells. The SP is also detected on spermatozoa collected from the epididymis and the deferent duct. Such a localization could, however, result from a contamination during preparation of the tissue. The fact that ejaculated spermatozoa are still coated with the SP is, then, of interest in this regard. The precise role of the SP in the process of fecundation is, however, still to be defined.

Acknowledgments. This work was supported by a grant from DGRST (No. 77.7.ORO6) and by WHO.

References

Acott, T.S., and Hoskins, D.D., 1977, Sperm bovine forward motility protein: Partial purification and characterization, *J. Biol. Chem.* **253:**6744–6750.

Acott, T.S., Johnson, D.J., Brandt, H., and Hoskins, D.D., 1979, Sperm forward motility protein: Tissue distribution and species cross reactivity, *Biol. Reprod.* **20:**247–252.

Amann, R.P., Killian, G.J., and Benton, A.W., 1973, Differences in the electrophoretic characteristics of bovine rete testis fluid and plasma from the cauda epididymidis, *J. Reprod. Fertil.* **35:**321–330.

Barker, L.D.S., and Amann, R.P., 1970, Epididymal physiology. I. Specificity of antisera against bull spermatozoa and reproductive fluids, *J. Reprod. Fertil.* **22:**441–452.

Bedford, J.M., 1975, Maturation, transport and fate of spermatozoa in the epididymis, in: *Handbook of Physiology,* Vol. V, *Male Reproductive System* (R.W. Greep and E.B. Astwood, eds.), pp. 303–317, American Physiological Society, Washington.

Cameo, M.S., and Blaquier, J.A., 1976, Androgen-controlled specific proteins in rat epididymis, *J. Endocrinol.* **69:**47–55.

Davis, B.J., 1964, Disc electrophoresis. II. Method and application to human serum proteins, *Ann. N. Y. Acad. Sci.* **121:**404–427.

Edelman, G.M., and Millette, C.F., 1974, Chemical dissections and surface mapping of spermatozoa, in: *Functional Anatomy of the Spermatozoan* (B.A. Afzelius, ed.), pp. 349–357, Pergamon Press, Oxford.

Einer-Jensen, N., 1974, Local recirculation of injected ^{3}H-testosterone from the testis to the epididymal fat pad and the corpus epididymidis in the rat, *J. Reprod. Fertil.* **37:**145–148.

Flechon, J.E., 1979, Sperm glycoproteins of the boar, bull, rabbit and ram. I. Acrosomal glycoproteins; II. Surface glycoproteins and free acidic groups, *Gamete Res.* **2:**43–64.
Flickinger, C.J., 1979, Synthesis, transport and secretion of protein in the initial segment of the mouse epididymis as studied by electron microscope radioautography, *Biol. Reprod.* **20:**1015–1030.
Fournier, S., 1968, Electrophorèse des protéines du tractus génital du rat. I. Présence dans le sperme épididymaire d'une glycoprotéine migrant vers l'anode à pH = 8.45, *C. R. Soc. Biol. (Paris)* **162:**568–571.
Fournier-Delpech, S., Bayard, F., and Boulard C., 1973, Isolement, extraction et caracterisation d'une sialoproteine du sperme épididymaire du rat par électrophorèse sur polyacrylamide, *C. R. Soc. Biol. (Paris)* **167:**543–546.
Fournier-Delpech, S., Danzo, B.J., and Orgebin-Crist, M.C., 1977, Extraction of concanavalin-A affinity material from rat testicular and epididymal spermatozoa, *Ann. Biol. Anim. Biochim. Biophys.* **17:**207–213.
Free, M.J., and Jaffe, R.A., 1978, Target organs for testosterone transferred from vein to artery in the pampiniform plexus: The epididymis *Biol. Reprod.* **18:**639–642.
Garberi, J.C., Kohane, A.C., Cameo, M.S., and Blaquier, J.A., 1979, Isolation and characterization of specific rat epididymal proteins, *Mol. Cell. Endocrinol.* **13:**73–82.
Gordon, M., Dandekar, P.V., and Bartoszewicz, W., 1975, The surface coat of epididymal, ejaculated and capacitated sperm, *J. Ultrastruct. Res.* **50:**199–207.
Graham, R.C., Jr., and Karnovsky, M.J., 1966, The early stages of absorption of injected horseradish peroxidase in the proximal tubules of mouse kidney: Ultrastructural biochemistry by a new technique, *J. Histochem. Cytochem.* **14:**291–302.
Gwatkin, R.B.L., 1976, Fertilization, in: *Cell Surface Reviews,* Vol. 1, *The Cell Surface in Aminal Embryogenesis and Development* (G. Poste and C.G. Nicolson, eds.), pp. 1–54, North-Holland, Amsterdam.
Hamilton, D.W., 1975, Structure and function of the epithelium lining the ductuli efferentes, ductus epididymidis and ductus deferens in the rat, in: *Handbook of Physiology,* Vol. V, *Male Reproductive System* (R.O. Greep and E.B. Astwood, eds.), pp. 259–301, American Physiological Society, Washington.
Hoffer, A.P., Hamilton, D.W., and Fawcet, D.W., 1973, The ultrastructure of the principal cells and intraepithelial leukocytes in the initial segment of the rat epididymis, *Anat. Rec.* **175:**169–202.
Huang, H.F.S., and Johnson, A.D., 1975, Comparative study of protein pattern of epididymal plasma of mouse, rat, rabbit and sheep, *Comp. Biochem. Physiol.* **51B:**337–341.
Knorr, D.W., Vanha-Perttula, T., and Lipsett, M.B., 1970, Structure and function of rat testis through pubescence, *Endocrinology* **86:**1298–1304.
Koehler, J.K., 1978, The mammalian sperm surface: Studies with specific labeling techniques, *Int. Rev. Cytol.* **54:**73–108.
Kohane, A.C., Garberi, J.C., Cameo, M.S., and Blaquier, J.A., 1979, Quantitative determination of specific proteins in rat epididymis, *J. Steroid Biochem.* **11:**671–674.
Koskimies, A.I., and Kormano, M., 1975, Proteins in fluids from different segments of the rat epididymis, *J. Reprod. Fertil.* **43:**345–348.
Lea, O.A., Petrusz, P., and French, F.S., 1978, Purification and localization of acidic epididymal glycoprotein (AEG): A sperm coating protein secreted by the rat epididymis, *Int. J. Androl., Suppl. 2,* pp. 592–607.
Loosfelt, P., Pavie-Fisher, J., and Kourilky, F.M., 1970, Technique d'immunofluorescence de membrane appliquée à la détection des antigènes tissulaires sur les cellules sanguines humaines, *Pathol. Biol.* **18:**795–800.
Midgley, A.R., Jr., 1966, Radioimmunoassay: A method for human chorionic gonadotropin and human luteinizing hormone, *Endocrinology* **79:**10–18.

Millette, C.F., 1977, Distribution and mobility of lectin binding sites on mammalian spermatozoa, in: *Immunobiology of Gametes* (M. Edidin and M.H. Johnson, eds.), pp. 51–71, Cambridge University Press, New York.

Nicholson, G.L., Usin, N., Yanagimachi, R., Yanagimachi, H., and Smith, J.R., 1977, Lectin binding sites on the plasma membranes of rabbit spermatozoa: Changes in surface receptor during epididymal maturation and after ejaculation, *J. Cell Biol.* **74:**950–962.

Orgebin-Crist, M.C., Danzo, B.J.. and Davies, J., 1975, Endocrine control of the development and maintenance of sperm fertilizing ability in the epididymis, in: *Handbook of Physiology,* Vol. V, *Male Reproductive System* (R.O. Greep and E.B. Astwood, eds.), pp. 319–338, American Physiological Society, Washington.

Pujol, A., and Bayard, F., 1979, Androgen receptors in the rat epididymis and their hormonal control, *J. Reprod. Fertil.* **56:**217–222.

Segrest, J.P., and Jackson, R.L., 1972, Molecular weight determination of glycoproteins by polyacrylamide gel electrophoresis in sodium dodecyl sulfate, in: *Methods in Enzymology,* Vol. XXVIII, Part B (V. Ginsburg, ed.), pp. 54–63, Academic Press, New York.

Vaitukaitis, J., Robbins, J.B., Nieschlag, E., and Ross, G.T., 1971, A method for producing specific antisera with small doses of immunogen, *J. Clin. Endocrinol. Metab.* **33:**988–991.

Weber, K., and Osborn, M., 1969, The reliability of molecular weight determination by sodium dodecyl sulfate–polyacrylamide gel electrophoresis, *J. Biol. Chem.* **244:**4406–4412.

Wetzel, B., Spicer, S., and Wollman, S., 1965, Changes in fine structure and acid phosphatase localization in rat thyroid cells following thyrotropin administration, *J. Cell Biol.* **25:**593–618.

VI

Control Mechanisms and Metabolic Regulation

18

The Use of Isopropyl-N-phenylcarbamate as a Potential Contraceptive

Control of Meiotic Maturation

Daniel Szöllösi and Nicole Crozet

1. Introduction

The most successful contraceptive practices, exclusive of the physical methods, affect the complex hypothalamic–hypophyseal–gonadal feedback systems. The intervention could theoretically occur at any point of this feedback loop. Steroid supplementation reduces the level of gonadotrophin-secretory activity and via it gonadal steroid production. The inconveniences of this method to the user and the user's undesirable physiological responses to this method have become known during recent decades of contraceptive practice (for a review, see *Régulation de la Fécondité,* Vol. 83, INSERM, Paris, 1979). Steroid-based contraceptives certainly influence gamete production and, as well, the hormone-secretory activity of the gonads themselves and various protein hormones of the central nervous system.

A direct action of contraceptive agents on production of mature gametes would be a more desirable and preferable method if it could be

Daniel Szöllösi and *Nicole Crozet* • INRA, Station Centrale de Physiologie Animale, 78350 Jouy-en-Josas, France.

specific. The cyclically occurring reactivation of the expended meiotic process, leading to the production of fertilizable oocytes, may lend itself to experimental approaches. In the experiments reported in this chapter, we have taken advantage of a unique feature of oocytes among mammalian cells: that the two spindles involved in the meiotic process lack centrioles (Szöllösi *et al.*, 1972). The herbicide isopropyl-*N*-phenylcarbamate (IPC), an antimitotic agent, apparently inhibits cell division only in cells in which centrioles are not part of the spindle apparatus. Apparently, this drug acts on the site of spindle microtubule assembly (Brown and Bouck, 1974; Coss and Pickett-Heaps, 1974; Crozet and Szöllösi, 1979; Hepler and Jackson, 1969; Nasta and Gunther, 1973).

Mature mammalian oocytes complete meiosis *in vivo* on a known time schedule after release of luteinizing hormone or *in vitro* after their removal from the vesicular Graafian follicle (Donahue, 1968; Sorensen, 1973). The *in vitro* system represents a suitable method for the study of the action of various chemical agents during the final maturation phases of the oocyte, referred to as meiotic maturation.

2. Materials and Methods

Fully grown mouse and rabbit oocytes were recovered from large, preovulatory Graafian follicles from young animals. The ovaries were placed directly into the respective incubation media, and the oocytes in them were removed from the follicles by puncture with sterile hypodermic needles. The harvested oocytes, possessing intact germinal vesicle (GVs), were washed and placed into small plastic wells containing 200 μl medium. Brinster's medium (Brinster, 1971) (Gibco, New York) was used for control cultures, while 2×10^{-4} M IPC (gift of Sigma Chemical Co., St. Louis, Missouri) was added to the experimental medium. The incubation was carried out under either parafilm or paraffin oil at 37°C in a humidified 5% CO_2–air atmosphere.

Most of the oocytes were mounted in a drop of culture medium on glass microscope slides, and coverslips were sealed with a Vaseline–paraffin mixture and observed with Nomarski differential-interference optics. Further oocytes were fixed for 40 min in a 2% glutaraldehyde–1% paraformaldehyde mixture in 0.075 M phosphate buffer containing 0.1–0.5% potassium ferricyanide (Elbers *et al.*, 1965) immediately after termination of culture. After a brief wash in the buffer, the fixed lots were postosmicated in 2% osmium tetroxide in distilled water for 1 hr. Subsequently, the oocytes were washed in distilled water, stained *in toto* for several hours to overnight in 0.5% aqueous uranyl acetate, dehydrated in ethyl alcohol, and embedded in Epon.

3. Results

In control cultures, meiotic maturation is initiated shortly after removal of the oocytes from the follicles. Following 5 hr of culture in Brinster's medium, both mouse and rabbit oocytes undergo germinal-vesicle breakdown (GVBD) (Figs. 1 and 2) corresponding well to the timing of nuclear progression published by other authors [mouse (Donahue, 1968; Sorensen, 1973), rabbit (Chang, 1955; Thibault, 1972)]. When the herbicide IPC is added to the culture medium and incubated for the same length of time, the GV remains intact, possessing one or two nucleoli (Fig. 3). The GVs are in fact very similar to oocytes at the time of their removal from the follicle.

The GV is nearly spherical at the time the oocytes are removed, slight undulations of the nuclear envelope being visible in electron micrographs. The nucleolus is composed of compacted thin, filamentous material exclusively, while at its periphery a ring of more looselypacked filaments is seen (Fig. 4). In the oocytes of both animals, the nucleoplasm contains clumps of chromatin. In mouse oocytes, in proximity to chromatin masses, large, electron-dense granules with a diameter of 50–60 nm form islets of different sizes. Nuclear bodies composed of a loosely packed filamentous matrix and granules with a diameter of about 15–20 nm characterize mouse GVs. At the periphery of the nuclear body, dense microspherules are seen, corresponding to heterochromatic knobs (Chouinard, 1973). In the GVs of rabbit oocytes, nuclear bodies and electron-dense, large granules are absent.

GVBD occurs in mouse oocytes in fact already following 3 hr of culture in Brinster's medium, while in the rabbit this may take 5 hr (see Fig. 2). Even though the nuclear envelope has broken down, its remnants are retained as quadruple membranes in the transitional region between nucleoplasm and cytoplasm while the chromosomes condense. In some areas, microtubule bundles are seen. After 4–6 hr of culture, GVBD has taken place in 86% of mouse oocytes. The quadruple membranes disappear and the first spindle is formed (see Fig. 1).

In contrast to the control cultures, when 2×10^{-4} M IPC is added to the culture medium for the duration of 5 hr, GVBD is inhibited. In the rabbit, the oocytes remain identical to those removed and fixed immediately (Fig. 5). Mouse oocytes after 5 hr of culture are similar in many respects to those fixed at the outset of culture (Fig. 6). Only the number and spatial relationship of the nuclear bodies are different. Serial sections through the GV confirm their numerical increase. Following culture, the nuclear bodies more frequently demonstrate continuity with the nucleolus and the islets of large, electron-dense granules (Fig. 7). If the oocytes are cultured for the first 30 min after their removal in Brinster's medium

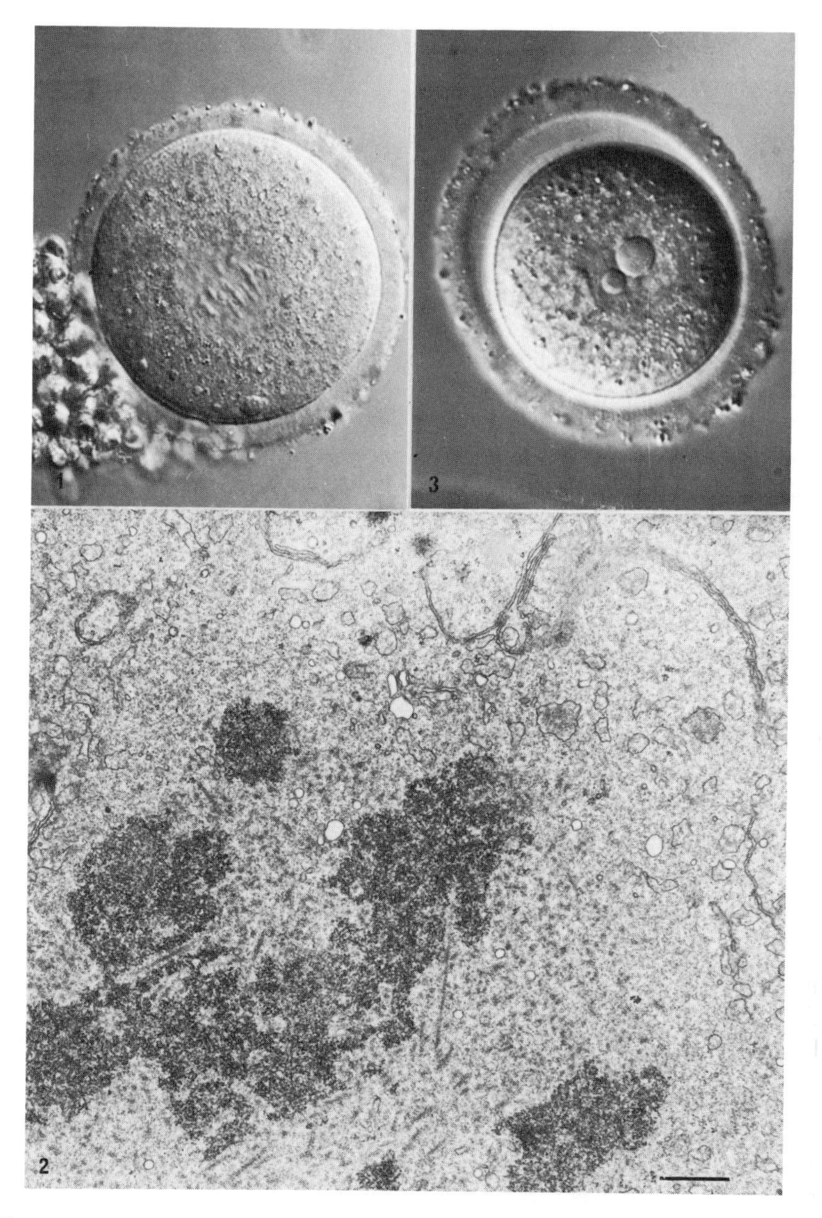

Figure 1. Occurrence of GVBD in mouse oocytes after 5 hr in culture and formation of the first meiotic spindle. Normarski differential-interference optics.

Figure 2. Broken nuclear envelope of a rabbit oocyte following 5 hr of culture in Brinster's medium. Quadruple-membrane remnants of the nuclear envelope, chromosomes, and spindle microtubules are seen. Scale bar: 1 μm.

Figure 3. Mouse oocyte cultured with IPC for 5 hr. The GV with two nucleoli remains intact. Nomarski differential-interference optics.

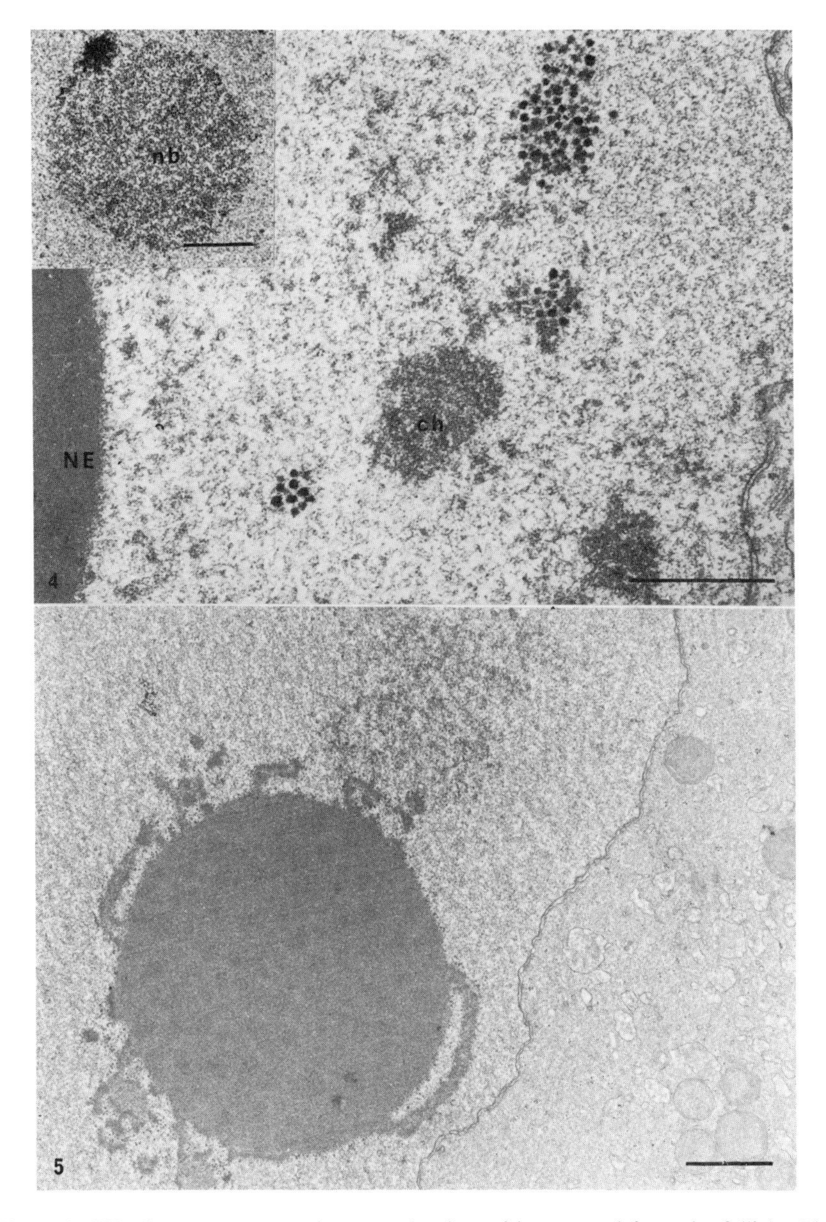

Figure 4. GV of a mouse oocyte intact at the time of its removal from the follicle. The nucleolus is composed of tightly packed thin filaments. In the nucleoplasm, electron-dense granules with a diameter of 50–60 nm form large aggregates. (ch) Condensing chromatin; (NE) nuclear envelope. *Inset:* A nuclear body (nb) composed of a thin, fibrillar matrix and granules with a diameter of 15–20 nm. At its periphery, a microspherule is visible. Scale bar: 1 μm.

Figure 5. Intact GV in a rabbit oocyte after 5 hr of culture with IPC. The nucleolus is unaltered. Scale bar: 1 μm.

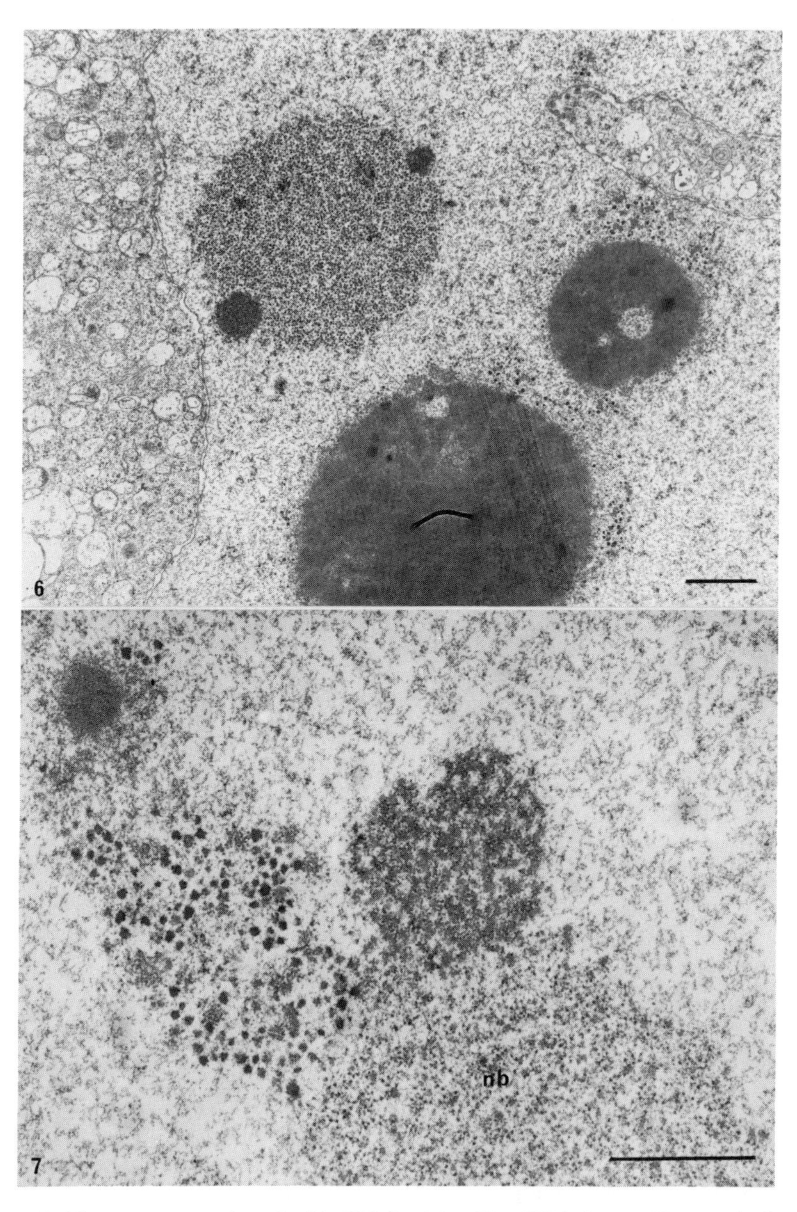

Figure 6. Mouse oocyte cultured with IPC for 9 hr. The GV is intact. Two nucleoli are present and are surrounded by condensing chromatin with associated dense granules. The nuclear body is associated with two microspherules. Scale bar: 1 μm. Reprinted from Crozet and Szöllösi (1979) with the permission of *Annales de Biologie Animale, Biochimie, Biophysique*.

Figure 7. Mouse oocyte cultured with IPC for 9 hr. The nuclear body (nb) often shows continuity with the nucleolus and with islets of large, electron-dense granules. Scale bar: 1 μm. Reprinted from Crozet and Szöllösi (1979) with the permission of *Annals de Biologie Animale, Biochimie, Biophysique*.

and then cultured further with IPC, the nuclear progression is still inhibited. The inhibitory effect of this herbicide is reversible, however, if the culture is continued in a medium lacking the drug. The nuclear envelope breaks down, microtubules form, the nucleolus disperses, and chromosomes condense. However, microtubules are randomly oriented (Fig. 8), the microtubule-organizing centers (MTOCs) are distributed around the chromosomes, and chromosomes are seen lagging outside the spindle area. Spindle formation is evidently disturbed. Nuclear bodies may remain present (Fig. 8). When, however, IPC is added to the culture medium after 3 hr of culture without it, after GVBD, the spindle organization is not affected (Fig. 9); it is very similar to that formed in control cultures. Still, some chromosomes may remain excluded from the spindle. IPC thus does not affect microtubules already formed.

4. Discussion

The herbicide IPC is apparently a specific antimitotic agent in cells in which centrioles do not form part of the spindle, e.g., higher plants and algae (Hepler and Jackson, 1969; Nasta and Gunther, 1973; Coss and Pickett-Heaps, 1974). This drug does not act in the same manner as the well-known antimitotic agents such as colchicine, vinblastine, and vincristine and its derivatives, because it does not bind with tubulin (Coss *et al.*, 1975). It seems to affect the site of microtubule assembly, the MTOC.

It was tempting, therefore, to study the effects of IPC *in vitro* on the meiotic maturation of mammalian oocytes after the discovery that centrioles are absent from both meiotic spindles (Szöllösi *et al.*, 1972). In the same study, GVBD was shown to be related to focal microtubule organization and their projections in small bundles toward the nuclear envelope. These microtubule bundles were always related to MTOCs.

Before the employment of IPC as a potential contraceptive agent, its possible toxic effects must be considered. In this regard, contradictory evidence is available. The mitotic index of lymphocytes was reduced by IPC following phytohemagglutinin stimulation (Timson, 1970). Reduction of growth rate and some morphological changes of mouse fibroblasts in culture with the chlorine derivative of IPC [isopropyl-*N*-(3-chlorophenyl)-carbamate] have also been reported more recently (Oliver *et al.*, 1978). In our own preliminary experiments, on the other hand, IPC had no apparent action on the mitotic rate of mouse L-cells (Laport and Szöllösi, unpublished). Neither were any significant effects of IPC detected on the proliferation of the Kc diploid *Drosophila* cell line at various concentrations of the herbicide and for a duration of several days (Delbec, personal communication). No toxic effects of IPC were ob-

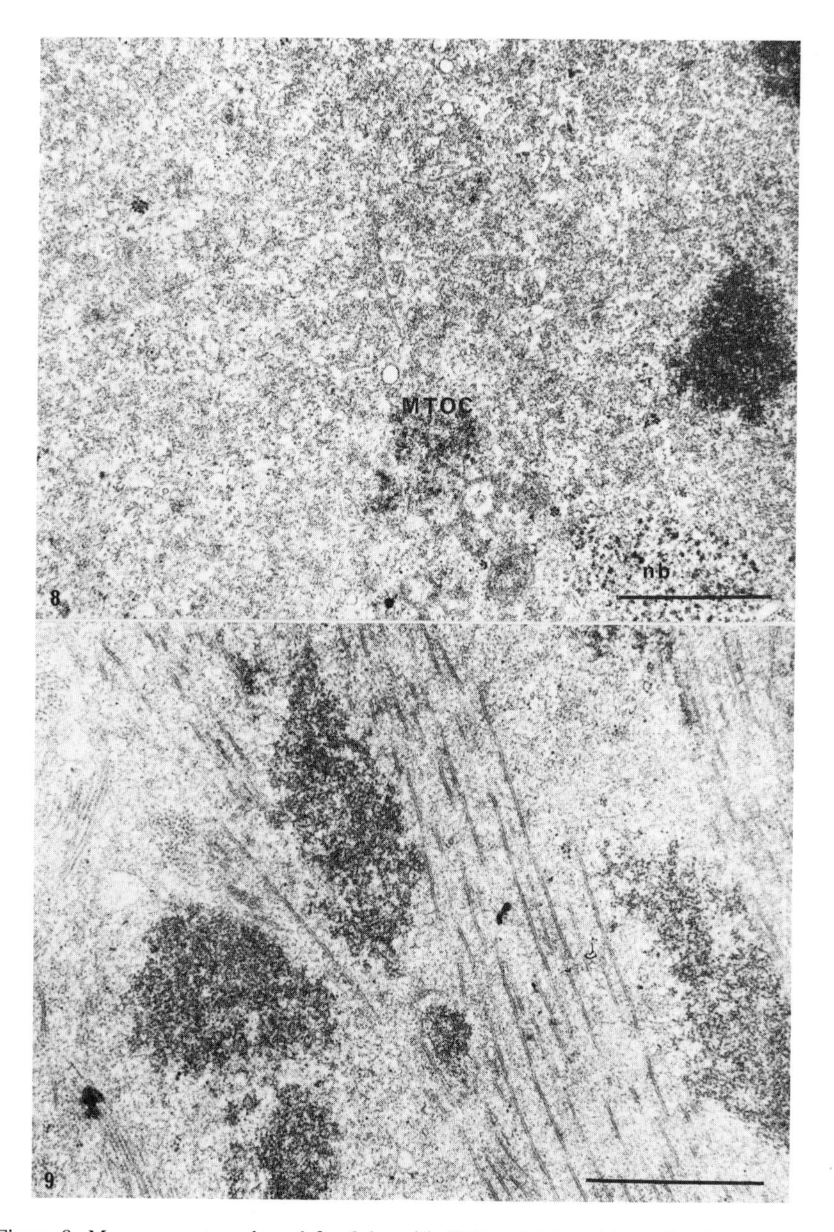

Figure 8. Mouse oocyte cultured for 3 hr with IPC and 3 hr without the drug. There is GVBD, chromosomes condense, and microtubules form but are irregularly oriented. A portion of a nuclear body (nb) is visible. (MTOC) Microtubule-organizing center. Scale bar: 1 μm.

Figure 9. Electron micrograph demonstrating that the first meiotic spindle, when formed in control culture medium, is little affected when cultured further with IPC. Scale bar: 1 μm.

served in mice injected intraperitoneally with various doses and for a duration of several days (up to 3 weeks). Unquestionably, the possible toxic effects of the drug must be thoroughly reexamined *in vivo* and *in vitro*.

In the experimental series presented herein, the results confirm the working hypothesis that IPC inhibits GVBD and formation of the first meiotic spindle, both of which depend on microtubule polymerization associated with MTOCs. Exposure for 30 min of oocytes containing intact GVs to culture media in which meiotic maturation usually progresses well preceding exposure to the drug is not sufficient to initiate irreversibly the nuclear events. The microtubules of the first meiotic spindle apparatus are formed and the spindle is organized within 3 hr of incubation in the control medium. The formed spindle is not markedly disturbed when the culture is continued with the herbicide. This observation is in contrast to the results obtained with algae and plants, in which the presence of IPC in the culture medium for a duration of 0.5–2 hr causes multipolar spindles and changes the disposition of the chromosomal microtubules in the proximity of the kinetochores (Hepler and Jackson, 1969). Because of the short time of exposure to the herbicide, the cells affected must have been either entering or already in mitosis at the time the culture was initiated. The drug must disturb the forming or already-formed spindle. The existing, individual microtubules, on the other hand, are not disturbed or depolymerized in either algae, flagellates, plant, or animal cells by the drug (Brown and Bouck, 1974; Coss and Pickett-Heaps, 1974; Hepler and Jackson, 1969; Mergulis and Banerjee, 1969).

More drastic disturbance of the spindle organization was observed in oocytes when the herbicide was washed out following 3 hr of exposure and the cells were further cultured in its absence. Following the removal of the drug, cytoplasmic microtubule organization, GVBD, and spindle formation take place. Lagging chromosomes are evidence of severe spindle malformation. The distribution of MTOCs is also altered. Similar delayed effects of IPC in algae and plants were also observed (Coss and Pickett-Heaps, 1974; Hepler and Jackson, 1969).

The increase in numbers of nuclear bodies in mouse oocytes may imply either a continuation or reinitiation of ribonucleoprotein (RNP) synthetis activity by the GVs. It may equally indicate the segregation and compaction of already-synthesized molecular complexes. The EDTA regressive-coloration technique with uranyl acetate on frozen sections suggests that nuclear bodies are RNP in nature (Palombi and Viron, 1977). It was suggested already from morphological studies that they represent accumulations of macromolecules in the nucleus (Chouinard, 1973, 1975) because they were present in larger numbers at the onset and during the oocyte growth phase, when they are associated with the condensing bivalents and when RNP synthesis is high. In preovulatory follicles, in contrast, nuclear bodies regress in size, decrease in number, and detach from the bivalents. RNP synthesis at that time is minimal or nonexistent (Moore *et al.*, 1974). IPC treatment thus apparently retains oocytes in an earlier stage, permitting further synthetic activity and accumulation of synthetic products. Preliminary studies on incorporation of [^{3}H] uridine are in agreement with this hypothesis (unpublished studies). When IPC is withdrawn, the maturation processes are initiated and nuclear synthetic activities cease.

The hypothesis discussed above is in apparent contradiction to the reported inhibition of enzyme synthesis by IPC during seed germination following gibberellin stimulation (Mann *et al.*, 1967). The evidence presented does not justify the assumption, however, that the inhibition is at the transcriptional or translational level, or both. The mechanism of action of IPC has to be studied more extensively to understand the various inhibitory effects of this herbicide.

ACKNOWLEDGMENT. The financial support of the Ford Foundation (Grant 770.0315) is gratefully acknowledged.

Discussion

CHANNING: Have you found an effect of your herbicide on naked vs. cumulus-enclosed vs. follicle-enclosed oocytes?

SZÖLLÖSI: IPC inhibits meiotic maturation in both denuded and cumulus-enclosed oocytes.

CHANNING: Could the lack of *in vivo* effect be due to inability to get into follicular fluid or to a short half-life due to liver metabolism?

SZÖLLÖSI: Our *in vivo* experiments were not extensive thus far. We may have to change the administered dose and the way of administration. We do not know if IPC actually enters the follicular fluid, nor do we know the half-life of the herbicide.

BOGDANOVE: In your toxicity studies in the mouse, did you have an opportunity to observe whether the animals were able to become pregnant?

SZÖLLÖSI: Yes, all females became pregnant when they were treated with IPC for 1 week before they were paired with males and then continued on the treatment while they were with males.

NOLIN: If intracytoplasmic oocyte (or a prolactin fragment) has meiotic maturation inhibitory activity (see Chapter 9), it may be that resumption of meiosis of naked oocytes *in vitro* involves "leakage" of prolactin into the medium, thus relieving the inhibition. Have you observed any changes in the volume with IPC which would support such an idea?

SZÖLLÖSI: No obvious changes of the cell-membrane structure were observed on the usual ultrastructural level or of the frequency of microvilli. We have not concentrated our efforts on these questions, however, and we should examine these parameters more closely.

C. A. VILLEE: I would like to follow up Dr. Channing's question. You said that the effective level of IPC *in vitro* was 2×10^{-4} M, a very substantial level. The compound is metabolized fairly rapidly, probably by the liver, and therefore it would take a fairly high amount of IPC given orally to attain a level of 2×10^{-4} M in the ovary *in vivo*. What amounts were you giving to the animals when you were testing the *in vivo* toxicity levels but which were not effective in preventing pregnancy?

SZÖLLÖSI: We have given a single or twice-daily intraperitoneal injection of 0.1 or 0.2 ml of a 10^{-3} M IPC solution. The dilution of the drug and its metabolism may have rendered it ineffective the way it was administered.

REFERENCES

Brinster, R.L., 1971, *In vitro* culture of the embryo, in: *Pathways to Conception* (A.I. Sherman, ed.), Charles C. Thomas, Springfield, Illinois.

Brown, D.L., and Bouck, G.B., 1974, Microtubule biogenesis and cell shape in *Ochromonas*. III. Effects of the herbicidal mitotic inhibitor isopropyl-*N*-phenylcarbamate on shape and flagellum regeneration, *J. Cell Biol.* **61:**514–536.

Chang, M.C., 1955, The maturation of rabbit oocytes in culture and their maturation, activation, fertilization and subsequent development in the fallopian tubes, *J. Exp. Zool.* **128:**379–405.

Chouinard, L.A., 1973, An electron-microscope study of the extranucleolar bodies during growth of the oocyte in the prebubertal mouse, *J. Cell Sci.* **12:**55–70.

Chouinard, L.A., 1975, A light- and electron-microscope study of the oocyte nucleus during development of the antral follicle in the prebubertal mouse, *J. Cell Sci.* **17:**589–615.

Coss, R.A., and Pickett-Heaps, J.D., 1974, Effects of isopropyl-*N*-phenylcarbamate on the green alga *Oedogonium cardiacum*. I. Cell division, *J. Cell Biol.* **63:**84–98.

Coss, R.A., Bloodgood, R.A., Brower, D.L., Pickett-Heaps, J.D., and McIntosh, J.R., 1975, Studies on the mechanism of action of isopropyl-*N*-phenylcarbamate, *Exp. Cell Res.* **92:**394–398.

Crozet, N., and Szöllösi, D., 1979, The effects of isopropyl-*N*-phenylcarbamate on meiotic maturation of mammalian oocytes, *Ann. Biol. Anim. Biochim. Biophys.* **19:**1131–1140.

Donahue, R.P., 1968, Maturation of the mouse oocyte *in vitro*. I. Sequence and timing of nuclear progression, *J. Exp. Zool.* **169:**237–250.

Elbers, P.J., Ververgaest, P.H.J.T., and Demel, R., 1965, Tricomplex fixation of phospholipids, *J. Cell Biol.* **23:**23–30.

Hepler, P.K., and Jackson, W.T., 1969, Isopropyl-*N*-phenylcarbamate affects spindle microtubule orientation in dividing endosperm cells of *Haemanthus Katherinae* Baker, *J. Cell Sci.* **5:**727–743.

Mann, J.D., Cota-Robles, E., Yung, K.H., and Haid, H., 1967, Phenylurethane herbicides: Inhibitors of changes in metabolic state. I. Botanical aspects. *Biochim. Biophys. Acta* **138:**133–139.

Mergulis, L., and Banerjee, D., 1969, Effects of mitotic spindle inhibitors on regenerating cilia of *Stentor coeruleus, J. Protozool.* **16:**75.

Moore, G.P.M., Lintern-Moore, S., Peters, H., and Faber, M., 1974, RNA synthesis in the mouse oocyte, *J. Cell Biol.* **60:**416–422.

Nasta, A., and Gunther E., 1973, Mitotic disturbances by application of carbamate herbicides on *Allium cepa* and *Hordeum vulgare, Biol. Zentralbl.* **92:**27–36.

Oliver, J.M., Krawiec, J.A., and Berlin R.D., 1978, A carbamate herbicide causes microtubule and microfilament disruption and nuclear fragmentation in fibroblasts, *Exp. Cell Res.* **116:**229–237.

Palombi, F., and Viron, A., 1977, Nuclear cytochemistry of mouse oogenesis. I. Changes in extranucleolar ribonucleoprotein components through meiotic prophase, *J. Ultrastruct. Res.* **61**:10–20.

Sorensen, R.A., 1973, Cinemicrography of mouse oocyte maturation utilizing Nomarski differential-interference microscopy, *Am. J. Anat.* **136**:265–276.

Szöllösi, D., Calarco, P., and Donahue, R.P., 1972, Absence of centrioles in the first and second meiotic spindles of mouse oocytes, *J. Cell Sci.* **11**:521–541.

Thibault, C., 1972, Final stages of mammalian oocyte maturation, in: *Oogenesis* (J.D. Biggers and A.W. Schuetz eds.), pp. 397–411, University Park Press, Baltimore.

Timson, J., 1970, Effect of the herbicides propham and chlorpropham on the rate of mitosis of human lymphocytes in culture, *Pestic. Sci.* **1**:191–192.

19

Interaction between Prolactin and Gonadotropin Secretion

J. A. F. Tresguerres, A. Esquifino, and A. Oriol-Bosch

1. Introduction

In a number of physiological or experimental reproductive situations as well as in many pathophysiological conditions and following pharmacological manipulations, an inverse relationship between the secretion of prolactin (PRL) and the gonadotropins emerges.

In this chapter, we aim to review, from animal experiments and human clinical experiences, the current information regarding the possible modulatory role of PRL in the secretion of gonadotropins and in their regulatory mechanisms.

For this purpose, we have organized the available information under the following headings: reciprocal relationship between basal or spontaneously secreted PRL and gonadotropins in various physiological, experimental, or clinical situations; modulatory effects of PRL on the negative- as well as the positive-feedback mechanisms of gonadal hormones regulating gonadotropin secretion; and PRL in relation to the pituitary gonadotropin secretory reserve as measurable by administration of gonadotropin-releasing hormone. Finally, we will attempt to summarize our present knowledge of the hypothalamic neurotransmitters involved in the regulation of PRL and gonadotropin secretion, attempting

J. A. F. Tresguerres, A. Esquifino, and *A. Oriol-Bosch* • Chair for Experimental Endocrinology, Department of Physiology, University Complutensis Medical School, Madrid-3, Spain.

to establish a unitarian hypothesis around the tuberoinfundibular dopaminergic (TIDA) system.

2. Apparent Reciprocity between Prolactin and Gonadotropin Secretion

Among the many situations in which an apparent reciprocity between the plasma levels of PRL and gonadotropins emerges, the most obvious one of physiological relevance is the postpartum lactational state in experimental animals and in women.

In the rat, during lactation, plasma PRL levels are elevated in proportion to the intensity of suckling (Amenori *et al.,* 1970). Hammons *et al.* (1973) showed that intact lactating rats maintained low plasma luteinizing hormone (LH) levels. Lu *et al.* (1976b) confirmed this finding and observed that follicle-stimulating hormone (FSH) was also decreased. Hypothalamic LH-releasing hormone (LH-RH) concentrations were lower in rats nursing eight pups than in those nursing only two pups (Smith, 1978). Muralidhar *et al.* (1977) were able to further suppress LH levels in rats nursing two pups by injecting PRL. This effect was also evident in ovariectomized–adrenalectomized animals, thus excluding the possibility that this is a steroid-mediated phenomenon. Nevertheless, Lu *et al.* (1976b) observed in lactating rats with reduced PRL levels due to ergocornine administration that while FSH increased, LH remained low until day 17 postpartum.

Kann *et al.* (1977) studied the gonadotropin secretion during the postpartum period in the ewe and observed that the reduction of PRL by bromocriptine administration shortens the physiological anestrous period. Plasma FSH increases more rapidly than LH, and secretory peaks of FSH appear after 4–5 days, leading to a preovulatory surge on day 25. LH remains at a low level, but the first peak also appears, on day 25.

Working with primates, Maneckjee *et al.* (1976) also observed that the lactational hyperprolactinemia is accompanied by low LH levels.

Nursing women have repeated bursts of PRL secretion related to the suckling stimulus and a higher mean plasma level of PRL (Noel *et al.,* 1974). Breast-feeding mothers can show delays up to 100 days or more between parturition and the next menstruation, while those who do not breast-feed their infants have the following menstruation, on the average, on day 35 postpartum. Plasma LH levels are diminished for longer periods in breast-feeding mothers, while those who do not breast-feed recover normal levels between days 16 and 23 postpartum. In lactating women, FSH remains in the upper normal range during the first year of lactation (Rolland *et al.,* 1975a,b). Delvoye *et al.* (1978) have studied the endocrine status of a population of Central African women among whom the custom is to continue lactation for 2 years. Of these women, 80% were still amenorrheic after the first postpartum year, and 20% of them still did not resume their menstruation at the end of the second postpartum year. In this prolonged-lactation situation, the women showed high PRL; moderately elevated FSH levels, higher in those with amenorrhea than in the menstruating women; and lower concentrations of plasma LH and estradiol in amenorrheic lactating mothers. The number of amenorrheic lactating mothers decreased postpartum in parallel with basal serum PRL.

Bohnet *et al.* (1975b) reported that during the puerperium, while PRL levels are elevated the spontaneously spiking LH plasma levels reflecting the secretory bursts of this hormone are either diminished or abolished.

On weaning, there is an immediate drop in blood concentration of PRL that is accompanied by an increase in concentration of LH and estradiol (McNeilly, 1979).

During recent years, it has been evident that at least in some 20% of secondary amenorrhoeas, the PRL concentrations are high, although galactorrhea may not be as frequently associated with amenorrhea as occurs with hyperprolactinemia. Although gonadotropins may be normal, quite often they have been shown to be diminished. LH is more often found subnormal (Bergh *et al.*, 1978; Bohnet *et al.*, 1975a; Corbey *et al.*, 1977; London *et al.*, 1977; Quigley *et al.*, 1979; Seki and Seki, 1974), while FSH is found at normal levels in some instances even when LH levels are depressed. Pulsatile LH fluctuations are absent in most cases of hyperprolactinemia (Bohnet and Schneider, 1977). The rise of PRL associated with sleep in hyperprolactinemic women is accompanied by a further fall of serum gonadotropins (Kappen *et al.*, 1975). The reduction of PRL either by surgical removal of a PRL-secreting adenoma or by treatment with bromocriptine leads to a normalization of the situation. The process requires a time interval, since Strauch *et al.* (1977) showed that treatment for 4 days with bromocriptine in amenorrheic–hyperprolactinemic patients did not modify LH, FSH, or estradiol plasma concentrations despite a prompt circulating PRL decrease. Shortly thereafter, though, the gonadotropins return to normal levels, spontaneous LH spiking is recovered (Tolis, 1977), and ovulation ocurs (Seki and Seki, 1974; Tyson *et al.*, 1975).

Several groups have recently studied an experimental model of hyperprolactinemia in the intact rat bearing a pituitary graft under the kidney capsule. The effects of hyperprolactinemia on gonadotropin secretion in this model have given contradictory results. Figure 1 shows the kinetics of plasma PRL and LH levels in 90-day-old female rats

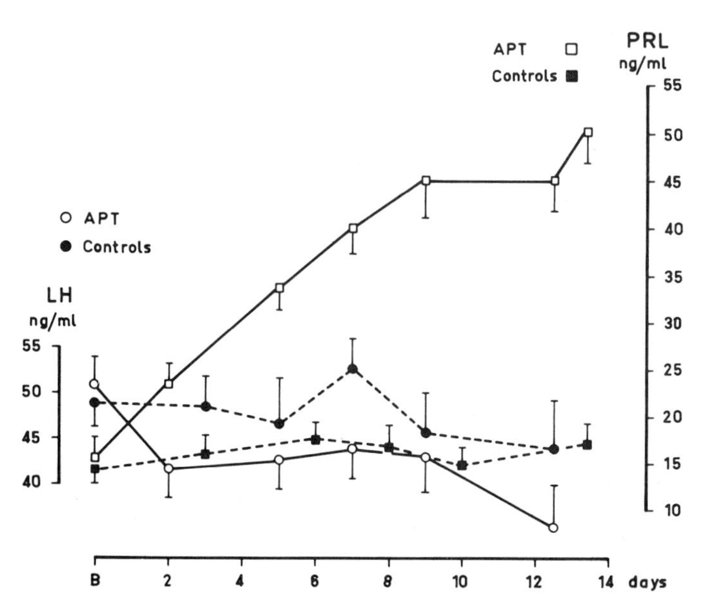

Figure 1. Evolution of plasma LH and PRL concentrations in female rats after an anterior-pituitary (APT) graft under the kidney capsule at 90 days of age. From Esquifino (1979).

during the first 14 days after they had received a pituitary transplant under the kidney capsule or had been sham-operated. Although PRL increases immediately, no apparent changes in LH concentration could be observed (Esquifino, 1979). Similar results were found in adult male rats by Winters and Loriaux (1978). On the other hand, young male rats bearing multiple pituitary grafts showed a concomitant 4-fold increase of plasma PRL and a significant decrease of LH after 30 days and of FSH 210 days after becoming hyperprolactinemic (Bartke *et al.,* 1977). Similarly, Greeley and Kizer (1979), using young adult male rats, could demonstrate that 9 days after a multiple pituitary graft, LH and FSH levels were decreased. Our group could also corroborate this finding when the animals received pituitary transplants prepubertally (30 days of age); 60 days later, they showed a decreased LH plasma level that was accompanied by a reduction in the percentage testicular weight and an increase in adrenal weight in males (Esquifino and Tresguerres, 1979) and delayed vaginal opening with alteration of the estrous cycle in females (Tresguerres and Esquifino, 1979). In both sexes, the administration of bromocriptine reduced PRL levels and allowed a recovery of the reduced LH levels. It is possible that some of the reported contradictory results could be due to the moment in the maturation of the gonadotropin-regulatory mechanism at which hyperprolactinemia was acting as well as the degree and duration of the elevated PRL levels (Esquifino, 1979).

Using a somewhat different model, Fang *et al.* (1974) showed that male rats transplanted with a PRL-producing tumor had moderately elevated estrogen and LH levels together with an extremely low testosterone concentration. Since the hyperprolactinemia obtained was very marked, the question arose about the source of the plasma LH found in these animals.

3. Prolactin and the Negative-Feedback Regulatory Mechanism of Gonadotropins by the Sex Hormones

Clinical and experimental evidence points to the possibility that PRL may play a modulatory role in the tonic secretion of gonadotropins regulated by the sex hormones through a well-known negative-feedback mechanism. The integrity of this regulatory mechanism can be explored by withdrawal of increase of the available sex hormones. The first situation can be achieved either by gonadectomy or by antihormone administration. Physiological or pathological situations such as menopause or primary ovarian failure can provide natural models that evidence that functioning of this regulatory mechanism.

In the rat, regardless of the cause of hyperprolactinemia, and independent of age, sex, or other circumstances, it has been shown that elevated PRL levels diminish the gonadotropin response to gonadectomy (Table I and Figs. 2 and 3). Hyperprolactinemic, neonatally estrogenized female rats (Table I) ovariectomized at 80 days of age showed a reduced LH response 10 days afterward in agreement with previously reported data (Neill, 1972; Turgeon and Barraclough, 1974). The same effect also appears in hyperprolactinemic, anovulatory rats subjected to a constant-illumination regime (Table I). In this case, with a less prominent increase of PRL, the response of LH to ovariectomy, although highly significant, is also less inhibited.

Figure 2 shows the LH and PRL levels obtained in hyperprolactinemic male rats 8 days after castration. Gonadectomy was performed 80 days after an anterior pituitary gland was transplanted under the kidney capsule in 90-day-old animals. The LH increase after castration in the hyperprolactinemic rats was significantly lower than in the control animals. Basically the same results were obtained in female animals of the same age or when the pituitary was transplanted at a different age (Esquifino, 1979; Esquifino and Tresguerres, 1979). Figure 3 shows the results obtained in female animals with a pituitary graft transplanted prepubertally and ovariectomized 60 days thereafter. It appears that the

Table I. Plasma LH and PRL Concentrations in Two Experimental Groups of Female Rats[a]

Group No.	Treatment	LH (ng/ml) Basal		LH (ng/ml) Post		PRL (ng/ml) Basal		PRL (ng/ml) Post	
1	Controls	24.5 ± 5.1	(8)	130.9 ± 12.0	(20)	45.6 ± 11.3	(7)	26.3 ± 2.8	(18
2	Estrogenized	33.7 ± 2.7	(24)	60.5 ± 6.1	(24)	403.4 ± 43.3	(21)	118.7 ± 15.7	(21
	P	0.01		0.01		0.01		0.01	
3	Controls (L/O)[b]	22.0 ± 3.7	(9)	159.0 ± 22.6	(9)	34.9 ± 10.2	(10)	20.6 ± 1.7	(8
4	Acyclic (L/L)[b]	24.0 ± 6.0	(8)	103 ± 20.3	(9)	121.2 ± 14.2	(10)	44.1 ± 7.2	(9
	P	N.S.		0.01		0.01		0.01	

[a] Data from Aguilar (1979). Basal samples were taken from 80-day-old females before ovariectomy. Postcastration samples were tak on day 90 (Groups 3 and 4) or day 94 (Groups 1 and 2). Estrogenized rats received 100 μg estradiol benzoate on day 5. Acylic ra were submitted to constant light from day 1. Results are means ± S.E. The numbers in parentheses are the numbers of rats.
[b] (L/O) Light–dark (obscurity) cycle; (L/L) constant illumination.

inhibition of the LH response to gonadectomy by hyperprolactinemia is more marked in young animals (Fig. 3) than in older ones (Fig. 2). Shaar *et al.* (1975) described their finding that old female rats, having higher plasma PRL levels, responded with a smaller increase of LH to ovariectomy then young females. Males, with no different PRL levels in relation to age, responded more similarly than the females, although the increases after castration were also diminished with age. Grandison *et al.* (1977) studied the effects of hyperprolactinemia induced by different procedures on the kinetics of the postcastration LH increases in male and female rats. Regardless of the procedure utilized to generate the hyperprolactinemia, the postcastration rise of LH was partially inhibited. The LH

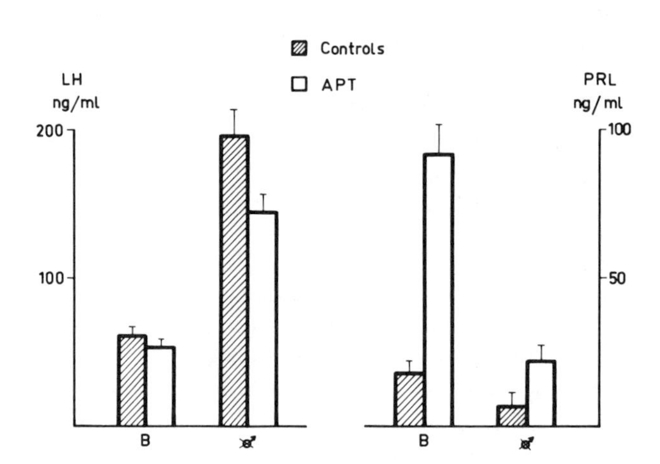

Figure 2. Plasma LH and PRL concentrations before and 8 days after castration of 170-day-old male rats. (APT) Animals received a pituitary graft under the kidney capsule at 90 days of age. From Esquifino (1979).

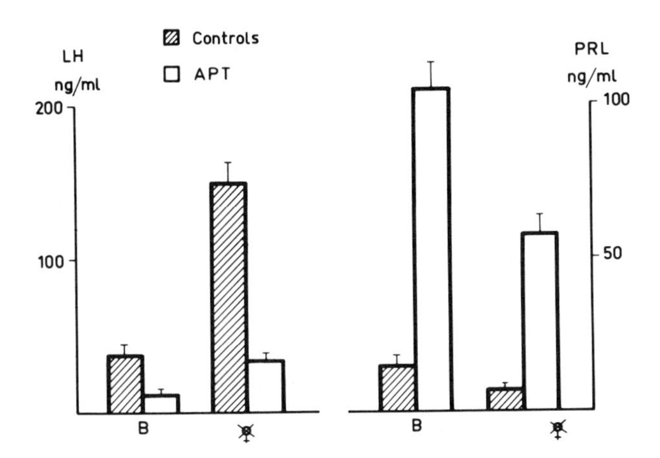

Figure 3. Plasma LH and PRL concentrations before and 8 days after ovariectomy of 90-day-old female rats. (APT) Animals received a pituitary graft under the kidney capsule at 30 days of age. From Esquifino (1979).

suppression was greater at the earlier phases following castration, the inhibitory effect of PRL diminishing with time. Basically the same results were obtained by Winters and Loriaux (1978), using pituitary-transplanted male rats. These authors studied the acute responses to castration and found a negative relationship between plasma PRL concentration and LH levels attained 24 hr after castration. Greeley and Kizer (1979) have recently reported that LH but not FSH levels 5–6 days after castration were diminished in hyperprolactinemic pituitary-bearing male rats and that adrenalectomy also diminished the FSH response to castration in the hyperprolactinemic animals.

The effects of the physiological PRL hypersecretion during lactation on the gonadotropin response to castration were studied by Ford and Melampy (1973) and Hammonns *et al.* (1973). The postcastration rise of LH and FSH can be diminished or even suppresed depending on the litter size and therefore on the magnitude of the suckling stimulation. On removal of the pups, the gonadotropin levels rapidly increased toward the high values of the control nonlactating ovariectomized mothers.

In postmenopausal women, this inhibitory effect of PRL can also be evidenced. Mancini *et al.* (1975) were able to demonstrate that the high plasma levels of LH and FSH due to the follicular failure in postmenopausal women could be diminished by producing hyperprolactinemia with sulpiride.

If PRL appears to block the release of gonadotropins when sex hormones are absent, no clear effect can be demonstrated on the capability of gonadal steroids to inhibit gonadotropin secretion. London *et al.* (1977) found that women with hyperprolactinemic amenorrhea had a normal negative-feedback mechanism evidenced by the acute suppression of circulating LH and FSH levels after estrogen administration and the increase of urinary estrogens after clomiphen. Furthermore, L'Hermite *et al.* (1978) have recently shown, by inducing hyperprolactinemia with sulpiride in 4 normally cycling women, that the estrogen negative-feedback mechanism regulating the tonic secretion of gonadotropins was not only intact but also seemed to be sensitized.

4. Prolactin and the Positive Feedback of Estrogens on Gonadotropin Secretion

The central reproductive feature in the female is the ovulation of fecundable ova. This phenomenon is caused by a preovulatory secretory burst of LH triggered by the increased follicular estrogen secretion. In the female, a preovulatory type of LH secretion can be experimentally induced by estrogen administration, and therefore the functioning of this positive-feedback mechanism can easily be explored. Since augmented PRL secretion is associated with anovulatory syndromes in experimental animals as well as women, it seems that PRL may be playing some modulatory role in the estrogen positive-feedback mechanism regulating the cyclic secretion of gonadotropins. Elevated basal PRL levels have been described in the anovulatory syndromes of most different origins in the rat, such as spontaneous age-dependent anovulation (Ratner and Peake, 1974), neonatal androgenization (Mallampatti and Johnson, 1974) or estrogenization (Aguilar, 1979) (see Table I), suprachiasmatic lesions (Bishop *et al.*, 1972), frontal hypothalamic deafferentation (Caligaris and Taleisnik, 1976), constant illumination (Aguilar, 1979) (see Table I), and so forth. Since in most of these anovulatory syndromes the animals are in a constant estrous situation, one could think of the hyperprolactinemia as being due to an estrogenic effect, but as a matter of fact, elevated estrogen levels cannot always be detected (Mallampatti and Johnson, 1974), while in some instances changes in hypothalamic neurotransmitter concentrations have been measured (Kledzik and Meites, 1974).

Figures 4 and 5 show LH and PRL response to estrogens in rats presenting an anovulatory syndrome of two different origins. The testing was done in 110-day-old animals, ovariectomized 10 days before, by administration of 75 μg estradiol benzoate at 10:00 hr. Normal rats first exhibit the negative-feedback mechanism by a lowering of LH level during the first 48 hr after estrogen administration, but they show an LH peak at 17:00 hr of the second day (55 hr post-estrogen injection). The

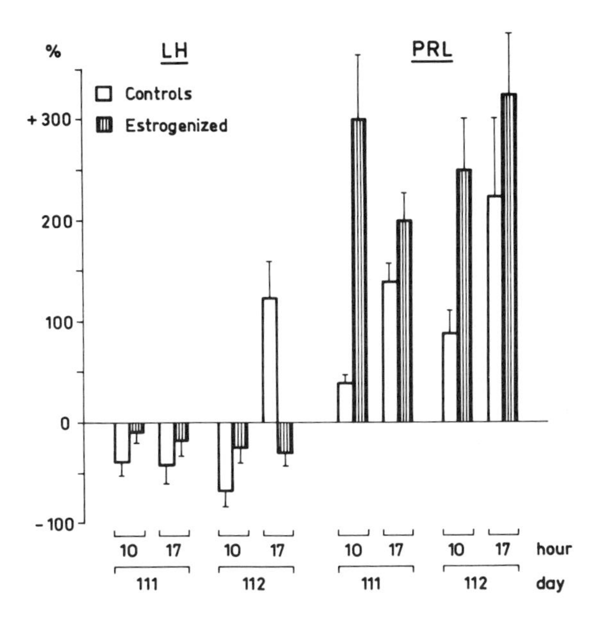

Figure 4. Changes of plasma LH and PRL concentrations expressed as percentages of the basal values after administration of 75 μg estradiol benzoate to 110-day-old female rats ovariectomized at 100 days of age. Estrogenized rats received 100 μg estradiol benzoate on day 5. From Vaticon *et al.* (1979).

PRL response to estrogen in these ovariectomized animals shows an increase with peaks at 17:00 hr on the first and second days. Although constant-illuminated rats develop hyperprolactinemia with their anovulatory syndrome, the ovariectomy decreases their basal PRL level (Table I). Estrogen challenge is followed by an increased PRL response in constant-illuminated acyclic rats and by an abolished LH response (Fig. 4). The phenomenon is reversible, since 15 days after the experimental rats had been returned to the normal light–dark regime, the augmented PRL response had begun to decrease and LH-positive response response reappeared, although with amplified magnitude and an advance of the response toward the first day. According to Meninin and Gorski (1975), the disappearance of the positive-feedback response of LH could be the cause of the anovulatory syndrome in constant-illuminated rats. Since the LH response recovers on return to the normal light–dark regime, it also explains the recovery of the estrous cycle (Hoffman and Cullin, 1975). Mann *et al.* (1976) also detected increased PRL responses to estrogen in this syndrome. The question that arises here is the causal role that hyperprolactinemia may play in this alteration of the LH cyclic regulatory mechanism.

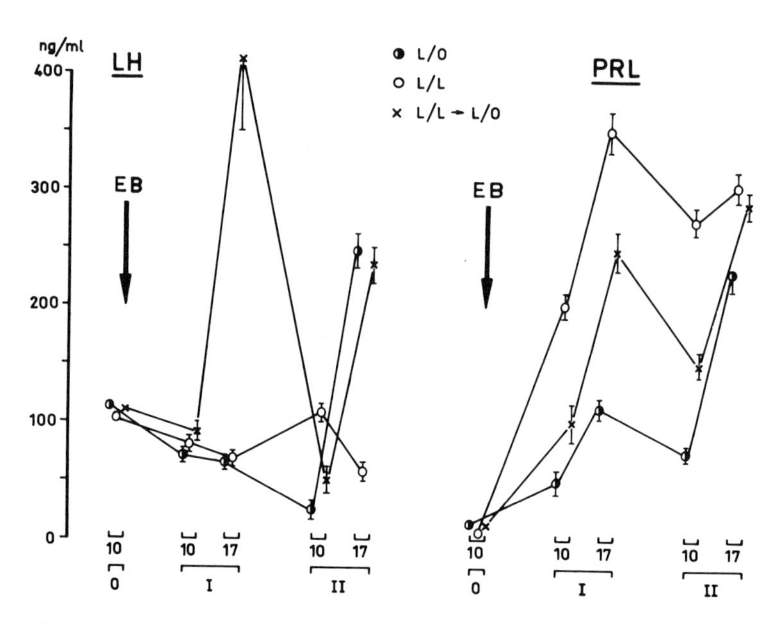

Figure 5. Variations of plasma LH and PRL concentrations in 110-day-old control female rats (L/O), in rats constant-illuminated from day 1 (L/L), or in rats constant-illuminated 15 days after being returned to a 12-hr light–dark (obscurity) cycle (L/L → L/O). From Fernandez-Galaz *et al.* (1979).

In Fig. 5, the results obtained in control and neonatally estrogenized female rats are presented. The LH and PRL plasma levels are expressed as percentages of the basal values before estrogen administration. The neonatally estrogenized females show an increased and erratic PRL response and an absent positive LH response. Basically the same results were obtained in hyperprolactinemic, pituitary-grafted female rats (Esquifino, 1979). In neonatally androgenized cyclic females, the LH response to estrogen also disappears, while PRL response is maintained (Puig-Duran and McKinnon, 1975). It will be of interest to observe whether correction of the hyperprolactinemia in this kind of animals can improve their responsiveness to estrogens and thereby restore their normal estrous cyclicity. Chatterton *et al.* (1975) were able to block ovulation in rats, presumably by interfering with the proestrous LH surge, by generating hyperprolactinemia with perphenazine. In lactating ewes, the positive gonadotropin response to an estrogen challenge can be elicited only if the hyperprolactinemia is corrected with bromocriptine (Kann *et al.*, 1977). In spayed ewes, the positive feedback can also be abolished if hyperprolactinemia is induced with thyrotropin-releasing hormone (TRH) (Kann *et al.*, 1977).

In hyperprolactinemic–amenorrheic women, Glass *et al.* (1975) reported that the gonadotropin rise induced by estrogen administration was absent. Aono *et al.* (1976a) confirmed this finding. Faglia *et al.* (1977) observed that surgical or pharmacological correction of hyperprolactinemia in six amenorrheic women allowed tbem to recover the previously absent positive feedback. Progesterone can also induce a positive gonadotropin response in estrogen-primed women during the early follicular phase. Rakoff *et al.* (1978) have shown that women with normogonadotropic hypothalamic chronic anovulation with hyperprolactinemia do not show the gonadotropin increase, but when the hyperprolactinemia is corrected by surgical removal of the pituitary microadenoma, this positive response is recovered.

Hyperprolactinemic–amenorrheic women with or without galactorrhea generally do not ovulate in response to clomiphen treatment (Thorner *et al.*, 1974; Tyson *et al.*, 1975; Jacobs *et al.*, 1976; Dickey and Stone, 1976; Seki *et al.*, 1976; Bohnet and Schneider, 1977). Since the women usually respond with increased urinary estrogen excretion, the failure must reside in the central mechanism capable of generating the increase in gonadotropin secretion (London *et al.*, 1977).

The administration of bromocriptine to lactating women shortens the amenorrheic period with an increase of the endogenous estrogen production that starts around day 12. LH levels begin to increase between days 16 and 23, pointing toward the recovery of the positive estrogenic feedback by the correction of the hyperprolactinemia (Rolland *et al.*, 1975a,b).

The induction of hyperprolactinemia in women with sulpiride diminishes the magnitude of the spontaneous LH ovulatory peak in normal volunteers (Robyn *et al.*, 1976) and abolishes gonadotropin response to estrogens (L'Hermite *et al.*, 1978). On the other hand, Weiss *et al.* (1977) induced moderate hyperprolactinemia with haloperidol, fluphenezine, or chlorpromazine without interfering with the positive response to estrogens.

5. *Prolactin and the Pituitary Responsiveness to LH-RH Administration*

To establish whether the pituitary could be the level at which PRL may modulate the gonadotropin-regulatory mechanism, a number of authors have studied the effects of hyperprolactinemic states in women and experimental animals on the so-called pituitary gonadotropin reserve, measured through the changes in plasma LH and FSH concentration after LH-RH administration.

Figures 6 and 8 show plasma LH responses to LH-RH in hyperprolactinemic rats, while Figs. 7 and 9 show the plasma PRL concentrations measured in the same samples. Anovulatory female rats (Fig. 6) previously ovariectomized show no impairment of their pituitary gonadotropin responsiveness to LH-RH. Neonatally estrogenized females, starting from a somewhat lower basal LH level, showed the same percentage increments as their controls. These results coincide with those of Uilenbroek and Gribbing-Hegge (1977). On the other hand, Debeljuk *et al.* (1975) and Castro-Vazquez and McCann (1975), using intact animals, observed increased LH response to LH-RH.

In ovariectomized female rats exposed to continuous illumination, the LH response to LH-RH was not significantly different from the response measured in the light–dark maintained controls.

On the contrary, intact adult male rats that bear a pituitary graft under the kidney capsule transplanted at 90 days of age show a diminished LH response to LH-RH administered after 40 or 80 days of hyperprolactinemia (Fig. 8). The same results were obtained when pituitaries were grafted prepubertally in male or female rats. Correction of hyperprolactinemia with bromocriptine administration, while it did not change the

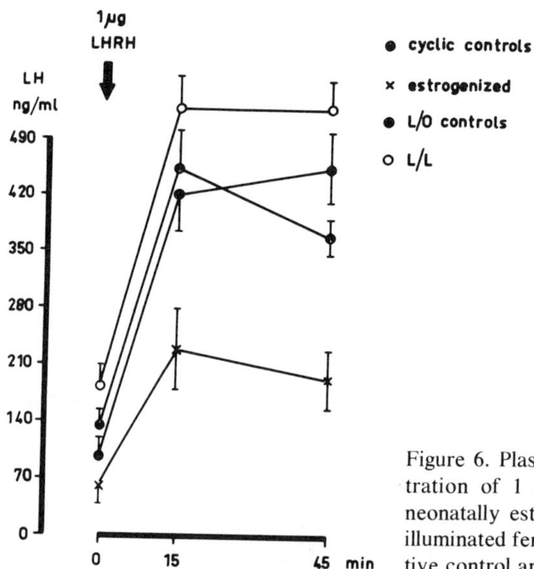

Figure 6. Plasma LH variations after administration of 1 μg LH-RH to ovariectomized, neonatally estrogenized female rats, constant-illuminated female rats (L/L), and their respective control animals. From Aguilar (1979).

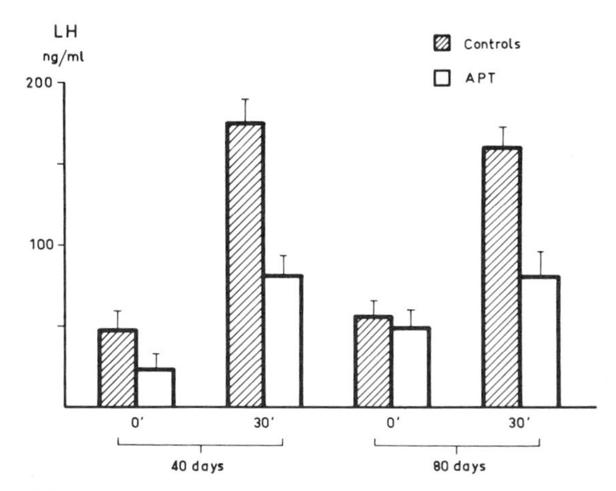

Figure 7. Plasma LH concentration before and 30 min after administration of 1 μg LH-RH to young male adult rats 40 and 80 days after they had received a pituitary graft under the kidney capsule on day 90. From Esquifino (1979).

LH response in control animals, improved it in the animals bearing a pituitary graft (Esquifino and Tresguerres, 1979; Tresguerres and Esquifino, 1979).

Winters and Loriaux (1978) and Greeley and Kizer (1979) also reported impaired LH responses to LH-RH in pituitary-grafted male rats. Greeley and Kizer (1979), administering PRL before the LH-RH test, also observed a reduction of the LH response that was not evident

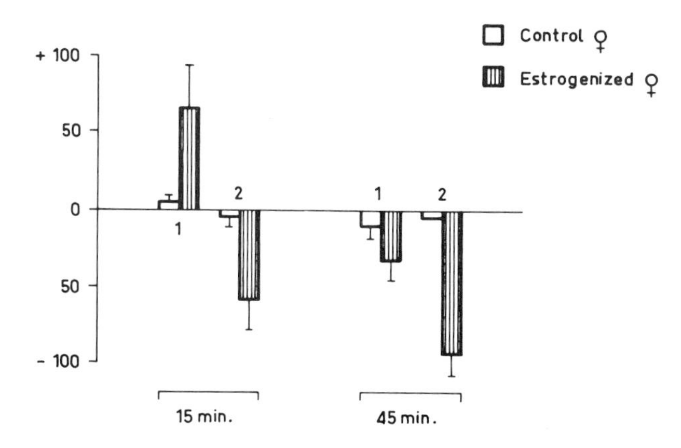

Figure 8. Variations in plasma PRL concentrations expressed as percentages of the basal values after administration of 1 μg LH-RH in control and neonatally estrogenized female rats. From Aguilar (1979).

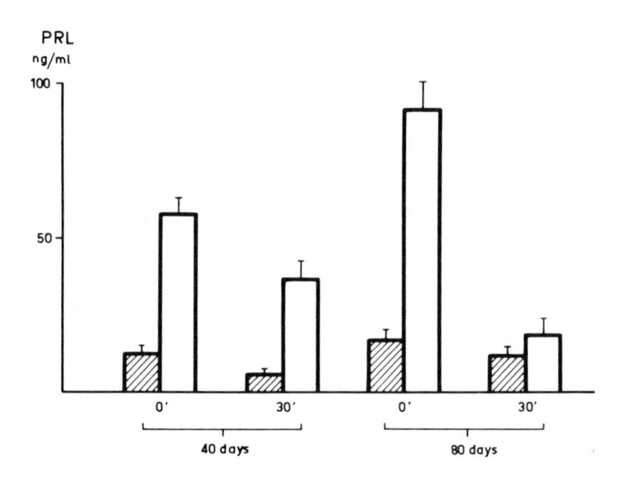

Figure 9. Plasma PRL concentration before and 30 min after administration of 1 μg LH-RH to young male adult rats 40 or 80 days after they had received a pituitary graft under the kidney capsule on day 90. From Esquifino (1979).

if the rats were previously adrenalectomized. On the other hand, Grandison *et al.* (1977) did not detect a decrease in pituitary sensitivity to LH-RH in ovariectomized, pituitary-grafted rats.

In control animals, plasma PRL levels are not modified by LH-RH administration (Ceballos and McCann, 1974), but we observed that in hyperprolactinemic rats, LH-RH induced a decrease in circulating PRL (Figs. 7 and 9).

In amenorrheic–hyperprolactinemic women with or without galactorrhea, the responses to LH-RH reported in the literature ranged from subnormal to elevated. Rjosk *et al.* (1976), Bergh *et al.* (1978), and Spellacy *et al.* (1978) predominantly found in their series normal gonadotropin responses to LH-RH.

Thorner *et al.* (1974), Jacobs *et al.* (1976), and Corbey *et al.* (1977), among others, found elevated LH and FSH responses in hyperprolactinemic–amenorrheic patients. Bergh *et al.* (1978) also observed higher responses, but only in those hyperprolactinemic women with normal sellae. Seppälä *et al.* (1975), Wiebe *et al.* (1976), and Chang *et al.* (1977) found increased LH responses, while the FSH responses were normal, and Lachelin *et al.* (1977b) reported increased FSH but normal LH responses in these patients. Thorner *et al.* (1974), Zarate *et al.* (1975), Glass *et al.* (1975), Tolis *et al.* (1975), Aono *et al.* (1976b), Wiebe *et al.* (1976), Van Campenhout *et al.* (1977), March *et al.* (1977), Lachelin *et al.* (1977b), and Spellacy *et al.* (1978) reported subnormal responses in some instances. Finally, while Agrain *et al.* (1977) reported a negative correlation between PRL and gonadotropin responses to LH-RH in hyperprolactinemic patients, Mortimer *et al.* (1973), Archer *et al.* (1976), Van Campenhout *et al.* (1977), and Wiebe *et al.* (1977) found no correlation whatsoever in their patients.

Since pituitary responsiveness to exogenous LH-RH is markedly modulated by sex steroids and LH-RH itself, there should be no surprise at the scattering of the data reported in the literature. The contradictory reported results probably reflect the differences among individual subjects presenting hyperprolactinemic amenorrhea of their endogenous estrogen production as well as their own LH-RH secretion. One is tempted to consider the human pituitary delicately modulated through the rate of endogenous LH-RH and estrogen secretion and by a mainly indirect hypothalamic PRL action. Judd *et al.* (1978), for example, reported that dopamine (DA) reduces the gonadotropin response to LH-RH the more, the higher the endogenous estrogen levels are. Acute hyperprolactinemia induced by sulpiride (L'Hermite *et al.*, 1978), TRH, hypoglycemic stress, or metoclopramide (Thorner *et al.*, 1977) does not modify the gonadotropin response to LH-RH.

The self-priming effect of LH-RH was studied by Lu *et al.* (1976a) in the rat, and it was observed that during the postpartum period, it was less evident than in the animals in diestrus. Muralidhar *et al.* (1977) observed an inverse relationship between the number of suckling pups and the response to LH-RH in the lactating mother. PRL administration prior to LH-RH in rats nursing only two pups inhibited the response to LH-RH. Smith (1978) described shorter gonadotropin-secretion periods after LH-RH in lactating rats than in diestrous day 1 and 2 cycling rats.

In lactating primates, the response of LH to LH-RH was improved when nursing was interrupted, and administration of PRL before LH-RH also reduced LH responses in lactating monkeys (Maneckjee *et al.*, 1976). In contrast, Kann *et al.* (1977) observed that in the postpartum ewe, there is a normal response to LH-RH, and neither bromocriptine nor TRH administration changes it in spayed animals.

In women during postpartum, there is a refractory period to LH-RH that has been interpreted by some as an indication of a central action of PRL (Andreassen and Tyson, 1976).

Keye and Jaffe (1976) studied the kinetics of the responses to LH-RH in postpartum nonnursing mothers and described a smaller response at the 2nd postpartum week and an exaggerated one from the 5th to the 8th weeks in comparison to the responses obtained in the early follicular phase. Since the postpartum pattern of response to LH-RH was essentially identical in these nonlactating mothers to those reported in lactating women (Lemaire *et al.*, 1974), the refractory and hyperresponsive periods described probably do not depend on the endogenous PRL levels.

6. Role of Hypothalamic Neurotransmitters in Prolactin and Gonadotropin Secretion

Since the pituitary responsiveness to LH-RH does not seem to be exclusively responsible for the PRL modulation of gonadotropin secretion, the hypothalamic level of regulatory integration has to be implicated in the process. The state of our knowledge of the neurotransmitter organization in the complex hypothalamic structures does not yet allow us to have a very clear picture of how it functions, since the available information up to now is full, not only of gaps, but also of apparently contradictory data. Although virtually all neurotransmitters have been implicated in the interrelationship between PRL and gonadotropin neuroregulation, much emphasis has been placed on DA. Indeed, it appears that the possible influences of other neurotransmitters are exerted through modulation of the dopaminergic systems. An attempt is made in this section to summarize the current information on the role of DA as a modulator of PRL and gonadotropin secretion and its physiological and clinical implications, with reference to other neurotransmitters only in regard to their effect on dopaminergic activity.

In the human, PRL levels decline during the few weeks after birth and remain fairly constant until the time of puberty, when the rise in estrogen secretion in the female generates an increase of PRL secretion. PRL secretion is predominantly inhibited by the CNS through one or more PRL-inhibiting factor(s) (PIF) (Boyd and Reichlin, 1978).

In the rat, PRL levels increase form day 0 until day 35, remaining constant thereafter (Ojeda and McCann, 1974). The gonadotropins, on the contrary, follow the opposite pattern, LH and FSH levels being higher in the first days of life than after puberty. This situation could be explained by the interactions of at least two modulatory systems: the positive feedback of estrogens to PRL and the negative feedback of the dopaminergic systems that mature at different rates.

The first PRL modulatory system starts its function at 11 days of age (Ojeda and McCann, 1974), the response being maximal at 35–37 days of age. This is in correspondence with the appearance of estrogen-binding sites of the hypothalamus, which increase dramatically between days 21 and 25 of life (Plapinger and McEwen, 1973) and could explain the increase in PRL secretion over this period. The gonadotropins, on the contrary, are subject mainly to negative feedback by the estrogens, which could explain the reduction of LH and FSH observed at the same time.

The second regulatory mechanism starts to work earlier, as reported by Ojeda and McCann (1973). At 3 days of age, a reduction in PRL levels was found in response to DA administration in the rat, and reached its maximal sensitivity again at 35 days of age.

This inhibitory effect of DA on prolactin has been clearly demonstrated by many authors with the administration of DA itself (Donoso *et al.*, 1971) or of DA precursors (Lu and Meites, 1972) or their agonists (Lachelin *et al.*, 1977a) or antagonists (Quigley *et al.*, 1979).

It has been demonstrated that section of the pituitary stalk increases PRL secretion in experimental animals and in humans (Kanematsu and Sawyer, 1973; Turkington *et al.*, 1971), suggesting the presence of a PIF at the suprasellar level. This PIF is identified by many researchers with DA, although brain extracts devoid of DA activity are also capable of reducing PRL secretion (Schally *et al.*, 1976). DA may serve as a common link in the regulation of PRL, LH, and FSH as suggested by Rakoff *et al.* (1978) through the tuberoinfundibular dopaminergic (TIDA) system described by Fuxi and Hökfelt (1969). DA terminal axons end in the perivascular spaces of the primary portal capillaries at the median eminence (ME), where DA is secreted into the portal blood and carried to the anterior pituitary gland, where specific DA receptors have been identified (Brown *et al.*, 1976; Caron *et al.*, 1978; McLeod *et al.*, 1976). On the other hand, the DA effects on gonadotropin secretion appear to be exerted centrally by means of axo–axonic contacts between DA and LH-RH terminals in the ME.

The inhibitory effect of DA on gonadotropin secretion has been found in several species. In the rat, Uemura and Kobayashi (1971) reported a reduction in gonadotropin secretion following implantation of DA in the ME. Sawyer *et al.* (1974) also found a decrease in gonadotropins on injecting DA intraventricularly into the rabbit. The same results were found in humans by Leblanc *et al.* (1976) after DA infusion and by Lachelin *et al.* (1977a) after the administration of a DA agonist. Recently, Quigley *et al.* (1979) observed a dose-related increase of LH and FSH after metoclopramide, a DA antagonist, in hyperprolactinemic, anovulatory women. This effect was interpreted by the authors as a demonstration that the hypogonadotropic status in these women was due to a high tonic inhibition by DA of LH-RH secretion. Vijayan and McCann (1978) reported a reduction in LH with high intraperitoneal doses of DA and an enhancement of LH after intravenous administration. This could be due to a modulatory effect of sex steroids on gonadotropin secretion, since the stimulation of LH occurs at low intravenous doses of DA in estrogen/progesterone-primed rats with reduced gonadotropin levels, while high DA doses reduced LH in nonprimed, ovariectomized rats showing elevated gonadotropins.

On the other hand, findings of Kamberi *et al.* (1970) showed that the intraventricular administration of DA increased plasma LH in the rat.

This effect was not present if the monoamine was administered directly into the pituitary or into the hypothalamic pituitary stalk. Rotszentejn *et al.* (1976) confirmed this finding, reporting an increase of *in vitro* release of LH-RH from rat medial basal hypothalamus (MBH), and later, Vernes and Telegdy (1979) and Ojeda *et al.* (1979) also reported increased LH-RH after DA *in vitro*, this effect being more marked in the ME than in the MBH. They postulated that this effect was mediated through prostaglandin E acting on the ME terminals of LH-RH-secreting neurons, since indomethacin, a very potent prostaglandin inhibitor, reduces the DA effect.

In terms of PRL, a short-loop feedback system may be operative where PRL excess promotes an increased DA turnover in the ME (Fuxe *et al.,* 1976) or an increased DA synthesis in the MBH and striatum, as Perkins *et al.* (1979) have shown in rats grafted with PRL-secreting tumors or treated with trifluoperazine. Perkins and Westfall (1978) also showed a marked enhancement by PRL of the electrically induced release of DA in MBH slices. Gudelski *et al.* (1977) showed an increase in DA content of the ME in the female rat 4 weeks after ovariectomy, a situation in which PRL levels are very low (Esquifino and Tresguerres, 1979), and the reversal of the enhanced DA levels after 7 days of treatment with estradiol benzoate, which elevates PRL levels. Taking into account that gonadotropin levels are high after ovariectomy and considering the experiments in which DA reduces gonadotropins (Uemura and Kobayashi, 1971; Sawyer *et al.,* 1974; Leblanc *et al.,* 1976), one could also speculate on the possibility of a gonadotropin–DA feedback.

The mode of action of the dopaminergic system on gonadotropin control nevertheless remains uncertain. From the data reported above showing the ineffectiveness of DA injected into the pituitary-stalk vessels or directly into the pituitary, a hypothalamic mechanism appears to be necessary (Kamberi *et al.,* 1970). On the other hand, a direct pituitary effect of DA on PRL seems well established, since L-dopa, a DA precursor, has been shown to be effective in suppressing PRL after pituitary-stalk section (Diefenbach *et al.,* 1976) and Lu and Meites (1972) demonstrated DA inhibition of PRL in pituitary-grafted, hypophysectomized rats. Bromocriptine stimulates DA receptors in the CNS, but also seems to act directly on the pituitary, since the stimulation of PRL by TRH, a direct action at the pituitary level, is also blocked by this drug (Del Pozo *et al.,* 1973).

Furthermore, bromocriptine is also able to inhibit PRL secretion of pituitary grafts located under the kidney capsule of rats, evidencing the existence of DA receptors in the pituitary, as has been experimentally supported also by Levin and Voogt (1978).

There is morphological and biochemical evidence in support of a double dopaminergic mechanism. The TIDA system has DA terminal axons ending at the perivascular spaces of the primary portal capillaries in the ME, where DA is secreted into the portal blood and carried down to the anterior pituitary gland, where specific DA receptors have also been identified (Brown *et al.,* 1976; Caron *et al.,* 1978; McLeod *et al.,* 1976). On the other hand, as noted above, the DA effects on gonadotropins could be exerted through the axo–axonic contacts between DA and LH-RH terminals present in the ME. Accordingly, situations are beginning to be described in which dissociation of DA effects at both levels, hypothalamic and pituitary, may occur (Quigley *et al.,* 1979; Scanlon *et al.,* 1979).

7. Conclusions

In many different circumstances, PRL and gonadotropin secretion seem to be organized by a reciprocal regulatory system. The data obtained cannot be explained solely by an effect of PRL on the pituitary responsiveness to LH-RH, and they point toward a suprapituitary inhibitory action on the tonic as well as cyclic regulatory mechanisms of gonadotropin secretion. On the other hand, hyperprolactinemic animals show a decrease of PRL secretion under stimulated gonadotropin secretion after LH-RH administration.

At present, we can best organize our knowledge and interpret our results as summarized in Fig. 10. The tuberoinfundibular dopaminergic (TIDA) system should inhibit LH-RH secretion as regulated by either the tonic or the cyclic centers where estrogens exert their negative- or

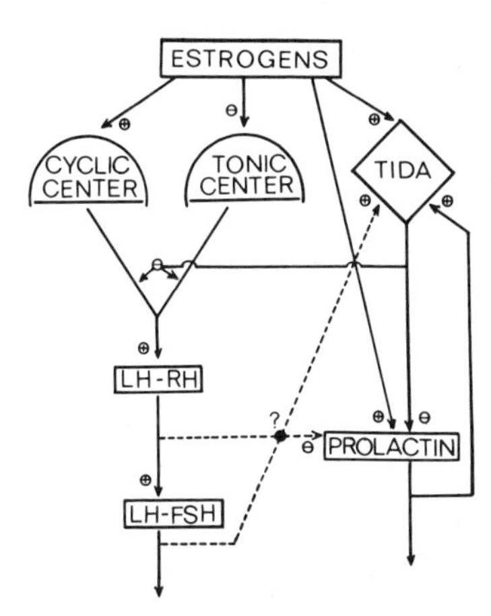

Figure 10. A proposed model for the interregulation of PRL and gonadotropins through the TIDA system.

positive-feedback regulation. LH-RH, either by itself or through the increased gonadotropin secretion it generates, inhibits PRL secretion either directly at the pituitary level or, more appealing in our view, through stimulation of the TIDA system. In the latter case, the gonadotropin autoregulatory short-loop feedback mechanism would be involving the TIDA system.

ACKNOWLEDGMENTS. This work has been made possible through Grant 2795 from the Comisión Asesora de Investigación Cientifica y Tecnica. One of the authors (A.E.) has been supported by the Caja de Ahorros y Monte de Piedad de Madrid. The National Pituitary Agency of the NIAMDD has supplied us with the RIA material for the PRL, LH, and FSH determinations. Dr. A. Gomez-Pan has kindly reviewed the manuscript and discussed it with us. We also thank our colleagues, E. Aguilar, A. Tejero, M. C. Fernandez-Galaz, and M. D. Vaticon, from the Department of Physiology, for their cooperation.

DISCUSSION

BOGDANOVE: Is there, in fact, good evidence that it is circulating PRL itself, and not the process of PRL hypersecretion, that "desensitizes" the LH cell's response to LH-RH? For example, can the LH-release response be dumped by injection of exogenous PRL? The reason I wonder lies in the anatomical intimacy of the gonadotroph and the PRL cell. As Nakare showed many years ago, the PRL cell not only makes a "maternal" hormone but also, itself, embraces the PRL cell in a more or less maternal fashion. Conceivably, it could thereby act as a barrier between the gonadotroph and the outside world and might not be "desensitizing" the cell but simply be denying it access to the impingement of LHRH. Is there evidence to refute this possibility?

Table II. Total Pituitary PRL and LH Content in Male and Female Rats[a]

Animal group	Number of rats	Total pituitary content (μg/hypophysis) PRL	LH
Male			
Experimental	5	2.5 ± 0.5	8 ± 0.3
Control	5	11 ± 0.5	13.5 ± 0.5
Female			
Experimental	4	1.6 ± 0.6	8.8 ± 0.6
Control	6	15.6 ± 0.7	15.8 ± 0.75

[a] The pituitary hormone content was measured in control and pituitary-grafted rats at 100 days of age, 70 days after the operation.

C. A. VILLEE: How do you visualize the effect of prolactin on the pituitary?

ORIOL-BOSCH: It has been shown that PRL administration prior to LH-RH in rats diminishes the gonadotropin response (Muralidhar *et al.,* 1977). Our hyperprolactinemic animals bearing a pituitary graft under the kidney capsule showed a reduction of their own *in situ* pituitary function regarding both PRL and gonadotropin as reflected by the diminished hormonal content (see Table II). Under these circumstances, although a direct action of PRL on the gonadotroph and lactotrophs from the *in situ* pituitary cannot be excluded, the situation is certainly different from the one in animals in which the hyperprolactinemia originates from the lactotrophs located near or around the gonadotrophs. For these reasons, we are inclined to prefer the alternative hypothesis of a common cause to explain the pituitary effects of the "extrapituitary" hyperprolactinemia. The only reasonable mechanism we could think of involved the suprahypophyseal TIDA system as we have postulated previously.

REFERENCES

Agrain, R., Russell, V., Toblowsky, P., and Vaitukaitis, J., 1977, New evidence for estradiol modulation of intergrafted pituitary gonadotropin and prolactin responses among women with hypothalamic amenorrhea, *Clin. Res. 25:289A.*

Aguilar, E., 1979, Regulación de la secreción de LH y prolactina en cuadros anovulatorios experimentales, *Fdn. J. March, Serie Univer.* 93.

Amenomori, Y., Chen, C.S., and Meites, J., 1970, Serum prolactin levels in rats during different reproductive states, *Endocrinology* **86:**506.

Andreassen, B., and Tyson, J.E., 1976, Role of hypothalamic–pituitary–ovarian axis in puerperal infertility, *J. Clin. Endocrionl. Metab.* **42:**696–702.

Aono, T., Miyake, A., Shioji, T., Kinugasa, T., Onishi, T., and Kurachi, K., 1976a, Impaired LH release following endogenous estrogen administration in patients with amenorrhea–galactorrhea syndrome, *J. Clin. Endocrinol. Metab.* **42:**696–702.

Aono, T., Shioji, T., Kohno, H., Veda, G., and Kurachi, K., 1976b, Pregnancy following 2-bromo-alpha-ergocryptine (CB-154) induced ovulation in an acromegalic patient with galactorrhea and amenorrhea, *Fertil. Steril.* **27:**341–344.

Archer, D.F., Sprung, J.W., Nankin, H.R., and Jojimovich, J.B., 1976, Pituitary gonadotropin response in women with idiopathic hyperprolactinemia, *Fertil. Steril.* **27:**1158.

Bartke, A., Smith, M.S., Michael, S.D., Peron, F.G., and Dalterio, S., 1977, Effects of experimentally induced chronic hyperprolactinemia on testosterone and gonadotropin levels in male rats and mice, *Endocrinology* **100:**182.

Bergh, T., Nillius, S.J., and Wide, L., 1978, Serum prolactin and gonadotrophin levels before and after luteinizing hormone-releasing hormone in the investigation of amenorrhea, *Br. J. Obstet. Gynaecol.* **85:**945–956.

Bishop, W., Kalra, P.S., Fawcett, C.P., Krulich, L., and McCann, S.M., 1972, The effects of hypothalamic lessions on the release of gonadotrophins and prolactin in response to estrogen and progesterone treatment in female rats, *Endocrinology* **91:**1404.

Bohnet, H.G., and Schneider, H.P.G., 1977, Prolactin as cause of anovulation, in: *Prolactin and Human Reproduction* (P.G. Crosignani and C. Robyn, eds.), pp. 153–159, Academic Press, New York.

Bohnet, H.G., Dahlen, H.G., and Schneider, H.P.G., 1975a, Influence of prolactin concentration on the tonic LH release (spiking), *Arch. Gynaekol.* **219:**592–593.

Bohnet, H.G., Wiest, H.J., Dahlen, H.G., and Schneider, H.P.G., 1975b, Pulsatile LH fluctuation (spiking) dependent on the circulating prolactin: Studies during physiological (puerperium), functional pathological and TRH induced hyperprolactinemia, *Endocrinology* **66:** 158–172.

Boyd, A.G., III, and Reichlin, S., 1978, Neural control of prolactin secretion in man, *Psychoneuroendocrinolgy* **3:** 113.

Brown, G.M., Seeman, P., and Lee, T., 1976, Dopamine-neuroleptic receptors in basal hypothalamus, *Endocrinology* **99:** 1407.

Caligaris, L., and Taleisnik, S., 1976, The role of hypothalamic afferents in the release of prolactin induced by ovarian steroids, *Neuroendocrinology* **21:** 139.

Caron, M.G., Beaulieu, M., Raymond, V., Gagne, B., Drovin, J., Lefkuwitz, R.J., and Labrie, F., 1978, Dopaminergic receptors in the anterior pituitary gland, *J. Biol. Chem.* **253:**2244.

Castro-Vazquez, A., and McCann, S.M., 1975, Responsiveness of the pituitary to LHRH induced by LHRH, *Endocrinology* **97:** 13.

Ceballos, G., and McCann, S.M., 1974, The effect of subcutaneous administration of luteinizing hormone releasing factor on plasma gonadotrophins and PRL in the rat, *Proc. Soc. Expt. Biol. Med.* **145:**415.

Chang, R.J., Keye, W.R., Jr., Young, J.R., Wilson, C.B., and Jaffe, R.B., 1977, Detection, evaluation and treatment of pituitary microadenomas in patients with galactorrhea and amenorrhea, *Am. J. Obstet. Gynecol.* **128:**356.

Chatterton, R.T., Chien, J.L., Ward, D.A., and Miller, T.T., 1975, Ovarian responses to perphenazine induced prolactin secretion in the rats: Effect of ovulation, stress and steroids, *Proc. Soc. Expt. Biol. Med.* **149:** 1027.

Corbey, R.S., Lequin, R.H., and Rolland, R., 1977, Hyperprolactinemia and secondary amenorrhea, in: *Prolactin and Human Reproduction* (P.G. Crosignani and C. Robyn, eds.) pp. 203–215, Academic Press, New York.

Debeljuk, L., Rozados, S., Daskal, H., and Villepas-Velez, C., 1975, Variation of the pituitary response to LHRH during a 24-hour period in male, diestrous female and androgenized female rats, *Neuroendocrinology* **17:**48.

Del Pozo, E., Friesen, H., and Burmeister, P., 1973, Endocrine profile of a specific prolactin inhibitor: BR-ergocriptine (CB-154): A preliminary report, *Schweiz. Med. Wochenschr.* **103:**847.

Delvoye, P., Demaegd, M., Uwayitu-Nyampeta., and Robyn, C., 1978, Serum prolactin, gonadotrophins and estradiol in menstruating and amenorrheic mothers during two year's lactation, *Am. J. Obstet. Gynecol.* **130:**635.

Dickey, R.P., and Stone, S.C., 1976, Effect of bromo-ergocriptine on serum hPRL, hLH, hFSH and estradiol-17-beta in women with galactorrhea–amenorrhea, *Obstet. Gynecol.* **48:**84–89.

Diefenbach, W.P., Carmel, P.W., Frantz, A., and Ferin, M., 1976, Suppression of PRL secretion by L-DOPA in the stalk sectioned rhesus monkey, *J. Clin. Endocrinol. Metab.* **43:**638.

Donoso, A.O., Bishop, W., Fawcett, C.P., Krulich, L., and McCann, S.M., 1971, Effects of drugs that modify brain monoamine concentrations on plasma gonadotrophin and prolactin levels in the rat, *Endocrinology* **89:**774.

Esquifino, A.I., 1979, Estudio de un modelo de hiperprolactinemia experimental, Doctoral thesis, Universidad Autonoma de Madrid.

Esquifno, A.I., and Tresguerres, J.A.F., 1979, Plasma LH responses in hyperprolactinemic male rats to LHRH before and after bromocriptine treatment. *Acta Endocrinol. Suppl.* **225:** 141.

Faglia, G., Beck-Peccoz, P., Travaglini, P., Ambrosi, B., Rondena, M., Parachi, A., Spada, A., Weber, G., Bara, R., and Bouzin, A., 1977, Functional studies in hyperprolactinemic states, in: *Prolactin and Human Reproduction* (P.G. Crosignani and C. Robyn, eds.), pp. 225–238, Academic Press, New York.

Fang, V.S., Refetoff, S., and Rosenfield, R.L., 1974, Hypogonadism induced by a transplantable, prolactin producing tumor in male rats: Hormonal and morphological studies, *Endocrinology* **95:**991.

Fernandez-Galaz, C., Tejero, A., Esquifino, A., Vaticon, M.D., and Aguilar, E., 1979, Influencia del regimen luminoso sobre los niveles plasmatico de LH y prolactina tras la administración de benzoato de estradiol en ratas hembra, *Endocrinologia* **26:**1.

Ford, J.J., and Melampy, R.M., 1973, Gonadotrophin levels in lactating rats: Effect of ovariectomy, *Endocrinology* **93:**540.

Fuxe, K., and Hökfelt, T., 1969, Catecholamines in the hypothalamus and pituitary gland, in: *Frontiers in Neuroendocrinology* (W. Ganong and L. Martini, eds.), pp. 47–90, Oxford University Press, New York.

Fuxe, K., Hökfelt, T., Agnati, L., Löfstrom, A., Everitt, B.J., Johansson, O., Jonsson, G., Wuttke, W., and Goldspein, M., 1976, in: *Neuroendocrine Regulation of Fertility* (T.C. Anaud-Kuman, ed.), p. 124, S. Karger, Basel.

Glass, M.R., Shaw, R.W., Butt, W.R., Edwards, R.L., and London, D.R., 1975, An abnormality of oestrogen feedback on amenorrhea–galatorrhea, *Br. Med. J.* **3:**274.

Grandison, L., Hodson, C., Chen, H.T., Advis, J., Simpkins, J., and Meites, J., 1977, Inhibition by prolactin of postcastration rise in LH, *Neuroendocrinology* **23:**312–322.

Greeley, G.H., and Kizer, J.S., 1979, Evidence for adrenal involvement in the modulatory role of prolactin in luteinizing hormone secretion in the male rat, *Endocrinology* **104:**948.

Gudelski, G.A., Annunziato, I., and Moore, K.E., 1977, Increase in dopamine content of the rat eminence after long-term ovariectomy and its reversal by estrogen replacement, *Endocrinology* **101:**1894.

Hammons, J.A., Velasco, H., and Rotchild, I., 1973, Effect of sudden withdrawal or increase of suckling on serum LH levels in ovariectomized post parturient rats, *Endocrinology* **92:**206.

Hoffman, J.C., and Cullin, A.M., 1975, Effects of pinealectomy and constant light on the estrous cycles of female rats, *Neuroendocrinology* **17:**167.

Jacobs, H.S., Franks, S., Murray, M.A.F., Hull, M.G.R., Steele, S.J., and Nabarro, J.D.N., 1976, Clinical and endocrine features of hyperprolactinemia, *Clin. Endocrinol.* **5:**439–454.

Judd, S.J., Kakuff, J.J., and Yenn, S.C., 1978, Inhibition of gonadotrophin and prolactin release by dopamine: Effect of endogenous estradiol levels, *J. Clin. Endocrinol. Metab.* **47:**494.

Kamberi, I.A., Mical, R.S., and Porter, J.C., 1970, Effect of anterior pituitary perfusion and intraventricular injection of catecholamines and indolamines on LH release, *Endocrinology* **87:**1.

Kanematsu, S., and Sawyer, C.H., 1973, Elevation of plasma PRL after hypophyseal stalk section in the rat, *Endocrinology* **93:**238.

Kann, G., Martinet, J., and Shirar, A., 1977, Modifications of gonadotrophin secretion during natural and artificial hyperprolactinemia in the ewe, in: *Prolactin and Human Reproduction* (P.G. Crosignani and C. Robyn, eds., pp. 47–59, Academic Press, New York.

Kappen, S., Boyar, R.M., Freeman, R., Frantz, A., Hellman, L., and Weitlman, E.D., 1975, Twenty-four-hour secretory patterns of gonadotrophins and prolactin in a case of Chiari–Fromel syndrome, *J. Clin. Endocrinol. Metab.* **40:**234.

Keye, W.R., and Jaffe, R.B., 1976, Changing patterns of FSH and LH response to gonadotrophin releasing hormone in the puerperium, *J. Clin. Endocrinol. Metab.* **42:**1133.

Kledzik, G.S., and Meites, J., 1974, Reinitiation of estrous cycles in light induced constant estrous female rats by drugs, *Proc. Soc. Expt. Biol. Med.* **146:**989.

Lachelin, G.C.L., Leblanc, H., and Yen, S.S.C., 1977a, The inhibitory effect of dopamine agonists on LH release in women, *J. Clin. Endocrinol. Metab.* **44:**728.

Lachelin, G.C.L., Abu-Fadil, S., and Yen, S.S.C., 1977b, Functional delineation of hyperprolactinemia–amenorrhea, *J. Clin. Endocrinol. Metab.* **44:**1163.

Leblanc, H., Lachelin, G.C.L., Abu-Fadil, S., and Yen, S.S.C., 1976, Effects of dopamine infusion on pituitary hormone secretion in humans, *J. Clin. Endocrinol. Metab.* **43:**668.

Lemaire, W.A., Shapiro, A.G., Rigall, F., and Yang, N.S.T., 1974, Temporary pituitary insensitivity to stimulation by synthetic LRF during the post partum period, *J. Clin. Endocrinol. Metab.* **38:**916.

Levin, J.I., and Voogt, J.L., 1978, Effects of dopamine, norepinephrine, and serotonin on plasma prolactin levels in ovariectomized, pituitary grafted rats, *Proc. Soc. Expt. Biol. Med.* **157:**576.

L'Hermite, M., Delogne-Desnoeck, J., Michaux-Duchene, A., and Robyn, C., 1978, Alteration of feedback mechanism of estrogen on gonadotrophin by sulpiride induced hyperprolactinemia, *J. Clin. Endocrinol. Metab.* **47:**1132.

London, D.R., Glass, M.R., Shaw, R.W., Butt, W.R., and Logan-Edwards, R., 1977, The modulation by ovarian hormone of gonadotrophin release in hyperprolactinemic women, in: *Prolactin and Human Reproduction* (P.G. Crosignani and C. Robyn, eds.), pp. 119–124, Academic Press, New York.

Lu, K.H., and Meites, J., 1972, Effect of L-DOPA on serum prolactin and PIF in intact and hypophysectomized, pituitary grafted rats, *Endocrinology* **91:**868.

Lu, K.H., Chen, H.T., Grandison, L., Huang, H.H.C., and Meites, J., 1976a, Reduced luteninizing hormone release by synthetic LHRH in post partum lactating rats, *Endocrinology* **98:**1235–1240.

Lu, K.H., Chen, K.T., Huang, H.H., Grandison, L., Marshall, S., and Meites, J., 1976b, Relation between prolactin and gonadotrophin secretion in postpartum lactating rats, *J. Endocrinol.* **68:**241–250.

Mallampatti, R.S., and Johnson, D.C., 1974, Gonadotrophins on female rats androgenized by various treatments: Prolactin as an index for hypothalamic damage, *Neuroendocrinology* **15:**255.

Mancini, A.M., Guitelman, A., Debeljuk, L., and Vargas, G.A., 1975, Effect of administration of sulpiride on serum follicle stimulating hormone and luteinizing hormone levels in a group of post menopausal women, *J. Endocrinol.* **67:**127–128.

Maneckjee, R., Srinath, B.R., and Mougdal, N.R., 1976, Prolactin suppresses release of luteinizing hormone during lactation in the monkey, *Nature (London)* **262:**507–508.

Mann, D.R., Korowitz, C.D., MacFarland, L.A., and Cost, G.M., 1976, Interactions of the light–dark cycle, and renal glands and time of steroid administration in determining the temporal sequence of LH and prolactin release in female rats, *Endocrinology* **99:**1252.

March, C.M., Kletzky, O.A., and Davaja, N.V., 1977, Clinical response to CB-154 and the pituitary response to thyrotrophin releasing hormone and gonadotrophin releasing hormone in patients with galactorrhea–amenorrhea, *Fertil. Steril.* **28:**521.

McLeod, R.M., Kimura, H., and Login, I., 1976, Inhibition of PRL secretion by dopamine and piribedil, in: *Growth Hormone and Related Peptides* (Pecile and Muller, eds.), p. 443, Excerpta Medica, Amsterdam.

McNeilly, A.S., 1979, Effects of lactation on fertility, *Br. Med. Bull.* **35:**151–154.

Meninin, S. P., and Gorski, R.A., 1975, Effects of ovarian steroids on plasma LH in normal and persistent estrous adult female rats, *Endocrinology* **96:**486.

Mortimer, C.H., Besser, G.M., McNeilly, A.S., Marshall, J.C., Tunbridge, W.M.G., Gomez-Pan, A., and Hall, R., 1973, Luteinizing hormone and follicle stimulating

hormone releasing hormone test in patients with hypothalamic–pituitary–gonadal dysfunction, *Br. Med. J.* **4:**73.
Muralidbar, K., Maneckjee, R., and Mougdal, N.R., 1977, Inhibition of *in vivo* pituitary release of luteinizing hormone in lactating rats by exogenous prolactin, *Endocrinology* **100:**1137.
Neill, J.D., 1972, Sexual differences in the hypothalamic regulation of prolactin secretion, *Endocrinology* **90:**1154.
Noel, G.L., Suh, H.K., and Frantz, A.G., 1974, Prolactin release during nursing and breast stimulation in post partum and non post partum subjects, *J. Clin. Endocrinol. Metab.* **38:**413.
Ojeda, S.R., and McCann, S.M., 1973, Evidence for participation of a catecholaminergic mechanism in the post castration rise in plasma gonadotrophins, *Neuroendocrinology* **12:**295–315.
Ojeda, S.R., and McCann, S.M., 1974, Development of dopaminergic and estrogenic control of prolactin release in the female rat, *Endocrinology* **95:**1499–1505.
Ojeda, S.R., Negro-Vilar, A., and McCann, S.M., 1979, Release of prostaglandins Es by hypothalamic tissues: Evidence for their involvement in catecholamine-induced luteinizing hormone release, *Endocrinology* **104:**617.
Perkins, N., and Westfall, T., 1978, Effect of prolactin on dopamine release from rat striatum and medial basal hypothalamus, *Neuroscience* **3:**59–63.
Perkins, N.A., Westfall, T.C., Paul, C.V., McLeod, R., and Rogol, A.D., 1979, Effect of prolactin on dopamine synthesis in medial basal hypothalamus: Evidence for a short loop feedback, *Brain Res.* **160:**431.
Plapinger, L., and McEwen, B.S., 1973, Ontogenicity of estradiol binding sites in rats brain. I. Appearance of presumptive adult receptors in cytosol and nuclei, *Endocrinology* **93:**1119.
Puig-Duran, E., and McKinnon, P.C.B., 1975, Differential output of luteinizing hormone and prolactin in response to oestrogen in rats of both sexes, *J. Endocrinol.* **67:**38.
Quigley, M.E., Judd, S.J., Gilliland, G.B., and Yen, S.S.G., 1979, Effects of a dopamine antagonist in normal women and women with hyperprolactinemic anovulation, *J. Clin. Endocrinol. Metab.* **48:**718.
Rakoff, J.S., Rigg, L.A., and Yen, S.S.C., 1978, The impairment of progesterone-induced pituitary release of prolactin and gonadotrophin with hypothalamic chronic anovulation, *Am. J. Obstet. Gynecol.* **130:**807–812.
Ratner, A., and Peake, G.T., 1974, Maintenance of hyperprolactinemia by gonadal steroids in androgen-sterilized and spontaneously constant-estrous rats, *Proc. Soc. Expt. Biol. Med.* **146:**680.
Rjosk, H.K., Von Werder, K., and Fahlbusch, R., 1976, Hyperprolactinemic amenorrhea: Clinical relevance, endocrine features, therapy, *Geburtshilfe Frauenheilkd.* **36:**575–587.
Robyn, C., Vekemans, M., Caufriez, A., and L'Hermite, M., 1976, Effects of sulpiride-induced hyperprolactinemia on circulating gonadotrophins and sex steroids during the menstrual cycle, *IRCS Med. Sci.* **4:**14.
Rolland, R., Lequin, R.M., Schellekens, L.A., and Dejong, F.H., 1975a, The role of prolactin in the restoration of ovarian function during the early post partum period in the human female. I. A study during physiological lactation, *Clin. Endocrinol.* **4:**15–25.
Rolland, R., De Jong, L.H., Schellekens, L.A., and Lequin, R.M., 1975b, The role of prolactin in the restoration of ovarian function during the early post partum period in the human female. II. A study during inhibition of lactation by bromoergocryptine, *Clin. Endocrinol.* **4:**27–38.
Rotszentejn, W.N., Chali, J.L., Pattou, E., Epelbaum, J., and Kordon, C., 1976, *In vitro*

release of luteinizing hormone-releasing hormone (LHRH) from rat mediobasal hypothalamus: Effects of potassium, calcium and dopamine, *Endocrinology* **99:**1663.

Sawyer, C.H., Hilliard, J., Kanematsu, S., Scaramuzzi, R., and Blake, C.A., 1974, Effects of intraventricular infusions of norepinephrine and dopamine on LH release and ovulation in the rabbit, *Neuroendocrinology* **15:**328–337.

Scanlon, M.F., Pourmand, M., McGregor, A.M., Rodriguez-Arnao, M.D., Hall, K., Gomez-Pan, A., and Hall, R., 1979, Some current aspects of clinical and experimental neuroendocrinology with particular reference to growth hormone, thyrotropin and prolactin, *J. Endocrinol. Invest.* **2:**307–331.

Schally, A.V., Dupont, A., Arimura, A., Takahara, J., Redding, T.N., Clemens, J., and Shaar, C., 1976, Purification of a catecholamine-rich fraction with PIF activity from porcine hypothalami, *Acta Endocrinol. (Copenhagen)* **82:**1.

Seki, K., and Seki, M., 1974, Successful ovulation and pregnancy achieved by CB-154 (2-Br alpha-ergocriptine) in a woman with Chiari–Frommel syndrome, *J. Clin. Endocrinol. Metab.* **38:**508–509.

Seki, K., Seki, M., Okumura, T., and Huang, K.E., 1976, Effect of clomiphene citrate on serum prolactin in infertile women with ovarian dysfunction, *Am. J. Obstet. Gynecol.* **124:**125–128.

Seppälä, M., Hirvonen, E., Ranta, T., Virkkunen, P., and Leppälouto, J., 1975, Raised serum prolactin levels in amenorrhea, *Br. Med. J.* **2:**305.

Shaar, C.J., Euker, J.S., Riegle, G.D., and Meites, J., 1975, Effects of castration and gonadal steroids on serum luteinizing hormone and prolactin in old and young rats, *J. Endocrinol.* **66:**45–51.

Smith, M.S., 1978, A comparison of pituitary responsiveness to luteinizing hormone-releasing hormone during lactation and the estrous cycle of the rat, *Endocrinology* **102:**114.

Spellacy, W.N., Cantor, B., Kalra, P.S., Buhi, W.C., and Birk, S.A., 1978, The effect of varying prolactin levels on pituitary luteinizing hormone and follicle-stimulating hormone response to gonadotrophin-releasing hormone, *Am. J. Obstet. Gynecol.* **132:**157–164.

Strauch, G., Valcke, J.C., Mahoudeau, J.A., and Bricaire, H., 1977, Hormone changes induced by bromocriptine (CB-154) at the early stage of treatment, *J. Clin. Endocrinol. Metab.* **44:**588.

Thorner, M.O., McNeilly, A.S., Hagan, C., and Besser, G.M., 1974, Long-term treatment of galactorrhea and hypogonadism with bromocriptine, *Br. Med. J.* **2:**419–422.

Thorner, M.O., Edwards, C.R.W., Hanker, J.P., Abraham, G., and Besser, G.H., 1977, Prolactin and gonadotrophin interaction in the male, in: *The Testis in Normal and Infertile Men* (P. Troen and H.R. Nankin, eds.), pp. 351–366, Raven Press, New York.

Tolis, G., 1977, Galactorrhea–amenorrhea and hyperprolactinemia: Pathophysiological aspects and diagnostic tests, *Clin. Endocrinol. Suppl.* **6:**81s.

Tolis, G.H., Friesen, H.G., and McKenzie, J.M., 1975, Altered pituitary responsiveness to gonadotrophin releasing hormone in hyperprolactinemic states, *Clin. Res.* **23:**390.

Tresguerres, J.A.F., and Esquifino, A.I., 1979, Influence of hyperprolactinemia in female rats, *Acta Endocrinol. Suppl.* **225:**176.

Turgeon, J.L., and Barraclough, C.A., 1974, Pulsatile plasma LH rhythms in normal and androgen-sterilized ovariectomized rats: Effects of estrogen treatment, *Proc. Soc. Exp. Biol. Med.* **195:**821.

Turkington, R.W., Underwood, L.E., and Van Wyk, J.I., 1971, Elevated serum PRL levels after pituitary stalk section in man, *N. Engl. J. Med.* **295:**707.

Tyson, J.E., Andreassen, B., Huth, J., Smith, B., and Zagur, H., 1975, Neuroendocrine dysfunction in galactorrhea–amenorrhea after oral contraceptive use, *Obstet. Gynecol.* **46:** 1–4.

Uemura, H., and Kobayashi, I., 1971, Effects of dopamine implanted in the median eminence on the estrous cycle of the rats, *Endocrinol. Jpn.* **18:**91.

Uilenbroek, J.T.J., and Gribling-Hegge, L.A., 1977, Pituitary responsiveness to LHRH in intact and ovariectomized androgen sterilized rats, *Neuroendocrinology* **23:**43.

Van Campenhout, J., Papas, S., Blanchet, P., Wyman, H., and Somma, M., 1977, Pituitary responses to synthetic luteinizing hormone-releasing hormone in thirty-four cases of amenorrhea and oligomenorrhea associated with galactorrhea, *Am. J. Obstet. Gynecol.* **127:**723.

Vaticon, M.D., Fernandez-Galaz, C., Aguilar, E., and Tejero, A., 1979, Variaciones de los niveles plasmaticos de LH y prolactina tras la administración de benzoato de estradiol en ratas hembras ciclicas y estrogenizadas postnatalmente, *Endocrinologia* **26:**9–12.

Vernes, I., and Telegdy, G., 1979, A cell suspension technique for study of hypothalamus–pituitary–testis function *in vitro*: Effects of neurotransmitters, *Acta Endocrinol. Suppl.* **225:**235.

Vijayan, E., and McCann, S.M., 1978, The effect of systemic administration of dopamine and apomorphine on plasma LH and prolactin concentrations, *Neuroendocrinology* **25:**221.

Weiss, G., Schmidt, C., Kleinberg, D.L., and Ganguly, M., 1977, Positive feedback of oestrogen on LH secretion in women on neuroleptic drugs, *Clin. Endocrinol.* **7:**423–427.

Wiebe, R.H., Hammond, C.B., and Borchert, L.G., 1976, Diagnosis of prolactin-secreting pituitary microadenoma, *Am. J. Obstet. Gynecol.* **126:**993.

Wiebe, R.H., Hammond, C.B., and Handwerger, S., 1977, Treatment of functional amenorrhea–galactorrhea with 2-ergocriptine, *Fertil. Steril.* **28:**426.

Winters, S., and Loriaux, D.L., 1978, Suppression of plasma luteinizing hormone by prolactin in the male rat, *Endocrinology* **102:**864.

Zarate, A., Canales, E.A., Villalobos, H., Soria, J., Jacobs, L.S., Kastin, A.J., and Schally, A.V., 1975, Pituitary hormonal reserve in patients presenting hyperprolactinemia, intrasellar masses and amenorrhea without galactorrhea, *J. Clin. Endocrinol. Metab.* **40:** 1034–1037.

20

Interactions among LH, FSH, PRL, LH-RH, and Sex Steroids in the Control of Testicular LH, FSH, and PRL Receptors in the Rat

P. A. Kelly, F. Labrie, C. Auclair, C. Séguin, A. Bélanger, and M. R. Sairam

1. Introduction

Within the last several years, improved methods to label highly purified polypeptide hormones to a high level of specific activity, with a retention of biological activity, along with advances in subcellular-fractionation techniques have allowed an important increase in our understanding of the mechanisms of action of peptide hormones. Binding of hormones to receptors located in the plasma membrane is the first event in the action of these hormones in their target tissues, and specific receptors for a large number of hormones have been identified (Kahn, 1976).

Recent studies have shown that testicular luteinizing hormone (LH)-receptor levels can be markedly influenced by LH itself. In fact, a marked loss of LH receptors is seen after changes of endogenous LH secretion induced by LH-releasing hormone (LH-RH) (Auclair *et al.*, 1978; Catt *et al.*, 1979) or its agonistic analogues (Auclair *et al.*, 1977a,b; Labrie *et al.*, 1978, 1980; Bélanger *et al.*, 1979; Rivier *et al.*, 1979; Kelly

P. A. Kelly, F. Labrie, C. Auclair, C. Séguin, and *A. Bélanger* • Department of Molecular Endocrinology, Le Centre Hospitalier de l'Université Laval, Quebec GIV 4G2, Canada. *M. R. Sairam* • Institut de Recherches Cliniques, Montreal H2W lR7, Canada.

et al., 1979), as well as after human chorionic gonadotropin (hCG) administration (Hsueh *et al.*, 1976; Sharpe, 1976; Auclair *et al.*, 1977a; Chen and Payne, 1977). On the other hand, a stimulatory effect of prolactin (PRL) on testicular LH-receptor levels has been demonstrated in intact (Bélanger *et al.*, 1979) as well as in hypophysectomized rats (Zipf *et al.*, 1978). A similar role of PRL has also been observed in immature rats (Aragona *et al.*, 1977), dwarf mice (Bohnet and Friesen, 1976), and light-deprived hamsters (Bex and Bartke, 1977).

Since not only LH and PRL but also follicle-stimulating hormone (FSH) is involved in the control of testicular function (Bartke *et al.*, 1978) and the action of these three pituitary hormones is likely to be dependent on the level of their corresponding specific receptors, the study presented in this chapter describes the interactions among LH, PRL, and FSH on the testicular levels of receptors for these three pituitary hormones in hypophysectomized adult rats. Moreover, since direct actions of estrogens and androgens on testicular steroidogenesis and spermatogenesis have been described (Kalla *et al.*, 1977), the effects of 17β-estradiol, testosterone, and dihydrotestosterone were studied on the same parameters.

2. Receptor Assays

For the measurement of PRL-, LH-, and FSH-receptor levels, ^{125}I-labeled hormones iodinated with chloramine T to a low specific activity (20–60 μCi/μg) were used. Testes were homogenized and centrifuged at 20,000*g* for 15 min. The pellet was recentrifuged once and resuspended in tris-HC1 buffer. Assay tubes contained 10 mg equivalent (wet weight) of testis. Saturating quantities of [^{125}I]-hCG, human FSH (hFSH), or ovine PRL (oPRL) were utilized (100,000–150,000 counts/min). Duplicate tubes were incubated in the absence or presence of 2 μg oLH or oFSH or 1 μg oPRL for the three corresponding labeled hormones. Hormone binding was represented as femtomoles of hormone bound per testis or per gram testis (Auclair *et al.*, 1977a,b; Bélanger *et al.*, 1979).

3. Plasma PRL and Androgen Assays and Calculations

Plasma concentrations of testosterone and dihydrotestosterone were measured by a specific double-antibody radioimmunoassay as described earlier (Bélanger *et al.*, 1980). Plasma PRL levels were measured by double-antibody radioimmunoassay using materials provided by Dr. A. F. Parlow of the National Institute of Arthritis, Metabolism and Digestive

Diseases. Radioimmunoassay data were calculated using a program based on Model II of Rodbard and Lewald (1970) using a Hewlett-Packard 9830A desk-top calculator. Statistical significance was determined using the Duncan–Kramer multiple range test (Kramer, 1956).

4. Hormones

Purified hCG (CR-119, 11,600 IU/mg) was generously supplied by the Center for Population Research of the NICHHD, NIH. oLH (NIH-LH-S19, 1.01 × NIH-LH-S1), oFSH (NIH-FSH-S12, 12.25 × NIH-FSH-S1), oPRL (NIH-P-S12, 35 IU/mg), rat PRL-I-3, rat PRL-RP-1, and anti-rat PRL-S6 were gifts of the National Pituitary Agency, NIH. For *in vivo* treatment, the following hormone preparations were used: hCG [2910 IU/mg (obtained from Dr. Jean-Pierre Raynaud, Roussel-UCLAF, France), oPRL (NIH-P-12, 35 IU/mg), 17β-estradiol [(E_2) Steraloids Inc.], testosterone [(T) Sigma], and dihydrotestosterone [(DHT) Sigma]. hFSH (S1194H2R) (120 × NIH-FSH-S10) was used for injection and iodination.

5. Down-Regulation of Testicular Gonadotropin Receptors by LH-RH and Its Agonists

Since LH-RH and its analogues are currently used in the treatment of oligospermia and male infertility and the success obtained has been limited (Schwartzstein, 1976), it became important to study in detail the effect of such treatment in experimental animals. The interest of this study was strengthened by the recent findings that a marked loss of LH receptors (Hsueh *et al.*, 1976, 1977) and steroidogenic response to gonadotropins (Hsueh *et al.*, 1977) occurs in the rat after systematic administration of oLH or hCG. This ability of gonadotropins to induce loss of their own receptors is analogous to the effect of insulin, thyrotropin-releasing hormone, growth hormone, and catecholamines on the level of their own receptors in respective target tissues (Gavin *et al.*, 1974; Hinkle and Tashjian, 1976; Lesniack and Roth, 1976; Mukherjee *et al.*, 1975). It thus seemed possible that changes of endogenous LH secretion induced by administration of LH-RH or its agonists could lead to a significant loss of testicular LH receptors.

We initially observed the effect of treatment of adult male rats with increasing doses (0.6, 3, 15, or 75 μg) of an LH-RH agonist, [D-Leu^6,des-Gly-NH_2^{10}]LH-RH ethylamide, hCG (50 IU), or the vehicle alone (0.15 M NaCl) three times a day (at 08:00, 16:00, and 24:00 hr) for 7 days. The LH-RH agonist resulted in a marked reduction of LH/hCG-binding sites in testicular tissue, while injection of hCG caused an almost complete loss of LH/hCG-binding sites. Testis weight was also significantly reduced by these treatments (Auclair *et al.*, 1977b).

When the time-course of action of the LH-RH agonist was studied, it was found that a single injection of 125 μg [D-Leu^6,des-Gly-NH_2^{10}]LH-RH ethylamide or 200 IU hCG induced a rapid decline in testicular LH/hCG-receptor levels. The effect was apparent at 12 hr, remained maximal up to 3 days, and decreased progressively thereafter. Testicular weight declined steadily and remained low until 8 days after injection. The changes of LH/hCG-receptor levels followed the same pattern when expressed per gram testis. After a small increase 12 hr after injection of [D-Leu^6,des-Gly-NH_2^{10}]LH-RH ethylamide, plasma testosterone levels were decreased at 3 days and returned to normal at later time intervals. A single injection of hCG led to a sustained elevation of plasma testosterone levels up to 3 days after its administration, with a small decrease at 5 and 6 days, and a return to normal levels at 8 days after injection (Auclair *et al.*, 1977b).

Since the lowest daily amount of the LH-RH analogue (1.8 μg) administered in the previous experiments was high, it was important to study the effect of lower and more physiological doses of the LH-RH agonist. Groups of rats were injected with saline or increasing doses (0.008, 0.04, 0.2, 1.0, or 5.0 μg) of [D-Leu^6,des-Gly-NH_2^{10}]LH-RH ethylamide once a day (08:00 hr) or three times a day (08:00, 16:00, and 24:00 hr). The last injection was at 08:00 hr on the 7th day, and the rats were sacrificed 24 hr later by decapitation.

As illustrated in Fig. 1A, a significant loss of testicular LH/hCG receptors (30%) is seen in animals treated with as little as 8 ng of the LH-RH agonist, with a maximal effect (80%) being observed between 40 and 200 ng. Interestingly, the desensitizing effect of the LH-RH agonist is more apparent in animals injected once rather than three times a day (Auclair *et al.*, 1977a). Figure 1B shows that a similar inhibitory effect is seen on testicular PRL-receptor levels, a maximal effect being observed at 200 ng of the LH-RH agonist. Since treatment with doses of 40 ng or higher of the LH-RH agonist leads to a reduction of testis weight, it is important to mention that a similar inhibitory pattern was observed when LH/hCG- and PRL-receptor levels were expressed per gram testis.

Although the total number of FSH receptors is decreased after treatment with [D-Leu^6,des-Gly-NH_2^{10}]LH-RH ethylamide, this effect is not significant when expressed per gram testis. It can also be seen that testis weight (Fig. 2A) and plasma testosterone levels (Fig. 2B) are reduced after treatment with doses of 40 ng or higher of the LH-RH agonist (Auclair *et al.*, 1977a).

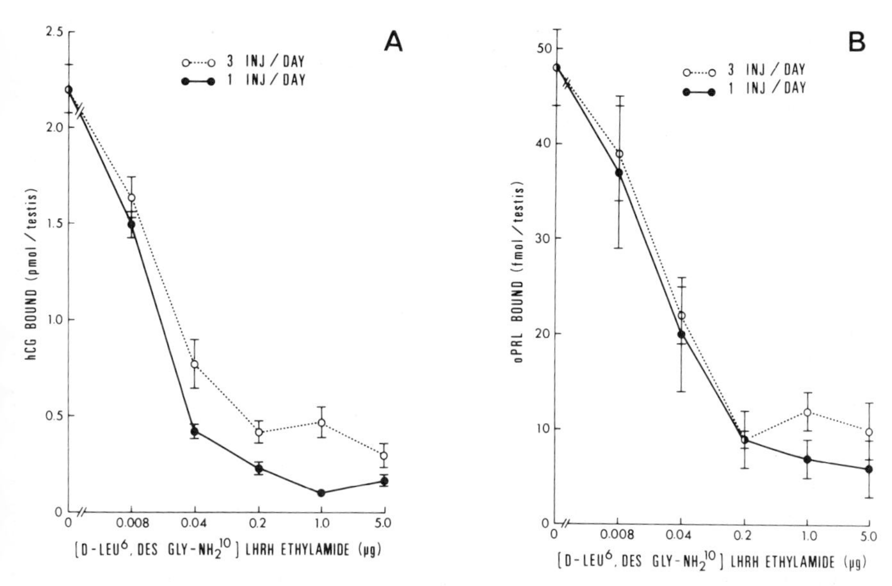

Figure 1. Effect of increasing doses of [D-Leu^6,des-Gly-NH_2^{10}]LH-RH ethylamide injected once or three times a day for 7 days on testicular LH/hCG-receptor (A) and PRL-receptor (B) levels. From Auclair *et al.* (1977a).

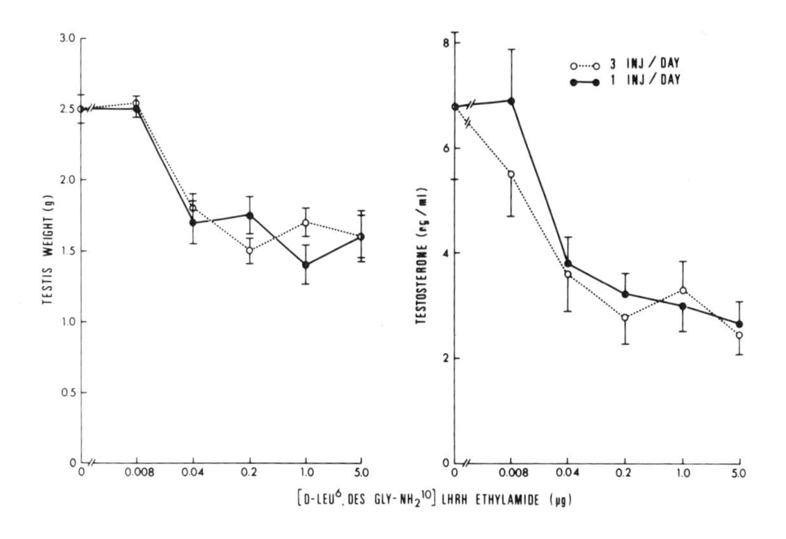

Figure 2. Effect of increasing doses of LH-RH agonist on testis weight *(left)* and plasma testosterone levels *(right)* After Auclair *et al.* (1977a).

Since the LH-RH agonist used previously, [D-Leu^6,des-Gly-NH$_2$^{10}]LH-RH ethylamide, is approximately 100 times more potent than LH-RH, and the natural decapeptide is used in many clinical studies for the treatment of oligospermia and male infertility (Schwartzstein, 1976), we studied the effect of treatment with LH-RH on testicular LH receptors and function in the rat (Auclair *et al.*, 1978). Adult male Sprague–Dawley rats were injected subcutaneously with a single dose of 5, 25, 100, or 500 μg LH-RH or 0.05 μg [D-Ala^6,des-Gly-NH$_2$^{10}]LH-RH ethylamide (an analogue approximately equipotent with [D-Leu^6,des-Gly-NH$_2$^{10}]LH-RH ethylamide), and the animals were killed 2 or 8 days later. It can be seen in Fig. 3 that a single injection of 5 μg LH-RH or 0.05 μg [D-Ala^6,des-Gly-NH$_2$^{10}]LH-RH ethylamide leads to approximately the same inhibition of testicular LH-receptor levels. It can also be noticed, as observed previously (Auclair *et al.*, 1977b), that the inhibitory effect is maximal at 2 days, with a 50–75% recovery of LH receptor levels at 8 days.

Treatment with LH-RH or its agonist also leads to a 50–75% inhibition of plasma testosterone levels at 2 days, with partial recovey at 8 days. Significant reduction of seminal-vesicle weight is observed at 8 days with the LH-RH agonist and at doses of LH-RH of 100 μg or higher (Auclair *et al.*, 1978),

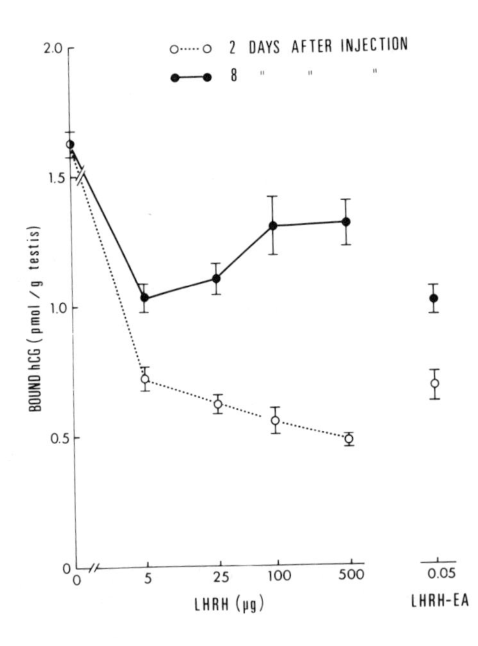

Figure 3. Effect of a single injection of increasing doses of native LH-RH (5, 25, 100, or 500 μg) or [D-Ala^6,des-Gly-NH$_2$^{10}]LH-RH ethylamide (0.05 μg) on binding of [^{125}I]-hCG to testicular LH/hCG receptors in adult male rats. Determinations were made 2 days (○) or 8 days (●) after administration of the peptide. From Auclair *et al.* (1978).

6. *Regulation of Testicular Gonadotropin-Receptor Levels in Hypophysectomized Animals*

In the first series of experiments, groups of hypophysectomized Sprague–Dawley rats, weighing 200–225 g, were injected subcutaneously twice a day for 5 days with either saline, hCG (2 IU), oPRL (250 μg), oFSH (1 μg), 17β-estradiol (0.5 μg), testosterone (200 μg), or dihydrotestosterone (200 μg) alone or in combination. Injections were started 8–10 days after hypophysectomy. All animals were sacrificed 16–18 hr after the last injection except in the hCG-treated groups, in which the last hCG injection was given 40–42 hr previously.

In a second series of experiments, hypophysectomy was performed under ether anesthesia by a parapharyngeal approach in intact rats (12–16 animals per group) or in animals bearing three pituitaries (obtained from adult female animals of the same strain) implanted under the left kidney capsule 5–7 hr earlier. Animals were killed by decapitation between 08:00 and 10:00 hr of the 5th day following hypophysectomy. Completeness of hypophysectomy was assessed by examination of the sella turcica. Testes, ventral prostate, and seminal vesicles were dissected immediately and weighed. Plasma was collected and kept at −20°C until assayed.

As illustrated in Fig. 4, following hypophysectomy, the total number of LH receptors is 90% reduced, from 2280 ± 180 to 255 ± 20 fmol/testis, while treatment with PRL for 5 days increases LH receptors 2.6-fold ($p < 0.01$). At the low dose used (2 IU), hCG further decreases LH-receptor levels in hypophysectomized animals to approximately 10% of the value found in intact animals (Table I). Treatment with PRL leads to a partial reversal of this inhibitory effect of hCG (Fig. 4 and Table I). Treatment with FSH, on the other hand, has little or no effect on testicular LH-receptor levels when administered alone or in combination with either PRL, hCG, or FSH.

When administered with PRL, E_2, T, and DHT lead to a 22–33% reduction ($p < 0.05$ for E_2) of testicular LH-receptor content. In the presence of FSH or in control hypophysectomized animals, only T and DHT cause a slight but nonsignificant inhibition. However, when data are expressed per gram testis, T and DHT lead to a significant inhibitory effect not only in animals treated with PRL but also in the control and FSH-treated groups (Table I). There is no effect of any steroid in hCG-treated animals.

To further examine the effect of steroids on testicular LH-receptor levels, Silastic implants of steroids were placed in hypophysectomized animals and LH-receptor levels measured. Figure 5 shows that E_2, T,

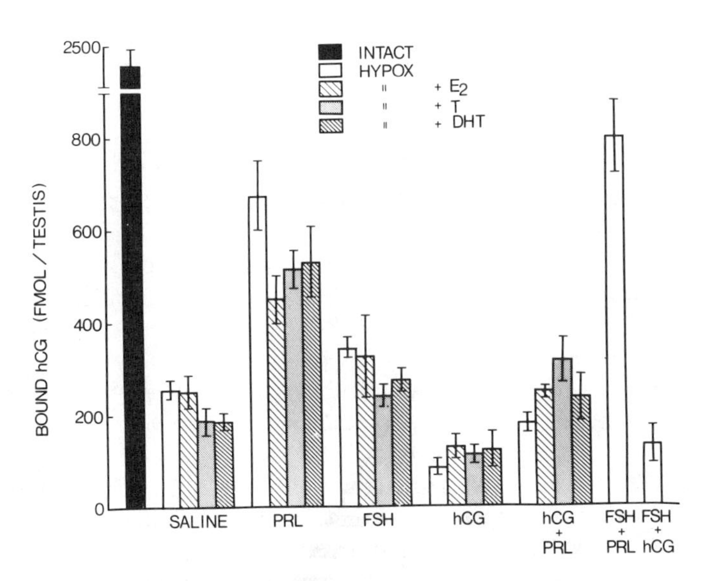

Figure 4. Effect of twice-daily injections of oPRL (250 μg), oFSH (1 μg), hCG (2 IU), 17β-estradiol [(E_2) 0.5 μg], testosterone [(T) 200 μg], or 5α-dihydrotestosterone [(DHT) 200 μg] alone or in combination on testicular LH-receptor levels in adult intact and hypophysectomized rats. Treatment was started 8–10 days after hypophysectomy and continued for 5 days.

and DHT, as well as progesterone (P), are all equally effective in reducing testicular LH-receptor levels compared to control hypophysectomized animals ($p < 0.01$ for all).

The effect of the various treatments on testicular FSH-receptor levels is shown in Fig. 6 and Table I. As can be seen in Fig. 6, hypophysectomy results in a marked loss of total FSH receptors, and none of the peptide hormone treatments, except FSH itself, has significant effects on the total number of FSH receptors. However, when expressed as femtomoles per gram testis (Table I), FSH-receptor concentrations are markedly increased by hypophysectomy, and FSH treatment reduces binding. However, this reduction in receptor concentration can be explained by the significant increase in testis weight induced by FSH (Table I).

As clearly illustrated in Fig. 7, treatment of hypophysectomized rats with hCG leads to a marked stimulation of the testicular PRL-receptor concentration from 10 ± 3 to 63 ± 6 fmol/testis, while a value of 125 ± 14 fmol/testis is found in intact animals. Treatment with PRL itself, FSH, or sex steroids has no significant effect on testicular PRL-receptor levels. As expected, treatment of hypophysectomized rats with hCG or hFSH

Table I. Effect of Twice-Daily Treatment of Hypophysectomized Adult Male Rats with hCG, oPRL, oFSH, E_2, T, or DHT Alone or in Combination on Testis, Ventral-Prostate, and Seminal-Vesicle Weight and on Testicular LH-, FSH-, and PRL-Receptor Levels[a]

Treatment[a]	Number of rats	Weight: Testis (g)	Weight: Ventral prostate (mg)	Weight: Seminal vesicle (mg)	Hormone-receptor levels (fmol/g testis): LH	FSH	PRL
Intact	9	2.64 ± 0.13[b]	242 ± 33[b]	179 ± 18[b]	1750 ± 160[b]	330 ± 14[b]	90 ± 10[b]
Hypox	9	0.55 ± 0.09	1 ± 14	40 ± 3	990 ± 105	675 ± 35	40 ± 16
Hypox + E_2	9	0.66 ± 0.05	9 ± 2	39 ± 1	750 ± 84[c]	637 ± 48	39 ± 10
Hypox + T	9	0.71 ± 0.06	50 ± 4[b]	94 ± 14[b]	508 ± 50[b]	620 ± 13	39 ± 10
Hypox + DHT	9	0.72 ± 0.08	54 ± 5[b]	110 ± 14[b]	571 ± 68[b]	625 ± 35	55 ± 9
Hypox + oPRL	9	0.81 ± 0.05[c]	15 ± 3	48 ± 6	1745 ± 65[b]	660 ± 19	32 ± 8
Hypox + oPRL + E_2	6	0.62 ± 0.05	11 ± 1	43 ± 4	1413 ± 116[b]	657 ± 54	45 ± 18
Hypox + oPRL + T	8	0.87 ± 0.07[c]	58 ± 2[b]	104 ± 10[b]	1230 ± 76[b]	596 ± 28	34 ± 8
Hypox + oPRL + DHT	8	0.74 ± 0.05	56 ± 3[b]	100 ± 8[b]	1438 ± 173[c]	708 ± 30	27 ± 6
Hypox + hCG	9	0.83 ± 0.05[c]	37 ± 3[b]	79 ± 4[b]	212 ± 42[b]	615 ± 32	150 ± 12[b]
Hypox + hCG + E_2	8	0.86 ± 0.06[b]	40 ± 6[b]	85 ± 5[b]	306 ± 59[b]	603 ± 17	155 ± 20[b]
Hypox + hCG + T	8	0.86 ± 0.14[b]	66 ± 1[b]	108 ± 8[b]	235 ± 41[b]	556 ± 45[c]	131 ± 13[b]
Hypox + hCG + DHT	9	0.79 ± 0.08[c]	53 ± 5[b]	89 ± 7[b]	324 ± 78[b]	552 ± 43[c]	157 ± 24[b]
Hypox + hCG + oPRL	8	0.86 ± 0.05[c]	53 ± 4[b]	86 ± 6[b]	440 ± 70[b]	635 ± 35	146 ± 15[b]
Hypox + hCG + oPRL + E_2	9	0.80 ± 0.05[c]	47 ± 4[b]	84 ± 10[b]	599 ± 90[b]	525 ± 32[b]	135 ± 16[b]
Hypox + hCG + oPRL + T	9	0.80 ± 0.07[c]	66 ± 5[b]	88 ± 6[b]	817 ± 134	613 ± 18	146 ± 22[b]
Hypox + hCG + oPRL + DHT	9	0.81 ± 0.07[c]	63 ± 7[b]	98 ± 10[b]	561 ± 88[b]	593 ± 37	120 ± 15[b]
Hypox + FSH	8	0.91 ± 0.06[b]	15 ± 2	43 ± 4	766 ± 53	386 ± 25[b]	69 ± 8[c]
Hypox + FSH + oPRL	7	0.89 ± 0.08[b]	22 ± 3[c]	41 ± 3	1780 ± 68[b]	450 ± 21[b]	33 ± 6
Hypox + FSH + E_2	9	0.80 ± 0.12[b]	15 ± 3	48 ± 6	716 ± 103[c]	450 ± 51[b]	25 ± 6
Hypox + FSH + T	9	0.98 ± 0.05[b]	41 ± 4[b]	89 ± 8[b]	475 ± 47[b]	416 ± 15[b]	16 ± 4
Hypox + FSH + DHT	9	1.02 ± 0.07[b]	50 ± 4[b]	99 ± 11[b]	515 ± 50[b]	436 ± 24[b]	69 ± 16
Hypox + FSH + hCG	9	1.12 ± 0.10[b]	48 ± 8[b]	85 ± 9[b]	215 ± 42[b]	387 ± 22[b]	138 ± 8[b]

[a] Abbreviations and doses: (DHT) 5α-dihydrotestosterone, 200 μg; (E_2) 17β-estradiol, 0.5 μg; (hCG) human chorionic gonadotropin; 2 IU; (LH) luteinizing hormone; (oFSH) ovine follicle-stimulating hormone, 1 μg; (oPRL) ovine prolactin, 250 μg; (T) testosterone, 200 μg. Treatment was started 8–10 days after hypophysectomy (Hypox) for 5 days.

[b] $p < 0.01$ (experimental vs. Hypox).

[c] $p < 0.05$ (experimental vs. Hypox).

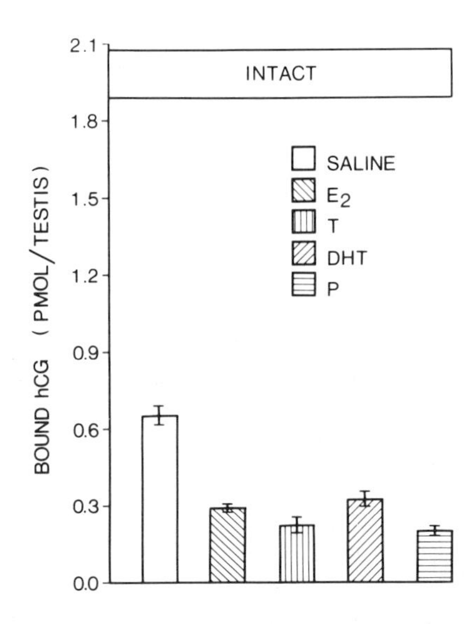

Figure 5. Effect of Silastic implants of 17β-estradiol (E_2), testosterone (T), 5α-dihydrotestosterone (DHT), or progesterone (P) on testicular LH-receptor levels in hypophysectomized rats. The horizontal bar at the top indicates the mean ± S.E.M. of testicular LH values seen in intact rats. Vertical columns represent LH-receptor levels in hypophysectomized rats. Silastic implants of the following length were implanted into male rats hypophysectomized 3 days previously: E_2, 2 cm; T, 2.5 cm; DHT, 2.5 cm; P, 10 cm. The animals were sacrificed 10 days after placement of the implants.

leads to increased testicular weight (Table I). The small inhibitory effect of androgens alone at the relatively low dose and for the short treatment period (5 days) is not significant.

Although measurements of plasma testosterone plus dihydrotestosterone concentrations at one time interval are of limited value, circulating levels of androgens were increased approximately to the same extent after injection of testosterone (3.4 ± 0.5 ng/ml) or hCG (5.2 ± 0.8 ng/ml). Changes of ventral-prostate and seminal-vesicle weight were those expected from changes of plasma androgen levels (Table I), increased weight of the accessory sex organs being seen after treatment with T, DHT, or hCG, while no effect or interaction of PRL or FSH could be observed under the experimental conditions used.

Since PRL has such dramatic effects on testicular LH-receptor levels, we next studied the effect of elevated endogenous PRL levels on the same parameters in intact and hypophysectomized rats. As can be seen in Table II, the implantation of three pituitaries (obtained from adult female rats of the same strain) leads to an approximate 20-fold increase of plasma PRL levels after 5 days in intact rats, and comparable levels are achieved in hypophysectomized animals. Hyperprolactinemia leads to a 75% increase of testicular LH-receptor levels in intact rats, while it completely prevents the 75% loss of LH receptors measured 5 days after hypophysectomy (Table II). Similar effects are found when data are expressed on a testicular-weight basis (Fig. 8). While hyperprolactinemia does not prevent the loss of testicular and ventral-prostate weight that

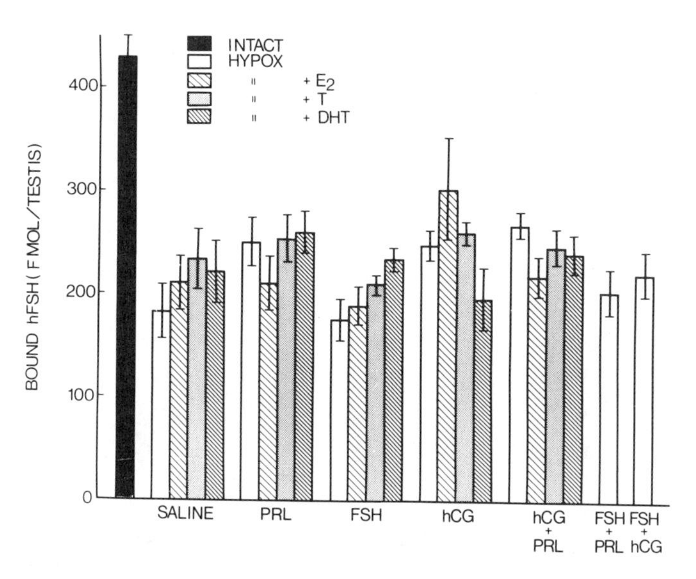

Figure 6. Effect of twice-daily injections of hCG, oPRL, oFSH, E_2, T, or DHT alone or in combination on testicular FSH-receptor levels in adult intact and hypophysectomized rats. Same experiment as described in the Fig. 4 caption.

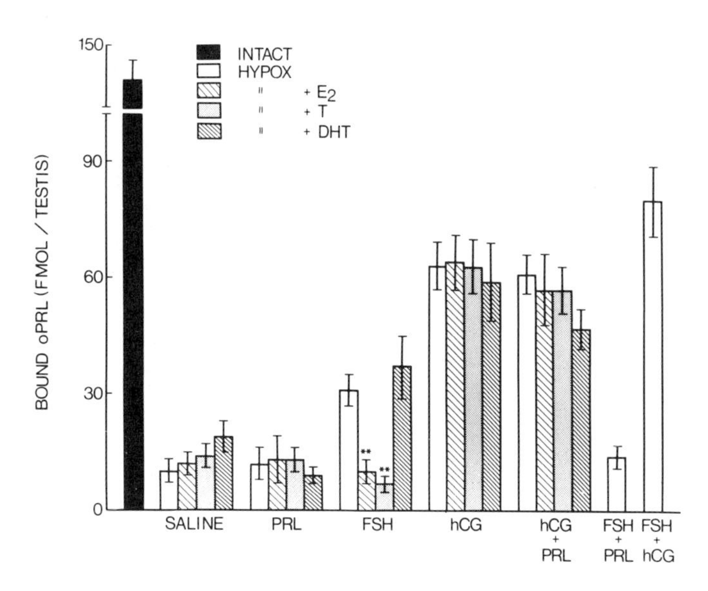

Figure 7. Effect of twice-daily injections of hCG, oPRL, oFSH, E_2, T, or DHT alone or in combination on testicular PRL-receptor levels in adult intact and hypophysectomized rats. Same experiment as described in the Fig. 4 caption.

Table II. Effects of Increased Circulating PRL Levels in Intact and Hypophysectomized Rats on Testis, Prostate, and Seminal-Vesicle Weight; Testicular LH-, FSH-, and PRL-Receptor Levels; and PRL and Testosterone Concentrations[a]

Treatment	Number of rats	Weight			Hormone-receptor levels (fmol/testis)			Plasma PRL (ng/ml)	Plasma testosterone (ng/ml)
		Testis (g)	Prostate (mg)	Seminal vesicle (mg)	LH	FSH	PRL		
Intact (control)	16	2.86 ± 0.05	213 ± 14	230 ± 8	1780 ± 90	323 ± 13	71 ± 6	12 ± 1	5.8 ± 0.7
Intact + implants	15	2.84 ± 0.08	223 ± 13	274 ± 14[b]	3110 ± 110[b]	329 ± 13	33 ± 6[b]	217 ± 25[b]	6.7 ± 1.3
Hypox (control)	13	2.35 ± 0.08	87 ± 9	131 ± 6	410 ± 60	349 ± 16	27 ± 5	1 ± 0.2	< 0.2
Hypox + implants	12	2.35 ± 0.05	91 ± 10	160 ± 12[c]	1820 ± 100[b]	285 ± 13	22 ± 6	178 ± 27[b]	< 0.2

[a] Animals were left intact or hypophysectomized 5–7 hr after pituitary implantation under the kidney capsule and killed 5 days later.
[b] $p < 0.01$.
[c] $p < 0.05$.

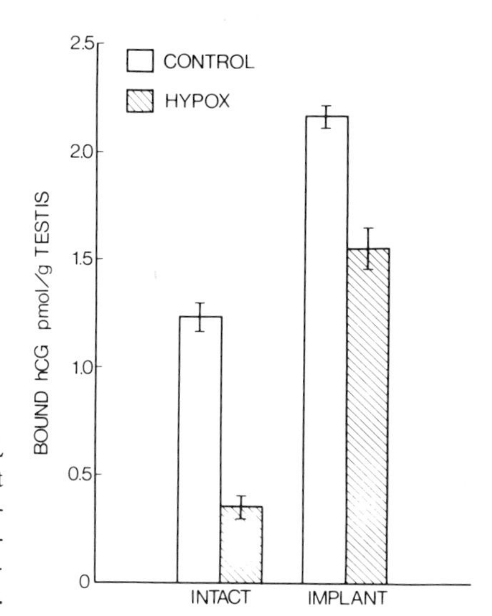

Figure 8. Effects of increased plasma PRL levels in intact and hypophysectomized adult rats on testicular LH-receptor levels. Hypophysectomy was performed 5–7 hr after pituitary implantation under the kidney capsule, and the animals were killed 5 days later.

followed hypophysectomy, it significantly increases seminal-vesicle weight in both intact and hypophysectomized animals (Table II). High circulating levels of PRL reduced testicular PRL-receptor levels in intact rats, such changes probably being due to occupancy of binding sites by endogenous PRL. Hyperprolactinemia has no effect on testicular FSH-receptor levels.

The stimulatory effect of LH or hCG on testicular PRL receptors is illustrated in Fig. 9. Hypophysectomized male rats were injected with 1 μg oLH or 10 IU hCG for 7 days. It can be seen that both treatments result in a marked stimulation, receptor levels approaching those found in intact animals.

While numerous observations indicate a major role of PRL in the control of testicular LH receptors and steroidogenesis in the rodent, few data are available on the control of testicular PRL-receptor levels (Auclair *et al.*, 1977a,b, 1978; Labrie *et al.*, 1978; Kelly *et al.*, 1979). PRL binding in rat Leydig cells has been described (Aragona *et al.*, 1977; Costlow and McGuire, 1977). As clearly illustrated in Figs. 7 and 9, treatment of hypophysectomized adult rats with hCG or LH stimulates testicular PRL receptors to about 50–80% of the level found in intact animals. Higher doses are even more stimulatory (Kelly *et al.*, manuscript submitted).

Following its binding to high-affinity membrane receptors on Leydig-cell membranes, LH exerts its stimulatory action on androgen biosynthesis through activation of the adenylyl cyclase–protein kinase enzy-

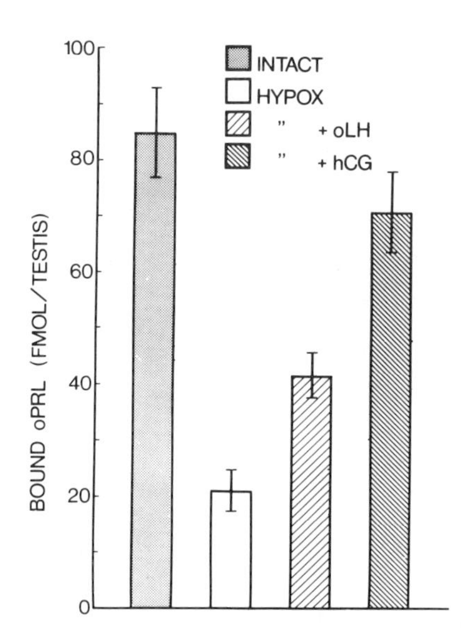

Figure 9. Effect of daily injection of hypophysectomized male rats with 1 μg oLH or 10 IU hCG for 7 days on testicular PRL-receptor levels.

matic system (Catt *et al.,* 1972; Mendelson *et al.,* 1975). Since treatment with T or DHT has no effect on PRL-receptor levels, it is likely that the potent stimulatory effect of hCG treatment on this parameter is mediated at a step independent of androgen formation, possibly by a specific cyclic-AMP-mediated pathway directly involved with the synthetis or processing, or both, of the PRL receptor. Since we have previously found that changes of endogenous gonadotropin secretion induced by treatment with LH-RH or its agonistic analogues inhibit testicular PRL-receptor levels in the intact animal (Auclair *et al.,* 1977a,b), it appears that although the effect of hCG treatment in the hypophysectomized animal leads to a marked stimulation of PRL-receptor levels, other important hormonal interactions having opposite effects on PRL-receptor levels are predominant during stimulation of Leydig-cell activity in the intact animal.

FSH has been found to increase LH-receptor levels in granulosa cells (Zeleznik *et al.,* 1974) as well as in immature rat testis (Odell and Swerdloff, 1975; Chen *et al.,* 1976). In agreement with previous observations (Hauger *et al.,* 1977), the data presented herein show that treatment with FSH does not prevent the marked loss of LH receptors seen after hypophysectomy in adult animals (see Fig. 1). Moreover, FSH has no effect on the stimulation and inhibition of LH-receptor levels induced by PRL and hCG, respectively. When LH-receptor levels are expressed per unit of testis weight, no effect of FSH treatment has been reported in immature Odell and Swerdloff, 1975; Chen *et al.,* 1976) or adult (Thanki and Sternberger, 1976) hypophysectomized male rats. However, more recently, it has been reported that treatment of immature rats with FSH led to an increase in testicular LH receptors and *in vitro* responses to hCG (Hsueh *et al.*, 1977). A stimulatory effect of FSH on LH-induced testosterone formation has also been reported in both immature (Chen *et al.*, 1976) and adult (ElSafoury and Bartke, 1974) hypophysectomized rats.

The stimulatory effect of PRL on testicular LH-receptor levels is well documented. Treatment with PRL increases testicular LH binding in golden hamsters with testicular regression induced by a short photoperiod (Bex and Bartke, 1977), in dwarf mice (Bohnet and Friesen, 1976), and in hypophysectomized rats (Zipf *et al.*, 1978). Moreover, treatment of immature (Aragona *et al.*, 1977) or mature (Bélanger *et al.*, 1979) rats with CB 154 (bromoeriptine), an inhibitor of PRL secretion, leads to reduced testicular LH-receptor levels. The data presented herein clearly show that treatment with oPRL can reverse the decrease of testicular LH-receptor levels seen after hypophysectomy in adult male rats. These findings are in agreement with the observation of a 25–45% reversal of the loss of testicular LH receptors in animals treated with a lower dose of PRL (Zipf *et al.*, 1978).

Although the loss of testicular LH receptors seen after hypophysectomy in the adult rat cannot be prevented by FSH, estradiol, or testosterone (Hauger *et al.*, 1977), it is of interest to study the possible interaction of these hormones with LH and PRL. Since estrogens have been found to have a direct inhibitory effect on testicular steroidogenesis (Samuels *et al.*, 1964; De Jong *et al.*, 1973; Tcholakian *et al.*, 1974; Desjardins *et al.*, 1975; Bartke *et al.*, 1977) and increased estrogen formation is seen in the testis during gonadotropin stimulation (De Jong *et al.*, 1973), it has been suggested that estrogens could be responsible for the blockage seen during desensitization. Our data suggest that estrogens can even have an inhibitory effect on testicular LH-receptor levels. We have carried out further studies using Silastic implants of estradiol, and with constant elevated levels of steroids, the reduction in testicular LH-receptor levels is more striking (see Fig. 5) (Bélanger, Kelly, Séguin, and Labrie, in prep.). It should be mentioned that the doses of estrogens needed to affect testicular steroidogenesis *in vitro* are relatively high, and the role of estrogens in the control of androgen biosynthesis *in vivo* remains to be determined.

Since the Leydig-cell number has been reported to remain unchanged until 2 weeks after hypophysectomy (Desjardins *et al.*, 1975), it is likely that the marked loss of testicular LH receptors seen within 48 hr after hypophysectomy in this chapter and a previous report (Hauger *et al.*, 1977) is due to a decrease of the number of LH receptors per Leydig cell. Although possible effects on Leydig-cell numbers will have to be assessed, it is likely that the effects of LH and PRL observed after 5 days of treatment are also due to corresponding changes of LH-receptor numbers per Leydig cell.

7. Possible Direct Testicular Effects of LH-RH Agonists

Although testicular desensitization (loss of LH- and PRL-receptor levels and blockage of testosterone biosynthesis) in male rats can be explained by the LH-RH-induced release of endogenous LH, recent data suggest an extrapituitary site of LH-RH action. Hsueh and Erickson (1979), Arimura *et al.* (1979), and Labrie *et al.* (1980) (also see Chapter 21), have found that treatment of hypophysectomized immature or adult male rats with LH-RH agonists decreases testicular LH-receptor levels. In an experiment aimed at dissociating the pituitary and testicular sites of LH-RH agonist action, male rats were injected with 1 μg [D-Ser(TBU)^6,des-GlyNH$_2$^{10}]LH-RH ethylamide 12 hr before and immediately after parapharyngeal hypophysectomy and sacrificed 2 days later. In control hypophysectomized rats, testicular LH receptors were 893 $\pm$ 57 fmol/g testis. Administration of the LH-RH analogue before hypophysectomy induces a 50% decline in receptor levels ($p < 0.01$) as compared to only a 22% ($p < 0.05$) decline in animals injected immediately after hypophysectomy.

Such observations of a decrease of testicular LH receptors in the absence of the pituitary gland clearly suggest a direct testicular site of action of LH-RH agonists.

8. Summary and Conclusions

The testicular LH, FSH, and PRL receptors are good examples of the multiplicity of factors involved in the regulation of hormone receptors. In this chapter, we showed that not only LH but also sex steroids and LH-RH agonists can cause a loss of testicular LH receptors. Moreover, PRL can play a major role in the maintenance of testicular LH receptors after hypophysectomy and can even prevent the inhibitory effect of LH-RH agonist on testicular down-regulation of LH receptors.

Discussion

SPILMAN: In your experiments where hCG was administered for several weeks, you attributed the rebound of LH receptors, after the initial down-regulation, to the formation of antibodies to hCG and the resultant decrease of the effectiveness of injected hCG. What would happen to LH receptors if increasing doses of hCG, LH, or LH-RH were administered during the rebound phase?

KELLY: This study has not been carried out, but I would assume the results would be another down-regulation of LH receptors.

C.A. VILLEE: Have you looked for direct effects of LH-RH on Leydig cells in culture?

KELLY: Yes, we are currently measuring LH-RH receptors in testicular membranes and Leydig cells grown in primary culture. I can sum up our results by saying that there appear to be specific LH-RH receptors (using ^{125}I-labeled LH-RH agonists) in Leydig cells.

BANIK: After 4 weeks of treatment by LH-RH, you are perhaps not able to induce decreased fertilizing ability of spermatozoa which were produced. Spermogenesis takes about 18 days in the rat and duration of sperm transport in the epididymis about 10 days. Possibly the effects should appear on fertilizing ability earlier than on sperm number in the ejaculate, and on testicular sperm production earlier than in ejaculate, due to the presence of a reserve of mature spermatozoa in the epididymis.

NOLIN: If, as appears to be the case for the testis, only a small fraction of total receptors need be occupied by hormone for maximal steroidogenic response, how can one reconcile the very nice correlations of decreased response with losses of large numbers of "unnecessary" receptors which you have demonstrated with what may be happening in individual target cells?

KELLY: Although only a small percentage of testicular receptors need to be occupied to induce an action, reducing the total number of available receptors would also reduce the chance of a sufficient number of receptors being occupied to induce a response. This block could be possibly overcome by using higher concentrations of hormone or one with a longer half-life (hCG vs. LH, for example).

NOLIN: The "model" for peptide hormone–target cell interaction you have just shown resembles very closely the one presented by Drs. Ordronneau and Petrusz earlier this week (see Chapter 10). Their model derives largely from their studies of gonadotropins other than prolactin and is based on ideas previously set forth by others, notably C. M. Szego. Would you please cite the evidence which supports this model for *prolactin?* And would you indicate which parts of the model still require substantiation?

KELLY: The model I showed in slide form was meant to be general for the action of peptide hormones with their receptors and the subsequent induction of action. Although not the subject of the presentation, I can summarize our current thoughts on this problem. Prolactin is a hormone for which the second messenger is not known. Therefore, a possible candidate is the hormone, receptor, or hormone–receptor complex itself. There is strong evidence that prolactin receptors are found within the cell, in the Golgi, lysosomes, and polysomes. Prolactin has been localized within the cell by you as well as others, with a recent report using ^{125}I-labeled PRL in rabbit mammary gland showing some labeled PRL even at the nuclear level. The concentration of PRL receptors is much higher in the Golgi than in the plasma membrane. Lysosomes appear to be involved in the "down-regulation" of prolactin receptors, and polysomes and perhaps Golgi are important in replenishing PRL receptors, since they appear to be turned over at a rapid rate.

HAOUR: Could you comment on the inhibiting effect of steroids on the level of hCG-binding sites in hypophysectomized rats and how you think that if any steroid is decreasing hCG-receptor levels, the effect of LH-RH could not be mediated by steroid production induced at the adrenal level, for example?

KELLY: The effect of hCG or LH-RH in reducing testicular LH-receptor levels is much more marked than that observed after estrogens, for example, and it is unlikely that LH-RH and its agonistic analogues exert their effect via the adrenal.

BOGDANOVE: Your diagram, and I think other, similar diagrams such as that shown previously at this meeting by Dr. Petrusz, represent entry of hormone (H) and receptor (R) exclusively as an H–R complex. However, the evidence I am aware of indicates only the internalization of H, with sustained conjunction of H and R, and consequent internalization of R, being only one of two ways H might enter the cell. Am I correct in believing that there is still no *direct* evidence that R is internalized in conjunction with H?

KELLY: Most of the evidence to date has demonstrated the entry of only the hormone into the cell, by either direct or indirect means. For example, Ira Pastan's group in Bethesda has shown the internalization of ferritin-labeled peptides into the cell using video intensification microscopy. In addition, peptide hormones such as prolactin have been reported to be inside mammary and adrenal cells by Dr. Nolin. There have been one or two studies with antiserum against receptors (for prolactin, for example) which have demonstrated receptors within the cell. In addition, intercellular hormone receptors (in Golgi, lysosomes, etc.) have been localized using labeled ligands. However, the actual process of internalization of receptor using techniques specific for receptors rather than hormones is yet to be reported.

BOGDANOVE: But that is an artificial molecule and not relevant to whether, physiologically, the hormone and receptor remain in conjunction during the process of hormone internalization and during its subsequent redistribution within the cell. It does seem that with antibodies to receptor as well as hormone now available, an effort should be made to see whether and when and where the H–R complex can be found in the cell.

KELLY: This is a research area in which a great deal of effort is currently being concentrated, and some positive results should soon be found.

REFERENCES

Aragona, C., Bohnet, H.G., and Friesen, H.G., 1977, Localization of prolactin binding in prostate and testis: The role of serum Prolactin concentrations on the testicular LH receptor, *Acta Endocrinol.* **84**:402.

Arimura, A., Serafini, P., Talbot, S., and Schally, A.V., 1979, Reduction of testicular luteinizing hormone/human chorionic gonadotropin receptors by [D-Trp^6]-luteinizing hormone-releasing hormone in hypophysectomized rats, *Biochem. Biophys. Res. Commun.* **90**:687.

Auclair, C., Kelly, P.A., Coy, D.H., Schally, A.V., and Labrie, F., 1977a, Potent inhibitory activity of [D-Leu^6,des-Gly-NH_2^{10}]LHRH ethylamide on LH/hCG and PRL testicular receptor levels in the rat, *Endocrinology* **101**:1890.

Auclair, C., Kelly, P.A., Labrie, F., Coy, D.H., and Schally, A.V., 1977b, Inhibition of testicular luteinizing hormone receptor level by treatment with a potent luteinizing hormone-releasing hormone agonist or human chorionic gonadotropin, *Biochem. Biophys. Res. Commun.* **76**:855.

Auclair, C., Ferland, L., Cusan, L., Kelly, P.A., Labrie, F., Azadian-Boulanger, G., and Raynaud, J.-P., 1978, Effet inhibiteur de la LHRH sur les récepteurs de la LH dans le testicule chez le rat, *C. R. Acad. Sci.* **286:**1305.

Bartke, A., Williams, K.I.H., and Dalterio, S., 1977, Effect of estrogens on testicular testosterone production *in vitro, Biol. Reprod.* **17:**645.

Bartke, A., Hafiez, A.A., Bex, F.J., and Dalterio, S., 1978, Hormonal interaction in regulation of androgen secretion, *Biol. Reprod.* **18:**44.

Bélanger, A., Auclair, C., S;ğuin, C., Kelly, P.A., and Labrie, F., 1979, Down-regulation of testicular androgen biosynthesis and LH receptor levels by an LHRH agonist: Role of prolactin, *Mol. Cell. Endocrinol.* **13:**47.

Bélanger, A., Caron, S., and Picard, V., 1980, Simultaneous radioimmunoassay of progestins, androgens and estrogens in rat testis, *J. Steroid Biochem.* **13:**123.

Bex, F.J., and Bartke, A., 1977, Testicular LH binding in the hamster: Modification by photoperiod and prolactin, *Endocrinology* **100:**1223.

Bohnet, H.G., and Friesen, H.G., 1976, Effect of prolactin and growth hormone on prolactin and LH receptors in the dwarf mouse, *J. Reprod. Fertil.* **48:**307.

Catt, K.J., Dufau, M.L., and Tsuruhara, T., 1972, Radioligand receptor assay of luteinizing hormone and chorionic gonadotropin, *J, Clin. Endocrinol. Metab.* **34:**123.

Catt, K.D., Bankal, A.J., Davies, T.F., and Dufau, M.L., 1979, Luteinizing hormone releasing hormone-induced regulation of gonadotropin and prolactin receptors in the rat testis, *Endocrinology* **104:**17.

Chen, Y.D.I., and Payne, A.J., 1977, Regulation of testicular LH receptors by homologous hormne: *In vitro* studies on receptor occupancy and receptor loss, *Biochem. Biophys. Res. Commun.* **74:**1589.

Chen, Y.D.I., Payne, A.H., and Kelch, R.P., 1976, Stimulation of Leydig cell function in the hypophysectomized immature rat, *Proc. Soc. Exp. Biol. Med.* **153:**473.

Costlow, M.E., and McGuire, W.L., 1977, Autoradiographic localization of binding of ^{125}I-labeled prolactin to rat tissues *in vitro, J. Endocrinol.* **75:**221.

De Jong, F.H., Hey, A.H., and Van der Molen, H.J., 1973, Effect of gonadotropins on the secretion of estradiol-17β and testosterone by rat testis, *J. Endocrinol.* **57:**277.

Desjardins, C., Zeleznik, A.J., and Midgley, A.R., 1975, Binding of radioiodinated gonadotropins to testes after induction of testicular dysfunction, *Endocrinology* **96:**145A.

ElSafoury, S., and Bartke, A., 1974, Effects of follicle-stimulating hormone and luteinizing hormone on plasma testosterone levels in hypophysectomized and in intact immature adult male rats, *J. Endocrinol.* **61:**193.

Gavin, P.A., Roth, J., Neville, D.M., De Meyts, P., and Buell, D.M., 1974, Insulin-dependent regulation of insulin receptor concentrations: A direct demonstration in cell culture, *Proc. Natl. Acad. Sci. U.S.A.* **71:**84.

Hauger, R.L., Chen, Y.D.I., Kelch, R.P., and Payne, A.H., 1977, Pituitary regulation of Leydig cell function in the adult male rat, *J. Endocrinol.* **74:**57.

Hinkle, P.M., and Tashjian, H.H., 1976, Thyrotropin-releasing hormone regulates the number of its own receptors in the GH_3 strain of pituitary cells in culture, *Biochemistry* **14:**3845.

Hsueh, A.J.W., and Erickson, G.F., 1979, Extrapituitary inhibition of testicular function by luteinizing hormone-releasing hormone, *Nature (London)* **281:**66.

Hsueh, A.J.W., and Dufau, M., and Catt, K.J., 1976, Regulation of luteinizing hormone receptors in testicular interstitial cells by gonadotropin, *Biochem. Biophys. Res. Commun.* **72:**1145.

Hsueh, A.J.W., Dufau, M.L., and Catt, K.J., 1977, Gonadotropin-induced regulation of luteinizing hormone receptors and desensitization of testicular 3′,5′-cyclic AMP and testosterone responses, *Proc. Natl. Acad. Sci. U.S.A.* **74:**592.

Kahn, C.R., 1976, Membrane receptors for hormones and neurotransmitter, *J. Cell Biol.* **70:**261.

Kalla, N.R., Nisula, B.C., Ménard, R.H., and Loriaux, D., 1977, Estrogen modulation of Leydig cell function, *Endocrinology* **100:**82A.

Kelly, P.A., De Léan, A., Auclair, C., Cusan, L., and Labrie, F., 1979, Regulation of peptide hormone receptors, in: *Clinical Endocrinology: A Pathological Approach* (G. Tolis, F. Labrie, J.B. Martin, and F. Naftolin, eds.), pp. 183–197, Raven Press, New York.

Kramer, C.Y., 1956, Extension of multiple range test to group means with unequal numbers of replications, *Biometrics* **12:**307.

Labrie, F., Auclair, C., Cusan, L., Kelly, P.A., Pelletier, G., and Ferland, L., 1978, Inhibitory effects of LHRH and its agonists on testicular gonadotropin receptors and spermatogenesis in the rat, *Int. J. Andrology,* Suppl. **2:**303.

Labrie, F., Bélanger, A., Pelletier, G., Séguin, C., Cusan, L., Kelly, P.A., Lemay, A., Auclair, C., and Raynaud, J.P., 1980, LHRH agonists: Inhibition of testicular functions and possible clinical applications, in: *Clinics in Andrology,* Vol. 5, *Regulation of Male Fertility* (G.R. Cunningham, W.B. Schill, and E.S.E. Hafez, eds.), pp. 64–75, Martinus Nijhoff, The Hague, The Netherlands.

Lesniak, M.A., and Roth, J., 1976, Regulation of receptor concentration by homologous hormone: Effects of human growth hormone on its receptor in IM-9 lymphocytes, *J. Biol. Chem.* **251:**3720.

Mendelson, C., Dufau, M.L., and Catt, K.J., 1975, Gonadotropin binding and stimulation of cyclic adenosine 3′,5′-monophosphate and testosterone production in isolated Leydig cells, *J. Biol. Chem.* **250:**8818.

Mukherjee, C., Caron, M.G., and Lefkowitz, R.L., 1975, Catecholamine-induced subsensitivity of adenylate cyclase associated with a loss of beta-adrenergic receptor binding sites, *Proc. Natl. Acad. Sci. U.S.A.* **72:**1945.

Odell, W.D., and Swerdloff, R.S., 1975, The role of testicular sensitivity to gonadotropins in sexual maturation of the male rat, *J. Steroid Biochem.* **6:**853.

Rivier, C., Rivier, J., and Vale, W., 1979, Chronic effects of [D-Trp^6,Pro^9-NET]luteinizing hormone-releasing factor in reproductive processes in the male rat, *Endocrinology* **105:**1191.

Rodbard, D., and Lewald, J.E., 1970, Computer analysis of radioligand assay and radioimmunoassay data, in: *Karolinska Symposia on Research Methods in Reproductive Endocrinology,* No. 2 (E. Diczfalusy, ed.), pp. 79–103, Bogtrykkeriet, Copenhagen.

Samuels, L.T., Short, J.G., and Huseby, 1964, The effect of diethylstilbestrol on testicular 17α-hydroxylase and 17-desmolase activities in BALB/C mice, *J. Endocrinol.* **63:**411.

Schwarzstein, L., 1976, Diagnostic and therapeutic use of LHRH in the infertile man, in: *Hypothalamus and Endocrine Functions* (F. Labrie, J. Meites, and G. Pelletier, eds.), pp. 73–91, Plenum Press, New York.

Sharpe, M., 1976, Relationships between testosterone, fluid content and luteinizing hormone receptors in rat testis, *Nature (London)* **264:**644.

Tcholakian, R.K., Chowdhury, M., and Steinberger, E., 1974, Time action of estradiol-17β on luteinizing hormone and testosterone, *J. Endocrinol.* **63:**411.

Thanki, K.H., and Sternberger, A., 1976, ^{125}I-LH binding to rat testes at various ages and post-hypophysectomy, *Endocr. Res. Commun.* **3:**49.

Zeleznik, A.J., Midgley, A.R., and Reichart, L.E., 1974, Granulosa cell maturation in the rat: Increased binding of human chorionic gonadotropin following treatment with follicle-stimulating hormone, *J. Biol. Chem.* **250:**8818.

Zipf, W.P., Payne, A.A., and Kelch, R.P., 1978, Prolactin, growth hormone and luteinizing hormone in the maintenance of testicular luteinizing hormone receptors, *Endocrinology* **103:**595.

21

Inhibition of Testicular Androgen Biosynthesis by Treatment with LH-RH Agonists

Alain Bélanger, Simon Caron, Lionel Cusan, Carl Séguin, Claude Auclair, and Fernand Labrie

1. Introduction

It is well known that testicular function is controlled by hormones of the anterior pituitary gland. The absolute requirement for these hormones is demonstrated by the rapid atrophy of the sex accessory organs and inhibition of spermatogenesis following hypophysectomy (Wood and Simpson, 1961). The production of testicular steroids, in particular testosterone, is under the control of luteinizing hormone (LH), which is able to stimulate steroidogenesis in a variety of *in vitro* systems such as perfused testes, incubation of decapsulated testes, and suspensions of purified Leydig cells (Tsuhuhara *et al.*, 1977). There is good evidence showing that the action of LH in Leydig cells involves binding to specific receptors located on the plasma membrane, stimulation of cyclic AMP formation, and activation of protein kinase. The acute steroidogenic response of the testis to LH depends primarily on the activation of the enzymes controlling cholesterol side-chain cleavage, although there is also evidence for control mechanisms at a later step in the steroid biosynthetic pathway (Chasalow, 1979).

Alain Bélanger, Simon Caron, Lionel Cusan, Carl Séguin, Claude Auclair, and *Fernand Labrie* • Department of Molecular Endocrinology, Le Centre Hospitalier de l'Université Laval, Quebec G1V 4G2, Canada.

The concept that several peptide hormones can specifically regulated their homologous binding sites in target tissues is now well established (Kahn *et al.,* 1973; Lesnick and Roth, 1976). This hormonal regulation of receptor sites is characterized in most cases by an inverse relationship between the concentration of the circulating hormone and the tissue level of receptors. This negative regulation of the receptor population has been especially well demonstrated to occur in ovarian and testicular tissue after administration of exogenous gonadotropins (Hsueh *et al.,* 1976; Sharpe, 1976; Auclair *et al.,* 1977a). Moreover, a marked loss of the androgen response to gonadotropins has been observed in the rat after systematic administration of ovine LH (oLH) or human chorionic gonadotropin (hCG) (Cigorraga *et al.,* 1978; Purvis *et al.,* 1977; Sharpe and McNeilly, 1977). It therefore occurred to us that the relative lack of therapeutic success of LH-releasing hormone (LH-RH) and its agonistic analogues in the treatment of oligospermia and male infertility in men could be attributed to desensitization of testicular receptors and decreased steroidogenesis following a marked increase of endogenous gonadotropin induced by administration of the gonadotropin-releasing peptides.

This chapter describes the inhibitory effect of an LH-RH agonist, [D-Ala^{6},des-Gly-NH_2^{10}]LH-RH ethylamide (LH-RH-A), on testicular gonadotropin receptors and steroidogenesis in the adult male rat. It also summarizes the comparative effect of LH-RH-A and hCG on testicular LH and prolactin receptors and testicular steroid concentrations. Following the observation of such dramatic effects of LH-RH and its agonists in the rat, we will describe the inhibitory effects of single intranasal administration of another LH-RH agonist, [D-Ser(TBU)^{6},des-Gly-NH_2^{10}]LH-RH ethylamide, on serum androgen levels in normal adult men. Finally, findings supporting a major role of prolactin in the control of testicular steroidogenesis will also be described.

2. Steroidogenesis in Adult Rat Testis

The conversion of pregnenolone to testosterone is well known to take place in testicular tissue from adult rats (Slaunwhite and Samuels, 1956), and upon *in vitro* incubation of tritiated pregnenolone or progesterone with testicular homogenate, several intermediates have been isolated, testosterone being the major metabolite. The transformation of pregnenolone into testosterone can occur through two biosynthetic routes, the Δ_5 (pregnenolone → 17-OH-pregnenolone → dehydroepiandrosterone → androst-5-ene-3β,17β-diol → testosterone) and Δ_4 (progesterone → 17-OH-progesterone → androstenedione → testosterone) pathways, in Leydig cells. The relative importance of the two pathways varies according to species, but the Δ_4 pathway appears to be the main biosynthetic route in the adult male rat (Fig. 1).

To study the testicular steroidogenic pathways in rats, a reliable method that permits the simultaneous measurement of pregnenolone, 17-OH-pregnenolone, progesterone, 17-OH-progesterone, androstenedione, testosterone, androst-5-ene-3β,17β-diol, dihydrotestosterone, and estradiol in extracts of rat testis or plasma has been developed. After extraction of the testicular homogenate with methanol, the method includes separation on an LH-20 column and steroid measurement by specific radioimmunoassays (Bélanger *et al.*, 1980a).

3. Down-Regulation of Testicular Androgen Biosynthesis and LH Receptors by an LH-RH Agonist

Our previous studies have shown that treatment of adult male rats with LH-RH or its agonistic analogues leads to a marked loss of testicular LH receptors accompanied by decreased testis, ventral-prostate, and

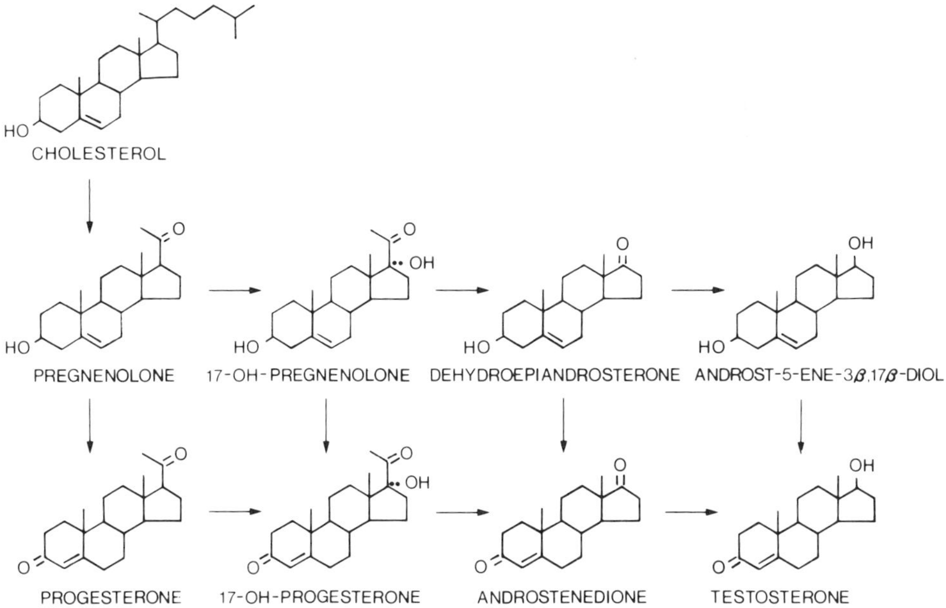

Figure 1. Biosynthesis of testosterone from cholesterol.

seminal-vesicle weight as well as lowered plasma testosterone concentration (Auclair *et al.*, 1977a,b; Labrie *et al.*, 1978).

Studies performed with rat Leydig cells desensitized by hCG treatment have shown that the hCG-induced desensitization process is accompanied by a defect of 17,20-desmolase activity (Cigorraga *et al.*, 1978). Moreover, it could be clearly seen that the extent and nature of the block in the androgen biosynthetic pathway are highly dependent on the dose of hCG used. To further analyze the steroidogenic pathway during desensitization induced by changes of endogenous gonadotropin secretion, we have studied the time-course of the effect of daily administration of 1 μg [D-Ala^6,des-Gly-$NH_2{}^{10}$]LH-RH ethylamide on testicular and plasma levels of steroid intermediates. Changes of testicular LH, prolactin, follicle-stimulating hormone (FSH) receptor levels have been correlated with changes of steroidogenesis.

In agreement with our previous observations (Auclair *et al.*, 1977a, b; Labrie *et al.*, 1978), daily treatment with a maximal dose of [D-Ala^6,des-Gly-$NH_2{}^{10}$]LH-RH ethylamide, a peptide approximately 200 times more potent than LH-RH itself on LH release, led to a progressive decrease of testicular LH receptors to 10–20% of control at 4 and 8 days (Fig. 2). No significant effect could be found at 24 hr. The inhibitory effect on testicular prolactin-receptor levels was more rapid, an inhibition to 40% of control being already seen 24 hr after the first injection of the

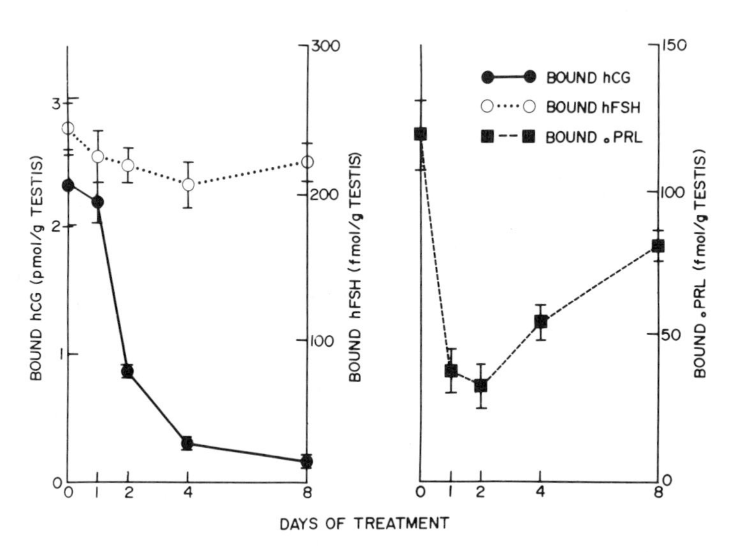

Figure 2. Time-course of the effect of daily subcutaneous injection of 1 μg [D-Ala^6,des-Gly-$NH_2{}^{10}$]LH-RH ethylamide on testicular LH, FSH *(left)*, and prolactin *(right)* receptor levels in adult rats. Data are presented as means ± S.E.M.

peptide. Testicular prolactin receptors decreased to 30% of control at 2 days, with a progressive return to 70% of control levels at 8 days. No significant effect was obtained on the testicular FSH-receptor concentration up to the last time interval studied (8 days).

As illustrated in Fig. 3, testicular pregnenolone levels did not change during treatment, while those of progesterone were markedly increased at days 1 and 2, with a return to normal at later time intervals. Testicular 17-OH-progesterone levels, on the other hand, decreased between days 2 and 4 to 30% of control at days 4 and 8. The data showing an accumulation of progesterone at early time intervals during treatment with the LH-RH agonist suggest that a blockage could exist between progesterone and 17-OH-progesterone at the level of the hydroxylase system.

The progressive and dramatic fall in testicular androstenedione and testosterone levels that follows treatment with the LH-RH agonist is illustrated in Fig. 4. It can be seen that 24 hr after the first injection of the LH-RH agonist, testicular levels of testosterone and androstenedione were already 25% reduced ($p < 0.05$) at a time when the concentration of LH receptors was still normal. The rapid decrease of androstenedione levels suggests that the enzyme responsible for the conversion of 17-OH-progesterone to androstenedione, the 17,20-desmolase, could also be affected by the LH-RH agonist. Following multiple injections of the LH-

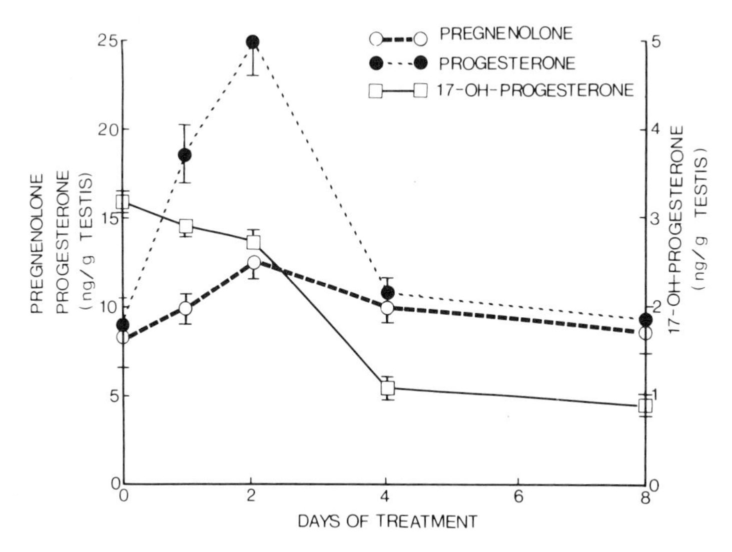

Figure 3. Time-course of the effect of daily subcutaneous injection of 1 μg [D-Ala^6,des-Gly-NH$_2$^{10}]LH-RH ethylamide on testicular pregnenolone (○), progesterone (●), and 17-OH-progesterone (□) levels in adult rats.

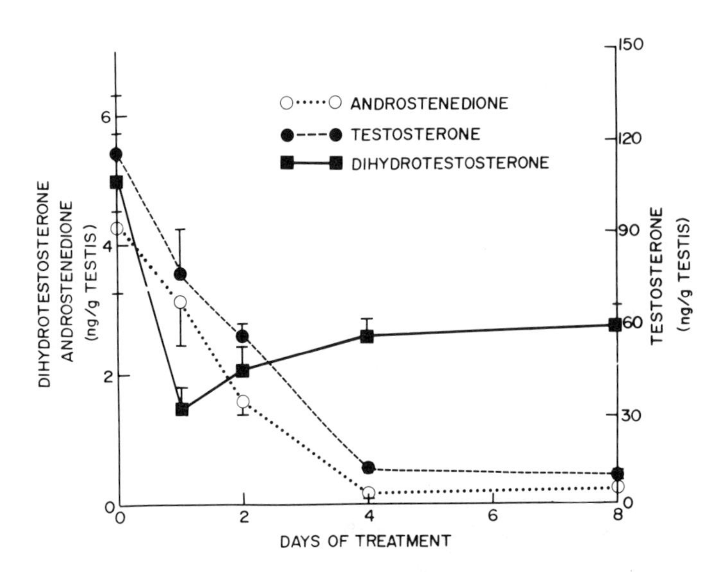

Figure 4. Time-course of the effect of daily subcutaneous injection of 1 μg [D-Ala^6,des-Gly-NH$_2$^{10}]LH-RH ethylamide on testicular androstenedione (○), testosterone (□), and dihydrotestosterone (■) in adult rats.

RH agonist, a progressive decrease to 5–15% of control was then observed for both androstenedione and testosterone. The inhibitory effect on 5α-dihydrotestosterone levels was, however, of much lower amplitude: the decrease to 30% of control observed at 1 day was followed by a slow increase reaching 50% of control values at 8 days.

Contrary to the marked changes observed in the androgen biosynthetic pathway after treatment with the LH-RH agonist, testicular estrone and 17β-estradiol levels remained unchanged during treatment with the peptide. It should also be mentioned that changes of plasma levels of progesterone, testosterone, and 17β-estradiol are parallel to those of the testicular levels of these steroids.

Although the present observations strongly suggest that LH-RH agonist treatment has inhibitory effects on testicular 17-hydroxylase and 17,20-desmolase activities and the 5α-reductase and aromatase activies appear to be increased, detailed analysis of these enzymatic activities remains to be performed. Preferential metabolism of the steroid intermediates through other biosynthetic routes could also explain some of the changes observed. Furthermore, possible changes of peripheral clearance rates of steroids could also be involved.

Since the study described above has shown that the nature and extent of blockage of the steroidogenic pathway are dependent on the duration of treatment with the LH-RH agonist, the purpose of the next study was to investigate the effect of increasing doses of the same LH-RH agonist, [D-Ala^6,des-Gly-$NH_2{}^{10}$]LH-RH ethylamide, on testicular gonadotropin receptors and steroid levels under basal conditions as well as after stimulation with oLH.

As previously observed with the LH-RH agonist [D-Leu^6,des-Gly-$NH_2{}^{10}$]LH-RH ethylamide (Auclair *et al.*, 1977a, b), daily injection of increasing doses of [D-Ala^6,des-Gly-$NH_2{}^{10}$]LH-RH ethylamide for 9 days induced a dose-dependent inhibition of testicular LH-receptor levels. While no significant effect was observed with the 1- and 5-ng doses, a 50% decrease in the number of LH receptors was found at the dose of 10 ng, while a maximal inhibition to 5–10% of control was obtained after administration of the 50- or 250-ng dose. A progressive decrease of testicular prolactin-receptor levels was also observed, a maximal effect being obtained with daily administration of 10 ng of the LH-RH agonist.

As shown in Fig. 5A, administration of the agonistic analogue at doses of 10 ng or higher caused a progressive decrease of basal testicular 17-OH-progesterone levels, while the progesterone concentration remained unchanged. When the testicular progesterone concentration was measured 2 hr after injection of 10 μg oLH (Fig. 5B), a 100% stimulation was observed at the 1-ng dose and no significant change was observed at higher doses. While administration of increasing doses of the LH-RH agonist led to a progressive inhibition of basal testicular 17-OH-progesterone levels to approximately 50% of control, the response to oLH was inhibited to 10–15% of control (Fig. 5B). The observation of such a marked inhibition of the 17-OH-progesterone response to oLH in the presence of unchanged testicular progesterone levels also suggests an enzymatic defect at the level of the 17-hydroxylase.

As illustrated in Fig. 6A, the basal levels of testicular testosterone and androstenedione were not affected after administration of 1 or 5 ng of the agonistic analogue, while the 50- and 250-ng doses induced a marked decrease of the level of these androgens to 5–10% of control. The 10-ng dose of the agonistic analogue induced a 50% decrease of basal testicular testosterone levels, while no significant effect was observed on androstenedione concentration. When the testicular testosterone and androstenedione responses to oLH were measured, a biphasic pattern was observed. At the lowest doses (1, 5, or 10 ng) of the agonistic analogue, the androstenedione and testosterone responses to oLH were increased (Fig. 6B), a maximal stimulatory effect being observed at the 5-ng dose. The androstenedione and testosterone responses to oLH were increased from 452 ± 22 and 815 ± 40 to 745 ± 22 and 1511 ± 150 ng/g testis, respectively. By contrast, an almost complete inhibition of the androgen response to oLH was seen with the 50- or 250-ng dose of the

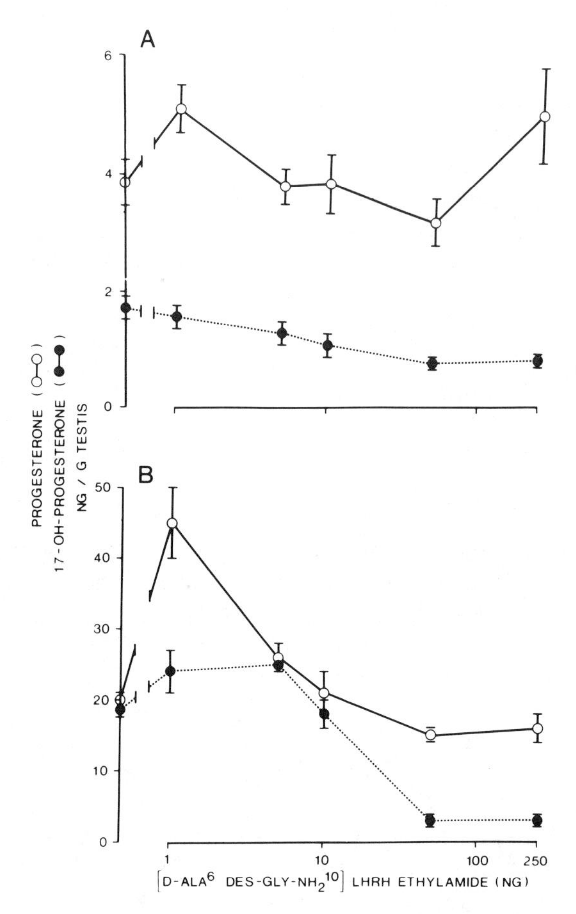

Figure 5. Effect of treatment with increasing doses of [D-Ala^6,des-Gly-NH_2^{10}]LH-RH ethylamide injected once daily for 9 days on testicular progesterone (○) and 17-OH-progesterone (●) concentrations. (A) Basal concentrations; (B) response to 10 μg oLH (2 hr after injection).

LH-RH agonist (Fig. 6B). It should be mentioned that slightly increased androstenedione and testosterone responses to oLH were obtained in animals treated with the 10-ng dose of the agonistic analogue when only 50% of LH receptors were present.

It is of great interest that treatment with low doses of the LH-RH agonist led to an increased steroidogenic response to oLH. In fact, after daily treatment with 1 or 5 ng [D-Ala^6,des-Gly-NH_2^{10}]LH-RH ethylamide for 9 days, the testicular response of progesterone, 17-OH-progesterone, androstenedione, and testosterone were increased in the presence of

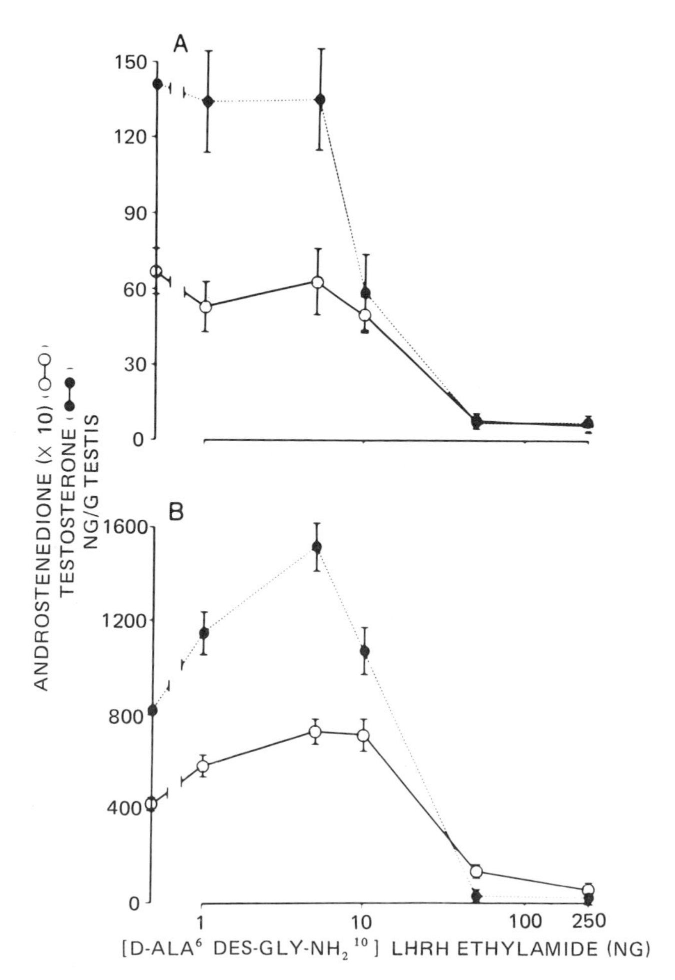

Figure 6. Effect of treatment with increasing doses of [D-Ala^6,des-Gly-NH_2^{10}]LH-RH ethylamide injection once daily for 9 days on testicular androstenedione (○) and testosterone (●) concentrations. (A) Basal testicular levels; (B) response to 10 μg oLH (2 hr after injection).

unchanged testicular LH-receptor levels. At the 10-ng dose of the LH-RH agonist, which led to a 50% loss of LH receptors, a significant increase of the testicular androstenedione and testosterone as well as plasma testosterone responses to oLH was also observed. Such findings raise the interesting possibility that treatment with low doses of LH-RH agonists could stimulate Leydig cell function, while high doses could be used to inhibit testicular functions (Labrie *et al.*, 1978; Pelletier *et al.*,

1978). In agreement with the aforementioned data describing the time-course of the effect of treatment with a daily dose of 1 μg of the LH-RH agonist [D-Ala^6,des-Gly-NH_2^{10}]-LH-RH ethylamide (Bélanger *et al.*, 1980b), the present data clearly show that a marked inhibition of basal testicular testosterone concentrations apparently due to a decrease of the enzymatic activities of 17-hydroxylase and 17,20-desmolase is obtained after treatment with daily doses of the LH-RH agonist above 10 ng.

4. *Comparative Effects of an LH-RH Agonist and hCG on Testicular Steroidogenesis*

Since a decline in LH-receptor levels and steroidogenesis induced by treatment with hCG is well described (Cigorraga *et al.*, 1978), we next compared the effect of treatment with hCG and an LH-RH agonist on basal and oLH-stimulated testicular steroid levels. A marked loss of testicular LH receptors was seen in animals treated with 1 μg of the LH-RH analogue (40% of control), while 30 IU hCG induced an even greater inhibition (8% of control). The testicular prolactin-receptor levels were also reduced to 65% of control after treatment with the LH-RH agonist, while hCG had no effect on this parameter.

A single injection of the LH-RH analogue led after 3 days to a 40–60% inhibition of testicular pregnenolone, 17-OH-pregnenolone, androst-5-ene-3β,17β-diol, 17-OH-progesterone, androstenedione, testosterone, and dihydrotestosterone levels, while the concentration of progesterone was stimulated to 225% of control and estradiol levels were unaffected (Fig. 7). In contrast, a single injection of hCG induced a significant increase of pregnenolone, 17-OH-pregnenolone, 17-OH-progesterone, androstenedione, testosterone, and estradiol to about 200–250% of control levels in the presence of normal concentrations of androst-5-ene-3β,17β-diol and dihydrotestosterone and a marked stimulation of the progesterone concentration to about 700% of control. It should be mentioned that the ratio of 17-OH-progesterone to progesterone was decreased to approximately the same extent after administration of either the LH-RH agonist or hCG, thus suggesting an inhibition of 17-hydroxylase activity in both cases.

At 2 hr after injection of 10 μg oLH into the LH-RH-agonist-treated rats (Fig. 8), the testicular progesterone levels were markedly stimulated (750% above the control value), while pregnenolone, androstenedione, and estradiol levels were within normal limits and 17-OH-pregnenolone and 17-OH-progesterone levels were 50–95% increased. The response of

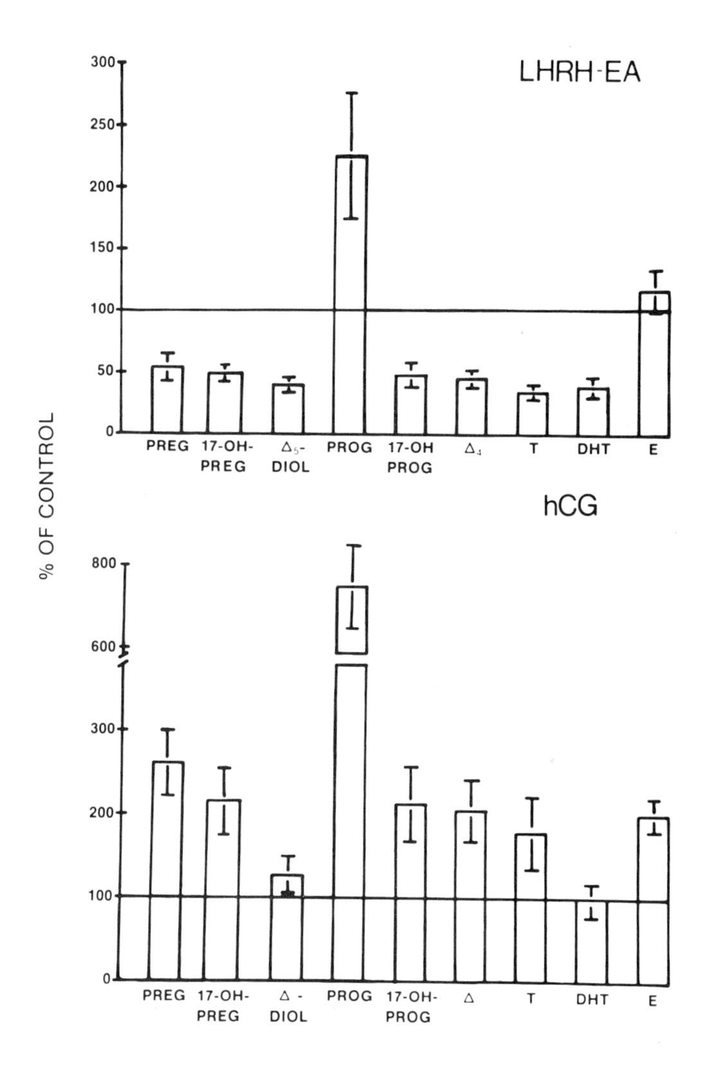

Figure 7. Effect of a single injection of 1 μg [D-Ala^6,des-Gly-NH$_2$^{10}]LH-RH ethylamide *(top panel)* or 30 IU hCG *(bottom panel)* on basal testicular pregnenolone (PREG), 17-OH-pregnenolone (17-OH-PREG), androst-5-ene-3β,17β-diol (Δ_5-DIOL), progesterone (PROG), 17-OH-progesterone (17-OH-PROG), androstenedione (Δ_4), testosterone (T), dihydrotestosterone (DHT), and estradiol (E_2) concentrations in adult male rats. The animals were sacrificed 3 days after injection. The control values were (in ng/g except for E_2): PREG, 20 $\pm$ 2; 17-OH-PREG, 4.9 $\pm$ 0.1; Δ_5-DIOL, 5.9 $\pm$ 0.4; PROG, 8.1 $\pm$ 0.8; 17-OH-PROG, 3.2 $\pm$ 0.4; Δ_4, 5.2 $\pm$ 0.2; T,165 $\pm$ 37; DHT,5.7 $\pm$ 0.3; and E_2 21 $\pm$ 3 pg/g.

the other androgens (androst-5-ene-3β,17β-diol, testosterone, and dihydrotestosterone) showed a significant reduction (Fig. 8). On the other hand, injection of oLH into hCG-treated rats led to a stimulation of progesterone (225%) and estradiol (100%) levels, while the concentration of the other steroids was reduced to about 25–50% of the response observed in saline-treated animals.

In *in vitro* studies performed with hCG-desensitized Leydig cells, it has been demonstrated that the steroidogenic pathway could be stimulated up to 17-OH-progesterone, thus suggesting that only 17,20-desmolase was affected (Cigorraga *et al.*, 1978). In the present *in vivo* study, it appears that following single administration of hCG, a further stimulation of the steroidogenic pathway by oLH can stimulate steroidogenesis up to progesterone and pregnenolone, while 17-OH-progesterone, testosterone, and the other androgen intermediates showed a marked inhibition of the response. Inhibition of both 17-hydroxylase and 17,20-desmolase activities after hCG treatment has also been described recently (Chasalow *et al.*, 1979).

The presence of high basal steroid levels 3 days after a single administration of hCG is probably due to the long half-life of the gonadotropin. It is in fact well known that high circulating hCG levels can lead to high plasma testosterone levels in the presence of desensitized Leydig cells (Auclair *et al.*, 1977a; Sharpe, 1977; Saez *et al.*, 1978).

Although the nature and extent of the enzymatic defects are dependent on the dose and time of administration, it appears that treatment with LH-RH agonists and hCG can lead to the same enzymatic defects at the level of 17-hydroxylase and 17,20-desmolase.

5. Recovery of Testicular Androgen Formation after 1 Month of Treatment with an LH-RH Agonist

The aforementioned data clearly show that the pituitary–testicular axis is highly sensitive to down-regulation by treatment with LH-RH and its agonists. These findings suggest the potential use of LH-RH agonists in the treatment of androgen-dependent pathologies and as a new approach to male contraception. It thus becomes important to gain precise knowledge of the recovery of testicular functions after cessation of such treatment. We will now describe the time-course of recovery (up to 4 months) of testicular gonadotropin-receptor levels and steroidogenesis after 4 weeks of treatment with the LH-RH agonist [D-Ala^6,des-Gly-NH$_2^{10}$]LH-RH ethylamide in adult male rats.

Testicular LH-receptor levels, which were reduced to 10% of control after 1 month of treatment with [D-Ala^6,des-Gly-NH$_2^{10}$]LH-RH ethylam-

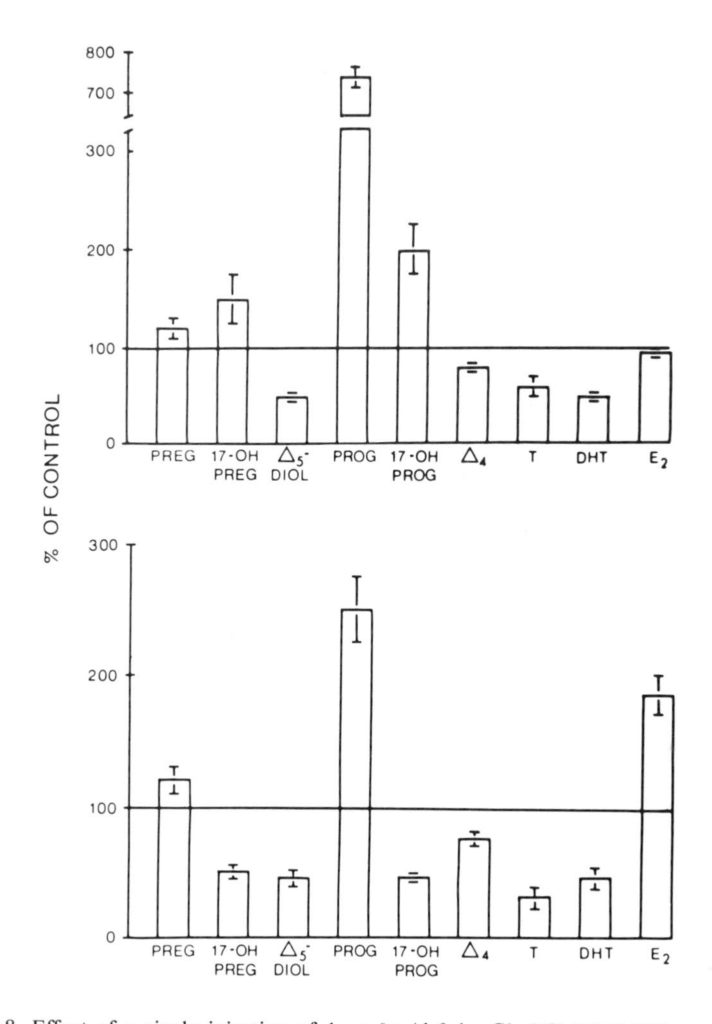

Figure 8. Effect of a single injection of 1 μg [D-Ala^6,des-Gly-NH$_2$^{10}]LH-RH ethylamide *(top panel)* or 30 IU hCG *(bottom panel)* on the testicular steroidogenic response to 10 μg oLH. Pregnenolone (PREG), 17-OH-pregnenolone (17-OH-PREG), androst-5-ene-3β,17β-diol (Δ_5-DIOL), progesterone (PROG), 17-OH-progesterone (17-OH-PROG), androstenedione (Δ_4), testosterone (T), dihydrotestosterone (DHT), and estradiol (E_2) concentrations in the adult male rat. The animals were sacrificed 3 days after injection of the LH-RH agonist or hCG and 2 hr after injection of oLH. The control values were (in ng/g except for E_2): PREG, 51.4 ± 4; 17-OH-PREG, 41 ± 10; Δ_5-DIOL, 26 ± 2; PROG, 24 ± 4; 17-OH-PROG, 17 ± 2; Δ_4, 16 ± 2; T, 852 ± 110; DHT, 13 ± 1; and E_2, 21 ± 3 pg/g.

ide (LH-RH-A) (1.0 μg, s.c., every second day for 4 weeks), had returned to normal values 1 month later. It has also been observed that the 35% loss of testicular prolactin receptors measured at the end of treatment with the LH-RH analogue had completely disappeared during the same recovery period. Both LH- and prolactin-receptor levels remained within normal values up to the last time interval studied (4 months after cessation of LH-RH-A administration).

In agreement with the changes of ventral-prostate and seminal-vesicle weight, it can be seen in Fig. 9 that testicular testosterone and dihydrotestosterone concentrations had returned to normal levels 1 month after cessation of LH-RH-A treatment. Parallel changes of plasma testosterone and dihydrotestosterone levels were observed.

The present day data clearly show that the marked reduction of the concentration of testicular LH and prolactin receptors, the lowering of plasma testosterone and dihydrotestosterone concentration, and the decrease of ventral-prostate and seminal-vesicle weight are reversible within 1 month after treatment of adult rats with the LH-RH agonistic analogue [D-Ala^6,des-Gly-NH_2^{10}]LH-RH ethylamide for the same period of time.

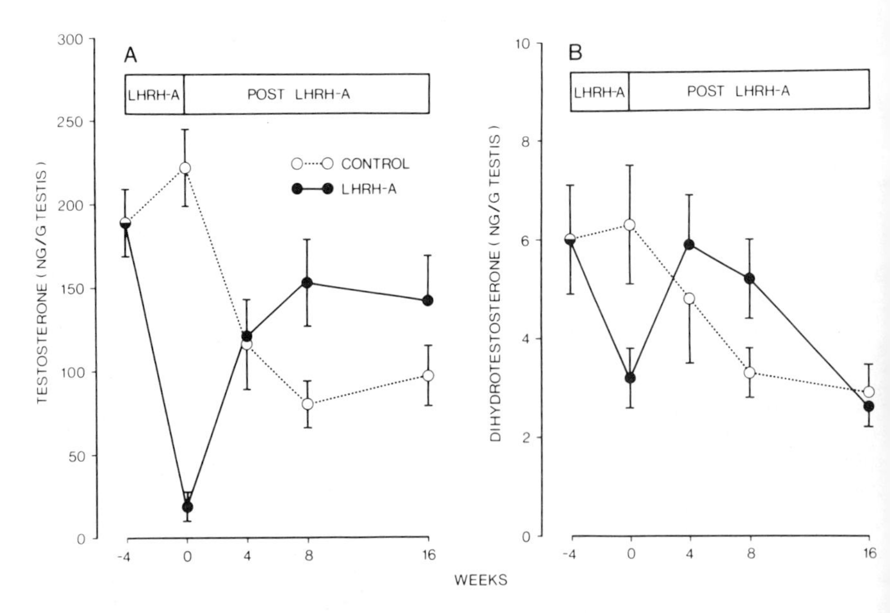

Figure 9. Recovery of testicular testosterone (A) and dihydrotestosterone (B) concentration following 1 month of treatment of adult rats with [D-Ala^6,des-Gly-NH_2^{10}]LH-RH ethylamide at a dose of 1 μg s.c. every second day.

6. Inhibitory Effects of a Single Intranasal Administration of [D-Ser(TBU)^6,des-Gly-NH_2^{10}]LH-RH Ethylamide on Serum Steroid Levels in Normal Adult Men

The relative lack of success of LH-RH and its agonistic analogues in the treatment of oligospermia and male infertility (Schwarzstein, 1976) might well be explained by the aforedescribed findings of a loss of testicular LH and prolactin receptors accompanied by decreased testis, seminal-vesicle, and ventral-prostate weight following treatment of adult male rats with gonadotropin-releasing peptides. This loss of receptors is accompanied by decreased androgen biosynthesis and accumulation of pregnenolone and progesterone, which suggest a blockage in the testicular steroidogenic pathway at the level of the 17-hydroxylase and 17,20-desmolase activities (Bélanger *et al.*, 1979, 1980b). It is thus of great interest to study the possibility of a similar inhibition of testicular steroidogenesis by treatment with LH-RH agonists in man.

Since the potent LH-RH agonist [D-Ser(TBU)^6,des-Gly-NH_2^{10}]LH-RH ethylamide (Hoe 766, Sandow *et al.*, 1978) is active by the intranasal route (Wiegelman *et al.*, 1978), we have studied the effect of a single intranasal administration of 500 μg of this peptide on plasma levels of pregnenolone, 17-OH-pregnenolone, progesterone, 17-OH-progesterone, androst-5-ene-3β, 17β-diol, testosterone, dihydrotestosterone, and 17β-estradiol during 7 consecutive days following treatment in six normal adult men.

In agreement with previous data (Sandow *et al.*, 1978; Wiegelman *et al.*, 1978), the intranasal application of 500 μg [D-Ser(TBU)^6,des-Gly-NH_2^{10}]LH-RH ethylamide led to a long-lasting increase of plasma LH and FSH levels, with a return to normal values 8–12 hr after administration of the peptide (data not shown).

It can be seen in Fig. 10 that the serum levels of all steroids measured were highest at 08:00 hr and lowest at 22:00 hr during the two control pretreatment days. Intranasal administration of 500 μg [D-Ser(TBU)^6,des-Gly-NH_2^{10}]LH-RH ethylamide led to a disturbance of this diurnal cyclicity of steroid levels: on the day of treatment, the serum concentrations of pregnenolone, 17-OH-pregnenolone, progesterone, 17-OH-progesterone, androst-5-ene-3β,17β-diol, and testosterone were increased, this elevation being followed by a loss of cyclicity and lowered serum levels for the next 3 days. The concentrations of 17-OH-progesterone and testosterone were decreased to, respectively, 40 and 50% of controls on day 3 following administration of the peptide. Normal cyclicity and serum steroid levels were observed on the 4th through 6th posttreatment days. The effect of treatment on the pattern of serum 17β-

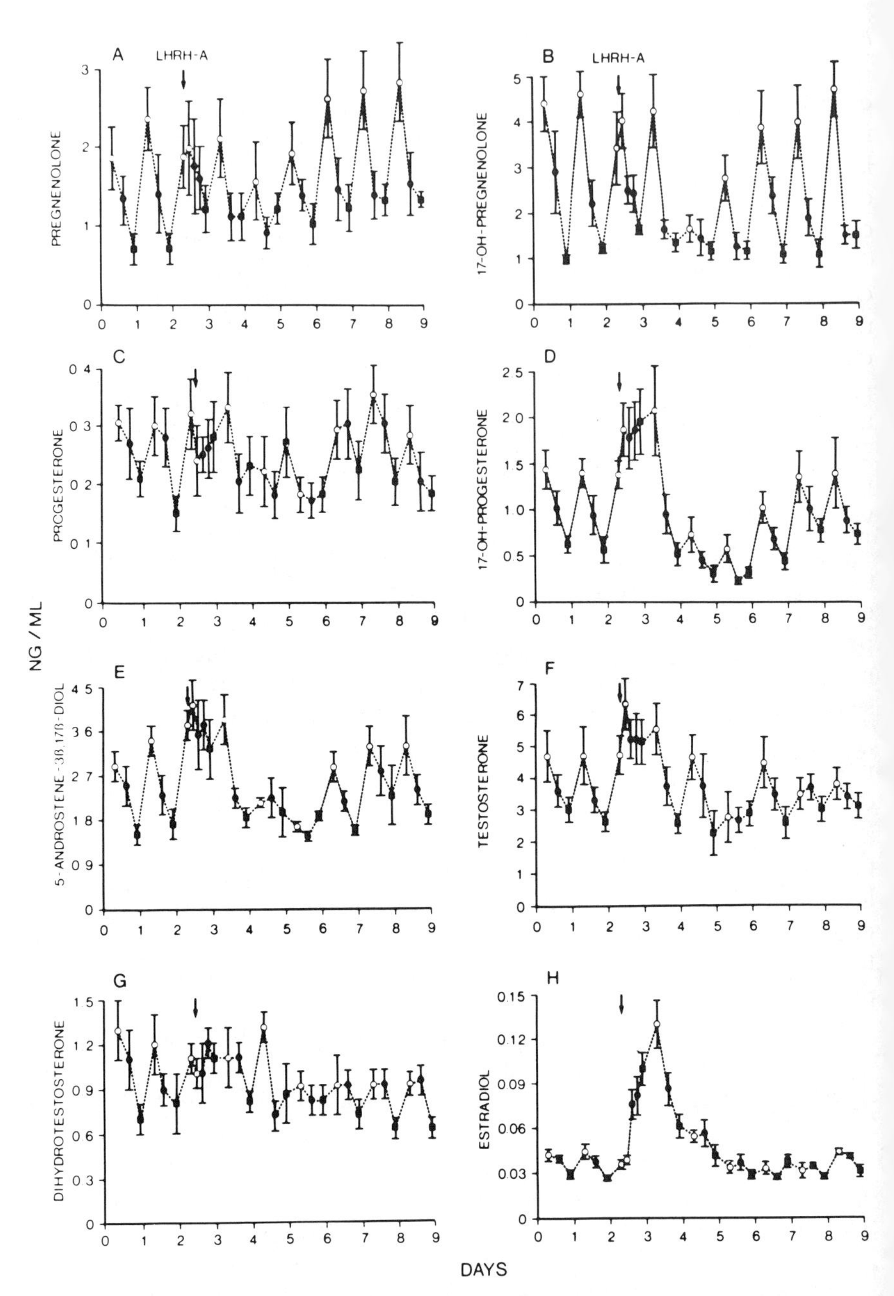

Figure 10. Effect of a single intranasal administration of 500 μg [D-Ser(TBU)^6,des-Gly-NH_2^{10}]LH-RH ethylamide on serum pregnenolone (A), 17-OH-pregnenolone (B), progesterone (C), 17-OH-progesterone (D), androst-5-ene-3β, 17β-diol (E), testosterone (F), dihydrotestosterone (G), and estradiol (H) in normal adult men. Steroid levels were measured on two pretreatments, one treatment, and six posttreatment days. Data are expressed as the means ± S.E.M. of values obtained from six normal adult men. (○) 08:00 hr and 11:00 hr; (●) 15:00 hr and 18:00 hr; (■) 22:00 hr.

estradiol concentrations was, however, completely different: a marked increase on the afternoon of [D-Ser(TBU)^6,des-Gly-NH_2^{10}]LH-RH ethylamide treatment lasted for 24 hr (from 20–35 to 100–250 pg/ml in different subjects) and was followed by a return to normal serum concentration at later time intervals.

In close agreement with the data obtained in the male rat, the present data clearly show that the administration of a potent LH-RH agonist in normal men leads to inhibition of testicular steroidogenesis. The loss of diurnal cyclicity and the decreased serum levels of testosterone and its precursors do in fact last for 3 days following a single intranasal administration of [D-Ser-(TBU)^6,des-Gly-NH_2^{10}]LH-RH ethylamide. These data agree with the recent findings of an inhibition of serum testosterone levels in normal men treated subcutaneously daily for 1 week with 5 μg of the same LH-RH analogue (Smith *et al.*, 1979).

The diurnal variation of serum testosterone in man is well known (Dray *et al.*, 1965). The present data show that this cyclicity can be extended to pregnenolone, 17-OH-pregnenolone, progesterone, 17-OH-progesterone, and androst-5-ene-3β,17β-diol. They also indicate the necessity of multiple daily blood sampling for proper assessment of the secretion of these steroids.

The present findings of a high sensitivity of testicular steroidogenesis to treatment with an LH-RH agonist in normal men and our observation of a complete inhibition of spermatogenesis after chronic treatment with a similar peptide in the male rat (Pelletier *et al.*, 1978) clearly suggest that chronic administration of LH-RH agonists in men could be used to lower circulating androgen levels and offer the possibility of a new contraceptive approach. Such an approach could possibly be useful in the treatment of androgen-dependent pathologies, such as cancer of the prostate and prostatic hyperplasia.

7. *Blockage of the Testicular Steroidogenic Pathway and Role of Prolactin*

Prolactin receptors have been localized on Leydig cells, and there is increasing evidence for a role of prolactin in the control of testicular function (Bartke *et al.*, 1978). These data pertain to the observations that prolactin can: (1) potentiate the stimulatory effect of LH on testosterone synthesis in hypophysectomized rats and spermatogenesis in hypophysectomized mice; (2) increase plasma testosterone levels in hamsters; (3) increase testicular LH binding in hamsters, rats, and dwarf mice; (4) stimulate the activity of testicular 3α- and 17α-hydroxy-steroid dehydrogenase; and (5) promote accumulation of testicular esterified cholesterol.

A role of prolactin in the stimulation of testosterone secretion has also been suggested in man.

Since we had recently found that treatment of adult male rats with LH-RH or its agonistic analogues leads to a marked reduction of testicular LH receptors accompanied by decreased testicular and androgen levels and accessory-sex-organ weight, we next studied the role of endogenous prolactin in this LH-RH-induced testicular desensitization.

The 7-fold increase of plasma prolactin concentration seen in pituitary-implanted animals was accompanied by a 20% rise of testicular LH/hCG-receptor levels. Interestingly, administration of bromocriptine (CB 154) at a dose that completely prevented the rise of plasma prolactin induced by the pituitary implants caused a 60% decrease of LH/hCG-binding sites in both intact and pituitary-implanted animals ($p < 0.01$). Injection of a low dose of [D-Ala^6,des-Gly-NH$_2$^{10}]LH-RH ethylamide (40 ng, LH-RH-EA) every third day led to an approximate 70% inhibition of testicular LH/hCG-receptor levels in both intact rats and animals bearing pituitary implants.

Administration of the LH-RH agonist to intact animals caused a 90 and 60% reduction of testicular testosterone and androstenedione concentrations, respectively ($p < 0.01$), this inhibitory effect being similar in all groups treated with the LH-RH agonist (Fig. 11). On the other hand, the testicular progesterone concentration increased 3.2-fold in intact rats treated with LH-RH ethylamide. This increase was even more dramatic (7-fold) in animals bearing pituitary transplants. That this marked increase of testicular progesterone is dependent on prolactin was further indicated by the finding of an almost complete suppression of the LH-RH agonist effect in animals treated with CB 154. It should be noted that CB 154 or pituitary transplants alone had no effect on testicular testosterone or progesterone levels. It has also been observed that the plasma content of testosterone and progesterone followed a pattern similar to that of the testicular levels of these two steroids.

The present data clearly demonstrate that endogenous plasma prolactin could play a role in the down-regulation of testicular LH receptors induced by treatment with an LH-RH agonist and that prolactin leads to an apparent accentuation of this blockage probably due to increased steroid-precursor formation.

References

Auclair, C., Kelly, P.A., Labrie, F., Coy, D.H., and Schally, A.V., 1977a, Inhibition of testicular luteinizing hormone receptor level by treatment with a potent luteinizing hormone-releasing hormone agonist or human chorionic gonadotropin, *Biochem. Biophys. Res. Commun.* **76:**855.

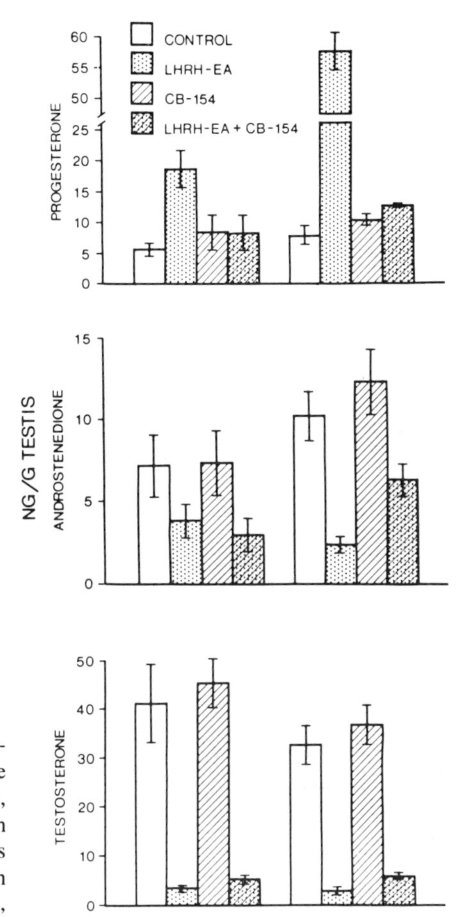

Figure 11. Effect of treatment with [D-Ala[6],des-Gly-NH_2[10]]LH-RH ethylamide (40 ng, every 3 days), CB 154 (500 μg, twice a day), or a combination of both treatments for 12 days in intact male rats or animals bearing pituitary transplants on testicular progesterone, androstenedione, and testosterone concentrations.

Auclair, C., Kelly, P.A., Coy, D.H., Schally, A.V., and Labrie, F., 1977b, Potent inhibitory activity of [D-Leu[6],des-Gly-NH_2[10]]LHRH ethylamide on LH/hCG and PRL testicular receptor levels in the rat, *Endocrinology* **101:** 1890.

Bartke, A., Hafiez, A.A., Bex, F.J., and Dalterio, S., 1978, Hormonal interactions in regulation of androgen secretion, *Biol. Reprod.* **18:**44.

Bélanger, A., Auclair, C., Seguin, C., Kelly, P.A., and Labrie, F., 1979, Down-regulation of testicular androgen biosynthesis and LH receptor levels by an LHRH agonist: Role of prolactin, *Mol. Cell. Endocrinol.* **13:**47.

Bélanger, A., Caron, S., and Picard, V., 1980a, Simultaneous radioimmunoassay of progestins, androgens and estrogens in rat testis, *J. Steroid Biochem.* **13:** 185.

Bélanger, A., Auclair, C., Ferland, L. and Labrie, F., 1980b, Time-course of the effects of treatment with a potent LHRH agonist on testicular steroidogenesis and gonadotropin receptor levels in the adult rat, *J. Steroid Biochem.* **13:**191.

Chasalow, F., 1979, Mechanism and control of rat testicular steroid synthesis, *J. Biol. Chem.* **254:**3000.

Chasalow, F., Marr, H., Haour, F., and Saez, J.M., 1979, Testicular steroidogenesis after human chorionic gonadotropin desensitization in rats, *J. Biol. Chem.* **254:**5613.

Cigorraga, S.B., Dufau, M.L., and Catt, K.J., 1978, Regulation of luteinizing hormone receptors and steroidogenesis in gonadotropin-desensitized Leydig cells, *J. Biol. Chem.* **254:**4297.

Dray, F., Reïnberg, A., and Sebaoun, J., 1965, Rythme biologique de la testostérone libre du plasma chez l'homme adult sain: Existence d'une variation circadienne, *C. R. Acad. Sci.* **261:**573.

Hsueh, A.J.W., Dufau, M.L., and Catt, K.J., 1976, Regulation of luteinizing hormone receptors in testicular cells by gonadotrophin, *Biochem. Biophys. Res. Commun.* **72:**1145.

Kahn, C.R., Neville, D.M., and Roth, J., 1973, Insulin–receptor interaction in the obese-hyperglycemic mouse: A model of insulin resistance, *J. Biol. Chem.* **248:**244.

Labrie, F., Auclair, C., Cusan, L., Kelly, P.A., Pelletier, G., and Ferland, L., 1978, Inhibitory effect of LHRH and its agonists on testicular gonadotropin receptors and spermatogenesis in the rat, Fifth Annual Workshop on the Testis: Endocrine Approach to Male Contraception, *Int. J. Androl. Suppl.* **2:**303.

Lesnick, M.A., and Roth, J., 1976, Regulation of receptor concentration by homologous hormone: Effect of human growth hormone on its receptors in IM-9 lymphocytes, *J. Biol. Chem.* **251:**3730.

Pelletier, G., Cusan, L., Auclair, C., Kelly, P.A., Désy, L., and Labrie, F., 1978, Inhibition of spermatogenesis in the rat by treatment with [D-Ala^{6},des-Gly-NH_2^{10}]LHRH ethylamide, *Endocrinology* **103:**641.

Purvis, K., Torgesen, P.A., Haug, E., and Hansson, V., 1977, hCG suppression of LH receptors and responsiveness of testicular tissue to hCG, *Mol. Cell. Endocrinol.* **8:**73.

Saez, J.M., Haour, F., Tell, G.P.E., Gallet, D., and Sanchez, P., 1978, Human chorionic gonadotropin-induced Leydig cell refractoriness to gonadotropin stimulation, *Mol. Pharmacol.* **14:**1054.

Sandow, J., Recherberg, W.V., Köning, W., Hahn, M., Jerzabek, G., and Frazer, H., 1978, Physiological studies with highly active analogues of LHRH, in: *Hypothalamic Hormone—Chemistry, Physiology and Clinical Applications* (Das Gupta and W. Voelter, eds), pp. 307 Weinheim, New York, Verlag Chemie.

Schwarzstein, L., 1976, Diagnostic and therapeutic use of LHRH in the infertile man, in: *Hypothalamus and Endocrine Functions* (F. Labrie, J. Meites, and G. Pelletier, eds.), p. 73, Plenum Press, New York.

Sharpe, R.M., 1976, hCG-induced decrease in availability of rat testis receptors, *Nature (London)* **264:**644.

Sharpe, R.M., 1977, Gonadotropin-induced reduction in the steroidogenic responsiveness of the immature rat testis, *Biochem. Biophys. Res. Commun.* **76:**957.

Slaunwhite, W.R., Jr., and Samuels, L.T., 1956, Progesterone as a precursor of testicular androgens, *J. Biol. Chem.* **220:**341.

Smith, R., Donald, R.A., Espiner, E.A., Stromach, S.G., and Edwards, I.A., 1979, Normal adults and subjects with hypogonatropin hypogonadism respond differently to [D-Ser(TBU)^{6},des-Gly-NH_2^{10}]LHRH-EA, *J. Clin. Endocrinol. Metab.* **48:**167.

Tsuruhara, T., Dufau, M.L., Cigorraga, S., and Catt, K.J., 1977, Hormonal regulation of testicular luteinizing hormone receptors, *J. Biol. Chem.* **252:**9002.

Wiegelman, W., Solbach, H.G., Kley, M.K., Nieschlag, E., Rudorff, K.H., and Kruskemper, M.L., 1978, A new LHRH analogue: [D-Ser(TBU)^{6},des-Gly-NH_2^{10}]LHRH-(1-9-nonapeptide ethylamide, in: *Hypothalamic Hormone-Chemistry Physiology and Clinical Applications* (Das Gupta and M. Voelter, eds.), p. 327, Verlag Chemie.

Woods, M.C., and Simpson, M.E., 1961, Pituitary control of the testis of the hypophysectomized rat, *Endocrinology* **69:**91.

22

Gonadotropic Stimulation of Enzymes Involved in Testicular Growth

Claude A. Villee and Janet M. Loring

The marked decrease in testicular size following hypophysectomy and its restoration toward normal following the injection of pituitary extracts have long been known. Analyses of the endocrine and molecular mechanisms involved in the growth response of the gonads to gonadotropins have been undertaken by many investigators. Gonadotropic hormones increase the production of cyclic AMP (cAMP) (Murad *et al.*, 1969; Hollinger, 1970; Kuehl *et al.*, 1970; Dorrington *et al.*, 1972; Braun and Sepsenwol, 1974) and increase the synthesis of RNA and protein in the testis of the rat (Means and Hall, 1967, 1968, 1969; Goswami *et al.*, 1968; D. Villee and Goswami, 1973). The increased synthesis of RNA and protein in response to gonadotropins is inhibited by pretreatment of the animals with actinomycin D (Means and Hall, 1969; Goswami *et al.*, 1968; Reddy and Villee, 1975a). The ubiquitous distribution of the polyamines, their responses to hormonal stimuli, and their possible involvement in both transcriptional and translational events of cellular growth processes have been reported repeatedly and reviewed extensively (Williams-Ashman *et al.*, 1972; Russell, 1973).

Claude A. Villee and *Janet M. Loring* • Department of Biological Chemistry and Laboratory of Human Reproduction and Reproductive Biology, Harvard Medical School, Boston, Massachusetts 02115.

The finding that an increased activity of ornithine decarboxylase is a feature of the early response of a great many target tissues to their respective hormones is especially intriguing. The ornithine decarboxylase activity of the liver is stimulated by growth hormone (Richman *et al.*, 1971) and that of the adrenal is stimulated by ACTH (Richman *et al.*, 1973). The activity of ovarian ornithine decarboxylase rises sharply during the preovulatory surge of luteinizing hormone (LH) (Kobayshi *et al.*, 1971); the rise can be prevented by injecting an anti-LH before proestros. Ovarian ornithine decarboxylase activity increases sharply 1 hr after the administration of LH. This increase appears to be due to the *de novo* synthesis of the enzyme, because actinomycin D or cyclohex-imide prevents the rise (Kaye *et al.* 1973). The ornithine decarboxylase activity of the immature rat testis is stimulated by both LH and follicle-stimulating hormone (FSH) (Reddy and Villee, 1975b; C.A. Villee and Loring, 1978). The ornithine decarboxylase activity of the prostate is stimulated by testosterone (Williams-Ashman *et al.*, 1969), and estradiol stimulates the ornithine decarboxylase activity of the immature rat uterus (Kaye *et al*, 1971; C.A. Villee and Loring, 1975) and of the chick oviduct (Cohen *et al.*, 1970).

Experiments in a variety of systems have provided evidence of a correlation between the concentration of polyamines and the rate of RNA synthesis. In addition, there is evidence that the amount of polyamines in a tissue can be affected by hormonal stimuli (Raina and Holtta, 1972). Using purified RNA polymerases I and II from pig kidney, Jänne *et al.* (1975) found that adding spermine or spermidine increased the rate of transcription *in vitro* of DNA or of native chromatin from pig kidney. The maximal stimulation of polymerase I occurred within the range of 0.5–2.0 mM spermine, and the optimal concentration for RNA polymerase II was 1–5 mM spermine. At higher concentrations of spermine, within the range of 5–10 mM, both polymerases were strongly inhibited. The experiments of Jänne and co-workers suggested that polyamines have little or no effect on chain initiation, but increase the rate of elongation of RNA chains some 2- or 3-fold. Spermine, at a concentration of 2 mM, increased the incorporation of tritium-labeled AMP, but had no effect on the incorporation of ^{32}P from γ-^{32}P-labeled ATP into nucleic acids. The authors calculated that the average size of the RNA made under control conditions was 570 nucleotides, whereas in the presence of 2 mM spermine it was 1330 nucleotides. Assuming that the rate of synthesis was constant for the 15 min of incubation, the rate of chain elongation was 0.64 nucleotide/sec under control conditions and 1.44 nucleotides/sec in the system stimulated by spermine.

The ornithine decarboxylase activity in the testis of an immature hypophysectomized rat decreased spectacularly to a level about 5% that in the testis of the intact control immature rat (C.A. Villee and Loring, 1978). The testis of the hypophysectomized rat responded to the injection of FSH with an increase in ornithine decarboxylase activity. At 4 hr after the injection of the hormone, the ornithine decarboxylase activity had returned to the level of the intact, immature rat not injected with the hormone.

The FSH used (NIH, FSH S-12) is known to be far from pure and to contain some LH. Through the kindness of Professor Leo Reichert of Emory University, we obtained a preparation of highly purified FSH (LER 1577) that contained 880 IU FSH and only 5.7 IU LH per milligram. As little as 0.01 μg of the highly purified FSH produced a significant increase in ornithine decarboxylase activity, and 0.5 μg of the highly purified FSH gave as great an increase as 20 μg of the NIH S-12 FSH (C.A. Villee and Loring, 1978).

To determine whether the increased RNA synthesis following the injection of FSH might be mediated by increased polyamines resulting from a stimulation of ornithine decarboxylase by the gonadotropin, immature rats were injected intratesticularly with spermine or spermidine and with [^{3}H]uridine to measure RNA synthesis (Tables I and II). The RNA was extracted as indicated and passed through an oligodeoxythymidylic acid [oligo(dT)] chromatography column to separate polyadenylic acid [poly(A)]-rich RNA from the bulk of the RNA. Spermine diphosphate or spermidine phosphate injected into the testis over the range of 0.5–2.0 μg/testis increased the incorporation of uridine into both bound

Table I. Specific Activity of RNA Fractions from Testes of Immature (21-Day) Rats Injected Intratesticularly with [^{3}H]Uridine and Spermidine Phosphate[a]

	Specific activity (cpm/μg)	
Fraction	Control	Spermidine phosphate (1 μg)
Total RNA	17.9	23.8
Unbound RNA	13.6	21.2
Bound RNA	61.6	112.0

[a] One testis was injected with spermidine phosphate; the contralateral testis was injected with saline. Both were injected with 2 μCi [^{3}H]uridine. Comparably treated testes were pooled and homogenized, and RNA was extracted with phenol–sodium dodecyl sulfate. The RNA was treated with DNAase, reextracted with phenol, precipitated with ethanol in the cold, and separated by binding to and eluting from an oligo(dT) cellulose column. The values are the means of four experiments.

Table II. Specific Activity of RNA Fractions from Testes of Immature (21-Day) Rats Injected Intratesticularly with [^{3}H]Uridine and Spermine Diphosphate.[a]

	Specific activity (cpm/μg)			
		Spermine diphosphate injected		
Fraction	Control	0.5 μg	1.0 μg	2.0 μg
Total RNA	1.87	2.90	4.51	2.30
Unbound RNA	1.90	2.38	2.97	2.10
Bound RNA	6.24	12.5	14.4	9.4

[a] One testis was injected with spermine diphospate; the contra-lateral testis was injected with saline. The rest of the treatment was as described in the Table I footnote.

and unbound RNA, with a greater effect on the incorporation into bound poly(A)-rich RNA. The maximum effect was observed at 1 μg spermine diphosphate/testis (Table II).

If the major effect of spermine or spermidine is on the elongation of nascent RNA chains, the incorporation of both [^{3}H]uridine and [α-^{32}P]-ATP should be increased comparably. The injection of spermine resulted in a 50% per cent increase in the incorporation of tritium-labeled uridine and a 47% increase in the incorporation of [α-^{32}P]-ATP into bound RNA (Table III). In further experiments, both [^{3}H]uridine and [γ-^{32}P]-ATP were injected intratesticularly along with 1 μg spermine. If the polyamine

Table III. Specific Activity of RNA Fractions from Testes of Immature (21-Day) Rats Injected Intratesticularly with [^{3}H]Uridine and [α-^{32}P]-ATP: Effect of Spermine Diphosphate.[a]

	Specific activity (cpm/μg)			
	Control		Spermine diphosphate (1 μg)	
Fraction	^{3}H	^{32}P	^{3}H	^{32}P
Total RNA	12.7	3.95	19.1	5.42
Unbound RNA	10.0	3.12	18.3	4.94
Bound RNA	58.6	15.2	88.1	22.4

[a] One testis was injected with spermine diphosphate; the contra-lateral testis was injected with saline. Both were injected with 2 μCi [^{3}H]uridine and 2 μCi [α-^{32}P]-ATP. The rest of the treatment was as described in the Table I footnote. The values are the means of three experiments.

increased the rate of chain initiation, the rate of incorporation of [γ-^{32}P]-ATP should be increased over the rate of incorporation of labeled uridine. In these experiments, the injection of spermine increased by 74% the rate of incorporation of tritium-labeled uridine into RNA and increased by 97% the incorporation of [γ-^{32}P]-ATP into bound RNA (Table IV). This raises the possibility that spermine may increase the rate of initiation as well as the rate of elongation of RNA chains, but the difference is too small to assert this with any degree of confidence.

Although there is a prompt and dramatic increase in ornithine decarboxylase activity in the testis following the injection of FSH or LH, the content of putrescine, spermine, and spermidine in the testis changed very little, if at all, either shortly after injection (4 or 6 hr) or 2 or 3 days after the injection of FSH, LH, or cAMP (Table V). This system appears to parallel many other systems in which a hormone increases the activity of adenylate cyclase and yet there is little or no demonstrable increase in the amount of AMP present in the tissue.

These experiments show clearly that the intratesticular injection of either spermine or spermidine will increase RNA synthesis, primarily by increasing the elongation of RNA chains. However, the question of whether the gonadotropins produce their effect on RNA synthesis in the testis by way of a stimulation of ornithine decarboxylase and an increased synthesis of polyamines remains open. During the hypertrophic response of the tissue to trophic hormones, a series of biochemical processes transfers the hormonal stimulus from the cell membrane through the

Table IV. Specific Activity of RNA Fractions from Testes of Immature (21-Day) Rats Injected Intratesticularly with [^{3}H]Uridine and [γ-^{32}P]-ATP: Effect of Spermine Diphosphate.[a]

	Specific activity (cpm/μg)			
	Control		Spermine diphosphate (1 μg)	
Fraction	^{3}H	^{32}P	^{3}H	^{32}P
Total RNA	17.3	7.1	21.6	9.1
Unbound RNA	16.7	6.2	20.9	8.1
Bound RNA	86.0	31.3	150.0	61.8

[a] One testis was injected with spermine diphosphate; the contralateral testis was injected with saline. Both were injected with 2 μCi [^{3}H]uridine and 2 μCi [γ-^{32}P]-ATP. The rest of the treatment was as described in the Table I footnote. The values are the means of six experiments.

Table V. Change in Content of Polyamines in Testes Injected with FSH, LH, or cAMP Compared to Contralateral Control Testes.[a]

Time after injection (hr)	FSH			LH			cAMP		
	Pu	Sd	Sp	Pu	Sd	Sp	Pu	Sd	Sp
4	−21	−8	−7	−8	+15	+28	−15	+7	+3
6	+39	−4	−7	+19	+34	−10	+3	+2	−2
12	−11	−2	−3	+5	−2	−3	+3	+6	+6
16	−19	−7	−12	+17	+17	+13	+29	+3	−1
24	−11	−28	−35	−7	+1	−2	0	+2	+6
48	0	−10	−5	—	—	—	+13	+12	+14
72	+5	−1	−2	+8	+5	+7	+9	+3	+5

[a] The values are expressed as percentage differences between the content in the hormone-injected testis and that in the contralateral, saline-injected testis. The mean concentrations in the control testes were (in nmol/g testis): putrescine (Pu), 45; spermidine (Sd), 502; and spermine (Sp), 613. Testes were removed, homogenized in 4% sulfosalicylic acid, and centrifuged. The supernatant fractions were separated and diluted 1:4 with buffer, and 100 μl was placed in a Beckman amino acid analyzer and measured by the method of Gehrke *et al.* (1977).

cytoplasm and results in the activation of certain nuclear genes, ultimately resulting in the synthesis of new species of RNA and of proteins. The hormone-sensitive adenylate cyclases and ornithine decarboxylase appear to be involved in this response in some way.

It is generally accepted that cAMP exerts its intracellular control primarily through its effect on cAMP-dependent protein kinases (Kuo and Greengard, 1969). Several laboratories have provided evidence that cAMP may stimulate protein kinase without a detectable increase in the concentration of cAMP in the tissue (Bylund and Krebs, 1975). Cyclic AMP can be rapidly degraded and no increase in the concentration of cAMP may be demonstrable in a tissue despite an increase in the activity of adenylate cyclase. In other systems in which an increase in cAMP can be detected, the increase is rapidly followed by an activation of cAMP-dependent protein kinase (Costa *et al.*, 1976; Keeley *et al.*, 1975; Langan, 1973).

The activity of the cAMP-dependent protein kinase in the testis and its response to gonadotropins were measured by the method of Witt and Roskoski (1975). Either histone or protamine sulfate was used as the phosphate acceptor and [γ-^{32}P]-ATP as the phosphate donor.

Using the purified FSH (LER 1577), a marked increase in the activity of the cAMP-dependent protein kinase in the testis could be demonstrated 15, 30, 60, or 120 min after the injection of FSH (Table VI). By 4 hr after the injection of gonadotropin, the cAMP-dependent protein kinase activity had decreased toward baseline. An increased cAMP-dependent protein kinase activity could be demonstrated 30 or 60 min after the intratesticular injection of LH, but not 2 or 4 hr after the injection of the hormone (Table VI). The concentration of cAMP in the testis was assayed by its ability to activate the cAMP-dependent protein kinase of the tissue using the method of Kuo and Greengard (1972). The administration of 1 μg LH intratesticularly nearly doubled the cAMP concentration within 30 min. The administration of 1 μg purified FSH (LER 1577) also increased the cAMP content of the tissue, with an increase detectable at 15 min. The peak in the content of cAMP was observed 60 min after the injection of the gonadotropin. With both FSH and LH, the cAMP content of the testis had decreased by 2 hr and was at control levels 4 hr after injection of the gonadotropin. This is very similar to the time-course of the response of cAMP content of the liver to injected aminophylline or methyl isobutyl xanthine (Costa *et al.*, 1975).

The effects of injected gonadotropins on the next enzyme in the sequence in the biosynthesis of polyamines, *S*-adenosyl-methionine decarboxylase, was tested in another series of experiments. *S*-Adenosyl-methionine decarboxylase activity was measured by the method of Feldman *et al.* (1972). ^{14}C-Carboxyl-labeled *S*-adenosyl-L-methionine was used as substrate. The incubations were carried out in stoppered test tubes fitted with a center well containing hyamine hydroxide to absorb the labeled carbon dioxide released in the reaction. These experiments

Table VI. Effects of FSH and LH Injected Intratesticularly on the Content of cAMP and on the Activity of cAMP-Dependent Protein Kinase of the Rat Testis.[a]

	FSH			LH		
		cAMP-dependent protein kinase			cAMP-dependent protein kinase	
Time (min)	cAMP (pmol/mg wet weight)	−cAMP / +cAMP	% of control	cAMP (pmol/mg wet weight)	−cAMP / +cAMP	% of Control
0	0.338	0.39	100	0.363	0.40	100
15	0.490	0.65	167	—	—	—
30	0.450	0.76	195	0.695	0.88	220
60	0.625	0.72	185	0.484	0.68	170
120	0.407	0.70	180	0.404	0.58	145
240	0.315	0.51	131	0.259	0.53	133

[a] Rats were injected intratesticularly with 1 μg FSH (LER 1577) or 1 μg LH (NIH S-20). At the times indicated, rats were killed, the testes were removed and homogenized, and the homogenate was centrifuged for 20 min at 50,000*g*. The cAMP-dependent protein kinase activity of the supernatant fraction was measured from the ratio of the cpm of [^{32}P]phosphate from [γ-^{32}P]-ATP incorporated into protamine sulfate in the absence and presence of cyclic AMP (−cAMP/+cAMP) during a 5-min reaction at 30°C. The reaction was terminated by transferring 50 μl of the reaction mixture to phosphocellulose paper (Whatman P81). The paper strips were washed with a large volume of tapwater, then acetone, and dried with ether. The samples on the paper strips were counted using a mixture of Omnifluor and toluene.

Table VII. *S*-Adenosyl-methionine Decarboxylase Activity in the Testis of the Immature Rat: Effect of Intratesticular Injection of FSH and LH[a]

Time (hr)	FSH	LH
0	1.23	1.23
1	1.07	—
2	1.88	1.00
4	1.75	1.33

[a] Values are expressed as nanomoles $^{14}CO_2$ released per milligram total protein per hour. Immature (21-day-old) rats were injected intratesticularly with 20 μg FSH or 20 μg LH (NIH S-20). The rats were killed at the times indicated, and the testes were removed and homogenized. The homogenate was centrifuged 20 min at 20,000*g*, and aliquots of the supernatant fraction were transferred to test tubes containing ^{14}C-carboxyl-labeled *S*-adenosyl-methionine. The test tubes were stoppered and fitted with a polyethylene center well containing hyamine hydroxide to absorb the labeled CO_2 released in the reaction. The reaction was stopped by the addition of 1 N H_2SO_4. After 30 min, the hyamine was transferred to a scintillation vial containing an Omnifluor–toluene mixture and counted.

showed (Table VII) that the injection of FSH increased *S*-adenosyl-methionine decarboxylase activity at 2 or 4 hr, but not at 1 hr. No effect of injected LH was observed at 2 hr, but the activity of the enzyme was increased 4 hr after the injection of LH.

These experiments show that the injection of gonadotropins leads to an increase in the activity of both ornithine decarboxylase and *S*-adenosyl-methionine decarboxylase without a demonstrable increase in the concentration of spermine or spermidine in the testis. The absence of an increase in the content of spermidine and spermine could reflect an increased rate of their degradation. Alternatively, it is possible that despite the increase in enzyme activity, there is no increase in the flux of substrates through the pathway, because of either a lack of precursors or a lack of some cofactor. To test this, a series of experiments was carried out in which [2-^{14}C]methionine was injected into the testis along with a gonadotropin or with saline as a control. The animals were sacrificed 1, 2, or 4 hr later. The testes were homogenized and deproteinized, and the methionine was separated from spermine and spermidine by passage through a Dowex 50W X-4H^+ column. The column was eluted first with water, than with 0.5 N HCl, which elutes methionine, and finally with 2.5 N HCl, which elutes spermidine and spermine separately. At 1, 2, or 4 hr, both FSH and LH increased the rate of disappearance of labeled methionine (Table VIII), but did not increase the rate of accumulation of spermidine and spermine. In both hormone-injected testes and control, saline-injected testes, the amount of radioactivity present in spermine and spermidine was greater at 2 hr than at 1 hr and less at 4 hr than at 2 hr. The interpretation of these unexpected results is not yet clear. The increased disappearance of methionine in the testes injected with gonadotropins is consistent with the effect on *S*-adenosyl-methionine decarboxylase; however, the amount of radioactivity in the spermine and spermidine was less rather than greater in the testes injected with FSH or LH. The finding that radioactivity from methionine is not incorporated more rapidly into spermine and spermidine in response to injected gonadotropin is consistent with our previous observation that the gonadotropins did not increase the content of polyamines in the tissue. Perhaps gonadotropins increase the rate of degradation of the polyamines as well as increase their rate of synthesis. However, there is no experimental evidence that bears on this point.

In previous experiments (Reddy and Villee, 1975a), the seminiferous tubules and interstitial cells were separated by treatment with collagenase by the method of Moyle and Ramachandran (1973). FSH treatment increased the incorporation of labeled uridine into bound poly(A)-rich messenger RNA (mRNA) only in the seminiferous tubules and not in Leydig cells, whereas human chorionic gonadotropin stimulated mRNA synthesis in the interstitial cells, but not in the seminiferous tubules. The intratesticular injection of cAMP increased the synthesis of mRNA in both interstitial-cell and seminiferous-tubule-cell preparations. Comparable experiments were carried out in which gonadotropins were injected intratesticularly and after 1, 2, or 4 hr the animal was killed and the testes were removed and separated into seminiferous-tubules- and Leydig-cell preparations by treatment with collagenase. The ornithine decarboxylase, *S*-adenosyl-methionine decarboxylase, and cAMP-dependent protein kinase activities in the two cell preparations were compared. The injection of LH into the testis resulted in increased ornithine decarbox-

Table VIII. Effects of FSH or LH on the Conversion of Methionine to Spermidine and Spermine by the Testes of the Immature Rat[a]

Time		Control testis			Gonadotropin-injected testis		
(hr)	Gonadotropin	Methionine	Spermidine	Spermine	Methionine	Spermidine	Spermine
1	FSH	49,800	7430	4950	15,850	1540	1570
2	FSH	5,320	9260	3540	5,900	7440	2020
4	FSH	300	7520	3560	180	1640	830
2	LH	1,760	8370	6000	1,230	5540	2830
4	LH	570	4190	2160	300	4200	2150

[a] [2-^{14}C]Methionine was injected, with or without FSH or LH, into the testes of immature rats. At 1, 2, or 4 hr later, the rats were killed and the testes were removed and homogenized. The homogenate was deproteinized with trichloroacetic acid (TCA) and centrifuged, the TCA was removed with ether, the pH was adjusted to pH 6, and the supernatant fluid was passed through a Dowex 50W X-4H$^+$ column. The column was eluted with water, 0 5 N HCl, and 2.5 N HCl. Values are expressed as cpm in the specific fractions isolated from the columns.

ylase activity in the Leydig-cell preparations 1, 2, or 4 hr after injection of the hormone, but was without effect on the ornithine decarboxylase activity in the seminiferous tubules (Table IX). The injection of 20 μg FSH S-12 into the testis increased ornithine decarboxylase activity in the seminiferous tubules 1, 2, or 4 hr later, but also increased the ornithine decarboxylase activity in Leydig cells 2 or 4 hr later. However, when 1 μg purified FSH (LER 1577) was injected, there was a clear effect at 2 or 4 hr on the ornithine decarboxylase activity in the seminiferous tubules and no increase in the activity of ornithine decarboxylase in the Leydig cells. This suggests strongly that the increased ornithine decarboxylase activity in the Leydig cells induced by the injection of 20 μg of the S-12 preparation of FSH was due to an LH contaminant in the preparation.

To measure the *S*-adenosyl-methionine decarboxylase activities of seminiferous-tubule-cell and Leydig-cell preparations, one testis of an immature rat was injected with 0.1 μg purified LER 1577 FSH and the contralateral testis was injected with saline. After 2 or 4 hr, the rats were killed and the testes removed, pooled, and separated into seminiferous tubules and Leydig cells by treatment with collagenase. The *S*-adenosyl-methionine decarboxylase activity of the seminiferous tubules of the FSH-injected testis was 65% greater than that of the contralateral control testis at 2 hr and 107% greater at 4 hr. In contrast, FSH had no effect on the *S*-adenosyl-methionine decarboxylase activity of the Leydig-cell

Table IX. Effects of LH and FSH on Ornithine Decarboxylase Activity in Seminiferous-Tubule-Cell Preparations and Leydig-Cell Preparations from the Testes of Immature Rats

Gonadotropin injected	Time (hr)	Seminiferous tubules[a]	Leydig cells[a]
LH (S-20) (20 μg)	0	1.09	0.23
	1	0.96	0.36
	2	1.15	0.35
	4	1.18	0.29
FSH (S-12) (20 μg)	0	1.06	0.22
	1	1.37	0.24
	2	2.02	0.39
	4	1.55	0.30
FSH (LER 1577) (1 μg)	0	0.88	0.19
	2	1.17	0.19
	4	1.43	0.18

[a] Values are expressed as nanomoles $^{14}CO_2$ released per milligram total protein per hour. There were 4–6 animals in each group.

Table X. Effects of FSH and LH on *S*-Adenosyl-methionine Decarboxylase in Seminiferous Tubules and Leydig Cells from the Testes of Immature Rats

			Seminiferous tubules[a]		Leydig cells[a]	
Gonadotropin injected	Amount (μg)	Time (hr)	Control	Gonadotropin-injected	Control	Gonadotropin-injected
FSH (LER 1577)	0.1	2	0.77	1.27	1.14	1.15
	0.1	4	1.18	2.43	1.66	1.67
LH (S-20) 20	20	2	1.57	1.50	1.36	1.85
	20	4	1.42	1.35	1.18	1.54

[a] Values are expressed as nmol $^{14}CO_2 \times$ mg protein$^{-1} \times$ hr^{-1}. For methods, see the Table VII footnote.

preparation (Table X). In other experiments, LH (NIH S-12), 20 μg/testis, was injected into one testis and saline into the contralateral one. Seminiferous-tubule-cell and Leydig-cell preparations were separated by collagenase treatment and tested for *S*-adenosyl-methionine decarboxylase activity. The injected LH had no effect on the *S*-adenosyl-methionine decarboxylase activity of the seminiferous tubules, but increased the activity of the enzyme in the Leydig cells 36% at 2 hr and 31% at 4 hr.

The cAMP contents and cAMP-dependent protein kinase activities of seminiferous-tubule-cell and Leydig-cell preparations were measured. One testis of an immature rat was injected with 0.1 μg purified LER 1577 FSH or with 1 μg LH NIH S-20 and the contralateral testis was injected with saline. After 30 or 60 min, the rats were killed and the testes removed, pooled, and separated into seminiferous tubules and Leydig cells by treatment with collagenase. Each was separately assayed for cAMP content (Kuo and Greengard, 1972) and for cAMP-dependent protein kinase (Witt and Roskoski, 1975). LH increased the cAMP content and cAMP-dependent protein kinase activity of the Leydig cells, but not of the seminiferous tubules, at both 30 and 60 min (Table XI). FSH increased the protein kinase activity of the seminiferous tubules at both 30 and 60 min and their cAMP content at 60 min; however, it was without effect on the Leydig cells (Table XI).

The present working hypothesis states that gonadotropins interact with specific receptors on the surface of the target cells and activate cyclase. This increases the intracellular concentrations of cAMP and activates the cAMP-dependent protein kinase. The increased RNA synthesis induced in the testis by the injection of gonadotropins can be mimicked by the injection of cAMP (Reddy and Villee, 1975a). The protein kinase or its catalytic subunit migrates into the nucleus, where it is believed to phosphorylate specific acidic nuclear proteins and stimulate

Table XI. Effects of FSH and LH on the Content of cAMP and on the Activity of cA Dependent Protein Kinase in Seminiferous Tubules and Leydig Cells from the Testes Immature Rats[a]

		Seminiferous tubules			Leydig cells		
Gonadotropin	Time (min)	cAMP (nmol/mg protein)	−cAMP / +cAMP	% of control	cAMP (nmol/mg protein)	−cAMP / +cAMP	% o contr
—	0	0.94	0.28	100	4.6	0.22	100
LH	30	0.57	0.25	90	6.1	0.44	200
	60	0.52	0.33	117	7.9	0.45	205
FSH	30	0.82	0.75	267	2.5	0.23	104
	60	1.70	0.71	253	1.5	0.20	91

[a] For methods, see the Table VI footnote.

the transcription of certain parts of the genome. This leads to the synthesis of new mRNAs coding for, among other enzymes, ornithine decarboxylase and *S*-adenosyl-methionine decarboxylase. Some of this RNA causes an early, and relatively short-lived, increased synthesis of ornithine decarboxylase and *S*-adenosyl-methionine decarboxylase on the ribosomes, and the enzymes produced increase the synthesis of spermine and spermidine. The effects of gonadotropins on RNA synthesis in the testis can be mimicked by the intratesticular injection of spermine or spermidine. The polyamines increase the rate of chain elongation during RNA synthesis. For reasons not yet clear, the injection of gonadotropins did not lead to an accumulation of polyamines in the testis in either the short term or the long term. Although the intratesticular injection of gonadotropin increased the rate of disappearance from the testis of labeled methionine, the precursor of *S*-adenosyl-methionine, it decreased the rate of incorporation of methionine into the polyamines.

These experiments add further evidence that FSH affects only the cells of the seminiferous tubules and not Leydig cells, whereas LH affects the Leydig cells but not the cells of the seminiferous tubules. Thus, FSH increases the synthesis of RNA and the activity of ornithine decarboxylase, *S*-adenosyl-methionine decarboxylase, and cAMP-dependent protein kinase in the seminiferous tubules, but is without effect on the interstitial cells of Leydig. The injection of LH stimulates RNA synthesis and orithine decarboxylase, *S*-adenosyl-methionine decarboxylase, and cAMP-dependent protein kinase activity in the interstitial cells of Leydig, but not in the cells of the seminiferous tubules. Other workers have shown that specific receptors for FSH are present in the seminiferous tubules but not in the Leydig cells, and receptors for LH are present in the Leydig cells but not in the seminiferous tubules. Thus, the

Leydig cells and seminiferous tubules exhibit parallel sequences of responses to their specific gonadotropins.

Discussion

PETRUSZ: Do bacterial toxins (e.g., cholera, diphtheria) affect ornithine decarboxylase levels in testis?

C.A. VILLEE: I don't know; we haven't tested them.

References

Braun, T., and Sepsenwol, S., 1974, Stimulation of ^{14}C-cyclic AMP accumulation by FSH and LH in testis from mature and immature rats, *Endocrinology* **94:**1028.

Bylund, D.B., and Krebs, E.G., 1975, Effect of denaturation on the susceptibility of proteins to enzymic phosphorylation, *J. Biol. Chem.* **250:**6355.

Cohen, S., O'Malley, B.W., and Stastny, M., 1970, Estrogen induction of ornithine decarboxylase *in vivo* and *in vitro, Science* **170:**336.

Costa, M., Manen, C.A., and Russell, D.H., 1975, *In vivo* activation of cAMP dependent protein kinase by aminophylline and 1-methyl—isobutyl xanthine, *Biochem. Biophys. Res. Commun.* **65:**75.

Costa, E., Kurasawa, A., and Guidotti, A., 1976, Activation and nuclear translocation of protein kinase during transsynaptic induction of tyrosine—monooxygenase, *Proc. Natl. Acad. Sci. U.S.A.* **73:**1058.

Dorrington, J.H., Vernon, R.G., and Fritz, I.B., 1972, The effect of gonadotropins on the 3′,5′-cAMP levels of seminiferous tubules, *Biochem. Biophys. Res. Commun.* **46:**1523.

Feldman, M.J., Levy, C.C., and Russell, D.H., 1972, Purification and characterization of *S*-adenosyl-L-methionine decarboxylase from rat liver, *Biochemistry* **11:**671.

Gehrke, C.W., Kuo, K.C.,and Ellis, R.L., 1977, Polyamines—an improved automated ion-exchange method, *J. Chromatogr.* **143:**345.

Goswami, A., Skipper, J.K., and Williams, W.L., 1968, Stimulation of fatty acid synthesis *in vitro* by gonadotropin-induced testicular ribonucleic acid, *Biochem. J.* **108:**147.

Hollinger, M.A., 1970, Studies on adenyl cyclase in rat testis, *Life Sci.* **9:**533.

Jänne, O., Bardin, C.W., and Jacob, S.T., 1975, DNA dependent RNA polymerases I and II from kidney: Effect of polyamines on the *in vitro* transcription of DNA and chromatin, *Biochemistry* **14:**3589.

Kaye, A.M., Icekson, I., and Lindner, H.R., 1971, Stimulation by estrogens of ornithine and *S*-adenosylmethionine decarboxylase in the immature rat uterus, *Biochim. Biophys. Acta* **252:**150.

Kaye, A.M., Icekson, I., Lamprecht, S.A., Gruss, R., Tsafriri, A., and Lindner, H.R., 1973, Stimulation of ornithine decarboxylase activity by luteinizing hormone in immature and adult rat ovaries, *Biochemistry* **12:**3072.

Keely, S.L., Jr., Corbin, J.D., and Park, C.R., 1975, On the question of translocation of heart cAMP-dependent protein kinase, *Proc. Nat. Acad. Sci.* **72:**1501.

Kobayashi, Y., Kupelian, J. and Maudsley, D.V., 1971, Ornithine decarboxylase stimulation in rat ovary by luteinizing hormone, *Science* **172:**379.

Kuehl, F.A., Patanelli, D. J., Tarnoff, J., and Humes, J.L., 1970, Testicular adenyl cyclase: Stimulation by the pituitary gonadotropins, *Biol. Reprod.* **2:**154.

Kuo, J.F., and Greengard, P., 1969, Cyclic nucleotide-dependent protein kinase. IV.

Widespread occurrence of adenosine 3′,5′-monophosphate dependent protein kinase in various tissues and phyla of the animal kingdom, *Proc. Natl. Acad. Sci. U.S.A.* **64:**1349.

Kuo, J., and Greengard, P., 1972, An assay method for cyclic AMP and cyclic GMP based upon their abilities to activate cyclic AMP-dependent protein kinase, in: *Advances in Cyclic Nucleotide Research,* Vol. II (P. Greengard and G.A. Robison, eds.), pp. 41–50, Raven Press, New York.

Langan, T.A., 1973, Protein kinases and protein kinase substrates, in: *Advances in Cyclic Nucleotide Research,* Vol. III (P. Greengard and G.A. Robison, eds.), pp. 99–153, Raven Press, New York.

Means, A.R., and Hall, P.F., 1967, Effect of FSH on protein biosynthesis in testes of the immature rat, *Endocrinology* **81:**1151.

Means, A.R., and Hall, P.F., 1968, Protein biosynthesis in the testis. I. Comparison between stimulation by FSH and glucose, *Endocrinology* **82:**597.

Means, A.R., and Hall, P.F., 1969, Protein biosynthesis in the testis. V. Concerning the mechanism of stimulation by follicle-stimulating hormone, *Biochemistry* **8:**4293.

Moyle, W.R., and Ramachandran, J., 1973, Effect of LH on steroidogenesis and cyclic AMP accumulation in rat Leydig cell preparations and mouse tumor Leydig cells, *J. Endocrinol.* **93:**127.

Murad, F., Strauch, S., and Vaughan, M., 1969, The effect of gonadotropins on testicular adenyl cyclase, *Biochim. Biophys. Acta* **177:**591.

Raina, A., and Holtta, E., 1972, The effect of growth hormone on the synthesis and accumulation of polyamines in mammalian tissues, in: *Growth and Growth Hormone, Excerpta Med. Int. Congr.* Ser., No. 244 (A. Pecile and E.E. Muller, eds.), pp. 143–149, Excerpta Medica, Amsterdam.

Reddy, P.R.K., and Villee, C.A., 1975a, Messenger RNA synthesis in the testis of immature rats: Effect of gonadotropins and cyclic AMP, *Biochem. Biophys. Res. Commun.* **63:**1063.

Reddy, P.R.K., and Villee, C.A., 1975b, Stimulation of ornithine decarboxylase activity by gonadotropic hormones and cyclic AMP in the testis of immature rats, *Biochem. Biophys. Res. Commun.* **65:**1350.

Richman, R.A., Underwood, L.E., Van Wyk, J.J., and Voina, S.J., 1971, Synergistic effect of cortisol and growth hormone on hepatic ornithine decarboxylase activity, *Proc. Soc. Exp. Biol. Med.* **138:**880.

Richman, R.A., Dobbins, D., Voina, S., Underwood, L.E., Mahaffee, D., Gitelman, H.S., Van Wyk, J.J., and Ney, R.L., 1973, Regulation of adrenal ornithine decarboxylase by adrenocorticotropic hormone and cyclic AMP, *J. Clin. Invest.* **52:**2007.

Russell, D.H. (ed.), 1973, *Polyamines in Normal and Neoplastic Growth,* Raven Press, New York.

Villee, C.A., and Loring, J.M., 1975, Estrogenic control of uterine enzymes, *Adv. Enzyme Regul.* **13:**137.

Villee, C.A., and Loring, J.M., 1978, Effects of FSH and LH on RNA synthesis in the testis: Role in ornithine decarboxylase, in: *Structure and Function of the Gonadotropins* (K. McKerns, ed.), pp. 295–313, Plenum Press, New York.

Villee, D., and Goswami, A., 1973, Effects of exogenous RNA on steroid synthesis in endocrine tissues, in: *The Role of RNA in Reproduction and Development* (M.C. Niu and S.J. Segal, eds.), pp. 73–85, American Elsevier, New York.

Williams-Ashman, H.G., Pegg, A.E., and Lockwood, D.H., 1969, Mechanism and regulation of polyamine and putrescine biosynthesis in male genital glands and other tissues of mammals, *Adv. Enzyme Regul.* **7:**291.

Williams-Ashman, H.G., Janne, T., Coppoc, G.L., Gerach, M.E., and Schienone, A., 1972, New aspects of polyamine biosynthesis in eukaryotic organisms, *Adv. Enzyme Regul.* **10**:225.

Witt, J.J., and Roskoski, R., 1975, Rapid protein kinase assay using phosphocellulose–paper absorption, *Anal. Biochem.* **66**:253.

23

Uterine Diamine Oxidase

A Marker for Progestin Action

C. H. Spilman, D. C. Beuving, and
K. K. Bergstrom

1. Introduction

The classic bioassays for progesterone and biologically active progestins are tedious and semiquantitative. These assays and their limitations were described recently (Glasser, 1975). The synthesis of RNA was suggested by Glasser (1975) to be a reasonably specific and reproducible quantitative biochemical endpoint for the action of progestins on the uterus. Another biochemical endpoint, the synthesis of uteroglobin (UTG), has been used by Jänne *et al.* (1978) as a measure of the progestin activity of natural and synthetic androgens. While there was good correlation between biological activity (induction of UTG synthesis) and binding affinity to the rabbit uterine progesterone receptor, the UTG bioassay is expensive to use routinely for the evaluation of a large number of compounds.

Diamine oxidase [(DAO) amine:oxygen oxidoreductase, deaminating, EC 1.4.3.6] activity was reported to be associated primarily with the maternal placenta in the rat (Guha and Jänne, 1976; Maudsley and Kobayashi, 1971, 1977). The appearance of DAO activity in the uterus of pregnant rats increased markedly on day 10 of pregnancy, reached a peak at about day 15, and then gradually declined until parturition (Guha and Jänne, 1976). Decidual tissue in pseudopregnant rats was also

C. H. Spilman, D. C. Beuving, and *K. K. Bergstrom* • Fertility Research, The Upjohn Company, Kalamazoo, Michigan 49001.

reported to contain high levels of DAO activity (Backus and Kim, 1970). Furthermore, it has been suggested that serum levels of DAO activity in pregnant women may be a reliable index of gestational age and fetal health (Beaven *et al:,* 1975). Taken together, all these studies indicate that DAO activity is associated with placental function.

Harris and Kim (1972) suggested that the measurement of DAO activity in uterine tissue of rats could be used as a sensitive bioassay for progestin as well as antiprogestin activity. Reported herein is some of our experience concerning the evaluation of DAO activity as a marker for progestin action in the mammalian uterus.

2. Materials and Methods

Bilaterally ovariectomized rats, 150–170 g, were purchased from Hormone Assay, Inc. Hamsters, 80–100 g, were purchased from Engle and bilaterally ovariectomized through a midventral laparotomy. Steroids, obtained from the references files of The Upjohn Company, were prepared in sesame oil and injected subcutaneously once daily in a volume of 0.2 ml. The first steroid injection was given at least 10 days after ovariectomy in both species. A 2-0 silk suture was placed in the entire length of both uterine horns on the 4th day of steroid treatment to induce a decidual reaction. Steroid injections were continued for 7 days in rats and for 5 days in hamsters.

The animals were sacrified by cervical dislocation on day 8. Both uterine horns were excised, the silk sutures removed, and the uteri trimmed and weighed. The tissue was homogenized in 9 volumes of ice-cold 0.25 M sucrose using a Polytron PT 10St homogenizer. The homogenate was centrifuged at 1450*g* for 20 min, and the supernatant was decanted for subsequent assay of DAO activity.

A DAO standard curve was prepared in 0.25 M sucrose using enzyme from porcine kidney (Sigma Chemical Company, Grade II, 0.06 U/mg). The levels of DAO ranged from 0 to 50 mU/200 μl sucrose. The putrescine substrate was prepared by dissolving 36.54 mg putrescine dihydrochloride (Sigma Chemical Company) in 0.2 M sodium phosphate buffer (pH 7.2), adding 200 μl [^{14}C]putrescine dihydrochloride (New England Nuclear, [1,4,-^{14}C], 96.4 mCi/mmol), and diluting to 100 ml with 0.2 M sodium phosphate buffer. The specific activity of this substrate solution is approximately 0.09 mCi/mmol.

The DAO assay was performed according to the method of Okuyama and Kobayashi (1961) with several modifications. To a 15-ml screw-cap centrifuge tube on ice were added 200 μl 0.2 M sodium phosphate buffer (pH 7.2), 400 μl distilled water, and 200 μl DAO standard or homogenized tissue supernatant. The tubes were incubated for 5 min in a shaking water bath at 37°C. Following this preincubation, 200 μl putrescine dihydrochloride substrate solution was added, and the incubation was continued for 30 min at 37°C with shaking. At the end of the incubation, the tubes were placed in an ice bath, and 100 μl 0.13 mM aminoguanidine sulfate (Eastman Kodak Company) was added to stop the reaction. Approximately 0.5 g sodium bicarbonate was added to adjust the pH. The end product, Δ^1-pyrroline, was extracted with 10 ml toluene by shaking for 5 min on a mechanical shaker. The tubes were centrifuged at 800 *g* for 8 min, and then placed in a 40% ethanol–dry ice bath to freeze the aqueous phase. The toluene phase was decanted into scintillation vials containing 400 μl Liquifluor (New England Nuclear). The samples were counted in a liquid scintillation counter for 10 min. The DAO activity in the tissue samples was calculated by comparison with the standard curve using a computer program developed for this assay.

3. Results

3.1. Rats

3.1.1. Assay Validation and Precision

Uterine horns from ovariectomized rats were pooled and homogenized in 8 volumes of 0.25 M sucrose. Known amounts of DAO prepared in 0.25 M sucrose were then added to aliquots of the homogenized-tissue supernatant. The final preparation was equivalent to having homogenized the tissue in 9 volumes of sucrose, similar to that for unknown samples. The recovery of DAO added to tissue homogenate supernatant ranged between 93 and 100% (Fig. 1). The sensitivity of the assay is 0.3 mU

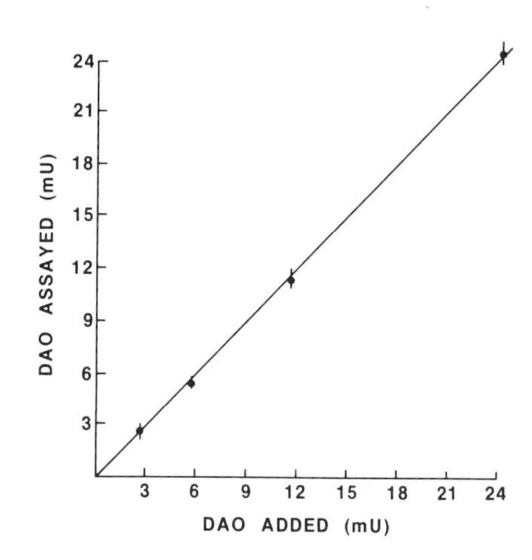

Figure 1. Quantitative recovery of DAO added to ovariectomized rat uterine tissue homogenate. Vertical bars are S.E.M.

DAO activity. The within- and between-assay coefficients of variation for uterine tissue pools assayed in duplicate in seven assays were 4.3 and 5.9%, respectively.

3.1.2. Progesterone Dose–Response

A typical dose–response curve for the induction of DAO activity by progesterone is shown in Fig. 2. Increasing doses of progesterone induced increasing amounts of DAO activity in rat uterine tissue in which a decidual reaction had been stimulated by a silk suture. The dose–response curve was reasonably linear between 0.2 and 0.8 mg progesterone daily. Over the dose range studied, there was an increase in DAO activity of approximately 25-fold. Although there was an increase

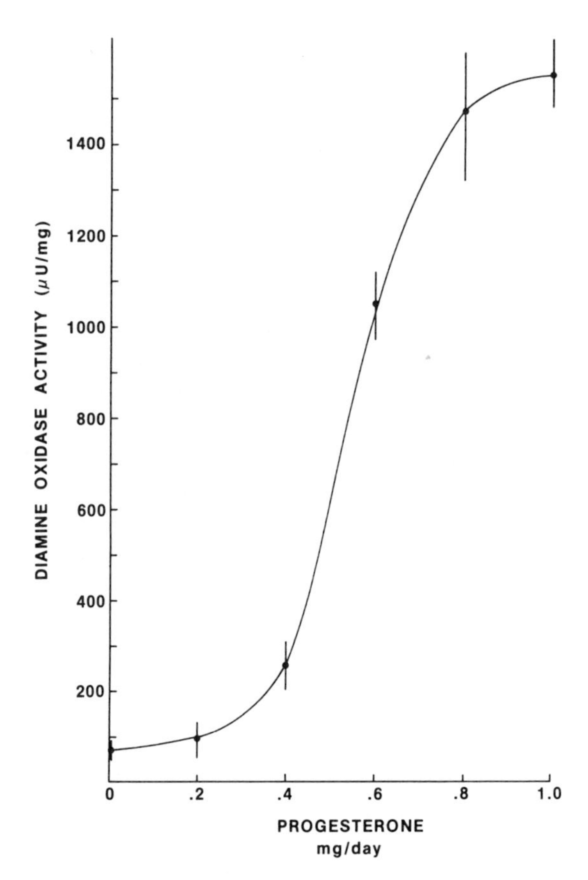

Figure 2. Dose–response curve for the induction of DAO activity in uteri of ovariectomized rats. Vertical bars are S.E.M.

Table I. Decline in Uterine DAO Activity in Ovariectomized Rats after Last Progesterone Injection

Uterine DAO activity (μU/mg)	Time after last injection (hr)			
	24	48	72	96
Mean (4 rats/group)	670	639	606	423
S.E.M.	126	222	124	78

in DAO activity in rats treated daily with only 0.2 mg progesterone, the response among rats was highly variable.

A study was performed to determine whether uterine DAO activity decreased rapidly during the 96 hr following the last progesterone injection. The data presented in Table I show that there was little difference in DAO activity in rats sacrificed from 24 to 72 hr after the last steroid administration. By 96 hr, DAO activity had declined, but this decline was not statistically significant ($P > 0.1$). This rather long lag time in the decline of DAO activity following the last progesterone injection is presumably related to the fact that the steroid was administered in an oil vehicle.

3.1.3. Specificity of DAO Induction

The data in Table II show that uterine DAO activity is induced by progesterone and medroxyprogesterone acetate, but not by estradiol-17β, testosterone propionate, norethisterone, or norethynodrel. Rats that were not treated with progestin but in which the uteri were traumatized with a silk suture had little detectable DAO activity (see Table II). Progestin treatment in the absence of uterine traumatization did not induce uterine DAO activity; less than 5 μU/mg DAO activity was measured in such animals.

3.1.4. Dose–Response of Additional Steroids

Several other steroids were tested in the DAO assay to determine whether they also increased enzyme activity (Table III). Steroids were administered at one of two doses, either 0.4 or 1.0 mg/day, and the levels of DAO activity were compared with those in rats treated with similar doses of progesterone. Of these additional steroids, two were inactive, two were more potent than progesterone, and one was approximately as potent as progesterone in the DAO assay.

Table II. Specificity of the Induction of Uterine DAO Activity in Ovariectomized Rats

Steroid treatment	Daily dose	Uterine DAO activity (μU/mg)[a]	
Sesame oil	0.2 ml	79 ± 15	(7)
Progesterone	1.5 mg	1187 ± 154	(6)
Medroxyprogesterone acetate	0.2 mg	1170 ± 83	(3)
Estradiol-17β	0.25 mg	18 ± 6	(3)
	1.0 mg	16 ± 12	(3)
Testosterone propionate	0.2 mg	107 ± 12	(3)
	3.0 mg	32 ± 6	(3)
Norethisterone	0.2 mg	29 ± 6	(3)
Norethynodrel	0.2 mg	41 ± 4	(3)

[a] Results are means ± S.E.M. The number in parentheses is the number of rats in the group.

3.1.5. *Inhibition of Progesterone-Induced DAO Activity*

The progesterone-induced increase in DAO activity measured in traumatized uteri can be inhibited by several steroids (Table IV). Estradiol-17β is particularly active in this regard. Additional experiments (data not shown) demonstrated that as little as 0.1 μg estradiol-17β was sufficient to cause a 90% inhibition of progesterone-induced DAO activity.

3.2. *Hamsters: Specificity of DAO Induction*

Uterine DAO activity was induced by progesterone, medroxyprogesterone acetate, and 19-norprogesterone, but not by estradiol-17β,

Table III. Dose–Response for the Induction of Uterine DAO Activity by Additional Steroids in Ovariectomized Rats

Steroid	Uterine DAO activity (μU/mg)[a]			
	0.4 mg/day		1.0 mg/day	
Progesterone	255 ± 93	(4)	1548 ± 96	(5)
5α-Pregnane-3,20-dione	68 ± 20	(5)	61 ± 3	(5)
6α-Methylprogesterone	1484 ± 242	(5)	1247 ± 198	(5)
19-Norprogesterone	1279 ± 183	(5)	962 ± 175	(5)
17-Methyltestosterone	70 ± 14	(5)	24 ± 6	(5)
17α-Ethynyl-17-methoxy-4-estrene-3-one	531 ± 218	(5)	1293 ± 195	(5)

[a] Results are means ± S.E.M. The number in parentheses is the number of rats in the group.

Table IV. Inhibition of Progesterone-Induced Uterine DAO Activity in Ovariectomized Rats

Treatment	Daily dose (mg)	Uterine DAO activity (μU/mg)[a]	
Progesterone	1.5	1187 ± 154	(6)
Progesterone (1.5 mg)			
+ Estradiol-17β	0.2	7 ± 6	(4)
+ Norethisterone	0.2	100 ± 54	(3)
+ Norethynodrel	0.2	28 ± 2	(2)
+ 7α-Methyl-3β,17β-dihydroxy-5-estrene	0.2	21 ± 4	(2)

[a] Results are means ± S.E.M. The number in parentheses is the number of rats in the group.

testosterone propionate, or 7α-methylprogesterone (Table V). Unlike the situation in the rat, the induction of DAO activity by progesterone could not be inhibited by as much as 200 μg estradiol-17β.

4. Discussion

A biochemical marker for the action of progestins on the mammalian uterus should possess several characteristics. Analytical methods for its detection should be sensitive, reliable, and facile. The marker should be present only in low amounts in unstimulated tissue, and increases in the marker should be dose-dependent and specific for progestins. Some aspect of uterine function that is known to be progestin-dependent should be associated with the biochemical marker. The half-life of the marker

Table V. Specificity of the Induction of Uterine DAO Activity in Ovariectomized Hamsters

Steroid treatment	Daily dose	Uterine DAO activity (μU/mg)[a]	
Sesame oil	0.2 ml	42 ± 3	(5)
Progesterone	0.1 mg	284 ± 65	(5)
	0.8 mg	2004 ± 138	(5)
Medroxyprogesterone acetate	0.5 mg	1790	(1)
19-Norprogesterone	0.1 mg	1183 ± 303	(5)
Estradiol-17β	0.4 mg	34 ± 2	(5)
Testosterone propionate	1.0 mg	61 ± 4	(4)
7α-Methylprogesterone	0.5 mg	0	(5)

[a] Results are means ± S.E.M. The number in parentheses is the number of hamsters in the group.

should be long enough that its degradation does not interfere with the measurement of its induction. All these criteria were enumerated by Lyttle and DeSombre (1977) with regard to the induction of uterine peroxidase activity by estrogens. In addition, the experimental model should be reasonably inexpensive, thus allowing its use for the evaluation of a large number of compounds. In the case of a specific progestin bioassay, the method should also be more quantitative than the classic histological methods, and the method should not be confounded by the necessity of pretreatment with estrogen.

The induction of DAO activity in the uteri of ovariectomized rats seems to satisfy all the characteristics listed above for a specific marker of progestin action. The assay is reliable and relatively easy to perform. Neufeld and Chayen (1979) recently compared three methods for the measurement of DAO activity. They concluded that the radiochemical method, as was used in the studies reported herein, was the most reliable. Using this method, the reaction was linear with time up to at least 60 min, and a lag phase in the reaction was not observed. The activity of DAO was independent of putrescine concentration in the range of 0.02–0.93 mM. The concentration of substrate in our assay system was 0.45 mM. Neufeld and Chayen (1979) also reported no significant difference in DAO activity at pH 6.8 or 7.4, a range that includes the pH used in our studies. It should be mentioned that Andersson *et al.* (1978) have suggested that the radiochemical method with toluene extraction of Δ^1-pyrroline may not be reliable for the determination of DAO activity in all tissues. When [^{14}C]putrescine was incubated with guinea pig liver homogenate, the major reaction products were gamma-aminobutyric acid and some unidentified compounds. The toluene-extractable Δ^1-pyrroline was only a minor reaction product. However, when the source of enzyme was maternal or fetal human placenta, the major reaction product was Δ^1-pyrroline. In the case of the maternal placenta, the amount of Δ^1-pyrroline formed was so high that it interfered with the determination of the other products. Over 60% of the reaction products were accounted for by Δ^1-pyrroline when fetal placenta homogenate was incubated with [^{14}C]putrescine. It has been reported that greater than 90% of the DAO activity of the placenta on day 18 of pregnancy in the rat is associated with the maternal part (Guha and Jänne, 1976; Maudsley and Kobayashi, 1977). Therefore, we believe that the radiochemical assay method based on the toluene extraction of Δ^1-pyrroline is a reliable method for the determination of DAO activity in uteri bearing decidual tissue.

DAO activity was detectable only in low amounts in rats not treated with progesterone, and detectable in even lower amounts in rats in which a decidual cell reaction had not been induced. Using an oil vehicle for the administration of the progestins resulted in a very slow decline in uterine levels of DAO activity following the last steroid injection (see Table I). Thus, the magnitude of the increase in DAO activity induced by progestins and the slow degradation of enzyme activity are sufficient for using this assay system as a sensitive marker for the action of progestins on the uterus.

Other proteins have been used recently as markers of progesterone action. Ornithine decarboxylase (ODC) activity was reported to be elevated in decidual tissue of the rat (Barkai and Kraicer, 1978; Guha and Jänne, 1976; Heald, 1979; Maudsley and Kobayashi, 1971, 1977; Saunderson and Heald, 1974) and mouse (Collawn and Baggett, 1977). The activity of ODC was higher in the fetal placenta than in the maternal placenta of rats (Guha and Jänne, 1976; Maudsley and Kobayashi, 1971, 1977). This distribution is the reverse of that for DAO activity in rat placentas as mentioned earlier. Leavitt *et al.* (1979) reported that ODC activity was induced in hamster decidual tissue by progestins, but not by cortisol, testosterone, or estradiol. Many investigators have studied the induction of uteroglobin (UTG) synthesis by rabbit uterus. Jänne *et al.* (1978) have shown a good correlation between the binding of natural and synthetic androgens to the rabbit uterine progesterone receptor and the ability of those steroids to induce UTG synthesis by rabbit uteri. Their data suggest that the action of these steroids is mediated through the progesterone receptor. Both these bioassays, the induction of ODC activity and the synthesis of UTG, seem to be specific for progestin action. We believe that the DAO assay is somewhat easier to perform than either the ODC or the UTG assay, and is considerably less expensive with regard to the animal used than the UTG assay.

On the basis of the results reported herein, the induction of uterine DAO activity appears to be specific for progestins. However, it should be mentioned that it is not known whether all the steroids that supported the induction of DAO activity in ovariectomized rats and hamsters would also maintain pregnancy in ovariectomized animals. There is ample evidence that norethisterone and norethynodrel will not maintain pregnancy in ovariectomized rats (Stucki, 1958); neither of these compounds supported the induction of DAO activity in the rat. It is unusual that norethisterone caused the same synthesis of UTG as did progesterone in the rabbit (Torkkeli *et al.*, 1977). Therefore, the induction of UTG by steroids may not be related to the most important function of progesterone, that of pregnancy maintenance. It should be determined which of the compounds that support DAO induction will also maintain pregnancy.

As suggested by Harris and Kim (1972), the DAO assay can also be used to detect antiprogestin activity. One potential disadvantage of using the rat DAO assay to detect antiprogestin activity is that the induction of enzyme activity is particularly sensitive to inhibition by estrogens. Therefore, a compound that inhibits the progesterone-induced increase in DAO activity may do so because of its intrinsic estrogenic activity and not because it inhibits progesterone binding to the uterine progesterone receptor. The induction of DAO activity in the hamster uterus appears not to be sensitive to inhibition by estrogens. From this point of view, the hamster may be a better animal model to use for the detection of antiprogestins. However, the relative binding affinities of many steroids for the hamster uterine progesterone receptor (Leavitt *et al.*, 1974; Wilks *et al.*, 1980) are very different from those for the human progesterone receptor (Kontula *et al.*, 1975). Thus, the hamster may not be the most reliable model to use for predicting activity in the human.

On the basis of the preliminary data reported herein, the DAO assay appears to have the necessary requirements of a specific, biochemical marker for the action of progesterone on the mammalian uterus. With the implementation of this assay by other laboratories, the apparent usefulness of this progestin bioassay may be validated.

Discussion

Pietras: Can your assay be modified for application to histologic or microfluorometric studies of progesterone-sensitive cells?

Spilman: I suspect that if you developed antibodies against DAO, you could localize the enzyme in the cells using immunohistochemical techniques.

Channing: (1.) What is the effect of 20α-OH-progesterone in your system? (2.) What cell type of the uterus makes DAO?

Spilman: (1) We have not tested this. In view of the low binding affinity of 20α-OH-progesterone for the progesterone receptor, I would expect that DAO activity would not be induced. (2) We have not done any work to identify the cell type. All we know is that DAO activity increases in parallel with decidualization. An increase in DAO activity cannot be detected in uterine tissue from progesterone-treated rats unless the uterus is traumatized.

Banik: You mentioned that the maternal part of the placenta contains more DAO than the fetal part. Do you know if anybody measured this enzyme from the maternal placenta following induced abortion in humans?

Spilman: To my knowledge, no one has done this. I am only aware of the measurement of DAO activity in human serum during pregnancy.

References

Andersson, A.-C., Henningsson, S., Persson, L., and Rosengren, E., 1978, Aspects on diamine oxidase activity and its determination, *Acta Physiol. Scand.* **102:**159.

Backus, B., and Kim, K.S., 1970, Subcellular distribution of diamine oxidase, *Comp. Gen. Pharmacol.* **1:**196.

Barkai, U., and Kraicer, P.F., 1978, Definition of period of induction of deciduoma in the rat using ornithine decarboxylase as a marker of growth onset, *Int. J. Fertil.* **23:**106.

Beaven, M.A., Marshall, J.R., Baylin, S.B., and Sjoerdsma, A., 1975, Changes in plasma histaminase activity during normal early human pregnancy and pregnancy disorders, *Am. J. Obstet. Gynecol.* **123:**605.

Collawn, S.S., and Baggett, B., 1977, Increased ornithine decarboxylase activity in the mouse uterus during early decidualization, *Fed. Proc. Fed. Am. Soc. Exp. Biol.* **36:**388 (Abstract 631).

Glasser, S.R., 1975, A molecular bioassay for progesterone and related compounds, in: *Hormone Action,* Part A, *Steroid Hormones, Methods in Enzymology,* Vol. XXXVI (B.W. O'Malley and J.G. Hardman, eds.), pp. 456–465, Academic Press, New York.

Guha, S.K., and Jänne, J., 1976, The synthesis and accumulation of polyamines in reproductive organs of the rat during pregnancy, *Biochim. Biophys. Acta* **437:**244.

Harris, M.E., and Kim, K.S., 1972, A new progestational activity assay: Uterine diamine oxidase (DAO), in: *IVth International Congress of Endocrinology, Excerpta Medica Int. Congr. Ser.,* No. 256, pp. 27–28 (Abstract 71), Excerpta Medica, Amsterdam.

Heald, P.J., 1979, Changes in ornithine decarboxylase during early implantation in the rat, *Biol. Reprod.* **20:**1195.

Jänne, O., Hemminki, S., Isomaa, V., Kokko, E., Torkkeli, T., and Vierikko, P., 1978, Progestational activity of natural and synthetic androgens, *Int. J. Androl. Suppl.* **2:**162.

Kontula, K., Jänne, O., Vihko, R., deJager, E., deVisser, J., and Zeelen, F., 1975, Progesterone-binding proteins: *In vitro* binding and biological activity of different steroidal ligands, *Acta Endocrinol.* **78:**574.

Leavitt, W.W., Toft, D.O., Strott, C.A., and O'Malley, B.W., 1974, A specific progesterone receptor in the hamster uterus: Physiologic properties and regulation during the estrous cycle, *Endocrinology* **94:**1041.

Leavitt, W.W., Chen, T.J., and Evans, R.W., 1979, Regulation and function of estrogen and progesterone receptor systems, in: *Steroid Hormone Receptor Systems, Adv. Exp. Med. Biol.,* Vol. 117 (W.W. Leavitt and J.H. Clark, eds.), pp. 197–222, Plenum Press, New York.

Lyttle, C.R., and DeSombre, E.R., 1977, Uterine peroxidase as a marker for estrogen action, *Proc. Natl. Acad. Sci. U.S.A.* **74:**3162.

Maudsley, D.V., and Kobayashi, Y., 1971, Diamine metabolism in the rat placenta, *Fed. Proc. Fed. Am. Soc. Exp. Biol.* **30:**204 (Abstract 59).

Maudsley, D.V., and Kobayashi, Y., 1977, Biosynthesis and metabolism of putrescine in the rat placenta, *Biochem. Pharmacol.* **26:**121.

Neufeld, E., and Chayen, R., 1979, An evaluation of three methods for the measurement of diamine oxidase (DAO) activity in amniotic fluid, *Anal. Biochem.* **96:**411.

Okuyama, T., and Kobayashi, Y., 1961, Determination of diamine oxidase activity by liquid scintillation counting, *Arch. Biochem. Biophys.* **95:**242.

Saunderson, R., and Heald, P.J., 1974, Ornithine decarboxylase activity in the uterus of the rat during early pregnancy, *J. Reprod. Fertil.* **39:**141.

Stucki, J.C., 1958, Maintenance of pregnancy in ovariectomized rats with some newer progestins, *Proc. Soc. Exp. Biol. Med.* **99:**500.

Torkkeli, T.K., Kontula, K.K., and Jänne, O.A., 1977, Hormonal regulation of uterine blastokinin synthesis and occurrence of blastokinin-like antigens in nonuterine tissues, *Mol. Cell. Endocrinol.* **9:**101.

Wilks, J.W., Spilman, C.H., and Campbell, J.A., 1980, Steroid binding specificity of the hamster uterine progesterone receptor, *Steroids* **35:**697.

VII

hCG Peptides and Antisera as Antifertility Agents

24

Specific Antisera to Human Choriogonadotropin

Steven Birken

1. Introduction

Since this monograph is concerned with reproductive processes and contraception, it is appropriate to direct our attention to measurement of the pregnancy hormone, human choriogonadotropin (hCG), a glycoprotein hormone secreted by the placenta that plays a crucial role in maintenance of pregnancy. Its function appears to be maintenance of the steroid secretions of the corpus luteum. The hormone binds to the same receptor as human luteinizing hormone (hLH) and inhibits the onset of menstruation after conception has taken place.

hCG is secreted in large quantities (10,000 mIU/ml in plasma) during the first trimester of pregnancy (Loraine, 1967; Kosasa *et al.*, 1973) and is excreted in the urine, from which it is usually purified. The hormone has also been reported as a secretory product of a variety of malignancies (Braunstein *et al.*, 1973), normal male genital tissue such as the testis (Braunstein *et al.*, 1975), normal nonpregnant urine, pituitary (Chen *et al.*, 1976) (Ayala *et al.*, 1978), and even normal nongenital tissues. hCG has recently been referred to as "human cellular gonadotropin" (Yoshi-

Steven Birken • Department of Medicine, Columbia University College of Physicians & Surgeons, New York, New York 10032.

moto *et al.*, 1979) because of its apparent ubiquitousness in all tissues examined. When extracted from ectopic tissue sources, the hormone appears to lack part or all of its carbohydrate moieties.

There are also reports of hCG-like immunoreactive materials in the cell walls of microbes (Livingston and Livingston, 1974; Slifkin *et al.*, 1979) (also see Chapter 26). Other than the urinary product from normal pregnancy or molar pregnancy, the identification of hCG in other tissues relies on the use of specific antisera for immunoassay or immunohistochemistry. Hence, identification of the unique antigenic determinants of hCG for use in generation of hCG-specific antisera is a task of crucial importance to confirm the apparent ubiquity of the hCG-like immunoreactive material. Specific antisera are also needed for measurement of hCG during therapy of certain hCG-secreting malignancies. Likewise, the development of a "contraceptive vaccine" is dependent on delineation of the unique sites of hCG so that the most appropriate immunogen may be selected.

2. *Measurement of hCG by Immunoassay with Anti-β-Subunit Antisera*

hCG and hLH are very similar in structure. They share a common α-subunit (Pierce, 1971), and their β-subunits differ by less than 20% in primary structure (Morgan *et al.*, 1973; Bellisario *et al.*, 1973; Carlsen *et al.*, 1973; Morgan *et al.*, 1975; Ward, 1979). This similarity of structure leads to difficulty in immunological measurement of hCG in the presence of hLH. If hCG is employed as immunogen in rabbits, the resultant antisera cross-react significantly with hLH. Several investigators have successfully employed purified hCG-β as immunogen to elicit sensitive hCG antisera of lesser hLH cross-reactivity (Vaitukaitis *et al.*, 1972; Swaminathan and Braunstein, 1978). Specific and sensitive anti-hCG-β antisera are relatively rare. One example is the SB-6 antiserum distributed by the NIH, which was selected from a large number of rabbits immunized by Dr. J. Vaitukaitis and colleagues (Vaitukaitis *et al.*, 1972) with an early preparation of hCG-β from this laboratory. Most hCG antisera cross-react with hLH to a significant degree in an immunoassay employing [^{125}I]-hCG as tracer or exhibit insufficient sensitivity to the native hormone, rather reacting far better with free β. One novel approach to attempt to overcome this difficulty is to use a hybrid gonadotropin immunogen composed of an animal pituitary α species (which does not cross-react with human α) and the hCG-β subunit. This approach was described by Lerario *et al.* (1978).

2.1. *Studies of Antisera Generated to hCG-β and to Asialo Reduced and Carboxymethylated β-Subunit*

In an effort to ascertain whether a specific β-antiserum such as SB-6 represented a unique pattern of antibody response of one rabbit, we performed a series of titrations using a variety of labeled hCG tracers and several hCG-β antisera. The tracers employed were hCG, hCG-β, and reduced and carboxymethylated (RCM)-β as well as the desialylated [asialo (As.)] forms of each. Thus, we sought to assess the importance of ligand conformation and sialic acid content by testing reactivity of disulfide-linked as well as disulfide-reduced tracers with or without sialic acid. Binding of [^{125}I]-hLH was also examined. This evaluation of ligand binding by titration reflects both molar quantity and affinity of the antibody populations in each antiserum for each of the labeled tracers. It provides a qualitative assessment of the binding preferences of each antiserum prior to confirmation by dose–response competition studies. Five anti-β sera were produced in our laboratory, and three of these were compared to SB-6 by the series of titrations shown in Fig. 1. The titration patterns of all the β-antisera were similar even though SB-6 displayed the greatest discrimination between hCG and hLH in competition assays. It was found that all the antisera are conformationally directed (i.e., they bind RCM tracers poorly), that desialylation of the ligand did not improve binding, and that all the anti-hCG-β antisera bound hLH tracer. Dose–response curves confirmed the very poor competition by RCM-β or AS.RCM-β in a radioimmunoassay (RIA) using labeled hCG as tracer. Therefore, it appears that the discriminating properties of the SB-6 antiserum do not result from an unusual response to the hCG-β immunogen, but rather lie in a subtle unique determinant recognized by a population of antibodies in SB-6. Such discriminating antibodies are presumably less numerous in those β-antisera that display more hLH cross-reactivity.

On the basis of reports of successful generation of hCG-specific antisera by use of RCM-β as immunogen (Pandian and Bahl, 1977), we examined a series of antisera from ten rabbits injected with As. RCM-β either as the free polypeptide or carrier-conjugated to ovalbumin by carbodiimide. Figure 2 shows that results of titrations of those four anti-As.RCM-β antisera with the same tracers employed in Fig. 1. Antisera R131 and R132 represent two of the five rabbits immunized with free As.RCM-β, while R137 and R138 are representative of rabbits immunized with As.RCM-β conjugated to ovalbumin. As is apparent in Fig. 2, none of these antisera is of use in measuring hCG, since they exhibit negligible binding to hCG and are all directed toward the unfolded polypeptide

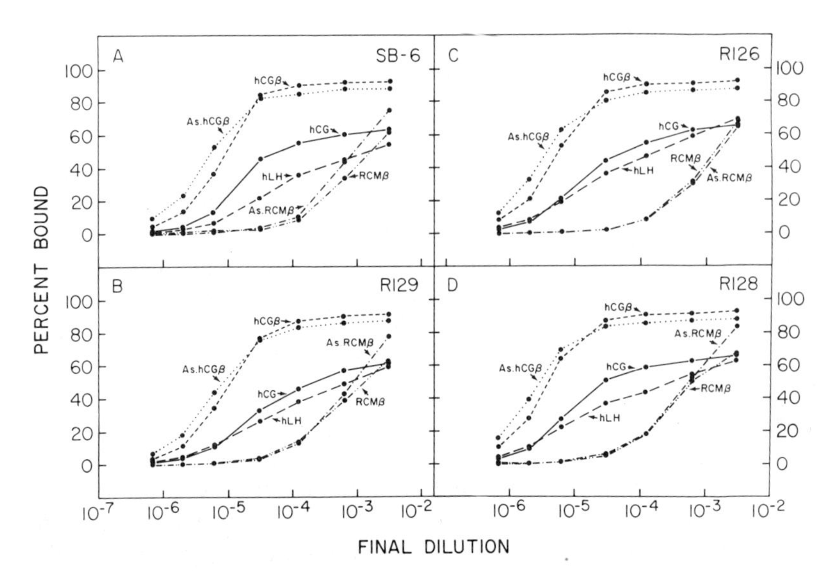

Figure 1. Titrations of hCG antisera with six ^{125}I-labeled tracers: Two tracers were desialylated: As.hCG-β and As.RCM-β. These were compared to their parent sialic-acid-containing tracers, hCG-β and RCM-β, as well as to the intact hormones, hCG and hLH. (A) Results obtained with a relatively specific hCG antiserum, the widely used SB-6; (B–D) results obtained with anti-β-antisera of lesser specificity. The titration patterns look very similar qualitatively, indicating that SB-6 is not a completely unique antiserum but exhibits a qualitative binding pattern similar to the other anti-β-antisera. The antisera are all conformationally directed (unfolded β-polypeptide binds less well) and are not sensitive to the presence or absence of sialic acid in the tracers.

structure. These antisera proved to be highly effective for immunoprecipitation of the cell-free translation product (hCG-β) of messenger RNA purified from first-trimester placentas (Daniels-McQueen *et al.,* 1978). Since the anti-RCM-β antisera were superior to anti-β antisera, we infer that such *in vitro* translated material is not likely to have assumed its native conformation.

2.2. Determinants Recognized by the SB-6 Antiserum

Since use of RCM-β as immunogen did not seem promising we returned to examination of the determinants recognized by SB-6 as being present in hCG but absent in hLH. We and others have shown that these determinants are not directed to the hCG-β COOH-terminus, but are based on subtle conformational differences between hLH and hCG (Birken and Canfield, 1979; Birken *et al.,* 1980).

2.2.1. *Peptides That Bind to SB-6*

Swaminathan and Braunstein (1978) performed a study of the relative immunopotency of several hCG-β disulfide-linked peptides in an SB-6 RIA. Although the peptides derived from partially reduced hCG-β had resulted from significant disulfide interchange and all exhibited great losses of immunopotency relative to whole β, several were shown to compete with [^{125}I]-hCG in the SB-6 RIA. These fragments were comprised of peptide 21–31 disulfide-bridged to peptide 106–114 and peptide 21–37 linked to 65–74. The investigators also added synthetic β COOH-terminal peptides of varying lengths and found no competition with [^{125}I]-hCG tracer and the SB-6 antiserum.

2.2.2. *An Asialo hCG-β Core That Contains All SB-6 Determinants*

We are exploring the determinants recognized by the SB-6 antiserum by an alternate approach. Using thermolysin, As.hCG-β was cleaved into a disulfide-linked "core" fragment that is completely free of the COOH-terminal peptide. The core fragment was found to retain full

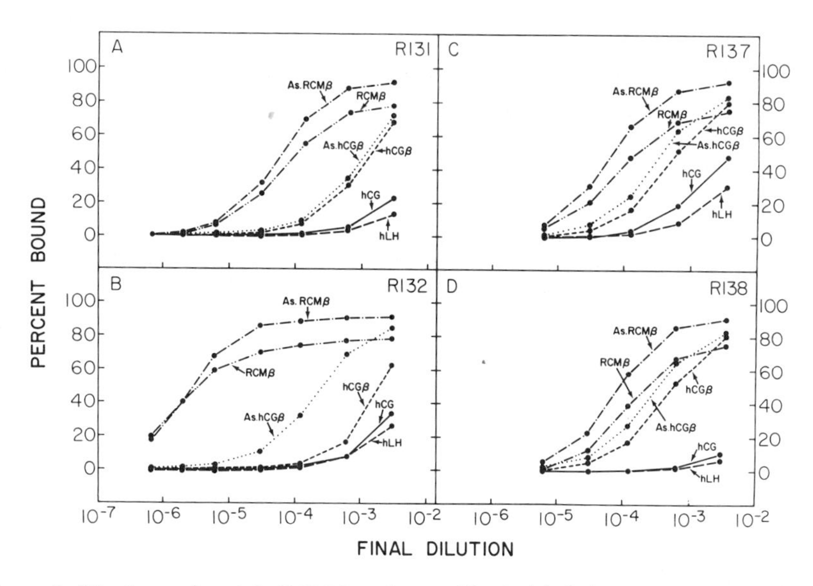

Figure 2. Titrations of anti-As.RCM-β antisera with six labeled tracers (see the Fig. 1 caption). It is noted that none of these antisera reacts well with hCG-β and that they are directed toward determinants masked by the conformation of the native hormone or disulfide-linked β-subunit.

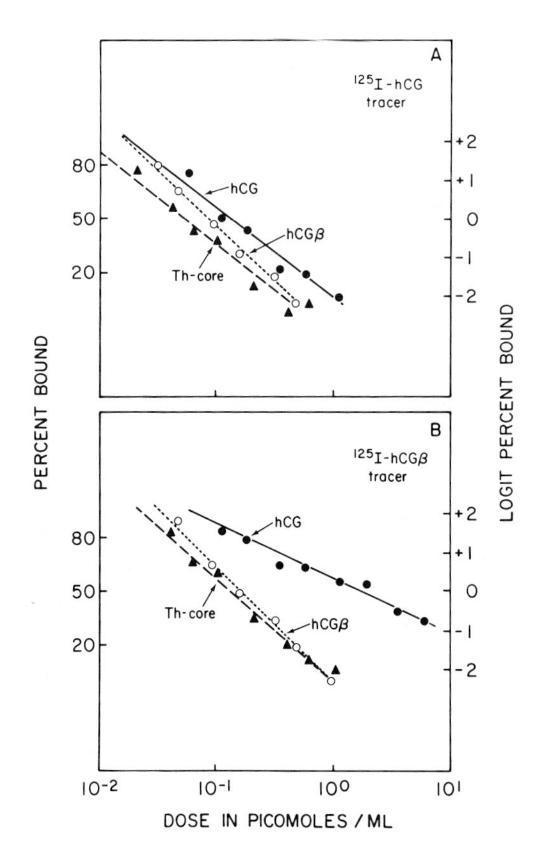

Figure 3. Comparison of immunoreactivity of hCG, hCG-β, and thermolytic core (Th-core) using SB-6 antiserum (1:120,000) in a series of dose–response curves using [^{125}I]-hCG tracer (A) or [^{125}I]-hCG-β tracer (B). The th-core, which is devoid of the COOH-terminal peptide (115–145), retains all its immunopotency relative to its parent hCG molecule in both RIA systems.

immunopotency relative to its parent As.hCG-β molecule in competition studies using the SB-6 antiserum and two tracers, [^{125}I]-hCG (Fig. 3A) and [^{125}I]-hCG-β (Fig. 3B).

This study indicates that: (1) The COOH-terminal peptide plays no role in the determinants recognized by SB-6. The Swaminathan and Braunstein (1978) study did not exclude participation of the COOH-terminal peptide in some conformational determinant located within the core. (2) The determinants for SB-6 lie within residues 4–114, since full immunopotency was conserved within this structure. The core material will be further cleaved and smaller "core" fragments purified to determine the smallest core unit that retains the determinants recognized by SB-6. (3) There are at least two determinants in hCG that are absent in hLH. One is the conformational determinant recognized by SB-6, and the second is the unique hCG-β COOH-terminal peptide that is present only in hCG.

3. *Measurement of hCG Using Anti-β COOH-Terminal Antisera*

This laboratory as well as other laboratories have generated antisera to the unique hCG-β COOH-terminal peptide in an effort to obtain hCG-specific antisera. We have employed an asialo (As.) tryptic peptide from this region, while others have used both natural and synthetic peptides (Louvet *et al.*, 1974; Chen *et al.*, 1976; Matsuura *et al.*, 1979; Stevens, 1976). The resultant antisera have proven to be specific for hCG with no hLH cross-reactivity. However, all are low in titer and sensitivity as compared to anti-hCG-β antisera. Furthermore, all the anti-As. tryptic peptide or anti-synthetic peptide antisera mapped by Drs. Matsuura and Chen (Matsuura *et al.*, 1978, 1979) appeared to be directed toward the same determinant. We undertook to characterize these antisera further to ascertain whether the conformation of hCG was masking other determinants present in the peptide immunogens, or whether the alteration or absence of carbohydrate in the immunogen was playing a role in the behavior of these antisera. If either of these masking effects were found, samples to be assayed could be treated to expose such "hidden" determinants and thus increase the effective sensitivity of the assay.

3.1. *Antisera to a Desialylated hCG-β COOH-Terminal Peptide*

In collaboration with colleagues at the NIH (Drs. G. Ross, G. Hodgen, H. Chen, and J. Louvet), we elicited a series of antisera to the As.hCG-β COOH-terminal tryptic peptide composed of residues 123–145 and conjugated by carbodiimide to bovine serum albumin (BSA). These antisera were characterized by colleagues (Louvet *et al.*, 1974; Chen *et al.*, 1976, Matsuura *et al.*, 1978). Five additional rabbits were immunized in New York with the same As.tryptic peptide conjugated to ovalbumin.

3.1.1. *Determinants Recognized by These Antisera*

Drs. Chen, Matsuura, and Hodgen have done extensive characterization of COOH-terminal antisera. Some of the antisera were generated in response to an As.tryptic COOH-terminal peptide isolated and conjugated to a carrier protein (BSA or ovalbumin) in our laboratory. A problem with all these antisera has been low titer, low affinity, and consequently low sensitivity to native hCG. Another observation has been that relatively few rabbits (i.e., 1–3 of 8 rabbits) produced an antiserum with a significant titer to native hCG (titer 1:1000). A number of these antisera, including some generated to synthetic peptides, which are easier and less expensive to produce than peptides from natural hCG, were mapped by a series of immunoassays using [^{125}I]-hCG as tracer and synthetic peptides of varying lengths as competitors (Matsuura *et al.*, 1978, 1979). It was observed that in all cases, the antisera exhibited binding to the same determinant within the last 15 amino acids of the molecule. This appears to be the only determinant available in synthetic or As.COOH-terminal peptides that is also exposed in native hCG.

3.1.2. Significance of Sialic Acid

Four antisera generated to an As.tryptic COOH-terminal peptide (residues 123–145) were compared by titration with six labeled hCG derivatives (Fig. 4) in a fashion similar to that performed with anti-β antisera in Section 2.1. The antisera in Fig. 4A and B are two produced in our laboratory in New York, while those in Fig. 4C and D were produced by our colleagues Drs. Louvet and Hodgen, using preparations

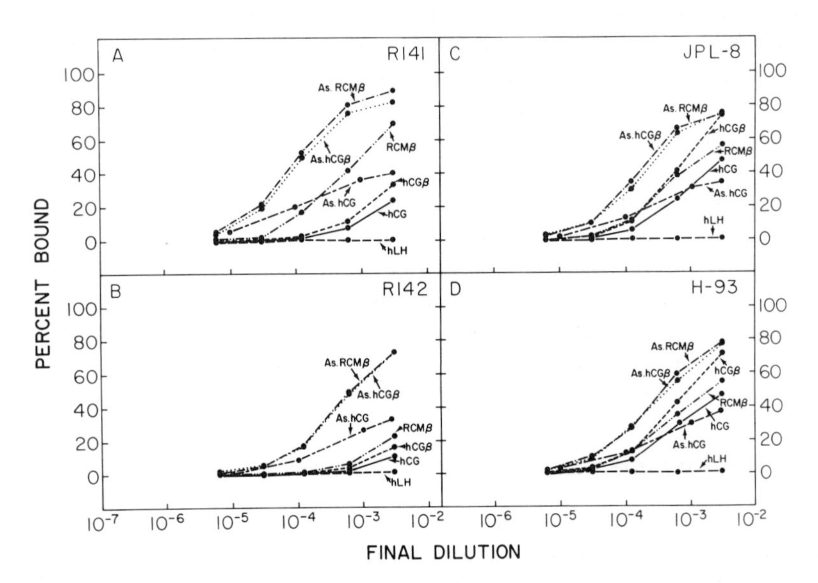

Figure 4. Titrations of antisera to the desialylated COOH-terminal tryptic peptide residues 123–145 with labeled tracers (see the Fig. 1 caption). Labeled desialylated hCG (As.hCG) was also used in these studies. It was noted that in contrast to the titrations shown in Figs. 2 and 3, sialic-acid-containing tracers bound very poorly to antisera R141 (A) and R142 (B). The other two antisera (C,D) were less affected by the presence of sialic acid in the tracer. Unfolding of the tracer (RCM-β) did not improve binding as significantly as desialylation; i.e., As.RCM-β bound to R141 as well as As.hCG-β did. The inference of a significant effect of sialic acid content on tracer binding was further explored by the dose–response curves shown in Figs. 5 and 6.

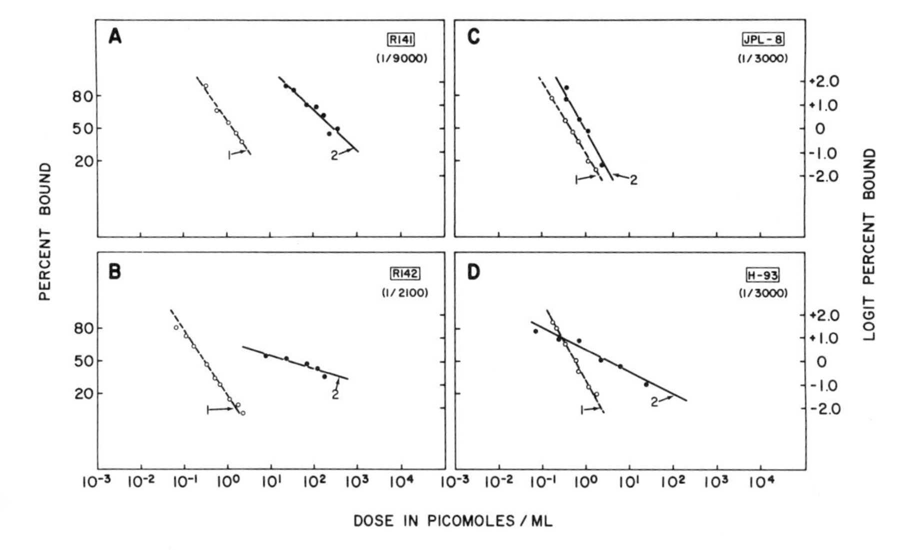

Figure 5. Immunoreactivity of four antisera to the desialylated tryptic peptide β 123–145 in a series of dose–response curves using [^{125}I]-As.RCM-β as tracer and As. 123–145 (1) or sialic-acid-containing 123–145 (2) as competitor. (A,B,D) Antisera show significant discrimination between the As. and the sialic-acid-containing peptides; (C) antiserum JPL-8 does not display such discrimination.

of the peptide immunogens made in our laboratory. The titrations indicated a pattern of response very different from that observed with antisera generated in response to hCG (Section 2.1). The desialylated tracers bound significantly better than ligands containing sialic acid; i.e., As. reduced and carboxymethylated (RCM)-β and As.hCG-β bound equally well, while RCM-β or hCG bound to a much lesser degree. This was the first indication of a significant effect of sialic acid on ligand binding to these COOH-terminal antisera, and it appeared that sialic acid rather than conformational factors was more crucial to these antisera. It was necessary to examine this premise by competition immunoassay to measure the significance of this sialic acid sensitivity.

One series of competition experiments was performed with [^{125}I]-As.RCM-β tracer, an unfolded and desialylated form of β (Fig. 5). This tracer was used because it most closely resembles the original As.tryptic peptide, which cannot itself be iodinated due to lack of tyrosine. Competitor 1 in Fig. 5 is the As.tryptic peptide 123–145, which is the tryptic peptide haptenic part of the immunogen, while competitor 2 is the same peptide with its sialic acid complement (10% wt./wt.) remaining intact. Clearly, R141, R142, and H-93 bind the sialic-acid-containing peptide poorly, while JPL-8 recognizes both peptides equally well. These

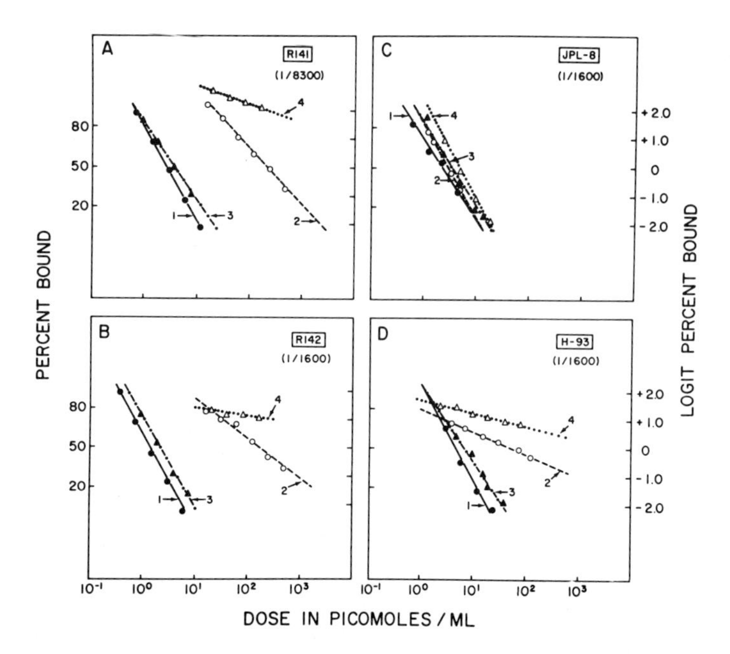

Figure 6. Immunoreactivity of four antisera to the desialylated tryptic peptide 123–145 in a series of dose–response curves using [^{125}I]-As.hCG as tracer. Four competitors were used: (1) As. 123–145; (2) 123–145; (3) As.hCG; (4) hCG. (A,B,D) It is significant that for the antisera shown, desialylation of hCG produces an As. hormone (3) with an immunopotency equal to that of the As. tryptic peptide (1) that was a part of the immunogen itself. (C) The curves show that JPL-8 antiserum does not discriminate between hCG and As.hCG.

results were further confirmed by a similar series of dose–response curves performed with [^{125}I]-As.hCG tracer. In this case, we are examining binding to the intact hormone that has been desialylated. The resultant dose–response curves are similar to those in Fig. 6. Competitor 1 is As.peptide 123–145, 2 is sialic-acid-containing 123–145, 3 is As.hCG, and 4 is native hCG. It can be inferred that R141, R142, and H-93 all contain a population of antibodies that bind poorly to native hCG or the sialic-acid-containing tryptic peptide. When hCG is treated with neuraminadase to remove its complement of sialic acid, it becomes as good a competitor as the As.tryptic peptide itself. This is interpreted to mean that all the determinants within the As.tryptic peptide are available in As.hCG and are masked by sialic acid in the native hormone.

The observation of sialic-acid-sensitive antibodies in the H-93 antiserum is apparent only when As.hCG is used as tracer. When labeled HCG is employed as tracer, H-93 does not discriminate between hCG and As.hCG (Fig. 7). Also, the titer of H-93 with [^{125}I]-hCG is higher

(×4) than the titer with [^{125}I]-As.hCG. In contrast, antiserum R141 has a titer (B/T = 0.3) of 1:8000 with As.hCG and only 1:300 with native hCG tracer. The differences between H-93 and R141 with respect to titers of hCG and As.hCG can be explained by two populations of antibodies with two different binding sites. R141 is likely to have one major population of antibodies to a site masked by sialic acid in hCG, whereas H-93 may have two populations of antibodies.

3.1.3. Problems Arising from Use of Synthetic or Desialylated Forms of hCG-β COOH-Terminal Peptides

In the light of the studies described above, it can be inferred that the use of a desialylated peptide as immunogen (and a synthetic peptide presumably would give the same results) invites production of antibodies to determinants within the hCG-β COOH-terminal region that are masked by sialic acid in the native hormone. This may provide part of the explanation for the low number of rabbits producing antibodies that react with native hCG (these antisera may contain a major population of the R141–R142 type of antibodies which bind preferentially to asialo hCG). Also, the use of peptides lacking sialic acid may be part of the reason for low sensitivity. Therefore, we embarked on a program to purify significant quantities of COOH-terminal fragments with their sialic acid content

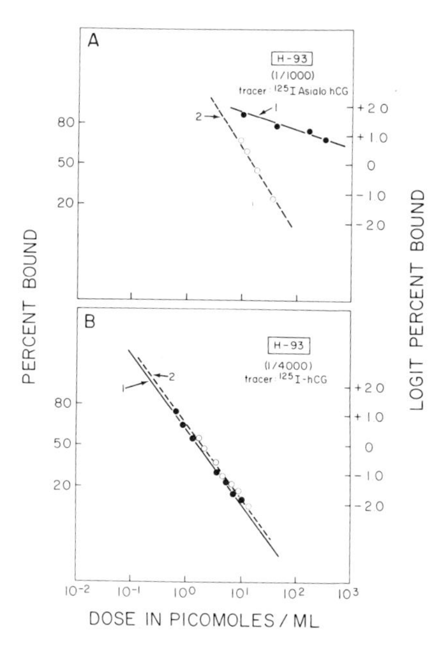

Figure 7. Comparison of immunoreactivity of hCG (1) and As. hCG (2) in RIAs using antiserum H-93 and [^{125}I]-As.hCG tracer (A) or [^{125}I]-hCG tracer (B). It is apparent that a population of antibodies that binds [^{125}I]-As.hCG tracer (B/T = 0.3) at a titer of 1:1000 does not cross-react well with hCG (1), but binds well to As.hCG (2). In contrast, when [^{125}I]-hCG is employed as tracer (B), the differential recognition between hCG and As.hCG disappears and both forms of hCG bind equally well in the competition assay vs. the native hCG tracer. The second population of hCG antibodies bind [^{125}I]-hCG tracer at a dilution of 1:4000 and appear to be the major antibody population present in the H-93 antiserum.

intact and to ascertain whether they would elicit antisera to the native hormone in a greater proportion of rabbits and result in antisera of greater sensitivity.

3.2. *Antisera to Sialic-Acid-Containing hCG-β COOH-Terminal Peptides*

The purification of a peptide from hCG-β containing residues 115–145 was accomplished by performing a limited tryptic digestion and a gel filtration on Sephadex G-100 followed by reduction, carboxymethylation, and refiltration on Sephadex G-75. The latter reduction and alkylation was required to remove residual disulfide-linked fragments that would be expected to generate antisera capable of cross-reacting with hLH. This peptide was next conjugated to thyroglobulin by carbodiimide, and the product was characterized by gel filtration and RIA with existent immunoassays to the COOH-terminal region. It was found that only about 10% of the peptide was coupled. The remainder appeared as polymers and modified free peptide. This conjugate mixture was employed as immunogen for eight rabbits by the multiple-site technique, which was performed by Dr. J. Lewis at Loma Linda University. In addition, eight rabbits were immunized with free peptide by the same technique. None of the rabbits immunized with the free peptide produced measurable titers to hCG when assayed with [^{125}I]-hCG tracer. Six of eight rabbits immunized with the conjugated peptide produced good titers to labeled hCG (1:2000 final dilution). Four of the best rabbits, R524, R525, R528, and R529, were characterized for binding to hCG and As.hCG by competitive-binding assays using both labeled hCG and labeled As.hCG as tracers in a series of dose–response curves.

It was observed that the new group of antisera generated to a sialic-acid-containing COOH-terminal peptide were 2–6 times more sensitive than the best antisera elicited earlier by the As.tryptic peptide (or reported for synthetic peptides). In addition, none of the antisera displayed antibodies binding better to As.hCG than to native hCG such as had occurred with R141, R142, and H-93. It appears that the native peptide is the immunogen of choice, since more sensitive antisera were generated and a greater percentage of rabbits responded to this immunogen than to the As. or synthetic peptide immunogens. The sensitivity of this new group of antisera is between 0.6 and 1.0 pmol/ml at the 50%-bound region. This is about 10% of the sensitivity of SB-6, the high-affinity hCG-β antiserum that exhibits 5–10% hLH cross-reactivity. All the COOH-terminal antisera exhibited no hLH cross-reactivity.

It is of interest that antisera generated to the native peptide appear less heterogeneous than those generated to peptides lacking sialic acid. Removal of sialic acid probably uncovers other determinants so that an antiserum such as H-93 appears to have one population of antibodies to a determinant in native hCG and a second smaller population that binds to another site available in As.hCG (see Section 3.1.2). R141 and R142 do not bind [^{125}I]-hCG except at 1:300. The binding may be due either to low affinity of a single set of antibodies to As.hCG or to a small population of anti-hCG antibodies. The R525–R529 series of rabbits, produced to a sialic-acid-containing hCG-β COOH-terminal peptide, have anti-native hCG antibodies that bind slightly less well to As.hCG. This latter group of antisera do not display any of the sialic-acid-sensitive antibodies elicited earlier by the As.tryptic peptide immunogen.

4. Areas that Require Additional Study

The studies described herein present a number of interesting questions worthy of further pursuit. Some of these questions are discussed below.

4.1. Peptide Conjugation

One area of interest is the conjugation of the COOH-terminal peptide. The sialic-acid-rich peptide appears to be difficult to conjugate, perhaps due to the presence of sialic acid and a consequent large radius due to hydration. Yet, good-quality antisera were generated. It is conceivable that only carbodiimide treatment of the peptide is required for immunogenicity. We have already shown that the free untreated peptide is not immunogenic; i.e., it does not produce antibodies of affinities useful in immunoassays (unpublished results). Carbodiimide is known to form stable isourea groups on the COOH groups of amino acids as an "undesirable" side reaction (Weare and Richert, 1979). These modified COOH groups (now with a positive charge), on the aspartic acids in this case, can conceivably interact electrostatically with the negatively charged sialic acid moieties. Since the free peptide is not immunogenic, carbodiimide-induced modifications may be of great importance for increasing immunogenicity of this peptide, since this material comprised 80–90% of the immunogen successfully employed. The coupling procedure and identification of the immunogenic components are therefore of keen interest.

4.2. Development of Immunoassay for Carbohydrate-Altered hCG

Another area of pursuit is the use of anti-asialo (As.) hCG antisera, such as R141 and R142, for measurement of As.hCG in clinical samples as compared to native hCG. This could be accomplished by assay of the same specimen with R141 and JPL-8 (or with the new anti-sialic-acid-containing peptide sera) using [^{125}I]-As.hCG tracer. The former sees only As.hCG, while the latter measures both native and As.hCG. It would be of interest to see whether patients with hCG-secreting tumors exhibited different carbohydrate-altered forms of hCG. There have been several reports that malignant tissues as well as normal tissues produce hCG without part of or all its carbohydrate component (Braunstein, 1979; Yoshimoto *et al.*, 1979). Failure of hCG-immunoactive material to absorb to concanavalin A–Sepharose indicates carbohydrate alteration. An immunoassay that could measure As.hCG would greatly facilitate such investigations, and we plan to develop such an assay using the reagents described above.

5. Conclusion

These findings are of importance to those investigating the use of the unique hCG-β COOH-terminal peptide as immunogen. It appears that use of synthetic peptide may not be capable of inducing adequate levels of sensitive anti-hCG antibodies. COOH-terminal peptides containing sialic acid are better immunogens and have elicited antisera of greater sensitivity to hCG in a larger number of rabbits.

ACKNOWLEDGMENT. This research was supported by Research Grants AM 09579 and HD 13496.

Discussion

PETRUSZ: (1) Antibodies raised against small peptides without using a carrier will tend to be of lower affinity, and represent a broader range of specificities, than antibodies raised to the carrier-coupled peptide. However, antisera to native peptides, although they may not show binding in radioimmunnoassay, could still be very useful for immunohistochemistry. (2) What is known about hCG prohormones?

BIRKEN: (1) It would be of interest to find useful low-affinity antibodies in the antisera made to the free COOH-terminal peptide. I will send you some of these antisera to test by immunohistochemical techniques. (2) In collaboration with the laboratory of Dr. I. Boime, we have determined most of the structure of the NH_2-terminal region of pre-hCG-α as translated from first-trimester placental mRNA. The precursor of hCG-α contains an amino-terminal precursor sequence of 24 amino acids with the following structure: Met-Asp-Tyr-Tyr-Arg-Lys-Tyr-Ala-Ala-Ile-Phe-Leu-X-Thr-Leu-X-X-Phe-Leu-X-X-Leu-X-X- (see also Birken *et al.*, 1978). The hCG-β subunit is apparently translated from a separate mRNA and has a leader peptide at its NH_2 terminus which we are in the process of sequencing (Daniels-McQueen *et al.*, 1978).

WILBUR: I have two questions in regard to the structural differences of the β-subunits of hTSH, hFSH, hLH, and hCG. (1) Can you identify these specific subunits based on the sialic acid content only? If so, what is the sialic acid content of these subunits? (2) Is the sialic acid moiety necessary for functional activity?

BIRKEN: (1) The antiserum R141 which distinguishes asialo from native hCG when using [^{125}I]-As.hCG tracer is an hCG-β COOH-terminal antiserum and cannot be used to measure the other glycoprotein hormones which lack this peptide region. (2) It is known that loss of sialic acid from hCG severely decreases its half-life *in vivo* because of removal of the As.hCG by a liver receptor which recognizes the newly exposed galactose sugar underlying sialic acid (Ashwell and Morell, 1974).

NOLIN: Have you studied antisera to prolactin along the lines you just described? If not, do you know of anyone who has?

BIRKEN: We have not studied prolactin. I do not know where a source of a variety of antiprolactin antisera can be obtained. With the advent of the technology of hybridoma-derived monoclonal antibodies, it is likely that antibodies specific to various determinants within many important hormones will soon be commercially available.

REFERENCES

Ashwell, G., and Morrell, A.G., 1974, The role of surface carbohydrates in the hepatic recognition and transport of circulating glycoproteins, in: *Advances in Enzymology*, Vol. 41 (A. Meister, ed.), pp. 99–128, John Wiley, New York.

Ayala, A.R., Nisula, B.C., Chen, H.C., Hodgen, G.D., and Ross, G.T., 1978, Highly sensitive radioimmunoassay for chorionic gonadotropin in human urine, *J. Clin Endocrinol. Metab.* **47**:767.

Bellisario, R., Carlsen, R.B., and Bahl, O.P., 1973, Human chorionic gonadotropin: Linear amino acid sequences of the subunit, *J. Biol. Chem.* **248**:6796.

Birken, S., and Canfield, R.E., 1979, Structural and biochemical properties of human choriogonadotropin, in: *Structure and Function of the Gonadotropins* (K.W. McKerns, ed.), pp. 47–80, Plenum Press, New York.

Birken, S., Fetherston, J., Desmond, J., Canfield, R., and Boime, I., 1978, Partial amino acid sequence of the preprotein form of the alpha subunit of human choriogonadotropin and identification of the site of subsequent proteolytic cleavage, *Biochem. Biophys. Res. Commun.* **85**:1247.

Birken, S., Canfield, R.E., Laver, R., Agosto, G., and Gabel, M., 1980, Immunochemical determinants unique to human chorionic gonadotropin: Importance of sialic acid for antisera generated to the human chorionic gonadotropinβ-subunit COOH-terminal peptide, *Endocrinology* **106**:1659.

Braunstein, G.D., 1979, Human chorionic gonadotropin in nontrophoblastic tumors and tissues, in: *Recent Advances in Reproduction and Regulation of Fertility* (G.P. Talwar, ed.), pp. 389–397, Elsevier/North Holland, Amsterdam.

Braunstein, G.D., Vaitukaitis, J.L., Carbone, P.O., and Ross, G.T., 1973, Ectopic production of human chorionic gonadotrophin by neoplasms, *Ann. Intern. Med.* **78**:39.

Braunstein, G.D., Rasor, J., and Wade, M.E., 1975, Presence in normal human testes of a chorionic-gonadotropin-like substance distinct from human luteinizing hormone, *N. Engl. J. Med.* **293**:1339.

Carlsen, R.B., Bahl, O.P., and Swaminathan, N., 1973, Human chorionic gonadotropin: Linear amino acid sequences of the subunit, *J. Biol. Chem.* **248:**6810.

Chen, H.-C., Hodgen, G.D., Matsuura, S., Lin, L.J., Gross, E., Reichert, L.E., Birken, S., Canfield, R.E., and Ross, G.T., 1976, Evidence for a gonadotropin from nonpregnant subjects that has physical, immunological, and biological similarities to human chorionic gonadotropin, *Proc. Natl. Acad. Sci. U.S.A.* **73:**2885.

Daniels-McQueen, S., McWilliams, D. Birken, S., Canfield, R.E., Landefeld, T., and Boime, I., 1978, Identification of mRNAs encoding the alpha and beta subunits of human choriogonadotropin, *J. Biol. Chem.* **253:**7109.

Kosasa, T.S., Taymor, M.L., Goldstein, D.P., and Levesque, A.L., 1973, Use of a radioimmunoassay specific for human chorionic gonadotropin in the diagnosis of early ectopic pregnancy, *Obstet. Gynecol.* **42:**868.

Lerario, A.C., Pierce, J.C., and Vaitukaitis, J.L., 1978, Effect of conformation of hCG on generation of hCG-specific antibody, *Endocr. Res. Commun.* **5:**43.

Livingston, V.W.-C., and Livingston, A.M., 1974, Some cultural, immunological, and biochemical properties of progenitor crytocides, *Trans. N. Y. Acad. Sci.* **36:**569.

Loraine, J.A., 1967, Human chorionic gonadotropin, in: *Hormones in Blood* (C.H. Gray and A.L. Bacharach, eds.), pp. 313–332, Academic Press, New York.

Louvet, J.-P., Ross, G.T., Birken, S., and Canfield, R.E., 1974, Absence of neutralizing effect of antisera to the unique structural region of human chorionic gonadotropin, *J. Clin. Endocrinol. Metab.* **39:**1155.

Matsuura, S.H., Chen, C., and Hodgen, G.D., 1978, Antibodies to the carboxyl-terminal fragment of human chorionic gonadotropin in β subunit: Characterization of antibody recognition sites using synthetic peptide analogues, *Biochemistry* **17:**575.

Matsuura, S., Ohashi, M., Chen, H.-C., and Hodgen, G.D., 1979, A human chorionic gonadotropin-specific antiserum against synthetic peptide analogs to the carboxylterminal peptide of its β-subunit, *Endocrinology* **104:**396.

Morgan, F.J., Birken, S., and Canfield, R.E., 1973, Human chorionic gonadotropin: A proposal for the amino acid sequence, *Mol. Cell. Biochem.* **2:**97–99.

Morgan, F.J., Birken, S., and Canfield, R.E., 1975, The amino acid sequence of human chorionic gonadotropin: The α subunit and the β subunit, *J. Biol. Chem.* **250:**5247.

Pandian, M.R., and Bahl, O.P., 1977, Immunological properties of the subunit of hCG: Preparation of specific antibodies and their properties, *Abstracts of the Endocrine Society Meeting* (59th), Abstract 12.

Pierce, J.C., 1971, The subunits of pituitary thyrotropin—their relationship to other glycoprotein hormones, *Endocrinology* **89:**1331.

Slifkin, M., Pardo, M., Pouchet-Melvin, G.R., and Acevedo, H.F., 1979, Immuno-electron microscopic localization of a choriogonadotropin-like antigen in cancer-associated bacteria, *Oncology* **36:**208.

Stevens, V.C., 1976, Antifertility effects from immunizations with intact subunits and fragments of hCG, in: *Physiological Effects of Immunity against Reproductive Hormones* (R.G. Edwards and M.H. Johnson, eds.) pp. 249–274, Cambridge University Press, Cambridge, Mass.

Swaminathan, N., and Braunstein, G.D., 1978, Location of major antigenic sites of the subunit of human chorionic gonadotropin, *Biochemistry* **17:**5832.

Vaitukaitis, J., Braunstein, G., and Ross, G., 1972, A radioimmunoassay which specifically measures human chorionic gonadotropin in the presence of human luteinizing hormone, *Am. J. Obstet. Gynecol.* **113:**751.

Ward, D.N., 1979, Chemical approaches to the structure–function relationships of luteinizing hormone (lutropin), in: *Structure and Function of the Gonadotropins* (K.W. McKerns, ed.), pp. 31–45, Plenum Press, New York.

Weare, J.A., and Reichert, L.E., 1979, Studies with carbodiimide-cross-linked derivatives of bovine lutropin, *J. Biol. Chem.* **254:**6964.

Yoshimoto, Y., Wolfsen, A.R., Hirose, F., and Odell, W.D., 1979, Human chorionic gonadotropin-like material: Presence in normal human tissues, *Am. J. Obstet. Gynecol.* **134:**729.

25

Development of Specific Antisera for Human Chorionic Gonadotropin

Om P. Bahl

1. Introduction

A highly sensitive and specific radioimmunoassay (RIA) for human chorionic gonadotropin (hCG) not only is of considerable value in the detection of normal (Goldstein *et al.*, 1968; Varma *et al.*, 1971; Vaitukaitis *et al.*, 1972) and ectopic pregnancies (Braunstein *et al.*, 1978; Franchimont *et al.*, 1978; Kosasa *et al.*, 1973; Rasor and Braunstein, 1977; Milwidsky *et al.*, 1978) but also is important in the diagnosis and management of hCG-producing neoplasms (Rutanen and Seppälä, 1978; Vaitukaitis, 1979; Ross, 1977; Jones *et al.*, 1975). Therefore, there have been continued efforts made toward the refinement of RIA for hCG by enhancing its sensitivity and specificity. Several basic advances including the purification and structural elucidation of various glycoprotein hormones and the techniques for radioisoptic labeling of proteins have contributed a great deal to the development of various assays for hCG. As the purified hCG (Bahl, 1969, 1973) became available, several assays based on the use of [^{125}I]-hCG and anti-hCG were developed. While these assays served a useful purpose, they suffer from the limitation of cross-reactivity with other gonadotropins, in particular with human luteinizing hormone (hLH). The next breakthrough came when hCG was first dissociated into subunits and the subunits were separated (Swami-

Om P. Bahl • Department of Biological Sciences, Division of Cell and Molecular Biology, State University of New York at Buffalo, Buffalo, New York 14260.

nathan and Bahl, 1970; Bahl, 1977; Morgan and Canfield, 1971). As a result, the structural elucidation of the subunits was accomplished (Bellisario *et al.*, 1973; Carlsen *et al.*, 1973; Kessler *et al.*, 1979a,b; Morgan *et al.*, 1975). It was recognized that whereas the α-subunits of all glycoprotein hormones were nearly identical (Bellisario *et al.*, 1973), the β-subunits had significant differences (Carlsen *et al.*, 1973; Giudice and Pierce, 1978). Also, the β-subunit of hCG (hCG-β) was found to have a unique feature of having at the carboxy terminus an additional 30-residue fragment that was missing from other hormones. This resulted in the use of anti-hCG-β or anti-carboxy terminal peptide antisera in place of anti-hCG antiserum in the RIA for hCG. Although anti-hCG-β antiserum has the ability to discriminate between hCG and hLH, it has some cross-reactivity with the latter. Nevertheless, the assays based on anti-hCG-β are relatively more specific than those involving anti-hCG. On the other hand, the assays employing anti-carboxy terminal peptide are highly specific but lack sensitivity due to poor affinity of the antibody for hCG. Thus, it is clear that while anti-hCG-β lacks specificity, the anti-carboxy terminus lacks sensitivity. We have therefore attempted to modify hCG-β so as to obtain a specific antigen and thereby a specific antibody for hCG RIA. This chapter describes the structure and function studies of hCG-β related to the development of the hCG-specific antigen. The properties of the antibody and its application to a specific RIA for hCG are also discussed.

2. *Characteristics of hCG and Its Subunits*

hCG is a glycoprotein hormone produced during normal and ectopic pregnancies and also during certain neoplasia. The hormone has a molecular weight of 39,000 computed from its complete covalent structure, which includes the structures of the polypeptide chains (Bellisario *et al.*, 1973; Carlsen *et al.*, 1973; Kessler *et al.*, 1979b) and the carbohydrate units (Kessler *et al.*, 1979a,b). The carbohydrate forms approximately 33% of the molecule. It consists of two noncovalently bonded dissimilar subunits designated α and β. The hormone can be dissociated into subunits that can be reassociated to form the reconstituted molecule (Pierce *et al.*, 1971). The subunits can be prepared simply by the dissociation of the hormone into subunits with urea followed by their separation by ion-exchange and gel-filtration chromatography. The molecular weights of the α-subunit and the β-subunit are 15,000 and 24,000, respectively, the carbohydrate again forming about 33% of each molecule. While hCG-α is interchangeable with the α-subunits of the other homologous pituitary glycoprotein hormones such as hLH, human follicle-stimulating hormone (hFSH), and human thyroid-stimulating hormone (hTSH), the β-subunit is hormone-specific. The covalent structures of both α- and β-subunits (Bellisario *et al.*, 1973; Carlsen *et al.*, 1973; Kessler *et al.*, 1979b), including the two polypeptide chains and the eight carbohydrate units (Kessler *et al.*, 1979a,b), have been completed. hCG-α has 89–92 amino acid residues and 18–22 sugar residues. The variation in the number of amino acid and carbohydrate residues is due to the microhetergeneity of the molecule. The microheterogeneity in the polypeptide chains is at the amino terminus and in the carbohydrate chains near the nonreducing termini. The amino acid sequence of hCG-α shown in Fig. 1 is identical to the amino acid sequences of the α-subunits of other human pituitary glycoprotein hormones. The β-subunit of hCG has 145 amino acid and 34–40 sugar residues. As noted in Section 1, a unique feature of hCG-β structure is the presence at the carboxy terminus of a 30-residue peptide that is not present in any other glycoprotein hormone. The structure of hCG-β is shown in Fig. 2. A comparison of the amino acid sequences of the β-subunits of hCG, hLH, hTSH, and hFSH shown in Fig. 3 clearly indicates that of all the β-subunits, hLH-β most closely resembles hCG-β. Despite extensive sequence homology between hLH-β and hCG-β, short stretches of variable sequences are also present across the entire polypeptide chains.

hCG has two types of carbohydrate units, asparagine-linked and serine-linked (Kessler *et al.*, 1979a,b). The asparagine-linked carbohydrate units are "complex"-type, two of which are present in the α-

```
                                          10
H-Ala-Pro-Asp-Val-Gln-Asp-Cys-Pro-Glu-Cys-Thr-

                                      20
Leu-Gln-Glu-Asp-Pro-Phe-Phe-Ser-Gln-Pro-Gly-Ala-

                              30
Pro-Ile-Leu-Gln-Cys-Met-Gly-Cys-Cys-Phe-Ser-Arg-

                      40
Ala-Tyr-Pro-Thr-Pro-Leu-Arg-Ser-Lys-Lys-Thr-Met-

              50
Leu-Val-Gln-Lys-[Asn(CHO)]-Val-Thr-Ser-Glu-Ser-

              60
Thr-Cys-Cys-Val-Ala-Lys-Ser-Tyr-Asn-Arg-Val-Thr-

70                                          80
Val-Met-Gly-Gly-Phe-Lys-Val-Glu-[Asn(CHO)]-His-Thr-

                                          90
Ala-Cys-His-Cys-Ser-Thr-Cys-Tyr-Tyr-His-Lys-Ser-OH
```

Figure 1. Linear amino acid sequence of the hCG α-subunit.

```
                                             10
H-Ser-Lys-Gln-Pro-Leu-Arg-Pro-Arg-Cys-Arg-Pro-Ile-Asn(CHO)-Ala-

                         20                            30
Thr-Leu-Ala-Val-Glu-Lys-Glu-Gly-Cys-Pro-Val-Cys-Ile-Thr-Val-Asn(CHO)-

                                     40
Thr-Thr-Ile-Cys-Ala-Gly-Tyr-Cys-Pro-Thr-Met-Thr-Arg-Val-Leu-Gln-Gly-

        50                                        60
Val-Leu-Pro-Ala-Leu-Pro-Glx-Leu-Val-Cys-Asn-Tyr-Arg-Asp-Val-Arg-Phe-

                        70                                    80
Glu-Ser-Ile-Arg-Leu-Pro-Gly-Cys-Pro-Gly-Val-Asn-Pro-Val-Val-Ser-Tyr-

                                90
Ala-Val-Ala-Leu-Ser-Cys-Gln-Cys-Ala-Leu-Cys-Arg-(Arg)-Ser-Thr-Thr-

   100                                     110
Asp-Cys-Gly-Gly-Pro-Lys-Asp-His-Pro-Leu-Thr-Cys-Asp-Asp-Pro-Arg-Phe-

                120
Gln-Asp-Ser-Ser-Ser-Ser(CHO)-Lys-Ala-Pro-Pro-Pro-Ser(CHO)-Leu-Pro-

130                                                  140
Ser-Pro-Ser(CHO)-Arg-Leu-Pro-Gly-Pro-Ser(CHO)-Asx-Thr-Pro-Ile-Leu-

    145
Pro-Gln-OH.
```

Figure 2. Linear amino acid sequence of the hCG β-subunit.

```
                                          10              *
HCG-β   Ser-Lys-Glu-Pro-Leu-Arg-Pro-Arg-Cys-Arg-Pro-Ile-Asn-Ala-Thr-Leu-Ala-Val-
HLH-β    -  Arg-Glu  -   -   -   -  Trp  -  His  -   -   -   -  Ile  -   -   -
HFSH-β                              Asn-Ser  -  Glu-Leu-Thr  -* Ile  -  Ile  -  Ile-
HTSH-β                                  Phe  -  Ile  -  Thr-Glx-Tyr-Met-Thr-His  -

          20                              30*
HCG-β   Glu-Lys-Glu-Gly-Cys-Pro-Val-Cys-Ile-Thr-Val-Asn-Thr-Thr-Ile-Cys-Ala-Gly-
HLH-β    -   -   -   -   -   -   -   -   -   -   -   -*  -   -   -   -   -   -
HFSH-β   -   -   -  Glu  -  Arg-Phe  -  Leu  -  Ile  -*  -   -  Trp  -   -   -
HTSH-β   -  Arg-Arg-Glx  -  Ala-Tyr  -  Leu  -  Ile  -*  -   -   -   -   -   -

                    40                                   50
HCG-β   Tyr-Cys-Pro-Thr-Met-Thr-Arg-Val-Leu-Gln-Gly-Val-Leu-Pro-Ala- X -Leu-Pro-
HLH-β    -   -   -   -   -   -  Arg(Met)Leu  -  Glx-Ala  -   -   -  Pro- X -Val  -
HFSH-β   -   -  Tyr  -  Arg-Asp-Leu  -  Tyr-Lys-Asn-Pro- X - X   -  Arg-Pro-Lys-
HTSH-β   -   -  Met-Thr-Arg-Asx-Ile-Asx-Gly-Lys-Leu-Phe  -   -  Lys-Tyr-Ala-Leu-

                        60                                       70
HCG-β   -X- Gln-Val-Val-Cys-Asn-Tyr-Arg-Asp-Val-Arg-Phe-Glu-Ser-Ile-Arg-Leu-Pro-
HLH-β   -X-  -  Pro  -   -  Thr  -   -  Asx  -   -   -  Glx  -   -   -   -   -
HFSH-β  Ile  -  Lys-Thr  -  Thr-Phe-Lys-Glu-Leu-Val-Tyr  -  Thr-Val  -  Val  -
HTSH-β  Ser  -  Asx  -   -  Thr  -   -   -  Phe-Ile-Tyr-Arg-Thr-Val-Glx-Ile  -

                                80                                       90
HCG-β   Gly-Cys-Pro-Arg-Gly-Val-Asn-Pro-Val-Val-Ser-Tyr-Ala-Val-Ala-Leu-Ser-Cys-
HLH-β    -   -   -   -   -   -  Asp  -   -   -   -  Phe-Pro  -   -   -   -   -
HFSH-β   -   -  Ala-His-His-Ala-Asp-Ser-Leu-Tyr-Thr  -  Pro  -   -  Thr-Gln  -
HTSH-β   -   -   -  Leu-His  - (Ala  -  Tyr)Phe  -   -  Pro  -   -   -   -   -

                                            100
HCG-β   Gln-Cys-Ala-Leu-Cys-Arg-Arg-Ser-Thr-Thr-Asp-Cys-Gly-Gly-Pro-Lys-Asp-His-
HLH-β   Arg  -  Gly-Pro  -   -   -   -   -  Ser  -   -   -   -   -   -  Asx  -
HFSH-β  His  -  Gly-Lys  -  Asp-Ser-Asp-Ser  -   -   -  Thr-Val-Arg-Gly-Leu-Gly-
HTSH-β  Lys  -  Gly-Lys  -  Asx-Thr-Asx-Tyr-Ser  -   -  Ile-His(Glu-Ala-Ile)Lys-

            100                                 120          *
HCG-β   Pro-Leu-Thr-Cys-Asp-Asp-Pro-Arg-Phe-Gln-Asp-Ser-Ser-Ser-Ser-Lys-Ala-Pro-
HLH-β    -   -   -   -   -  Asx(Glx-Asx-Ser-Lys)Gly-COOH
HFSH-β   -  Ser-Tyr  -  Ser-Phe-Gly-Glu-Met-Lys-Gln-Tyr-Pro-Thr-Ala-Leu-Ser-Tyr-
HTSH-β  Thr-Asx-Tyr  -  Thr-Lys  -  Glx-Lys-Ser-Tyr-COOH

                *                   *                   140*
HCG-β   Pro-Pro-Ser-Leu-Pro-Ser-Pro-Ser-Arg-Leu-Pro-Gly-Pro-Ser-Asp-Thr-Pro-Ile-
HFSH-β  -COOH
HCG-β   Leu-Pro-Gln-COOH
```

Figure 3. Comparison of the linear amino acid sequences of hCG-β, hLH-β, hFSH-β, and TSH-β.

subunit at positions 52 and 78 and two in the β-subunit at positions 13 and 30. In addition, there are four serine-linked carbohydrate units located in the carboxy terminus of hCG-β at positions 121, 127, 132, and 138. All "complex"-type carbohydrate units are identical in structure with the exception that the two in hCG-β have an additional 1 residue each of L-fucose. The structure of the "complex"-type unit is shown in Fig. 4. In this structure, all the chains are shown to be complete. In fact, as with other glycoproteins, one also encounters in hCG microheterogeneity that is due to the variation in the number of sialic acid, fucose, and galactose residues, i.e., the peripheral sugar residues. The reason for assigning a structure with completed branches is that we have been able to isolate hCG, hCG-α, and hCG-β in which the number of sialic acid residues has been found to be approximately 16, 4, and 12, respectively, indicating the presence of complete chains. Similarly, the structure of the *O*-glycosidic carbohydrate units in Fig. 4 also represents the complete structure assigned on similar grounds as described above (Kessler *et al.*, 1979a,b).

Whereas the amino acid sequences of the polypeptide chains of several glycoprotein hormones from various species have been determined, the elucididation of their carbohydrate units, with the exception of hCG, has not received as much attention (Kennedy *et al.*, 1974). Only recently, a tentative proposal for the carbohydrate units of the α-subunits of hLH, hFSH, and hTSH based on enzymatic degradation and periodate oxidation has been made (Hara *et al.*, 1978). The pituitary glycoprotein hormones, hLH, and hTSH consist of only three "complex"-type asparagine-linked carbohydrate units, two in the α-subunit and one in the β-

```
        α2,3        β1,4          β1,2
NANA ——— Gal ——— GlcNAc ——— Man
                                   \α1,6
                                    Man ——— GlcNAc ——— GlcNAc ——— (Asn)
                                   /α1,3  β1,4      β1,4   |α1,6
NANA ——— Gal ——— GlcNAc ——— Man                          (Fuc) 0,1
   α2,3      β1,4         β1,2

                               α
NANA ——— Gal ——— GalNAc ——— SER
   α2,3     β1,3     |α2,6
                    NANA
```

Figure 4. Structures of asparagine- and serine-linked carbohydrate units of hCG-α and hCG-β. hCG-α has two asparagine-linked carbohydrate units, while hCG-β has two asparagine and four serine-linked carbohydrate units. The carbohydrate units of hCG-α lack fucose residue. All chains are shown complete, but microheterogeneity has been found in sialic acid, galactose, and fucose residues, particularly in the asparagine-linked carbohydrate units.

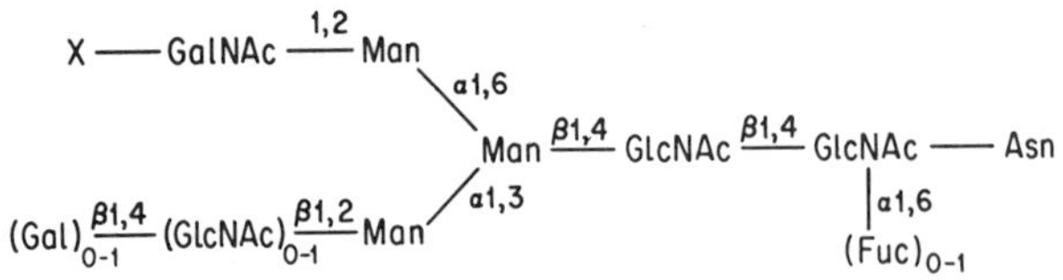

```
                         1,2
        X —— GalNAc ——— Man
                            \ α1,6
                             Man —β1,4— GlcNAc —β1,4— GlcNAc —— Asn
                            / α1,3                      | α1,6
(Gal)0-1 —β1,4— (GlcNAc)0-1 —β1,2— Man                 (Fuc)0-1
```

Figure 5. A proposal for the structure of carbohydrate units of bLH and oLH.

subunit. hFSH contains four such carbohydrate units, two in the α- and two in the β subunit. One unique feature of these carbohydrate units is the presence of *N*-acetylgalactosamine. So far, this is the first example of asparagine-linked carbohydrate units containing *N*-acetylgalactosamine as an integral part of the structure. Generally, *N*-acetylgalactosamine is found as a component of the *O*-glycosidic carbohydrate chains. We have investigated the detailed structures of the carbohydrate parts of bovine LH (bLH) and ovine LH (oLH). These investigations have led to the discovery of a novel structure hitherto unknown in any other glycoprotein. The proposal for the structure of the carbohydrate units of bLH and oLH is presented in Fig. 5. This structure is derived from extensive investigations involving periodate oxidation, methylation, and deamination techniques, and by enzymatic hydrolysis using exo- and endoglycosidases. The details of these studies are described elsewhere (Bahl *et al.*, 1980). The main structural feature of the carbohydrate units is the presence of a substituted *N*-acetylgalactosaminyl branch at one of the residues of the common core structure (Fig. 5). It would be interesting to uncover any special role that *N*-acetylgalactosamine may play in the function of these hormones.

3. Development of Specific Antisera for hCG

3.1. Immunological Properties of hCG and Its Subunits

hCG and hLH, because of their close structural similarity, show immunological cross-reactivity; i.e., the antibody against either of the two antigens reacts with the other. On the other hand, anti-hCG-β has the ability to discriminate between hCG and hLH, although it still shows appreciable cross-reactivity with the latter. Thus, the RIAs based on the use of anti-hCG or anti-hCG-β are not specific. Two approaches have been pursued to enhance the specificity of the anti-hCG-β assay. One approach involves the use of the carboxy-terminal peptide of hCG-β, and the second approach has been to modify the native hCG-β so as to eliminate or reduce cross-reactivity of hCG-β with anti-hLH. It was expected that the modified hCG-β would elicit antibody with little or reduced cross-reactivity with hLH. In pursuit of the first approach, several natural and synthetic peptides ranging in size from 23 to 35 amino acid residues of the carboxy terminus of hCG-β after conjugation with a carrier protein have been used for antibody production in animals (Chen *et al.*, 1976; Matsuura *et al.*, 1979; Ayala *et al.*, 1978; Stevens, 1980). The antibodies to these peptides were highly specific and did not cross-react with other glycoprotein hormones, particularly hLH. However, the affinity of the antibody to the hormone was weak, and therefore the sensitivity of the RIA was low, rendering the currently available carboxy-terminal peptide antibodies unsuitable for an RIA for hCG. Furthermore, antibody to a tryptic peptide (123–145 residues) has been found to react with urinary extract from a normal postmenopausal woman or from a patient with Klinefelter's syndrome (Chen *et al.*, 1976).

3.2. Modifications of hCG-β

The second approach to obtain a specific antigen was by enzymatic and chemical modifications of hCG-β. hCG-β was modified in both the carbohydrate and protein parts of the molecule. The carbohydrate was modified by sequential degradation with exoglycosidases such as *Vibrio cholerae* neuraminidase, *Aspergillus niger* β-D-galactosidase, *A. niger* β-*N*-acetylglucosaminidase, and *A. niger* α-D-mannosidase. The protein part was modified by specific blocking groups. The resulting derivatives were evaluated for their immunological activity in the [^{125}I]-hCG-β–anti-hCG-β system (hCG-β system), and the extent of their cross-reactivity with hLH was measured in the [^{125}I]-hLH–anti-hLH system (hLH system).

3.2.1. Enzymatic Modifications

Several glycosidase-treated derivatives of hCG-β designated as N-hCG-β, NG-hCG-β, NGA-hCG-β, and NGAM-hCG-β were obtained, the prefixes N, G, A, and M denoting, respectively, neuraminidase, β-D-galactosidase, β-*N*-acetylglucosaminidase, and α-D-mannosidase. The amounts of sialic acid, galactose, *N*-acetylglucosamine, and mannose removed by enzymatic hydrolysis of hCG-β in N-hCG-β, NG-hCG-β, NGA-hCG-β, and NGAM-hCG-β were approximately 100, 63, 61, and 40%, respectively. The immunological properties of these derivatives in hCG-β and hLH RIAs are given in Fig. 6. It is clear that the glycosidase treatment affected neither the immunological activity in the hCG-β RIA

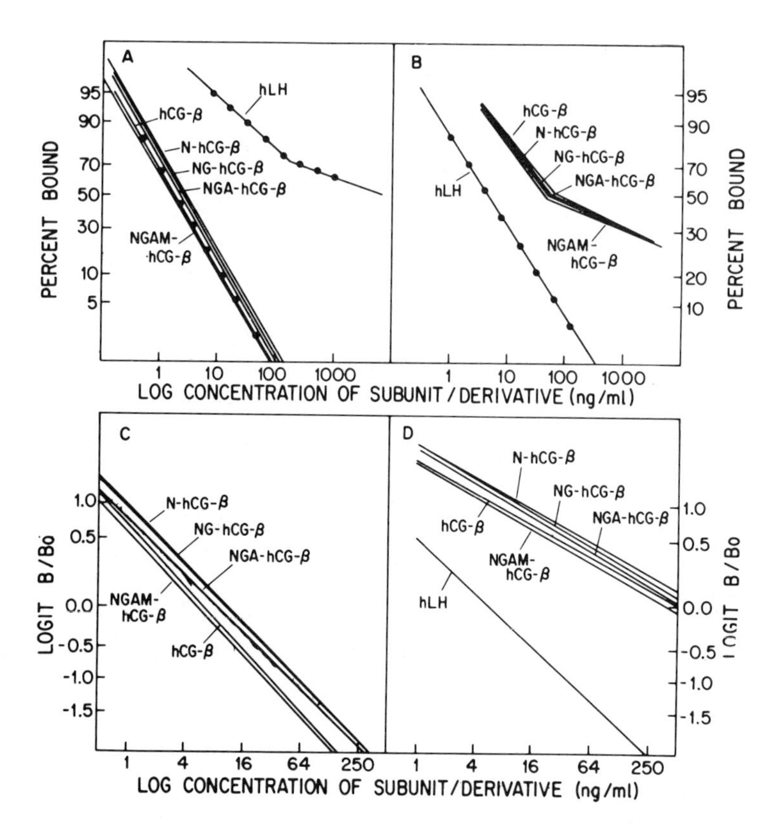

Figure 6. Immunological activity of glycosidase-treated derivatives of hCG-β by RIA: (A) in the [^{125}I]-hCG-β–anti-hCG-β system (hCG-β system); (B) in the [^{125}I]-hLH–anti-hLH-system (hLH system); (C) computer-generated logit-log plots of the data in (A) (hCG-β system); (D) computer-generated logit-log plots of the data in (B) (hLH system).

system nor the cross-reactivity with hLH in the hLH RIA system. This strongly suggests that the antigenic determinants are not associated with the carbohydrate of hCG-β.

3.2.2. Chemical Modifications

The following modifications of hCG-β in the protein part were carried out (Ghai *et al.*, 1980; Bahl and Muralidhar, 1980): The amino groups were modified by succinylation (Suc-hCG-β) and by polymerization with glutaraldehyde (hCG-β_n). The number of amino groups modified by succinylation was 4, and hCG-β_n was found to be a dimer from its molecular weight by gel filtration. Arginyl residues of hCG-β were modified with 1,2-cyclohexanedione to form N^7,N^8-(1,2-dihydroxycyclohex-1,2-ylene)-L-arginine (Dhc-hCG-β). Thus, 7 of the 12 arginyl residues were modified. This was determined from the number of unreacted arginine residues obtained from the amino acid analysis. The carboxyl groups in hCG-β were protected by coupling with glycyl ethyl ester (Gee) or tyrosyl ethyl ester (Tee) in the presence of a water-soluble carbodiimide. Thus, all the carboxyl groups in Gee conjugate (Gee-hCG-β) and 5 carboxyl groups in Tee conjugate (Tee-hCG-β) were modified as found by the number of additional glycyl and tyrosyl residues on amino acid analysis.

The tyrosyl residues of hCG-β were modified with both tetranitromethane (NO_2-hCG-β) and *N*-bromosuccinimide (NB_s-hCG-β). Under the conditions employed, only one half the tyrosyl residues were nitrated. The extent of reaction with *N*-bromosuccinimide was also about 50% as indicated by the number of unreacted tyrosyl residues. In addition, *N*-bromosuccinimide oxidized the only histidine residue in hCG-β. All the derivatives thus obtained were evaluated by the hCG-β and hLH RIA systems. The results are summarized in Table I and Fig. 7. It is clear from Table I that none of the modifications resulted in a specific antigen as evidenced by the ratio of hCG-β and hLH activities in the two RIAs.

Since all six disulfide bonds of hCG-β are present in most hLH-homologous regions, it was thought that the reduction and *S*-alkylation or oxidation of cystinyl residues should give rise to a specific antigen.

Table I. Effect of Chemical Modification on the Immunological Activity of hCG-β

		Activity (%)[a]				
		RIA system				
G-β/ vative	Modifying reagent	hCG-β[b]	hLH[b,c]		hCG-β–anti-hLH	Activity ratio (hCG-β/hLH)[d]
β	—	100	100	(4.50)	100	1.0
hCG-β	Succinic anhydride	70.0 ± 1.9	88.0 ± 3.3	(4.00)	81.0 ± 1.0	0.8
$β_n$	Glutaraldehyde	47.1 ± 3.4	52.3 ± 5.3	(2.30)	42.0 ± 3.0	0.9
CG-β	Glutaraldehyde and tetanus toxoid	4.0 ± 0.2	4.0 ± 0.25	(0.30)	3.8 ± 0.2	1.0
hCG-β	Tetranitromethane	60.0 ± 5.1	56.0 ± 3.1	(2.50)	51.0 ± 4.0	1.1
hCG-β	*N*-Bromosuccinimide	25.0 ± 2.8	50.0 ± 4.0	(2.20)	54.0 ± 3.5	0.5
hCG-β	1,2-Cyclohexanedione	33.3 ± 1.0	80.0 ± 5.8	(3.60)	19.0 ± 6.0	0.4
hCG-β	Glycine ethyl ester	N.C.	N.C.		—	—
CG-β	Tyrosine ethyl ester	N.C.	N.C.		—	—

ues represent mean percentage activity ± S.E. of 2–6 experiments in duplicate and were calculated as follows: concentration of hCG-β and its derivatives that inhibited binding of [¹²⁵I]-hCG-β or [¹²⁵I]-hLH to anti-hCG-β anti-hLH by 50% was obtained from the weighted regression of computer analysis. Assuming the concentration hCG-β (6–9 ng/ml) and of hLH (80–130 ng/ml) in hCG-β and hLH systems representing 50% inhibition of ding to represent 100% activity, the percentage activity of the hormone or the derivative was calculated in erent RIA systems.

C.) Not calculated, since the dose–response curves of hCG-β derivatives were nonparallel to the standard ve.

ues in parentheses represent the actual hLH activity of hCG-β or its derivatives in the hLH system.

ios of the percentage activity of hCG-β and hLH in the hCG-β and hLH systems.

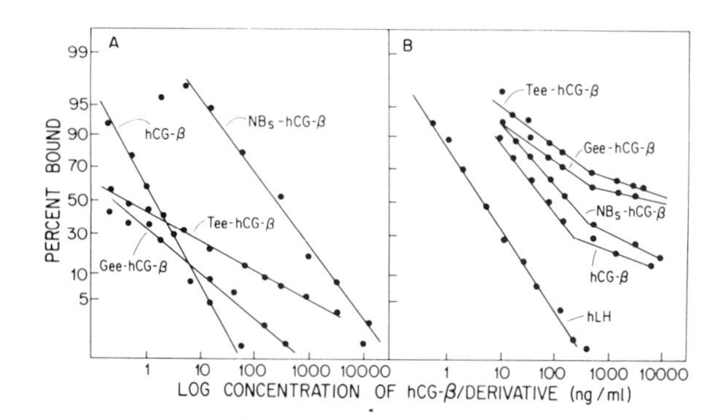

Figure 7. Inhibition of the binding of [^{125}I]-hCG-β or [^{125}I]-hLH by various analogues of hCG-β obtained by modifications of the polypeptide chain. (A) RIA in the hCG-β system. (B) RIA in the hLH system. (NB_s-hCG-β) *N*-bromosuccinimide-treated hCG-β; (Tee-hCG-β) hCG-β conjugated with tyrosine ethyl ester; (Gee-hCG-β) hCG-β conjugated with glycine ethyl ester.

Consequently, when all six disulfide bonds in hCG-β were cleaved, the resulting derivative did not show any immunological activity in the hCG-β RIA system, indicating that the antigenic determinants were conformational rather than sequential in nature. Therefore, a systematic attempt was made to prepare several derivatives of hCG-β with varying numbers of cystinyl residues modified.

hCG-β was modified in the cystinyl residues by reduction with dithioerythritol followed by *S*-alkylation with iodoacetamide or iodoacetic acid or aminoethylation with ethylenimine and by oxidation with performic acid. The reduction and *S*-carboxamidomethylation was carried out under controlled conditions so as to obtain several derivatives with varying numbers of disulfide bonds cleaved. The various derivatives thus obtained were evaluated as described above for immunological activity in the two RIA systems. There was a progressive decrease in the reactivity of the derivatives with anti-hCG-β with the increase in the number of disulfide bonds cleaved (Fig. 8A and B). Table II shows a decrease in hCG-β activity from 73.6 to 6.5% when the number of disulfide bonds modified in the molecule are increased from 2 (DS_2-hCG-β) to 5 (DS_5-hCG-β). A similar loss in activity was also noted in the hLH system (Fig. 8B). However, the decrease in activity was not proportionate in the two systems. As can be seen from Table II, the ratio of hCG-β to hLH activity of the derivatives was more than 1, and ranged from 1.3 to infinity, indicating that as the number of disulfide bonds cleaved increased, the hLH activity was lost progressively and preferentially over hCG-β activity. These changes in the immunological activity cannot be attributed to the charge introduced into the molecule, since all three

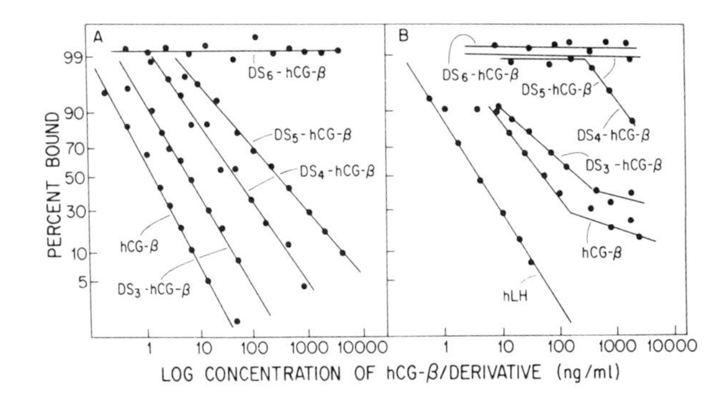

Figure 8. Immunological activity of the various reduced and carboxamidomethylated derivatives of hCG-β. (A) RIA in the hCG-β system; (B) RIA in the hLH system. hCG-β was reduced and alkylated at various levels using dithioerythritol and iodoacetamide. The derivatives thus obtained were designated as DS_3-hCG-β, DS_4-hCG-β, DS_5-hCG-β, and DS_6-hCG-β; the subscript number denoting the number of disulfide bonds modified.

derivatives prepared by *S*-carboxamidomethylation, *S*-carboxymethylation, and *S*-aminoethylation of the reduced hCG-β resulted in a similar change (Fig. 9A and B). The *S*-carboxymethylated (DS_6-C_m-hCG-β) and aminoethylated (DS_5-Ae-hCG-β) hCG-β derivatives showed higher activity in both the systems than did the corresponding *S*-carboxamidomethylated derivatives. However, the ratio of hCG-β activity to hLH activity varied from 3.3 to 7.6 depending on the number of disulfide bonds modified (Table II). When all the disulfide bonds in hCG-β were reduced and alkylated, with either iodoacetamide or iodoacetic acid, the deriva-

ble II. Effect of Reduction and Alkylation and Oxidation of Disulfide Bonds on the Immunological Activity of hCG-β

G-β/derivative[a]	Number of disulfide bonds cleaved	Alkylating reagent	Activity (%)[b] hCG-β system	hLH system		Activity ratio (hCG-β/hLH)
G-β	0.0	—	100	100	(4.5)	1.0
$_2$-hCG-β	2.3	Iodoacetamide	73.6	55	(2.50)	1.3
$_3$-hCG-β	3.1	Iodoacetamide	35.0	16	(0.72)	2.0
$_4$-hCG-β	4.1	Iodoacetamide	8.0	0.2	(0.01)	40.0
$_5$-hCG-β	5.1	Iodoacetamide	6.5	0		∞
$_6$-hCG-β	6.1	Iodoacetamide	0	0		—
$_4$-C_m-hCG-β	4.5	Iodoacetic acid	42.8	5.6	(0.25)	7.6
$_6$-C_m-hCG-β	6.0	Iodoacetic acid	0	0		—
$_5$-Ae-hCG-β	5.2	Ethylenimine	6.6	2.0	(0.09)	3.3
$_5$-Sul-hCG-β	4.8	Performic acid	0.3	0.03	(0.001)	10.0

he subscript number denotes the number of disulfide bonds modified; calculated from the *S*-carboxymethylcysteine, *S*-aminoethylcysteine, and cysteic acid residues obtained on amino acid analysis.

alues in parentheses represent the actual hLH activity of hCG-β or its derivatives in the hLH system.

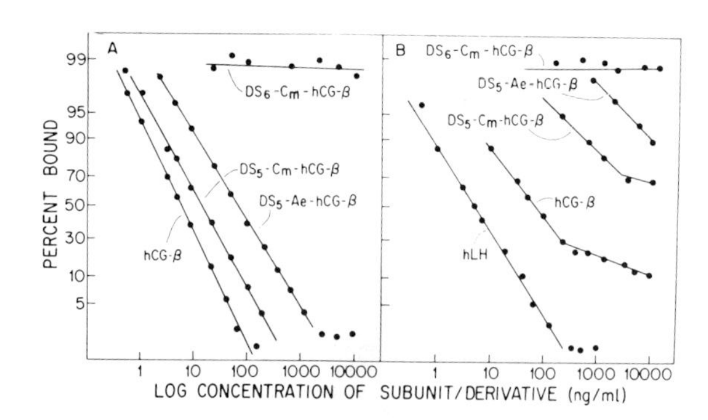

Figure 9. Immunological activity of the reduced and alkylated derivatives of hCG-β. (A) RIA in the hCG-β system; (B) RIA in the hLH system. (DS_5-C_m-hCG-β, DS_6-C_m-hCG-β) Carboxymethyl-hCG-β; (DS_5-Ae-hCG-β) aminoethyl-hCG-β. The subscript number denotes the number of disulfide bonds modified.

tives failed to inhibit the binding of the labeled hCG-β and hLH with their respective antibodies (Figs. 8 and 9). It can be concluded from this result that the antigenic determinants in hCG-β are conformational rather than sequential in nature (Bahl *et al.*, 1976).

It may be pointed out that since the carboxy-terminal region of hCG-β, which is unique to hCG, is composed predominantly of proline, serine, and leucine residues, most of the chemical modifications described above probably occurred in the main core of the hCG-β polypeptide chain. On the basis of immunological behavior, these derivatives can be divided into three classes: (1) derivatives with a proportionate loss in hCG-β and hLH activities, i.e., those with an hCG-β/hLH activity ratio of approximately 1, assuming the activities of hCG-β in the hCG-β and hLH RIA systems to be 1, e.g., Suc-hCG-β, hCG-β_n (Table I); (2) derivatives with a preferential loss in the hLH activity, i.e., those with an hCG/hLH activity ratio of more than 1, e.g., DS_3-hCG-β (Table II); (3) derivatives with a preferential loss in the hCG-β activity, i.e., those with an hCG-β/hLH activity ratio of less than 1, e.g., NB_s-hCG-β, Dhc-hCG-β (Table I). These data would imply the existence of two different types of antigenic determinants, those that are specific to hCG-β and those that are common to hCG-β and hLH. The concept of two types of antigenic determinants in hCG-β is further supported by the fact that modifications of hCG-β result in derivatives that, while they retain significant amounts of reactivity with anti-hCG-β, concomitantly lose cross-reactivity with hLH. The hCG-β/hLH activity ratio in these derivatives as DS_5-hCG-β and DS_6-hCG-β was close to infinity (Table II), indicating the presence of only hCG-specific determinants.

3.3. Properties of the Antibodies against the Modified Derivatives of hCG-β: DS_5-hCG-β and DS_6-hCG-β

Among the various chemically and enzymatically modified derivatives of hCG-β tested for cross-reactivity with anti-hLH, DS_5-hCG-β and DS_6-hCG-β were found to be devoid of any detectable cross-reactivity with hLH in the RIA system. Consequently, they were administered to rabbits after being coupled with hemocyanin, to determine whether they also produced specific antibodies. An outline of the procedure used for the preparation of DS_5-hCG-β-heme or DS_6-hCG-β-heme is given in Fig. 10. The antibodies raised against DS_5-hCG-β-heme and DS_6-hCG-β-heme were highly specific for hCG or hCG-β. However, only anti-DS_5-hCG-β-heme was suitable for an RIA, since it not only gave a high titer of 1:8000 for 30% binding of the tracer but also showed high avidity for the hormone. On the other hand, although the anti-DS_6-hCG-β was very specific for hCG or hCG-β, it was not suitable for a sensitive RIA because of its low titer and weak avidity for the hormone. The titer of the anti-DS_5-hCG-β-heme and its specificity could be maintained by periodic injections of the rabbits with DS_6-hCG-β-heme or a mixture of DS_5-hCG-β-heme and DS_6-hCG-β-heme. It is worth noting that while the continuous injections with hCG-β caused an increase in the titer as well as the cross-reactivity of anti-hCG-β with hLH, the injections with a mixture of DS_5-hCG-β-heme and DS_6-hCG-β-heme resulted in the maintenance of high titer with a gradual decrease in the cross-reactivity. This antibody thus obtained is referred to as anti-DS_5,DS_6-hCG-β. Any ob-

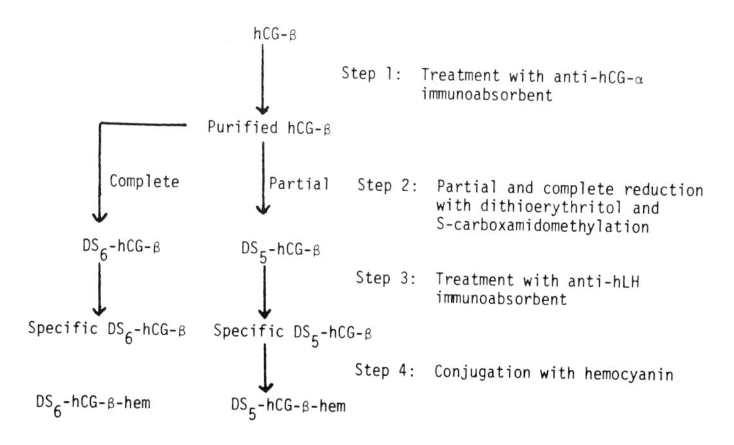

Figure 10. Outline of the procedure for the preparation of DS_5-hCG-β-heme and DS_6-hCG-β-heme. Subscripts 5 and 6 denote 5 and 6 disulfide bonds modified.

served cross-reactivity of the antiserum with hLH is probably due to some discrepancy in the preparation of the antigens. It may be pointed out that the product at each step of the preparation, such as treatment with anti-hCG-α immunoabsorbent reduction and *S*-alkylation and treatment with anti-hLH immunoabsorbent, must be carefully monitored. Furthermore, it is advisable to have highly purified hCG-β with less than 0.1% hCG activity and also to modify no fewer than 5 or preferably slightly more than 5 disulfide bonds in DS_5-hCG-β.

The RIA using anti-DS_5,DS_6-hCG-β-heme is highly specific for hCG-β and hCG and has no detectable cross-reactivity with other hormones. The sensitivity of the assay is about 0.1–1 ng/ml and can be further increased by increasing the specific activity of the tracer hormone (Pandian *et al.*, 1980). It is interesting to note that the dose–response curves for hCG and hCG-β are parallel and almost overlapping, indicating that the antibody may be directed against a common hCG-specific antigenic site (Fig. 11). Furthermore, since the carboxy-terminal 37-residue chymotryptic peptide from the reduced and carboxamidomethylated hCG-β failed to compete with [^{125}I]-hCG or [^{125}I]-hCG-β for binding to the antibody, the antibody seems to be directed against some site in the main polypeptide chain rather than the unique carboxy-terminal peptide (Fig. 12). This is further supported by the fact that hCG-des-C-terminus competed as well as hCG-β or hCG in binding of [^{125}I]-hCG to the antibody (Bahl and Muralidhar, 1980). Our preliminary studies also suggest that the specific site may be in the vicinity of the disulfide bond involving cysteinyl residues 9 and 90. We have also found that the

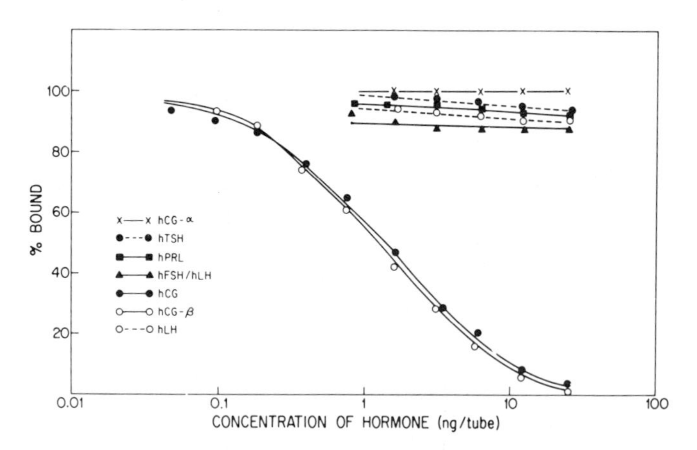

Figure 11. RIA using [^{125}I]-hCG, anti-DS_5,DS_6-hCG-β-heme antiserum, and the various competing antigens as indicated. Results are expressed as $B/B_0 \times 100$ vs. log dose of the antigen. The value of B_0 ranged from 25 to 30% ($\approx$ 8000–10000 cpm).

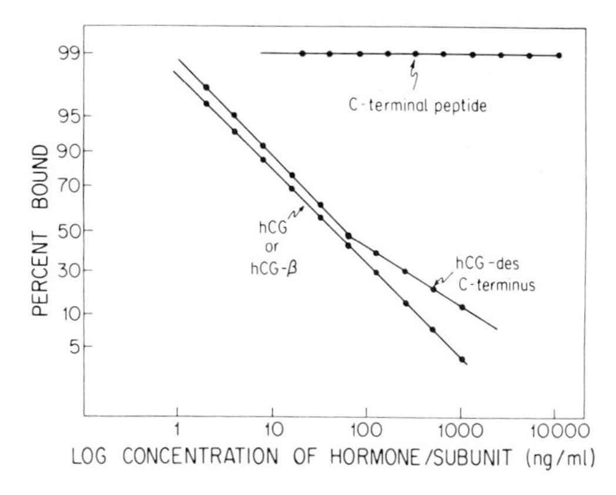

Figure 12. Inhibition of binding of [^{125}I]-hCG to anti-DS$_5$,DS$_6$-hCG-β-heme in an RIA by hCG-β, hCG-des-C-terminus, and carboxy-terminal 37-residue peptide of hCG-β.

controlled reduction and *S*-alkylation of hCG-β probably results in the preferential cleavage of disulfide bonds between residues 93 and 100 and 26 and 110 (Giudice and Pierce, 1978). This unfolding of the tertiary structure may result in the specificity of the DS$_5$-hCG-β by the destruction of the hCG–hLH common antigenic sites. Finally, coupling with hemocyanin, which involves the lysyl residues present mostly in the hLH-β

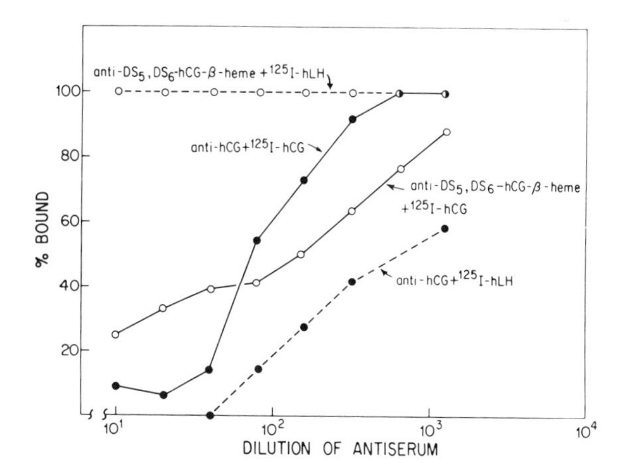

Figure 13. *In vitro* effect of anti-DS$_5$,DS$_6$-hCG-β-heme serum on the binding of [^{125}I]-hCG and [^{125}I]-hLH to ovarian receptors. Equal aliquots of a homogenate of ovaries from superovulated immature rats were incubated with either [^{125}I]-hCG or [^{125}I]-hLH ($\approx 1.0 \times 10^5$ cpm/tube) each in the presence of the anti-hCG serum or anti-DS$_5$,DS$_6$-hCG-β-heme serum at the dilutions indicated in the figure. The receptor-bound [^{125}I]hormone was subsequently measured.

homologous regions, could conceivably mask the hCG–hLH common sequences in hCG-β and may further add to the specificity of the antigen.

Finally, the anti-DS_5,DS_6-hCG-β-heme is capable of neutralizing the *in vitro* and *in vivo* biological activity of hCG without any detectable effect on the hLH activity. For example, the binding of [^{125}I]-hCG but not of [^{125}I]-hLH to rat ovarian receptors is inhibited by the anti-DS_5,DS_6-hCG-β-heme (Fig. 13). Similarly, the anti-DS_5,-DS_6-hCG-β-heme *in vivo* is able to inhibit the effect of hCG but not of hLH on ovarin ascorbic acid content (Table III and Fig. 14) and on rat uterine weight (Table IV and Fig. 15). The studies discussed above therefore clearly demonstrate the biological specificity of the antibody. In contrast, the antibody against the carboxy terminus lacks the ability to neutralize the biological activity of hCG (Chen *et al.,* 1976).

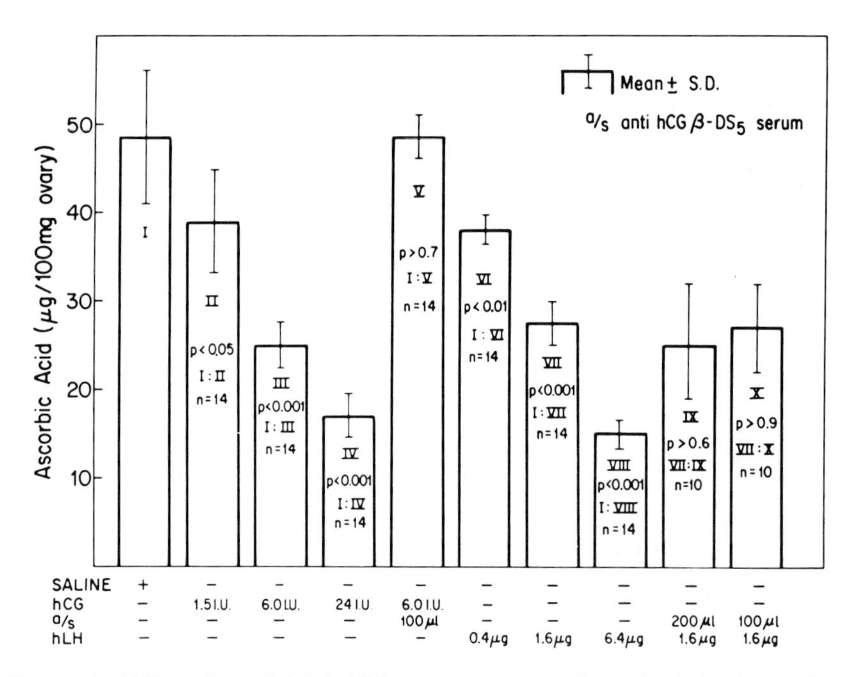

Figure 14. Ability of anti-DS_5,DS_6-hCG-β serum to neutralize selectively the ovarian-ascorbic-acid-depleting activity of hCG but not that of hLH. Groups of female 26-day-old immature rats (3–5 per group) were made pseudopregnant by subcutaneous injections of pregnant mare serum gonadotropin (50 IU/0.2 ml saline per rat) followed 56 hr later by hCG (25 IU/0.2 ml saline per rat). On the 5th day after hCG injections, they were administered intraperitoneally either 0.9% saline or hormone or a hormone and antiserum (a/s) as indicated in the figure. At 4 hr later, they were sacrificed and the ascorbic acid content in the ovaries measured. The mean values for ovarian ascorbic acid content were compared by Student's *t* test for test significance.

Table III. Ability of DS_5, DS_6-hCG-β-heme Antiserum to Neutralize Biological Activity[a] of hCG but Not of hLH

Group	Treatment	Antiserum (μl)	Ascorbic acid (μg ± S.D./100 mg ovary)	Depletion (100% − % of control)	Remarks[b]
I	0.9% Saline control	None	48.48 ± 7.9	0.00	—
II	1.5 IU hCG	None	39.00 ± 5.84	19.50	$p < 0.05$ (I:II) ($n = 14$)
III	6.0 IU hCG	None	24.27 ± 2.34	49.80	$p < 0.001$ (I:III) ($n = 14$)
IV	24.0 IU hCG	None	16.93 ± 2.5	65.10	$p < 0.001$ (I:IV) ($n = 14$)
V	6.0 IU hCG	100	48.47 ± 4.6	0.00	$p > 0.7$ (I:V) ($n = 14$)
VI	0.4 μg hLH	None	37.7 ± 3.1	22.30	$p < 0.01$ (I:VI) ($n = 14$)
VII	1.6 μg hLH	None	27.51 ± 2.6	43.50	$p < 0.001$ (I:VII) ($n = 14$)
VIII	6.4 μg hLH	None	15.13 ± 1.77	68.80	$p < 0.001$ (I:VIII) ($n = 14$)
IX	1.6 μg hLH	200	25.8 ± 7.77	46.80	$p > 0.6$ (VII:IX) ($n = 10$)
X	1.6 μg hLH	100	27.33 ± 5.3	43.50	$p > 0.9$ (VII:X) ($n = 10$)

[a] Ovarian-ascorbic-acid-depletion assay.
[b] (n) Number of independent determinations.

Table IV. Ability of Anti-hCG-β Derivatives to Neutralize the Biological Activity[a] of hCG but Not of hLH

Treatment	Antiserum or normal serum (ml)	Weight of uterus (mg/100 g body wt.)
Saline (0.4 ml)	—	67.8 ± 8.2
hCG		
0.6 IU	0.2 (normal	224.5 ± 20.6
0.6 IU	0.05 (1:5)	79.8 ± 5.0
0.6 IU	0.2 (1:5)	86.9 ± 10.3
hLH		
3.0 IU	0.2 (normal)	136.5 ± 6.0
3.0 IU	0.2 (1:5)	149.8 ± 15.9
3.0 IU	0.2 (1.2)	146.7 ± 16.6

[a] Rat uterine-weight assay.

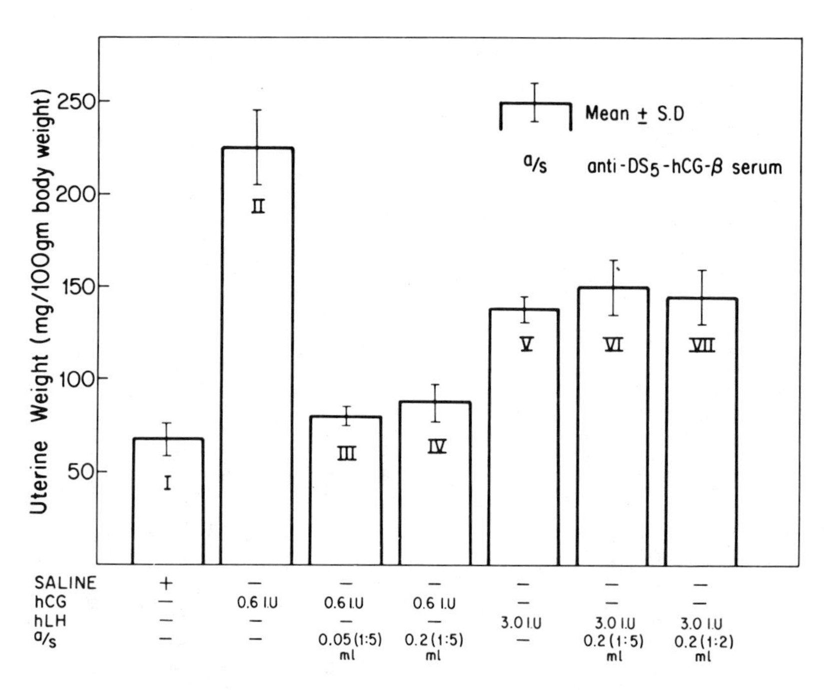

Figure 15. Ability of anti-DS_5,DS_6-hCG-β serum to discriminate between hCG and hLH in rat uterine-weight assay for hCG. Groups of female immature rats (18–21 days old) were administered either saline or hCG or hLH or antiserum plus hormone as indicated in the figure. All injections were intraperitoneal. Antiserum at the indicated dilution and volume was mixed with the hormone before injection. After three daily injections, the uterine weights of these animals were measured on the 4th day.

4. Conclusions

In summary, we have been able to obtain a specific antiserum for hCG or hCG-β by using chemically modified hCG-β derivatives. The antibody possesses both immunological and biological specificity, and it seems to be directed against an hCG-specific site(s) in the main polypeptide chain rather than the unique carboxy-terminal peptide. Furthermore, the antibody is highly suitable for a specific and sensitive RIA for hCG or hCG-β. When the assay was applied to the clinical serum samples obtained from pregnant women at various stages of gestation, the results were compatible with the known secretory pattern of hCG (Fig. 16). The interassay variation (as indicated by coefficient of variation) for these samples ranged from 6.7 to 16.6%. Thus, the RIA based on the use of anti-DS_5,DS_6-hCG-β is applicable to serum samples from normal pregnancies. Furthermore, such an assay should prove to be highly useful in the early detection of normal and ectopic pregnancies and certain types of hCG-producing neoplasms. Last but not least, the assay should also facilitate the monitoring of the treatment of these tumors.

Acknowledgments. The research reported herein was supported by United States Public Health Service Grant HD-08766. The author wishes to acknowledge the scientific contributions of Drs. M. R. Pandian, R. D.

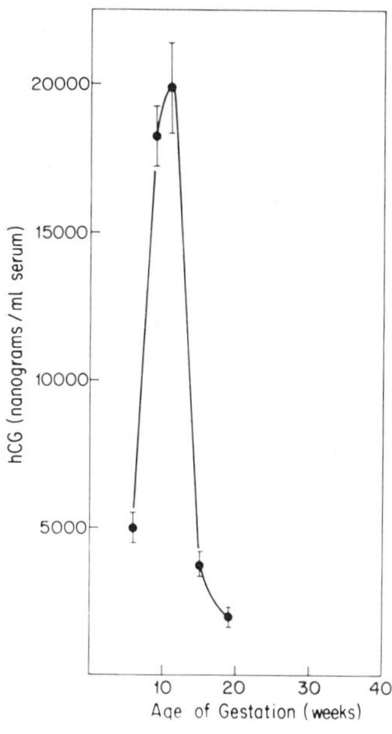

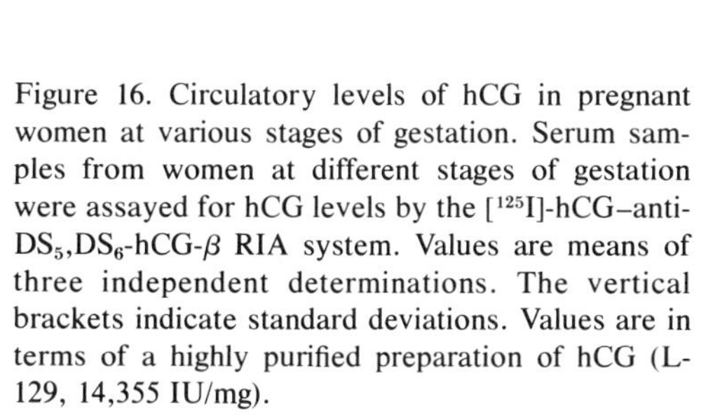
Figure 16. Circulatory levels of hCG in pregnant women at various stages of gestation. Serum samples from women at different stages of gestation were assayed for hCG levels by the [^{125}I]-hCG–anti-DS_5,DS_6-hCG-β RIA system. Values are means of three independent determinations. The vertical brackets indicate standard deviations. Values are in terms of a highly purified preparation of hCG (L-129, 14,355 IU/mg).

Ghai, T. Mise, R. Mitra, and K. Muralidhar to this project. All those investigators who have made significant contributions to the development of this field, and in particular those whose work has not been cited due to space limitations, are also acknowledged.

References

Ayala, A.R., Nisula, B.C., Chen, H.C., Hodgen, G.D., and Ross, G.T., 1978, Highly sensitive radioimmunoassay for chorionic gonadotropin in human urine, *J. Clin. Endocrinol. Metab.* **47:**767–773.

Bahl, O.P., 1969, Human chorionic gonadotropin. I. Purification and physicochemical properties, *J. Biol. Chem.* **244:**567–574.

Bahl, O.P., 1973, Chemistry of human chorionic gonadotropin, in: *Hormonal Proteins and Peptides,* Vol. 1 (C.H. Li, ed.), pp. 171–199, Academic Press, New York.

Bahl, O.P., 1977, Chemistry and biology of human chorionic gonadotropin and its subunits, in: *Frontiers in Reproduction and Fertility,* Part II (R.O. Greep, ed.), pp. 11–24, MIT Press, Cambridge.

Bahl, O.P., and Muralidhar, K., 1980, Current status of antifertility vaccines, in: *Immunological Aspects of Infertility and Fertility Regulation* (G.B.F. Schumacher and D.S. Dhindsa, eds.), pp. 225–257, Elsevier/North-Holland, Amsterdam (in press).

Bahl, O.P., Pandian, M.R., and Ghai, R.D., 1976, Immunological properties of the β-subunit of human chorionic gonadotropin, *Biochem. Biophys. Res. Commun.* **70:**525–532.

Bahl, O.P., Reddy, M., and Bedi, G., 1980, A novel carbohydrate structure in bovine and ovine luteinizing hormone, *Biochem. Biophys. Res. Commun.* **96:**1192–1199.

Bellisario, R., Carlsen, R.B., and Bahl, O.P., 1973, Human chorionic gonadotropin: Linear amino acid sequence of the α subunit, *J. Biol. Chem.* **248:**6796–6809.

Braunstein, G.D., Karow, W.G., Gentry, W.C., Rasor, J., and Wade, M.E., 1978, First-trimester chorionic gonadotropin measurements as an aid in the diagnosis of early pregnancy disorders, *Am. J. Obstet. Gynecol.* **131:**25–32.

Carlsen, R.B., Bahl, O.P., and Swaminathan, N., 1973, Human chorionic gonadotropin: Linear amino acid sequence of the β subunit, *J. Biol. Chem.* **248:**6810–6827.

Chen, H.C., Hodgen, G.D., Matsuura, S., Lin, L.J., Gross, E., Reichert, L.E., Birken, S., Canfield, R.E., and Ross, G.T., 1976, Evidence for a gonadotropin from nonpregnant subjects that has physical, immunological, and biological similarities to human chorionic gonadotropin, *Proc. Natl. Acad. Sci. U.S.A.* **73:**2885–2889.

Franchimont, P., Reuter, A., and Gaspard, U., 1978, Ectopic production of human chorionic gonadotropin and its α- and β-subunits, in: *Current Topics in Experimental Endocrinology,* Vol. 3 (L. Martinini and V.H.T. James, eds.), pp. 201–216, Academic Press, New York.

Ghai, R.D., Mise, T., Pandian, M.R., and Bahl, O.P., 1980, Immunological properties of the β-subunit of human chorionic gonadotropin. I. Effect of chemical and enzymatic modifications, *Endocrinology* **107:**1556–1563.

Giudice, L.C., and Pierce, J.G., 1978, Glycoprotein hormones: Some aspects of studies of secondary and tertiary structure, in: *Structure and Function of the Gonadotropins* (K.W. McKerns, ed.), pp. 81–110, Plenum Press, New York.

Goldstein, D.P., Aono, T., Taymor, M.L., Jochelson, K., Todd, R., and Hines, E., 1968, Radioimmunoassay of serum chorionic gonadotropin activity in normal pregnancy, *Am. J. Obstet. and Gynecol.* **102:**110–114.

Hara, K., Rathnam, P., and Saxena, B.B., 1978, Structure of the carbohydrate moieties of

α subunits of human follitropin, lutropin, and thyrotropin, *J. Biol. Chem.* **253:**1582–1591.

Jones, W.B., Lewis, J.L., and Lehr, M., 1975, Monitor of chemotherapy in gestational trophoblastic neoplasm by radioimmunoassay of the β-subunit of human chorionic gonadotropin, *Am. J. Obstet. Gynecol.* **121:**669–673.

Kennedy, J.F., Chaplin, M.F., and Stacey, M., 1974, Periodate oxidation, acid hydrolysis, and structure–activity relationships of human-pituitary, follicle-stimulating hormone and human chorionic gonadotropin, *Carbohydr. Res.* **36:**369–377.

Kessler, M.J., Reddy, M.S., Shah, R.H., and Bahl, O.P., 1979a, Structures of *N*-glycosidic carbohydrate units of human chorionic gonadotropin, *J. Biol. Chem.* **254:**7901–7908.

Kessler, M.J., Mise, T., Ghai, R.D., and Bahl, O.P., 1979b, Structures and location of the *O*-glycosidic carbohydrate units of human chorionic gonadotropin, *J. Biol. Chem.* **254:**7909–7914.

Kosasa, T.S., Taymor, M.L., Goldstein, D.P., and Levesque, L., 1973, Use of a radioimmunoassay specific for human chorionic gonadotropin in the diagnosis of early ectopic pregnancy, *Obstet. Gynecol.* **42:**868–871.

Matsuura, S., Ohashi, M., Chen, H.C., and Hodgen, G.D., 1979, A human chorionic gonadotropin-specific antiserum against synthetic peptide analogs of the carboxyl-terminal peptide of its β-subunit, *Endocrinology* **104:**396–401.

Milwidsky, A., Adoni, A., Miodovnik, M., Segal, S., and Palti, Z., 1978, Human chorionic gonadotropin (β-subunit) in the early diagnosis of ectopic pregnancy, *Obstet. Gynecol.* **51:**725–726.

Morgan, F.J., and Canfield, R.E., 1971, Nature of the subunits of human chorionic gonadotropin, *Endocrinology* **88:**1045–1053.

Morgan, F.J., Birken, S., and Canfield, R.E., 1975, The amino acid sequence of human chorionic gonadotropins: The α subunit and β subunit, *J. Biol. Chem.* **250:**5247–5258.

Pandian, M.R., Mitra, R., and Bahl, O.P., 1980, Immunological properties of the β-subunit of human chorionic gonadotropin: (ii) Properties of an hCG specific antibody prepared against a chemical analog of the β-subunit, *Endocrinology* **107:**1564–1571.

Pierce, J.G., Bahl, O.P., Cornell, J.S., and Swaminathan, N., 1971, Biologically active hormones prepared by recombination of the α chain of human chorionic gonadotropin and the hormone-specific chain of bovin thyrotropin or of bovine luteinizing hormone, *J. Biol. Chem.* **246:**2321–2324.

Rasor, J.L., and Braunstein, G.D., 1977, A rapid modification of the beta-hCG radioimmunoassay, *Obstet. Gynecol.* **50:**553–558.

Ross, G.T., 1977, Clinical relevance of research on the structures of human chorionic gonadotropin, *Am. J. Obstet. Gynecol.* **129:**795–808.

Rutanen, E.M., and Sepp lä, M., 1978, The hCG-beta subunit radioimmunoassay in nontrophoblastic gynecologic tumors, *Cancer* **41:**692–696.

Stevens, V.C., 1980, The current status of anti-pregnancy vaccines based on synthetic fraction of hCG, in: *Immunological Aspects of Reproduction and Fertility* (J. Hearn, ed.), pp. 203–216, MTP Press, London.

Swaminathan, N., and Bahl, O.P., 1970, Dissociation and recombination of the subunits of human chorionic gonadotropin, *Biochem. Biophys. Res. Commun.* **40:**422–427.

Vaitukaitis, J.L., 1979, Human chorionic gonadotropin—a hormone secreted for many reasons, *N. Engl. J. Med.* **301:**324–326.

Vaitukaitis, J.L., Braunstein, G.D., and Ross, G.T., 1972, A radioimmunoassay which specifically measures human chorionic gonadotropin in the presence of human luteinizing hormone, *Am. J. Obstet. Gynecol.* **113:**751–758.

Varma, K., Larraga, L., and Selenkow, H.A., 1971, Radioimmunoassay of serum human chorionic gonadotropin during normal pregnancy, *Obstet. Gynecol.* **37:**10–18.

26

Isolation and Characterization of an "hCG"-like Protein from Bacteria

Takeshi Maruo, Herman Cohen, Sheldon J. Segal, and S. S. Koide

1. Introduction

Human choriogonadotropin (hCG) is a glycoprotein hormone normally produced by the trophoblastic tissue of the placenta. However, it has also been reported that nontrophoblastic malignant tumors (Rabson et al., 1973; Vaitukaitis *et al.,* 1976) and normal human tissues (Yoshimoto *et al.,* 1977) may produce hCG or an hCG-like substance. All these descriptions of extra placental production of hCG deal with tissues of mammalian origin.

Several publications describing the presence of hCG-like substances ("hCG") in microorganisms (Livingston and Livingston, 1974; Cohen and Strampp, 1976) have also been presented, but these papers have lacked data obtained with the purified materials. Our interest in this source of gonadotropin was stimulated by these reports, and in the study we have purified the "hCG" material from bacteria and established its physicochemical as well as its immunological and biological similarity to hCG. We have also demonstrated that its activity in radioisotope assays,

Takeshi Maruo and *S.S. Koide* • Center for Biomedical Research, The Population Council, The Rockefeller University, New York, New York 10021. *Herman Cohen* • Carter-Wallace, Inc., Cranbury, New Jersey 08512. *Sheldon J. Segal* • The Rockefeller Foundation, New York, New York 10036.

i.e., radioimmunoassay (RIA) and radioreceptor assay (RRA), is not due to contamination with proteases as has been described by Maruo *et al.* (1979) for another nonmammalian source of "hCG."

2. Materials and Methods

An acetone-dried preparation of the bacterium identified as *"Progenitor cryptocides"** was obtained from Mr. John Majnarich and Dr. V. W. C. Livingston (Livingston and Livingston, 1974). A 10-g sample of this acetone powder was suspended and stirred in 50 ml 0.01 M tris-HCl, pH 7.4, for 6 hr at 4°C. The undissolved material was removed by centrifugation at 10,000*g* for 20 min, and the supernatant was dialyzed against distilled water and lyophilized. The weight of the lyophilized powder was 2.5 g. A measured amount of the lyophilized powder was reconstituted in 0.01 M phosphate-buffered saline (PBS) (pH 7.4), and the presence of "hCG" was determined by RIA and RRA with appropriate protease inhibitors in the assay medium.

Streptococcus faecalis (ATCC 12818) was purchased from American Type Culture Collection (ATCC) and grown in trypticase soy broth at 37°C for 24 hr. Acetone powder of the culture contents was treated according to the same method (Livingston and Livingston, 1974) by which the lyophilized extract described above was prepared.

Highly purified hCG (CR-119), hCG-α (CR-119), and hCG-β (CR-119-2) used as standard preparations in each assay were gifts from the Center for Population Research of NICHHD and Dr. R. Canfield, Columbia University. Specific antisera to hCG, hCG-α, and hCG-β were generated by Dr. C. C. Chang in this laboratory using the multiple intradermal injection technique as described by Vaitukaitis *et al.* (1971). Highly purified hCG (12,000 IU equivalent to Second International Standard hCG/mg by the mouse uterine-weight method), HCG-α and hCG-β used as antigens to produce the antisera were prepared by Dr. Y. Y. Tsong in this laboratory (from commercial hCG, 2700–3000 IU/mg, Organon) by a modified method described by Canfield and Morgan (1973). Specific antisera to hCG-β COOH-terminal peptide (H93) was a generous gift of Dr. H. C. Chen and Dr. G. D. Hodgen of NICHHD.

N-α-*p*-tosyl-L-lysine chloromethyl ketone (TLCK), L-1-tosylamide-2-phenylethylchloromethyl ketone (TPCK), soybean trypsin inhibitor (type 1-S), and methyl-α-D-glucopyranoside were purchased from the

* This is the classification according to Dr. Livingston. No such classification is noted in Bergey's manual.

Sigma Chemical Company. Sephadex G-100, concanavalin A (Con A)–Sepharose, and diethylaminoethyl (DEAE)–Sephadex were obtained from Pharmacia Fine Chemicals, and Na[^{125}I] from New England Nuclear. Leupeptin and antipain were donated by the U.S.–Japan Cooperative Medical Science Program.

2.1. Radioimmunoassay and Radioreceptor Assay

Radioiodination of hCG (CR-119), hCG-α (CR-119), and hCG-β (CR-119-2) with Na[^{125}I] was performed using the lactoperoxidase method as described in a previous report (Ashitaka and Koide, 1974). All RIAs were carried out by the double-antibody technique as described previously (Maruo *et al.*, 1979). hCG (CR-119) showed 1.4 and 9.0% cross-reactivity in the homologous RIA for hCG-α and hCG-β, respectively. RRA was performed by the method described by Saxena *et al.* (1974), using plasma membranes from bovine corpora lutea.

2.2. Sephadex Column Chromatography

A sample of 100 mg of the lyophilized extracts of the bacterial culture was dissolved in 2.0 ml 0.01 M tris-HCl buffer (pH 7.4) and 0.15 M NaCl and the mixture applied to a column (1.5 × 85 cm) of Sephadex G-100, equilibrated at 4°C with 0.01 M tris-HCl and 0.15 M NaCl. The eluate was collected in 2.0-ml fractions and the activity determined by RIA and RRA. The void volume was measured with blue dextran. The column was standardized on separate days with [^{125}I]-hCG, [^{125}I]-hCG-α and [^{125}I]-hCG-β.

2.3. Concanavalin A–Sepharose Column Chromatography

Columns (9.0 × 130 mm) of Con A–Sepharose were prepared and extensively washed at 4°C with 0.01 M PBS (pH 7.4) containing 1% BSA. The extracts were dissolved in 0.01 M PBS (pH 7.4). The unadsorbed substances were eluted with 0.01 M PBS (pH 7.4). Con-A-adsorbed glycoproteins were eluted with 0.01 M PBS containing 0.2 M methyl-α-D-glucopyranoside. Fractions of 1.0 ml were collected and activity measured by RIA and RRA.

2.4. Ion-Exchange Chromatography

Fractions containing hCG activity obtained from the Con A–Sepharose column were pooled, dialyzed, lyophilized, and applied to a column (1.5 × 40 cm) of DEAE–Sephadex A-50, equilibrated at 4°C with 0.01 M tris-HCl (pH 7.4). The adsorbed protein on the column was eluted by a linear gradient to 0.5 M NaC1 in 0.01 M tris-HCl buffer (pH 7.4). Fractions of 2.0 ml were collected. Activity was measured by RIA.

2.5. Polyacrylamide Gel Electrophoresis

Electrophoresis was performed utilizing 7.5% polyacrylamide gels containing 0.1% sodium dodecyl sulfate (SDS) as described by Weber and Osborn (1969). The gels were stained with 0.2% Coomassie blue.

2.6. Bioassay

Biological activity of the purified bacterial "hCG" factor was determined by measuring the increase in uterine weight and ovarian weight of immature female rats as described by Diczfalusy and Loraine (1955). The Second International Standard hCG was used as reference. Bovine serum albumin (BSA) was added as a stabilizing agent to all test solutions to a final concentration of 0.1%. The assay was run at three dose levels for both the standard and the unknown.

3. Results

Serial dilutions of the lyophilized extract derived from the acetone-dried preparation of "*Progenitor cryptocides*" demonstrated dose–response curves parallel to that of hCG (CR-119) in RRA and RIA using a homologous hCG assay system (Fig. 1) and RIA utilizing antiserum (H93) to hCG-β COOH-terminal peptide (Fig. 2). To exclude the influence of protease in the assays, TLCK (1 mM), TPCK (1 mM), leupeptin (1 mM), antipain (1 mM), and soybean trypsin inhibitor (0.1%) were added to the RRA and RIA systems. The addition of these protease inhibitors did not affect the results as described above, indicating that the results were not due to proteolytic activity. Extracts derived from the cultures of *Streptococcus faecalis* (ATCC 12818) showed no displacement of [^{125}I]-hCG in the RRA and RIA assays.

To verify that the putative "hCG" factor present in the bacterial culture extract is similar to hCG, the factor was isolated and purified by a combination of gel filtration on Sephadex G-100 and chromatography with Con A–Sepharose and DEAE–Sephadex A-50. When the extract was filtered through a column of Sephadex G-100, hCG-like factor determined by both RRA and RIA was found to be eluted at the position corresponding to that of [^{125}I]-hCG used as a marker, suggesting that the

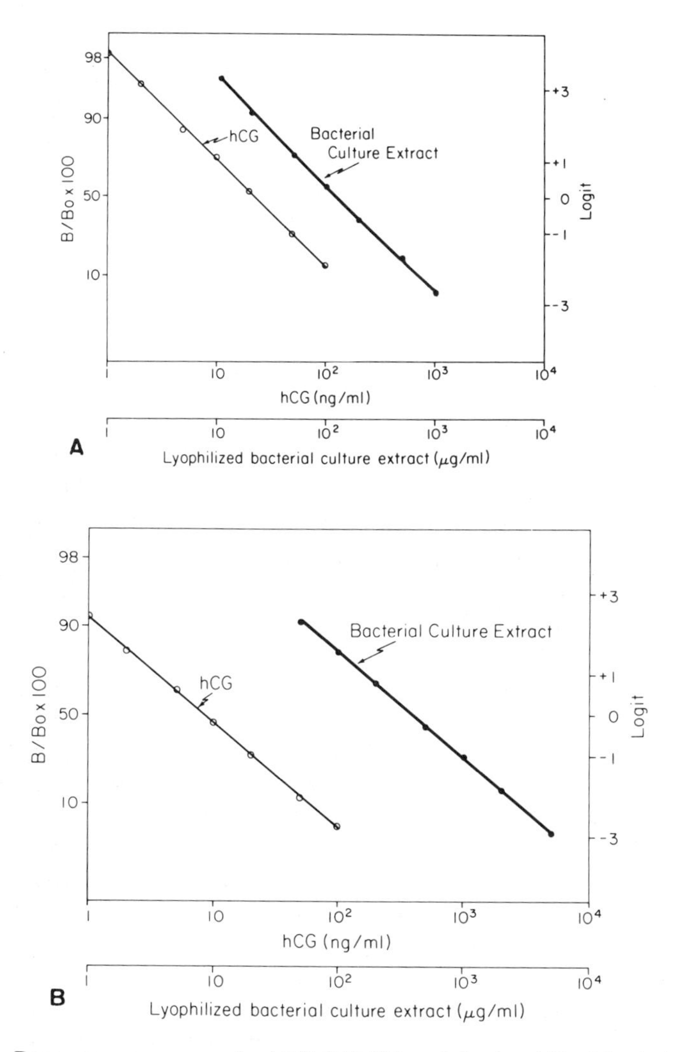

Figure 1. Dose–response curves for hCG (CR-119) and the lyophilized bacterial culture extract in the RRA (A) and RIA (B). Results are shown as the log dose of the hCG reference preparation (ng/ml) or of the lyophilized extract (μg/ml) plotted against the logit transformation of the response. B, cpm bound in the presence of [^{125}I]-hCG and unlabeled ligand; B^0, cpm bound in the presence of [^{125}I]-hCG alone.

apparent molecular weight of the "hCG" factor is similar to that of hCG. A noteworthy finding was that an immunoreactive peak that reacted only with antisera to hCG-β or to hCG-β COOH-terminal peptide was eluted. This peak was considerably retarded when compared to [^{125}I]-hCG-β, suggesting that the apparent molecular weight of this immunoreactive

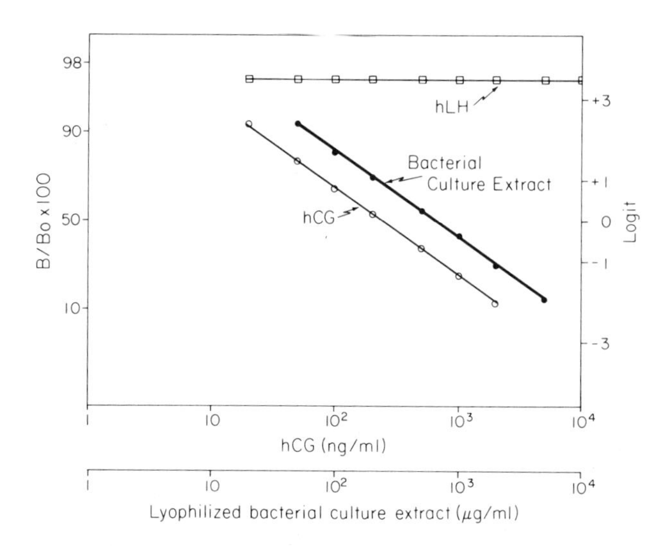

Figure 2. Dose–response curves for hCG (CR-119) and the lyophilized bacterial culture extract in the RIA using antiserum to hCG-β COOH-terminal peptide (H93).

fragment is much smaller than that of hCG-β (CR-119-2). However, no free α-subunit was detected. Following gel filtration of the lyophilized extract (Fig. 3), the fractions containing "hCG" (tubes 21–27) were pooled, dialyzed against distilled water, and lyophilized. The yield of the "hCG" fraction obtained at this step was 31.2 mg.

To confirm that the "hCG" fraction obtained from Sephadex G-100 did not contain proteolytic activity, [^{125}I]-hCG (500,000 cpm), used as a tracer in the RRA and RIA, was incubated with the "hCG" fraction (1 mg) at 37°C for 60 min. The reaction mixture was analyzed by gel filtration on Sephadex G-100. The radiolabeled hCG was recovered intact, indicating that the "hCG" fraction does not contain any proteases.

The aforegoing results indicated that the bacterial factor is similar to hCG and that interference from proteases in the RIA and RRA is unlikely.

To ascertain whether or not the bacterial "hCG" factor contains sugar moieties, 6-mg aliquots of the "hCG" fraction obtained from Sephadex G-100 fractionation were reconstituted in 2.0 ml 0.01 M PBS (pH 7.4) and applied to a column of Con A–Sepharose. The "hCG" factor was adsorbed to the Con A–Sepharose column and eluted with 0.2 M methyl-α-D-glucopyranoside (Fig. 4), indicating that the "hCG" contains glucose or mannose moieties.

After the total "hCG" fraction was chromatographed on Con A–Sepharose, the immunoreactive fractions were pooled, dialyzed against distilled water, and lyophilized. The yield of "hCG" at this step

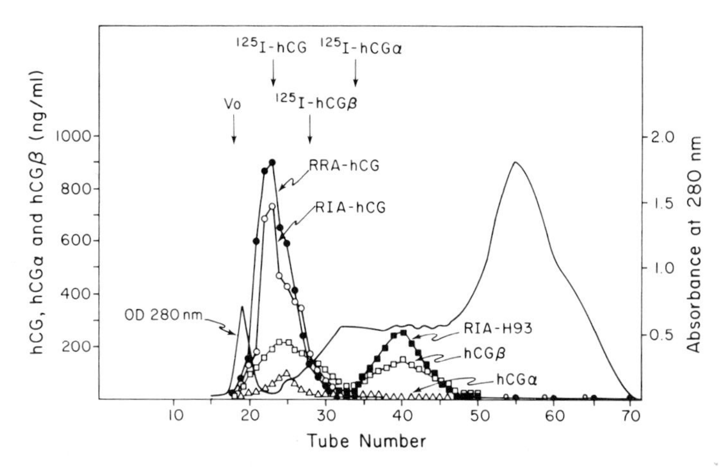

Figure 3. Sephadex G-100 elution profile of the bacterial culture extract. Each 2.0-ml fraction was assayed by the RRA and by the RIA system for hCG, hCG-α, hCG-β, and hCG-β COOH-terminal peptide. The protein content of the eluted fractions was measured by absorbance at 280 nm. (V_0) Void volume determined by elution of blue dextran. The vertical arrows indicate the elution positions of [^{125}I]-hCG, [^{125}I]-hCG-α, and [^{125}I]-hCG-β.

of purification was 6.4 mg. The lyophilized material was reconstituted in 0.01 M tris-HCl buffer (pH 7.4) and applied to a column of DEAE–Sephadex A-50. When this column was eluted with a linear gradient, the "hCG" was eluted in a single peak at a conductivity of 4.0–6.0 mmho (Fig. 5). This value is similar to that found for hCG from

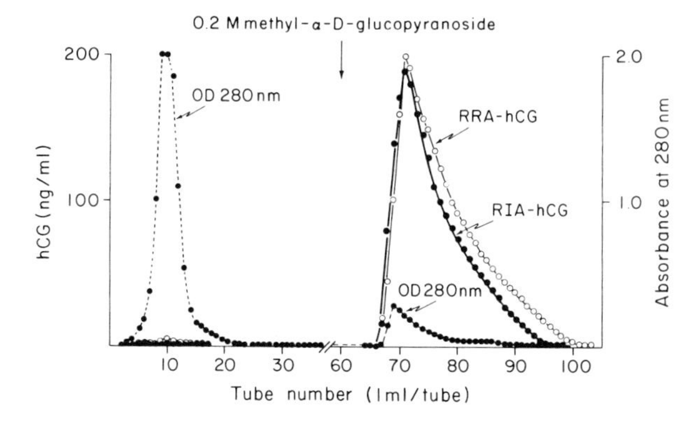

Figure 4. Elution profile on Con A–Sepharose column of the "hCG" fraction obtained by chromatography on a Sephadex G-100 column. Fractions 1–59 were eluted with PBS and followed by a second elution with PBS containing 0.2 M methyl-α-D-glucopyranoside. Each 1-ml fraction was assayed for hCG by the RRA and the RIA system. The protein content was measured by absorbance at 280 nm.

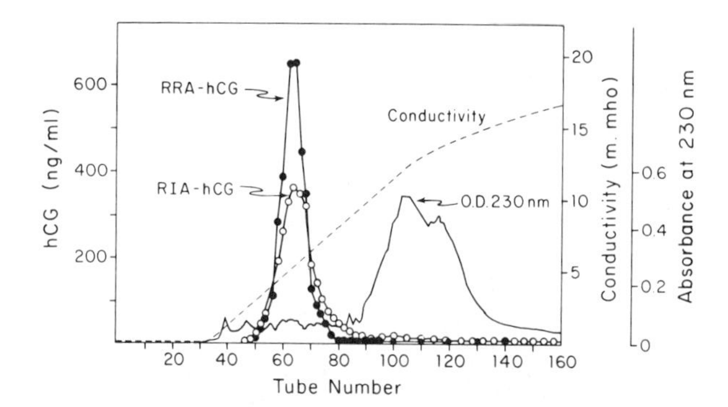

Figure 5. Elution profile of the Con A-adsorbed "hCG" fraction on DEAE–Sephadex A-50 column chromatography. The adsorbed protein on the column was eluted with a linear 300-ml gradient from 0 to 0.5 M NaCl in 0.01 M tris-HCl buffer (pH 7.4). Each 2.0-ml fraction was assayed for hCG by RRA and RIA. The protein content was measured by absorbance at 230 nm.

placental extract (Maruo *et al.*, 1974). The fractions containing "hCG" activity were combined, dialyzed against distilled water, and lyophilized. The final yield was 570 μg, and was equivalent to 2260 IU/mg when assayed in a homologous hCG RIA and 3580 IU/mg in the RRA.

Table I summarizes the yield and potency of the fractions containing "hCG" activity obtained at each step of the purification procedure.

To establish that the purified bacterial "hCG" is composed of subunits, the factor was analyzed by electrophoresis on SDS–polyacrylamide gels (Fig. 6). The bacterial factor separated into two major bands with mobility rates corresponding to α- and β-subunits dissociated from hCG (CR-119) and a minor band that was retarded. To identify each band, the gel was sliced into 2.2-mm segments. Each segment was homogenized in 500 μl PBS (pH 7.4) and centrifuged at

Table I. Yield and Relative hCG Potency at Each Purification Step

Fraction	Yield (mg)	hCG potency (IU/mg)[a] RRA	RIA
Lyophilized extract	2500[b]	3.4 ± 0.2	2.1 ± 0.1
Sephadex G-100	31.2	231 ± 11	150 ± 7
Con A–Sepharose	6.4	880 ± 45	542 ± 27
DEAE–Sephadex A-50	0.57	3580 ± 160	2260 ± 102

[a] The relative hCG potency (mean ± S.D.) was determined by the RRA and homologous hCG RIA using the Second International Standard hCG as a reference preparation. The yield is expressed as amounts of protein, except that of the starting extract.
[b] Weight of dry material.

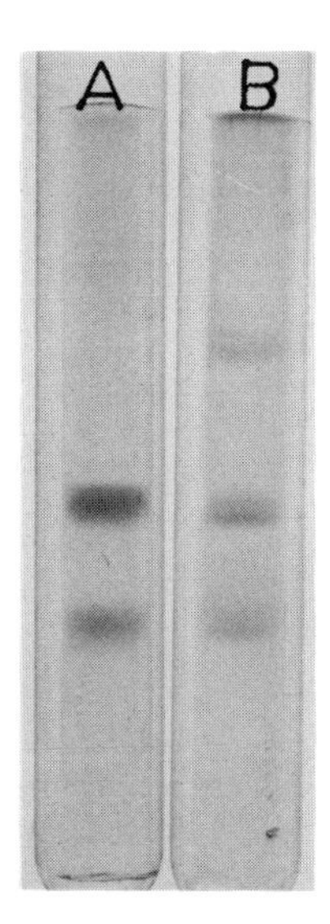

Figure 6. SDS–polyacrylamide gel electrophoresis. (A) Electrophoretic pattern of hCG (CR-119) (20 μg), which dissociated in SDS–gel into α- and β-subunits; (B) electrophoretic pattern of the purified bacterial hCG factor (20 μg) obtained from DEAE–Sephadex A-50 chromatography. The migration is toward the anode at the bottom of the figure.

3000*g* for 20 min, and the supernatant was assayed for hCG, hCG-α, and hCG-β using the respective homologous RIA systems. The peaks designated as segments 28, 22, and 11 showed immunoreactivities with anti-hCG-α, hCG-β, and hCG, respectively (Fig. 7). Hence, the immunoreactivity and protein peaks of the bacterial "hCG" subunits coincided with the migration bands of α- and β-subunits dissociated from hCG (CR-119), and the retarded minor band corresponds to the undissociated form of the bacterial "hCG" factor. These results demonstrate that the bacterial "hCG" is composed of two subunits with electrophoretic mobilities corresponding to those of the subunits of authentic hCG. The biological potency of the purified bacterial "hCG" factor determined by the rat uterine-weight assay and by the ovarian weight assay at three dose levels (Fig. 8) was found to be 380 IU/mg and 880 IU/mg, respectively, although by RIA in a homologous hCG system its activity is 2260 IU/mg and by RRA, 3580 IU/mg.

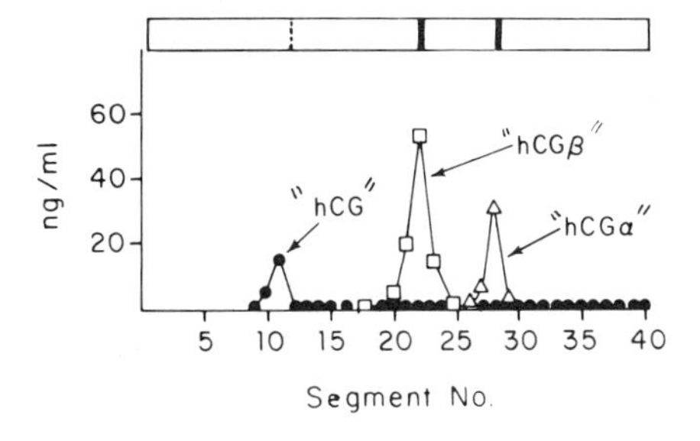

Figure 7. Localization of the immunoreactivity of hCG-α and hCG-β in the gel prepared in the same manner as gel B in Fig. 6. Each 2.2-mm segment of the gel was homogenized in PBS, and the supernatant was assayed for hCG, hCG-α, and hCG-β by the respective homologous RIAs.

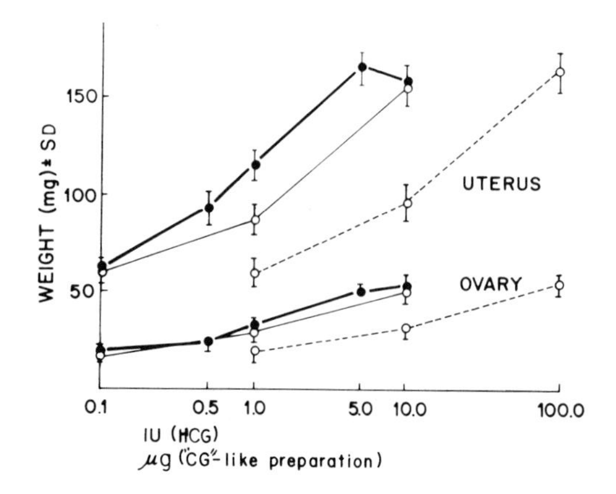

Figure 8. Dose–response curves of standard hCG and bacterial hCG-like preparations determined by measuring the uterine and ovarian weights in immature female rats. (●—●)Second International Standard hCG; (○--○) hCG-like preparation obtained from Sephadex G-100; (○—○) hCG-like preparation from DEAE–Sephadex A-50.

4. Discussion

On the basis of the aforegoing data, it may be concluded that a protein similar to hCG is produced by the bacterium "*Progenitor cryptocides*." First, the bacterial extracts show dose–response curves that parallel those of hCG (CR-119) in the RRA and homologous hCG RIA systems as well as in the RIA system using antiserum (H93) generated against hCG-β COOH-terminal peptide, which interacts specifically with the unique carboxyl-terminal peptide sequence of hCG-β (Chen *et al.*, 1976). Interference by proteases has been excluded by the findings that inhibitors added to the RRA and RIA systems did not influence the results and that radiolabeled hCG incubated with the partially purified bacterial fraction was not hydrolyzed. Second, the elution profiles of the bacterial "hCG" on Sephadex G-100, Con A–Sepharose, and DEAE–Sephadex A-50 were consistent with those of hCG. These data indicate that the bacterial "hCG" factor is a glycoprotein containing glucose or mannose moieties, or both, and that the apparent molecular weight and electrical charge of the glycoprotein are similar to those of hCG. Third, after extensive purification, the bacterial "hCG" factor may be dissociated into two subunits with mobility rates corresponding to those of hCG-α and hCG-β as determined by electrophoresis on SDS–polyacrylamide gel. Fourth, the bacterial factor was biologically active as determined by the rat uterine-weight bioassay and the ovarian weight bioassay, albeit much less so than purified hCG.

The biological potency of the purified bacterial factor (380 IU/mg) is extremely low compared to its immunological potency as well as to the biological activity of purified hCG (Morgan *et al.,* 1974). The basis for the reduced biological activity has not been clarified. However, it may be due to a low sialic acid content, since it is well known that desialylated hCG possesses extremely low biological activity *in vivo* due to an enhancement of its metabolic clearance (Van Hall *et al.,* 1971; Tsuruhara *et al.,* 1972), and yet it retains full activity in the RRA and RIA systems.

An interesting finding is the observation that significant amounts of free "hCG"-β immunoreactive fragments are obtained following gel filtration on Sephadex G-100 (see Fig. 3). On the other hand, no detectable free α-subunit was found in the bacterial culture extract. It has been frequently reported that substantial amounts of free α-subunit are found in sera of pregnant women, while minimal or no free β-subunit is detectable (Ashitaka *et al.,* 1974). Moreover, placental-tissue extracts contain substantial amounts of free hCG-α without accompanying free hCG-β (Vaitukaitis, 1974; Maruo, 1976), suggesting that the production of hCG-β might be a rate-limited step in the biosynthesis of hCG in human placenta. Hence, the unbalanced formation of significant quantities of free "hCG"-β immunoreactive subunits with no detectable free α-subunit in the bacterial preparation observed in this study suggests that the regulatory mechanism controlling the biosynthesis of "hCG" factor in bacteria may differ from that taking place in human placenta.

Recently, with the use of indirect fluorescein-labeled and peroxidase-labeled antibody techniques, Acevedo *et al.* (1978) reported that hCG-like immunoreactive protein could be demonstrated in several microorganisms that were isolated from patients bearing malignant neoplasms. Richert and Ryan (1977) also reported that [^{125}I]-hCG binds to *Pseudomonas maltophilia* and that the culture media may contain a protein molecule that cross-reacts with antisera to hCG as well as to hCG-β. However, in a subsequent publication, they interpret this data as being due to artifacts, namely, a protease and a soluble hCG-binding substance (Richert *et al.,* 1978).

The production of a mammalian hormone by a microorganism is an unusual phenomenon; hence, the expression of this mammalian gene in bacteria should be further investigated. We have also recently isolated from the urine of a patient (A.K.) with adenoid cystic carcinoma of the ethmoid sinus a strain of *Streptococcus faecalis* that produces only an hCG-β-like immunoreactive fragment. This finding of "hCG" production by a strain of microorganism that is a normal inhabitant of the human intestinal tract is a novel demonstration of an expression of a mammalian gene. To explain this observation, two alternative hypotheses are offered. One is that the hCG gene has a prehistoric origin and the other is that it is a consequence of natural recombinant DNA.

ACKNOWLEDGMENTS. We are grateful to Drs. V. W. C. Livingston, J. Majnarich, and also Dr. A. Strampp for making available to us acetone-powder preparations of the bacteria. This work was performed as part of the Contraceptive Development Program of the International Committee for Contraceptive Research of the Population Council. T. M. is supported by a Rockefeller Foundation Fellowship Award in Reproductive Biology and by a Biomedical Fellowship of the Population Council, Rockefeller University.

DISCUSSION

BOGDANOVE: One thing I noted was that the bacterial extract and authentic hCG seemed to have about the same potency ratio by RRA and by RIA with antibody to authentic hCG, but a very different ratio by the RIA with H93 antibody. Thus, an index of discrimination would suggest some subtle difference between the bacterial "hCG" (at least in the crude extract) and authentic mammalian hCG.

COHEN: RIA-H93 is an antiserum raised against the carboxyl-terminal peptide of hCG by Chen *et al.* (1976). The RIA-H93 was not used to monitor the initial peak (hCG). The delayed peak had a greater immunoreactivity with H93 than with anti-hCG-β, suggesting that the material is a segment of hCG-β and contains the carboxyl-terminal peptide. Part of the amino-terminal segment might be missing. An alternative explanation is that the hCG-β is aglycosylated and the H93 antiserum may be more active. To interpret a result with the crude fraction is "risky," and conclusion should be based on results obtained with purified product.

HAOUR: What is the quantity of hCG-like material produced, compared to the total protein synthesized?

COHEN: Less than 0.8% of total protein in the active precipitate contained hCG-like material by RIA. This amount is comparable to that produced by first-trimester placenta, which was estimated to be about 0.2–0.3% of the total protein.

PIETRAS: It is noteworthy to add that Watanabe and colleagues have presented evidence for the occurrence of specific, high-affinity receptors for testosterone in the periplasmic membranes of *Pseudomonas*.

COHEN: Richert and Ryan (1977) reported that *Pseudomonas maltophilia* possesses receptors for hCG. We did not detect any hCG-like factor in a culture of this organism.

CHANNING: What was the carbohydrate composition of your bacterial hCG? Dr. Bahl and I have found that (*in vitro*) the carbohydrate residues internal to sialic acid are important in biological activity of hCG. hCG with galactose or *N*-acetyl glucosamine removed can bind to the receptor (granulosa and Leydig cell), but did not have ability to stimulate AMP or progesterone secretion. They also can inhibit hCG action and have potential as contraceptives.

COHEN: The carbohydrate content has not been determined directly as yet because of the small quantity of purified material. The finding that the factor does adsorb to Con A–Sepharose suggests that it contains mannose or glucose, or both, indicating that it is a glycoprotein.

C. A. VILLEE: There would appear to be three possible explanations for the presence of hCG in microbial extracts. One, that it is some kind of hoax, but the fact that this has been found in extracts in several laboratories argues against that. Two, that several kinds of bacteria have the genes to make both the α- and β-subunits of hCG, which seems rather unlikely. Or three, that these are examples of natural recombinant DNA phenomena, in which the genes from some human cell have been incorporated into an appropriate plasmid and have been incorporated into the bacterial genome and are now being transcribed and translated.

COHEN: To exclude possible contamination of the samples by exogenous hCG, the organisms were cultured at Memorial Hospital and Rockefeller University. It is possible that the hCG gene(s) resides in the chromosomal DNA of the bacteria because the data presented by Boime and associates indicate that the two genes are involved. The probability of insertion of two genes in plasmid is rather remote. However, we favor the notion that the genes were inserted into the organism possibly by a process of natural recombination, since the ability to produce hCG-like material appears to be a transient state. In other words, this capability is lost after storage or after a period of repeated culture. Furthermore, the possibility that the genes might exist in close proximity to each other in the same chromosomes has not been excluded.

BIRKEN: (1) Could you please comment as to how you exclude the possibilities of a protease-type artifactual immunoreactivity? (2) In reference to the gel-filtration figure of the bacterial "hCG" which was analyzed by RIA with anti-α, anti-β, and anti-COOH-terminal antisera, the pattern of reactivity can be partially explained by proteolysis of hCG with a trypsinlike enzyme which generates a free COOH-terminal peptide and would result in two peaks of activity with hCG (i.e., a peak with whole hormone and a peak with the COOH-terminal portion). The hCG without the COOH-terminal may still bind to receptor and would result in a split or spread-out receptor-assay peak. What is unclear is that the small-molecular-weight peak also reacts with anti-hCG-β antisera, which is unusual, since such antisera do not generally react with this peptide.

COHEN: (1) Protease activity in the samples was excluded by the addition of the following inhibitors to the assay system: TPCK, soybean trypsin inhibitor, ovalbumin, and TLCK. The addition of the inhibitors did not influence the results obtained with RIA or RRA for hCG. An additional test was performed by heating the sample at 90°C for 30 min, which denatures proteases but does not affect hCG. Treatment of the bacterial material did not influence its immunoreactivity. Furthermore, partially purified bacterial hCG material did not hydrolyze [^{125}I]-hCG, in contrast to an apparent hCG-like factor in the crab stomach. (2) The delayed peak on gel filtration probably contained an hCG-β-like fragment with an intact carboxyl terminal and substantially part of or all of the amino terminal. Although this fraction was not characterized, it might be an aglycosylated hCG-β molecule, hence eluted after hCG-β. The fact that its inactivity with H93 antiserum was greater than that with anti-hCG-β suggests that the carboxyl-terminal segment is intact. Deficiency or absence of carbohydrate moieties might influence its reactivity with anti-hCG-β.

References

Acevedo, H.F., Slifkin, M., Pouchet, G.R., and Pardo, M., 1978, Immunohistochemical localization of a choriogonadotropin-like protein in bacteria isolated from cancer patients, *Cancer* **41:**1217–1229.

Ashitaka, Y., and Koide, S.S., 1974, Interaction of human chorionic gonadotropin, *Fertil. Steril.* **25:**177–184.

Ashitaka, Y., Nishimura, R., Endoh, Y., and Tojo, S., 1974, Subunits of human chorionic gonadotropin and their radioimmunoassays, *Endocrinol. Jpn.* **21:**429–435.

Canfield, R.E., and Morgan, F.J., 1973, Human chorionic gonadotropin (hCG). 1. Purification and biochemical characterization, in: *Methods in Investigative and Diagnostic Endocrinology* (S.A. Berson and R.S. Yallow, eds.), pp. 727–742, North-Holland, Amsterdam.

Chen, H.C., Hodgen, G.D., Matsuura, S., Lin, L.J., Gross, E., Reichert, L.E., Jr., Birkin, S., Canfield, R.E., and Ross, G.T., 1976, Evidence for a gonadotropin from nonpregnant subjects that has physical, immunological, and biological similarities to human chorionic gonadotropin, *Proc. Natl. Acad. Sci. U.S.A.* **73:**2885–2889.

Cohen, H., and Strampp, A., 1976, Bacterial synthesis of substance similar to human chorionic gonadotropin, *Proc. Soc. Exp. Biol. Med.* **152:**408–410.

Diczfalusy, E., and Loraine, J.A., 1955, Sources of error in clinical bioassays of serum chorionic gonadotropin, *J. Clin. Endocrinol. Metab.* **15:**424–434.

Livingston, V.W.C., and Livingston, A.M., 1974, Some cultural, immunological, and biochemical properties of progenitor cryptocides, *Trans. N. Y. Acad. Sci.* **36**(2):569–582.

Maruo, T., 1976, Studies on *in vitro* synthesis and secretion of human chorionic gonadotropin and its subunits, *Endocrinol. Jpn.* **23:**119–128.

Maruo, T., Ashitaka, Y., Mochizuki, M., and Tojo, S., 1974, Chorionic gonadotropin synthesized in cultivated trophoblast, *Endocrinol. Jpn.* **21:**499–505.

Maruo, T., Segal, S., and Koide, S.S., 1979, Studies on the apparent human chorionic gonadotropin-like factor in the crab *Ovalipes ocellatus*, *Endocrinology* **104:**932–939.

Morgan, F.J., Canfield, R.E., Vaitukaitis, J.L., and Ross, G.T., 1974, Properties of the subunits of human chorionic gonadotropin, *Endocrinology* **94:**1601–1606.

Rabson, A.S., Rosen, S.W., Tashjian, A.H., Jr., and Weintraub, B.D., 1973, Production of human chorionic gonadotropin *in vitro* by a cell line derived from a carcinoma of the lung, *J. Natl. Cancer Inst.* **50:**669–674.

Richert, N.D., and Ryan, R.J., 1977, Specific gonadotropin binding to *Pseudomonas maltophilia*, *Proc. Natl. Acad. Sci. U.S.A.* **74:**878–882.

Richert, N.D., Bramley, G.A., and Ryan, R.J., 1978, Hormone binding, proteases and the regulation of adenylate cyclase activity, in: *Novel Aspects of Reproductive Physiology* (C.H. Spilman and J.W. Wilks, eds.), pp. 81–106, SP Medical and Scientific Books, New York.

Saxena, B.B., Hasen, S.H., Haour, F., and Schmidt-Gollwitzer, M., 1974, Radioreceptor assay of human chorionic gonadotropin: Detection of early pregnancy, *Science* **184:**793–795.

Tsuruhara, T., Dafau, M.L., and Hickman, J., 1972, Biological properties of hCG after removal of terminal sialic acid and galactose residues, *Endocrinology* **91:**296–301.

Vaitukaitis, J.L., 1974, Changing placental concentrations of human chorionic gonadotropin and its subunits during gestation, *J. Clin. Endocrinol. Metab.* **38:**755–759.

Vaitukaitis, J.L., Robbins, J.B., Nieschlag, E., and Ross, G.T., 1971, A method for producing specific antisera with small doses of immunogen, *J. Clin. Endocrinol. Metab.* **33:**988–991.

Vaitukaitis, J.L., Ross, G.T., Braunstein, G.D., and Rayford, P.L., 1976, Gonadotropins and their subunits: Basic and clinical studies, *Recent Prog. Horm. Res.* **32:**289–331.
Van Hall, E.V., Vaitukaitis, J.L., and Ross, G.T., 1971, Immunological and biological activity of HCG following progressive desialylation, *Endocrinology* **88:**456–464.
Weber, K., and Osborn, M., 1969, The reliability of molecular weight determinations by dodecyl sulfate–polyacrylamide gel electrophoresis, *J. Biol. Chem.* **244:**4406–4412.
Yoshimoto, Y., Wolfson, A.R., and Odell, W.D., 1977, Human chorionic gonadotropin-like substance in nonendocrine tissues of normal subjects, *Science* **197:**575–577.

VIII

Luteinization, Oocyte Maturation, and Early Pregnancy

27

Comparative Approach to Mechanisms in the Maintenance of Early Pregnancy

Fuller W. Bazer, D. C. Sharp III, W. W. Thatcher, and R. Michael Roberts

1. Endocrine Changes during the Estrous Cycle and Early Pregnancy

1.1. The Pig

The sexually mature female pig (*Sus scrofa, domestica*) has recurring estrous cycles of 18–21 days that are not affected by season. The onset of behavioral estrus is characterized by elevated plasma levels of luteinizing hormone (LH surge) and elevated plasma estrogen (estradiol and estrone) concentrations that precede onset of estrus by 24–48 hr. Ovulation occurs 38–42 hr after onset of estrus, with gilts (virgin females) releasing 10–16 ova. Corpora lutea (CL) are well formed by day 4 or 5 of the estrous cycle, and plasma progesterone concentrations increase from about 5 ng/ml on day 4 to 30–40 ng/ml between days 12 and 14. Luteal regression begins on about day 15 in nonpregnant females, and plasma progesterone concentrations decline to basal levels (1 ng/ml or less) by day 17–18. This cycle repeats itself unless interrupted by

Fuller W. Bazer and *D. C. Sharp III* • Department of Animal Science, University of Florida, Gainesville, Florida 32611. *W. W. Thatcher* • Department of Dairy Science, University of Florida, Gainesville, Florida 32611. *R. Michael Roberts* • Department of Biochemistry and Molecular Biology, University of Florida, Gainesville, Florida 32611.

pregnancy or events leading to endocrine dysfunction. These general comments concerning the porcine estrous cycle are based on data presented by Guthrie *et al.* (1972) and Anderson (1974).

Mating occurs during the period of estrus, which lasts for 24–72 hr, and if the female becomes pregnant, a gestation period of 114 days follows. The hormonal events associated with estrus and the first 14 days of the estrous cycle and pregnancy are essentially identical. After that time, however, functional CL must be maintained for the duration of pregnancy. Loss of CL function at any stage of gestation leads to abortion within 24–36 hr (Belt *et al.*, 1971). Plasma progesterone levels of 30–40 ng/ml on days 12–14 of pregnancy decrease to 20–25 ng/ml by day 25 pregnancy (Guthrie *et al.*, 1974). Other research has indicated plasma progesterone concentrations of 10–20 ng/ml by day 20 of gestation (Robertson and King, 1974; Knight *et al.*, 1977), which then remain fairly constant until about day 100 of gestation, when they begin to decline slowly to parturition (day 114–115), when they decrease abruptly to less than 1 ng/ml.

Prolonged CL maintenance occurs in nonpregnant pigs if they are bilaterally hysterectomized by day 12 of the estrous cycle (Anderson *et al.*, 1969) or if they are injected with estradiol valerate (E_2V) (5 mg/day) on days 11 through 15 of the estrous cycle (Frank *et al.*, 1977). The source of the uterine luteolysin in pigs, presumably prostaglandin $F_{2\alpha}$, is removed in hysterectomized females, and its release into the uterine venous drainage appears to be blocked by E_2V (Frank *et al.*, 1977).

Plasma estradiol and estrone, both free and conjugated forms, change dramatically during the course of pregnancy in pigs. Estrone sulfate is, however, the predominant plasma form of estrogen during pregnancy. This hormone increases in plasma from day 16 (60 pg/ml) to day 30 (>3 ng/ml), decreases to day 46 (35 pg/ml), increases slightly to day 60, and then increases steadily to day of parturition (>3 ng/ml) (Robertson and King, 1974). Knight *et al.* (1977) reported the same pattern for unconjugated estrone and estradiol in uterine-vein and peripheral plasma and amniotic and allantoic fluids. It seems clear that these free and conjugated estrogens are of conceptus origin (see Knight *et al.*, 1977). The time of initiation of estrogen production by the pig blastocyst is discussed in more detail in Section 2.1.2.

There are several key events in conceptus development during pregnancy that must be mentioned to put subsequent discussion into proper perspective. The pig embryo moves from the oviduct into the uterus at about the 4-cell stage, i.e., 60–72 hr after onset of estrus. The embryo reaches the blastocyst stage by day 5, and the zona pellucida is shed (hatching) between days 6 and 7. The exposed blastocyst expands from about 0.5–1 mm in diameter at hatching to 2–6 mm in diameter on day 10 and then elongates rapidly to a threadlike organism of 700–1000 mm in length by days 14–16 of pregnancy (Anderson, 1978). These long, filamentous blastocysts lie end to end and follow the contour of the deep uterine endometrial folds. Rapid expansion and development of the allantois occurs between days 18 and 30 of gestation. Fusion of the chorion and allantois takes place between days 30 and 60 of pregnancy, and by days 60–70, placental development is complete. The pig blastocyst is said to undergo a central type of implantation; however, there is no invasion of the maternal uterine endometrium and the term "placentation" is more appropriate. Placentation in the pig involves interdigitation of microvilli on the surface of the trophoblast and epithelial cells lining the uterine endometrium. Since placentation is superficial in the pig, as it is in the cow, ewe, and mare, the direct transfer of nutrients or histotroph from endometrial surface and glandular epithelium to the chorioallantois occurs at least through the second trimester of gestation. The role of the blastocyst in the establishment and maintenance of pregnancy is discussed in Section 2.1.2.

1.2. The Mare

The sexually mature mare (*Equus caballus*) is seasonally polyestrous, with recurring estrous cycles during the spring, summer, and early fall. The anestrous season is from about mid-October through March or April. The length of the estrous cycle is 21–24 days, with estrus lasting 5–7 days. Mares ovulate about 2 days prior to the end of the estrus. Estrogen concentrations in urine of nonpregnant mares increase for 6–7 days before ovulation to peak levels 1–2 days before to the day of ovulation and then decline (Hillman and Loy, 1975; Palmer and Terqui, 1977). Plasma LH concentrations of 10–15 ng/ml during diestrus increase to 30–40 ng/ml on the day of ovulation and days 1–2 postovulation (Evans and Irvine, 1975). Plasma progesterone concentrations increase rapidly from day of ovulation ($\approx$1 ng/ml) to days 6–12 postovulation (8–10 ng/ml) and then decline to basal levels ($<$1 ng/ml) if pregnancy is not established (Sharp and Black, 1973).

Burns and Fleeger (1975) reported that equine plasma progestin concentrations increased from about 1 ng/ml at ovulation to 10–15 ng/ml by day 30 of pregnancy and remained at about that concentration until days 50–70 of pregnancy, when values increased to 20–30 ng/ml. The increase in plasma progestin concentrations between days 30 and 50 coincides with the formation of accessory CL on the mare ovary. Data of Burns and Fleeger (1975) were extended by Ganjam *et al.* (1975) to indicate that plasma progestin concentrations decreased from about day 70 (10–20 ng/ml) to days 140–260 (5–10 ng/ml), increased to 15–20 ng/ml from day 300 to term, and fell precipitously to about 1 ng/ml by 24 hr postpartum. Burns and Fleeger (1975) and Lovell *et al.* (1975) found that progesterone accounted for only a portion (40–60%) of the plasma progestins. Atkins *et al.* (1974) suggested that 5 α-dihydroprogesterone is a metabolite of progesterone from the fetal–placental unit. Young Lai (1971) reported that 17α-hydroxyprogesterone, 20α-dihydroprogesterone, and pregnenolone were present in equine luteal tissue. One or more of these metabolites probably accounts for the discrepancy in values when progestins and progesterone are reported (Burns and Fleeger, 1975). Nett *et al.* (1975) reported plasma estrone values of 10–20 pg/ml between days 0 and 90 of gestation and then a steady increase to days 210–240 (600–800 pg/ml), followed by a decrease to term. Estradiol followed the same pattern, but maximum plasma concentrations of only 200 pg/ml were measured. Data are not available to characterize changes in plasma concentrations of conjugated estrogens during the course of gestation in the mare. In addition to estrone and estradiol, equilin and equilenin are found in the plasma of pregnant mares and follow the same temporal changes of estrone and estradiol (Nett *et al.,* 1975).

Pregnant mare serum gonadotropin is produced by the chorionic cells of the endometrial cups between days 30 and 150 of gestation (Allen, 1975). This is the only well-characterized gonadotropinlike hormone known to be produced by the trophoblast or chorion, or both, of domestic animals.

1.3. The Cow

The sexually mature bovine female (*Bos taurus*) exhibits recurring estrous cycles of 20–21 days that are not affected by the season of the year. Onset of estrus is associated with plasma LH concentrations of 10–25 ng/ml above basal levels of about 1 ng/ml. Plasma estrogens increase from about 3 days before onset of estrus to peak levels on the day of onset of estrus ($\approx$10 pg/ml), and then decline to basal levels of 1–3 pg/ml. Ovulation occurs about 30 hr after onset of estrus, and then plasma progesterone concentrations begin to increase from less than 1 ng/ml to 6–10 ng/ml on days 12–15 of the estrous cycle and then decline to 1 ng/ml or less about 2 days before onset of the next estrus. These general comments are based on data reported by Foote (1974) and Chenault *et al.* (1976). In the pregnant cow, plasma progesterone concentrations of about 10 ng/ml are maintained from around days 14–16 of gestation to term. Estrone sulfate, the major form of estrogen in the pregnant cow, is present in highest concentrations in allantoic fluid. Eley *et al.* (1979a) reported allantoic-fluid estrone sulfate concentrations ($\overline{X} \pm$ S.E.M.) to increase from day 33 (0.21 $\pm$ 0.03 ng/ml) to days 70 (6.65 $\pm$ 2.90 ng/ml) and 111 (220.3 $\pm$ 2.07 ng/ml) of pregnancy. Robertson and King (1979) extended these observations. They indicated an initial increase in estrone sulfate ($\overline{X}$) in allantoic fluid from day 41 (0.4 ng/ml) to day 132 (475 ng/ml), followed by a sharp decrease to about 50 ng/ml on day 170. Then a second increase occurred between days 200 and 250 (150–200 ng/ml), followed by a decrease to term (≈ 100 ng/ml). These data indicate that estrogen production by the fetal–placental unit undergoes marked changes during the course of gestation.

Bovine placental lactogen (bPL) has been isolated from placental cotyledons (Bolander and Fellows, 1976; Buttle and Forsyth, 1976), but not amnion, allantochorion, endometrium, or fetal skin (Buttle and Forsyth, 1976). Serum levels of bPL in dairy and beef cows appear to be less than 50 ng/ml prior to day 160 of pregnancy, but increase to an average of 1103 ± 342 ng/ml ($\overline{X}$ ± S.D.) by day 200 of gestation for dairy cattle and remain at that level until term. Peak levels in beef cattle were only 650 ± 37 ng/ml, but temporal changes in serum bPL were parallel in beef and dairy breeds of cattle (Bolander *et al.*, 1976). Recently, Flint *et al.* (1979) detected bPL in the preattachment blastocyst of cattle.

1.4. The Ewe

The sexually mature female sheep (*Ovis aries*) is seasonally polyestrous during late summer, fall, and, in some breeds, early winter. Estrous-cycle length ranges from 14 to 19 days, with an average length being 17 days. Estrus lasts 24–48 hr and is preceded by a rise in peripheral plasma total estrogens from basal levels (20–50 pg/ml) to peak levels of 600–700 pg/ml the day preceding and the day of onset of estrus. Peak plasma levels of LH (20–100 ng/ml) occur near the time of onset of estrus and persist for only 6–9 hr. Basal LH concentrations (about 1 ng/ml) are found during the remainder of the cycle. Plasma progesterone concentrations begin to exceed basal levels (<1 ng/ml) on day 3–4 of the estrous cycle, increase to about 2 ng/ml by day 12, and then rapidly decrease to near basal levels again by day 15–16. These general observations relative to the estrous cycle are based on data presented by Terrill (1974) and Hansel *et al.* (1973).

During the course of pregnancy, plasma progesterone concentrations are similar to those found during the luteal phase of the estrous cycle, i.e., about 2 ng/ml, up to about day 60 of gestation. The levels of progesterone then increase to 4–8 ng/ml by days 60–100. A further increase to 16–20 ng/ml occurs between days 60 and 140 due to placental progesterone production. After day 140, plasma progesterone levels decline to time of parturition (145–150 days) according to Kelly *et al.* (1974). Estrone concentrations in peripheral plasma are low (10–20 pg/ml) from days 20 to 110 of gestation and then increase to 40–60 pg/ml by term (Carnegie and Robertson, 1978; M. Arriojas, D. H. Clark, F. W. Bazer, W. W. Thatcher, and D. H. Barron, unpublished data). Estradiol values follow the same pattern, but are about 50% lower than those for estrone.

Estrone sulfate is the major form of estrogen in the pregnant ewe according to Carnegie and Robertson (1978) and M. Arriojas, D. H. Clark, F. W. Bazer, W. W. Thatcher, and D. H. Barron (unpublished data), and the highest concentrations are present in allantoic fluid. Estrone sulfate concentrations in allantoic fluid increase from day 20 (30–50 pg/ml) to a first peak on day 46 (14,200 pg/ml), and then decrease to day 55 (850 pg/ml). A second estrone sulfate peak occurs between days 100 and 120 (10,000–15,000 pg/ml), which is followed by a decline to day 140 (1000–2000 pg/ml). Data are not available to indicate whether or not estrone sulfate in allantoic fluid increases again near time of parturition as do plasma estrogen levels (Challis, 1971; Robertson and Smeaton, 1973).

Ovine placental lactogen (oPL) is known to be produced by the ovine chorion (Martal and Djiane, 1977; Carnegie *et al.*, 1977). This hormone has prolactin and growth-hormone activity (Martal, 1978; Reddy and Watkins, 1978). Martal and Djiane (1977) and Martal and Lacroix (1978) have suggested that oPL may be a component of the "luteotrophic" complex in pregnant ewes.

oPL concentrations in maternal plasma reach levels of 2 ng/ml by day 48 and increase steadily to 800–1000 ng/ml between days 130 and 140 and then decrease to term. Although undetectable in plasma before days 45–50 with current assay techniques, oPL has been extracted from trophoblast tissue of days 16 and 17 blastocysts (Martal and Djiane, 1977) and detected in day 16 trophoblast tissue by immunofluorescent microscopy (Carnegie *et al.*, 1977).

Martal *et al.* (1979) reported the presence in sheep of a trophoblast factor, "trophoblastin," that has antiluteolytic properties when infused into the uterine lumen of nonpregnant ewes. The nature of this proposed protein and its mode of action are not known.

2. The Uterus, Corpus Luteum Maintenance, and Maternal Recognition of Pregnancy Signals in Domestic Animals

2.1. The Pig

2.1.1. The Uterus and Corpus Luteum Maintenance

Loeb (1923) first noted that hysterectomy of guinea pigs during the luteal phase of the estrous cycle allowed prolonged CL maintenance. This was later demonstrated in pigs by Spies *et al.* (1958), du Mesnil du Buisson and Dauzier (1959), and Anderson *et al.* (1961), who found that bilateral hysterectomy in the early to midluteal phase of the estrous cycle resulted in CL maintenance for periods equal to or longer than the 114 days of a normal pregnancy. Corrosive and toxic agents that destroy the integrity of the endometrium or congenital absence of the uterine endometrium also leads to prolonged CL maintenance (see the review by Anderson *et al.*, 1969). Thus, the pig uterine endometrium was identified as the source of a substance that would cause morphological regression of CL and cessation of progesterone secretion. This substance was designated the uterine luteolysin.

Schomberg (1967) initially reported that a component of uterine flushings from pigs had a luteolytic effect on cultured granulosa cells, but this effect was later shown to be cytolytic (Schomberg, 1969).

Pharriss and Wyngarden (1969) proposed that prostaglandin $F_{2\alpha}$ ($PGF_{2\alpha}$) was luteolytic in rats, and subsequently, $PGF_{2\alpha}$ was evaluated as a luteolytic agent in swine. Diehl and Day (1973) found 2–5 mg $PGF_{2\alpha}$ to be ineffective in shortening the estrous cycle when administered into the uterus on day 10 or intramuscularly (i.m.) on day 12 of the cycle. Hallford *et al.* (1974) injected gilts with 20 mg $PGF_{2\alpha}$ i.m. at 12 hr intervals on days 4 and 5 (a total dose of 80 mg $PGF_{2\alpha}$) with no effect on the length of the estrous cycle. However, when the same injection protocol and dosage were used on days 12 and 13, the length of the estrous cycle was reduced in treated (17.2 days) as compared to control (18.6 days) gilts. Douglass and Ginther (1975a) found that injections of 10 or 20 mg $PGF_{2\alpha}$ i.m. on day 12 of the estrous cycle reduced CL weight, but neither estrus nor ovulation occurred by day 16 of the cycle. In contrast to these results, injection of 5 mg $PGF_{2\alpha}$ i.m. into early pregnant (day 25–30) gilts caused plasma progesterone levels to decrease from 10.2 to 2.5 ng/ml within 12 hr, and abortion and estrus occurred within the next 72 hr (Diehl and Day, 1973). Furthermore, Kraeling *et al.* (1975) reported that CL maintained in intact nonpregnant gilts, by administration of 5 mg estradiol valerate daily from days 10 to 15 of the estrous cycle, were susceptible to luteolysis (10–12 gilts) when 10 mg PGF was administered on day 20.

The results of these initial studies on the effect of $PGF_{2\alpha}$ on porcine CL must now be considered in light of some previous work. du Mesnil du Buisson and Leglise (1963) reported that CL develop normally in pigs hypophysectomized within 3 hr after onset of estrus. This was confirmed by Anderson and Melampy (1967). They reported that CL of pigs hypophysectomized on the first day of estrus were apparently normal up to day 12 of the cycle, but were undergoing regression by day 16. From these data, it was concluded that pig CL require only initial pituitary support at or about the time of onset of estrus for formation and function for the first 12 days of the estrous cycle. In subsequent studies, it was shown that CL maintenance could be extended beyond 12 days in hysterectomized pigs that were later hypophysectomized and given either bovine LH (5 mg/day), human chorionic gonadotropin (hCG) at 1000 IU/day, or 250 mg/day of an acid–acetone extract of ovine pituitary (du Mesnil du Buisson and Leglise, 1963; du Mesnil du Buisson, 1966; Anderson *et al.*, 1965). The gonadotropin preparations were not effective in gilts that had intact uteri, which indicated that the uterine luteolytic factor could "override" the pituitary gonadotropin (luteotropin) after day 12 of the estrous cycle.

Since the porcine CL appears to be "autonomous" for the first 12 days of the estrous cycle, relative to a need for luteotrophic support, the absence of a marked luteolytic effect of $PGF_{2\alpha}$ observed by Diehl and Day (1973), Hallford *et al.* (1974), and Douglass and Ginther (1975a) could be due to refractoriness of the CL to $PGF_{2\alpha}$ through day 12. That is, the pig CL become susceptible to exogenous $PGF_{2\alpha}$ only at about the time they would begin to regress normally, and the only effect of exogenous $PGF_{2\alpha}$ is to cause a slight reduction in the length of the estrous cycle. Moeljono *et al.* (1976) therefore used the hysterectomized gilt as a model for assessing the luteolytic effect of $PGF_{2\alpha}$ on CL that would normally be regressing, i.e., day 17 after onset of estrus.

Moeljono *et al.* (1976) reported that $PGF_{2\alpha}$ administered i.m. to hysterectomized gilts at 08:00 hr (10 mg) and 20:00 hr (10 mg) on either day 17 or 38 postestrus resulted in expression of estrus at an average of 88.0 ± 13.5 hr after first injection. Saline-treated controls did not exhibit estrus for at least 60 days posttreatment. Two gilts were bilaterally hysterectomized and the saphenous artery catheterized on day 7 after onset of estrus, and $PGF_{2\alpha}$ was administered, as previously described, on day 17. Plasma progesterone concentrations fell precipitously and the gilts exhibited estrus between 90 and 110 hr after the first $PGF_{2\alpha}$ injection. An additional gilt was bilaterally hysterectomized and the CL marked with India ink. Treatment with $PGF_{2\alpha}$, as before, resulted in estrus 92 hr after first $PGF_{2\alpha}$ injection, and on day 4 of the treatment cycle, regression of marked CL to corpora albicantia and presence of newly formed CL were confirmed. These data indicate that $PGF_{2\alpha}$ is, by definition, luteolytic in swine, since it causes CL regression and a decline in progesterone secretion followed by onset of behavioral estrus.

Henderson and McNatty (1975) suggested that CL may remain refractory to $PGF_{2\alpha}$ until such time as LH begins to dissociate from luteal membrane receptors. At this time, conformational changes within the membranes facilitate $PGF_{2\alpha}$ binding. The $PGF_{2\alpha}$, in turn, alters the adenylate cyclase system to inhibit progesterone secretion and activates lysosomal enzymes that lead to lysing of the luteal cells. Support for such a mechanism in swine is based on research that indicated that CL induced with hCG on days 6, 8, 10, and 16 of the estrous cycle had a functional life-span of 12.5, 14.5, 15.0, and 19.3 days, respectively, from the day they were induced (Caldwell *et al.*, 1969). Therefore, only those CL induced on day 6 were old enough to be susceptible to endogenous luteolytic agents and undergo simultaneous regression with the natural CL.

Moeljono *et al.* (1976), again using bilaterally hysterectomized gilts, administered $PGF_{2\alpha}$ at 08:00 hr (10 mg) and 20:00 hr (10 mg) on either day 8, 11, 14, or 17 after onset of estrus. None of the gilts treated on days 8 and 11 exhibited estrus for at least 60 days thereafter. Two of three gilts treated on day 14 and all three gilts treated on day 17 exhibited estrus ($\bar{X} \pm$ S.E.M.) at 116.0 ± 9.8 hr posttreatment. Krzymowski *et al.* (1976) infused a total of 2 mg $PGF_{2\alpha}$ into the anterior uterine vein of sows over a 10-hr period on either day 6, 8, 10, 12, 14, or 15 of the estrous cycle. On the basis of CL weight and plasma progesterone concentration, $PGF_{2\alpha}$ had no luteolytic effect when given on day 6, 8, or 10, but caused luteolysis when given on day 12, 14, or 15. These data clearly support the concept that porcine CL are refractory to the luteolytic effect of $PGF_{2\alpha}$ up to day 12 of the estrous cycle. This refractoriness may be associated with LH binding to the luteal cells.

2.1.2. Maternal Recognition of Pregnancy Signals

The CL of pigs must be maintained and continue to secrete progesterone if pregnancy is to be established and maintained to term (du Mesnil du Buisson and Dauzier, 1957). In the pig, therefore, "maternal recognition of pregnancy" signals are those, presumably from the conceptus, that act to prevent the uterus from exerting a luteolytic effect on the CL. On the basis of work from our laboratory, a theory has been developed relative to a mechanism whereby an agent (estrogen) produced by the embryo prevents $PGF_{2\alpha}$ secretion from the uterine endometrium into the uterine venous drainage, where it could gain access to the CL and cause luteolysis (Bazer and Thatcher, 1977).

The following assumptions were made and may help to put the following discussion into proper context: (1) $PGF_{2\alpha}$ is synthesized and secreted by the epithelial cells of the uterine endometrium; (2) $PGF_{2\alpha}$ can move either into the uterine lumen (exocrine direction) or toward the endometrial stroma and associated vasculature (endocrine secretion); (3) the direction of movement of $PGF_{2\alpha}$ is determined by the local concentration of estrogen established by the trophoblast (chorion) in the pregnant pig; and (4) $PGF_{2\alpha}$ is the uterine luteolysin in swine.

Moeljono *et al.* (1976) established that $PGF_{2\alpha}$ is luteolytic in pigs, and Patek and Watson (1976) demonstrated that the uterine endometrium produces $PGF_{2\alpha}$.

Studies of uteroovarian-vein plasma immunoreactive prostaglandin F (PGF) concentrations indicated that elevated levels of PGF were temporally related to CL regression and declining plasma progesterone levels in nonpregnant gilts (Gleeson *et al.*, 1974; Moeljono *et al.* 1977). This relationship was not observed in pregnant gilts. Moeljono *et al.* (1977) found that basal and peak PGF concentrations in uteroovarian-vein plasma were significantly lower in pregnant than in nonpregnant gilts and that plasma progesterone levels in pregnant gilts were maintained. Uteroovarian-vein plasma estradiol and estrone concentrations were also studied, and a transient rise in one or both of these estrogens between days 13 and 15 was detected in all pregnant gilts, but in none of the nonpregnant gilts. This observation is consistent with evidence that the pig blastocyst begins to produce estrogen by day 12 of pregnancy (Heap *et al.*, 1979) and that plasma conjugated estrogens increase steadily from days 14 to 30 of gestation (Robertson and King, 1974). Heap *et al.* (1979) suggested that estrogen produced by the porcine blastocyst is the luteotrophic agent in swine. Available evidence makes it clear that plasma and urinary estrogens of conceptus origin are increasing between days 12 and 30 of gestation in pigs (Lunaas, 1962; Bowerman *et al.*, 1964; Robertson and King, 1974; Knight *et al.*, 1977).

Kidder *et al.* (1955) first reported that injection of diethylstilbestrol (DES) on day 11 of the estrous cycle consistently led to prolonged CL maintenance. Gardener *et al.* (1963) also found that DES, estradiol-17β, or estrone injection on day 11 of the estrous cycle would prolong CL maintenance. Anderson *et al.* (1965) demonstrated that LH (5 mg/day) allows for luteal maintenance in hypophysectomized sows only if the uterus is removed or LH is given concurrently with estrogen. du Mesnil du Buisson (1966) studied the effect of estrogen in nonpregnant, hypophysectomized gilts in which the uterus was intact. Injection of 5 mg estradiol valerate (E_2V) and 5 mg LH daily from days 12 to 20 allowed maintenance of CL of normal size. He concluded that since the pituitary is absent, estrogen is not acting by modifying a gonadotropic factor, but is inhibiting the luteolytic action of the uterus by blocking secretion or excretion of the uterine luteolysin.

Data from our laboratory (Frank *et al.*, 1977) indicate that uteroovarian-vein basal PGF concentrations, PGF peak concentrations ($> \overline{X} + 2$ S.D.), and number of PGF peaks were significantly lower between days 12 and 20 after onset of estrus in gilts treated with E_2V (5 mg/day) on days 11 through 15 after onset of estrus as compared to control gilts receiving injections of corn oil. The reduced uteroovarian-vein PGF levels in E_2V-treated gilts were associated with an interestrus interval of 146.5 days as compared to 19.0 days for control gilts. This study raised the question as to whether estrogen influenced the production of PGF or its direction of secretion.

Before we discuss the next point, an observation based on studies of the pattern of synthesis, secretion, and direction of movement of a progesterone-induced purple protein (uteroferrin) by the pig uterus is in order. Uteroferrin is synthesized and secreted by the surface and glandular epithelium of the uterine endometrium (Chen *et al.*, 1975). In nonpregnant gilts, this protein is secreted into the lumen of uterine glands through day 13 of the estrous cycle. Beginning on day 14, however, uteroferrin begins to become localized in the endometrial stroma surrounding the basement membrane of the uterine glands. This protein could then enter the interstitial fluid and vascular system within the endometrial stroma. In contrast to this pattern of movement of uteroferrin in nonpregnant females, uteroferrin was not found in the endometrial stroma at any of the stages of pregnancy studied, i.e., day 6, 8, 12, 14, 16, 18, 30, 50, 70, or 90 of gestation. In the pregnant animal, therefore, uteroferrin, a component of uterine histotroph, continues to be secreted into the uterine lumen, where it is available to the developing conceptus. These observations by Chen *et al.* (1975) provided the basis for the question of whether or not estrogen might prevent CL regression by affecting the direction of PGF secretion.

Changes in tissue localization of PGF, specifically, during the menstrual cycle were reported by Ogra *et al.* (1974). They used immunofluorescent antibody procedures to localize PGF in oviductal epithelium of women. PGF was localized in the mucosa during the preovulatory phase of the menstrual cycle, but during the postovulatory phase, PGF was localized in the lamina propria.

A study was conducted by Frank *et al.* (1978) to determine total protein, uteroferrin (acid phosphatase activity), and PGF in uterine flushings obtained from gilts during a control (corn-oil-treated) and E_2V-treatment cycle on days 11, 13, 15, 17, and 19 after onset of estrus. The same gilts were assigned to serve on a given day during both the control and treatment periods and therefore served as their own controls. Total recoverable PGF per uterine horn during control vs. treatment periods, respectively, was: day 11, 2.0 vs. 1.9 ng; day 13, 27.1 vs. 18.4 ng; day 15, 31.3 vs. 2022.6 ng; day 17, 210.2 vs. 4144.3 ng; day 19, 66.2 vs. 4646.7 ng. Uteroferrin was also present in greater amounts on days 17 and 19 of the treatment period. As in the previous study, the length ($\overline{X} \pm$ S.E.M.) of the control estrous cycle was only 19.4 ± 0.3 days, whereas the interestrus interval after E_2V treatment was 92.0 ± 11.2 days. In the previous study, Frank *et al.* (1977) found markedly reduced uteroovarian-vein basal PGF concentrations and numbers of PGF peaks after E_2V treatment, while this study indicated a linear 2000-fold increase in total recoverable PGF per uterine horn between days 11 and 19 after E_2V treatment. These data are consistent with the theory that E_2V treatment does not inhibit PGF synthesis and secretion into the uterine lumen, but does reduce PGF secretion into the uterine venous drainage. Therefore, the CL are "protected" from the uterine luteolysin, as suggested by du Mesnil du Buisson (1966), since it becomes sequestered in the uterine lumen.

The next experiment was designed to determine whether or not PGF accumulated within the uterine lumen of pregnant gilts (Zavy *et al.*, 1979b). In this study, control uterine flushings were obtained from four gilts each on days 6, 8, 10, 12, 14, 15, 16, and 18 of the estrous cycle as described by Bazer *et al.* (1978). Gilts were allowed one estrous cycle for recovery and then mated at 12 and 24 hr after onset of the next estrus. Uterine flushings were then obtained from the same gilts representing their respective days during pregnancy as confirmed by the presence of blastocysts in uterine flushings. Concentrations of estrone, estradiol, and PGF in the uterine flushings ($\approx$ 40 ml of total fluid from the two uterine horns) are summarized in Table I. For the purpose of comparing total recoverable PGF values ($\bar{X} \pm$ S.E.M.) with those from E_2V-treated gilts, in this experiment, total recoverable PGF per uterus increased from day 6 (67.6 ± 27.3 ng) to day 16 (827.6 ± 347.9 ng) and decreased slightly to day 18 (820.9 ± 126.6 ng) of the estrous cycle. In the pregnant gilts, total recoverable PGF increased slightly from day 6 (13.7 ± 3.3 ng) to day 10 (16.7 ± 4.7 ng) and then steadily to day 18 (22,466.3 ± 1761.1 ng). These values should be divided by two for comparison with the data of Frank *et al.* (1978), who expressed the data on a per-uterine-horn basis. In the pregnant gilt on day 18, there is about 11 μg PGF per uterine horn as compared with about 5 μg per uterine horn in the nonpregnant E_2V-treated gilts. This difference may be attributed to differences among animals, experiments, and other possible factors. However, it is also possible that the porcine conceptuses are

Table I. Prostaglandin F and Estrone Concentrations in Uterine Flushings from Nonpregnant and Pregnant Gilts

	PGF (ng/ml)[a]		Estrone (pg/ml)[a]	
Day[b]	NP	P	NP	P
6	1.4 ± 0.5	0.3 ± 0.1	10.9 ± 3.6	6.0 ± 0.7
8	0.6 ± 0.1	0.5 ± 0.1	7.8 ± 2.1	36.2 ± 17.7
10	0.5 ± 0.1	0.4 ± 0.1	7.2 ± 1.3	6.8 ± 0.3
12	0.9 ± 0.2	2.6 ± 1.2	5.0 ± 0.1	22.3 ± 10.4
14	4.5 ± 1.3	60.2 ± 31.3	5.2 ± 0.2	47.6 ± 22.0
15	15.0 ± 11.3	40.2 ± 18.7	8.2 ± 2.0	46.5 ± 9.6
16	18.6 ± 7.0	357.6 ± 148.1	5.9 ± 0.7	74.3 ± 33.8
18	16.8 ± 2.3	464.5 ± 31.5	22.4 ± 4.4	76.3 ± 14.4

[a] Flushing volume recovered was about 40 ml regardless of day or status of female. Results are means ± S.E.M. (NP) Nonpregnant; (P) pregnant.
[b] Days after onset of estrus (day of onset of estrus = day 0).

synthesizing and secreting PGF in addition to steroid. Lewis *et al.* (1979) reported synthesis of prostaglandins from arachidonic acid by bovine blastocysts.

The theory of maternal recognition of pregnancy is based on available evidence to indicate that estrogen, produced by the blastocyst from about day 12 to day 30 of pregnancy or injected as E_2V, reduces the release of the porcine uterine luteolysin, $PGF_{2\alpha}$, in an endocrine direction. Significantly lower basal PGF levels, PGF peak concentrations, and frequencies of PGF peaks in pregnant and E_2V-treated gilts support this point. The rate of secretion of PGF by the porcine uterine endometrium does not appear to be inhibited in the pregnant or E_2V-treated gilt; rather, the PGF secreted appears to be sequestered within the uterine lumen.

The mechanism whereby PGF can be sequestered within the uterine lumen is not known. However, evidence that PGF can be sequestered is available. Data from our laboratory have been presented to indicate that total PGF is markedly elevated within the uterine lumen of E_2V-treated gilts (Frank *et al.*, 1978) and pregnant gilts (Zavy *et al.*, 1979b). Harrison *et al.* (1972) reported recovering from 150 ng to 7.6 mg of authentic $PGF_{2\alpha}$ (analysis by gas chromatography–mass spectrometry) from fluid within the uterine horn of sheep after autotransplantation of the ovary to the neck. This finding was confirmed with ewes in which a uterine pouch was surgically constructed before breeding (Harrison *et al.*, 1976). Uterine fluid from these ewes had from 100 to 313,700 ng total $PGF_{2\alpha}$ sequestered. Bazer *et al.* (1979) reported mean (± S.E.M.) content of 222.8 ± 68 μg in uterine fluid of unilaterally pregnant ewes at day 140 of gestation. Harrison *et al.* (1976) were unable to detect $PGF_{2\alpha}$ (<2 ng/ml) in jugular-vein or uterine-vein plasma. Thatcher *et al.* (1979) reported that total recoverable PGF in bovine uterine flushings increased from 13.9 ng on day 8 to 111.0 ng on day 19 of the estrous cycle. However, total recoverable PGF increased from 17 ng on day 8 to 1188 ng on day 19 of pregnancy, i.e., a 10-fold higher quantity on day 19 of pregnancy as compared to day 19 of the estrous cycle.

In the pig, allantoic fluid appears to sequester substantial amounts of PGF (M. P. E. Moeljono, F. W. Bazer, and W. W. Thatcher, unpublished data). Based on six to eight samples from various days of pregnancy, PGF content (concentration × allantoic fluid volume/conceptus, $\bar{X}$ ± S.E.M.) increased from day 20 (0.4 ± 0.04 μg) to day 35 (98.4 ± 4.5 μg), decreased to day 40 (46.1 ± 1.9 μg), increased to day 60 (76.0 ± 4.5 μg), and then decreased steadily to day 100 (8.1 ± 0.2 μg). Bovine allantoic fluid total PGF has been shown to increase steadily from days 27 (48.9 ± 12.0 ng) to 111 (4228.9 ± 177.2 ng), and total PGF (ng) in bovine amniotic fluid also increases from days 60 (206.9 ± 109.9) to 111 (8111.2 ± 1918.9) according to Eley *et al.* (1979a).

Available data clearly indicate that significant quantities of PGF can be sequestered or compartmentalized, or both, within the female reproductive tract that has been subjected to the proper endocrine environment. As pointed out by Bito *et al.* (1976), "Physiologists who study membrane transport accept that prostaglandins (PGs) are carrier-mediated while physiologists studying the role of PGs have been skeptical and, in fact, assume that PGs freely penetrate cell membranes."

The assumption of "free movement" of prostaglandins appears incorrect. Facilitated transport of $PGF_{2\alpha}$ from the lumen of the rabbit vagina has been demonstrated (Bito and Spellane, 1974). Bito (1972) provided evidence that rabbit and rat uterine tissue accumulated PGE_1, $PGF_{1\alpha}$, and PGA_1, and that tissue from the gravid rat uterus accumulated more $PGF_{2\alpha}$ than that from the nongravid uterus.

The precise mechanism whereby estrogen of blastocyst origin or injected as E_2V prevents regression is unclear. We have proposed that secretion of $PGF_{2\alpha}$ in nonpregnant pigs is primarily in an endocrine direction, i.e., toward the uterine vasculature, whereby it reaches the CL and leads to their demise. In the pregnant gilt, however, $PGF_{2\alpha}$ secretion is in an exocrine direction; it is sequestered within the uterine lumen and is thereby prevented from entering the uterine venous drainage and ultimately affecting the CL. That the luteostatic effect of estrogen is at the level of the uterus in swine is supported by several lines of evidence: (1) it is well established that simple removal of the uterus from otherwise intact gilts, without hormone therapy, allows for prolonged CL maintenance; (2) CL of pregnant gilts (Diehl and Day, 1973) and E_2V-treated gilts (Kraeling *et al.*, 1975) are susceptible to the luteolytic effect of $PGF_{2\alpha}$; (3) production of $PGF_{2\alpha}$ by endometrium from pregnant gilts is equal to or greater than that for nonpregnant gilts (Zavy *et al.*, 1979b); and (4) LaMotte (1977) has concluded that $PGF_{2\alpha}$ is secreted into the uterine lumen and transferred from there to uterine venous circulation.

The fate and role of $PGF_{2\alpha}$ sequestered within the uterine lumen of pregnant gilts have not been extensively studied. Walker *et al.* (1977) have shown that the conceptus membranes, especially the amnion, have a high capacity for metabolizing $PGF_{2\alpha}$ to 13,14-dihydro-15-keto-$PGF_{2\alpha}$, which is believed to be biologically inactive. On the other hand, Saksena *et al.* (1976) suggest that PGF is required for implantation in mice.

2.2. The Mare

2.2.1. The Uterus and Corpus Luteum Maintenance

Ginther and First (1971), Stabenfeldt *et al.* (1972), and Squires *et al.* (1974) provided data to indicate that the primary CL is maintained for up to 140 days in bilaterally hysterectomized mares. In pregnant mares, the primary CL and accessory CL, i.e., CL formed between days 40 and 50, regress between days 140 and 210 of gestation, and primary CL weight at day 140 is greater for pregnant than for hysterectomized mares (Squires *et al.*, 1974). The gestation period in mares is about 335 days. Progesterone of placental origin is adequate for pregnancy maintenance after the CL regression.

$PGF_{2\alpha}$ causes luteolysis when injected into mares (Douglas and Ginther, 1972; Noden *et al.*, 1973; Allen and Rowson, 1973) in doses as small as 1.25 mg (Douglas and Ginther, 1975b). A 1.25-mg dose of $PGF_{2\alpha}$ given on day 32 of pregnancy will cause termination of pregnancy and pseudopregnancy (induced by embryo removal on day 24) (Kooistra and Ginther, 1976). It has been postulated that $PGF_{2\alpha}$ is the equine luteolysin (Douglas and Ginther, 1976).

Douglas and Ginther (1976) reported that PGF concentration ($\overline{X}$ ± S.E.M.) in uterine venous blood increased from 2.4 ± 1.0 ng/ml on day 2 of diestrus to 14.9 ± 0.5 ng/ml on day 14 and then decreased to day 18 (6.4 ± 1.3 ng/ml). Furthermore, on days 10 and 14, uterine-vein PGF levels ($\overline{X}$ ± S.E.M.) were higher for diestrous (8.2 ± 0.8 and 14.9 ± 0.5 ng/ml) than for pregnant (4.3 ± 0.9 and 9.3 ± 2.0 ng/ml) mares. Zavy *et al.* (1978b) found that total recoverable PGF ($\overline{X}$ ± S.E.M.) in uterine flushings of cycling mares followed a similar pattern, i.e., increased steadily from day 4 (52.2 ± 19.6 ng) to day 14 (1133.8 ± 377.4 ng) and then declined to day 20 (255.1 ± 69.0 ng) postovulation. Production of PGF (expressed as ng/mg endometrium per 2 hr) and PGF content of endometrium (ng), respectimely, from nonpregnant mares increased ($\overline{X}$ ± S.E.M.) from day 4 (3.5 ± 2.3 and 1.5 ± 0.6) to day 16 (41.2 ± 2.3 and 6.9 ± 0.6) and then decreased to day 18 (15.5 ± 2.1 and 2.6 ± 0.6) postovulation according to Vernon (1979). These three independent studies indicate that endometrial production, uterine-fluid content, and uterine-vein plasma concentrations of PGF are significantly elevated at a time when degeneration of the equine CL is occurring (Van Niekerk *et al.*, 1975) and plasma progesterone levels are declining (Sharp and Black, 1973; Douglass and Ginther, 1976).

2.2.2. *Maternal Recognition of Pregnancy Signals*

Maternal recognition of pregnancy signals is poorly understood in the mare, but some data are available to suggest some of the conditions associated with CL maintenance. As with the pig, it appears that the primary CL of the mare must be "protected" from the luteolytic effect of the uterine endometrium if pregnancy is to be established.

Kooistra and Ginther (1976) demonstrated that removal of the equine conceptus, i.e., embryo, placental membranes, and fluids, on day 24 of pregnancy resulted in an extended interestrus interval. Ovulation did not occur for 34.8 ± 5.8 days following conceptus removal. This period of prolonged CL maintenance of pseudopregnancy of 50–60 days was shorter than CL life-span for hysterectomized (70–140 days) or pregnant (140–210 days) mares. It is possible that the equine conceptus protects the CL from a uterine luteolysin.

A synthetic estrogen, diethylstilbestrol (DES), was administered daily to three mares by Nishikawa (1959) for 13, 15, or 20 days beginning on day 7, 8, or 9 postovulation. Ovaries were removed on day 24, 32, or 47, and all contained the CL believed to have been present at the time of initiation of treatment. Similarly, prior mares given 5.0 mg DES, but not 0.5 mg estradiol-17β or vehicle alone, had prolonged interovulatory intervals (35 ± 3.5 vs. 27 ± 1.7 and 24.6 ± 0.9 days, respectively) (Berg and Ginther, 1978). Attempts to confirm these observations have been unsuccessful (D. C. Sharp, unpublished data).

Zavy *et al.* (1979c) incubated conceptus membranes *in vitro* for 120 min in a chemically defined medium to assess estrone and estradiol production capabilities on days 8, 12, 14, 16, 18, and 20 of pregnancy. Estradiol production (pg/5 ml medium per hr) increased from day 8 (243 pg) to day 20 (128,763 pg). A similar trend, but of higher magnitude, was found for estrone. It is apparent from these high rates of estrogen production by equine blastocysts that the possible effect of these estrogens on the uterine luteolytic agent is worthy of further study. An additional, although parodoxical, effect of the equine conceptus is its apparent stimulatory ($P < 0.05$) effect on *in vitro* PGF production by endometrium from the gravid uterine horn (42.2 ± 2.5 ng PGF/mg tissue) vs. the nongravid uterine horn (19.1 ± 1.0 ng PGF/mg tissue) (Vernon, 1979). This worker also reported that PGF production was maximal from endometrium from ovariectomized mares treated systemically with progesterone (*in vivo*) and then exposed to estrogen *in vitro*.

Inferential evidence suggests that the embryonic signal for CL maintenance probably acts at the level of the uterus. First, endometrial production of PGF by the pregnant uterus increases steadily between days 8 and 20 of gestation, and therefore PGF must be prevented from reaching the CL. Second, luteal-cell-membrane $PGF_{2\alpha}$-binding capacity between days 14 and 18 of pregnancy is equal to or greater than that on day 12 of the estrous cycle, which also suggests that the uterine luteolysin must be prevented from reaching the pregnant CL. These two points are based on data presented by Vernon (1979). Douglass and Ginther (1976) reported that uterine venous blood from pregnant mares does, in fact, have lower PGF concentrations.

Pregnant mare serum gonadotropin (PMSG) is produced by the chorionic cells of the endometrial cups. It is detectable in maternal plasma by day 37, reaches peak values around day 60, and then declines to undetectable levels by day 140 of pregnancy. Kooistra and Ginther (1976) suggests that the longer period of CL maintenance in pregnant (140–210 days) as compared to hysterectomized (70–140 days) and pseudopregnant mares (about 50 days) may be due to effects of PMSG. However, data are not available to suggest that PMSG is present between days 12 and 14 of pregnancy when CL maintenance must be established.

2.3. The Cow

2.3.1. The Uterus and Corpus Luteum Maintenance

The uterus has been shown to control the life-span of the bovine CL (Wiltbank and Casida, 1956; Malven and Hansel, 1964), since hysterectomy leads to prolonged CL maintenance ($\approx$ 150 days). Estradiol injection between days 2 and 12 of the estrous cycle causes premature CL regression in cows with intact uteri (Greenstein *et al.*, 1958; Wiltbank *et al.*, 1961), but estradiol is not effective in hysterectomized heifers (Brunner *et al.*, 1969). Oxytocin injections on days 1 through 7 of the bovine estrous cycle also cause premature luteal regression, but only if the uterus is present (Anderson and Bowerman, 1963). The luteolytic effect of a nongravid uterine horn is unilateral. If the nongravid uterine horn of a nonpregnant heifer is located contralateral to the ovary with the CL, CL maintenance is prolonged. However, CL life-span is of normal duration if the nongravid uterine horn of a nonpregnant heifer is ipsilateral to the ovary with the CL.

The bovine luteolytic agent appears to be produced by the uterine glandular endometrium. Hansel *et al.* (1973) reported CL maintenance of 30 days or more for cows with congenital absence of endometrial glandular epithelium. Distention or irritation of the uterine horn adjacent to the ovary containing the CL shortens CL life-span (Hansel and Wagner, 1960; Ginther *et al.*, 1960; Mapletoft *et al.*, 1976).

The effects of estrogen, oxytocin, intrauterine devices, and irritants are believed to act through a common mechanism. That is, they may cause the premature release of PGF from the uterine endometrium, which in turn causes luteal regression.

Exogenous $PGF_{2\alpha}$ clearly exerts a luteolytic effect in the cow (Hafs *et al.*, 1974; Lauderdale, 1974; Thatcher and Chenault, 1976). Hansel *et al.* (1975) reported that arachidonic acid was the active uterine endometrial luteolytic agent, since it served as precursor for $PGF_{2\alpha}$ synthesis and its activity is regulated by phospholipase A (Kunze and Vogt, 1971). Evidence for elevated endometrial and uterine-vein plasma $PGF_{2\alpha}$ concentrations between days 15 and 20 of the estrous cycle (Shemesh and Hansel, 1975; Nancarrow, 1972) and preferential transfer of $PGF_{2\alpha}$ from uterine vein to ovarian artery (Hixon and Hansel, 1974) supports the notion that the bovine uterine luteolysin is $PGF_{2\alpha}$.

After comparing peripheral plasma concentrations of 13,14-dihydro-15-keto-$PGF_{2\alpha}$ (PGF-M), the inactive metabolite of $PGF_{2\alpha}$, in nonpregnant and pregnant heifers, Kindahl *et al.* (1976) reported (1) a temporal association between elevated PGF-M concentrations and declining plasma progesterone levels and (2) the absence of elevated PGF-M concentrations in the pregnant female.

Thatcher *et al.* (1979) reported that estradiol-17β given intravenously on day 13 of the estrous cycle caused an acute increase in plasma PGF-M levels at 6 hr postinjection, and complete luteolysis occurred earlier in treated vs. control heifers (96.9 vs. 153.6 hr postinjection). The mechanism whereby the bovine conceptus "protects" the CL is not known; however, Bartol *et al.* (1977) reported that total PGF is greater in uterine flushings from pregnant than nonpregnant cows between days 16 and 19 after onset of estrus. This may indicate a mechanism for sequestering PGF produced by the endometrium or conceptus, or both, in the pregnant female to prevent PGF release in the vascular system. From days 27 to 111 of pregnancy, appreciable amounts of PGF are sequestered within allantoic and amniotic fluids of the conceptus (Eley *et al.*, 1979a).

Binding sites for $PGF_{2\alpha}$ on bovine-CL-cell-membrane preparations have been reported (Kimball and Lauderdale, 1975; Rao, 1974). However, data comparing $PGF_{2\alpha}$ binding by bovine-luteal-cell membranes from pregnant and nonpregnant females are not available.

2.3.2. Maternal Recognition of Pregnancy Signals

The presence of a developing embryo in the bovine uterus allows for CL maintenance for the duration of pregnancy; however, the mechanism involved is not known. Northey and French (1978) reported that pregnant cows from which embryos were removed on days 17 and 19 had interestrus intervals of 25.0 $\pm$ 1.2 and 26.2 $\pm$ 0.6 days as compared to those having embryos removed on day 13 (20.2 $\pm$ 0.8 days) or not mated (20–21 days). Again, no explanation for the effect of the bovine conceptus on the interovulatory interval is available.

Amoroso (1952) indicated that prior to attachment of the conceptus, there is increased edema, vascularity, and glandular hypertrophy of the uterine endometrium. Fusion of the trophoblast and endometrium is followed by a gradual erosion of the endometrial surface epithelium

(Amoroso, 1952; Melton *et al.*, 1951). Elongation of the bovine blastocyst is most rapid between days 16 and 20 of gestation, and the trophoblast cells are either columnar or giant cells. The columnar cells accumulate lipid material between days 16 and 25 of gestation (Greenstein *et al.*, 1958), which coincides with increasing amounts of PGF in the uterine lumen (Bartol *et al.*, 1977). Lewis *et al.* (1979) demonstrated synthesis of $PGF_{2\alpha}$ from arachidonic acid by day 16 and day 19 bovine blastocysts.

Bovine blastocysts recovered on days 15, 16, and 17 of pregnancy can convert androstenedione to estradiol *in vitro* (Eley *et al.*, 1979c). The physiological importance of this observation is not known. As indicated earlier, estrogens have a luteolytic action in the cow whether administered in the unconjugated (Greenstein *et al.*, 1958; Wiltbank *et al.*, 1961) or conjugated (estrone sulfate) form (Eley *et al.*, 1979b). Extensive metabolism of androstenedione to 5β-reduced compounds by the conceptus, and the role of these compounds in regulating the bovine endometrium, warrants further study (Eley *et al.*, 1979c).

2.4. The Ewe

2.4.1. The Uterus and Corpus Luteum Maintenance

Available data on the ovine uterus–CL–hypothalamic–pituitary axis is extensive and has been reviewed recently by Horton and Poyser (1976). As with other farm animals, the ovine uterine endometrium appears to control CL life-span in nonpregnant females (Wiltbank and Casida, 1956).

McCracken (1971) and McCracken *et al.* (1972) obtained evidence that $PGF_{2\alpha}$ was released into the uterine vein of the ewe and elevated levels were temporally associated with CL regression. Goding (1974) reviewed evidence for and against the proposition that $PGF_{2\alpha}$ is the ovine uterine luteolysin and concluded that $PGF_{2\alpha}$ serves that role.

2.4.2. Maternal Recognition of Pregnancy Signals

The question of how the embryo prevents luteal regression is applicable to the ewe. A functional CL is required for at least 50–60 days of the 145- to 150-day gestation period. After that time, placental progesterone production is adequate to allow pregnancy maintenance (Casida and Warwick, 1945).

Moor and Rowson (1966a,b) removed embryos from mated ewes on either day 5, 7, 9, 12, 13, 14, or 15 of pregnancy. No significant effect of the embryo on the interestrus interval was detected until day 13 of gestation. About one third of the ewes from which embryos were removed between days 13 and 15 had interestrus intervals of greater than 25 days. Using the opposite approach, the pregnancy rate for nonmated ewes to which embryos were transferred on either day 12, 13, or 14 was 60–67, 22, and 0%, respectively. These data indicated that the conceptus influences CL maintenance between days 12 and 15 of pregnancy. Continuous intrauterine infusion of homogenates of 14- to 15-day ovine embryos also extends CL life-span in sheep (Moor and Rowson, 1967; Ellinwood, 1978); however, infusion of the homogenate into the uterine vein was not effective (Ellinwood, 1978).

In contrasting uteroovarian-vein and endometrial concentrations of $PGF_{2\alpha}$, available data provide no evidence for differences between nonpregnant and pregnant ewes. Thorborn *et al.* (1973) reported the absence of PGF peaks in uterine-vein plasma of two pregnant ewes between days 15 and 16 after onset of estrus. Barcikowski *et al.* (1974), on the basis of data from a single pregnant ewe, also reported the absence of PGF peaks in uterine-vein plasma after day 14, but the PGF patterns were similar in the pregnant ewe and a nonpregnant ewe on days 13 and 14.

Nett *et al.* (1976) reported that mean levels and peak concentrations of $PGF_{2\alpha}$ in ovine uteroovarian-vein plasma were not different in nonpregnant and pregnant ewes between days 11 and 17 after onset of estrus, but the frequency of $PGF_{2\alpha}$ peaks was lower in pregnant ewes. Pexton *et al.* (1975) and Ellinwood (1978) also failed to detect differences between uteroovarian-vein and uterine-vein concentrations of $PGF_{2\alpha}$ in nonpregnant and pregnant ewes between days 11 and 17 after onset of estrus.

L. Wilson *et al.* (1972) reported that endometrial content and concentration and uterine-vein concentration of $PGF_{2\alpha}$ were higher for pregnant than nonpregnant ewes on day 13 after the onset of estrus. Pexton *et al.* (1975) sampled ovarian vein, ovarian artery, and uterine vein and concluded that there were no significant differences in plasma PGF concentrations between pregnant and nonpregnant females on day 15 after the onset of estrus, regardless of the sampling site. In support of these findings, Lewis *et al.* (1977) found endometrial content and concentration of $PGF_{2\alpha}$ to be greater in pregnant than nonpregnant ewes, but no differences in uterine-vein PGF concentrations were detected between days 11 and 18 after onset of estrus. Uterine-vein plasma $PGF_{2\alpha}$ concentrations on day 15 and $PGF_{2\alpha}$ concentrations in endometrium, ovarian artery, and uterine vein on day 16 were not different in pregnant and nonpregnant ewes. Ellinwood (1978) reported that *in vitro* production of $PGF_{2\alpha}$ by ovine endometrium (pg/ml endometrium per 80 min) was not different in pregnant and nonpregnant ewes on days 13 and 17;

however, production was significantly higher for pregnant ewes on day 15. Endometrial concentration of $PGF_{2\alpha}$ showed the same pattern. Uteroovarian-vein and uterine-fluid concentrations of $PGF_{2\alpha}$ were significantly higher for pregnant than nonpregnant ewes on days 13, 15, and 17 after onset of estrus.

Available data indicate that findings of Thorburn *et al.* (1973) with two pregnant ewes and Barcikowski *et al.* (1974) with one pregnant ewe are exceptions to the findings from a number of other studies involving much larger numbers of ewes.

Ellinwood (1978) presented preliminary evidence for a protein from the pregnant ovine endometrium that bound $PGF_{2\alpha}$ preferentially and, to a lesser extent, PGE_2. He proposed the theory that the blastocyst induces the endometrium to secrete a $PGF_{2\alpha}$-binding protein into the uterine lumen and uterine vasculature and that there may be a shift in $PGF_{2\alpha}$ secretion toward the uterine lumen. The $PGF_{2\alpha}$-binding protein, if present, would antagonize the transfer of $PGF_{2\alpha}$ from uterine vein to ovarian artery and therefore protect the CL. Confirmation of the observations of Ellinwood (1978) is needed.

Peterson *et al.* (1976) reported peaks of 13,14-dihydro-15-keto-$PGF_{2\alpha}$ (PGF-M) on days 14 through 17 of the estrous cycle that were temporally related to declining plasma progesterone levels. However, no PGF-M peaks were detected in pregnant ewes. Lewis *et al.* (1977), on the other hand, reported no differences in jugular-vein plasma PGF-M concentrations between days 11 and 16 of the estrous cycle and pregnancy. Peterson *et al.* (1976) sampled every 2 hr for 3 days and then twice daily for the next 4 days beginning on either day 12 or 13 of the estrous cycle or pregnancy; however, Lewis *et al.* (1977) sampled only once daily for PGF-M analysis. Because of the well-documented episodic type of release of $PGF_{2\alpha}$, it would be expected that frequent sampling would favor detection of differences in PGF-M.

The mechanism whereby the conceptus directly or indirectly provides the signal for maternal recognition of pregnancy in the ewe is not known. Ellinwood (1978) found that the presence of embryos within the uterine lumen, up to day 14 or 15 of pregnancy when they were removed, extended the interestrus interval from 16.9 ± 0.4 days to 29.1 ± 1.4 days. Homogenates of days 14 and 15 ovine conceptuses did not exhibit LH- or prolactinlike activity in radioreceptor assays, nor did they stimulate progesterone or cAMP synthesis in isolated ovine luteal cells. These results are not consistent with other data. Godkin *et al.* (1978) reported that extracts of ovine embryos from days 13 to 15 of pregnancy stimulated progesterone production by CL slices obtained from nonpregnant ewes between days 12 and 14 after onset of estrus and from cows between days 14 and 15 of the cycle. The nature of the "stimulatory" agent is not known, but preliminary data indicate that it is nondialyzable and heat-labile. Godkin *et al.* (1978) suggested that the sheep conceptus produces a factor that acts at the CL to allow CL maintenance. Ellinwood (1978), on the other hand, concludes that homogenates of ovine embryos are effective in allowing CL maintenance only when infused into the uterine lumen; i.e., the effect of the conceptus is at the uterine endometrium.

In considering potential agents that might act on the uterus or CL or both to allow CL maintenance, several general comments are in order. First, Martal and Djiane (1977) and Carnegie *et al.* (1977) reported the presence of ovine placental lactogen (oPL) in ovine conceptuses between days 16 and 17 of pregnancy, but data on effects of oPL on CL function are not available. Steroid production by the ovine blastocysts between days 12 and 15 of pregnancy has not been established. Ellinwood (1978) measured estradiol-17β and estrone on days 13, 15, and 17 of the estrous cycle and pregnancy in uteroovarian-vein plasma. For pregnant and nonpregnant ewes, estradiol-17β concentrations ($\bar{X} \pm$ S.E.) were 4.2 $\pm$ 0.5 and 4.4 $\pm$ 1.3, 3.7 $\pm$ 0.7 and 4.3 $\pm$ 0.8, and 4.1 $\pm$ 1.3 and 4.0 $\pm$ 0.4 pg/ml on days 13, 15, and 17, respectively, while estrone levels were 41.2 $\pm$ 7.3 and 37.9 $\pm$ 20.4, 30.1 $\pm$ 5.7 and 24.3 $\pm$ 4.7, and 41.6 $\pm$ 10.2 and 20.0 $\pm$ 5.5 pg/ml for the same respective days. These estrogen values were not significantly different, but estrone concentrations were, on the average, consistently higher for pregnant ewes. Ellinwood (1978) speculates that estrogen, of conceptus origin, may stimulate $PGF_{2\alpha}$-binding protein synthesis or secretion or both.

The most convincing evidence for a factor emanating from the gravid uterus that affects the CL directly was reported by Mapletoft *et al.* (1975). As pointed out earlier, CL regression occurs in sheep if a conceptus is not present in the uterine horn ipsilateral to the CL. If a ewe has one CL on each ovary and the uterine horns are surgically separated so that an embryo is present on only one side, the CL ipsilateral to the conceptus is maintained while the CL on the contralateral ovary regresses. Mapletoft *et al.* (1975) anastomosed the main uterine veins from the gravid and nongravid uterine horns of unilaterally pregnant ewes having one CL on each ovary. This resulted in bilateral CL maintenance. They proposed that venous blood from the gravid uterine horn contains a "luteotrophic" factor that results in CL maintenance. They also found that CL regressed when the uterine vein ipsilateral to the conceptus received blood only from the nongravid horn, indicating the presence of a uterine luteolysin. But if the anastomosis was made so that the ipsilateral uterine vein contained blood from the gravid and nongravid uterine horn, CL maintenance occurred, indicating the presence of a "luteotrophic" agent. The nature of the "luteotrophic" agent has not been directly determined.

Inskeep *et al.* (1975), Henderson *et al.* (1977), Mapletoft *et al.* (1977), Colcord *et al.* (1978), and Magness *et al.* (1978) provided evidence for an "antiluteolytic" effect of PGE_2, and Hoyer *et al.* (1978) reported a similar effect for PGE_1. If PGE_2 or PGE_1 or both are exerting a "luteotrophic" effect on CL maintenance and if there is an endometrial $PGF_{2\alpha}$-binding protein, one can speculate, as Ellinwood (1978) has, that the uterus, under an appropriate signal from the conceptus, and the conceptus itself act in concert to allow for CL maintenance and the establishment of pregnancy in sheep.

Finally, it should be noted that estradiol may have either a "luteotrophic" effect (Piper and Foote, 1965; Denamur *et al.,* 1970) or a luteolytic effect (Stormshak *et al.,* 1969; Warren *et al.,* 1973) when injected. The effect is influenced by the stage of the estrous cycle when injected and the dosage. Further research is necessary to clarify these observations.

3. Conceptus–Endometrial Steroid Metabolism

3.1. The Pig

Perry *et al.* (1973) first demonstrated conversion of tritium-labeled dehydroepiandrosterone (DHEA), androstenedione, progesterone, and estrone sulfate to unconjugated estradiol and estrone by days 14–16 pig blastocysts. Later, Perry *et al.* (1976) confirmed that DHEA, androstenedione, and estrone sulfate could be converted to estradiol and estrone as early as day 12 of pregnancy. Conversion of progesterone, pregnenolone, and testosterone to estrogens was found to be less efficient. More recently (Heap *et al.,* 1979), evidence for conversion of acetate and cholesterol to estrogens by pig blastocysts has been provided. The source of these estrogen precursors was initially assumed to be maternal (Heap and Perry, 1974). Heap and Perry (1974) proposed that androgens and conjugated estrogens of maternal origin were available to the porcine blastocyst, which would convert them to free estrodiol and estrone. These free estrogens would result from the action of aromatase and sulfatase enzymatic activities within the blastocyst. Data from our laboratory (Valdivia, 1977) confirm estrone and estradiol production *in vitro* by day 20 porcine conceptuses incubated *in vitro* in modified Medium 199. The data of Valdivia (1977) differ from those summarized by Heap *et al.* (1979) in that estradiol was the predominant estrogen in the medium after a 4-hr incubation [estradiol ($\overline{X}$) = 16,782 pg/5 ml medium per g conceptus; estrone ($\overline{X}$) = 6292 pg/5 ml medium per g conceptus]. However, estrone ($\overline{X}$ = 9293 pg/5 ml medium per g conceptus) was present in higher concentrations than estradiol (5954 pg/5 ml medium per g conceptus) in conceptuses that represented nonincubated controls.

Heap and Perry (1974) suggest that the uterine enzymes, 17β-hydroxysteroid dehydrogenase and sulfotransferase, result in estrone sulfate's being the primary form of estrogen to enter the maternal circulation. It is known that estrone sulfate is the primary estrogen in maternal plasma (Robertson and King, 1974) and allantoic fluid (Dueben *et al.,* 1979) of pigs. The concept is very appealing from a biological standpoint, since it allows one to explain very high levels of estrogen production by the porcine placenta. These free estrogens may exert a local effect on such things as uterine blood flow, water and electrolyte movement, maternal recognition of pregnancy phenomena, uterine secretory activity, and other events at the site of placentation. But before leaving the uterus, these estrogens are conjugated and therefore enter the maternal circulation in a biologically inactive form. Local effects of estrogens are thus restricted to tissues having sulfatase enzymatic activity, e.g., the mammary gland and blastocyst, while systemic effects associated with estrogen are minimized.

The possibility that the uterine endometrium of pregnant gilts can metabolize progesterone was suggested by two lines of evidence from our laboratory. Knight *et al.* (1977) observed that uterine artery–uterine vein (A-V) differences in plasma progestins were positive at all stages of gestation studied between days 20 and 100. Conversely, the A-V differences in estradiol and estrone were negative at all stages of gestation studied. This indicated that progesterone was being "taken up" and possibly metabolized by the pregnant uterus. The trophoblast (chorion) of the porcine conceptus is known to be a source of estrogens (Velle, 1960; Raeside, 1963; Molokwu and Wagner, 1973; Choong and Raeside, 1974; Robertson and King, 1974; Wetteman *et al.,* 1974; Perry *et al.,* 1976; Knight *et al.,* 1977).

Data from our laboratory also indicated that gilts bilaterally ovariectomized on day 4 of pregnancy and treated with 25, 50, 100, or 200 mg progesterone per day to maintain pregnancy had plasma progestin concentrations which were markedly affected by pregnancy status on day 60 after mating. Gilts that were pregnant (viable conceptuses) or had been pregnant (resorbing conceptuses) had plasma progestin concentrations ranging from 7.0 to 26.5 ng/ml. However, gilts for which there was no evidence that pregnancy had been established had plasma progestin concentrations of 163.4–428.0 ng/ml. These data suggest that the uterus or fetal–placental unit, or both, or the uterus alone, if pregnancy had been temporarily established, had a marked effect on plasma progestin concentrations.

In the normal course of pregnancy in pigs, there is a 30–70% decrease in maternal plasma progestin concentrations between days 14 and 30 of gestation (Guthrie *et al.*, 1972; Robertson and King, 1974). This decrease in plasma progestins has been attributed to partial CL regression during the period of "maternal recognition" of pregnancy. The decrease may, however, reflect onset of progesterone metabolism by the pregnant uterine endometrium or the conceptus or both.

It was learned, on the basis of work by Henricks and Tindall (1971), that endometrial tissue from nonpregnant gilts metabolizes [^{3}H]progesterone. They noted ten metabolites of progesterone and identified 5 α-dihydroprogesterone and its 3β-hydroxylated counterpart. They also found 36% of the radioactivity in the water-soluble fraction after extraction of the endometrial incubates and suggested that steroid sulfates and glucuronides were also products.

In an attempt to determine whether or not the porcine uterus metabolized progesterone to steroids known to be precursors for estrogen production (Perry *et al.*, 1976), studies were initiated in which endometrial tissue from nonpregnant and pregnant gilts was incubated *in vitro* with [^{3}H]progesterone (about 30 ng) for 15, 30, 60, 120, or 180 min.

Dueben *et al.* (1977, 1979) have incubated uterine endometrial tissue from nonpregnant and pregnant gilts with [^{3}H]progesterone. Androstenedione, testosterone, 5α-dihydroprogesterone, estrone, and estradiol have been identified by Sephadex LH 20 chromatography followed by recrystallization to constant specific activity. Furthermore, a major portion of the radioactivity is, as reported by Henricks and Tindall (1971), present in the aqueous fraction after extraction of the endometrial homogenate three times with diethyl ether.

Estriol, in addition to estrone and estradiol, has been found in experiments wherein day 14, 16, or 18 pregnant endometrium was incubated with [^{3}H]progesterone for 3 hr. After that time, the endometrium was removed and 14, 16, or 18 blastocysts placed in the medium and incubated for an additional 3 hr. This suggested that an interaction between the endometrium and blastocyst was necessary for conversion of [^{3}H]progesterone to estriol. To confirm this observation, allantoic fluid was obtained from litters of three gilts on day 30 of pregnancy. The allantoic fluid was assayed for unconjugated estrone, estradiol, and estriol and conjugated (sulfates and glucuronides) estrone, estradiol, and estriol. Mean concentrations ($\bar{X} \pm$ S.E., ng/ml) were 3.0 ± 5.0, 0.6 ± 0.5, and 2.2 ± 1.8 for unconjugated estrone, estradiol, and estriol and 28.1 ± 13.6, 2.3 ± 0.5, and 2.1 ± 1.7 for conjugated estrone, estradiol, and estriol, respectively.

Collectively, these data indicate that the pig endometrium can convert progesterone to androstenedione, testosterone, estrone, estradiol, and conjugated estrone and estradiol. The percentage conversions of conjugated estrogens, e.g., estrone sulfate to estrone (82%) and estradiol (16%), by pig blastocysts is very efficient as compared with conversion of androstenedione to estrone (13.6%) and estradiol (2.4%) according to Perry *et al.* (1973, 1976). The fact that conjugated estrogens are major products of endometrial metabolism of progesterone and may be efficiently converted to free estrogens by the blastocyst may be only a partial explanation for their significance. The pig endometrium is known to have 5α-reductase enzymatic activity (Hendricks and Tindall, 1971), which would result in the 5α-reduction of progestins and androgens. Such steroids cannot be aromatized to estrogens (J. D. Wilson, 1972). Metabolism of progesterone to conjugated estrogens would circumvent the possibility of estrogen precursor pools (androgens) that could be reduced and rendered nonaromatizable and, at the same time, provide the most efficient form of unconjugated estrogen precursor to the blastocysts, which can initiate key events associated with the "maternal recognition of pregnancy" in swine.

Perry *et al.* (1976) reported that endometrium from pregnant sows can convert androstenedione and DHEA to unconjugated estrogens to a limited extent. Renfree and Heap (1977) found that endometrium of the marsupial, *Macropus eugenii,* metabolized androstenedione to phenolic steroids, presumably estrogens.

3.2. The Ewe

Willis *et al.* (1979a,b) incubated ovine endometrium obtained on days 20, 60, and 140 of gestation with [^{3}H]progesterone for 15, 30, 60, 120, or 240 min. Androstenedione, testosterone, estrone, and estradiol were identified by recrystallization to constant specific activity. No differences in progesterone metabolism between endometrium from pregnant and nonpregnant uterine horn, caruncle vs. intercaruncular endometrium, or presence or absence of chorioallantois in medium with endometrium was detected. No estriol was detected in any of the incubations.

3.3. The Mare

Incubation of endometrium from nonpregnant and pregnant mares with [^{3}H]progesterone resulted in minimal conversion ($<1\%$) to unconjugated estrogens. However, conceptus membranes (trophectoderm and yolk sac) alone were capable of converting [^{3}H]progesterone to estrogens ($>7\%$), especially estrone as identified by recrystallization to constant specific activity (Seamans *et al.*, 1979). Estrone and estradiol were monitored by radioimmunoassay following *in vitro* incubation of conceptus membranes during early pregnancy. Similar to the results of Valdivia (1977), estradiol was the primary estrogen and was produced in relatively large quantity (>6000 pg/5 ml incubation medium) as early as day 12 of pregnancy (Zavy *et al.*, 1979c). The lack of significant endometrial aromatase activity compared to prolific free-estrogen production by the conceptus membranes suggests that the conceptus is the primary source of estrogen at the time of maternal recognition of pregnancy in the mare.

3.4. The Cow

The bovine conceptus can convert [^{3}H]androstenedione to estrone, estradiol-17β, androstanediol, etiocholanolone, and androstanedione, i.e., 5β-reduced and hydroxylated androgens, from as early as day 15 to day 27 of pregnancy (Eley *et al.*, 1979c). Other unidentified metabolites were also present, but no estradiol-17α or estriol was detectable. The bovine conceptus was shown to convert [^{3}H]progesterone to pregnanediol, pregnanedione, and pregnan-one-ol, i.e., 5β-reduced and hydroxylated progestins, but not estrogens, by the same researchers.

The bovine endometrium from pregnant females was found to convert [^{3}H]progesterone and [^{3}H]androstenedione to their 5α-reduced and hydroxylated metabolites (Eley *et al.*, 1979c). These workers also observed that (1) the endometrium metabolized progesterone more rapidly than androstenedione; (2) the endometrial metabolites of either progesterone or androstendione did not include free or conjugated estrogens; (3) the endometrium from day 27 of pregnancy metabolized progesterone and androstenedione more actively than that obtained on days 16 and 19; and (4) the metabolites of androstenedione and progesterone appeared to be the same when pregnant and nonpregnant endometrium from days 16 and 19 of gestation were compared.

Discussion

NOLIN: What is all that prostaglandin doing in the uterine lumen?

BAZER: I am not certain of its precise role or value in the domestic animals. However, it has been implicated in effecting changes in blood flow, vascular permeability, water and electrolyte transport, and stimulation of steroidogenesis. All of these responses have been shown to increase in one or more species during early pregnancy.

PIETRAS: Do you know if prostaglandins are transported into the lumen or the endometrium *per se* if they are generated by enhanced activity of an appropriate phospholipase?

BAZER: I assume that synthesis of prostaglandins is enhanced by phospholipase A2 and that the prostaglandins are secreted from the apical portion of the endometrial epithelial cells into the uterine lumen.

BOGDANOVE: I was wondering, with regard to the 7-day period of estrus in the mare, whether conception can occur at any time during this period, or whether mating during the early part of estrus is—like most human mating—a delightful but unproductive activity. Do horse spermatozoa have an unusually long period of intrauterine viability, or does the horse share, with man, the capability of using sex for recreation rather than (or in addition to) procreation?

BAZER: It is my impression that conception takes place after ovulation, which is late evening or early morning of the 5th day of an "average" 7-day estrus period. The horses certainly appear to enjoy this long estrus period.

Although there is one report of a mare conceiving to a breeding some 6 days before ovulation, the experience of my colleague, Dr. D. C. Sharp, and the equine veterinarians with whom he has contact, indicates that sperm longevity in the reproductive tract of the mare is 18–24 hr. However, conception to breeding 48 hr preovulation occasionally results, but is uncommon.

There is some tentative evidence, according to Dr. Sharp and his colleagues, that the refusal activity of mares and stallions intensifies near the time of ovulation. If so, this suggests that under natural conditions, breeding during the early part of the estrus period may be relatively inefficient.

References

Allen, W.R., 1975, The influence of fetal genotype upon endometrial cup development and PMSG and progestagen production in equids, *J. Reprod. Fertil. Suppl.* **23:**405.

Allen, W.R., and Rowson, L.E., 1973, Control of the mare's estrous cycle by prostaglandins, *J. Reprod. Fertil.* **33:**359.

Amoroso, E.C., 1952, Placentation, in: *Marshall's Physiology of Reproduction* (A.S. Parkes, ed.), pp. 127–311, Little, Brown, Boston.

Anderson, L.L., 1974, Reproduction cycle in pigs, in: *Reproduction in Farm Animals* (E.S.E. Hafez, ed.), pp. 275–287, Lea and Febiger, Philadelphia.

Anderson, L.L., 1978, Growth, protein content and distribution of early pig embryos, *Anat. Rec.* **190:**143.

Anderson, L.L., and Bowerman, A.M., 1963, Utero-ovarian function of oxytocin-treated heifers, *J. Anim. Sci.* **22:**1136 (abstract).

Anderson, L.L., and Melampy, R.M., 1967, Hypophysial and uterine influences on pig luteal function, in: *Reproduction in the Female Mammal* (G.E. Lamming and E.C. Amoroso, eds.), pp. 285–316, Plenum Press, New York.

Anderson, L.L., Butcher, R.L., and Melampy, R.M., 1961, Subtotal hysterectomy and ovarian function in gilts, *Endocrinology* **69:**571.

Anderson, L.L., Leglise, P.C., du Mesnil du Buisson, F., and Rombauts, P., 1965, Interaction des hormones gonadotropes et l'uterus dans le maintien du tissu luteal ovarien de la Truie, *C. R. Acad. Sci.* **261:**3675.

Anderson, L.L., Bland, K.P., and Melampy, R.M., 1969, Comparative aspects of uterine–luteal relationships, *Recent Prog. Horm. Res.* **25:**27.

Atkins, D.T., Sorenson, A.M., and Fleeger, J.L., 1974, 5α-Dihydroprogesterone in the pregnant mare, *J. Anim. Sci.* **39:**196 (abstract).

Barcikowski, B., Carlson, J.C., Wilson, L., and McCracken, J.M., 1974, The effect of endogenous and exogenous estradiol-17β on the release of prostaglandin $F_{2\alpha}$ from the ovine uterus, *Endocrinology* **95:**1340.

Bartol, F.F., Kimball, F.A., Thatcher, W.W., Bazer, F.W., Chenault, J.R., Wilcox, C.J., and Kittok, R.J., 1977, Follicle, luteal and uterine parameters during the bovine estrous cycle, *Proc. Am. Soc. Anim. Sci.,* p. 134, Madison, Wisconsin.

Bazer, F.W., and Thatcher, W.W., 1977, Theory of maternal recognition of pregnancy in swine based on estrogen controlled endocrine versus exocrine secretion of prostaglandin $F_{2\alpha}$ by the uterine endometrium, *Prostaglandins* **14:**397.

Bazer, F.W., Roberts, R.M., and Sharp, D.C., 1978, Collection and analysis of female genital tract secretions, in: *Methods in Mammalian Reproduction* (J.C. Daniels, Jr., ed.), pp. 503–528, Academic Press, New York.

Bazer, F.W., Roberts, R.M., Basha, S.M.M., Zavy, M.T., Caton, D., and Barron, D.H., 1979, Method for obtaining ovine uterine secretions from unilaterally pregnant ewes, *J. Anim. Sci.* **49:**1522.

Belt, W.D., Anderson, L.L., Cavazos, L.F., and Melampy, R.M., 1971, Cytoplasmic granules and relaxin levels in porcine corpora lutea, *Endocrinology* **89:**10.

Berg, S.L., and Ginther, O.J., 1978, Effect of estrogens on uterine tone and life span of the corpus luteum in mares, *J. Anim. Sci.* **47:**203.

Bito, L.Z., 1972, Comparative study of concentrative prostaglandin accumulation by various tissues of mammals and marine vertebrates and invertebrates, *Comp. Biochem. Physiol.* **43A:**65.

Bito, L.Z., and Spellane, P.J., 1974, Saturable, "carrier-mediated," absorption of prostaglandin $F_{2\alpha}$ from the *in vivo* rabbit vagina and its inhibition by prostaglandin F_2B, *Prostaglandins* **8:**345.

Bito, L.Z., Wallenstein, M., and Baroody, R., 1976, The role of transport processes in the distribution and disposition of prostaglandins, *Adv. Prostaglandin Thromboxane Res.* **1:**297.

Bolander, F.F., and Fellows, R.E., 1976, Purification and characterization of bovine placental lactogen, *J. Biol. Chem.* **251:**2703.

Bolander, F.F., Ulberg, L.C., and Fellows, R.E., 1976, Circulating placental lactogen levels in dairy and beef cattle, *Endocrinology* **99:**1273.

Bowerman, A.M., Anderson, L.L., and Melampy, R.M., 1964, Urinary estrogens in cycling, pregnant, ovariectomized and hysterectomized gilts, *Iowa State J. Sci.* **38:**437.

Brunner, M.A., Donaldson, L.E., and Hansel, W., 1969, Exogenous hormones and luteal function in hysterectomized and intact heifers, *J. Dairy Sci.* **52:**1849.

Burns, S.J., and Fleeger, J.L., 1975, Plasma progestogens in the pregnant mare in the first and last 90 days of gestation, *J. Reprod. Fertil. Suppl.* **23:**435.

Buttle, H.L., and Forsyth, I.A., 1976, Placental lactogen in the cow, *J. Endocrinol.* **68:**141.

Caldwell, B.V., Moor, R.M., Wilmut, I., Polge, C., and Rowson, L.E.A., 1969, The relationship between day of formation and functional lifespan of induced corpora lutea in the pig, *J. Reprod. Fertil.* **18:**107.

Carnegie, J.A., and Robertson, H.A., 1978, Conjugated and unconjugated estrogens in fetal and maternal fluids of the pregnant ewe: A possible role for estrone sulfate during early pregnancy, *Biol. Reprod.* **19:**202.

Carnegie, J.A., Chan, J.S.D., Robertson, H.A., Friesen, H.G., and McCully, M.E., 1977, Placental lactogen in the preattachment sheep embryo, *Proc. Am. Soc. Anim. Sci.*, p. 142, Madison, Wisconsin.

Casida, L.E., and Warwick, E.J., 1945, The necessity of the corpus luteum for maintenance of pregnancy in the ewe, *J. Anim. Sci.* **4:**34.

Challis, J.R.G., 1971, A sharp increase in free circulating estrogens immediately before parturition in sheep, *Nature (London)* **229:**208.

Chen, T.T., Bazer, F.W., Gebhardt, B.M., and Roberts, R.M., 1975, Uterine secretion in mammals: Synthesis and placental transport of a purple acid phosphatase in pigs, *Biol. Reprod.* **13:**304.

Chenault, J.R., Thatcher, W.W., Kalra, P.S., Abrams, R.M., and Wilcox, C.J., 1976, Plasma progestins, estradiol and luteinizing hormone following prostaglandin $F_{2\alpha}$ injection, *J. Dairy Sci.* **59:**1342.

Choong, C.H., and Raeside, J.I., 1974, Chemical determination of estrogen distribution in the fetus and placenta of the domestic pig, *Acta Endocrinol.* **77;**171.

Colcord, M.L., Hayer, G.L., and Weems, C.W., 1978, Effect of prostaglandin E_2 (PGE_2) as an antiluteolysin on estrogen-induced luteolysis in ewes, *Proc. Am. Soc. Anim. Sci.*, p. 352, East Lansing, Michigan.

Denamur, R., Martinet, J., and Short, R.V., 1970, Mode of action of estrogen in maintaining the functional life of the corpora lutea in the sheep, *J. Reprod. Fertil.* **23:**109.

Diehl, J.R., and Day, B.N., 1973, Effect of prostaglandin $F_{2\alpha}$ on luteal function in swine, *J. Anim. Sci.* **37:**307.

Douglas, R.H., and Ginther, O.J., 1972, Effect of prostaglandin $F_{2\alpha}$ on length of diestrous in mares, *Prostaglandins* **2:**265.

Douglas, R.H., and Ginther, O.J., 1975a, Effect of prostaglandin $F_{2\alpha}$ on estrous cycles or corpus luteum in mares and gilts, *J. Anim. Sci.* **40:**518.

Douglas, R.H., and Ginther, O.J., 1975b, Route of prostaglandin $F_{2\alpha}$ injection and luteolysis in mares, *Proc. Soc. Exp. Biol. Med.* **145:**263.

Douglas, R.H., and Ginther, O.J., 1976, Concentration of prostaglandin F in uterine vein plasma of anesthetized mares during the estrous cycle and early pregnancy, *Prostaglandins* **11:**251.

Dueben, B.D., Wise, T.H., Bazer, F.W., and Fields, M.J., 1977, Metabolism of H^3-progesterone to androgens by pregnant gilt endometrium, *Proc. Am. Soc. Anim. Sci.*, p. 153, Madison, Wisconsin.

Dueben, B.D., Wise, T.H., Bazer, F.W., Fields, M.J., and Kalra, P.S. 1979, Metabolism of H^3-progesterone to estrogens by pregnant gilt endometrium and conceptus, *Proc. Am. Soc. Anim. Sci.*, p. 293, Tuscon, Arizona.

du Mesnil du Buisson, F., 1966, Contribution a l'étude du maintien du corps jaune de la Truie, Ph.D. dissertation, Institut National de la Recherche Agronomique, Nouzilly, France.

du Mesnil du Buisson, F., and Dauzier, L., 1957, Controle mutuel de l'uterus et de l'ovaire chez la Truie, *Ann. Zootechn. Serie Suppl.* **D:**147.

du Mesnil du Buisson, F., and Leglise, P.C., 1963, Effect de l'hypophysectomie sur les corps jaunes de Truie: Resultats preliminaires, *C. R. Acad. Sci.* **257:**261.

Eley, R.M., Thatcher, W.W., and Bazer, F.W., 1979a, Hormonal and physical changes associated with bovine conceptus development, *J. Reprod. Fertil.* **55:**181.

Eley, R.M., Thatcher, W.W., and Bazer, F.W., 1979b, Luteolytic effect of estrone sulphate on cyclic beef heifers, *J. Reprod. Fertil.* **55:**191.

Eley, R.M., Thatcher, W.W., Bazer, F.W., and Fields, M.J., 1979c, Metabolism of progesterone and androstenedione *in vitro* by bovine endometrium and conceptus, *Proc. Am. Soc. Anim. Sci.*, p. 294, Tuscon, Arizona.

Ellinwood, W.E., 1978, Maternal recognition of pregnancy in the ewe and the rabbit, Ph.D. dissertation, Colorado State University, Fort Collins.

Evans, M.J., and Irvine, C.H.G., 1975, Serum concentrations of FSH, LH and progesterone during the estrous cycle and early pregnancy in the mare, *J. Reprod. Fertil. Suppl.* **23:**193.

Flint, A.P.F., Henville, A., and Christie, W.B., 1979, Presence of placental lactogen in bovine conceptuses before attachment, *J. Reprod. Fertil.* **56:**305.

Foote, W.D., 1974, Reproductive cycles in cattle, in: *Reproduction in Farm Animals* (E.S.E. Hafez, ed.), pp. 257–264, Lea and Febiger, Philadelphia.

Frank, M., Bazer, F.W., Thatcher, W.W., and Wilcox, C.J., 1977, A study of prostaglandin $F_{2\alpha}$ as the luteolysin in swine. III. Effects of estradiol valerate on prostaglandin F, progestins, estrone and estradiol concentrations in the utero-ovarian vein of nonpregnant gilts, *Prostaglandins* **14:**1183.

Frank, M., Bazer, F.W., Thatcher, W.W., and Wilcox, C.J., 1978, A study of prostaglandin $F_{2\alpha}$ as the luteolysin in swine. IV. An explanation for the luteotrophic effect of estradiol, *Prostaglandins* **15:**151.

Ganjam, V.K., Kenney, R.M., and Flickinger, G., 1975, Plasma progestogens in cyclic, pregnant and post-partum mares, *J. Reprod. Fertil. Suppl.* **23:**441.

Gardener, M.L., First, N.L., and Casida, L.E., 1963, Effect of exogenous estrogens on corpus luteum maintenance in gilts. *J. Anim. Sci.* **22:**132.

Ginther, O.J., and First, N.L., 1971, Maintenance of the corpus luteum in hysterectomized mares, *Am. J. Vet. Res.* **32:**1687.

Ginther, O.J., Woody, C.O., Janakiraman, K., and Casida, L.E., 1960, Effect of an intrauterine plastic coil on the estrous cycle of the heifers, *J. Reprod. Fertil.* **12:**193.

Gleeson, A.R., Thorburn, G.D., and Cow, R.I., 1974, Prostaglandin F concentrations in the utero-ovarian vein plasma of the sow during the late luteal phase of the estrous cycle, *Prostaglandins* **5:**521.

Goding, J.R., 1974, The demonstration that $PGF_{2\alpha}$ is the uterine luteolysin in the ewe, *J. Reprod. Fertil.* **38:**261.

Godkin, J.D., Cote, C., and Duby, R.T., 1978, Embryonic stimulation of ovine and bovine corpora lutea, *J. Reprod. Fertil.* **54:**375.

Greenstein, J.S., Murray, R.W., and Foley, R.C., 1958, Effect of exogenous hormones on the reproductive processes of the cycling dairy heifer, *J. Dairy Sci.* **41:**1834 (abstract).

Guthrie, H.D., Henricks, D.M., and Handlin, D.L., 1972, Plasma estrogen, progesterone and luteinizing hormone prior to estrus and during early pregnancy in pigs, *Endocrinology* **91:**675.

Guthrie, H.D., Henricks, D.M., and Handling, D.L., 1974, Plasma hormone levels and fertility in pigs induced to superovulate with PMSG, *J. Reprod. Fertil.* **41:**361.

Hafs, H.D., Louis, T.M., Noden, P.A., and Oxender, W.D., 1974, Control of the estrous cycle with prostaglandin $F_{2\alpha}$ in cattle and horses, *J. Anim. Sci. Suppl. I* **38:**10.

Hallford, D.M., Wettemann, R.P., Turman, E.J., and Omtvedt, I.T., 1974, Luteal function in gilts after prostaglandin $F_{2\alpha}$, *J. Anim. Sci.* **38:**213 (abstract).

Hansel, W., and Wagner, W.C., 1960, Luteal inhibition in the bovine as a result of oxytocin injections, uterine dilation, and intrauterine infusions of seminal and preputial fluids, *J. Dairy Sci.* **43:**796.

Hansel, W., Concannon, P.W., and Lukaszewska, J.H., 1973, Corpora lutea of large domestic animals, *Biol. Reprod.* **8:**222.

Hansel, W., Shemesh, M., Hixon, J., and Lukaszewska, J., 1975, Extraction, isolation and identification of a luteolytic substance from bovine endometrium, *Biol. Reprod.* **13:**30.

Harrison, F.A., Heap, R.B., Horton, E.W., and Poyser, N.L., 1972, Identification of prostaglandin $F_{2\alpha}$ in uterine fluid from the nonpregnant sheep with an autotransplanted ovary, *J. Endocrinol.* **53:**215.

Harrison, F.A., Heap, R.B., and Poyser, N.L., 1976, Production, chemical composition and prostaglandin $F_{2\alpha}$ content of uterine fluid in pregnant sheep, *J. Reprod. Fertil.* **48:**61.

Heap, R.B., and Perry, J.S., 1974, The maternal recognition of pregnancy, *Br. J. Hosp. Med.* **12:**8.

Heap, R.B., Flint, A.P.F., Gadsby, J.E., and Rice, C., 1979, Hormones, the early embryo and the uterine environment, *J. Reprod. Fertil.* **55:**267.

Henderson, K.M., and McNatty, K.P., 1975, A biochemical hypothesis to explain the mechanism of luteal regression, *Prostaglandins* **9:**779.

Henderson, K.M., Scaramuzzi, R.J., and Baird, D.T., 1977, Simultaneous infusion of prostaglandin E_2 antagonizes the luteolytic effect of prostaglandin $F_{2\alpha}$ *in vivo, J. Endocrinol.* **73:**379.

Henricks, D.M., and Tindall, D.J., 1971, Metabolism of progesterone-4-C^{14} in porcine uterine endometrium. *Endocrinology* **89:**920.

Hillman, R.B., and Loy, R.G., 1975, Estrogen excretion in mares in relation to various reproductive states, *J. Reprod. Fertil. Suppl.* **23:**223.

Hixon, J., and Hansel, W., 1974, Evidence for preferential transfer of prostaglandin $F_{2\alpha}$ to the ovarian artery following intrauterine administration in cattle, *Biol. Reprod.* **11:**543.

Horton, E.W., and Poyser, N.L., 1976, Uterine luteolytic hormone: A physiological role for prostaglandin $F_{2\alpha}$, *Physiol. Rev.* **56:**595.

Hoyer, G.L., Colcord, M.L., and Weems, C.W., 1978, Effect of prostaglandin E_1 (PGE_1) on estrogen-induced luteolysis in ewes, *Proc. Am. Soc. Anim. Sci.*, p. 367, East Lansing, Michigan.

Inskeep, E.K., Smutny, W.J., Butcher, R.L., and Pexton, J.E., 1975, Effects of intrafollicular injections of prostaglandins in nonpregnant and pregnant ewes, *J. Anim. Sci.* **41:**1098.

Kelly, P.A., Robertson, H.A., and Friesen, H.G., 1974, Temporal pattern of placental lactogen and progesterone secretion in sheep, *Nature (London)* **248:**435.

Kidder, H.E., Casida, L.E., and Grummer, R.H., 1955, Some effects of estrogen injections on the estrual cycle of gilts, *J. Anim. Sci.* **14:**470.

Kimball, F.A., and Lauderdale, J.W., 1975, Prostaglandin E_1 and $F_{2\alpha}$ specific binding in bovine corpora lutea: Comparison with luteolytic effects, *Prostaglandins* **10:**313.

Kindahl, H., Edquist, L.E., Bane, A., and Granström, E., 1976, Blood levels of progesterone and 15-keto-13,14-dihydro-prostaglandin $F_{2\alpha}$ during the normal estrous cycle and early pregnancy in heifers, *Acta Endocrinol.* **82:**134.

Knight, J.W., Bazer, F.W., Thatcher, W.W., Franke, D.E., and Wallace, H.D., 1977, Conceptus development in intact and unilaterally hysterectomized–ovariectomized gilts: Interrelations between hormonal status, placental development, fetal fluids and fetal growth, *J. Anim. Sci* **44:**620.

Kooistra, L.H., and Ginther, O.J., 1976, Termination of pseudopregnancy by administration of prostaglandin $F_{2\alpha}$ and termination of early pregnancy by administration of prostaglandin $F_{2\alpha}$, colchicine or by removal of embryo in mares, *Am. J. Vet. Res.* **37:**35.

Kraeling, R.R., Barb, C.R., and Davis, B.J., 1975, Prostaglandin induced regression of porcine corpora lutea maintained by estrogen, *Prostaglandins* **9:**459.

Krzymowski, T., Kotwica, J., Okrasa, S., and Doboszynska, T., 1976, The function and regression of corpora lutea during the sow's estrous cycle after 10 hrs of prostaglandin $F_{2\alpha}$ infusion into the anterior uterine vein, Proceedings of the VIIIth International Congress on Animal Reproduction and Artificial Insemination, p. 143, Krakow.

Kunze, H., and Vogt, W., 1971, Significance of phospholipase A from prostaglandin formation, *Ann. N. Y. Acad. Sci.* **180:**123.

LaMotte, J.O., 1977, Prostaglandins. I. Presence of a prostaglandin like substance in porcine uterine flushings; II. A new chemical assay for PGF_2 alpha, *Diss. Abstr. Int. B* **37:**3798.

Lauderdale, J.W., 1974, Distribution and biological effects of prostaglandins, *J. Anim. Sci. Suppl. 1* **38:**22.

Lewis, G.S., Wilson, L., Wilks, J.W., Pexton, J.E., Fogwell, R.L., Ford, S.P., Butcher, R.L., Thayne, W.V., and Inskeep, E.K., 1977, $PGF_{2\alpha}$ and its metabolites in uterine and jugular venous plasma and endometrium of ewes during early pregnancy, *J. Anim. Sci.* **45:**320.

Lewis, G.S., Thatcher, W.W., Bazer, F.W., Roberts, R.M., and Williams, W.F., 1979, Metabolism of arachidonic acid by bovine blastocysts and endometrium, *Proc. Am. Soc. Anim. Sci.,* pp. 313–314, Tucson, Arizona.

Loeb, L., 1923, The effect of extirpation of the uterus on the life and function of the corpus luteum in the guinea pig, *Proc. Soc. Exp. Biol. Med.* **20:**441.

Lovell, J.D., Stabenfeldt, G.H., Hughes, J.P., and Evans, J.W., 1975, Endocrine patterns of the mare at term, *J. Reprod. Fertil. Suppl.* **23:**449.

Lunaas, T., 1962, Urinary estrogen levels in the sow during estrous cycle and early pregnancy, *J. Reprod. Fertil.* **4:**13.

Magness, R.R., Huie, J.M., and Weems, C.W., 1978, Effect of contralateral and ipsilateral infusion of prostaglandin E_2 (PGE_2) on luteal function in the nonpregnant ewe, *Proc. Am. Soc. Anim. Sci.,* p. 376, East Lansing, Michigan.

Malven, P.V., and Hansel, W., 1964, Ovarian function in dairy heifers following hysterectomy, *J. Dairy Sci.* **47:**1388.

Mapletoft, R.J., Del Campo, M.R., and Ginther, O.J., 1975, Unilateral luteotropic effect of uterine venous effluent of a gravid uterine horn in sheep, *Proc. Soc. Exp. Biol. Med.* **150:**129.

Mapletoft, R.J., Del Campo, M.R., and Ginther, O.J., 1976, Local venoarterial pathway for uterine-induced luteolysis in cows, *Proc. Soc. Exp. Biol. Med.* **153:**289.

Mapletoft, R.J., Miller, K.F., and Ginther, O.J., 1977, Effects of PGF_2 and PGE_2 on corpora lutea in ewes, *Proc. Am. Soc. Anim. Sci.,* p. 185, Madison, Wisconsin.

Martal, J., 1978, Placental growth hormone in sheep: Purification, properties and variations, *Ann. Biol. Anim. Biochem. Biophys.* **18:**45.

Martal, J., and Djiane, J., 1977, The production of chorionic somatomammotropin in sheep, *J. Reprod. Fertil.* **49:**285.

Martal, J., and Lacroix, M.C., 1978, Production of chorionic somatomammotropin (oCS), fetal growth and growth of the placenta and corpus luteum in ewes treated with 2-bromo-α-ergocryptin, *Endocrinology* **103:**193.

Martal, J., Lacroix, C., Loudes, C., Saunier, M., and Wintenberger-Torres, S., 1979, Trophoblastin, an antiluteolytic protein present in early pregnancy in sheep, *J. Reprod. Fertil.* **56:**63.

McCracken, J.A., 1971, Prostaglandin $F_{2\alpha}$ and corpora luteal regression, *Ann. N. Y. Acad. Sci.* **180:**456.

McCracken, J.A., Carlson, J.C., Glew, M.E., Goding, J.R., Baird, D.T., Green, K., and

Samuelsson, B., 1972, Prostaglandin $F_{2\alpha}$ identified as a luteolytic hormone in sheep, *Nature (London) New Biol.* **238:**129.

Melton, A.A., Berry, R.O., and Butter, O.D., 1951, The interval between time of ovulation and attachment of the bovine embryo, *J. Anim. Sci.* **10:**993.

Moeljono, M.P.E., Bazer, F.W., and Thatcher, W.W., 1976, A study of prostaglandin $F_{2\alpha}$ as the luteolysin in swine. I. Effect of prostaglandin $F_{2\alpha}$ in hysterectomized gilts, *Prostaglandins* **11:**737.

Moeljono, M.P.E., Thatcher, W.W., Bazer, F.W., Frank, M., Owens, L.J., and Wilcox, C.J., 1977, A study of prostaglandin $F_{2\alpha}$ as the luteolysin in swine. II. Characterization and comparison of prostaglandin F, estrogens and progestin concentrations in utero-ovarian vein plasma of nonpregnant and pregnant gilts, *Prostaglandins* **14:**543.

Molokwu, E.C., and Wagner, W.C., 1973, Endocrine physiology of the puerperal sow, *J. Anim. Sci.* **36:**1158.

Moor, R.M., and Rowson, L.E.A., 1966a, The corpus luteum of the sheep: Effect of removal of embryos on luteal function, *J. Endocrinol.* **34:**497.

Moor, R.M., and Rowson, L.E.A., 1966b, Local maintenance of the corpus luteum in sheep with embryos transferred to various isolated protions of the uterus, *J. Reprod. Fertil.* **12:**539.

Moor, R.M., and Rowson, L.E.A., 1967, The influence of embryonic tissue homogenate infused into the uterus on the lifespan of the corpus luteum in sheep, *J. Reprod. Fertil.* **13:**511.

Nancarrow, C.D., 1972, Hormonal changes around estrous in the cow, *Proc. Aust. Soc. Reprod. Biol.* **4:**21.

Nett, T.M., Holtan, D.W., and Estergreen, L., 1975, Oestrogens, LH, PMSG and prolactin in serum of pregnant mares, *J. Reprod. Fertil. Suppl.* **23:**457.

Nett, T.M., Staigmiller, R.B., Akbar, A.M., Diekman, M.A., Ellinwood, W.E., and Niswender, G.D., 1976, Secretion of prostaglandin $F_{2\alpha}$ in cycling and pregnant ewes, *J. Anim. Sci.* **42:**876.

Nishikawa, Y., 1959, *Studies in Reproduction in Horses,* Shiba Tumuracho Minatoku, Tokyo.

Noden, P.A., Hafs, H.D., and Oxender, W.W., 1973, Progesterone, estrus and ovulation after prostaglandin $F_{2\alpha}$ in horses, *Fed Proc. Fed. Am. Soc. Exp. Biol.* **32:**299.

Northey, D.L., and French, L.R., 1978, Effect of embryo removal on bovine interestrus interval, *Proc. Am. Soc. Anim. Sci.,* p. 380, East Lansing, Michigan.

Ogra, S.S., Kirton, K.T., Tomasi, T.B., and Lippes, J., 1974, Prostaglandins in the human fallopian tube, *Fertil. Steril.* **25:**250.

Palmer, E., and Terqui, M., 1977, The measurement of total plasma estrogens during the follicular phase of the mare's estrous cycle, *Theriogenology* **7:**331.

Patek, C.E., and Watson, J., 1976, Prostaglandin F and progesterone secretion by porcine endometrium and corpus luteum *in vitro, Prostaglandins* **12:**97.

Perry, J.S., Heap, R.B., and Amoroso, E.C., 1973, Steroid hormone production by pig blastocysts, *Nature (London)* **245:**45.

Perry, J.S., Heap, R.B., Burton, R.D., and Gadsby, J.E., 1976, Endocrinology of the blastocyst and its role in the establishment of pregnancy, *J. Reprod. Fertil. Suppl.* **25:**85.

Peterson, A.J., Tervit, H.R., Fairclough, R.J., Havik, P.G., and Smith, J.F., 1976, Jugular levels of 13,14-dihydro-15-keto-prostaglandin F and progesterone around luteolysis and early pregnancy in the ewe, *Prostaglandins* **12:**551.

Pexton, J.E., Weems, C.W., and Inskeep, E.K., 1975, Prostaglandins F in uterine and ovarian venous plasma from nonpregnant and pregnant ewes collected by cannulation, *Prostaglandins* **9:**501.

Pharriss, B.B., and Wyngarden, I.J., 1969, The effect of prostaglandin $F_{2\alpha}$ on the progesterone content of ovaries from pseudopregnant rats, *Proc. Soc. Exp. Biol. Med.* **130:**92.

Piper, E.L., and Foote, W.C., 1965, A luteotrophic effect of estradiol in the ewe, *J. Anim. Sci.* **24:**927 (abstract).

Raeside, J.I., 1963, Urinary estrogen excretion in the pig during pregnancy and parturition, *J. Reprod. Fertil.* **6:**427.

Rao, C.V., 1974, Characterization of prostaglandin receptors in the bovine corpus luteum cell membranes, *J. Biol. Chem.* **249:**7203.

Reddy, S., and Watkins, W.B., 1978, Purification and properties of ovine placental lactogen, *J. Endocrinol.* **78:**59.

Renfree, M., and Heap, R.B., 1977, Steroid metabolism by the placenta, corpus luteum and endometrium during pregnancy in the marsupial, *Macropus eugenii, Theriogenology* **8:**164.

Robertson, H.A., and King, G.J., 1974, Plasma concentrations of progesterone, estrone, estradiol-17β and of estrone sulphate in the pig at implantation during pregnancy and at parturition, *J. Reprod. Fertil.* **40:**133.

Robertson, H.A., and King, G.J., 1979, Conjugated and unconjugated estrogens in fetal and maternal fluids of the cow throughout pregnancy, *J. Reprod. Fertil.* **55:**453.

Robertson, H.A., and Smeaton, T., 1973, The concentration of unconjugated estrone, estradiol-17α and estradiol-17β in the maternal plasma of the pregnant ewe in relation to the initiation of parturition and lactation, *J. Reprod. Fertil.* **35:**461.

Saksena, S.K., Lau, I.F., and Chang, M.C., 1976, Relationship between estrogen, prostaglandin $F_{2\alpha}$ and histamine in delayed implantation in the mouse, *Acta Endocrinol.* **81:**801.

Schomberg, D.W., 1967, A demonstration *in vitro* of luteolytic activity in pig uterine flushings, *J. Endocrinol.* **38:**359.

Schomberg, D.W., 1969, The concept of a uterine luteolytic hormone, in: *The Gonads* (K.W. McKerns, ed.), pp. 383–397, Appleton-Century-Crofts, New York.

Seamans, K.W., Fields, M.J., Bazer, F.W., Vernon, M.W., and Sharp, D.C., 1979, *In vitro* aromatase activity of pregnant and nonpregnant equine endometrium and conceptus membranes, *Proc. Am. Soc. Anim. Sci.*, p. 335, Tuscon, Arizona.

Sharp, D.C., and Black, D.L., 1973, Changes in peripheral plasma progesterone throughout the estrous cycle of the pony mare, *J. Reprod. Fertil.* **33:**535.

Shemesh, M., and Hansel, W., 1975, Levels of prostaglandin F (PGF) in the bovine endometrium, uterine venous, ovarian artery and jugular plasma during the estrous cycle, *Proc. Soc. Exp. Biol. Med.* **148:**123.

Spies, H.G., Zimmerman, D.R., Self, H.L., and Casida, L.E., 1958, Influence of hysterectomy and exogenous progesterone and size and progesterone contact of the corpora lutea in gilts, *J. Anim. Sci.* **17:**1234 (abstract).

Squires, E.L., Wentworth, B.L., and Ginther, O.J., 1974, Progesterone concentration in the blood of mares during the estrous cycle, pregnancy and after hysterectomy, *J. Anim. Sci.* **39:**759.

Stabenfeldt, G.H., Hughes, J.P., and Evans, J.W., 1972, Ovarian activity during the estrous cycle of the mare, *Endocrinology* **90:**1379.

Stormshak, F., Kelley, H.E., and Hawk, H.W., 1969, Suppression of ovine luteal function by 17-β estradiol, *J. Anim. Sci.* **29:**476.

Terrill, C.E., 1974, Reproductive cycle in sheep, in: *Reproduction in Farm Animals* (E.S.E. Hafez, ed.), pp. 265–274, Lea and Febiger, Philadelphia.

Thatcher, W.W., and Chenault, J.R., 1976, Reproductive physiological responses of cattle to exogenous prostaglandin $F_{2\alpha}$, *J. Dairy Sci.* **59:**1366.

Thatcher, W.W., Wilcox, C.J., Bazer, F.W., Collier, R.J., Eley, R.M., Stover, D.G., and Bartol, F.F., 1979, Bovine conceptus effects prepartum and potential carryover effects postpartum, Proceedings of the Beltsville Symposium on Animal Reproduction, May 14–17, 1978, Beltsville, Maryland.

Thornburn, G.D., Cox, R.I., Currie, W.B., Restall, B.J., and Schnieder, W., 1973, Prostaglandin F and progesterone concentrations in the utero-ovarian venous plasma of the ewe during the estrous cycle and early pregnancy, *J. Reprod. Fertil. Suppl.* **18:**151.

Valdivia, E.O., 1977, Estrogen secretion by the procine conceptus, M.S. thesis, University of Florida, Gainesville.

Van Niekerk, C.H., Morgenthal, J.C., and Gerneke, W.H., 1975, Relationship between the morphology of and progesterone production by the corpus luteum of the mare, *J. Reprod. Fertil. Suppl.* **23:**171.

Velle, W., 1960, Early pregnancy diagnosis in the sow, *Vet. Rec.* **72:**116.

Vernon, M.W., 1979, The role of prostaglandins in the utero-ovarian axis of the cycling and early pregnant mare, Ph.D. dissertation, University of Florida, Gainesville.

Walker, F.M.M., Patek, C.E., Leaf, C.F., and Watson, J., 1977, The metabolism of prostaglandins $F_{2\alpha}$ and E_2 by nonpregnant porcine endometrial and luteal tissue and early pregnant porcine endometrial tissue, luteal tissue and conceptuses *in vitro, Prostaglandins* **14:**557.

Warren, J.E., Hawk, H.W., and Bolt, D.J., 1973, Evidence for progestational priming of estradiol induced luteal regression in the ewe, *Biol. Reprod.* **8:**435.

Wettemann, R.P., Hallford, D.M., Kreider, D.L., and Turman, E.J., 1974, Parturition in swine after prostaglandin $F_{2\alpha}$, *J. Anim. Sci.* **39:**228 (abstract).

Willis, M.C., Fields, M.J., Wise, T.H., Dueben, B.D., and Bazer, F.W., 1979a, *In vitro* metabolism of H^3-progesterone to androgens and estrogens by pregnant sheep endometrium, *Proc. Am. Soc. Anim. Sci.*, p. 350, Tuscon, Arizona.

Willis, M.C., Fields, M.J., Wise, T.H., Dueben, B.D., and Bazer, F.W., 1979b, *In vitro* metabolism of H^3-progesterone to androgens and estrogens by pregnant sheep endometrium and conceptus membranes, *Proc. Am. Soc. Anim. Sci.*, p. 350, Tuscon, Arizona.

Wilson, J.D., 1972, Recent studies on the mechanism of action of testosterone, *N. Engl. J. Med.* **287:**1284.

Wilson, L., Butcher, R.L., and Inskeep, E.K., 1972, Prostaglandin $F_{2\alpha}$ in the uterus of ewes during early pregnancy, *Prostaglandins* **1:**479.

Wiltbank, J.N., and Casida, L.E., 1956, Alteration of ovarian activity by hysterectomy, *J. Anim. Sci.* **15:**134.

Wiltbank, J.N., Ingalls, J.E., and Rowden, W.W., 1961, Effects of various forms and levels of estrogen alone or in combinations with gonadotropins on the estrous cycle of beef heifers, *J. Anim. Sci.* **20:**341.

Young Lai, E.V., 1971, Steroid content of the equine ovary during the reproductive cycle, *J. Endocrinol.* **50:**589.

Zavy, M.T., Bazer, F.W., and Sharp, D.C., 1978a, A nonsurgical technique for the collection of uterine fluid from the mare, *J. Anim. Sci.* **47:**672.

Zavy, M.T., Bazer, F.W., Sharp, D.C., Frank, M., and Thatcher, W.W., 1978b, Uterine luminal prostaglandin F in cycling mares, *Prostaglandins* **16:**643.

Zavy, M.T., Bazer, F.W., Sharp, D.C., and Wilcox, C.J., 1979a, Uterine luminal proteins in the cycling mare, *Biol. Reprod.* **20:**689.

Zavy, M.T., Bazer, F.W., Sharp, D.C., and Thatcher, W.W., 1979b, Endocrine aspects of the uterine environment in nonpregnant and pregnant gilts, *Proc. Am. Soc. Anim. Sci.*, pp. 351–352, Tuscon, Arizona.

Zavy, M.T., Mayer, R., Vernon, M.W., Bazer, R.W., and Sharp, D.C., 1979c, An investigation of the uterine luminal environment of nonpregnant and pregnant pony mares, *J. Reprod. Fertil. Suppl.* **27:**403.

28

Porcine and Human Ovarian Nonsteroidal Follicular Regulators

Oocyte-Maturation Inhibitor, Luteinization Inhibitor, Luteinizing-Hormone-Receptor-Binding Inhibitor, Follicle-Stimulating-Hormone-Binding Inhibitor, and Inhibin F

Cornelia P. Channing, L. D. Anderson, Sarah Lipford Stone, and Satish Batta

1. Introduction

The ovarian follicle and the oocyte mature at a well-controlled rate in response to a constant as well as to changing levels of pituitary gonadotropins. In the primate, only one follicle grows to maturity and ovulates an ovum each menstrual cycle, while at least 200 other follicles undergo atresia (for reviews, see Greenwald, 1978; Byskov, 1978). The mechanism for choice of one follicle over the other is not known. Since all follicles are exposed to the same level of luteinizing hormone (LH) and follicle-stimulating hormone (FSH) (Sakai and Channing, 1979a,b), local ovarian factors rather than systemic hormones should be responsible for choice

Cornelia P. Channing, Sarah Lipford Stone, and *Satish Batta* • Department of Physiology, University of Maryland School of Medicine, Baltimore, Maryland 21201. *L.D. Anderson* • Department of Anatomy, University of Maryland School of Medicine, Baltimore, Maryland 21201.

of follicle. Local factors controlling follicular maturation include steroids as well as nonsteroidal factors. The local steroids primarily secreted by the ovary are: (1) estrogen, which stimulates granulosa-cell growth (Bradbury, 1961; Rao *et al.,* 1978; Harman *et al.,* 1975; (2) androgen, which may play a role in atresia in the rat (Louvet *et al.,* 1975), since administration of testosterone antiserum diminishes follicular atresia in the rat; and (3) progesterone, which may partially regulate the initiation of new follicular growth, since DeZerga and Hodgen (1979), personal communication) found that the insertion of Silastic wafers impregnated with progesterone into the ovary immediately after removal of its corpus luteum can lead to a delay in follicular growth. Nonsteroidal factors include: an *LH-receptor-binding inhibitor* (LHRBI), which is a small peptide present in aqueous extracts of corpora lutea from rat (Yang *et al.,* 1976, 1978, 1979), pig (Sakai *et al.,* 1977; Kumari *et al.,* 1979; Tucker *et al.,* 1979), and human ovaries (Kumari *et al.,* 1979); *luteinization inhibitor* (LI), a polypeptide present in follicular fluid from small antral follicles of the pig (Ledwitz-Rigby *et al.,* 1977) and human (Lunenfeld *et al.,* 1976); an *FSH-binding inhibitor* (FSHBI), present in bovine and porcine follicular fluid (Darga and Reichert, 1978); an *oocyte-maturation inhibitor* (OMI), a polypeptide of 2000 daltons present in human (Hillensjö *et al.,* 1978) and porcine (Tsafriri *et al.,* 1974; Hillensjö *et al.,* 1979a,b; Stone *et al.,* 1978) follicular fluid; and *inhibin F,* a polypeptide of greater than 10,000 daltons from follicular fluid. Inhibin-F activity has been found in follicular fluid of the human (Chari *et al.,* 1979) and the pig (Schwartz and Channing, 1977; Marder *et al.,* 1978). However, it does not act directly as a local regulator of follicular growth; rather, it acts indirectly to control follicular growth by inhibiting pituitary FSH secretion.

This chapter will summarize some recent studies on the role of nonsteroidal follicular regulators in porcine and human follicular function. The reader is referred to several excellent reviews on the role of steroids in control of follicular function (Schomberg, 1979; Hillier *et al.,* 1977).

2. Materials and Methods

2.1. Oocyte-Maturation-Inhibitor Assay

OMI obtained from medium-sized follicles using cultured porcine cumulus-enclosed oocytes was assayed as detailed previously (Tsafriri and Channing, 1975a,b; Stone *et al.,* 1978). In brief, fractions of follicular fluid were added in various dilutions to 4 replicate culture wells of 12–15 oocytes per well followed by incubation of the cumulus-enclosed oocytes for 2 days under 5% CO_2 in air at 37°C. At the end of the culture period, the oocytes were examined for nuclear maturation after being stained with aceto-orcein and classified according to the criteria of Hunter and Polge (1966) (Fig. 1A). The percentage inhibition of maturation was estimated for each (Fig. 1B). The culture media obtained from each well were assayed for progesterone by radioimmunoassay (RIA) (using specific 11-OH-antiprogesterone–bovine serum albumin conjugate kindly provided by G. Niswender) without prior extraction using methods described by Hillensjö *et al.*, (1978). We have found that fractions containing OMI activity also inhibit cumulus-cell progesterone secretion as well as oocyte maturation.

2.2. Human Ovarian-Cell and Follicular-Fluid Harvest

Follicular fluid was aspirated from follicles of two groups of patients, normal and those with cystic follicles. The average age of the patients was 35 ± 8 years (range 29–48). Follicular fluid was pooled from normal subjects who had at least one regular menstrual cycle prior to biopsy. A total of 92 subjects were used in this study. Normal follicles ranging from 5 to 15 mm were used as a source of follicular fluid. The fluids (cystic and normal) were divided in half, and half of each aliquot was subjected to charcoal stripping by mixing with 2% activated charcoal (wt./vol.) for 1 hr at 4°C. The charcoal was removed by centrifugation at 1000 rpm for 1 hr and at 106,000 *g* for 1 hr. Previously, this treatment has been shown to remove radiolabeled steroids and prostaglandins added to follicular fluid from 97.6 to 99.9%, respectively (Batta and Channing, 1980).

The other half of the two pools of human follicular fluid was filtered through a PM-10 membrane (as detailed by Hillensjö *et al.*, 1978), with the resulting PM-10 retentate (>10,000 daltons) being half the original fluid volume, while the filtrate (<10,000-dalton fraction) was lyophilized to dryness and reconstituted with distilled water to give a 5-fold concentrate. Serum obtained from 23 of these patients (chosen at random) was pooled and added directly or after PM-10 filtration. A volume of cystic follicular fluid was treated similarly.

2.3. *In Vivo* Bioassay for Inhibin F

The test system used to assay for inhibin-F activity was the bilaterally ovariectomized female rat in the metestrous phase of the estrous cycle. In these rats: FSH levels begin rising approximately 4 hr after surgery, and by 8 hr the plasma FSH levels reach 6- to 8-fold the

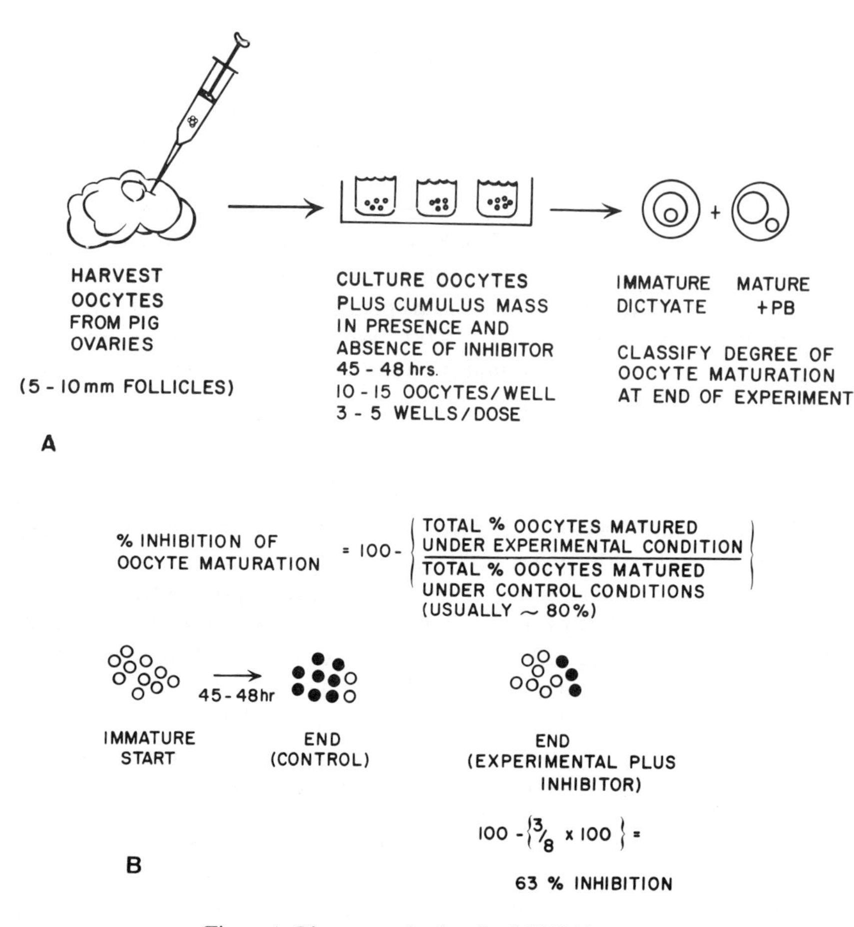

Figure 1. Diagrammatic sketch of OMI bioassay.

presurgery values. The postcastration rise in FSH is not significantly suppressed by the injection of large doses of estrogen and progesterone (Campbell and Schwartz, 1977). However, two injections of porcine follicular fluid (pFFl) given in doses of 500 μl or less, one within 30 min of surgery and the other at 3 hr postsurgery, have been shown to significantly supress the postcastration rise in plasma FSH (Marder *et al.,* 1978; Batta and Channing, unpublished data). In the present experiment, human follicular fluid (hFFl) was given in a volume of 500 μl at 30 min and 3 hr after bilateral castration. Blood samples were taken prior to and after surgery, and every hour thereafter to 8 hr, when the experiment was terminated. The plasma FSH was determined by RIA using RP-I (supplied by the NIAMDD through Dr. A. F. Parlow) as reference.

2.4. *Inhibin-F Bioassay in Cultured Rat Anterior-Pituitary Cells*

For inhibin-F studies with monkey ovarian-vein blood samples and hFFl, rat anterior-pituitary-cell cultures were used. Pituitary glands were dissected from adult female rats after decapitation and kept in Hanks' salts without Ca^{2+} and Mg^{2+} until 40 anterior pituitary glands were accumulated. Subsequently, the cells were enzymatically dispersed using collagenase, hyaluronidase, and Viokase according to the method of Vale *et al.* (1972). The pituitary cells were counted using 0.06% trypan blue dye exclusion and inoculated at a density of 150,000 viable cells per culture well and incubated for 3 days in Dulbecco's Modified Eagle's Medium (DMEM) containing 10% rat, 3% horse, and 2.5% fetal calf serum. The fluids to be tested were added for a 24-hr period from days 3–4 of culture after the cells were washed twice with DMEM. The test incubation medium consisted of DMEM plus 10^{-8} M estradiol and 10^{-7} M progesterone plus 10% rat, 3% horse, and 2.5% fetal calf serum. Each fluid was added to three replicate wells in 0.5 ml medium, while a standard pool of charcoal-treated pFFl was run in each assay at doses of 2 and 0.2%. Each anterior pituitary gland yielded about 6 wells of cultured cells.

FSH and LH in the conditioned culture medium were measured by RIA using an NIAMDD kit generously provided by Dr. A. F. Parlow.

For studies on hFFl, the pituitary cells were cultured in TC 199 plus 15% fetal calf serum rather than the DMEM growth media as described above.

2.5. *Collection of Monkey Ovarian-Vein Blood*

Adult female, cycling rhesus monkeys were observed for at least three normal cycles prior to blood sampling. In the cycle chosen for sampling, blood was collected for days 4–5 through days 9–12 of the cycle. Serum estrogen and progesterone were measured by RIA as detailed previously (Channing and Coudert, 1976).

On the day serum estrogen levels exceeded 200 pg/ml, the monkey was subjected to laparotomy and bilateral samples of ovarian-vein blood were obtained as described previously (Channing and Coudert, 1976). Peripheral samples were obtained prior to and after the ovarian-vein samples. In some instances, the preovulatory follicle was removed and the theca and granulosa cells cultured and later classified using the criteria for atretic and viable follicles that have been summarized elsewhere (Channing, 1980). The fluid was aspirated from the follicle using a 25-gauge needle attached to PE50 tubing and diluted with DMEM.

3. Results and Discussion

3.1. Oocyte-Maturation Inhibitor

OMI has been isolated from pFFl according to the purification scheme outlined in Fig. 2 (Pomerantz *et al.,* 1979). Results shown in Table I show that after Sephadex G25 column chromatography or carboxymethyl (CM)–and diethylaminoethyl (DEAE)–Sephadex, there was about a 125-fold purification compared to the PM-10 low-molecular-weight fraction of follicular fluid, while there was about a 15,000-fold purification compared to native follicular fluid.

The inhibitory effects of the Sephadex Peak A fraction of OMI on oocyte maturation were reversible, demonstrating that they did not cause

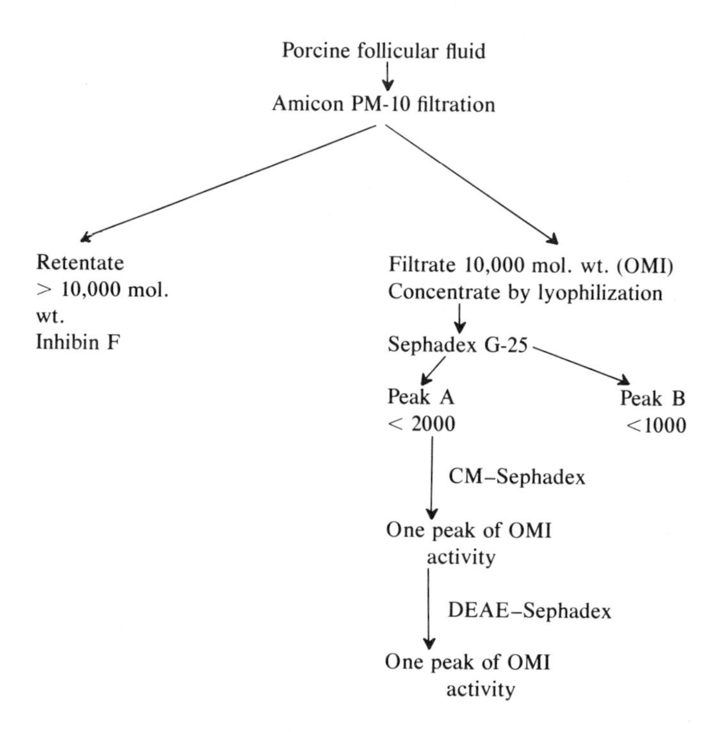

Figure 2. Purification scheme for isolation of OMI from pFFl.

Table I. Purification of OMI from pFFI[a]

Fraction	Volume (ml)	Peptide (mg/ml)[b]	Units/mg	Total units	Fold purification[b]
Follicular fluid	900	—	—	1800[c]	—
Amicon PM-10 filtrate	24.5	105	0.48	1225	1
Sephadex G-25 Peak A	20	43.4	1.38	1100	2.9
CM–Sephadex	15	29	5.17	2200[d]	10.8
DEAE–Sephadex (on 1100 units of CM–Sephadex fraction)	2	1.67	60	200	125

[a] Adapted from Pomerantz *et al.* (1979).
[b] Based on peptide determination by the fluorescamine method (Udenfriend *et al.*, 1972) using β-MSH as a standard.
[c] Estimated; 1:1 dilution inhibited maturation by approximately 50%.
[d] The observed increase in activity was probably due to removal of stimulator by purification.

nonspecific necrosis of the oocyte. The OMI fractions eluted from each column also inhibited cumulus-cell progesterone secretion in a reversible fashion (Hillensjö *et al.*, 1978) (Fig. 3). The inhibitory actions of the OMI on oocyte maturation appear to be mediated by the cumulus cells, since mechanically denuded oocytes were not inhibited from maturing (Hillensjö *et al.*, 1979a,b) (Fig. 4).

Viable (Hillensjö *et al.*, 1978) (Fig. 5) as well as cystic (Fig. 6) hFFl contained OMI in the low-molecular-weight PM-10 filtrate. However, the pool of cystic fluid contained less OMI activity as compared to the pool of viable follicular fluid, since a 1:20 dilution of the viable fluid, but not the cystic fluid, exerted a significant ($P < 0.01$) inhibition of oocyte maturation.

The cystic fluid inhibited progesterone secretion by the cultured cumulus-enclosed oocytes in a dose-dependent manner (Fig. 6 B). This inhibition of progesterone secretion was observable at lower doses than was the inhibition of oocyte maturation. Both the hFFl and the pFFl (low-molecular-weight fraction) also inhibited cumulus-cell outgrowth (Fig. 7).

The significance of observation of less OMI activity in cystic compared to viable hFFl is of interest to speculate on, even though more observations are needed to confirm this finding. Since granulosa cells are the source of porcine follicular OMI (Centola and Anderson, 1979; Tsafriri and Channing, 1975a), it is possible that the loss of granulosa cells in some types of cystic follicles results in decreased OMI levels in granulosa-cell-poor "cystic" follicles. This could lead to premature

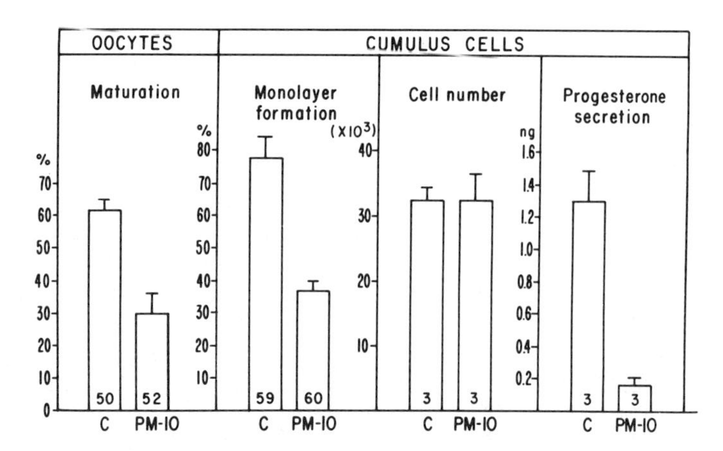

Figure 3. Actions of low-molecular-weight fraction of pFFl. Cumulus-enclosed oocytes were cultured in 199A in the absence and presence of the PM-10 filtrate of pFFl in 1:50 dilution. Values are means ± S.E. The numbers at the bottom of the bars in the two left most panels indicate the total number of oocytes examined for maturation or cumuli examined for monolayer formation. Cell number and progesterone secretion (the two right most panels) are expressed per culture ($N = 3$), each consisting of 15 oocyte–cumulus complexes. The effect of the inhibitor was significant ($P < 0.01$) except on the cell number. Reproduced from Hillensjö *et al.* (1979a) with permission.

maturation of the oocyte followed by oocyte degeneration. This working hypothesis is supported by the finding of no viable oocytes from human cystic follicles (Channing, unpublished observation). This may be a physiological protective mechanism to hasten degeneration of an abnormal oocyte.

3.2. Luteinizing-Hormone-Receptor-Binding Inhibitor

In 1976 Yang *et al.* (1976) reported that extracts of pseudopregnant and pregnant but not immature rat ovaries contained a low-molecular-weight polypeptide LHRBI.

We have observed that extracts of old and midluteal phase porcine corpus luteum tissue contained more LHRBI activity compared to extracts of young corpus luteum (Tucker *et al.,* 1979) (Fig. 8). It is possible that LHRBI may enter the follicle antrum and control the binding of LH to its receptor. Recently, Pomerantz, Anderson, and Channing (unpublished observations) found variable amounts of LHRBI activity in some pools of pFFl, suggesting the concept that it may diffuse from the corpus luteum to the follicle.

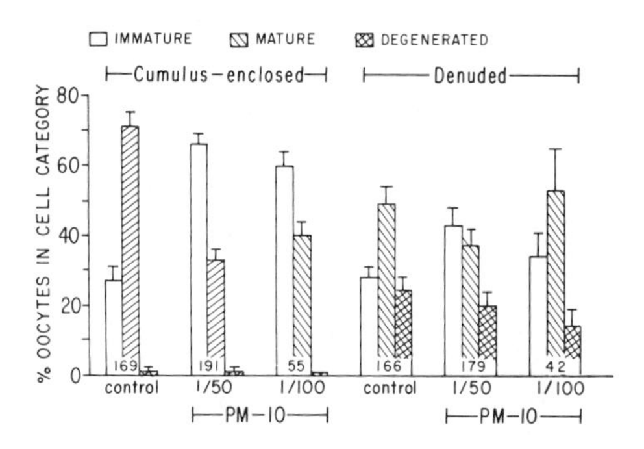

Figure 4. Role of cumulus cells in the inhibitory action of OMI on porcine oocyte maturation. Reproduced from Hillensjö *et al.* (1979a) with permission.

3.3. *Inhibin-F Activity*

3.3.1. *Studies in the Rat*

To determine whether pFFl can exert an inhibitory action on serum FSH, oocyte maturation, and follicular steroidogenesis, 1 ml charcoal-treated pFFl from medium and large follicles was administered to 26-day-old rats 24, 48, and 52 hr after injection of 20 IU pregnant mare serum gonadotropin (PMSG). Control rats were given charcoal-treated porcine serum. Starting 40 hr after PMSG injection, rats (N = 7–11) were sacrificed at 4-hr intervals, and the reproductive organs were weighed. Serum LH, FSH, estrogen, and progesterone were measured by RIA. Follicular fluid was collected from 0.5-mm follicles. Oocytes were examined for maturation (germinal-vesicle breakdown) under a

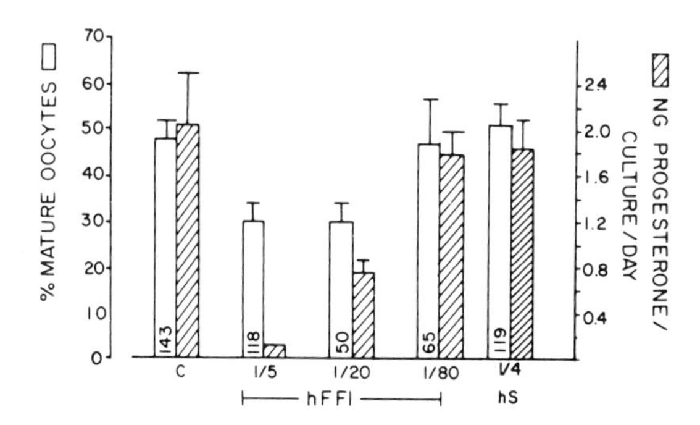

Figure 5. Effect of normal hFFl PM-10 filtrate on maturation of porcine oocytes (□) and cumulus-cell progesterone secretion (▨).

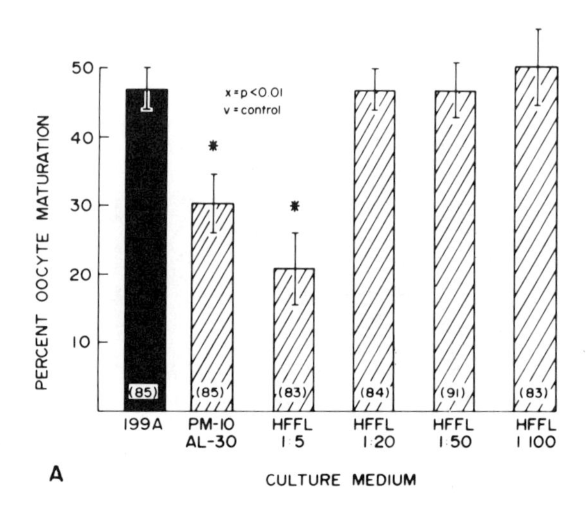

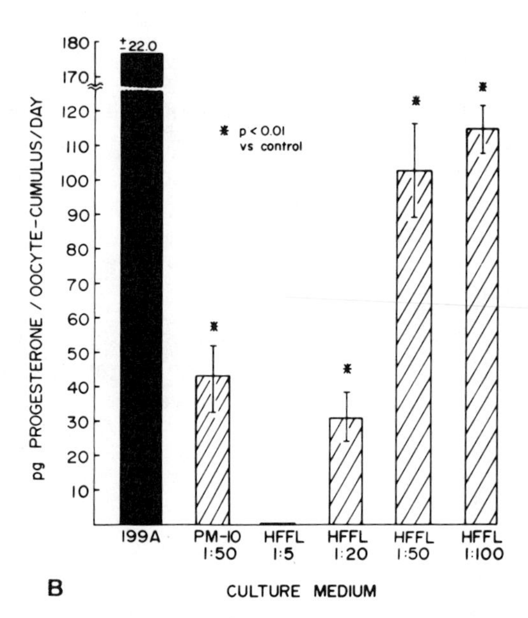

Figure 6. Effect of PM-10 filtrate of hFFl on porcine oocyte maturation (A) and cumulus-cell progesterone secretion (B).

stereomicroscope. The results demonstrated that pFFl caused a decrease in serum FSH levels (ng/ml, means ± S.E.M., RP-I from 127.2 ± 16.1, 1013.7 ± 374.0, 1281.0 ± 164.9, 1416.1 ± 170.4, 412.3 ± 37.9, and 252.3 ± 30.5 at 48, 52, 56, 60, 64, and 68 hr after PMSG in control rats to less than 50 ng/ml at all time intervals in treated rats (Figs. 9 and 10).

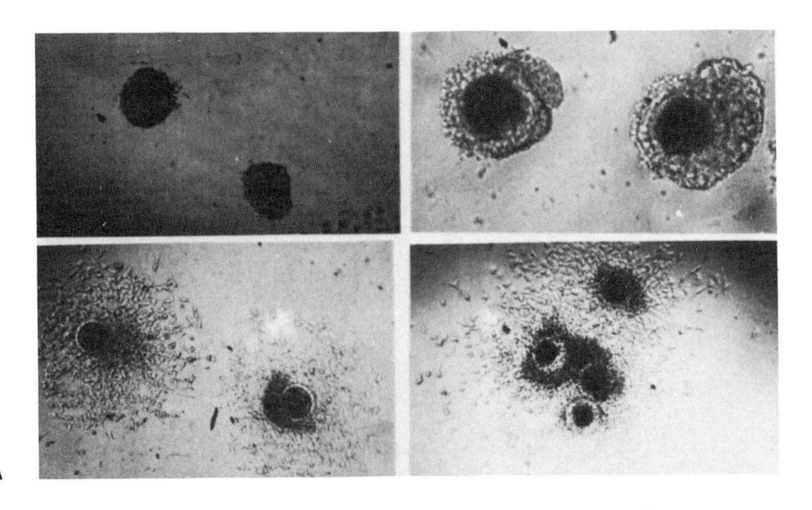

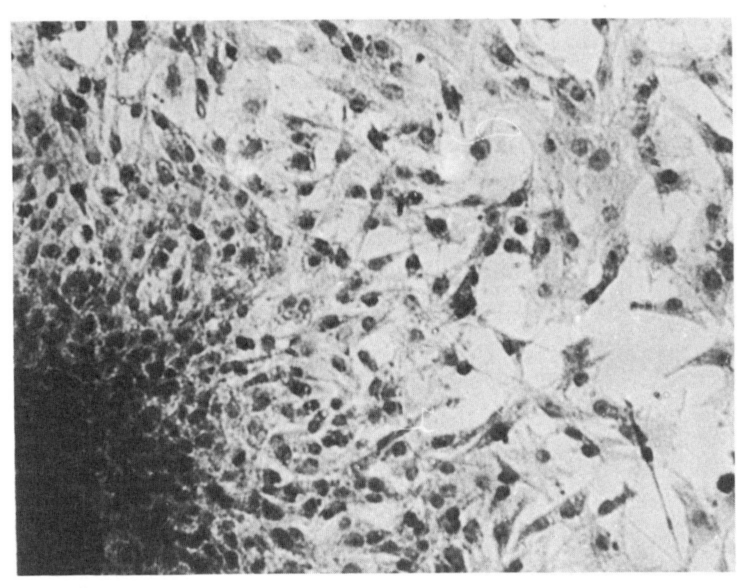

Figure 7. Effects of low-molecular-weight fraction of pFFl on cumulus-cell morphology in culture. (A) *Top right:* Two isolated cumulus–oocyte complexes prior to culture. × 100. *Bottom left:* Two cumulus–oocyte complexes after culture in control medium for 48 hr. × 50. *Top left:* Two cumulus–oocyte complexes after culture for 48 hr in the presence of a 1:50 dilution of PM-10, the low-molecular-weight concentrate of pFFl. × 50. *Bottom right:* Four cumulus–oocyte complexes cultured for 24 hr in the presence of a 1:50 dilution of the low-molecular-weight concentrate of pFFl followed by culture in control medium for an additional 24 hr × 50. (B) A higher-power picture of a cumulus–oocyte complex fixed and stained after culture for 48 hr under control conditions. × 200.

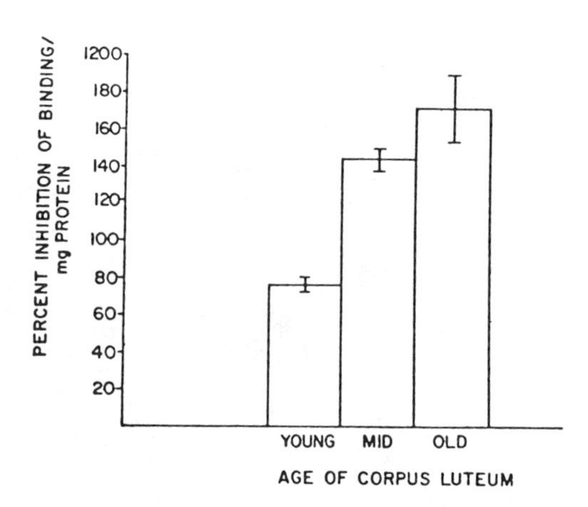

Figure 8. LHRBI levels in extracts of young, middle, and old porcine corpus luteum. Adapted from Tucker *et al.* (1979) with permission.

The pFFl did not alter serum LH levels, the degree of oocyte maturation, or the rate of ovulation. Estrogen and progesterone patterns in follicular fluid followed those in serum, and both were not altered by pFFl (Fig. 10). These studies showed that the pFFl could selectively inhibit FSH but not LH secretion and that the endogenous preovulatory FSH surge was not required for oocyte maturation, ovulation, and follicular estrogen and progesterone secretion in PMSG-treated immature rats (Figs. 9 and 10). The possibility exists that the PMSG itself brought about these ovarian changes and that abolition of the endogenous FSH would not have made any difference.

3.3.2. *Inhibin-F levels in Cystic and Normal hFFl*

3.3.2a. Effect of Normal hFFl on Rat Plasma in Vivo Bioassay. Two injections of a pool of normal hFFl (charcoal-treated or nontreated) given at 30 min and at 3 hr postsurgery reduced the plasma FSH rise by 33% compared to control rats receiving serum, whereas pFFl decreased serum FSH levels by 55% (Fig. 11). This indicates that hFFl has inhibinlike activity, and this activity was not removed by charcoal treatment, which would remove steroids, prostaglandins, thymidine, and uridine (Batta and Channing, 1979). In addition, hFFl was found to be about 40% less active, as compared to pFFl, in its FSH-suppressing activity using this test system. Efforts were made to partially purify the hFFl inhibin and determine its molecular weight. The hFFl was passed through a PM-10 membrane in an Amicon cell under nitrogen at 65 psi with filtrate and retentate recovered and tested for inhibinlike activity. The PM-10 membrane filter allows low-molecular-weight substances (≤ 10,000 daltons) to pass through, whereas substances larger than 10,000 daltons remain in the retentate. Aliquots of filtrate and retentate were injected twice into bilaterally ovariectomized female rats. The data on FSH determinations of periodic blood samples indicated that

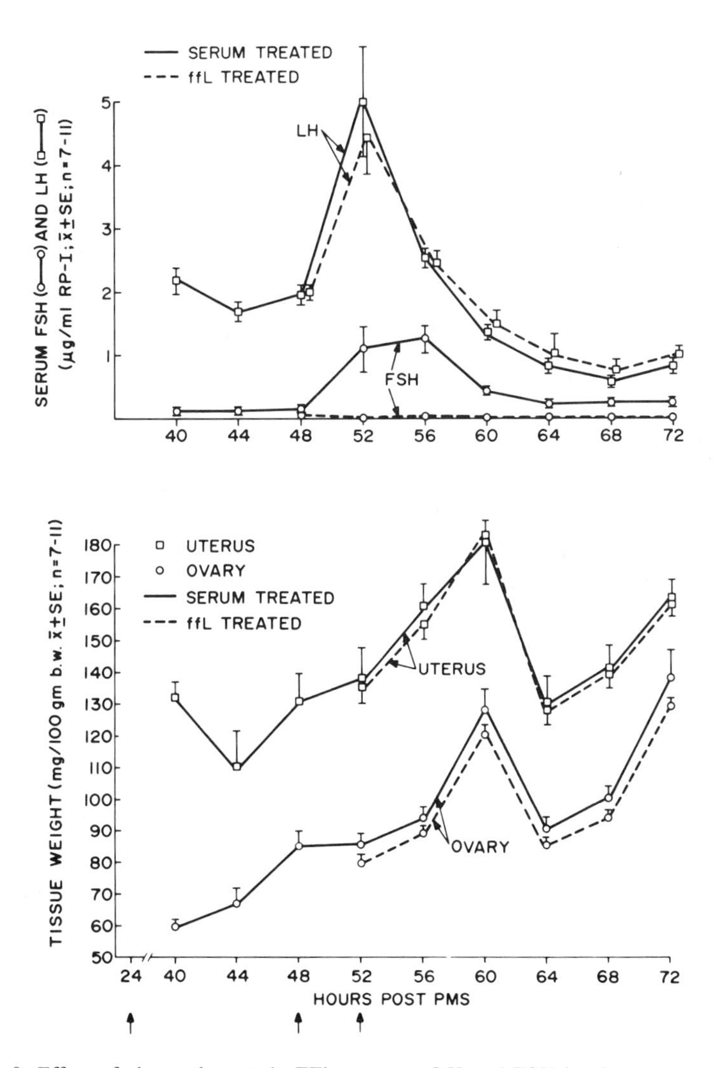

Figure 9. Effect of charcoal-treated pFFl on serum LH and FSH levels *(top)* and uterine and ovarian weight *(bottom)* in the PMSG-treated immature rat. Injections (1 ml) of pFFl were given at the times indicated by the arrows at the bottom of the graph. From Batta, Luchansky, Knudsen, and Channing (unpublished observations).

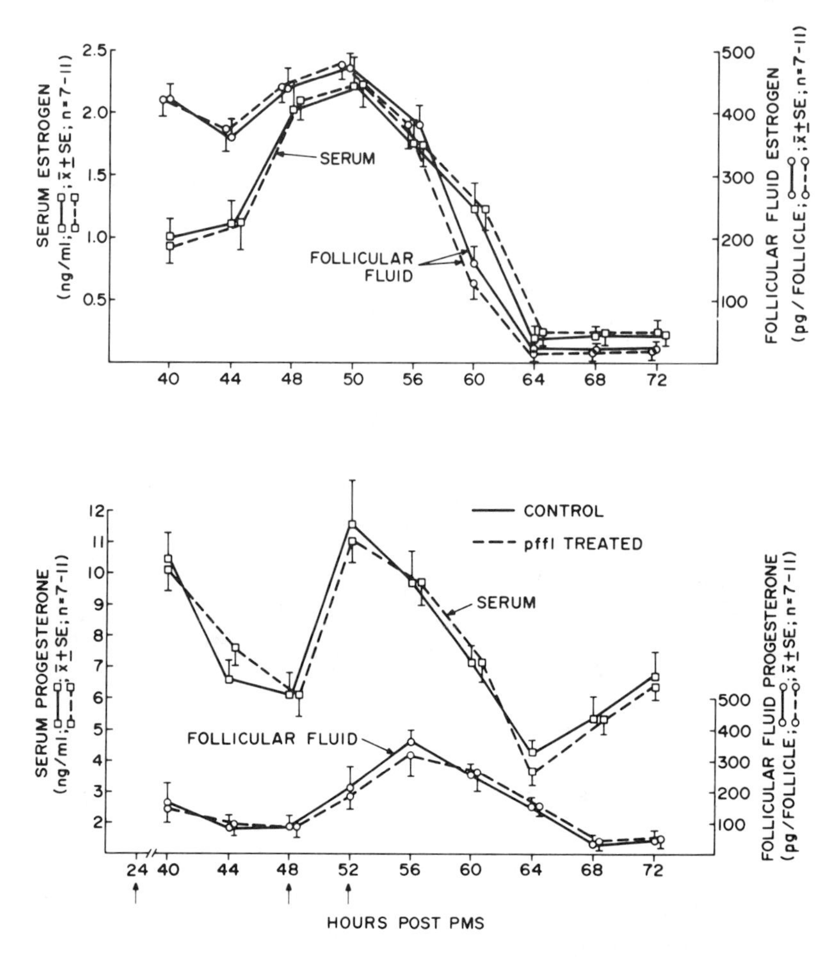

Figure 10. Effect of charcoal-treated pFFl on serum and follicular fluid estrogen *(top)* and progesterone *(bottom)* levels in the PMSG-treated immature rat. Intraperitoneal injections (1 ml) of pFFl were administered at the times designated by the arrows at the bottom of the graph. From Batta, Luchansky, Knudsen, and Channing (unpublished observations).

inhibin activity resided in PM-10 retentate, which contains substances of more than 10,000 daltons, whereas PM-10 filtrate was devoid of inhibin activity (Fig. 12).

3.3.2b. Effect of hFFl on FSH Secretion by Rat Pituitary Cells in Culture. Since the *in vivo* method required large volumes of hFFl for assay of inhibinlike activity, the effect of hFFl on rat pituitary-cell secretion of FSH in culture was studied, since it required much less material (Table II). During a 24-hr culture period, hFFl was effective in depressing the basal FSH secretion by direct action on the pituitary cells. The hFFl exerted a dose-related decrease in FSH secretion into the culture medium. In addition, when comparison of hFFl with pFFl (vol./vol.) was made on the monolayer culture of rat anterior pituitary, the hFFl was about 5 times less active than pFFl (Table II).

3.3.3. Inhibin Secretion by Cultured Normal Human Ovarian Cell Types

To determine whether human ovarian cells secrete an inhibin-F substance *in vitro* that could suppress FSH secretion, a 2% concentration of ovarian-cell medium was added to pituitary-cell cultures followed by incubation for 24 hr (Fig. 13). The data indicated that media from granulosa cells and theca (combined) produced a maximum amount of inhibinlike activity, with lesser activity found in the media from granulosa cells alone or a combined culture of granulosa, theca, and stroma media. In contrast, conditioned media from the culture of theca or stroma alone did not produce any significant decrease in FSH secretion from the cultured pituitary cells. Addition of testosterone to the cultures of granulosa cells decreased the secretion of inhibin by the granulosa cells. In contrast, in the case of stroma, which did not produce any significant amounts of inhibin activity when cultured alone, the addition of testos-

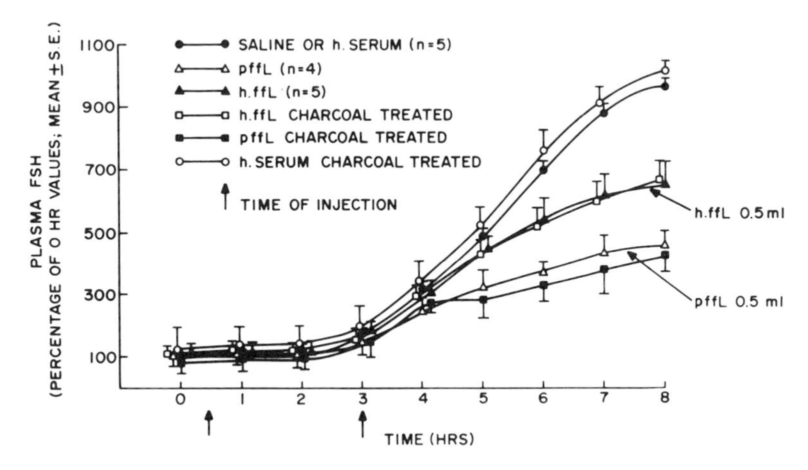

Figure 11. Effect of hFFl, human serum, pFFl with or without charcoal treatment, and saline on plasma FSH in bilaterally ovariectomized female rats. The fluids were administered intraperitoneally at the times indicated by the arrows at the bottom of the graph. Reproduced from Batta and Channing (1980) with permission.

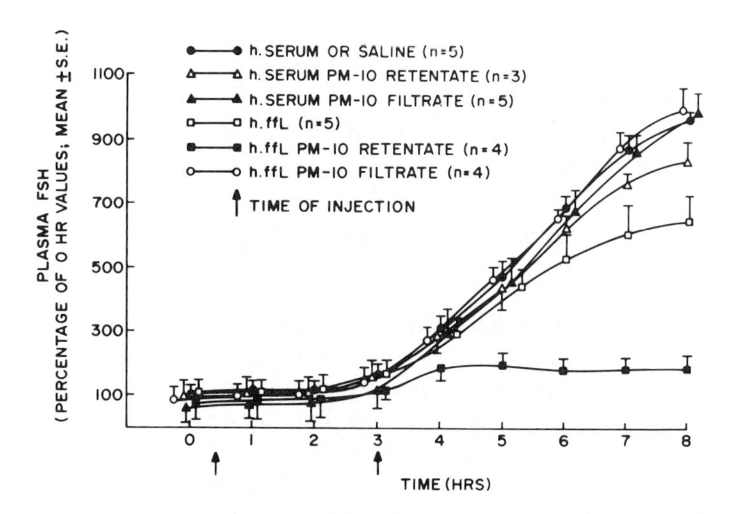

Figure 12. Effect of PM-10 retentate and filtrate of human serum and follicular fluid and saline on plasma FSH in bilaterally ovariectomized female rats. The fluids were given at 0.5-ml doses intraperitoneally at the times indicated by the arrows at the bottom of the graph. Reproduced from Batta and Channing (1980) with permission.

terone led to production of some substance(s) that when added to the pituitary cells resulted in a stimulation of FSH secretion. Addition of testosterone alone to cultured pituitary cells produced little or no effect on FSH secretion (Shander *et al.*, 1980). The data obtained thus far indicate that: (1) Granulosa cells were able to secrete inhibin, and addition of testosterone led to a decrease in activity. In another series of

Table II. Effect of pFFl and Various Dilutions of hFFl on 24-Hour FSH Secretion by Monolayer Cultures of Rat Anterior-Pituitary Cells

Treatment	FSH (ng/ml)	Percent of control
Control	1285.3 ± 30	100.0
hFFL[a]		
2%	1207.5 ± 156	93.3
5%	1059.3 ± 80	82.4
10%	892.5 ± 97	69.5
pFFl (2%)	872.1 ± 52	67.8

[a] Each dose of hFFl was added to 3 replicate culture wells and the culture incubated for 24 hr.

experiments, testosterone was found to inhibit progesterone secretion by granulosa alone and granulosa and theca. (2) The combined culture of granulosa and theca secreted more inhibin, which not only affected the secretion of FSH by the pituitary cells but also caused the cells to morphologically round up and reverse their explantation. At this time, we do not know which effect was first, reversal of explantation or a strong inhibition of FSH secretion. (3) The combined culture of granulosa, theca, and stroma also produced inhibin activity, but less than that found with the granulosa and theca combined culture and similar to granulosa-cell inhibin activity. (4) Stroma culture without testosterone added does not appear to secrete inhibin activity, but when cultured in the presence of testosterone, stroma secretes a substance that may

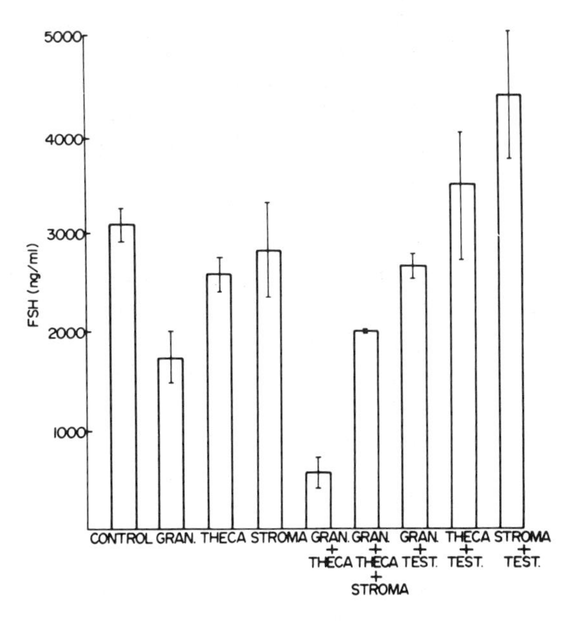

Figure 13. Effect of 2% conditioned culture medium from cultured human ovarian cell types on rat pituitary FSH secretion. Each point is the mean ± S.E. of 4 replicate pituitary cultures. Statistical analysis using the Duncan multiple-range test indicated that FSH inhibition by media conditioned by granulosa, granulosa plus theca, and theca plus stroma was significant vs. control. Addition of testosterone to granulosa caused a significant ($P < 0.05$) reversal of the inhibitory effect. Various human ovarian cell types were cultured for 2 days and the conditioned medium was at a concentration of 2% to rat anterior pituitary cultures and the mixture incubated for 24 hr. Cell types from 6 pairs of ovaries were grown for 2–8 days and the conditioned media pooled. Ovarian cell types were cultured as detailed previously (Batta *et al.*, 1980).

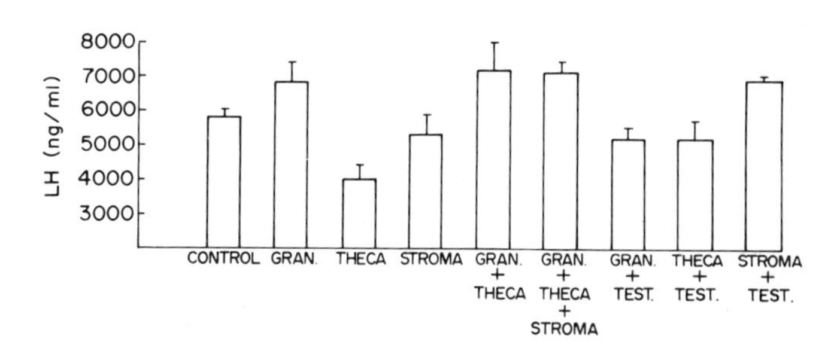

Figure 14. Effect of 2% conditioned culture medium from cultured human ovarian cell types on rat pituitary LH secretion.

significantly increase the secretion of FSH by the pituitary cells in culture.

The effect of culture medium conditioned by various ovarian-cell types on LH secretion by cultured pituitary cells was also studied. The data indicated (Fig. 14) that culture medium conditioned by theca caused a significant diminution of LH secretion, whereas granulosa-cell-conditioned culture medium caused a minor increase in LH secretion by the cells in culture. However, when granulosa and theca were cultured in combination, the conditioned culture medium increased the LH secretion significantly over that of control pituitary-cell cultures. The combined granulosa and theca culture in the presence of testosterone did not elicit any significant stimulatory response of LH secretion. To our surprise, culture medium conditioned by stroma grown in the presence of testosterone stimulated LH secretion, whereas medium conditioned by stroma cultured alone had no effect on LH secretion.

3.3.4. *Inhibitory Effect of Charcoal-Treated pFFI on Serum FSH Levels and Follicular Development in the Rhesus Monkey*

The ultimate demonstration of a significant role of inhibin F in control of follicular maturation is to prove that changes in blood levels of inhibin are temporally related to changes in follicular maturation. Such a study is best done in a subhuman primate that has one dominant follicle.

The rhesus monkey was chosen. First, it was necessary to determine whether exogenous inhibin does indeed suppress FSH in this species as well as suppress follicular maturation. This is summarized below.

In addition, it is necessary to measure the blood levels of inhibin throughout the menstrual cycle. Since RIA of inhibin F is not available, ovarian-vein levels of inhibin were measured using the rat anterior-pituitary monolayer system. These measurements are discussed in the next section.

In a series of experiments in collaboration with Gary Hodgen (Channing *et al.*, 1979), we observed that injection of 4 ml charcoal-treated pFFl led to a significant decrease in serum FSH but not LH (Fig. 15), if administered early in the menstrual cycle or at midcyle. If administered early in the menstrual cycle, the expected dominant preovulatory follicle present on days 11–13 of the cycle was either absent or atretic, i.e., contained few or no viable granulosa cells. Administration of either 5 or 15 ml pFFl doses to a long-term castrate female rhesus monkey also decreased serum FSH but not LH levels. (Fig. 16).

3.3.5. *Observation of Inhibin Activity in Rhesus Monkey Follicular Fluid and Ovarian-Vein Blood*

Since DePaolo *et al.*, (1979) observed changes in ovarian-vein inhibin levels occurring throughout the rat estrous cycle, it was appropriate to look for comparable changes in monkey ovarian-vein blood throughout

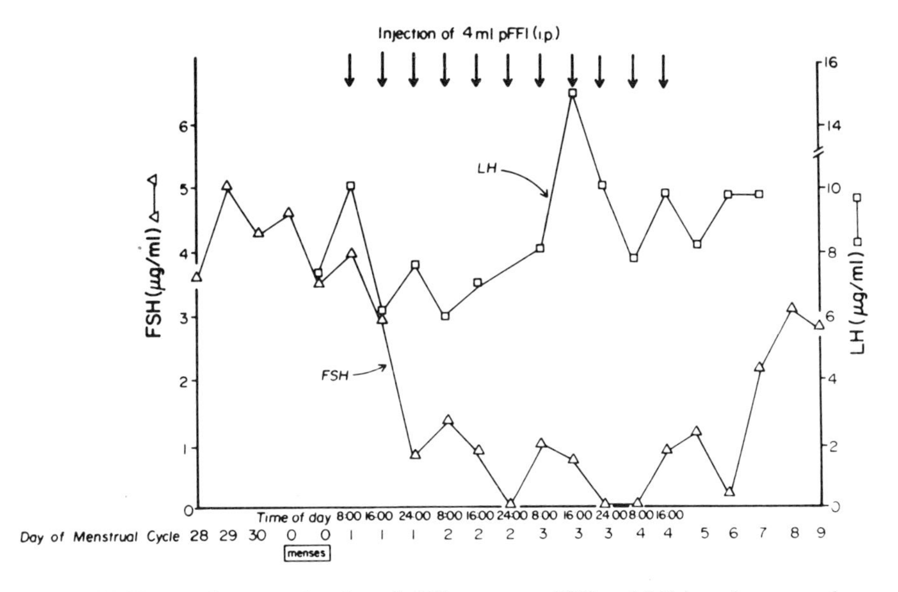

Figure 15. Temporal course of action of pFFl on serum FSH and LH in a rhesus monkey given pFFl every 8 hr for 4 days during the early follicular phase of the menstrual cycle. Reproduced from Channing *et al.* (1979) with permission.

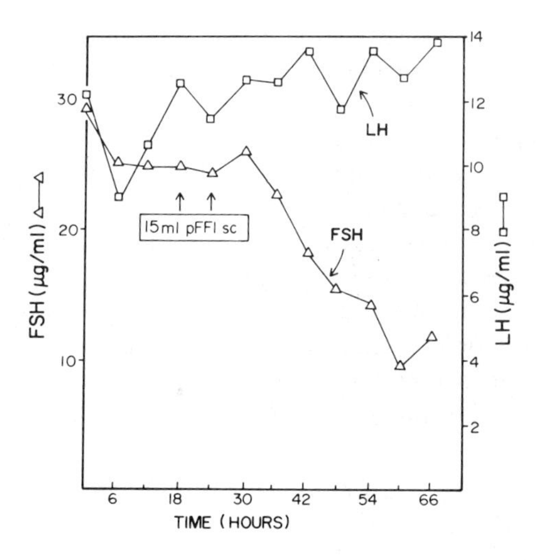

Figure 16. Temporal course of action of pFFl on serum FSH and LH in a long-term castrate female monkey. The fluid was given subcutaneously in two 5 or 15 ml doses 6 hr apart and blood samples were taken every 3 days. Serum FSH and LH levels were measured in each blood sample. Reproduced from Channing *et al.* (1979) with permission.

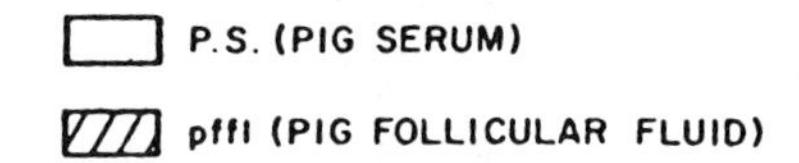

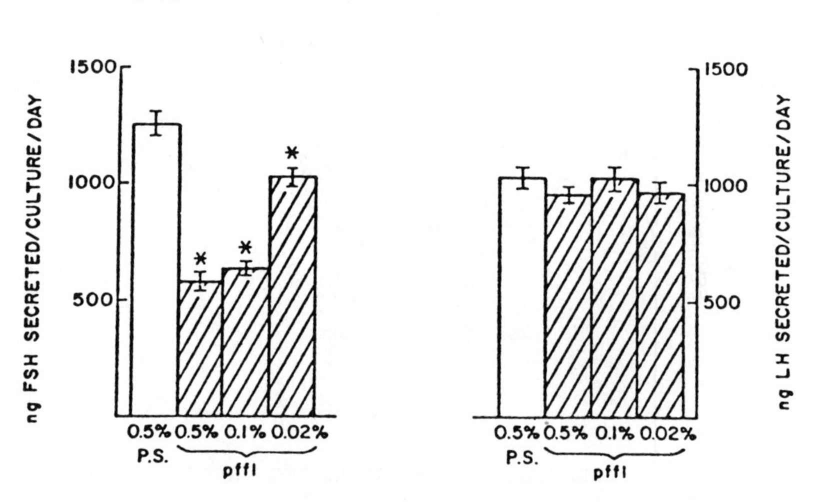

Figure 17. Effect of various doses of charcoal-treated pFFl on 24-hr secretion of FSH and LH by female rat pituitary monolayer culture. $^*P < 0.01$ as control. From Shander, Anderson, Barraclough, and Channing (unpublished observations).

the menstrual cycle. This study is in a very preliminary stage, but nevertheless worthy of mention here. The rat anterior-pituitary bioassay system has been validated by us (Shander *et al.*, 1979a,b); in each assay, various doses of pFFl were added and run as standard (Fig. 17). During the 24-hr test incubation, as little as 0.02% charcoal-treated pFFl produced a significant ($P < 0.01$) inhibition of FSH secretion. A dose of 0.5–2% pFFl produced a maximal 50–70% inhibition of FSH secretion.

To assay for inhibin activity, ovarian-vein and peripheral sera were added in triplicate at a dose of 20% to rat pituitary-cell monolayers. pFFl was added at a final dose of 0.2%. The pituitary cells were inoculated at a density of 150,000 viable cells/culture well and cultured for 3 days in DMEM containing 10% rat, 3% horse, and 2.5% fetal calf serum. The sera and fluid were added at day 3 in DMEM plus 10^{-8} M estradiol and 10^{-7} M progesterone. The incubation was continued for 24 hr, followed by RIA of FSH and LH in the culture medium. In each instance, 0.2–1% monkey follicular fluid exerted a highly significant dose-dependent 70–80% ($P < 0.001$) inhibition of FSH secretion. Ovarian-vein blood from the ovary containing the dominant follicle exerted a significant ($P < 0.01$) 30–40% inhibition of FSH secretion, Analogous peripheral serum failed to alter FSH secretion in the pituitary cultures (Fig. 18)

None of these fluids altered LH secretion. It is possible that the inhibitory effects on FSH secretion could be due to steroidal as well as nonsteroidal compounds. However, the secretion of LH was unaltered by addition of the same samples that suppressed FSH, suggesting a selective action by inhibin constituents of monkey ovarian-vein blood and follicular fluid. We are examining the steroidal and nonsteroidal effects of these fluids. Additional evidence that inhibin activity originates from the dominant follicle was obtained in the finding that 4 days after removal of the large preovulatory follicle, there was a significant ($P < 0.01$) 160% rise in serum FSH compared to a small rise in serum LH (Channing, Anderson, and Hodgen, unpublished observation).

In future studies, ovarian-vein samples will be taken throughout the menstrual cycle and inhibin measured to see whether changes in inhibin levels can be correlated with serum FSH levels and follicular development.

4. Summary

At least five nonsteroidal follicular regulators (?) exist that, along with gonadotropins and ovarian steroids, regulate follicular maturation

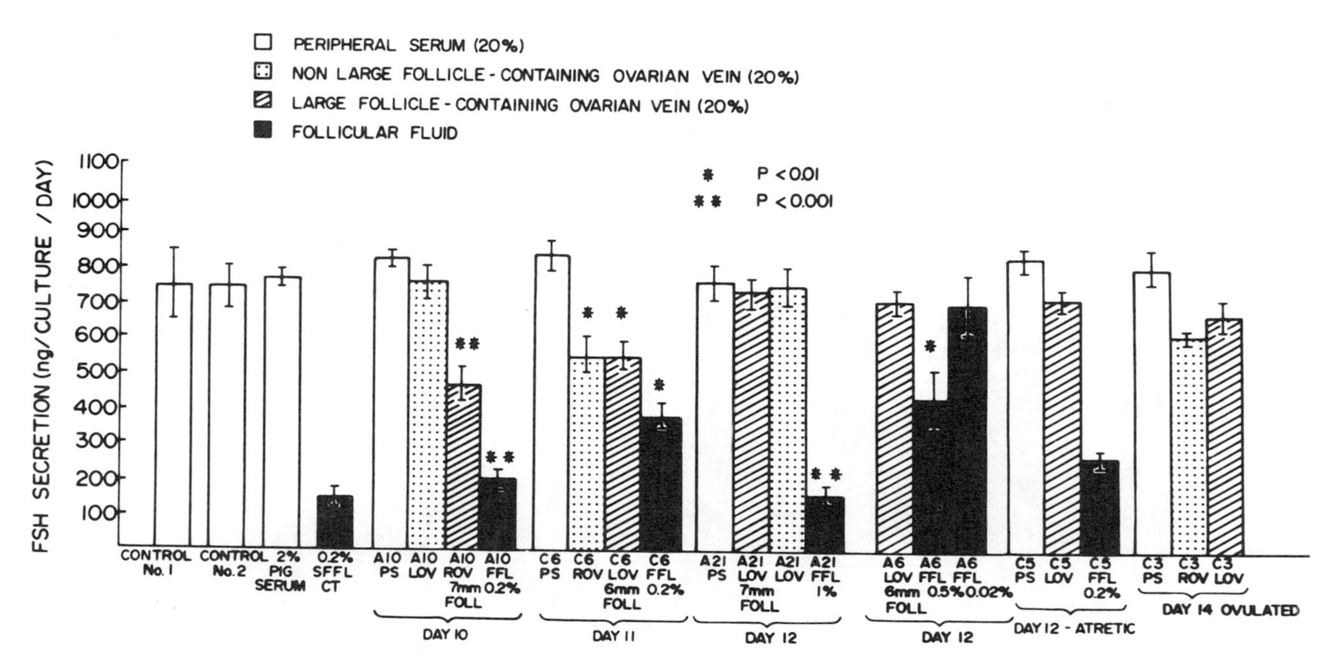

Figure 18. Inhibin activity in monkey ovarian-vein blood and follicular fluid. Serum and ovarian-venous serum were added at a dose of 20% and follicular fluid was added at a dose of 0.2 or 1% to rat anterior-pituitary monolayer cultures, and the cultures were incubated for 24 hr. A standard inhibin preparation consisting of 0.2% small pFFl charcoal-treated (four from left) produced a maximal inhibition (80%) of FSH. LH secretion was not altered by the fluids and is not shown. From Channing, Anderson, and Hodgen (unpublished observations).

and function. These include oocyte-maturation inhibitor (OMI), luteinizing-hormone-receptor-binding inhibitor (LHRBI), FSH-binding inhibitor (FSHBI), luteinization inhibitior (LI), and inhibin F.

A pool of human cystic follicular fluid contained less OMI compared to a pool of normal follicular fluid; both contained inhibin activity. Cultured human granulosa cells or theca plus granulosa secreted inhibin activity into culture medium, while human thecal tissue alone secreted no inhibin activity.

Inhibin activity has been demonstrated in monkey follicular fluid and ovarian venous blood. Porcine follicular fluid, containing crude inhibin, inhibits serum FSH but not LH in cycling and castrate female monkeys.

Since removal of the preovulatory follicle of the monkey causes a selective rise in serum FSH, and venous blood draining the dominant-follicle-containing ovary contains inhibin activity, it is most likely that the dominant monkey follicle is capable of secreting inhibin F into the blood.

ACKNOWLEDGMENTS. The financial aid in the form of grants from the NIH (HD08834), The Ford Foundation (760-0530), and the World Health Organization (Project 6008) is gratefully acknowledged. The expert technical assistance of Ms. Alison Schwartz-Kripner, Ms. Pat Gagliano, and Ms. Sandy Fowler is acknowledged with thanks. The generous cooperation of the following clinicians in obtaining the human ovarian follicular fluid and follicular biopsies is acknowledged: Dr. Joan Sulewski, Dr. N. Rezai, Dr. Luigi Mastroianni, Dr. Anne Wentz, and Dr. H. Taavon.

DISCUSSION

SPILMAN: You mentioned that LH inhibits OMI production. Do you know what stimulates the production of OMI, luteinization inhibitor or luteinization stimulator?

CHANNING: We are currently examining what controls OMI biosynthesis in cultured porcine granulosa cells. They are being cultured with estrogen, FSH, LH, testosterone, and other agents. Extracts of granulosa cells obtained from large porcine follicles (6–12 mm) contain less OMI (per mg protein or DNA) compared to extract of granulosa cells from small (1–2 mm) follicles (Centola and Anderson, 1979).

BANIK: Have you tested OMI in any *in vivo* system?

CHANNING: Not yet. That is an important experiment that remains to be done and will be done!

BOGDANOVE: Are inhibin levels higher in males or in females?

CHANNING: Three lines of evidence to date indicate that there is more inhibin present in and secreted by the ovary compared to the testis: (1) Pig follicular fluid contains at least 100-fold more inhibin compared to pig testis extract. (2) Ovarian-vein blood (pig, rat, monkey) contains measurable inhibin activity, whereas testicular vein blood does not contain much inhibin. (3) Pig and rat granulosa cells (Channing *et al.,* 1980; Erickson and Hseuh, 1978) in culture secrete manyfold more inhibin compared to Sertoli-cell cultures (Steinberger and Steinberger, 1976).

DE-REVIERS: (1) Do you look at the ovary content of OMI, luteinizing-inhibitor factor, and inhibin during pregnancy or after hypopophysectomy in the sow? (2) I would like to comment on Channing's answer to Bogdanove's question. In testis extract, we have to pay attention to the presence of trypsinlike ensyme, which can destroy the activity.

CHANNING: We have not examined OMI and inhibin F after hypophysectomy and during pregnancy.

REFERENCES

Batta, S.K. and Channing, C.P. 1980, Demonstration of inhibin activity in human follicular fluid and its possible role in human ovarian dysfunction, in: *Advances en Obstetricia y Ginecologia,* Vol. 6 (J. Gonzélez-Merlo and J. Guiu, eds), pp. 125–134, Salvat, Barcelona.

Batta, S.K., Wentz, A., and Channing, C.P., 1980, Steroidogenesis by human ovarian cell types in culture: Influences of mixing of cell types and effect of added testosterone, *J. Clin. Endocrinol.* **50:**274–279.

Bradbury, J., 1961, Direct action of estrogen on the ovary of the immature rat, *Endocrinology* **68:**115–120.

Byskov, A., 1978, Follicular atresia, in: *The Vertebrate Ovary* (R.E. Jones, ed.), pp. 533–562, Plenum Press, New York.

Campbell, C.S., and Schwartz, N.B., 1977 Steroid feedback regulation of LH and FSH secretion rates in male and female rats, *J. Toxicol. Environ. Health* **3:**61.

Centola, G., and Anderson, L., 1979, Porcine granulosa cells as a possible source of oocyte maturation inhibitor (OMI), 12th Annual Meeting of the Society for the Study of Reproduction, August 21–24, 1979, Quebec, Canada.

Channing, C.P. 1980, Oocyte maturation, progesterone and estrogen secretion by cultured monkey ovarian cell types: Influences of follicular size, serum luteinizing hormone levels and follicular fluid estrogen levels, *Endocrinology* **107:**342–352.

Channing, C.P., and Coudert, S.P., 1976, Contribution of granulosa cells and follicular fluids to ovarian estrogen secretion in the rhesus monkey *in vivo, J. Endocrinol.* **98:**590–597.

Channing, C.P., Anderson, L.D., and Hodgen, G.D., 1979, Inhibitory effect of charcoal-treated porcine follicular fluid upon serum FSH levels and follicular development in the rhesus monkey, in: *Ovarian Follicular and Corpus Luteum Function* (C.P. Channing, J. Marsh, and W.D. Sadler, eds.), pp. 407–415, Plenum Press, New York.

Channing, C.P., Schaerf, F.W., Anderson, L.D., and Tsafrire, A., 1980, Ovarian follicular and luteal physiology, in: *Reproductive Physiology,* Vol. III, *International Review of Physiology,* Vol 22 (R.D. Greer, ed.), pp. 117–1201, University Park Press, Baltimore.

Chari, S., Hopkinson, C.R.N., Daume, E., and Strum, G., 1979, Purification of "inhibin" from human ovarian follicular fluid, *Acta Endocrinol.* **90:**157.

Darga, N.C., and Reichert, L.E., 1978, Some properties of the interaction of follicle stimulating hormone with bovine granulosa cells and its inhibition by follicular fluid, *Biol. Reprod.* **19:**235–241.

DePaolo, L.V., Shander, D., Wise, P.M., Channing, C.P., and Barraclough, C.A., 1979, Evidence for ovarian secretion of "inhibin," *Endocrinology* **104:**402.

Erickson, G.F., and Hseuh, A.J.W., 1978, Secretion of "inhibin" by rat granulosa cells *in vitro, Endocrinology* **103:**1960.

Greenwald, G., 1978, Follicular activity in the mammalian ovary, in: *The Vertebrate Ovary* (R.E. Jones, ed.), pp. 639–689, Plenum Press, New York.

Harman, S.M., Louvet, J.P., and Ross, G.T., 1975, Interaction of estrogen and gonadotropins on follicular atresia, *Endocrinology* **96:**1145.

Hillensjö, T., Batta, S.K., Schwartz-Kripner, A., Wentz, A.C., Sulewski, J., and Channing, C.P., 1978, Inhibitory effect of human follicular fluid upon the maturation of porcine oocytes in culture, *J. Clin Endocrinol. Metab.* **47:**1332.

Hillensjö, T., Channing, C.P., Pomerantz, S.H., and Schwartz-Kripner, A., 1979a, Intrafollicular control of oocyte maturation in the pig, *In Vitro* **15:**32–39.

Hillensjö, T., Schwartz-Kripner, A., Pomerantz, S.H., and Channing, C.P., 1979b, Action of porcine follicular fluid oocyte maturation inhibitor *in vitro:* Possible role of cumulus cells, in: *Ovarian Follicular and Corpus Luteum Function* (C.P. Channing, J.M. Marsh, and W.D. Sadler, eds.), p. 283, Plenum Press, New York.

Hillier, S.G., Knazek, R.A., and Ross, G.T., 1977, Androgenic stimulation of progesterone production by granulosa cells from pre-antral ovarian follicles: Further *in vitro* studies using replicate cell cultures, *Endocrinology* **100:**1539.

Hunter, R.H.F., and C. Polge, 1966, Maturation of follicular oocytes in the pig aftr injection of human chorionic gonadotrophin, *J. Reprod. Fertil.* **12:**525–531.

Kumari, L., Tucker, S., and Channing, C.P., 1979, Demonstration of a larger amount of inhibitor of binding of labelled human chorionic gonadotropin and of progesterone secretion by cultured porcine granulosa cells in aqueous extracts of old compared to young porcine corpus luteum, *Biol. Reprod.* **21:**1043.

Ledwitz-Rigby, F., Rigby, B.W., Gay, V.L., Stetson, M., Young, J., and Channing, C.P., 1977, Inhibitory action of porcine follicular fluid upon granulosa cell luteinization *in vitro:* Assay and influence of follicular maturation, *J. Endocrinol.* **74:**175–184.

Louvet, J.-P., Harman, S.M., Schreiber, J.R., and Ross, G.T., 1975, Evidence for a role of androgens in follicular maturation, *Endocrinology* **97:**366.

Lunenfeld, B., Ben-Aderet, N., Ben-Michael, R., Grunslein, S., Kraiem, Z., Potashnik, G., Rofch, C., Shalit, A., and Tikotsky, D.T., 1976, Temporal relationship between hormonal profile and ovarian morphology during the preovulatory period in humans, in: *Endocrinology of the Ovary* (Rischoller, ed.), pp. 87–202, Edition Sepe, Paris.

Marder, M.L., Channing, C.P., and Schwartz, N.B., 1978, Suppression of serum follicle stimulating hormone in intact and acutely ovariectomized rats by porcine follicular fluid, *Endocrinology* **101:**1639.

Pomerantz, S.H., Tsafriri, A., and Channing, C.P., 1979, Partial purification and action of an oocyte maturation isolated from porcine follicular fluid, in: *Peptides: Structure and Biological Function,* Proceedings of the Sixth American Peptide Symposium (E. Grass and J. Meinhafer, eds.), pp. 765–774, Pierce Chemical Co.

Rao, M.C., Midgley, A.R., Jr., and Richards, J.S., 1978, Hormonal regulation of ovarian cellular proliferation, *Cell* **14:**71.

Sakai, C.N., and Channing, C.P., 1979a, Evidence for qualitative changes in luteinizing hormone secreted in monkeys with inadequate luteal phase, *Endocrinology* **104:**1217.

Sakai, C.N., and Channing, C.P., 1979b, Evidence for uptake of biologically active luteinizing hormone (LH) into the preovulatory monkey follicle *in vivo, Endocrinology* **104:** 1226.

Sakai, C.N., Channing, C.P., and Engle, B., 1977, Ability of an extract of pig corpus luteum to inhibit binding of ^{125}I-labeled human chorionic gonadotropin to porcine granulosa cells, *Proc. Soc. Exp. Biol. Med.* **155:**373.

Schomberg, D., 1979, Steroidal modulation of steroid secretion *in vitro:* An experimental approach to intra-follicular regulatory mechanisms, in: *Ovarian Follicular and Corpus Luteum Function* (C.P. Channing, J. Marsh, and W.D. Sadler, eds.), pp. 155–165, Plenum Press, New York.

Schwartz, N.B., and Channing, C.P., 1977, Evidence for ovarian inhibin: Suppression of the secondary rise in serum follicle stimulating hormone levels in proestrous rats by injection of porcine follicular fluid, *Proc. Natl. Acad. Sci. U.S.A.* **74:**5721.

Shander, D., Anderson, L.D., Barraclough, C.A., and Channing, C.P., 1979, Modulation of pituitary responsiveness to LHRH by porcine follicular fluid: Time and dose dependent effects, in: *Ovarian Follicular and Corpus Luteum Function* (C.P. Channing, J. Marsh, and W.D. Sadler, eds.), pp. 155–165, Plenum Press, New York.

Shander, P., Anderson, L.D., Barraclough, C., and Channing, C.P., 1980, Interactions of porcine follicular fluid with ovarian steroids and luteinizing hormone-releasing hormone on the secretion of luteinizing hormone and follicle-stimulating hormone by cultured pituitary cells, *Endocrinology* **106:**237–242.

Steinberger, A., and Steinberger, E., 1976, Secretion of an FSH-inhibiting factor by cultured Sertoli cells, *Endocrinology* **98:**918.

Stone, S.L., Pomerantz, S.H., Schwartz-Kripner, A. and Channing, C.P., 1978, Inhibition of oocyte maturation from porcine follicular fluid: Further purification and evidence for reversible action, *Biol. Reprod.* **19:**585.

Tsafriri, A., and Channing, C.P. 1975a, An inhibitory influence of granulosa cells and follicular fluid upon porcine oocyte meiosis *in vitro, Endocrinology* **96:**922.

Tsafriri, A., and Channing, C.P., 1975b, Influence of follicular maturation and culture conditions on the meiosis of pig oocytes *in vitro, J. Reprod. Fertil.* **43:** 149.

Tsafriri, A., Pomerantz, S.H., and Channing, C.P., 1974, Inhibition of oocyte maturation by porcine follicular fluid; Partial characterization of the inhibitor, *Biol. Reprod.* **14:** 511.

Tucker, S., Kumari, L., and Channing, C.P., 1979, Observation of greater LH/hCG binding inhibitor activity in aqueous extracts of old compared to young porcine corpus luteum, in: *Ovarian Follicular and Corpus Luteum Function* (C.P. Channing, J. Marsh, and W.D. Sadler, eds.), pp. 723–728, Plenum Press, New York.

Udenfriend, S., Stein, S., Bohlen, P., Dairman, W., Leingruber, W., and Weigle, N., 1972, Fluorexamine: A reagent for assay of amino acids, peptides, proteins, and primary amines, *Science* **178:**871.

Vale, W., Grant, G., Amoss, M., Blackwell, R., and Guillemin, R., 1972, Culture of enzymatically dispersed anterior pituitary cells: Functional validation method, *Endocrinology* **61:**562.

Yang, K., Samaan, N., and Ward, D.N., 1976, Characterization of an inhibitor for luteinizing hormone receptor site binding, *Endocrinology* **98:**233.

Yang, K.P., Gray, K.N., Jardine, J.H., Yen, H.L.N., Samaan, N.A., and Ward, D.N., 1978, LH-RBI—An inhibitor of *in vitro* luteinizing hormone binding to ovarian receptors and LH-stimulated progesterone synthesis by ovary, in: *Novel Aspects of*

Reproductive Physiology (C.H. Spilman and J. Wilks, eds.), p. 61, Spectrum, New York.

Yang, K., Samaan, N.A., and Ward, D.N., 1979, Effects of luteinizing hormone receptor-binding inhibitor on the *in vitro* steroidogenesis by rat ovary and testis, *Endocrinology* **104:**552.

IX

Steroids and Cell Growth

29

Estrogen-Induced Growth of Uterine Cells

Evidence for Involvement of Surface Membranes, Calcium, and Proteinase Activity

Richard J. Pietras and Clara M. Szego

1. Introduction

A general hypothesis has been developed in recent years to describe the mechanism of action of steriod hormones. Circulating steroid, having dissociated from serum carrier proteins, is considered to diffuse passively into most tissue cells (Peck *et al.,* 1973; Gorski and Gannon, 1976). The hormones are retained and accumulated only in responsive cells by interactions with extranuclear macromolecules that possess high affinity and specificity for hormone (Jensen and Jacobson, 1962; Talwar *et al.,* 1964; Gorski *et al.,* 1968; Jensen *et al.,* 1974; King and Mainwaring, 1974). In the chick, labeled hormone is detected bound to its receptor in the oviduct within 1–2 min following injection (O'Malley *et al.,* 1970). Once bound to hormone, the receptor is transformed or activated (Jensen *et al.,* 1968; Puca *et al.,* 1972) and thereby rendered capable, within 2–4 min after injection (O'Malley *et al.,* 1970), of transport to the nucleus. Within the nucleus, the steroid–receptor complex interacts with unspecified "acceptor sites" in the chromosomal material and thereby elicits

---

Richard J. Pietras and *Clara M. Szego* • Department of Biology and the Molecular Biology Institute, University of California, Los Angeles, California 90024.

the synthesis of specific RNA species that are believed to underlie expression of the phenotypic effects (Hamilton, 1968; Gorski and Gannon, 1976; Buller and O'Malley, 1976; Thrall *et al.*, 1978).

The focus of this chapter will be on the primary interactions of estradiol-17β ($E_2\beta$) with membranes and extranuclear receptors in uterine target cells. The relationship of these early interactions to hormone-induced cell proliferation will be considered.

2. Background

2.1. Estrogen-Induced Membrane Alterations and Growth in Uterine Cells

Direct mitogenic effects of estrogens on target cells *in vitro* have been demonstrated only in recent investigations (Gerschenson *et al.*, 1974; Lippman *et al.*, 1976; Gerschenson and Berliner, 1976; Pietras and Szego, 1979b). Nevertheless, it is well established that the proliferative response of endometrial cells to estrogen is preceded by alterations in membrane transport and ionic flux (Astwood, 1939; Roberts and Szego, 1953; Spaziani and Szego, 1959; Riggs, 1970; Spaziani, 1975; Pietras and Szego, 1975a). These studies are paralleled by other reports of estradiol effects on Na^+, K^+-activated ATPase (Karmakar, 1969; cf. Spaziani, 1975), adenylate cyclase activity (Szego and Davis, 1967; Rosenfeld and O'Malley, 1970), electrical activity of the endometrial membrane (Levin and Edwards, 1968; Levin and Pawlowski, 1973), and alterations in Ca^{2+} binding and flux in endometrial-cell suspensions (Pietras and Szego, 1975b). Recent studies also show an acute increase in the lateral mobility of lectin-binding components in the plane of the surface membrane of endometrial cells exposed to estrogen *in vitro* (Pietras and Szego, 1975c, 1979b). It has been widely suggested that reorganization of the cell surface is an obligatory step in the normal progression of cells to mitosis (Burger, 1973; DeTerra, 1974; Shodell, 1975; Nicolson, 1976; Noonan, 1978).

It is important to know whether such alterations in endometrial-cell membranes occur concomitantly with, or as a consequence of, the primary interaction of estrogen with specific receptors. New evidence indicates that early changes induced by $E_2\beta$ at the surfaces of endometrial cells, as well as enhanced growth evoked by the hormone in primary culture, are both associated with increased availability of an endogenous leupeptin-sensitive proteinase (Pietras and Szego, 1979b). The cathepsin-B-like proteinase, sequestered in lysosomes in the resting cell (Dean and Barrett, 1976), shows a moderately acid pH optimum with synthetic substrates (cf. Szego *et al.*, 1976) and is activated by calcium ion (cf. Szego *et al.*, 1976; Quinn and Judah, 1978). The enzyme is strongly inhibited by leupeptin, a peptide aldehyde (cf. Umezawa and Aoyagi, 1977), and by thiol-blocking compounds such as iodoacetic acid. Treatment of cells with $E_2\beta$, but not the relatively inert congener $E_2\alpha$, evokes a rapid rise in the access of cathepsin-B-like activity to the external cell surfaces (Pietras and Szego, 1979b). A striking localization of cathepsin B at or near the surfaces of transformed and neoplastic cells has also been decribed by immunohistochemical methods (Sylvén *et al.*, 1974). Cathepsin-B-like activity is also markedly elevated in sera of patients with cancer (Pietras *et al.*, 1978b, 1979). Several lines of evidence indicate that cathepsin-mediated alterations in the enzymatic activity or in the integrity of membrane-associated proteins by limited proteolysis may contribute to changes in the composition or distribution of membrane components (Szego, 1974; Pietras and Szego, 1975c, 1979b; Szego *et al.*, 1976; Seetharam *et al.*, 1976; Pietras, 1978). Mild treatment of normal cells with exogenous proteinases is known to lead to alterations in the cell surface (cf. Burger, 1973; Nicolson, 1976; Noonan, 1978) and a level of cell growth (cf. Burger, 1973; Carney and Cunningham, 1977; Noonan, 1978) usually seen only in transformed or tumor cells. Recent data show that treatment of endometrial cells with leupeptin-loaded liposomes evokes a selective reduction in endogenous cathepsin-B activity and marked suppression of the anticipated growth response after $E_2\beta$ exposure (Pietras and Szego, 1979b). Likewise, subcutaneous administration of milligram amounts of leupeptin to intact cyclic mice elicits a dramatic decrease in uterine weight and DNA content and in fertility (Katz *et al.*, 1977). Collectively, such observations lend further support to the concept that a leupeptin-sensitive proteinase may promote both membrane alterations and enhanced genic expression induced by estrogen (cf. Szego, 1975; Szego *et al.*, 1976; Pietras and Szego, 1979b).

2.2. Properties of Estrogen-Receptor Molecules

The pioneering studies of Gorski and associates (cf. Gorski *et al.*, 1968; Gorski and Gannon, 1976) indicated that the bulk of $E_2\beta$ that became specifically bound when uterine segments were incubated with hormone at 0–4°C occurred in association with 100,000 *g* supernatant fractions (i.e., cytosol) obtained after homogenization. These cellular extracts were prepared by extensive disruption of tissues in hypotonic media with ethylenediamine tetraacetic acid (EDTA). Most properties of steroid receptors have been described after further purification of hormone-binding macromolecules from cytosol. Toft and Gorski (1966) found that the molecule was probably a protein, and it exhibited a sedimentation coefficient of about 8 S when subjected to ultracentrifugation in low-salt sucrose density gradients (cf. also Erdos, 1968; Baulieu, 1975). High-ionic-strength solutions (e.g., 0.4 M KCl) elicited a hormone-binding component with a sedimentation coefficient of about 4 S (Gorski *et al.*, 1968; Gorski and Gannon, 1976). Though noncovalent, the binding of $E_2\beta$ to receptor proteins is very strong, with association constants ranging from 10^9 to 10^{12} M^{-1} (cf. Jensen and DeSombre, 1973). The tendency of the receptor molecule to aggregate in low-salt solutions may be attributable to its hydrophobic properties (cf. Baulieu, 1975). Preliminary characterization of partially purified material indicates that these receptors are acidic proteins (Puca *et al.*, 1971; King and Mainwaring, 1974; Thrall *et al.*, 1978). The presence of a lipid moiety in cytosol receptor is indicated by transformation of 8 S complex to a 4 S form after addition of lipase (Erdos, 1968) and by phospholipase-induced reduction of $E_2\beta$ binding to the receptor (Paton, 1969; King *et al.*, 1971; Hähnel *et al.*, 1974). The occurrence of phosphorous and carbohydrate moieties in association with cytosol receptors has also been reported (Jensen and DeSombre, 1972; Hähnel *et al.*, 1974).

Purification of the cytosol estrogen receptor from calf uterus 15,000-fold to apparent homogeneity has been reported in recent studies by Sica and Bresciani (1979). The product was prepared by sequential affinity chromatography and found to consist of a single subunit of about 70,000 daltons. On sucrose gradients in low-salt buffer, the binding component sedimented at 8 S.

2.3. Evidence for Activation of Estrogen-Bound Receptor

On heating of cytosol preparations to 25–30°C in the presence of estrogen, the receptor is transformed or activated, as manifested in a shift in the sedimentation coefficient from 4 S to 5 S (Jensen and DeSombre, 1972). The 5 S form binds to nuclei and chromatin, whereas the 4 S form does not (Jensen and DeSombre, 1972). Recently, antibodies raised against a homogeneous preparation of nuclear 5 S estradiol receptor from calf uterus were found to cross-react with cytosol 4 S receptor, indicating that, as suspected previously (cf. Jensen and DeSombre, 1972; Thrall *et al.*, 1978), these two forms represent interconversions of the same molecules (Greene *et al.*, 1977).

The nature of the receptor-activation process remains unknown. Puca *et al.* (1977) have proposed that a Ca^{2+}-activated enzyme may trigger the transformation of the cytosol estradiol–receptor complex into the modified form capable of binding to nuclear acceptor sites. A crude preparation of an intracellular Ca^{2+}-dependent proteinase that exhibits maximal activity at pH 8.5 and that possesses relatively high affinity for cytosol estradiol–receptor from calf uterus has recently been described (Puca *et al.,* 1977). Intracellular pH determined in a variety of different cells and tissues is normally in the range of 7.1–7.4 and reportedly is reduced to about 6.8 in chemically transformed, and thus rapidly metabolizing, liver cells (cf. Kitagawa and Kuroiwa, 1976). Consequently, the physiological significance of proteinase with such a strongly alkaline pH optimum for transformation of cytosol receptor remains to be demonstrated. Estrogen-induced translocation of Ca^{2+}-dependent cathepsin-B-like (Szego *et al.,* 1976) and of trypsin-like (Katz *et al.,* 1976) proteinase activities to the nuclear compartment of target cells has been reported to occur concomitantly with the formation of the "mobile" $E_2\beta$-binding derivative that enters the nucleus. Sherman *et al.* (1978) have also described the occurrence in breast-tumor cytosol of a Ca^{2+}-dependent and leupeptin-sensitive proteinase that acts on cytosol estrogen receptor at pH 7.4. The products of such limited proteolysis include a globular steroid-binding domain and an asymmetric nuclear-binding domain. Which, if any, of these several proteinases may be involved in receptor activation remains to be determined (cf. Szego and Pietras, 1981).

3. New Evidence on Interactions of Estrogen with Target-Cell Membranes

3.1. Estrogen-Binding Properties of Endometrial Cells Capable of Association with Immobilized Estradiol

Previous investigations have provided evidence for the availability of specific estrogen-binding sites at the surface membranes of target cells (Pietras and Szego, 1977, 1979a,c; Pietras *et al.,* 1978a). Significant numbers of hormone-responsive endometrial and liver cells were found to adhere to 17β-estradiol-17-hemisuccinyl–albumin–nylon fibers, but the binding of nontarget cells from intestine was negligible (Pietras and Szego, 1977). Target cells bound to immobilized hormone at an effective molar concentration of 5×10^{-10} M were displaced by incubation at 22°C in 150 m0smol saline containing a 400-fold molar excess of free $E_2\beta$ or diethylstilbestrol (DES) (Pietras and Szego, 1979a). Similar concentrations of other ligands such as $E_2\alpha$, cortisol, progesterone, or testosterone were ineffective. Such stereospecific binding to immobilized estrogen by target cells exclusively must be attributable to the occurrence of recognition sites for hormone at the surface membranes of responsive cells.

Specific accumulation of free [^{3}H]-$E_2\beta$ by subpopulations of endometrial cells with diverse affinities for binding to immobilized estrogen was evaluated in equilibrium binding experiments. In these and the further experiments reported below, endometrial cells bound to immobilized estrogen (i.e., 42% of total cell population) were dislodged from the fibers in the presence of excess $E_2\beta$ and recovered intact by methods described previously (Pietras and Szego, 1979a). Corresponding cells that had not become bound to the derivatized fibers (i.e., 58% of total cell population) were processed and recovered under parallel conditions. Both cell groups were then incubated at 37°C in steroid-free chemically defined medium (CDM) in an atmosphere of 5% CO_2 in air. CDM consisted of Earle's balanced salt solution enriched with 1 mM dextrose, 1×10^{-8} M insulin, 1×10^{-9} M cortisol, 0.5 μg transferrin/ml, 0.1% (wt./vol.) albumin, Modified Eagle's Medium amino acids, and 50 μg Gentamicin/ml. After 72-hr incubation to minimize residual estrogen, the interaction of cells with free $E_2\beta$ was determined.

The results shown in Fig. 1 indicate that binding of hormone by both fiber-binding and non-fiber-binding cells is a saturable process. To establish the concentration of specific binding sites for $E_2\beta$ and the equilibrium constant for the binding reaction in each group of cells, data in Fig. 1 were analyzed further (see Fig. 2) by the notation of Scatchard (1949). Mathematical resolution of the binding data by the method of

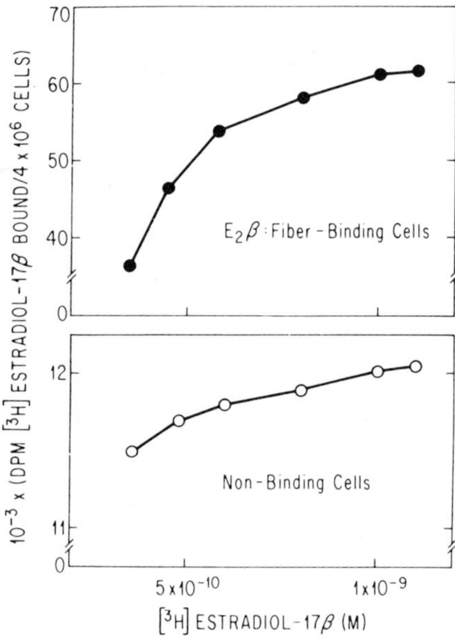

Figure 1. Specific binding of free [^{3}H]-$E_2\beta$ by subpopulations of endometrial cells with diverse affinities for binding to immobilized estrogen. Isolated endometrial cells that bind ($E_2\beta$:Fiber-Binding Cells) and those that do not bind (Non-Binding Cells) to estrogen-derivatized fibers were recovered under parallel conditions as described previously (Pietras and Szego, 1977, 1979a). After 72 hr in primary culture with steroid- and serum-free medium (see the text), the two groups of cells were suspended in 0.5 ml Ringer solution (8×10^6 cells/ml) and exposed for 30 min at 22°C to a series of [^{3}H]-$E_2\beta$ (99 Ci/mmol) concentrations ranging from 3.5×10^{-10} to 2.0×10^{-9} M as indicated on the abscissa (cf. Pietras and Szego, 1979b). Only specific binding (defined as the difference in bound steroid between paired tubes, one of which contained a 200-fold molar excess of unlabeled $E_2\beta$ throughout the experiment) is shown (cf. Williams and Gorski, 1973). Each point represents the mean of values obtained in two independent experiments.

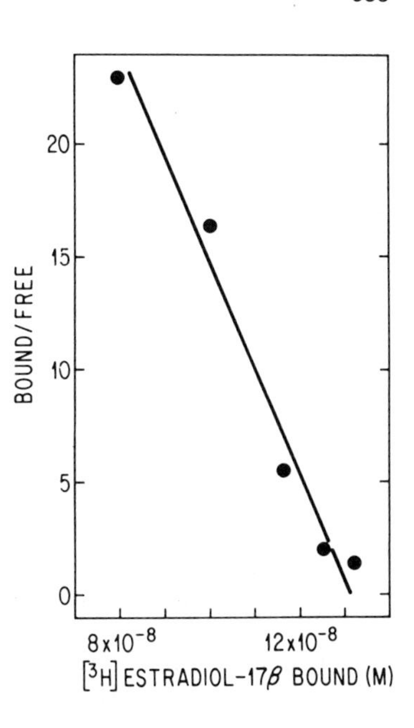

Figure 2. Scatchard analyses of specific [^{3}H]-$E_2\beta$ binding in estrogen–fiber-binding populations of endometrial cells. Calculation of these data is based on the results of equilibrium binding studies presented in Fig. 2. The mass of bound estradiol is obtained from the specific radioactivity of the hormone associated with the cells. The concentration of $E_2\beta$ so bound is obtained by estimating a volume of 1×10^{-6} liter per 2×10^6 endometrial cells (i.e., 2×10^9 packed cells/ml). The free estradiol (unbound) at the end of the incubation is calculated from the difference between the total amount of $E_2\beta$ added initially to the incubation medium and the amount of bound hormone.

least squares yields an apparent molar dissociation constant for the binding process of 2.1×10^{-9} among $E_2\beta$–fiber-binding cells, as compared to 3.1×10^{-9} in populations of nonbinding cells. The total number of $E_2\beta$-binding sites per cell at saturation corresponds to 41,527 in fiber-binding cells (see Fig 2) and to 8305 in nonbinding cells (not shown). Thus, endometrial cells that bind immobilized estradiol are also capable of retaining 5 times more $E_2\beta$ during subsequent exposure to physiological concentrations of hormone than those cells that do not bind. The presently reported binding capacity of $E_2\beta$–fiber-binding cells also represents a 2-fold enrichment over that reported previously for unfractionated endometrial cells [e.g., 21,110 $E_2\beta$-binding sites/cell with $K_D = 1.9 \times 10^{-9}$ M (cf. Pietras and Szego, 1979b)].

3.2. Effect of Estrogen on the Calcium Content of Endometrial Cells Fractionated by Affinity Binding to Immobilized Hormone

The physiological responsiveness of fiber-separated populations of endometrial cells to free $E_2\beta$ was evaluated in further experiments. Previous work has demonstrated a pronounced alteration in Ca^{2+} exchange by unfractionated endometrial cells within minutes after exposure to $E_2\beta$ *in vitro* (Pietras and Szego, 1975b). Determinations of the calcium contents of $E_2\beta$–fiber-binding and nonbinding endometrial cells incubated with or without estradiol for 15 min are shown in Table I. These preliminary results indicate that $E_2\beta$–fiber-bindings cells, but not non-binding cells, respond to hormone with a significant increase in the level of intracellular calcium to 1.2 times that of paired controls. This increase is in the range of that predicted from earlier studies on Ca^{2+} flux in hormone-stimulated endometrial cells. Considerable evidence indicates that increments in cell calcium accumulation precede or trigger, or both, division and growth of other cell types by a mechanism that remains unclear (Wasserman and Corradino, 1973; Whitfield *et al.*, 1973; Baulieu *et al.*, 1978). Puca *et al.* (1977) have hypothesized that estrogen-induced increase in Ca^{2+} may stimulate the activity of a receptor-transforming proteinase in calf uterus.

3.3. Effect of Estrogen on the Proliferative Activity of Endometrial Cells Fractionated by Affinity Binding to Immobilized Estradiol

The proliferative response of fiber-separated endometrial cells in the presence of $E_2\beta$ was investigated as shown in Fig. 3. No significant mitotic activity of cells was found in the absence of hormone. In contrast, significant increases in cell numbers were observed in both fiber-binding and non-binding populations at both 24 and 48 hr after exposure of cells to 2×10^{-9} M $E_2\beta$ [all at $P < 0.05$ (see Fig. 3)]. However, hormone-stimulated proliferation of fiber-binding cells was 1.6 times that of

Table I. Effect of Estradiol-17β on Intracellular Calcium Content of $E_2\beta$–Fiber-Binding and Nonbinding Endometrial Cells[a]

Group	Treatment	Cell calcium (mmol/kg intracellular water)
Nonbinding cells	Control (0.02% ethanol)	2.3 ± 0.1 (3)
	$E_2\beta$ (2×10^{-9} M)	2.1 ± 0.2 (3)
$E_2\beta$–fiber-binding cells	Control (0.02% ethanol)	1.9 ± 0.2 (3)
	$E_2\beta$ (2×10^{-9} M)	2.3 ± 0.2 (3)[b]

[a] Endometrial cells capable of binding to immobilized estrogen and nonbinding cells were collected as described in the text. Cells were incubated 15 min at 22°C in the presence and absence of $E_2\beta$. Determinations of cell calcium were carried out as reported previously (Pietras *et al.*, 1976).
[b] Value significantly different from paired control at $P < 0.05$.

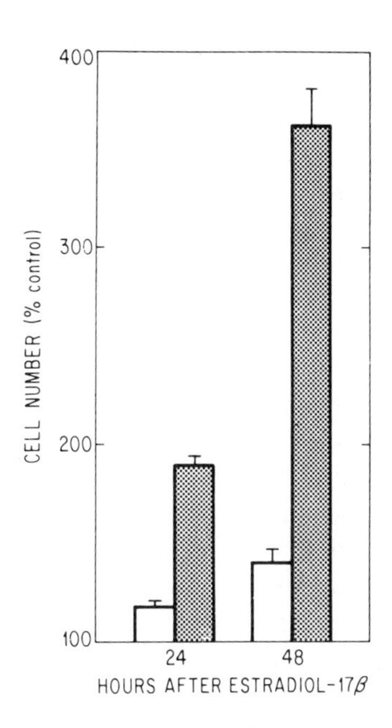

Figure 3. Influence of $E_2\beta$ on proliferation of endometrial cells with diverse affinities for binding to immobilized estrogen. $E_2\beta$–fiber-binding cells (stippled bars) and nonbinding cells (clear bars) were selected and then processed under parallel conditions (cf. Pietras and Szego, 1979a). After 72 hr in primary culture with $E_2\beta$- and serum-free medium (see the text), proliferation of endometrial cells after 24- or 48-hr incubation with or without 2×10^{-9} M $E_2\beta$ was determined from cell numbers as described previously (Pietras and Szego, 1979b). Values are the means ± S.E.M. of results obtained in three independent experiments.

nonbinding cells after 24 hr ($P < 0.01$) and increased to 2.6 times after 48 hr ($P < 0.001$). Thus, cells with capacity for interaction with fiber-immobilized estrogen exhibit a preferential growth response to $E_2\beta$. Such observations (cf. also Pietras and Szego, 1979a) indicate that the availability or concentration, or both, of plasmalemmal receptors for $E_2\beta$ may govern the relative responsiveness of a given cell to hormone.

3.4. *Inhibition by Liposome-Entrapped Leupeptin of Cellular Cathepsin-B Activity and Estrogen-Induced Growth in Endometrial Cells Capable of Association with Fiber-Immobilized Estradiol*

In previous experiments with unfractionated endometrial cells, prior treatment of cells with liposome-entrapped leupeptin was found to abolish the stimulation by $E_2\beta$ of cell proliferation and [^{3}H]thymidine incorporation into macromolecular form that was evident in paired preparations (Pietras and Szego, 1979b). To test the effect of leupeptin on the $E_2\beta$-stimulated proliferation of fiber-binding cells, cationic liposomes were prepared by the methods of Magee *et al.* (1974). The procedure for entrapment of leupeptin in liposomes was presented previously (Pietras, 1978; Pietras and Szego, 1979b).

Cell cultures were incubated with liposomes containing either leupeptin at a final concentration of 7×10^{-8} M or control vehicle. After 30 min, the incubation medium was removed and replaced with fresh medium before assessment was made of $E_2\beta$-induced proliferative responses in cells with or without prior exposure to liposome-entrapped leupeptin. Such brief treatment reduced the total cellular cathepsin-B activity to 27% that in control cells that were exposed to liposomes without leupeptin [$P < 0.001$ (Fig. 4A)]. That this procedure did not lead to generalized inhibition of enzymic function was demonstrated by lack of significant alteration in cellular activity of acid phosphatase, alkaline phosphatase, 5′-nucleotidase, or succinate dehydrogenase (see Table II). Under these conditions, the profound stimulation of cell proliferation normally elicited by $E_2\beta$ was markedly reduced, amounting to only 24% that among cells with prior exposure to control liposomes [$P < 0.001$ (Fig. 4B)]. Cells incubated with control liposomes before $E_2\beta$ treatment showed the anticipated increase in cell proliferation to 98% of the level found in cells exposed only to hormone [$P > 0.90$ (Fig. 4b)]. Thus, reduction of the endogenous activity of cathepsin B in fiber-binding endometrial cells elicits a corresponding loss in the capacity of the cells to respond to $E_2\beta$.

4. Conclusions

4.1. Reevaluation of the Native State of Unoccupied Estrogen Receptors

The studies reported herein provide additional evidence that binding components with specificity for $E_2\beta$ are present in plasma membranes of estrogen-responsive cells. The early studies of Noteboom and Gorski (1965) also identified a small fraction of specific $E_2\beta$ binding in ''mitochondrial'' (i.e., mitochondria and lysosome) and ''microsomal'' (i.e., microsomes and plasma-membrane vesicles) fractions of uterine homogenates. In more recent work, additional evidence for the occurrence of binding components with high affinity and specificity for $E_2\beta$ in particulate fractions of target cells has been presented (Blyth *et al.*, 1971; Little *et al.*, 1972; Milgrom *et al.*, 1973; Jackson and Chalkley, 1974; Hirsch and Szego, 1974; Sen *et al.*, 1975; Pietras and Szego, 1979c).

There is now substantial evidence demonstrating that binding sites with high affinity and specificity for $E_2\beta$ occur at the plasma membranes of responsive cells (Pietras *et al.*, 1978a; Pietras and Szego, 1979c) and may mediate the uptake of hormone (Milgrom *et al.*, 1973; Szego, 1975;

Rao *et al.*, 1977) as well as modulate its local effects (Dufy *et al.*, 1979). Likewise, the work of Jackson and Chalkley (1974) indicated that more than 70% of native estradiol-receptor components possess high affinity for target-cell membranes and are sequestered in discrete extranuclear compartments before interaction with hormone.

Further studies provide evidence that about 27% of total cellular binding sites with high affinity and ligand specificity for $E_2\beta$ are concentrated in plasma membranes purified from uterine cells (Pietras and

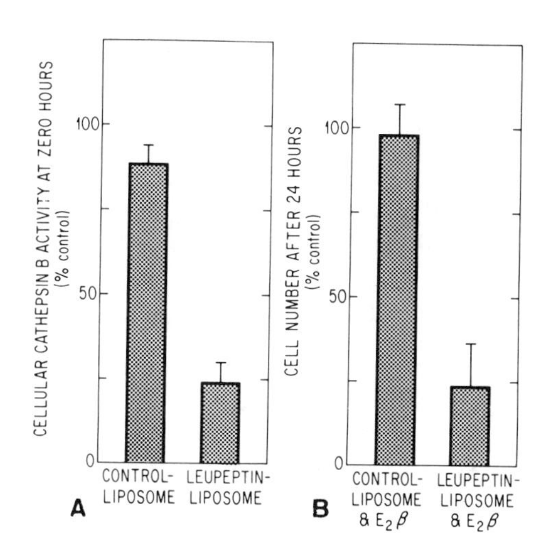

Figure 4. Influence of liposome-entrapped leupeptin on cellular cathepsin-B activity (A) and estrogen-stimulated proliferation of $E_2\beta$–fiber-binding populations of endometrial cells (B). Cells were selected and then processed in primary culture without $E_2\beta$ or serum for 72 hr (see the text). (A) The fiber-binding cells were then incubated with cationic liposomes containing either 7×10^{-8} M leupeptin (LEUPEPTIN-LIPOSOME) or control vehicle (CONTROL-LIPOSOME). After 30 min, cells in both groups were sedimented and washed once by centrifugation at 400*g*. Sedimented cells were sonicated and then suspended in 0.1% (vol./vol.) Triton X-100 at 4°C for 2 hr. Portions of the sediments thus solubilized were analyzed for protein and cathespin-B activity as described previously (Pietras and Szego, 1979b). Basal activity of cathepsin B in control cells not exposed to liposomes averaged 49 pmol substrate degraded/min/mg cell protein. (B) Proliferation of $E_2\beta$–fiber-binding endometrial cells was determined after 30-min exposure to either control- or leupeptin-liposomes. Cells were incubated with or without 2×10^{-9} M $E_2\beta$ for 24 hr. Cell numbers were then determined as described elsewhere (Pietras and Szego, 1979b). The control group represents fiber-binding cells treated with $E_2\beta$ but not treated with liposomes; cell numbers in the presence of hormone for 24 hr increased by 1.9 times that determined in the absence of $E_2\beta$. Throughout these experiments, more than 94% of liposome-exposed cells retained the capacity to exclude vital dyes such as nigrosin and trypan blue. Values in (A) and (B) are the means ± S.E.M. of results obtained in three independent experiments.

Table II. Effect of Brief Incubation of Fiber-Binding Endometrial Cells with Liposome-Entrapped Leupeptin on Total Activities of Various Hydrolytic Enzymes[a]

Treatment	Cellular enzyme activity (nmol/min/mg protein)			
	Acid phosphatase	Alkaline phosphatase	5'-Nucleotidase	Succinate dehydrogenase
Control vehicle	14 ± 3 (3)	57 ± 4 (3)	112 ± 2 (3)	21 ± 2 (3)
Control liposomes	13 ± 2 (3)	52 ± 3 (3)	106 ± 4 (3)	20 (2)
Leupeptin liposomes	12 ± 3 (3)	53 ± 2 (3)	104 ± 4 (3)	23 (2)

[a] Fiber-binding cells were treated for 30 min as indicated in the table and then sedimented and washed once with centrifugation at 400*g*. Sedimented cells were sonicated and then suspended in 0.1% (vol./vol.) Triton X-100 at 4°C for 2 hr. Portions of the sediment thus solubilized were analyzed for protein and enzyme activities as described previously (Pietras and Szego, 1979b).

Szego, 1979c). The analytical cell-fractionation scheme used in the latter study differed in the following significant ways from schemes in general use for the preparation of cell fractions enriched in $E_2\beta$-binding components (cf. Raspé, 1971; O'Malley and Hardman, 1975):

1. Isolated cell preparations, rather than organ segments, were used as starting material to reduce the shearing and grinding forces required to break cells. Previous workers have reported that the drastic homogenization necessary to disrupt the tough muscular and connective-tissue components of uterus results in very small fragments of plasma membrane, most of which is lost in postnuclear supernatant fractions at the first step of low-speed centrifugation (Kidwai *et al.*, 1971).

2. Disruption of a maximum of only 98% of isolated cells in a Teflon–glass homogenizer was used to prevent the aberrant redistribution of macromolecules known to be elicited by excessive cell homogenization (Plagemann, 1969).

3. Cells were disrupted in buffered isotonic sucrose containing $CaCl_2$ and fractionated using isotonic media throughout, conditions known to promote the maintenance of the integrity of cell structures (Berman *et al.*, 1969; DePierre and Karnovsky, 1973). Appropriate control experiments indicated that no more than 5% of the high, though saturable and specific, binding capacity of the plasma membranes for $E_2\beta$ may be attributable to cytosol protein entrapment or adsorption artifacts arising during cell disruption and fractionation (Pietras and Szego, 1979c). In paired experiments, alternative cell-disruption procedures utilizing glass–glass homogenization with hypotonic buffer and EDTA were tested. Under these conditions, binding sites for $E_2\beta$ were found predom-

inantly in cytosol. However, activity of 5′-nucleotidase, a plasma-membrane marker enzyme (Kidwai *et al.*, 1971; Pietras and Szego, 1979c; Matlib *et al.*, 1979), showed a highly aberrant redistribution from particulate to cytosol fractions (Pietras and Szego, 1979c). Despite the use of isotonic buffer, homogenization of uterine segments with a Polytron (cf. Müller *et al.*, 1979) also elicits the solubilization of 43% of total cellular 5′-nucleotidase activity (cf. Matlib *et al.*, 1979). Such results have raised the question of whether the widely reported predominance of receptors for $E_2\beta$ in cytosol (cf. Raspé, 1971; O'Malley and Hardman, 1975) might have resulted from inadvertent extraction of native hormone receptors by homogenization procedures that resulted in extensive damage to cellular structures (cf. Little *et al.*, 1972; Jackson and Chalkley, 1974; Hirsch and Szego, 1974; Szego, 1974; Pietras and Szego, 1979c). Moreover, considerable evidence discussed elsewhere (Szego and Pietras, 1981) also implicates membrane functions in the binding and responsiveness of target cells to additional steroid hormones including androgens (Giorgi *et al.*, 1973; Rao *et al*, 1977; Watanabe *et al.*, 1979), glucocorticoids (Harrison *et al.*, 1974, 1979; Suyemitsu and Terayama, 1975; Rao *et al.*, 1976; Koch *et al.*, 1978; Fant *et al.*, 1979), mineralocorticoids (Ožegović *et al.*, 1977), and progestogens (L. D. Smith and Ecker, 1971; Ishikawa *et al.*, 1977; Godeau *et al.*, 1978).

4.2. Consideration of the Role of Leupeptin-Sensitive Proteinase in the Stimulation of Cell Growth by Estrogen

Lysosomes of responsive cells are known to concentrate a wide variety of mitogens (cf. Allison, 1974) and possess sites specific for binding $E_2\beta$ (Hirsch and Szego, 1974; Szego, 1975; Pietras and Szego, 1979c). The redistribution of limited amounts of lysosomal hydrolases (Szego *et al.*, 1976) and antigens (Szego *et al.*, 1977) from this organelle in a dose-dependent manner to the nucleus and external surfaces of target cells after exposure to $E_2\beta$ has also been reported. These cellular responses to hormone may correspond to the coupled processes of endocytosis and lysosomal fusion (cf. Cohn, 1975; Szego, 1975), processes that are being exploited as a means of introduction of antimitotic drugs into the cell interior (deDuve *et al.*, 1974). Similarly, the present data show that treatment of endometrial cells with leupeptin-loaded liposomes evokes a virtual suppression of the anticipated growth response after $E_2\beta$ exposure. Such results lend support to the concept that a leupeptin-sensitive proteinase may promote enhanced genic expression elicited by estrogens (Szego, 1974; Szego *et al.*, 1976; Katz *et al.*, 1977; Pietras and Szego, 1979c). Additional evidence for the involvement of proteinase activity in estrogen action is reviewed elsewhere (Szego and Pietras, 1981).

Recent studies provide evidence that lysosomal proteinases are primarily responsible for the turnover of intracellular proteins (Dean, 1975; Ward *et al.*, 1977; Lasch *et al.*, 1977; Segal *et al.*, 1978; MacGregor *et al.*, 1979). The functions of such proteinases are far more diverse and exhibit greater substrate specificity than had previously been recognized (cf. Barrett, 1977). In this regard, there are strong indications that cathepsin B may serve in vital cell functions by means of specific, limited proteolytic steps. Such specialized actions may include the activation and inactivation of enzymes (cf. Barrett, 1977) and the conversion of precursor proteins, such as proinsulin (R. E. Smith and Van Frank, 1975) and proalbumin (Quinn and Judah, 1978), to their active states. The precise functions of enhanced cathepsin-B activity at the surface membrane (Pietras and Szego, 1975a, 1979b) and in nuclei (Szego *et al.*, 1976) of estrogen-stimulated cells remain to be determined.

The effects of inhibitors for cathepsin B (e.g., iodoacetate, leupeptin) in living cells point to an important role for this thiol proteinase in estrogen action. Previous experiments have shown that leupeptin specifically binds to and inhibits cathepsin B in estradiol target tissues (cf. Szego *et al.*, 1976; Seglen *et al.*, 1979). Moreover, the tetrapeptide aldehyde appears to be a relatively selective inhibitor of the lysosomal pathway of protein turnover (Seglen *et al.*, 1979).

Incorporation of proteinase inhibitors in biodegradable lipid vesicles offers a new strategy for the intracellular delivery of such compouunds to the lysosomal compartment (cf. Magee *et al.*, 1974; Lasch *et al.*, 1977; Pietras, 1978; Pietras and Szego, 1979b). Liposomes are taken up into cells both by endocytosis and by fusion of the lipid vesicles with plasma membranes (Poste and Papahadjopoulos, 1976). However, addition of significant amounts of cholesterol to the mixture of lipids elicits a solidifying effect and suppresses fusion of liposomes with surface membranes, thereby favoring uptake by endocytosis (Papahadjopoulos *et al.*, 1974). One recent finding indicates that a small amount of the content of cationic liposomes may be discharged into cytosol on initial interaction of the lipid vesicles with plasma membrane (Steger and Desnick, 1977). Nevertheless, it is considered that leupeptin, once oriented toward lysosomes by its liposomal carrier, is unable to diffuse out and elicits its primary inhibitory action within the lysosomal compartment (cf. Lasch *et al.*, 1977).

The studies described in this chapter suggest that a leupeptin-sensitive lysosomal proteinase is involved in the expression of uterine functions elicited by exposure to $E_2\beta$. The inhibition of endometrial-cell responses to $E_2\beta$ by leupeptin may be attributable, in part, to reduction of hormone accumulation by intact cells after exposure to the proteinase inhibitor (Pietras and Szego, 1979b). Various sulfhydryl-group blocking reagents, including iodoacetate and iodoacetamide, potential inhibitors of cathepsin-B activity (Otto, 1971; Barrett, 1977), restrict the entry of estrogen into uterine cells while exerting markedly less inhibition of steroid binding by extracted cytosol receptors (Terenius, 1967; Milgrom *et al.*, 1973). Iodoacetamide is also known to suppress the nuclear accumulation of $E_2\beta$–receptor complexes in uterus (Milgrom *et al.*, 1973). Thus, it is conceivable that enhanced availability of a Ca^{2+}- and leupeptin-sensitive thiol proteinase in target cells after $E_2\beta$ exposure may promote cleavage of an estrogen-binding fragment with high nuclear affinity from a membrane-localized binding component (cf. Pietras and Szego, 1979c; Szego and Pietras, 1981). The control of zymogen activation by limited proteolysis (cf. Neurath and Walsh, 1976) and the demonstrated capability of cathepsin B to convert precursor protein to active form (R. E. Smith and Van Frank, 1975; Quinn and Judah, 1978) offer ample precedent for this hypothesis. Clearly, the involvement of proteinase activity in the action of estrogen and possibly other steroid hormones (Baker and Fanestil, 1977; Wrange and Gustafsson, 1978; Sherman *et al.*, 1978; Carlstedt-Duke *et al.*, 1979; Vedeckis *et al.*, 1980) requires intensive investigation in future studies.

ACKNOWLEDGMENTS. Reaction product for standardization of cathepsin-B activity was kindly provided by Dr. E. Smithwick, Jr. Highly purified insulin was donated by Dr. W. W. Bromer. B. J. Seeler carried out analyses for cathepsin-B activity. This investigation was aided by U.S. Public Health Service (USPHS) Postdoctoral Fellowship CA-5176 (to R. J. Pietras) and research grants HD 4354 and FR 7009 (USPHS), PCM 78-22489 (National Science Foundation), a grant from the Eli Lilly Research Laboratories, and general research funds of the University of California.

DISCUSSION

MATHER: Have you looked at cells bound to estradiol–fibers to determine if short-term estrogenic responses such as amino acid transport can be elicited?

PIETRAS: No, but we do plan to investigate this important question in future experiments. However, we shall need to proceed with some caution, because such an approach is subject to several major pitfalls; (1) The association constant for binding of 17β-estradiol-17-hemisuccinyl–albumin to target-cell receptors from calf uterus is approximately 100 times less than that for free 17β-estradiol (cf. Sica *et al.*, 1976). On this basis, we estimate that the association constant for specific binding of the $E_2\beta$–fiber complex to rat uterine-membrane receptor is approximately 10^9 M^{-1} (cf. Pietras and Szego, 1979c). This suggests that the estradiol–fiber complex may have only a fraction of the biologic activity of native estrogen. Consequently, the determination of relatively small alterations in membrane transfer of ions or nonelectrolytes might not be possible with methods currently available (2) Release of free steroid from the estradiol–fiber complex may occur when the immobilized hormone comes into contact with isolated cells. Secretion of several hydrolytic enzymes by endometrial cells has previously been demonstrated (cf. Pietras and Szego, 1979b). The concept of insulin acting exclusively at the plasma membrane was first supported by experiments in which insulin was covalently coupled to beads of insoluble agarose and found to have biologic potency close to that of native hormone (Cuatrecasas, 1969). However, subsequent studies indicated that much of the biologic activity of insulin–agarose could be accounted for by the solubilization of free insulin from the hormone–agarose complex after association with biologic materials (Davidson *et al.*, 1973; Garwin and Gelehrter, 1974; Kolb *et al.*, 1975; Goldfine, 1977).

In related investigations, Ishikawa *et al.* (1977) and Godeau *et al.* (1978) have reported that meiotic maturation (i.e., germinal-vesicle breakdown) of amphibian oocytes is induced by exposure to immobilized steroid derivatives. Release of free steroid from deoxycorticosterone (DOC)–agarose beads (Ishikawa *et al.*, 1977) or from DOC–polyethylene oxide complexes (Godeau *et al.*, 1978) into the cellular incubation medium appeared to be insignificant. After 20-hr incubation, the concentration of free steriod in the medium was found to be less than 0.2% that of the initial steroid–polymer complex (Godeau *et al.*, 1978). Ishikawa *et al.* (1977) also reported that little to no oocyte maturation occurred when DOC–agarose was added to oocytes covered with follicle cells or when DOC–agarose was physically separated from oocytes by a porous nylon net. In both investigations, a relatively high concentration (i.e., 10^{-6} to 10^{-5} M) of the steroid derivative was necessary to elicit the maturation response. This may be attributable to the lower receptor-association constant of such derivatives [see above] or to the limited area for contact between the bound steroid and the oocyte membrane.

MATHER: Did you treat the nonbinding fraction of cells with the same high levels of estrogen required to release the binding fraction?

PIETRAS: Yes. As described in Section 3.1 of this chapter and in previous investigations (Pietras and Szego, 1979a), both $E_2\beta$–fiber-binding and nonbinding cells were incubated at 22°C in 150 mOsmol saline containing 2×10^{-7} M $E_2\beta$ and processed subsequently in parallel. The latter concentration of hormone represents a 400-fold molar excess over that bound covalently to albumin–nylon fibers (i.e., effective molar concentration of 5×10^{-10} M). It should be emphasized that only physiologically active estrogens such as $E_2\beta$ and DES are effective in displacing target cells from the estradiol–fibers at 22°C. $E_2\beta$ and a variety of nonestrogenic steroids are ineffective (cf. Pietras and Szego, 1979a).

MATHER: Have you ever tried to keep the nonbinding fraction in culture and see if they can develop into "binders," possibly with an estrogen treatment?

PIETRAS: No, but we have found that prior incubation of freshly isolated endometrial cells with 1×10^{-9} M $E_2\beta$, but not $E_2\beta$, elicits a marked reduction in the subsequent binding of washed cells to immobilized estradiol (Pietras and Szego, 1977, 1979a). This may be attributable to masking of the surface-membrane sites by bound estrogen or to a hormone-induced change in the location or orientation of the estrogen-binding components. Other studies have shown that target-selective uptake of specific antigens is associated with a localized disordering of membrane structure and subsequent internalization, presumably of those regions of the surface membrane equipped with, or immediately adjacent to, specific recognition sites (cf. Cohn, 1975; Szego, 1978). A corresponding sequence of events may occur at the surfaces of endometrial cells in response to estrogen, as is evident from reports of altered membrane functions (Szego and Davis, 1967; Levin and Pawlowski, 1973; Spaziani, 1975; Pietras and Szego, 1975b,c, 1979b) and micropinocytotic vacuolation of the plasmalemma (Szego, 1975) after brief exposure to hormone.

The results of the present study indicate that nonbinding cells are not devoid of estrogen-receptor activity but are markedly deficient in the concentration of receptor sites per cell and in their response to free hormone as compared to $E_2\beta$–fiber-binding cells. In independent studies in which endometrial cells were isolated from uteri of ovariectomized rodents by comparable procedures (Alberga and Baulieu, 1968; J. A. Smith *et al.*, 1970), it was found that 75–85% of the cell population was composed of luminal and glandular epithelial cells, with the remainder being cells of the connective tissue stroma. Autoradiographic studies of [^{3}H]thymidine incorporation in thin uterine sections show that 70–80% of luminal epithelial cells, 20–30% of glandular epithelial cells, and essentially no stromal cells are labeled with the nucleotide after 15-hr treatment with $E_2\beta$ *in vivo* (Martin and Finn, 1971). Thus, the present results correspond with the latter observation that the response of endometrial cells of $E_2\beta$ is not uniform among the several cell types (cf. Pietras and Szego, 1979a,b). The present work indicates that cells preferentially responsive to free hormone (i.e., $E_2\beta$–fiber-binding cells) possess a high concentration of receptors capable of interaction with immobilized estradiol.

The inordinate degree of $E_2\beta$ retention in fiber-binding cells is consistent with the results of other studies which show that the number of glucocorticoid receptors per lymphoid cell increases severalfold after mitogenic stimulation (K. A. Smith *et al.*, 1977).

Kelly: Have you examined other estrogen-responsive cells, such as mammary cells or human breast-cancer cells, for their ability to bind to the fiber-linked estrogen?

Pietras: We have found significant binding of hormone-responsive hepatocytes to immobilized estradiol (Pietras and Szego, 1977, 1979a). Fiber-bound cells could be dislodged by brief incubation with 2×10^{-7} M $E_2\beta$ or DES, but not $E_2\alpha$, cortisol, progesterone, or testosterone, and recovered intact. As in the present experiments, fiber-binding cells showed a higher concentration of specific $E_2\beta$-binding sites per cell and a greater response to free hormone than corresponding cells which had not bound the immobilized estradiol (Pietras and Szego, 1979a).

Epithelial cells isolated from rat intestine have no specific binding sites for $E_2\beta$ (Toft and Gorski, 1966). In fiber-binding experiments, such epithelial cells likewise show no binding to immobilized estradiol (Pietras and Szego, 1977).

We have not yet attempted any experiments with mammary cells or human breast-cancer cells.

Kelly: Did you look for physiological regulation of membrane-bound receptors as compared to cytosol estrogen receptors? For example, can you see replenishment of estrogen receptors in membranes as well as cytosol?

Pietras: We have not yet had an opportunity to study the physiological regulation of membrane binding sites for $E_2\beta$. As I mentioned earlier in response to Dr. Mather, we have found evidence for an apparent depletion of plasmalemmal receptors after *in vitro* exposure of cells to free $E_2\beta$ (Pietras and Szego, 1977, 1979a).

Aakvaag: Do you have any idea whether the membrane receptor and the cytosol receptor are the same? Could they serve different functions in the two locations and in some ways move back and forth to control the functions of the cell?

PIETRAS: Your questions are very provocative, but I cannot provide you with adequate answers at the present time. In contrast to numerous reports devoted to the identification and properties of membrane-associated receptors for polypeptide hormones, membrane binding sites with specificity and high affinity for steroid hormones have scarcely been investigated (cf. Szego and Pietras, 1981). Evidence was provided in the pioneering studies of Noteboom and Gorski (1965) that specific binding sites for $E_2\beta$ are present in cell membranes from uterus, but such data have been largely ignored. As reviewed in this chapter and elsewhere (Szego and Pietras, 1981), new data from several laboratories indicate that a significant portion of steroid-hormone receptors found in cytosol of target tissues are subject to extraction from particulate fractions during the process of tissue disruption and fractionation. This important question on the origin of cytosol receptors must be resolved before your questions can be answered.

BAZER: Have you determined whether or not endometrial epithelial cells respond to progesterone *in vitro* by making protease inhibitor? Debbie Mullins (1979) has presented data which indicate that the progesterone-dominated pig uterus (e.g., days 12–18 of pregnancy) produces proteinase inhibitors. Could these factors inhibit translocation of estrogen–receptor complexes into the nucleus?

PIETRAS: We have not investigated the response of endometrial cells to treatment with progesterone. However, others have reported that relatively high doses of progesterone inhibit the subsequent mitotic response of uterine epithelial cells to $E_2\beta$ (J. A. Smith *et al.*, 1970; Martin and Finn, 1971). In cultures of rabbit endometrial cells, DES and progesterone were found to have antagonistic effects on the rate of cell growth (Gerschenson *et al.*, 1974). Since estrogen-stimulated uterine cells show an enhanced availability of lysosomal proteinase activity both at the cell surface (Pietras and Szego, 1975c, 1979b) and in ultrapurified nuclei (Szego, 1975; Szego *et al.*, 1976), it would certainly be worthwhile to consider whether the antagonistic action of progesterone may be attributable to the enhanced activity of a proteinase inhibitor.

BAZER: Is it possible that estrogen receptors on the plasma membrane may mediate effects that do not require nuclear translocation? For example, are there enzymes and/or transport-related proteins that are "activated" by estrogens or affected only by the plasma-membrane receptor?

PIETRAS: As reviewed in this chapter, there are numerous reports on changes in the activity of membrane-associated enzymes and transport functions early after exposure of target cells to $E_2\beta$. It is not known if such changes result from a direct action of estradiol or via the mediation of a membrane or cytosol receptor. At pharmacologic concentrations, some hormonal steroids are known to exert an antihemolytic effect on human red-blood-cell membranes (DeVenuto *et al.*, 1969). This action is probably a direct physical effect, since no specific, high-affinity binding sites for progesterone, testosterone, or estradiol are found in erythrocyte membranes (DeVenuto *et al.*, 1969; Brinkmann *et al.*, 1970; Brinkmann and van der Molen, 1972). The effect of 1×10^{-9} M $E_2\beta$, but not $E_2\alpha$, on Ca^{2+}-dependent changes in the electrical potential of pituitary-cell membranes shows a latency of less than 1 min (Dufy *et al.*, 1979). Such responses are considered to be too rapid for genic mediation (cf. Dufy *et al.*, 1979; Szego and Pietras, 1981). Spaziani (1975) has also reviewed several other effects of gonadal steroid hormones that are not likely to be explained as dependent on transcription of mRNA followed by assembly and positioning of specialized proteins. Such findings indicate that the assumption that gene activation accounts for all known effects of steroid hormones (cf. Hamilton, 1968) is premature. The potential role of a membrane receptor in mediating such responses to hormone must await further investigation on the nature of these binding sites and their relation to previously identified receptors for steroids in cytosol extracts (cf. Szego and Pietras, 1981).

References

Alberga, A., and Baulieu, E.-E., 1968, Binding of estradiol in castrated rat endometrium *in vivo* and *in vitro, Mol. Pharmacol.* **4:**311.

Allison, A.C., 1974, Lysosomes in cancer cells, *J. Clin. Pathol.* **27**(*Suppl. Roy. Coll. Pathol.* 7):43.

Astwood, E.B., 1939, Changes in the weight and water content of the uterus of the normal adult rat, *Am. J. Physiol.* **126:**162.

Baker, M.E., and Fanestil, D.D., 1977, Effect of proteinase inhibitors and substrates on deoxycorticosterone binding to its receptor in dog MDCK kidney cells, *Nature (London)* **269:**810.

Barrett, A.J., 1977, The diversity of cellular proteinases in physiology and pathology, *Acta Biol. Med. Ger.* **36:**1959.

Baulieu, E.-E., 1975, The mode of action of steriod hormones: Some recent findings, in: *Pathobiology Annual* (H.L. Ioachim, ed.), pp. 105–132, Appleton-Century-Crofts, New York.

Baulieu, E.-E., Godeau, F., Schorderet, M., and Schorderet-Slatkine, S., 1978, Steroid-induced meiotic division in *Xenopus laevis* oocytes: Surface and calcium, *Nature (London)* **275:**593.

Berman, H.M., Gram, W., and Spirtes, M.A., 1969, An improved, reproducible method of preparing rat liver cell plasma membranes in buffered isotonic sucrose, *Biochim. Biophys. Acta* **183:**10.

Blyth, C.A., Freedman, R.B., and Rabin, B.R., 1971, Sex specific binding of steroid hormones to microsomal membranes of rat liver, *Nature (London) New Biol.* **230:**137.

Brinkmann, A.O., and van der Molen, H.J., 1972, Localization and characterization of steroid binding sites of human red blood cells, *Biochim. Biophys. Acta* **274:**370.

Brinkmann, A.O., Mulder, E., and van der Molen, H.J., 1970, Interaction of steroids with human red cells, *Ann. Endocrinol.* **31:**789.

Buller, R.E., and O'Malley, B.W., 1976, The biology and mechanism of steroid hormone receptor interaction with the eukaryotic nucleus, *Biochem. Pharmacol.* **25:**1.

Burger, M.M., 1973, Surface changes in transformed cells detected by lectins, *Fed. Proc. Fed. Am. Soc. Exp. Biol.* **32:**91.

Carlstedt-Duke, J., Wrange, Ö., Dahlberg, E., Gustafsson, J.-Å., and Högberg, B., 1979, Transformation of the glucocorticoid receptor in rat liver cytosol by lysosomal enzymes, *J. Biol. Chem.* **254:**1537.

Carney, D.H., and Cunningham, D.D., 1977, Initiation of chick cell division by trypsin action at the cell surface, *Nature (London)* **268:**602.

Cohn, Z.A., 1975, Macrophage physiology, *Fed. Proc. Fed. Am. Soc. Exp. Biol.* **34:**1725.

Cuatrecasas, P., 1969, Interaction of insulin with the cell membrane: The primary action of insulin, *Proc. Natl. Acad. Sci. U.S.A.* **63:**450.

Davidson, M.B., van Herle, A.J., and Gerschenson, L.E., 1973, Insulin and sepharose-insulin effects on tyrosine transaminase levels in cultured rat liver cells, *Endocrinology* **92:**1442.

Dean, R.T., 1975, Direct evidence of importance of lysosomes in degradation of intracellular proteins, *Nature (London)* **257:**414.

Dean, R.T., and Barrett, A.J., 1976, Lysosomes, in: *Essays in Biochemistry,* Vol. 12 (P.N. Campbell and W.N. Aldridge, eds.), pp. 1–40, Academic Press, London.

deDuve, C., deBarsey, L., Poole, B., Trouet, A., Tulkens, P., and van Hoof, F., 1974, Lysosomotropic agents, *Biochem. Pharmacol.* **23:**2495.

DePierre, J.W., and Karnovsky, M.L., 1973, Plasma membranes of mammalian cells: A review of methods for their characterization and isolation, *J. Cell Biol.* **56:**275.

DeTerra, N., 1974, Cortical control of cell division, *Science* **184:**530.

DeVenuto, F., Ligon, D.F., Friedrichsen, D.H., and Wilson, H.L., 1969, Human erythrocyte membrane uptake of progesterone and chemical alterations, *Biochim. Biophys. Acta* **193:**36.

Dufy, B., Vincent, J.-D., Fleury, H., Du Pasquier, P., Gourdji, D., and Tixier-Vidal, A., 1979, Membrane effects of thyrotropin-releasing hormone and estrogen shown by intracellular recording from pituitary cells, *Science* **204:**509.

Erdos, T., 1968, Properties of a uterine oestradiol receptor, *Biochem. Biophys. Res. Commun.* **32:**338.

Fant, M.E., Harbison, R.D., and Harrison, R.W., 1979, Glucocorticoid uptake into human placental membrane vesicles, *J. Biol. Chem.* **254:**6218.

Garwin, J.L., and Gelehrter, T.D., 1974, Induction of tyrosine aminotransferase by sepharose-insulin, *Arch. Biochem. Biophys.* **164:**52.

Gerschenson, L.E., and Berliner, J.A., 1976, Further studies on the regulation of cultured rabbit endometrial cells by diethylstilbestrol and progesterone, *J. Steroid Biochem.* **7:**159.

Gerschenson, L.E., Berliner, J., and Yang, J.-J., 1974, Diethylstilbestrol and progesterone regulation of cultured rabbit endometrial cell growth, *Cancer Res.* **34:**2873.

Giorgi, E.P., Shirley, I.M., Grant, J.K., and Sewart, J.C., 1973, Androgen dynamics *in vitro* in the human prostate gland: Effect of cyproterone and cyproterone acetate, *Biochem. J.* **132:**465.

Godeau, J.F., Schorderet-Slatkine, S., Hubert, P., and Baulieu, E.-E., 1978, Induction of maturation in *Xenopus laevis* oocytes by a steriod linked to a polymer, *Proc. Natl. Acad. Sci. U.S.A.* **75:**2353.

Goldfine, I.D., 1977, Does insulin need a second messenger?, *Diabetes* **26:**148.

Gorski, J., and Gannon, F., 1976, Current models of steroid hormone action: A critique, *Annu. Rev. Physiol.* **38:**45.

Gorski, J., Toft, D., Shymala, G., Smith, D., and Notides, A., 1968, Hormone receptors: Studies on the interaction of estrogen with the uterus, *Recent Prog. Horm. Res.* **24:**45.

Greene, G.L., Closs, L.E., Fleming, H., DeSombre, E.R., and Jensen, E.V., 1977, Antibodies to estrogen receptor: Immunochemical similarity of estrophilin from various mammalian species, *Proc. Natl. Acad. Sci. U.S.A.* **74:**3681.

Hähnel, R., Twaddle, E., and Brindle, L, 1974, The influence of enzymes on the estrogen receptors of human uterus and breast carcinoma, *Steroids* **24:**489.

Hamilton, T., 1968, Control by estrogen of genetic transcriptions and translation, *Science* **161:**649.

Harrison, R.W., Fairfield, S., and Orth, D.N., 1974, Evidence for glucocorticoid transport through the target cell membrane, *Biochem. Biophys. Res. Commun.* **61:**1262.

Harrison, R.W., Balasubramanian, K., Yeakley, J., Fant, M., Svec, F., and Fairfield, S., 1979, Heterogeneity of AtT-20 cell glucocorticoid binding sites: Evidence for a membrane receptor, in: *Steroid Hormone Receptor Systems* (W.W. Leavitt and J.H. Clark, eds.), pp. 423–440, Plenum Press, New York.

Hirsch, P.C., and Szego, C.M., 1974, Estradiol receptor functions of soluble proteins from target-specific lysosomes, *J. Steroid Biochem.* **5:**533.

Ishikawa, K., Hanaoka, Y., Kondo, Y., and Imai, K., 1977, Primary action of steroid hormone at the surface of amphibian oocyte in the induction of germinal vesicle breakdown, *Mol. Cell. Endocrinol.* **9:**91.

Jackson, V., and Chalkley, R., 1974, The cytoplasmic estradiol receptors of bovine uterus: Their occurrence, interconversion, and binding properties, *J. Biol. Chem.* **249:**1627.

Jensen, E.V., and DeSombre, E.R., 1972, Estrogens and progestins, in: *Biochemical Actions of Hormones* (G. Litwack, ed.), pp. 215–253, Academic Press, New York.

Jensen, E.V., and DeSombre, E.R., 1973, Estrogen–receptor interaction, *Science* **182:**126.

Jensen, E.V., and Jacobson, H.I., 1962, Basic guides to the mechanism of estrogen action, *Recent Prog. Horm. Res.* **18:**387.

Jensen, E.V., Suzuki, T., Kawashima, T., Stumpf, W.E., Jungbult, P.W., and DeSombre, E.R., 1968, A two-step mechanism for the interaction of estradiol with rat uterus, *Proc. Natl. Acad. Sci. U.S.A.* **59:**632.

Jensen, E.V., Mohla, S., Gorell, T.A., and DeSombre, E.R., 1974, The role of estrophilin in estrogen action, *Vitam. Horm.* **32:**89.

Karmakar, P.K., 1969, The influence of oestradiol-17β on the rat uterine Na^{+}, K^{+}-Mg^{++} activated adenosine triphosphatase activity, *Experientia* **25:**319.

Katz, J., Troll, W., Levy, M., Filkins, K., Russo, J., and Levitz, M., 1976, Estrogen-dependent trypsin-like activity in the rat uterus, *Arch. Biochem. Biophys.* **173:**347.

Katz, J., Troll, W., Adler, S.W., and Levitz, M., 1977, Antipain and leupeptin restrict uterine DNA synthesis and function in mice, *Proc. Natl. Acad. Sci. U.S.A.* **74:**3754.

Kidwai, A.M., Radcliffe, M.A., and Daniel, E.E., 1971, Studies on smooth muscle plasma membrane. I. Isolation and characterization of plasma membrane from rat myometrium, *Biochim. Biophys. Acta* **233:**538.

King, R.J.B., and Mainwaring, W.I.P., 1974, *Steriod-Cell Interactions,* University Park Press, Baltimore.

King, R.J.B., Gordon, J., Marx, J., and Steggles, A.W., 1971, Localization and nature of sex steroid receptors within the cell, in: *Basic Actions of Sex Steroids on Target Organs* (P.O. Hubinont, F. Leroy, and P. Galand, eds.), pp. 21–43, S. Karger, New York.

Kitagawa, Y., and Kuroiwa, Y., 1976, Change in intracellular pH of rat liver during azo-dye carcinogenesis, *Life Sci.* **18:**441.

Koch, B., Lutz-Bucher, B., Briaud, B., and Mialhe, C., 1978, Specific interaction of corticosteroids with binding sites in the plasma membranes of the rat anterior pituitary gland, *J. Endocrinol.* **79:**215.

Kolb, H.J., Renner, R., Hepp, K.D., Weiss, L., and Wieland, O.H., 1975, Re-evaluation of sepharose-insulin as a tool for the study of insulin action, *Proc. Natl. Acad. Sci. U.S.A.* **72:**248.

Lasch, J., Koelsch, R., Brezesinski, G., Hermann, V., Riemann, S., and Bohley, P., 1977, Pepstatin- and leupeptin-loaded liposomes: A tool in protein breakdown studies, *Acta Biol. Med. Ger.* **36:**1829.

Levin, R.J., and Edwards, F., 1968, The transuterine endometrial potential difference, its variation during the oestrous cycle and its relation to uterine secretion, *Life Sci.* **7:**1019.

Levin, R.J., and Pawlowski, J., 1973, Microelectrode studies on endometrial cells of rat uterus, *J. Physiol.* **237:**27P.

Lippman, M.E., Bolan, G., and Huff, K., 1976, The effects of estrogens and antiestrogens on hormone-responsive human breast cancer in long-term tissue culture, *Cancer Res.* **35:**4595.

Little, M., Rosenfeld, G.C., and Jungblut, P.W., 1972, Cytoplasmic estradiol "receptors" associated with the "microsomal" fraction of pig uterus, *Hoppe-Seyler's Z. Physiol. Chem.* **353:**231.

MacGregor, R.R., Hamilton, J.W., Kent, G.N., Shofstall, R.E., and Cohn, D.V., 1979, The degradation of proparathormone and parathormone by parathyroid and liver cathepsin B, *J. Biol. Chem.* **254:**4428.

Magee, W.E., Goff, C.W., Schoknecht, J., Smith, M.D., and Cheriah, K., 1974, The interaction of cationic liposomes containing entrapped horseradish peroxidase with cells in culture, *J. Cell Biol.* **63:**492.

Martin, L, and Finn, C.A., 1971, Oestrogen–gestagen interactions on mitosis in target tissues, in: *Basic Actions of Sex Steroids on Target Organs* (P.O. Hubinont, F. Leroy, and P. Galand, eds.), pp. 172–197, S. Karger, New York.

Matlib, M.A., Crankshaw, J., Garfield, R.E., Crankshaw, D.J., Kwan, C.-Y., Branda, L.A., and Daniel, E.E., 1979, Characterization of membrane fractions and isolation of purified plasma membranes from rat myometrium, *J. Biol. Chem.* **254:**1834.

Milgrom, E., Atger, M., and Baulieu, E.-E., 1973, Studies on estrogen entry into uterine cells and on estradiol–receptor complex attachment to the nucleus—Is the entry of estrogen into uterine cells a protein-mediated process?, *Biochim. Biophys. Acta* **320:**267.

Müller, R.E., Johnston, T.C., and Wotiz, H.H., 1979, Binding of estradiol to purified uterine plasma membranes, *J. Biol. Chem.* **254:**7895.

Mullins, D.E., 1979, Characterization of the surface epithelium of the porcine uterus, Ph.D. dissertation, University of Florida, Gainesville.

Neurath, H., and Walsh, K.A., 1976, Role of proteolytic enzymes in biological regulation (a review), *Proc. Natl. Acad. Sci. U.S.A.* **73:**3825.

Nicolson, G.L., 1976, Trans-membrane control of the receptors on normal and tumor cells. II. Surface changes associated with transformation and malignancy, *Biochim. Biophys. Acta* **458:**1.

Noonan, K.D., 1978, Proteolytic modification of cell surface macromolecules: Mode of action in stimulating cell growth, *Curr. Top. Membr. Transp.* **11:**397.

Noteboom, W.D., and Gorski, J., 1965, Stereospecific binding of estrogens in the rat uterus, *Arch. Bichem. Biophys.* **111:**559.

O'Malley, B.W., and Hardman, J.G. (eds)., 1975, Cytoplasmic receptors for steroid hormones, *Methods Enzymol. Pt. A* **36:**156.

O'Malley, B.W., Sherman, M.R., and Toft, D.O., 1979, Progesterone "receptors" in the cytoplasm and nucleus of chick oviduct target tissue, *Proc. Natl. Acad. Sci. U.S.A.* **67:**501.

Otto, K., 1971, Cathepsins B1 and B2, in: *Tissue Proteinases* (A.J. Barrett and J.J. Dingle, eds.), pp. 1–28, North-Holland, Amsterdam.

Ožegović, B., Schön, E., and Milković, S., 1977, Interaction of [^{3}H]aldosterone with rat kidney plasma membranes, *J. Steroid Biochem.* **8:**815.

Papahadjopoulos, D., Poste, G., Schaeffer, B.E., and Vail, W.J., 1974, Membrane fusion and molecular segregation in phospholipid vesicles, *Biochim. Biophys. Acta* **352:**10.

Paton, D.M., 1969, Factors affecting the retention of [^{3}H]oestradiol by rat uterus and diaphragm, *Arch. Int. Pharmacodyn.* **181:**118.

Peck, E.J., Jr., Burgner, J., and Clark, J.H., 1973, Estrophilic binding sites of the uterus: Relation to uptake and retention of estradiol *in vitro, Biochemistry* **12:**4596.

Pietras, R.J., 1978, Heritable membrane alterations and growth associated with enhanced leupeptin-sensitive proteinase activity in epithelial cells exposed to dibutylnitrosamine *in vitro, Cancer Res.* **38:**1019.

Pietras, R.J., and Szego, C.M., 1975a, Steroid hormone-responsive, isolated endometrial cells, *Endocrinology* **96:**946.

Pietras, R.J., and Szego, C.M., 1975b, Endometrial cell calcium and oestrogen action, *Nature (London)* **253:**357.

Pietras, R.J., and Szego, C.M., 1975c, Surface modifications evoked by estradiol and diethylstilbestrol in isolated endometrial cells: Evidence from lectin probes and extracellular release of lysosomal protease, *Endocrinology* **97:**1445.

Pietras, R.J., and Szego, C.M., 1977, Specific binding sites for oestrogen at the outer surfaces of isolated endometrial cells, *Nature (London)* **265:**69.

Pietras, R.J., and Szego, C.M., 1979a, Metabolic and proliferative responses to estrogen by hepatocytes selected for plasma membrane binding-sites specific for estradiol-17β, *J. Cell. Physiol.* **98:**145.

Pietras, R.J., and Szego, C.M., 1979b, Estrogen-induced membrane alterations and growth associated with proteinase activity in endometrial cells, *J. Cell Biol.* **81:**649.

Pietras, R.J., and Szego, C.M., 1979c, Estrogen receptors in uterine plasma membranes, *J. Steroid Biochem.* **11:**1471.

Pietras, R.J., Naujokaitis, P.J., and Szego, C.M., 1976, Differential effects of vasopressin on the water, calcium and lysosomal enzyme contents of mitochondria-rich and lysosome-rich (granular) epithelial cells isolated from bullfrog urinary bladder, *Mol. Cell. Endocrinol.* **4:**89.

Pietras, R.J., Hutchens, T.W., and Szego, C.M., 1978a, Hepatocyte plasma membrane subfractions enriched in high affinity, low-capacity binding sites specific for estradiol-17β, *Endocrinology* **102:**[76].

Pietras, R.J., Szego, C.M., Mangan, C.E., Seeler, B.J., Burtnett, M.M., and Orevi, M., 1978b, Elevated serum cathepsin B1 and vaginal pathology after prenatal DES exposure, *Obstet. Gynecol.* **52:**321.

Pietras, R.J., Szego, C.M., Mangan, C.E., Seeler, B.J., and Burtnett, M.M., 1979, Elevated serum cathepsin B1-like activity in women with neoplastic disease, *Gynecol. Oncol.* **7;**1.

Plagemann, P.G.W., 1969, RNA synthesis in exponentially growing rat hepatoma cells. I. A caution in equating pulse-labeled polyribosomal RNA with messenger RNA, *Biochim. Biophys. Acta* **182:**46.

Poste, G., and Papahadjopoulos, D., 1976, Lipid vesicles as carriers for introducing materials into cultured cells: Influence of vesicle lipid composition on mechanism(s) of vesicle incorporation into cells, *Proc. Natl. Acad. Sci. U.S.A.* **73:**1603.

Puca, G.A., Nola, E., Sica, V., and Bresciani, F., 1971, Estrogen-binding proteins of calf uterus: Partial purification and preliminary characterization of two cytoplasmic proteins, *Biochemistry* **10:**3769.

Puca, G.A., Nola, E., Sica, V., and Bresciani, F., 1972, Estrogen-binding proteins of calf uterus: Interrelationship between various forms and identification of a receptor-transforming factor, *Biochemistry* **11:**4157.

Puca, G.A., Nola, E., Sica, V., and Bresciani, F., 1977, Estrogen binding proteins of calf uterus: Molecular and functional characterization of the receptor transforming factor: A Ca^{2+}-activated protease, *J. Biol. Chem.* **252:**1358.

Quinn, P.S., and Judah, J.D., 1978, Calcium-dependent Golgi-vesicle fusion and cathepsin B in the conversion of proalbumin into albumin in rat liver, *Biochem. J.* **172:**301.

Rao, M.L., Rao, G.S., Höller, M., Breuer, H., Schattenberg, P.J., and Stein, W.D., 1976, Uptake of cortisol by isolated rat liver cells: A phenomenon indicative of carrier-mediation and simple diffusion, *Hoppe-Seyler's Z. Physiol. Chem.* **357:**573.

Rao, M.L., Rao, G.S., and Breuer, H., 1977, Uptake of estrone, estradiol-17β and testosterone by isolated rat liver cells, *Biochem. Biophys. Res. Commun.* **77:**566.

Raspé, G. (ed.), 1971, Estrogen receptors, *Adv. Biosci.* **7:**5.

Riggs, T.R., 1970, Hormones and transport across cell membranes, in: *Biochemical Actions of Hormones*, Vol. 1 (G. Litwack, ed.), pp. 157–208, Academic Press, New York.

Roberts, S., and Szego, C.M., 1953, Steroid interaction in the metabolism of reproductive target organs, *Physiol. Rev.* **33:**593.

Rosenfeld, M.G., and O'Malley, B.W., 1970, Steroid hormones: Effects on adenyl cyclase activity and adenosine 3′,5′-monophosphate in target tissues, *Science* **168:**253.

Scatchard, G., 1949, The attraction of proteins for small molecules and ions, *Ann. N.Y. Acad. Sci.* **51:**660.

Seetharam, B., Grimme, N., Goodwin, C., and Alpers, D.H., 1976, Differential sensitivity of intestinal brush border enzymes to pancreatic and lysosomal proteases, *Life Sci.* **18:**89.

Segal, H.L., Brown, J.A., Dunaway, G.A., Jr., Winkler, J.R., Madnick, H.M., and Rothstein, D.M., 1978, Factors involved in the regulation of protein turnover, in: *Protein Turnover and Lysosome Function* (H.L. Segal and D.J. Doyle, eds.), pp. 9–28, Academic Press, New York.

Seglen, P.O., Grinde, B., and Solheim, A.E., 1979, Inhibition of the lysosomal pathway of protein degradation in isolated rat hepatocytes by ammonia, methylamine, chloroquine and leupeptin, *Eur. J. Biochem.* **95:**215.

Sen, K.K., Gupta, P.D., and Talwar, G.P., 1975, Intracellular localization of estrogens in chick liver: Increase of the binding sites for the hormone on repeated treatment of the birds with the hormone, *J. Steroid Biochem.* **6:**1223.

Sherman, M.R., Pickering, L.A., Rollwagen, F.M., and Miller, L.K., 1978, Mero-receptors: Proteolytic fragments of receptors containing the steroid-binding site, *Fed. Proc. Fed. Am. Soc. Exp. Biol.* **37:**167.

Shodell, M., 1975, Reversible arrest of mouse 3T3 cells in G_2 phase of growth by manipulation of a membrane-mediated G_2 function, *Nature (London)* **256:**578.

Sica, V., and Bresciani, F., 1979, Estrogen-binding proteins of calf uterus: Purification to homogeneity of receptor from cytosol by affinity chromatography, *Biochemistry* **18:**2369.

Sica, V., Nola, E., Puca, G.A., Cuatrecasas, P., and Parikh, I., 1976, Purification of estrogen receptor by affinity chromatography, in: *Receptors and Mechanism of Action of Steroid Hormones*, Part 1 (J.R. Pasqualini, ed.), pp. 85–90, Marcel Dekker, New York.

Smith, J.A., Martin, L., King, R.J.B., and Vertes, M., 1970, Effects of oestradiol-17β and progesterone on total and nuclear-protein synthesis in epithelial and stromal tissues of the mouse uterus, and of progesterone on the ability of these tissues to bind oestradiol-17β, *Biochem. J.* **119:**773.

Smith, K.A., Crabtree, G.R., Kennedy, S.J., and Munck, A.U., 1977, Glucocorticoid receptors and glucocorticoid sensitivity of mitogen stimulated and unstimulated human lymphocytes, *Nature (London)* **267:**523.

Smith, L.D., and Ecker, R.E., 1971, The interaction of steroids with *Rana pipiens* oocytes in the induction of maturation, *Dev. Biol.* **25:**232.

Smith, R.E., and Van Frank, R.M., 1975, the use of amino acid derivatives of 4-methoxy-β-naphthylamine for the assay and subcellular localization of tissue proteinases, in: *Lysosomes in Biology and Pathology* (J.T. Dingle and R.T. Dean, eds.), pp. 193–249, North-Holland, Amsterdam.

Spaziani, E., 1975, Accessory reproductive organs in mammals: Control of cell and tissue transport by sex hormones, *Pharmacol. Rev.* **27:**207.

Spaziani, E., and Szego, C.M., 1959, Early effects of estradiol and cortisol on water and electrolyte shifts in the uterus of the immature rat, *Am. J. Physiol.* **197:**355.

Steger, L.D., and Desnick, R.J., 1977, Enzyme therapy. VI. Comparative *in vivo* fates and effects on lysosomal integrity of enzyme entrapped in negatively and positively charged liposomes, *Biochim. Biophys. Acta* **464:**530.

Suyemitsu, T., and Terayama, H., 1975, Specific binding sites for natural glucocorticoids in plasma membranes of rat liver, *Endocrinology* **96:**1499.

Sylvén, B., Snellman, O., and Sträuli, P., 1974, Immunofluorescent studies on the occurrence of cathepsin B1 at tumor cell surfaces, *Virchows Arch. B* **17:**97.

Szego, C.M., 1974, The lysosome as a mediator of hormone action, *Recent Prog. Horm. Res.* **30:**171.

Szego, C.M., 1975, Lysosomal function in nucleocytoplasmic communication, in: *Lysosomes in Biology and Pathology,* Vol 4 (J.T. Dingle and R.T. Dean, eds.), pp. 385–477, North-Holland, Amsterdam.

Szego, C.M., 1978, Parallels in the modes of action of peptide and steroid hormones: Membrane effects and cellular entry, in: *Structure and Function of Gonadotropins* (K.W. McKerns, ed.), pp. 431–472, Plenum Press, New York.

Szego, C.M., and Davis, J.S., 1967, Adenosine 3′,5′-monophosphate in rat uterus: Acute elevation by estrogen, *Proc. Natl. Acad. Sci. U.S.A.* **58:**1711.

Szego, C.M., and Pietras, R.J., 1981, Membrane recognition and effector sites in steroid hormone action, in: *Biochemical Actions of Hormones,* Vol. 8 (G. Litwack, ed.), pp. 307–463, Academic Press, New York.

Szego, C.M., Seeler, B.J., and Smith, R.E., 1976, Lysosomal cathepsin B1: Partial characterization in rat preputial-gland and recompartmentation in response to estradiol-17β, *Eur. J. Biochem.* **69:**463.

Szego, C.M., Nazareno, M.B., and Porter, D.D., 1977, Estradiol-induced redistribution of lysosomal proteins in rat preputial gland: Evidence from immunologic probes, *J. Cell Biol.* **73:**354.

Talwar, G.P., Segal, J.J., Evans, A., and Davidson, O.W., 1964, The binding of estradiol in the uterus: A mechanism for derepression of RNA synthesis, *Proc. Natl. Acad. Sci. U.S.A.* **52:**1059.

Terenius, L., 1967, SH-groups essential for estrogen uptake and retention in the mouse uterus, *Mol. Pharmacol.* **3:**423.

Thrall, C.L., Webster, R.A., and Spelsberg, T.C., 1978, Steroid receptor interaction with chromatin, in: *The Cell Nucleus, Chromatin,* Part C, Vol. VI (H. Busch, ed.), pp. 461–529, Academic Press, New York.

Toft, D., and Gorski, J., 1966, A receptor molecule for estrogen: Isolation from the rat uterus and preliminary characterization, *Proc. Natl. Acad.Sci. U.S.A.* **55:**1574.

Umezawa, H., and Aoyagi, T., 1977, Activities of proteinase inhibitors of microbial origin, in: *Proteinases in Mammalian Cells and Tissues* (A.J. Barrett, ed.), pp. 637–662, North-Holland, New York.

Vedeckis, W.V., Freeman, M.R., Schrader, W.T., and O'Malley, B.W., 1980, Progesterone-binding components of chick oviduct: Partial purification and characterization of a calcium-activated protease which hydrolyzes the progesterone receptor, *Biochemistry* **19:**343.

Ward, W.F., Cox, J.R., and Mortimore, G.E., 1977, Lysosomal sequestration of intracellular protein as a regulatory step in hepatic proteolysis, *J. Biol. Chem.* **252:**6955.

Wasserman, R.H., and Corradino, R.A., 1973, Vitamin D, calcium, and protein synthesis, *Vitam. Horm.* **31:**43.

Watanabe, M., Sy, L.P., Hunt, D., and Lefebvre, Y., 1979, Binding of steroids by a partially purified periplasmic protein from *Pseudomonas testosteroni, J. Steroid Biochem.* **10:**207.

Whitfield, J.F., Rixon, R.H., MacManus, J.P., and Balk, S.D., 1973, Calcium, cyclic adenosine 3′,5′-monophosphate, and the control of cell proliferation: A review, *In Vitro* **8:**257.

Williams, D., and Gorski, J., 1973, Preparation and characterization of free cell suspensions from the immature rat uterus, *Biochemistry* **12:**297.

Wrange, Ö., and Gustafsson, J.-Å., 1978, Separation of the hormone- and DNA-binding sites of the hepatic glucocorticoid receptor by means of proteolysis, *J. Biol. Chem.* **253:**856.

X

Prostaglandins and Cell Function

30

Prostaglandin-Induced Luteolysis in the Superluteinized Rat Ovary

A. Aakvaag and P. A. Torjesen

1. Introduction

The history of prostaglandin (PG) research in reproductive biology is characterized by discrepancies; completely different effects may be observed, and these seem to be related to numerous factors: the doses used, experimental design, *in vivo,* vs. *in vitro,* and others. It is now, however, firmly established that $PGF_{2\alpha}$ is the luteolysin from the uterus in a number of animal species, first established in the sheep by McCracken *et al.* (1972) (for a review, see Horton and Poyser, 1976).

The corpus luteum (CL) is necessary for the maintenance of early pregnancy in all mammals, and for some species, luteal function is required throughout pregnancy. By interfering with the function of the CL, possible means of controling fertility may be obtained.

Luteolysis has two aspects, anatomical–structural and functional. This chapter will deal with functional luteolysis, which in the rat is characterized by the following phenomena: (1) a fall in progesterone production; (2) increased 20α-dihydroprogesterone (20α-DHP) production; and (3) loss of luteinizing-hormone (LH) receptor in the CL.

Although the luteolytic effect of $PGF_{2\alpha}$ on the rat CL is firmly established, the mechanism of its action is still unknown. Continued

A. Aakvaag and *P. A. Torjesen* • Hormone and Isotope Laboratory, University of Oslo Hospitals, Aker Hospital, Oslo 5, Norway.

research on this subject is therefore needed—basic research at the cellular and molecular level and clinical physiological studies for the possible application of such compounds in fertility regulation.

2. Animal Model

The superluteinized rat ovary (Parlow, 1961) has certain distinct advantages in the study of luteal function and luteolysis. Immature female rats 25 days old are injected with 50 IU pregnant mare serum gonadotropin, followed by 25 IU human chorionic gonadotropin (hCG) 65 hr later. By this treatment, superluteinization is achieved; 95% of the ovary is composed of luteal tissue that is all of the same generation and at the same stage of development.

With this animal model, we have followed spontaneous luteolysis and compared it to PG-induced luteolysis.

Ovarian function has been monitored by the estimation of progesterone and 20α-DHP in serum (Torjesen and Aakvaag, 1976, 1977), secretion of the same progestins by ovarian tissue *in vitro,* adenylate cyclase activity and the quantity of LH receptor in a membrane fraction of the CL (Birnbaumer *et al.,* 1976; Torjesen and Aakvaag, 1976, 1977), and response to hCG *in vivo* and *in vitro*.

3. Luteinizing-Hormone Receptors in the Superluteinized Rat Ovary and Luteolysis

We have previously reported on the progesterone level in serum throughout pseudopregnancy induced as described (Torjesen and Aakvaag, 1976). It is evident that the serum progesterone level reaches a plateau followed by a decline, with a pattern that follows closely the concentration of LH receptor in a membrane fraction (Fig. 1). There is a good correlation ($r = 0.92$) between serum progesterone and LH-receptor concentration in the ovary (Fig. 2). This suggests that there may be a causal relationship between loss of LH receptor and luteolysis, if we can assume that the plasma progesterone level reflects the ovarian production of progesterone.

In support of this concept was the observation by Grinwich *et al.* (1976) using an identical animal model. They induced luteolysis with $PGF_{2\alpha}$ and found reduced serum progesterone and reduced capacity of luteal tissue to bind hCG within 30 hr of $PGF_{2\alpha}$ administration.

The use of natural $PGF_{2\alpha}$ to induce luteolysis has certain disadvantages. The general toxicity severely affects the animal, with watery diarrhea and reduced motility, and the substance may even be lethal.

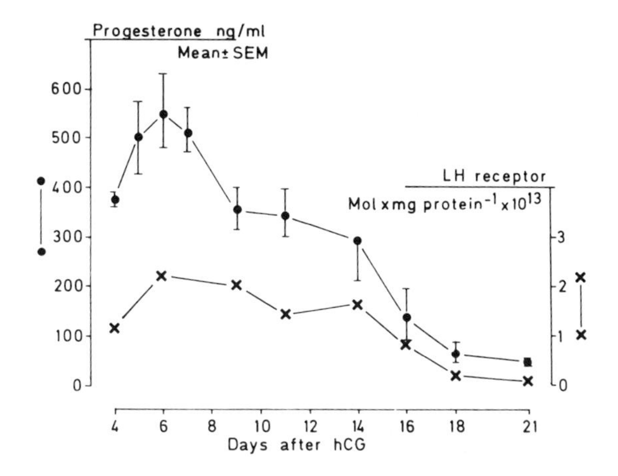

Figure 1. Serum progesterone levels and LH-receptor concentrations of a luteal-cell fraction of rats with superluteinized ovaries.

The effects observed in the ovary may therefore, in addition to a specific effect (Lahav *et al.,* 1976), also be a consequence of a general toxic effect. The development of PG analogues without significant side effects was therefore a great advantage (Dukes *et al.,* 1974). We have used substance ICI 80996 (cloprostenol, Estrumate®) (Fig. 3). This substance is very potent as a luteolytic substance, approximately 500 times more on a weight basis than $PGF_{2\alpha}$, and is without significant side effects.

Cloprostenol was injected into rats with superluteinized ovaries on day 8 after hCG administration, and the animals were sacrificed 21 hr later (Torjesen and Aakvaag, 1977). A dose–response curve for progesterone and 20α-DHP in serum and the luteal LH-receptor concentration (Fig. 4) shows a clear relationship among the three, and high correlation coefficients were found (between 0.81 and 0.92). This further supports the concept of the causal relationship between loss of LH receptor and reduced progesterone secretion. For the subsequent *in vivo* experiment, a dose of 5 μg cloprostenol was used.

A time study of the events following PG administration made it clear that the drop in the serum level of progesterone occurred almost instantaneously (Fig. 5), whereas the drop in LH receptors was not apparent for several hours (Torjesen *et al.,* 1978), suggesting that the drop in LH receptors could not be the cause of the progesterone reduction. Behrman *et al.* (1978) found in pseudopregnant rats on day 6 of pseudopregnancy that although luteal-membrane binding capacity for hCG *in vitro* remained constant within 2 hr of $PGF_{2\alpha}$ treatment, a significant decrease in the luteal accumulation of $[^{125}I]$-hCG *in vivo* occurred within 2 hr. The basis for this discrepancy of hCG binding *in*

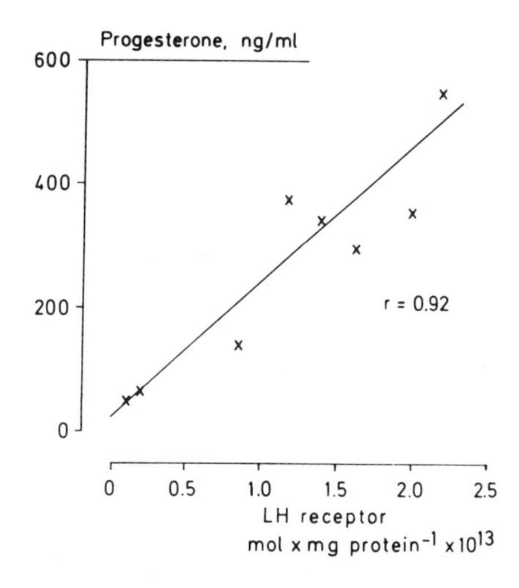

Figure 2. Correlation between serum progesterone and LH-receptor concentrations of corpus luteum in rats with superluteinized ovaries. Data from Torjesen and Aakvaag (1976).

vivo and *in vitro* is at present obscure and deserves some future investigation, and the validity of the measurement of LH receptors in luteal-cell membranes may be questionable in a physiological context.

At 24 hr after administration of the PG analogue, all signs of luteolysis—reduced progesterone and increased 20α-DHP secretion and loss of luteal LH receptor—are present, regardless of the method used to quantitate the receptor. We wanted to study the effect of cloprostenol in this situation, and gave rats another injection of the PG analogue, 19 hr after the first; control animals received saline. Groups of animals were decapitated at various time intervals, and serum progesterone and 20α-DHP were measured (Table I). A very rapid reduction in the serum level

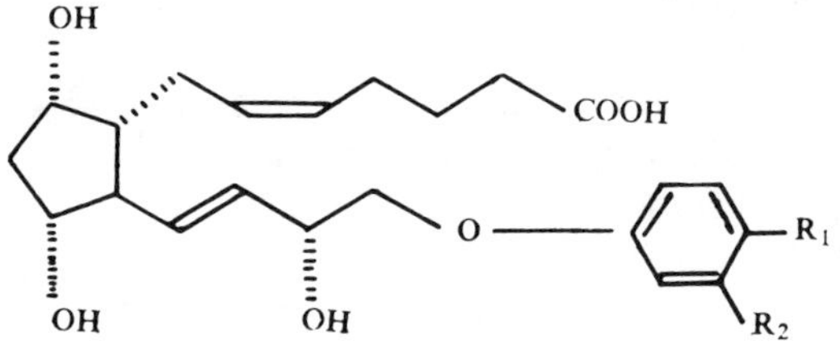

I

ICI 79,939 $R_1 = F; R_2 = H$ (racemic)

ICI 80,996 $R_1 = H; R_2 = Cl$ (racemic)

ICI 81,008 $R_1 = H; R_2 = CF_3$ (racemic)

Figure 3. Formulas of some PG analogues.

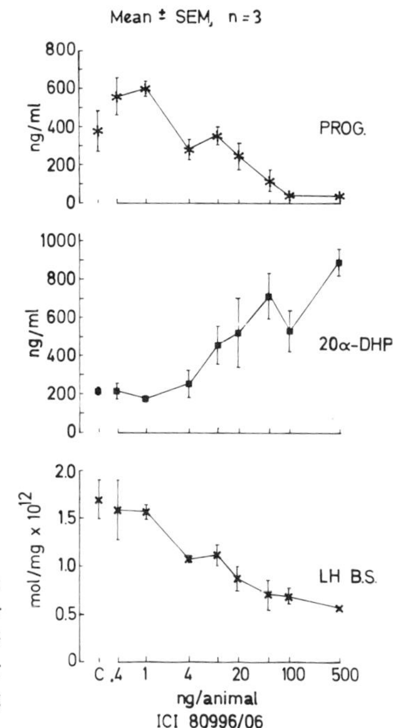

Figure 4. Serum progesterone and 20α-DHP levels and LH binding sites (LH B.S.) before and 20 hr after treatment of superluteinized rats with the PG analogue cloprostenol (ICI 80996) on day 8 after hCG. Reproduced from Torjesen and Aakvaag (1977) with permission.

of 20α-DHP, to less than 25% within 2 hr ($p < 0.01$), was observed, and a further reduction occurred during the next 4 hr ($p < 0.02$), very similar to the progesterone pattern after the first injection (see Fig. 5). During the next 20 hr, however, an increase in the serum concentration of 20α-DHP was observed, to a level not significantly different from that in the controls. Thus, the PG analogue has the same immediate effect on luteal progestin secretion regardless of which progestin is being secreted, and the amount of LH receptor in the tissue seems to be without any importance, suggesting that the early PG effects may be mediated via a system different from the LH receptor.

4. *Human Chorionic Gonadotropin and Prolactin in Vivo in Prostaglandin-Induced Luteolysis*

In agreement with the hypothesis that PG does exert its luteolytic effect via the LH receptor is the observation by Fuchs *et al.* (1974). These workers found that administration of LH concomitant with $PGF_{2\alpha}$ prevented luteolysis in pregnant rats. The same effect was observed in the animal model (see Section 2) (Table II), in which pretreatment with

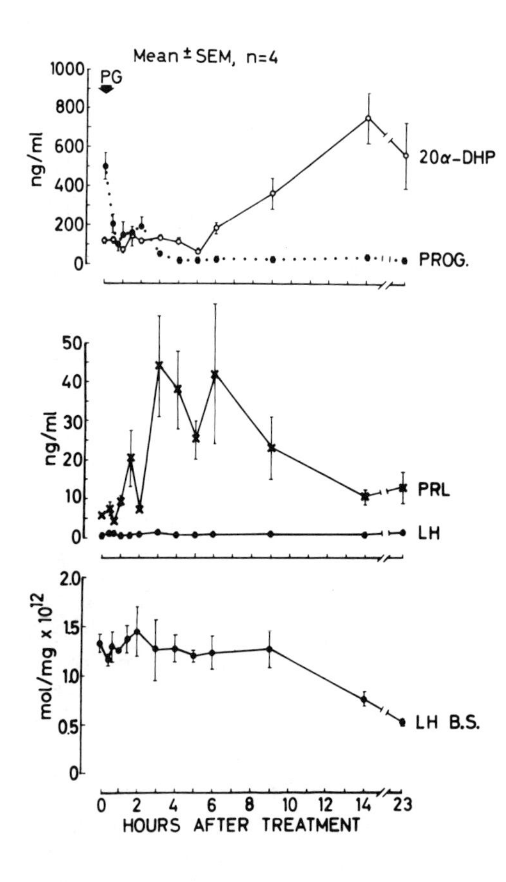

Figure 5. Serum progesterone, 20α-DHP, prolactin (PRL), and ovarian LH binding sites (LH B.S.) in rats with superluteinized ovaries before and after treatment with 5 μg of the PG analogue cloprostenol on day 8 after hCG. Reproduced from Torjesen *et al.* (1978) with permission.

25 IU hCG 2 hr prior to administration of the PG analogue totally protected the CL from the luteolytic effect of PG (Fig. 6).

A similar type of protection was observed for prolactin (PRL) by Behrman *et al.* (1978) in pseudopregnant rats on the 6th day of pseudopregnancy. The luteolytic effect of $PGF_{2\alpha}$ was prevented in animals treated with 2 IU PRL 6 hr before $PGF_{2\alpha}$ injection (Behrman *et al.*, 1978).

We have, however, been unable to demonstrate such an effect of PRL in the present animal model. When 30 μg ovine PRL (oPRL) (NIH) was given 2 hr before the PG analogue, the same serum progesterone level was observed as in controls receiving cloprostenol only, 2 hr before sacrifice (Table III). Fuchs *et al.* (1974), who gave $PGF_{2\alpha}$ to pregnant rats on day 4 and day 10 of pregnancy, observed only a marginally detectable protective effect of PRL in the former group and no effect in the latter.

Table I. Effect of the Prostaglandin Analogue Cloprostenol (5 μg s.c.) on Progestin Levels in Rats with Superluteinized Ovaries

Treatment		Progestins in serum (ng/ml)[a]					
		21 hr		25 hr		44 hr	
0 hr	19 hr	P	20α-DHP	P	20α-DHP	P	20α-DHP
Cloprostenol	Saline	55 ± 6	964 ± 86	—	—	37 ± 11	689 ± 183
		(N = 5)				(N = 4)	
Cloprostenol	Cloprostenol	16 ± 1	225 ± 17	6 ± 1	129 ± 18	13 ± 2	434 ± 30
		(N = 4)		(N = 4)		(N = 5)	

[a] Results are means ± S.E.M. (P) Progesterone; (20α-DHP) 20α-dihydroprogesterone.

Therefore, at present, the effect of PRL on $PGF_{2\alpha}$-induced luteolysis is still an open question. The discrepancies observed may be due to the use of different animal models or to the dosage of PRL in relation to PG. It should be pointed out that the dose of the PG analogue we have used is very high in terms of luteolytic effect compared to the dosages of $PGF_{2\alpha}$ used by other investigators, due to the toxic effect of the latter.

When the luteolysis has been initiated with PG analogue, hCG stimulation *in vivo* has only moderate effects. Administration of 25 IU hCG 1 hr after cloprostenol stimulated the progesterone level in serum 4 hr later [$p < 0.02$ (Fig. 7)], whereas no significant change was found for 20α-DHP (Fig. 8). hCG had no effect on the serum level of either progestin when given 24 hr after the PG analogue. In the early phase of PG-induced luteolysis, the CL has retained some ability to respond to hCG, an ability that is lost during the next 23 hr. These data must be seen in relation to results on the effect of LH on progesterone secretion by the ovary *in vitro* (see Section 3) and on adenylate cyclase in a membrane fraction of the luteal cells (see Section 5).

Table II. Effect of Pretreatment with 25 IU hCG on the Luteolytic Effect of 5 μg Cloprostenol on the Luteal Function of the Superluteinized Rat Ovary

Treatment		Progestins in serum (ng/ml)[a]					
		¾ hr		6 hr		26 hr	
hCG (−2 hr)	Cloprostenol (0 hr)	P	20α-DHP	P	20α-DHP	P	20α-DHP
		(N = 5)		(N = 5)		(N = 4)	
+	−	725 ± 89	266 ± 15	462 ± 37	203 ± 46	400 ± 47	176 ± 12
+	+	686 ± 86	277 ± 34	420 ± 25	138 ± 15	340 ± 45	174 ± 24

Results are means ± S.E.M. (P) Progesterone; (20α-DHP) 20α-dihydroprogesterone.

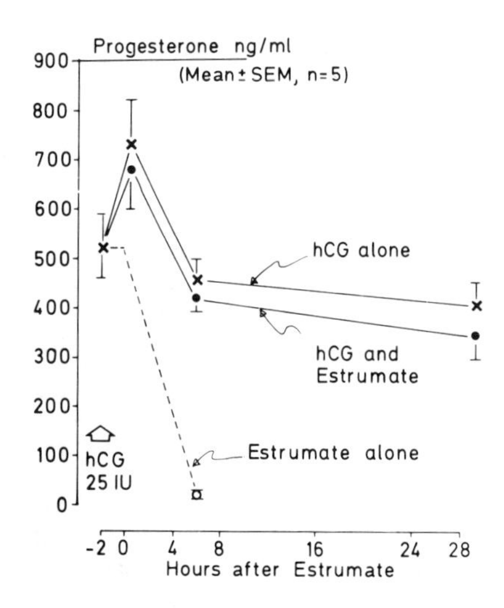

Figure 6. Effect of pretreatment with 25 IU hCG 2 hr before 5 μg of the PG analogue Estrumate® (cloprostenol) on serum progesterone level in rats with superluteinized ovaries on day 8 after hCG.

5. *Corpus Luteum Function in Vitro Following Prostaglandin Treatment in Vivo*

Since the effect of PG on the corpus luteum is a very rapid one, it seemed necessary to study the early changes in CL function. In light of the reported loss of LH binding *in vivo* following $PGF_{2\alpha}$, and the subsequent loss of luteal-cell-membrane receptor for LH, it was important to further characterize the CL function following PG treatment.

Ovaries removed at different time intervals from cloprostenol-treated and control animals were cut in four pieces and distributed so that each incubation flask contained one ovary from four different animals of the same treatment group. Incubations were carried out in 3 ml Krebs–Ringer bicarbonate buffer with 0.3% bovine serum albumin and 2 mM ATP in the presence or absence of 50 IU hCG. Aliquots of 100 μl medium were removed at various time intervals for the determination of progesterone and 20α-DHP. The secretion into the medium was linear up to 5 hr, and the data are presented as the amount of progesterone and 20α-DHP secreted into the medium per hour per ovary.

The tissue concentration of progesterone and 20α-DHP was measured before and after incubation, and it was evident that the steroids in the medium represented new synthesis and not merely secretion of preformed steroids.

Within 1 hr of the PG treatment, there was no change in unstimulated progesterone secretion, and the tissue was able to respond to hCG,

Table III. Effect of Ovine PRL on Prostaglandin-Induced Luteolysis in Superluteinized Rat Ovaries on Day 8 after hCG

Treatment		Progestins (ng/ml)[a]	
oPRL (30 μg, −2 hr)	Cloprostenol (5 μg, 0 hr)	P (2 hr)	20α-DHP (2 hr)
+	+	127 ± 4	63 ± 12
−	+	111 ± 21	49 ± 6

[a] Results are means ± S.E.M. ($N = 3$). (P) Progesterone; (20α-DHP) 20α-dihydroprogesterone.

although not quite as strongly as the saline control [$p < 0.05$ (Table IV)]. After 24 hr, progesterone secretion was reduced to about 10% of controls ($p < 0.01$) in accordance with the *in vivo* data, but still with retained ability to respond to hCG. At this time, there was an increase in 20α-DHP secretion that responded to hCG.

It should be recalled that no response to hCG was observed *in vivo* 24 hr after PG administration (see Fig. 7).

In another set of experiments, a luteal-cell-membrane fraction was prepared following the PG analogue treatment, and adenylate cyclase was determined according to Birnbaumer *et al.* (1976). PG treatment had no effect on unstimulated adenylate cyclase activity either 3 or 17 hr after treatment (Table V); also, the tissue content of cylic AMP (cAMP) was unaffected (data not shown). It was, however, evident that PG treatment completely abolished the adenylate cyclase response to hCG

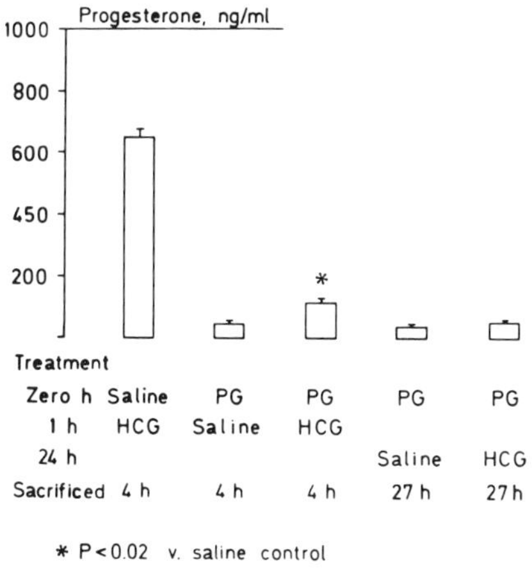

Figure 7. Effect of 25 IU hCG after 5 μg of the PG analogue cloprostenol on serum progesterone levels in rats with superluteinized ovaries on day 8 after hCG. The vertical bars indicate S.E.M.

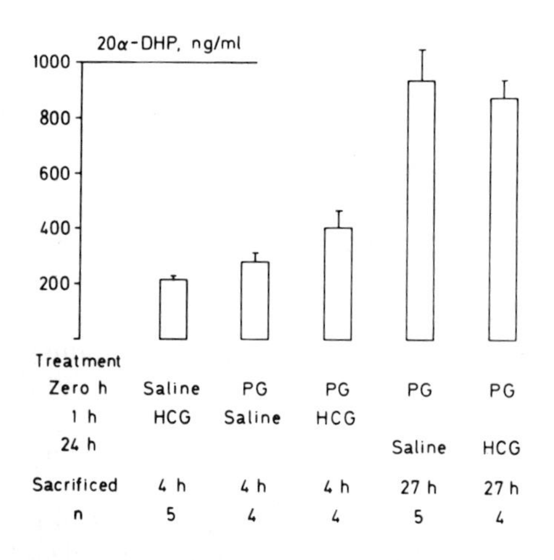

Figure 8. Effect of 25 IU hCG after 5 μg of the PG analogue cloprostenol on serum 20α-DHP levels in rats with superluteinized ovaries on day 9 after hCG. The vertical bars indicate S.E.M.

at both time intervals after the treatment. The response to isoproterenol remained intact following treatment with PG.

6. 20α-Hydroxysteroid Dehydrogenase Activity of the Ovary Following Prostaglandin

One of the characteristic features of luteolysis in the rat is the increased serum level of 20α-DHP. In the last experiment to be reported, we measured 20α-hydroxysteroid dehydrogenase (20α-HSD) in a 20,000*g* supernatant of a homogenate according to Eckstein *et al.* (1977). Animals were killed at various time intervals after an injection of 5 μg cloprostenol; control animals received saline.

It can be seen in Fig. 9 that as early as 2½ hr after administration of the PG analogue, there was a significant increase in 20α-HSD ($p <$ 0.001), and there was a continued increase during the next 20 hr. Also, the control showed some increase, which probably reflects an early phase of spontaneous luteolysis. Despite the early increase in enzyme activity, there was no increase in serum 20α-DHP at this time, 2½ hr after PG, which probably reflects the reduced steroidogenic activity *in vivo* of the ovary at this time, an activity that is regained during the next 10 hr.

This experiment does not allow any conclusions as to the underlying mechanism for this increase in 20α-HSD. The fact that it has more than

Table IV. Effect of hCG on Progestin Secretion by Ovarian Tissue *in Vitro* after Treatment with Cloprostenol *in Vivo*

	Progestins (ng/ovary per hr)[a]			
	Progesterone		20α-DHP	
Treatment[b]	Control	hCG	Control	hCG
Saline (1 hr)	550 ± 110	2000 ± 90[c]	1630 ± 150	1910 ± 100
Cloprostenol (1 hr)	640 ± 110	1530 ± 100[c,d]	1300 ± 160	2000 ± 50[c]
Saline (24 hr)	805 ± 55	1410 ± 120[c]	1680 ± 80	1920 ± 230
Cloprostenol (24 hr)	78 ± 11[e]	152 ± 23[c,e]	1950 ± 180	3100 ± 230[c]

[a] Results are means ± S.E.M. ($N = 6$). (20α-DHP) 20α-Dihydroprogesterone. For experimental details, see the text.
[b] Hours between treatment and sacrifice are indicated in parentheses.
[c] Significantly different from control, $P < 0.01$.
[d] Significantly different from saline-treated, $P < 0.01$.
[e] Significantly different from saline-treated, $P < 0.05$.

doubled within 2½ hr makes a hypothesis of a direct PG effect on the luteal cell an attractive one.

7. *Concluding Remarks and Summary*

PG induces a very rapid luteolytic process in the superluteinized rat ovary, the half-life of progesterone being of the order of 15–20 min. Within 1–3 hr of PG administration, the adenylate cyclase of a membrane fraction has lost its ability to respond to hCG, giving support to the concept that blocking the LH receptor may be the mechanism of action of PG. The fact that hCG *in vivo* and *in vitro* will stimulate progesterone

Table V. Adenylate Cyclase in Membranes of Corpora Lutea from Superluteinized Rat Ovaries

	cAMP (pmol)/min per mg protein)[a]		
Treatment	Unstimulated	hCG	Isoproterenol
Control			
day 8	24.8 ± 4.2	54.5 ± 16.8	83.5 ± 16.0
day 9	17.3 ± 1.8	47.0 ± 9.1	75.8 ± 10.0
Prostaglandin[b]			
3 hours earlier	24.3 ± 1.7	25.5 ± 2.8	52.3 ± 5.8
17 hours earlier	13.3 ± 1.5	14.8 ± 1.8	56.0 ± 4.6

[a] Each figure is the mean ± S.E.M. of four estimations.
[b] The prostaglandin analogue cloprostenol, 5 μg, was used.

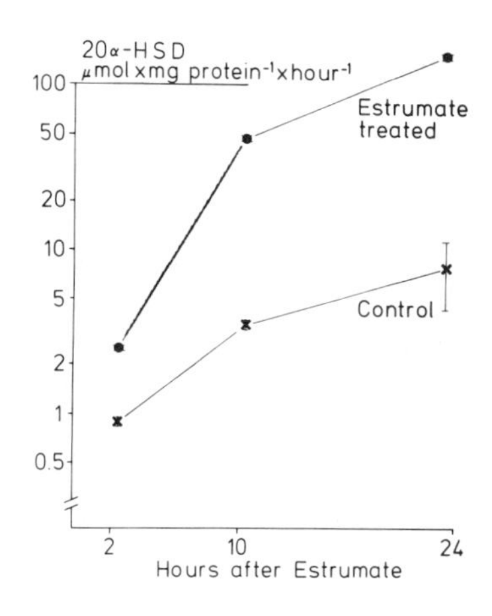

Figure 9. Ovarian 20α-HSD activity from superluteinized rat ovaries after 5 μg of the PG analogue Estrumate® (cloprostenol) on day 8 after hCG. Each point is the mean of six estimations ± S.E.M.

secretion at this early stage of luteolysis is, however, not consistent with this hypothesis. At 24 hr after treatment with PG, hCG had no effect *in vivo* and no effect on adenylate cyclase *in vitro,* but still, hCG clearly stimulated the secretion of both progesterone and 20α-DHP *in vitro.* Again, this discrepancy is difficult to explain.

In some way, the LH receptor does seem to be involved; the administration of hCG prior to PG completely protects the ovary from the luteolytic effect of the latter. A possible mechanism may be that binding of hCG to the LH receptor prevents the subsequent binding of PG to its receptor, which fits in nicely with the model of Henderson and McNatty (1975). This might possibly be the mechanism by which the human CL of pregnancy resists the luteolytic effect of $PGF_{2\alpha}$; hCG from the placenta might protect the CL from the PG effect. If PG acts solely on the LH receptor, one might expect only a moderate effect of a repeated PG dose 24 hr after the first. At this time a dramatic reduction in LH receptor has occurred, and there is no hCG-responsive adenylate cyclase and no response to hCG *in vivo.* Nevertheless, the drop in serum 20α-DHP resembles that of progesterone after the first injection. This finding may give some support to alternative theories, i.e., the vascular theory (Thorburn and Hales, 1972). It seems difficult to explain the data presented herein on the basis of just the one effect of PG on the superluteinized rat ovary—blocking the LH receptor.

Acknowledgments. Cloprostenol (ICI 80996) was kindly supplied by Dr. Bolton of Imperial Clinical Industries Limited, England. We are grateful to Ms. Lotte Aakvaag, Ms. Ruth Dahlin, and Ms. Sidsel Bugge for expert technical help and to Ms. Kjersti Gunneng for secretarial help.

Discussion

HAOUR: hCG is known to stimulate prostaglandin synthesis. Do you think it is possible that the protective effect of hCG in relation to the prostaglandin-induced luteolysis might be mediated via local prostaglandin synthesis?

AAKVAAG: I do not quite see how locally synthesized prostaglandins in response to hCG might protect the luteal tissue against the luteolytic effect of the subsequent exogenous prostaglandins.

BAZER: Relative to the question by Dr. Haour, the endogenous prostaglandins, e.g., PGE_2, may protect the corpora lutea by occupying the same receptor required for $PGF_{2\alpha}$ to exert its effect as proposed by the model of Henderson and McNatty (1975), which you mentioned. Data from Dr. Keith Inspeek of the University of West Virginia and Dr. C. W. Weems of the University of Arizona with sheep suggest that prostaglandins of the E series may play a role in protection of the corpus luteum during the period of establishment of the corpus luteum of pregnancy.

BANIK: Have you or anybody tried to induce luteolysis in pregnant or pseudopregnant rats with this prostaglandin analogue?

AAKVAAG: We have worked with this animal model only, and we have found that the sensitivity to prostaglandins increases from total resistance shortly after the induction of superluteinization to the very sensitive situation at this stage, day 8 after hCG injection. Similar data have been obtained when inducing luteolysis in pregnant or pseudopregnant animals. Dukes *et al.* (1974) showed that cloprostenol, the substance we have been using, would interrupt pregnancy in rats given as early as day 5 of pregnancy, and Salazar *et al.* (1976) showed that this was due to induction of luteolysis.

Russell (1975) induced early menstruation in nonpregnant monkeys, and Csapo and Mocsary (1976) were able to terminate human pregnancy at 4 weeks, with a very similar analogue, ICI 81008, probably by inducing luteolysis.

KELLY: Your failure to prevent the luteolytic effect of ICI 80996 by prolactin may be due to the low dose (30 mg) utilized and the very short period (2 hr) examined.

References

Behrman, H.R., Grinwich, D.L., Hichens, M., and MacDonald, G.J., 1978, Effect of hypophysectomy, prolactin, and prostaglandin $F_{2\alpha}$ on gonadotropin binding *in vivo* and *in vitro* in the corpus luteum, *Endocrinology* **103:**349.

Birnbaumer, L., Young, P.C., Hunzicker-Dunn, M., Bockaert, J., and Duran, J.M., 1976, Adenyl cyclase activities in ovarian tissues. I. Homogenization and conditions of assay in Graafian follicles and corpora lutea of rabbits, rats and pigs, *Endocrinology* **99:**163.

Csapo, A.I., and Mocsary, P., 1976, Menstrual induction with the PGF2α-analogue ICI 81008, *Prostaglandins* **11:**155.

Dukes, M., Russell, W., and Walpole, A.L., 1974, Potent luteolytic agents related to prostaglandin $F_{2\alpha}$, *Nature (London)* **250:**330.

Eckstein, B., Raanan, M., Lerner, N., Cohen, S.: and Nimrod, A., 1977, The appearance of 20α-hydroxysteroid dehydrogenase activity in preovulatory follicles of immature rats treated with pregnant mare serum gonadotropin, *J. Steroid Biochem.* **8**:213.

Fuchs, A.-R., Mok, E., and Sundaram, K., 1974, Luteolytic effects of prostaglandins in rat pregnancy, and reversal by luteinizing hormone, *Acta Endocrinol. (Copenhagen)* **76**:583.

Grinwich, D.L., Ham, E.A., Hichens, M., and Behrman, H.R., 1976, Binding of human chorionic gonadotropin and response of cyclic nucleotides to luteinizing hormone in luteal tissue from rats treated with prostaglandin $F_{2\alpha}$, *Endocrinology* **98**:146.

Henderson, K.M., and McNatty, K.P., 1975, A biological hypothesis to explain the mechanism of luteal regression, *Prostaglandins* **9**:779.

Horton, E.W., and Poyser, N.L., 1976, Uterine luteolytic hormone: A physiological role for prostaglandin $F_{2\alpha}$, *Physiol. Rev.* **56**:595.

Lahav, M., Freud, A., and Lindner, H.R., 1976, Abrogation by prostaglandin $F_{2\alpha}$ of LH-stimulated cyclic AMP accumulation in isolated rat corpora lutea of pregnancy, *Biochem. Biophys. Res. Commun.* **68**:1294.

McCracken, J.A., Carlson, J.C., Glew, M.E., Goding, J.R., Baird, D.T., Green, K., and Samuelsson, B., 1972, Prostaglandin $F_{2\alpha}$ identified as a luteolytic hormone in sheep, *Nature (London) New Biol.* **238**:129.

Parlow, A.F., 1961, Bioassay of luteinizing hormone by ovarian ascorbic acid depletion, in: *Human Pituitary Gonadotrophins* (A. Albert, ed.), pp. 300–310, Charles C. Thomas, Springfield, Illinois.

Russell, W., 1975, Luteolysis induced in pigtail monkeys *(Macaca nemestrina)* with prostaglandin $F_{2\alpha}$, ICI 80996 and ICI 81008, *Prostaglandins* **10**:163.

Salazar, H., Furr, B.J.A., Smith, G.K., Bentley, M., and Gonzalez-Angulo, A., 1976, Luteolytic effects of a prostaglandin analogue, cloprostenol (ICI 80,996), in rats: Ultrastructural and biochemical observations, *Biol. Reprod.* **14**:458.

Thorburn, G.D., and Hales, J.R.S., 1972, Selective reduction in blood flow to the ovine corpus luteum after infusion of prostaglandin $F_{2\alpha}$ into the uterine vein, *Proc. Aust. Physiol. Pharmacol. Soc.* **3**:145.

Torjesen, P.A., and Aakvaag, A., 1976, The affinity and capacity of the LH receptor of the superluteinized rat ovary in relation to progesterone production, *Eur. J. Obstet. Gynecol. Reprod. Biol.* **6**:181.

Torjesen, P.A., and Aakvaag, A., 1977, The serum levels of progesterone and 20α-dihydroprogesterone, and the ovarian LH, FSH and PRL binding during luteolysis of the superluteinized rat ovary, *Acta Endocrinol. (Copenhagen)* **86**:162.

Torjesen, P.A., Dahlin, R., Haug, E., and Aakvaag, A., 1978, The sequence of hormonal changes during prostaglandin induced luteolysis of the superluteinized rat ovary, *Acta Endocrinol. (Copenhagen)* **87**:617.

31

Role of Prostaglandins in Leydig-Cell Stimulation by hCG and Leydig-Cell Function

F. Haour, J. Mather, B. Bizzini-Kouznetzova, and F. Dray

1. Introduction

In many different cell types, prostaglandins have the ability to modify hormone action (Shio *et al.*, 1971), an effect known to be correlated with changes in cyclic AMP (cAMP) levels. In the restricted case of steroid-producing cells, as found in ovarian and testicular tissues, numerous data indicate that prostaglandins are directly involved in the mechanism of luteinizing hormone (LH)–human chorionic gondadotropin (hCG) action and in post-cAMP events (Phariss and Behrman, 1973; Kuehl *et al.*, 1970; Kuehl, 1974; Cenedella, 1975; Samuelson *et al.*, 1978).

The ability of prostaglandins to influence steroidogenesis was first recognized from investigations of the action of prostaglandins on the corpus luteum. A large number of publications demonstrate that prostaglandin $F_{2\alpha}$ ($PGF_{2\alpha}$) administered *in vivo* induces luteolysis of the corpus luteum through an inhibition of ovarian steroidogenesis (for reviews, see

F. Haour • INSERM U 162, Hôpital Debrousse, 69322 Lyon, France. *J. Mather* • Center for Biomedical Research, The Population Council, The Rockefeller University, New York, New York 10021. *B. Bizzini-Kouznetzova* and *F. Dray* • INSERM FRA 8, Institut Pasteur, 75724 Paris, France.

Phariss and Behrman, 1973; Cenedella, 1975). The mechanism of this inhibition involves interference with cholesterol ester synthetase activity (Behrman *et al.*, 1971), modification of the number of LH–hCG receptors (Grinwich *et al.*, 1976; Behrmann and Hickens, 1976), or inhibition of the LH–hCG-dependent cAMP accumulation (McNatty *et al.*, 1975; Thomas *et al.*, 1978).

However, the *in vitro* data indicate that $PGF_{2\alpha}$, as well as PGE_1 and PGE_2, mimic LH–hCG action and induce an increase in progesterone secretion (for a review, see Phariss and Behrman, 1973). This can be related to the fact that prostaglandin receptors linked to functional adenylate cyclase are present in the corpus luteum (Kuehl *et al.*, 1970; Rao, 1976) and might also induce steroidogenesis through cAMP accumulation (Shio *et al.*, 1971; Samuelson *et al.*, 1978). Since it has been demonstrated that prostaglandin biosynthesis is increased under LH–hCG stimulation (Chasalow and Pharriss, 1972; Marsh and LeMaire, 1974; Marsh *et al.*, 1974; Clark *et al.*, 1978 a,b; Zor *et al.*, 1977 a,b), a positive-feedback system could take place. However, the existence of one agonist-specific desensitization of the adenylate cyclase system for either LH–hCG or prostaglandin receptor (Bockaert *et al.*, 1976; Lamprecht *et al.*, 1977; Tell *et al.*, 1978) should provide a stop signal to this positive effect.

In the case of the testicular Leydig cell, though the amount of information is limited, the reported effect of PGE, PGF, and PGA in the male rodent on the synthesis of testosterone, *in vivo,* is a pronounced decrease in plasma testosterone (Bartke *et al.*, 1973; Saksena *et al.*, 1978; Kimball *et al.*, 1978). In contrast, in the perfused dog testis, testosterone production is stimulated (Eik-Nes, 1969). In *in vitro* studies, PGA_1, PGA_2, and PGE_1 inhibit testosterone production in mice (Bartke *et al.*, 1976) and in rat (Grotjan *et al.*, 1978). PGE_1 has also been shown to increase cAMP production while testosterone production was unchanged in a rat Leydig-cell preparation (Cooke *et al.*, 1974).

PGE and PGF are present and produced in the testis (Carpenter, 1974; Ellis *et al.*, 1972; Gerozissis and Dray, 1977a,b). Leydig cells themselves produce PGE and PGF (Carpenter *et al.*, 1978; Haour *et al.*, 1979), and this production is enhanced under LH–hCG stimulation (Haour *et al.*, 1979).

Since prostaglandin receptors are present in Leydig cells (Haour *et al.*, 1979), a local feedback regulation of Leydig-cell function is possible. Since it was known that prostaglandins have an inhibitory effect on testosterone production, it was of special interest to investigate the possible role of prostaglandins during the period of desensitization of LH–hCG stimulation and "down-regulation" of LH–hCG receptor (Sharpe, 1976; Haour and Saez, 1977, 1978; Hsueh *et al.*, 1977; Saez *et al.*, 1978a,b; Chasalow *et al.*, 1979). It has previously been shown (Haour *et al.*, 1979) that the period of acute increases in prostaglandin production corresponds to the period of intense LH–hCG-receptor down-regulation. Moreover, high levels of prostaglandins are observed following LH stimulation for 48 hr, corresponding to the refractory or desensitized period.

This chapter provides indications, obtained *in vitro* in the rat, suggesting that PGE_2 and $PGF_{2\alpha}$ are involved in the modulation of LH and hCG receptors and in the modulation of the cell sensitivity to hCG stimulation. *In vitro* studies on Leydig cells in primary culture have made possible a more detailed study of the importance of prostaglandins in the regulation of Leydig-cell function.

2. Role of Prostaglandins during LH–hCG Stimulation of Leydig Cells in the Adult Rat

2.1. Production of Prostaglandins during hCG Stimulation

The testicular content of prostaglandins increases drastically following hCG stimulation. Figure 1 indicates the pattern of PGE_2 and $PGF_{2\alpha}$ compared to the testicular testosterone (T) content and hCG binding sites. T production increases sharply immediately after the injection, but

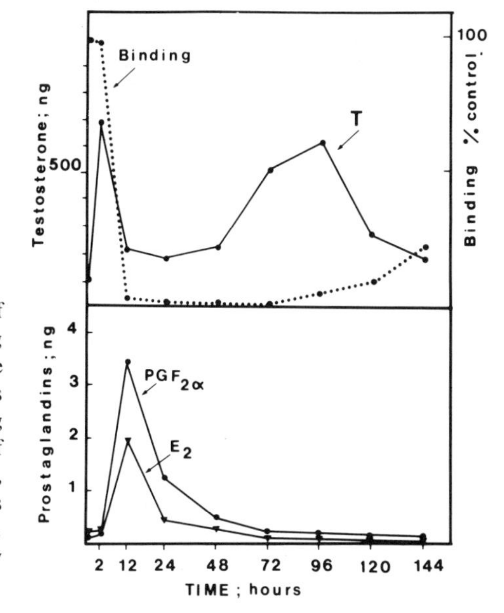

Figure 1. Pattern of the concentration of T, PGE_2, and $PGF_{2\alpha}$ and hCG binding sites in testicular tissue following one hCG injection (30 μg). Prostaglandins and T are expressed in ng/testis. Binding sites are expressed as percentages of control values (from Haour and Saez, 1977; Haour *et al.*, 1979). Prostaglandins were assayed according to Dray *et al.*, (1975) and Gerozissis and Dray (1977a,b).

this increase is transient, since 12 hr later the testicular concentration is about 2-fold the control value. Up to 48 hr following the injection, this plateau level is maintained and corresponds to the period of so-called desensitization. The spontaneous T peak observed between 48 and 144 hr is interpreted as the resensitization period (Haour and Saez, 1978).

PGE_2 and $PGF_{2\alpha}$ do not increase before 2 hr following hCG injection at a time when T concentration is at its peak value. Levels higher than control are maintained for 48 hr, for both PGE_2 and $PGF_{2\alpha}$. During the same period, PGE_1 is low and does not increase under hCG stimulation.

hCG binding sites, as previously shown (Haour and Saez, 1977, 1978), decrease sharply from 2 to 12 hr following hCG stimulation. They are almost undetectable from 12 to 96 hr following hCG injection. A slow rise in the binding-site number is observed after 72 hr.

This pattern of prostaglandin and androgen secretion and binding-site fluctuation suggests several comments. First, the acute increase in prostaglandin concentration corresponds to the acute disappearance in hCG binding sites. Second, elevated prostaglandin levels are observed for 48 hr, which corresponds to the period of refractoriness to hCG. Third, prostaglandin and T production are not directly linked and seem to alternate: T production is high when prostaglandin levels are still basal, T is lower than expected during the period of high prostaglandin levels, and rebound of T production is observed when prostaglandin levels are back to control values.

The same conclusions can be derived from the results obtained in hypophysectomized animals (Fig. 2). Despite the lack of an immediate T peak following hCG injection, PGE_2 and $PGF_{2\alpha}$ increase sharply at 2 hr following hCG injection and elevated T values are not observed before both prostaglandins reach control level.

These patterns are very suggestive of a direct role of prostaglandins in the regulation of steroidogenesis and hCG binding sites.

2.2. *Origin of the Prostaglandin Production*

Increase in Prostaglandin production might take place in Leydig cells but might also be a secondary phenomenon, for example, a consequence of the increase in T production. The time lag of 2 hr before the increase in prostaglandin concentrations would be in favor of the second hypothesis. However, it is likely that most of the prostaglandins assayed are produced by the Leydig cells themselves. As shown in Fig. 3, Leydig cells isolated (Haour and Saez, 1977; Saez *et al.*, 1978a) at different times following hCG injection secrete prostaglandins *in vitro*. The pattern of secretion and the relative proportions of PGE_2 and $PGF_{2\alpha}$ are very similar to what is observed in the total testicular tissue. Despite

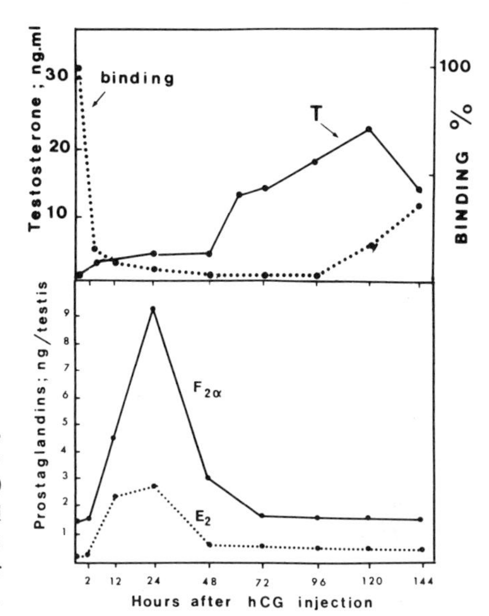

Figure 2. Hypophysectomized animals: Pattern of the concentrations of T, PGE_2, and $PGF_{2\alpha}$ and hCG binding sites in testicular tissue following one hCG injection (30 μg). Prostaglandins and T are expressed in ng/testis. Binding sites are expressed as percentages of control values (from Haour and Saez, 1978; Haour *et al.*, 1978, 1979).

the lack of careful purification of the Leydig cells, this is suggestive that the hCG-induced prostaglandin synthesis takes place in the Leydig cells.

2.3. Prostaglandin Receptors

Receptors for prostaglandins are present in the Leydig cell (Haour *et al.*, 1979). PGE_1 and PGE_2 seem to bind to the same receptor with different affinities. $PGF_{2\alpha}$ is bound by a different receptor than the E series.

The presence of prostaglandin binding sites in Leydig cells confirms the idea that prostaglandins might contribute to the regulation of Leydig-cell function. Autofeedback regulations are possible, which would explain a role of the prostaglandins during the desensitization process and the down-regulation of the hCG receptor.

2.4. Prostaglandin Production and Testosterone Production

T Production can occur without any detectable changes in prostaglandin levels. This is the case during the first 2 hr following hCG stimulation (see Fig. 1) and during the delayed T Peak after 48 hr (see Fig. 1). Similarly, under indomethacin treatment (Fig. 4) sufficient to cause a total blockage of prostaglandin production, there are subnormal T levels (Fig. 5).

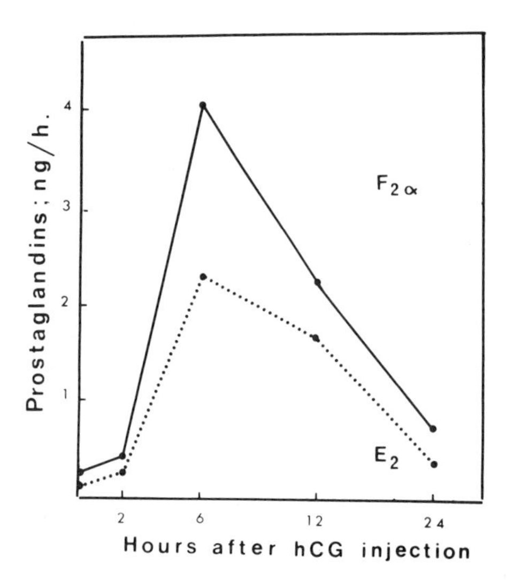

Figure 3. PGE_2 and $PGF_{2\alpha}$ production (ng/5 million cells in 1 hr at 34°C in Modified Eagle's Medium by isolated Leydig cells prepared at 2, 6, 12, or 24 hr following one hCG injection (from Haour *et al.*, 1979).

Alternatively, prostaglandins can be produced in great amount while T production is minimal. This is the case in the hypophysectommized rat during the first 48 hr following hCG injection (see Fig. 2). However, the increase in prostaglandin levels is not under the control of T itself, since high doses of T (T 1 mg plus T propionate 10 mg/rat) do not modify prostaglandin levels in the testis (data not shown). Similarly, the delayed T peak (72–144 hr) in either intact or hypophysectomized rats occurs without detectable changes in prostaglandin levels. The two biosynthetic pathways seem to be either controlled by different factors or triggered at different levels of cell stimulation by LH. The last hypothesis is more likely, since there is a definite correlation between the respective levels of both. T seems to be lower when prostaglandins are high and vice versa. Prostaglandins would consequently be one of the factors that contribute to the decrease in T production during the desensitized period. This is in agreement with the reports of inhibition of androgen levels *in vivo* (Saksena *et al.,* 1978) and *in vitro* (Grotjan *et al.,* 1978) in adult male rats by PGE_2 and $PGF_{2\alpha}$.

2.5. Relationship among Prostaglandins, cAMP, and Testosterone Production: Effect of Indomethacin

Indomethacin blocks the prostaglandin synthestase (or cyclooxygenase) that transforms arachidonic acid into endoperoxides (PGG_2, PGH_2) that are precursors to PGE_2, $PGF_{2\alpha}$, thromboxane, and other substances.

In our experiment, indomethacin treatment blocked prostaglandin production either under basal conditions or under hCG stimulation (see Fig. 4). During the same time, hCG-induced cAMP production is increased by simultaneous injection of indomethacin plus hCG compared to hCG alone (Fig. 6) due to an inhibitory effect of indomethacin on phosphodiesterase activity. T production (see Fig. 5) seems to be somewhat reduced by indomethacin treatment.

Since cAMP increase occurs before prostaglandin release, it is possible that the increase in prostaglandin production could be mediated through cAMP accumulation. However, the role of cAMP in the regulation of prostaglandin synthesis is unclear. In the platelet, cAMP has been shown to inhibit the cyclooxygenase activity (Malmsten *et al.*, 1976) and also to inhibit the phospholipase A_2 (Lapetina *et al.*, 1977) that transforms phospholipids into arachidonic acid. Both effects decrease prostaglandin production. In contrast, in the preovulatory rat follicle, cAMP enhances PGE production with a lag-time of about 3 hr (Clark *et al.*, 1978a,b).

Another point to be taken into consideration is the specific biological

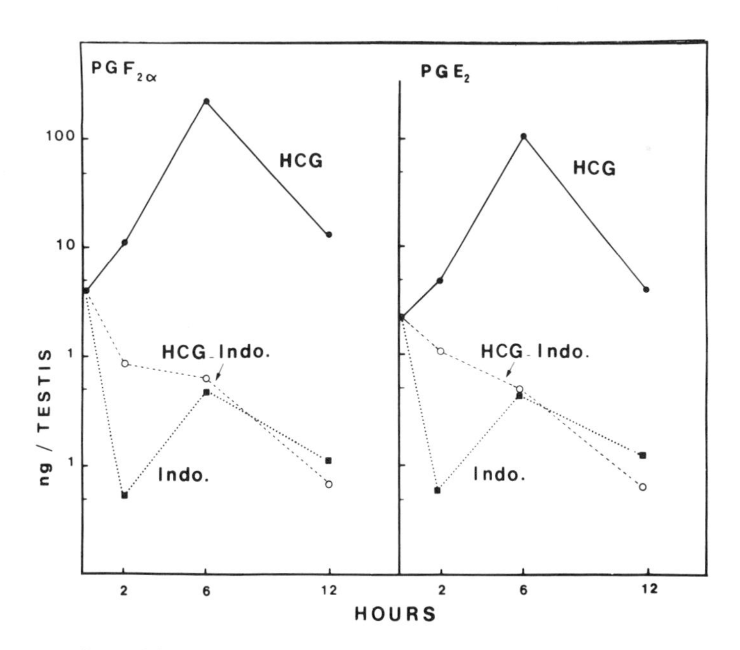

Figure 4. Effect of indomethacin (12 mg/kg) and hCG treatment on testicular levels of PGE_2 and $PGF_{2\alpha}$. Indomethacin was injected either alone or with hCG (30 μg); in the latter case, indomethacine was administered 2 hr before hCG injection.

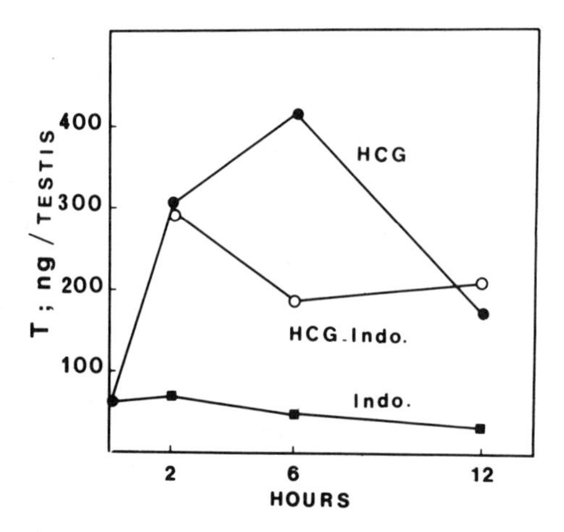

Figure 5. Effect of hCG and indomethacin (as indicated in Fig. 4) on T content in testicular tissue.

action of each prostaglandin. It has been shown in adrenal cells, in which prostaglandin receptors are present (Dazord *et al.*, 1974) that PGE_1 and PGE_2 increase cAMP accumulation and cortisol output, while $PGF_{1\alpha}$ and $PGF_{2\alpha}$ decrease both parameters (Honn and Chavin, 1976). Since in Leydig cells both E and F series are synthesized in similar amounts under hCG stimulation, the combined effect might be difficult to interpret.

2.6. *Prostaglandins and the Down-Regulation of the LH–hCG Receptors*

Two facts are in favor of a direct role of prostaglandins in the disappearance of the LH–hCG receptors. The first is temporal relationship between the two phenomena. Figure 7 indicates in detail $PGF_{2\alpha}$ levels and LH–hCG binding sites for 8 hr following hCG injection. It is striking that the acute increase of prostaglandin synthesis occurs between 2 and 6 hr, precisely at the time of acute disappearance of the LG–hCG receptors. Second, indomethacin treatment prevents hCG-receptor down-regulation for a certain period of time (Fig. 8). It is worth noting that this inhibition is observed between 3 and 6 hr and the receptors disappear eventually despite the apparent blockage of prostaglandin PGE_2 and $PGF_{2\alpha}$ synthesis (see Fig. 4). It is clear that other mechanisms of receptor regulation are also involved. We will see in Section 3.3 that a direct effect of $PGF_{2\alpha}$ on hCG binding sites can be obtained *in vitro*, confirming the probable role of prostaglandins in hCG-receptor regulation.

3. Role of Prostaglandins in Immature Porcine Leydig Cells in Primary Culture

An adaptation (Mather and Sato, 1978) of a serum-free, hormone-supplemented culture system originally developed for mouse Leydig cells (Mather and Sato, 1979) was used to culture interstitial cells from 3- to 4-week-old piglets. There are several advantages to using the pig as the experimental animal in this instance. At this age, porcine interstitial tissue is 80% Leydig cells, allowing for a relatively pure Leydig-cell population without extensive cell purification before culturing. The number of hCG receptors per cell is 10- to 20-fold higher than in rodents (Peyrat *et al.*, 1980), and a large number of cells can be obtained from a few animals.

It has been found that Leydig cells in this culture system maintain receptors for and responsiveness to hCG throughout the life-span of the cells in culture. This has allowed studies of cell regulation by various

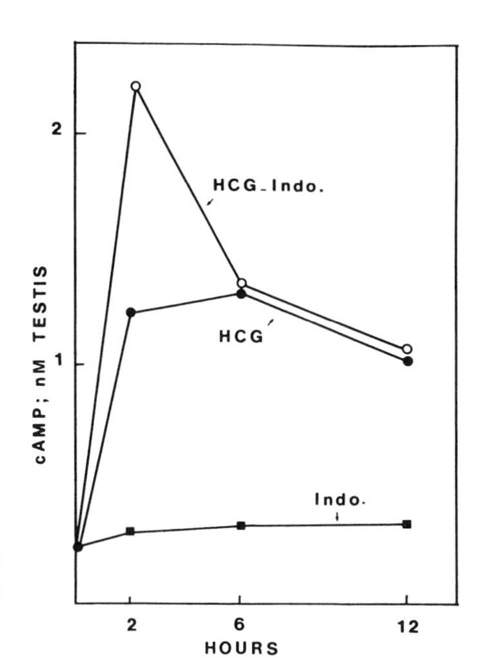

Figure 6. Effect of hCG and indomethacin (as indicated in Fig. 4) on cAMP content in the testis. cAMP was assayed as previously indicated (Saez *et al.*, 1978a,b).

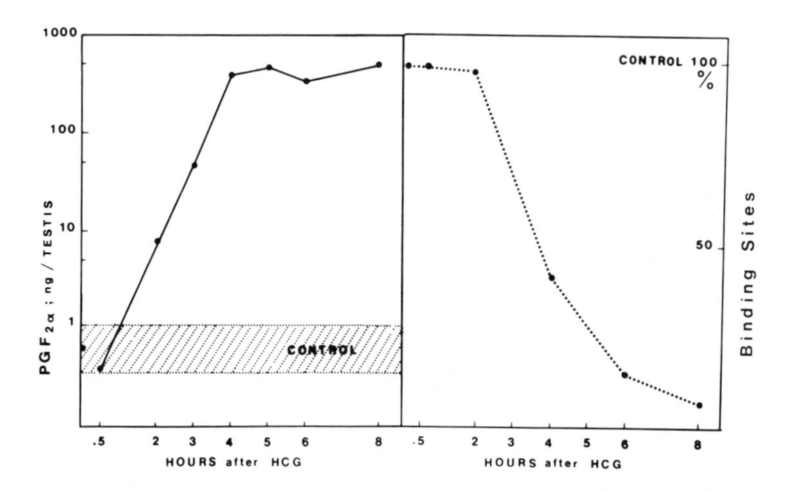

Figure 7. Variation of $PGF_{2\alpha}$ content in testis and hCG binding sites as a function of time following hCG injection (techniques as indicated in Fig. 1).

hormones in a more defined system over extended periods of time in culture.

3.1. Prostaglandin Production under hCG Stimulation

Cultured porcine Leydig cells are capable of producing PGE_2 and $PGF_{2\alpha}$ when stimulated *in vitro* with hCG (Fig. 9). This is in agreement with the *in vivo* data shown for the rat. However, the *in vitro* production corresponds to a summation of the 4 hr following hCG stimulation and

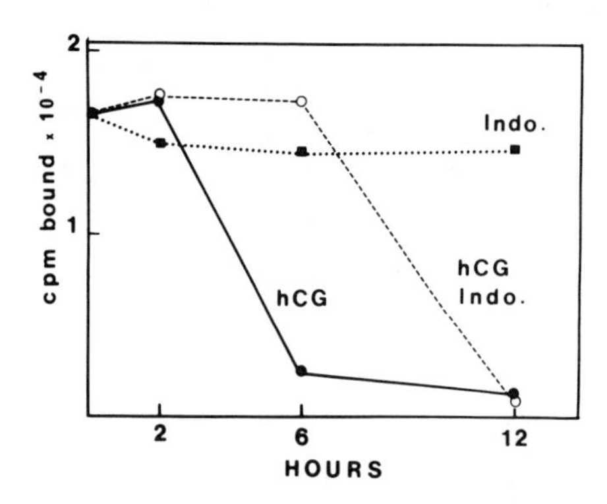

Figure 8. Effect of hCG and indomethacin (as indicated in Fig. 4) on hCG binding sites.

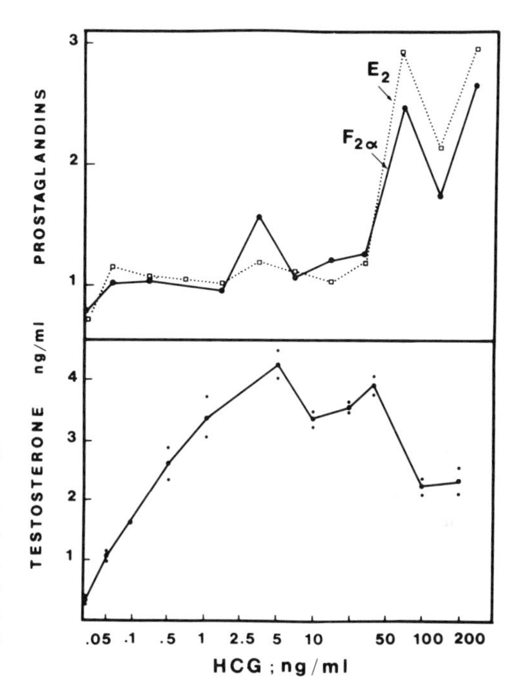

Figure 9. Production of T, PGE_2, and $PGF_{2\alpha}$ by porcine interstitial cells *in vitro* as a function of hCG concentration. Fresh medium (1.5 ml) containing the indicated hCG was added after 2 days in culture. Medium was collected and assayed at 4 hr. Results are expressed as ng/ml per 4 hr (2.0 $\times$ 10^6 cells/culture).

cannot be compared directly to *in vivo* concentrations. A clonal mouse Leydig-cell line has also been shown to produce PGE_2 and $PGF_{2\alpha}$ (unpublished results), lending further support to the hypothesis that Leydig cells are the source of prostaglandin production in the testis.

It can be seen that the concentration of hCG required for maximal stimulation of PGE_2 and $PGF_{2\alpha}$ is the same. Maximal stimulation of T is seen at concentrations of hCG 10-fold lower than that required for maximal stimulation of prostaglandin production. At concentrations (50–200 ng/ml hCG) where PGE_2 and $PGF_{2\alpha}$ concentrations are maximal, T production is decreasing. This pattern suggests that prostaglandins might be one of the mediators of the relative decrease in T production for high hCG concentrations.

3.2. *Role of Prostaglandins in Cell Survival*

Porcine Leydig cells survive well in culture for 3–4 days, thereafter showing a sharp decline in cell number. The addition of $PGF_{2\alpha}$ significantly extends the life-span of the cultured cells over that of the control

(Fig. 10). $PGF_{2\alpha}$ seems not only to prevent cell death but also to stimulate cell growth, with the cell number doubling during the first 5 days in culture.

The effect of prostaglandins on cell survival is dose-dependent (Fig. 11), with $PGF_{2\alpha}$ being effective at doses as low as 10 ng/ml.

3.3. Role of Prostaglandins in hCG-Receptor Regulation

$PGF_{2\alpha}$ can be shown to have a direct effect on hCG-receptor levels *in vitro* (Fig. 12). The hCG receptors are maintained throughout the culture period at or above the levels seen immediately after isolation. The addition of hCG causes the receptor levels to decrease sharply and remain low as long as hCG is maintained in the culture. The addition of low levels (25 ng/ml) of $PGF_{2\alpha}$ causes a gradual decline in hCG-receptor levels. The reduction of hCG receptors is first seen 48 hr after addition of $PGF_{2\alpha}$, with an 80% decrease in receptors by the 6th day. These results lend support to the idea that prostaglandins may play a direct role *in vivo* in the dow-regulation of hCG receptors during acute hCG stimulation.

3.4. Role of Prostaglandins in Steroidogenesis

Basal T production *in vitro* is unaffected by $PGF_{2\alpha}$ (Fig. 13, left). The hCG-stimulated production of T, however, is increased in the presence of $PGF_{2\alpha}$ (Fig. 13, right). This increase is seen both in contin-

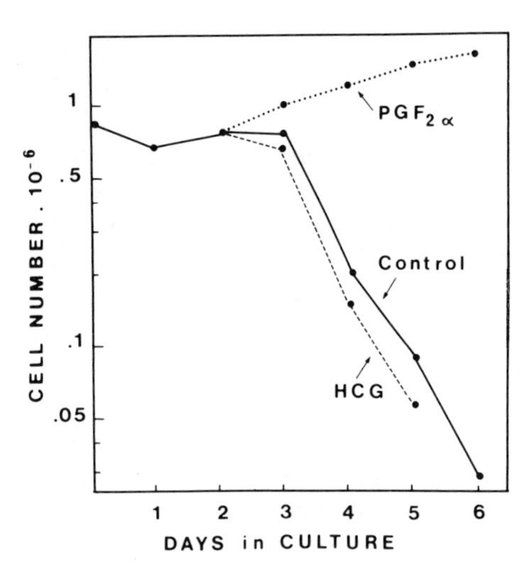

Figure 10. Effect of $PGF_{2\alpha}$ and hCG on porcine interstitial cell survival. Cells were plated in medium containing insulin, transferrin, epidermal growth factor, and 0.11% calf serum (control), with the addition of $pGF_{2\alpha}$ (25 ng/ml) or hCG (50 ng/ml) (Mather *et al.*, 1980).

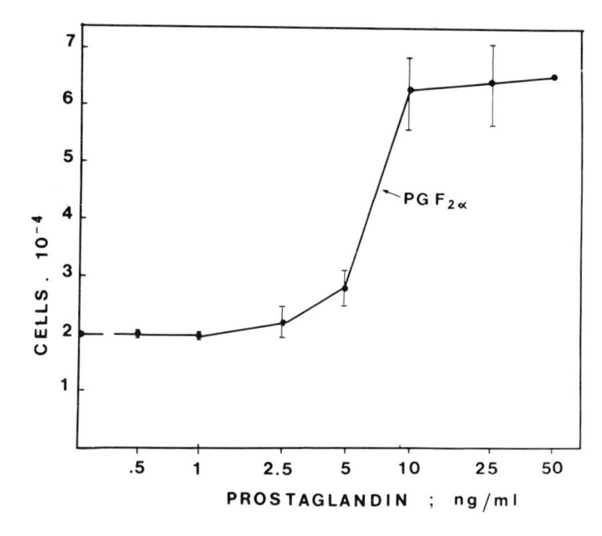

Figure 11. Porcine interstitial cell survival as a function of $PGF_{2\alpha}$ concentration. Cells were plated as in Fig. 10, with the indicated concentrations of $PGF_{2\alpha}$ added on day 3. Cells were counted on day 7.

uous hCG stimulation and when $PGF_{2\alpha}$ is present continuously and hCG is added 24 hr before collection of the medium. The cells require the continual presence of both $PGF_{2\alpha}$ and hCG for maximal response on days 5 and 6 of culture.

4. Discussion

Among the metabolic changes induced in the Leydig cell by hCG stimulation, the increase in prostaglandin synthesis is now well established. Our results show that this increase in prostaglandin production can be achieved *in vivo* as well as *in vitro*. These results are comparable to those previously observed in the ovary (Marsh *et al.*, 1974; Zor *et al.*, 1977a, b; Clark *et al.*, 1978a, b). Therefore, it seems that enhanced prostaglandin synthesis is part of the consequences of hCG stimulation of the target cells. The results obtained *in vitro* indicate that prostaglandin production might be triggered only at high concentrations of hCG. These levels being far above the concentrations that provided maximal steroidogenic response, it can be suggested that prostaglandin production is not involved in steroidogenesis itself, but on the contrary is one of the factors that reduce the capacity for steroid production during desensiti-

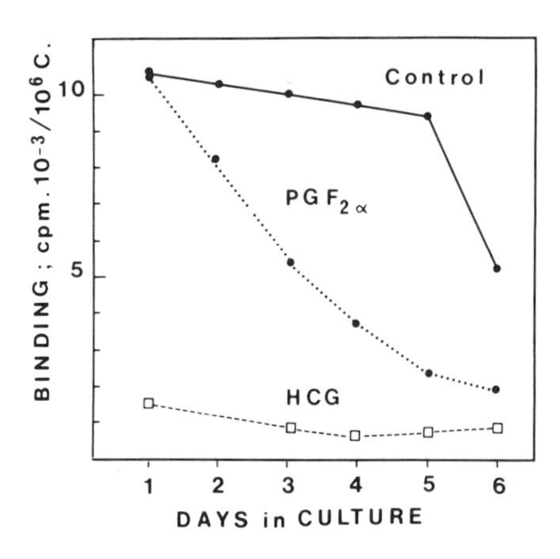

Figure 12. Effect of $PGF_{2\alpha}$ (25 ng/ml) on hCG receptors in cultured porcine interstitial cells. Cells were plated as indicated in Fig. 10. [^{125}I]-hCG binding sites are expressed in cpm/10^6 cells.

zation to hCG. This hypothesis is sustained by the incidence of high prostaglandin levels when T production is reduced either *in vivo* (see Fig. 1) or *in vitro* (see Fig. 9). It is also in accordance with the reduction of T production observed following prostaglandin stimulation either *in vivo* (Bartke *et al.*, 1973; Saksena *et al.*, 1978) or *in vitro* (Bartke *et al.*, 1976; Grotjan *et al.*, 1978). However, this is in contradiction to the fact that $PGF_{2\alpha}$ does not significantly decrease basal T level (see Fig. 13) and, on the contrary, potentiates the action of hCG. These contradictions can be partially resolved by considering that prostaglandins are involved at several levels of Leydig-cell function. The marked role of $PGF_{2\alpha}$ in increasing cell survival (see Figs. 10 and 11) indicates that prostaglandins are probably playing a major role in Leydig-cell regulation. Moreover, both E and F series are synthesized in similar amount in response to hCG stimulation. Since their biological action is often opposite (Phariss and Behrman, 1973; Honn and Chavin, 1976), the combined effects are diifficult to foresee.

The trigger for the increase in prostaglandin synthesis is not yet determined. It is not T itself, since elevated T levels do not modify prostaglandin production. The trigger could be cAMP accumulation. It has been clearly demonstrated in the rat follicle (Clark *et al.*, 1978a) that high doses of dibutyryl cAMP (DbcAMP) mimic the effect of LH and induce prostaglandin synthesis. Since it is well known that cAMP production at high hCG concentrations greatly exceeds the amount necessary for maximal T production, it is possible that above a certain level cAMP accumulation might trigger other metabolic functions, particularly prostaglandin biosynthetic pathways.

It is of interest to note that in similar endocrine systems, such as thyroid (Haye and Jacquemin, 1977) or adrenal (Laychock *et al.*, 1977), TSH and ACTH stimulation results in an increase in prostaglandin production. Since both hormonal activities are mediated by cAMP, this supports the idea that prostaglandin accumulation is a widespread result of polypeptide-hormone stimulation and cAMP accumulation.

The time course of prostaglandin biosynthesis in response to hCG exhibits a lag-time of about 2 hr, which is very similar to that observed in the ovary (Clark *et al.*, 1978a,b). Both results suggest that this is the time necessary for the synthesis of one or several enzymes in the prostaglandin biosynthetic pathway. Clark *et al.* (1978b) have confirmed this hypothesis by demonstrating an increase in prostaglandin synthetase activity following LH stimulation in the ovary.

The presence of prostaglandin receptors on the Leydig-cell membrane confirms the possibility of an autofeedback regulation of the Leydig cell by the prostaglandin produced *in situ*. This result can be compared with the results obtained on granulosa cells and the corpus luteum (Rao, 1976) and offers another example of similarities of regulation in both

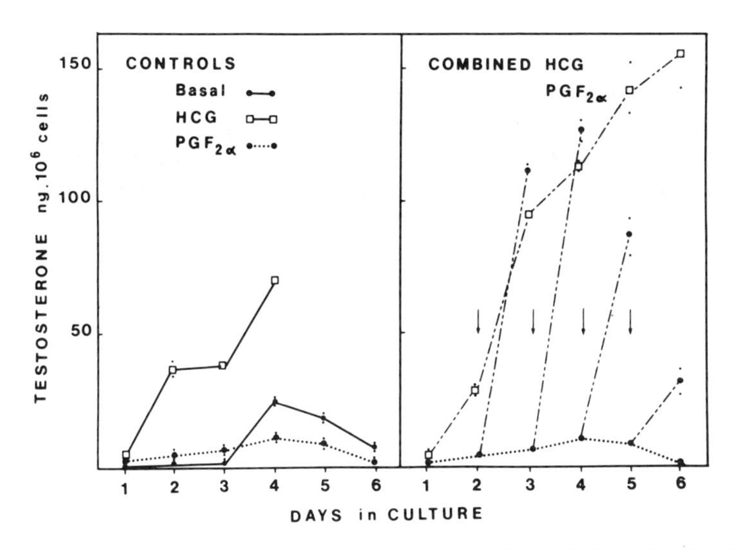

Figure 13. T Production by cultured porcine interstitial cells. *Left:* Basal levels with and without $PGF_{2\alpha}$ (25 ng/ml) and hCG (50 ng/ml) stimulated production in the absence of $PGF_{2\alpha}$. *Right:* T production when hCG and $PGF_{2\alpha}$ are both present throughout the culture period (□) or $PGF_{2\alpha}$ is present continually and hCG is added (↓) 24 hr before collection of the medium (●).

tissues. Endogenous and exogenous hormones may alter Leydig-cell function by binding to the membrane receptor; however, it is not known in this case whether endogenous prostaglandin may act directly on Leydig-cell metabolism without the mediation of the prostaglandin receptors.

The role of prostaglandins in LH–hCG-receptor regulation has been demonstrated *in vivo* as well as *in vitro*. *In vivo*, the acute increase in prostaglandin production corresponds to the disappearance of the receptor, and receptor loss can be antagonized by indomethacin. However, other factors are probably involved, since indomethacin alone cannot maintain LH–hCG receptors. The *in vitro* experiment indicates that $PGF_{2\alpha}$ is able to induce the disappearance of the gonadotropin receptor. The mechanism of this regulation in the absence of hCG is unkown. However, it provides an argument in favor of a role of prostaglandins in hCG-induced down-regulation of hCG receptors.

In summary, despite minor differences, the *in vivo* and *in vitro* systems provide new evidence for a major role of prostaglandins in the regulation of Leydig-cell function and particularly in the phenomenon of desensitization to hCG and down-regulation of the gonadotropin receptors.

ACKNOWLEDGMENTS. The Upjohn Company generously donated prostaglandins used in this study. hCG was kindly provided by Dr. Bosch (Oss, The Netherlands) and by Dr. Canfield (New York). Ms. P. Sanchez, O. Joannequin, and N. Virard-Micewicz provided skillful technical assistance. Special thanks are due Ms. J. Bois for typing the manuscript and to the Oberlin's farm for providing pig testis. This work was supported by grants INSERM CRL 78.5.028.4 and DGRST 78.7.0361.

DISCUSSION

WILBUR: Would you please comment on your very interesting finding that $PGF_{2\alpha}$ stimulated the longivity of the Leydig cells?

MATHER: These cells were isolated from testis from 3- to 4-week-old animals. At this time, the testes are growing rapidly. I would not say that $PGF_{2\alpha}$ is mitogenic per se in this system, but is perhaps allowing the cells to divide normally by allowing survival.

KELLY: Do receptors go into the cell with the hormone?

MATHER: Dr. Fox has photochemically cross-linked PGF and its receptor. The hormone–receptor complex is internalized and found in lysosomes, and eventually degraded. Dr. Roth's laboratory has detected insulin receptor inside the cell using antireceptor antibody. These data would indicate that the receptor moves into the cell with the hormone.

I might add that I have tested the effect of various hormones on the growth of two clonal cell lines isolated from immature mouse testis. These cell lines have been tentatively identified as a Leydig- and a Sertoli-cell line. The Leydig-cell line does not have a growth response to either LH or gonadotropin-releasing hormone (GnRH) in a serum-free, hormone-supplemented culture system.

Using a similar culture system, Dr. Haour and I have studied hormones affecting hCG-receptor regulation and steroid responsiveness in primary cultures of porcine Leydig cells. In this system, there was no direct effect of prolactin on hCG-receptor levels. The effect of GnRH has yet to be tested.

This approach should be of help in differentiating which hormone effects seen *in vivo* are primary effects of the hormone on Leydig cells and which are secondary effects mediated via other cells or organs.

McKERNS: It would be interesting to see the response with time in your *in vivo* system or in your *in vitro* cell cultures to cAMP, hCG, and both together using the maximum-response dose of each. This may give some information as to whether hCG acts independently of cAMP.

HAOUR: We tried to stimulate the cells with DbcAMP (1 mm), but this concentration is too high and kills the cells. The maximum dose response in this case might be difficult to establish.

C.A. VILLEE: In both Dr. Aakvaag's system and Dr. Haour's system, we see that prostaglandins seem to be involved in the down-regulation of LH receptors, and it would seem a reasonable inference to suppose that the molecular mechanism involved may be similar in the two systems. Do we have any idea as to what that mechanism of down-regulation might be?

HAOUR: Down-regulation implies probably internalization of membrane components and profound changes in the membrane structure. Whether prostaglandin release is the cause or the consequence of such changes remains still an open question. Mechanism or more probably mechanisms remain to be fully investigated.

KELLY: Could you discuss the mechanism involved in the short-term action of indomethacin you observed *in vivo?* Have you tried PRL in your incubation medium to protect against down-regulation?

MATHER: Prolactin does not seem to protect against down-regulation.

HAOUR: With respect to the indomethacin effect, we can just say that prostaglandin production is part of the acute response of the Leydig cells to hCG stimulation and participates in the down-regulation of the hCG receptors. However, as would be expected, other factors are involved and down-regulation occurs in spite of the lack of prostaglandin production.

BANIK: Have you measured estrogen level after administration of 500 IU hCG? Also, have you checked the mating behavior of these animals, especially during the down-regulation process?

HAOUR: No.

BANIK: At the beginning, you used 500 IU hCG, and at the end, you injected 0.6–30 mg hCG. What was the purity of this second product of hCG?

HAOUR: In both cases, the hCG injected is a commercial progestin [Pregnyl-Organon (3000 IU/mg)].

NOLIN: One aspect of hormone–target cell interaction we ought to look out for, now that we know that peptide hormones may be retained by their targets for long periods, is that the length of this retention may influence experimental results and their interpretations considerably.

HAOUR: This a very interesting comment, which relates directly to desensitization. Desensitization occurs probably when the cell cannot eliminate the stimulatory hormone fast enough. This has to be kept in mind for the interpretation of phenomena involving pulsatile release. It might just be the way to let the target cells eliminate what they have incorporated during the previous pulse.

BAZER: Was the slide or cell number you discussed relative to an actual increase in number of cells above the number you started with or an increase in the survival of the cells incubated with $PGF_{2\alpha}$?

HAOUR: Yes, the number of cells increased in culture.

SPILMAN: Your data showed that the basal levels of PGE_2 and $PGF_{2\alpha}$ in the testes were not very different between intact and hypophysectomized rats. Yet hCG caused a large increase in PGE_2 and $PGF_{2\alpha}$ in the testes. Wouldn't you expect PG levels to be lower in hypox. animals, in the absence of LH?

HAOUR: LH triggers prostaglandin release, but is certainly not the unique trigger, and production can be maintained in the absence of LH either in hypox animals or *in vitro*.

REFERENCES

Bartke, A., Musto, N., Caldwell, B.V., and Behrman, H.R., 1973, Effects of a cholesterol inhibitor and of prostaglandin $F_{2\alpha}$ on testis cholesterol and on plasma testosterone in mice, *Prostaglandins* **3**:97.

Bartke, A., Kupfer, D., and S. Dalterio, 1976, Prostaglandins inhibit testosterone secretion by mouse testes *in vitro, Steroids* **28**:81.

Behrman, H.R., and Hickens, M., 1976, Rapid measurement of gonadotropin uptake by corpora lutea *in vivo* induced by prostaglandins $F_{2\alpha}$, *Prostaglandins* **12**:83.

Behrman, H.R., MacDonald, G.J., and Greep, R.O., 1971, Regulation of ovarian cholesterol esters: Evidence for the enzymatic sites of prostaglandin-induced loss of corpus luteum function, *Lipids* **6**:791.

Bockaert, J., Hunzicker-Dunn, M., Biznbaumer, L., 1976, Hormone stimulation of hormone dependent adenylate cyclase. Dual action of luteinizing hormone on pig Graafian follicle membranes, *J. Biol. Chem* **251**:2653.

Carpenter, M.P., 1974, Prostaglandins of rat testis, *Lipids* **9**:397.

Carpenter, M.P., Manning, L.M., Robinson, R.D., and To, D., 1978, Prostaglandin synthesis by interstitial tissue of rat testis, *Prostaglandins* **15**:711.

Cenedella, R.J., 1975, Prostaglandins and male reproduction physiology, in: *Advances in*

Sex Hormone Research, Vol. 1, *Molecular Mechanism of Gonadal Hormone Action* (J.A. Thomas and R.L. Singhal, eds.), pp. 327–360, University Park Press, Baltimore.

Chasalow, F., and Pharriss, B.B., 1972, Luteinizing hormone stimulation of ovarian prostaglandin biosynthesis, *Prostaglandins* **1:**105.

Chasalow, F., Marr, H., Haour, F., and Saez, J.M., 1979, Testicular steroidogenesis after hCG desensitization in rats, *J. Biol. Chem.* **254:**5613.

Clark, M.R., Marsh, J.M., and Lemaire, W.J., 1978a, Stimulation of prostaglandin accumulation in preovulatory rat follicles by adenosine 3′,5′-monophosphate, *Endocrinology* **102:**39.

Clark, M.R., Marsh, J.M., and Le Maire, W.J., 1978a,b, Mechanism of luteinizing hormone regulation of prostaglandin synthesis in rat granulosa cells, *J. Biol. Chem.* **253:**7757.

Cooke, B.A., Rommerts, F.F.G., van der Kemp, J.W.C.M., and van der Molen, H.J., 1974, Effects of luteinizing hormone, follicle stimulating hormone, prostaglandin E_1 and other hormones on adenosine-3′:5′-cyclic monophosphate and testosterone production in rats testis tissues, *Mol. Cell Endocrinol.* **1:**99.

Dazord, A., Morera, A.M., Bertrand, J., and Saez, J.M., 1974, Prostaglandin receptors in human and ovine adrenal glands: Binding and stimulation of adenyl cyclase in subcellular fractions, *Endocrinology* **95:**352.

Dray, F., Charbonnel, B., and Maclouf, J., 1975, Radioimmunoassay of prostaglandins $F_{2\alpha}$, E_1 and F_2 in human plasma, *Eur. J. Clin. Invest.* **5:**311.

Dray, F., Charbonnel, B., and Maclouf, J., 1976, Primary prostaglandins in human peripheral plasma by radioimmunoasssy, in: *Advances in Prostaglandin and Thromboxane Research*, Vol. 1 (B. Samuelson and R. Paoletti, eds.), p. 93, Raven Press, New York.

Eik-Nes, K.B., 1969, Pattern of steroidogenesis in the vertebrate gonad, *Gen. Comp. Edocrinol. Suppl.* **2:**87.

Ellis, L.C., Johnson, J.M., and Hargrove, J.L., 1972, Cellular aspect of prostaglandin synthesis and testicular functions, in: *Prostaglandins in Cellular Biology* (P.W. Ramwell and B.P. Phariss, eds.), p. 385, Plenum Press, New York.

Gerozissis, K., and Dray, F., 1977a, Prostaglandins in the isolated testicular capsule of immature and young adult rats, *Prostaglandins* **13:**177.

Gerozissis, K., and Dray, F., 1977b, Selective and age dependent changes of prostaglandins E_2 in the epididymis and vas deferens of the rat, *J. Reprod. Fertil.* **50:**113.

Grinwich, D.L., Hichens, M., and Behrman, H.R., 1976, Control of the LH receptor by prolactin and prostaglandin $F_{2\alpha}$ in rat corpora lutea, *Biol. Reprod.* **14:**212.

Grotjan, H.E., Heindel, J.S., and Steinberger, E., 1978, Prostaglandin inhibition of testosterone production induced by luteinizing hormone, DbcAMP or 3-isobutyl-1-methylxanthine in dispersed rat testicular interstitial cells, *Steroids* **32:**307.

Haour, F., and Saez, J.M., 1977, hCG-dependent regulation of gonadotropin receptor sites: Negative control in testicular Leydig cells, *Mol, Cell Endocrinol.* **7:**17.

Haour, F., and Saez, J.M., 1978, Leydig cell responsiveness to LH–hCG stimulation: Mechanisms of hCG and steroid induced refractoriness, in: *Structure and Function of the Gonadotropins* (K.W. McKerns, ed.), pp. 497–516, Plenum Press, New York.

Haour, F., Sanchez, P., Cathiard, A.M., and Saez, J.M., 1978, Gonadotropin receptor regulation in hypophysectomized rat Leydig cells, *Biochem. Biophys. Res. Commun.* **81:**547.

Haour, F., Kouznetzova, B., Dray, F., and Saez, J.M., 1979, hCG-induced prostaglandin E_2 and F_2 release in adult rat testis: Role in Leydig cells desensitization to hCG, *Life Sci.* **24:**2151.

Haye, B., and Jacquemin, C., 1977, Incorporation of ^{14}C arachidonate in pig thyroid lipids and prostaglandins, *Biochim. Biophys. Acta* **687:**231.

Honn, K.J., and Chavin, W., 1976, Prostaglandin modulation of the mechanism of ACTH action in the human adrenal, *Biochem. Biophys. Res. Commun.* **73:**164.

Hsueh, A.J.W., Dufau, M.L., and Catt, K.J., 1977, Gonadotropin induced regulation of luteinizing hormone receptors and desensitization of testicular 3′5′-cAMP and testosterone response, *Proc. Natl. Acad. Sci. U.S.A.* **74:**592.

Kimball, F.A., Frielink, R.D., and Porteus, S.E., 1978, Effect of 15 (S)-15 methylprostaglandin $F_{2\alpha}$ methyl ester containing Silastic discs in male rats, *Fertil. Steril.* **29:**103.

Kuehl, F.A., Jr., Humes, J.L., Tarnoff, J., Cirillo, C.J., and Ham, E.A., 1970, Prostaglandin receptor site: Evidence for an essential role in the action of luteinizing hormone, *Science 169:*883.

Kuehl, F.A., Jr., 1974, Prostaglandins, cyclic nucleotides and cell function, *Prostaglandins* **5:**325.

Lamprecht, S.A., Zor, U., Solomon, Y., Kock, Y., Ahren, K., and Lindner, H.R., 1977, Mechanism of hormonally induced refractoriness of ovarian adenylate cyclase in luteinizing hormone and prostaglandin E_2, *J. Cyclic Nucleotide Res.* **3:**69.

Lapetina, E.G., Schmitges, C.J., Chandrabose, K., and Cuatrecasas, P., 1977, Cyclic adenosine 3′5′-monophosphate and prostacyclin inhibit membrane phospholipase activity in platelets, *Biochem. Biophys. Res. Commun.* **76:**828.

Laychock, S.G., Warner, W., and Rubin, R.P., 1977, Further studies on the mechanism controlling prostaglandin biosynthesis in the cat adrenal cortex: Role of calcium and cAMP, *Endocrinology* **100:**74.

Malmsten, C., Granström, E., And Samuelson, B., 1976, Cyclic AMP inhibits synthesis of prostaglandin endoperoxide (PGG_2) in human platelets, *Biochem. Biophys. Res. Commun.* **68:**569.

Marsh, J.M., Yang, N.S.T., and Lemaire, W.J., 1974, Prostaglandin synthesis in rabbit Graafian follicle *in vitro:* Effect of luteinizing hormone and cyclic AMP, *Prostaglandins* **7:**269.

Marsh, J.M., and Le Maire, W.J., 1974, Cyclic AMP accumulation and steroidogenesis in the human corpus luteum: Effect of gonadotropins and prostaglandins, *J. Clin. Endocrinol. Metab.* **38:**99.

Mather, J., and Sato, G.H., 1978, Hormones and growth factors in cell cultures: Problems and perspectives, in: *Cell Tissue and Organ Cultures in Neurobiology* (S. Federoff and L. Hertz, eds.), pp. 619–630, Academic Press, New York.

Mather, J.P., and Sato, G.H., 1979, Use of hormone-supplemented serum-free media in primary culture, *Exp. Cell Res.* **124:**215.

Mather, J.P., Saez, J.M., and Haour, F., 1980, *Life Sci.* (submitted).

McNatty, K.P., Henderson, K.M., and Sawers, R.S., 1975, Effects of prostaglandin $F_{2\alpha}$ and E_2 on the production of progesterone by human granulosa cells in tissue culture, *J. Endocrinol.* **67:**231.

Peyrat, J.P., Meusy-Dessolle, N., and Garnier, J., 1980, Properties of luteinizing hormone receptors in porcine testis: Variations with age and interstitial tissue development, *Endocrinology* (submitted).

Phariss, B.B., and Behrman, H.R., 1973, in: *Gonadal Function in the Prostaglandins,* Vol. 1 (P.W. Ramwell, ed.), pp. 347–363, Plenum Press, New York.

Rao, C.V., 1976, Discrete prostaglandin receptors in the outer cell membrane of bovine corpora lutea, in: *Advances in Prostaglandin and Thromboxane Research,* Vol. 1 (B. Samuelson and R. Paoletti, eds.), p. 267, Raven Press, New York.

Saez, J.M., Haour, F., Gallet, D., and Sanchez, P., 1978a, hCG-induced Leydig cells refractoriness to gonadotropin stimulation, *Mol. Pharmacol.* **14:**1054.

Saez, J.M., Haour, F., and Cathiard, A.M., 1978b, Early hCG-induced desensitization in Leydig cells, *Biochem. Biophys, Res. Commun.* **81:**552.

Saksena, S.K., Lan, I.F., and Chang, M.C., 1978, Effect of intrascrotal implants of prostaglandins E_2 or $F_{2\alpha}$ on blood steroids in the adult male rat, *Int. J. Androl.* **1:**180.

Samuelson, B., Goldyne, M., Granstrom, E., Hamberg, M., Hammaastrom, S., and Malmsten, C., 1978, Prostaglandins and thromboxane, *Annu. Rev. Biochem.* **47:**997.

Sharpe, R.M., 1976, hCG-induced decrease in availability of rat testis receptors, *Nature (London)* **264:**644.

Shio, H., Shaw, J., and Ramwell, P., 1971, Relation of cyclic AMP to the release and action of prostaglandins, *Ann. N. Y. Acad. Sci.* **185:**326.

Tell, G.P., Haour, F., and Saez, J.M., 1978, Hormonal regulation of membrane receptors and cell responsiveness: A review, *Metabolism* **27:**1566.

Thomas, J.P., Dorflinger, L.J., and Behrman, H.R., 1978, Mechanism of the rapid antigonadotropic action of prostsaglandins in cultured luteal cells, *proc. Natl. Acad. Sci. U.S.A.* **75:**1344.

Zor, U., Strulovici, B., and Lindner, H.R., 1977a, Stimulation by cyclic AMP of prostaglandin E production in isolated Graafian follicles, *Biochem. Biophys. Res. Commun.* **76:**1086.

Zor, U., Strulovici, B., and Lindner, H.R., 1977b, Stimulation by cyclic nucleotides of prostaglandin E production in isolated Graafian follicles, *Prostaglandins* **14:**947.

Index